# HILBERT SPACE METHODS IN QUANTUM MECHANICS

FUNDAMENTAL SCIENCES *Physics*

# HILBERT SPACE METHODS IN QUANTUM MECHANICS

Werner O. Amrein

EPFL Press
A Swiss academic publisher distributed by CRC Press

CRC Press
Taylor & Francis Group

Taylor and Francis Group, LLC
6000 Broken Sound Parkway, NW, Suite 300,
Boca Raton, FL 33487

Distribution and Customer Service
orders@crcpress.com

**www.crcpress.com**

Library of Congress Cataloging-in-Publication Data
A catalog record for this book is available from the Library of Congress.

This book is published under the editorial direction of
Professor Philippe-André Martin (EPFL, Lausanne).

EPFL Press

is an imprint owned by Presses polytechniques et universitaires romandes, a Swiss academic publishing company whose main purpose is to publish the teaching and research works of the Ecole polytechnique fédérale de Lausanne.

Presses polytechniques et universitaires romandes
EPFL – Centre Midi
Post office box 119
CH-1015 Lausanne, Switzerland
E-Mail: ppur@epfl.ch
Phone: 021 / 693 21 30
Fax: 021 / 693 40 27

**www.epflpress.org**


ISBN 978-2-940222-35-3 (EPFL Press)
ISBN 978-1-4200-6681-4 (CRC Press)

Printed in Spain

# Preface

This text is based on lectures given at the University of Geneva during the period 1994 - 2005. These courses were intended for advanced undergraduate students who had completed a first course in quantum mechanics and were thus expected to be familiar with the physical aspects and the basic mathematical formalism of quantum theory. Partly due to lack of time, quantum mechanics is often taught with no or little exposition of the mathematical questions arising through the introduction of infinite-dimensional Hilbert spaces. I hope that the present volume will prove useful, especially to somewhat theoretically minded students, for deepening their knowledge and understanding of the Hilbert space aspects of quantum mechanics, and prepare them for reading research papers.

Mostly the lectures were organised as one-year courses (80 hours plus 25 hours of problem sessions) and covered essentially the contents of Chapters 1 - 5 and parts of Chapters 6 or 7. To offer a few more applications, some material has been added to the original lecture notes (Sections 5.8, 6.6, 6.7, 7.2 and 7.5). Of course a strict selection of topics and applications to be treated had to be made from the outset. The emphasis is placed on a certain number of basic mathematical techniques, usually without striving for the most general results. For example there is no discussion of quadratic forms, and we have avoided the use of techniques from stochastic analysis. However we give essentially complete proofs for all results involving Hilbert space objects. Some of these proofs are collected in appendices to the various chapters. Some acquaintance with measure theory is required; the essential facts are explained without detailed proofs.

Chapter 1 gives the basic properties of Hilbert space and a description of the necessary material from measure theory. In Chapter 2 we present various classes of bounded linear operators and general notions on unbounded operators, including the invariance of self-adjointness under a class of perturbations. The problem of self-adjointness is further investigated in Chapter 3 which contains the theory of extensions of symmetric operators, with applications to Sturm-Liouville and Schrödinger operators. Chapter 4 deals with the spectral theory of self-adjoint operators, in particular with the spectral theorem and with the various spectral types. In Chapter 5 we prove Stone's Theorem and then discuss the fundamental aspects of scattering theory: scattering states, asymptotic condition and wave operators, $S$-matrix, scattering cross sections. Chapter 6 is devoted to the Mourre method for controlling the resolvent of self-adjoint operators near the real axis and to implications for their spectrum. In the final Chapter 7 we present stationary-state scattering theory and various applications of the results of Chapter 6: asymptotic completeness, properties of the $S$-matrix, time delay and the Flux-Across-Surfaces Theorem.

Each chapter ends with some bibliographical notes and a selection of problems. The bibliography consists mostly of books. They are referred to for alternative or more advanced presentations of some material or for certain points not treated in the present text. A very small number of original and review papers are cited for those interested in a deeper understanding of certain topics treated in the text. In combination with modern electronic means, these papers can also be useful as a basis for searching in the vast literature.

The majority of the problems have been tested in class-room sessions. Many of them are meant to help students to become familiar with the concepts discussed in the main body; some require a more detailed study of technical aspects of the text. A few of the more difficult problems are provided with a hint for the solution.

As regards notations, it should be pointed out that some symbols have more than one meaning in the text. In particular: the symbol $\|\cdot\|$ is used for various norms, the Greek letter $\sigma$ for spectra and for scattering cross sections, the letter $P$ for projections and for momentum, and the letter $R$ for resolvents and for $S - I$, where $S$ is the scattering operator. For the convenience of the reader we have included a Notation Index (page 385). Constants are often denoted generically by $c$, but in some proofs different constants are numbered as $c_1, c_2$, etc.

It is a great pleasure to thank all those who helped me in one way or another during the preparation of the lectures or of this volume. The problems in the text are mostly due to Marius Mantoiu, Joachim Stubbe and Rafael Tiedra de Aldecoa; they assumed the responsibility for the problem sessions with much competence and devotion. The continual interest of my students and their constructive comments on earlier versions of the text have influenced a considerable number of details. I received precious support from Philippe Jacquet, Andreas Malaspinas, Peter Wittwer and Luis Zuleta for coming to grips with TEX-related difficulties. I thank Philippe Martin for proposing the publication of my lectures, the referee for pointing out some errors and for useful suggestions, and Fred Fenter from EPFL Press for advice and his very efficient management of the publishing process. Finally I am indebted to the Physics Department of the University of Geneva for its kind hospitality after my retirement.

*Werner Amrein*
Geneva, Switzerland
December 2008

# Contents

# CHAPTER 1

# Hilbert Spaces

Hilbert space sets the stage for standard quantum theory: the pure states of a physical system are identified with the unit rays of a Hilbert space $\mathcal{H}$ and observables with self-adjoint operators acting in $\mathcal{H}$. In this initial chapter we present the essential concepts and prove the basic results concerning separable Hilbert spaces (Sections 1.1 - 1.3). In Section 1.4 we then introduce $L^2$ spaces, which are of special importance for quantum mechanics. This requires some familiarity with measure theory, and we include a short description of the necessary concepts from this theory.

## 1.1 Definition and elementary properties

**1.1.1.** Throughout this text a ***Hilbert space*** means a *complex linear vector space*, equipped with a Hermitian *scalar product*, which is *complete* and admits a *countable basis*. More precisely a (separable) Hilbert space $\mathcal{H}$ is defined by the four postulates (H1) - (H4) stated below:

**(H1)** *$\mathcal{H}$ is a linear vector space over the field $\mathbb{C}$ of complex numbers:*

With each couple $\{f, g\}$ of elements of $\mathcal{H}$ there is associated another element of $\mathcal{H}$, denoted $f + g$, and with each couple $\{\alpha, f\}$, $\alpha \in \mathbb{C}$, $f \in \mathcal{H}$, there is associated an element $\alpha f$ of $\mathcal{H}$, and these associations have the following properties (where $f, g, h \in \mathcal{H}$ and $\alpha, \beta \in \mathbb{C}$):

$$f + g = g + f \qquad f + (g + h) = (f + g) + h \tag{1.1}$$

$$\alpha(f + g) = \alpha f + \alpha g \qquad (\alpha + \beta)f = \alpha f + \beta f \tag{1.2}$$

$$\alpha(\beta f) = (\alpha\beta)f \qquad 1f = f. \tag{1.3}$$

Furthermore there exists a unique element $\mathbf{0} \in \mathcal{H}$ (called the ***zero vector***) such that [1]

$$\mathbf{0} + f = f \quad , \quad 0f = \mathbf{0} \qquad \forall f \in \mathcal{H}. \tag{1.4}$$

[1] Here 0 denotes the complex number $\alpha = 0$.

**(H2)** *$\mathcal{H}$ is equipped with a strictly positive scalar product*[2]*:*

With each couple $\{f, g\}$ of elements of $\mathcal{H}$ there is associated a complex number $\langle f, g\rangle$, and this association has the following properties[3]:

$$\langle g, f\rangle = \overline{\langle f, g\rangle} \qquad \forall f, g \in \mathcal{H} \tag{1.5}$$

$$\langle f, g + \alpha h\rangle = \langle f, g\rangle + \alpha\langle f, h\rangle \qquad \forall \alpha \in \mathbb{C},\ \forall f, g, h \in \mathcal{H} \tag{1.6}$$

$$\langle f, f\rangle > 0 \qquad \text{except for } f = \mathbf{0}. \tag{1.7}$$

One then defines

$$\|f\| := [\langle f, f\rangle]^{1/2}. \tag{1.8}$$

**(H3)** *$\mathcal{H}$ is complete:*

Each Cauchy sequence in $\mathcal{H}$ has a limit in $\mathcal{H}$. In other terms, if $\{f_n\}_{n\in\mathbb{N}}$ is a sequence of elements of $\mathcal{H}$ satisfying[4]

$$\lim_{m,n\to\infty} \|f_n - f_m\| = 0,^{5} \tag{1.9}$$

then there exists an element $f$ of $\mathcal{H}$ such that $\lim_{n\to\infty} \|f - f_n\| = 0$.

**(H4)** *$\mathcal{H}$ is separable:*

$\mathcal{H}$ has a countable orthonormal basis, i.e. there exists a sequence $\{e_1, e_2, e_3, \ldots\}$ of elements of $\mathcal{H}$ such that

$$\langle e_j, e_k\rangle = \delta_{jk} \tag{1.10}$$

and such that each element $f$ of $\mathcal{H}$ can be expressed as a linear combination of $e_1, e_2, e_3, \ldots$

The ***dimension*** of a Hilbert space $\mathcal{H}$ is defined as the number of elements in an orthonormal basis. This number may be finite or infinite, and it is independent of the chosen basis. In an infinite-dimensional space the postulate (H4) involves infinite linear combinations. These are interpreted as follows: if $\{e_1, e_2, e_3, \ldots\}$ is an orthonormal basis of $\mathcal{H}$ and $f$ an arbitrary element of $\mathcal{H}$, there exists a sequence $\{\alpha_1, \alpha_2, \alpha_3, \ldots\}$ of complex numbers such that $f = \sum_k \alpha_k e_k$ and $\|f\|^2 = \sum_k |\alpha_k|^2$. The coefficient $\alpha_k$ in this development of $f$ is given by $\alpha_k = \langle e_k, f\rangle$, and the series $\sum_k \alpha_k e_k$ converges to $f$ in the sense that $\|f - \sum_{k=1}^{N} \alpha_k e_k\| \to 0$ as $N \to \infty$.

The elements of $\mathcal{H}$ are called ***vectors***; they will be denoted by $f, g$ or $h$ (sometimes by $e$ for vectors satisfying $\|e\| = 1$). The letters $\alpha$ and $\beta$ will stand for complex numbers. The reader should be familiar with finite-dimensional Hilbert spaces from Linear Algebra. However many problems in quantum mechanics involve Hilbert spaces of infinite dimension. We shall see that, when suitably interpreted, some properties of finite-dimensional spaces have analogues in infinite-dimensional spaces (for example the spectral properties of compact operators are similar to those of matrices), while

[2] Also called "inner product" by some authors.

[3] $\overline{\alpha}$ denotes the complex conjugate of the complex number $\alpha$.

[4] $\mathbb{N} = \{1, 2, 3, \ldots\}$.

[5] i.e. for each $\varepsilon > 0$ there exists a number $N = N(\varepsilon) > 0$ such that $\|f_n - f_m\| < \varepsilon\ \forall n, m > N$.

other aspects of infinite-dimensional spaces do not appear in finite-dimensional ones (for example in finite-dimensional spaces there is no distinction between weak and strong convergence, and matrices do not have continuous spectrum).

Observe that, contrary to the convention adopted by mathematicians, the scalar product is linear in the second entry and anti-linear in the first one. The non-negative number $\|f\|$ defined by (1.8) is called the ***norm*** of the vector $f$. This norm function has the properties of a metric. The quantity $\|f-g\|$ can be interpreted as the distance from $f$ to $g$, and one has $\|\alpha f\| = |\alpha|\|f\|$, in particular $\|-f\| \equiv \|(-1)f\| = \|f\|$. The scalar product can be expressed in terms of the norm of $\mathcal{H}$; this is the content of the ***polarisation identity***:

$$4\langle f, g\rangle = \|f+g\|^2 - \|f-g\|^2 - i\|f+ig\|^2 + i\|f-ig\|^2. \tag{1.11}$$

This equation is easily checked by using the definition of the norm and the linearity of the scalar product [as in (1.16) below].

The following four inequalities are simple consequences of (H1) and (H2)[6]:

**Schwarz inequality**:
$$|\langle f, g\rangle| \le \|f\| \cdot \|g\| \tag{1.12}$$

**Triangle inequality**:
$$\|f+g\| \le \|f\| + \|g\| \tag{1.13}$$

$$\|f+g\|^2 \le 2\|f\|^2 + 2\|g\|^2 \tag{1.14}$$

$$\big|\|f\| - \|g\|\big| \le \|f-g\|. \tag{1.15}$$

PROOF. If $f = g$, (1.12) is evident: $\langle f, f\rangle = \|f\|^2$. If $f \neq g$, assume for example that $g \neq \mathbf{0}$. Then, for any $\alpha \in \mathbb{C}$:

$$0 \le \|f+\alpha g\|^2 = \langle f+\alpha g, f+\alpha g\rangle = \|f\|^2 + \alpha\langle f, g\rangle + \overline{\alpha}\langle g, f\rangle + |\alpha|^2\|g\|^2. \tag{1.16}$$

By taking $\alpha = -\overline{\langle f, g\rangle}/\|g\|^2$ one obtains

$$0 \le \|f\|^2 - \frac{1}{\|g\|^2}|\langle f, g\rangle|^2,$$

which implies (1.12) upon multiplication by $\|g\|^2$. For (1.13) we take $\alpha = 1$ in (1.16):

$$\begin{aligned}\|f+g\|^2 &= \|f\|^2 + \langle f, g\rangle + \langle g, f\rangle + \|g\|^2 \\ &\le \|f\|^2 + |\langle f, g\rangle| + |\langle g, f\rangle| + \|g\|^2 \\ &\le \|f\|^2 + 2\|f\| \cdot \|g\| + \|g\|^2 = (\|f\| + \|g\|)^2.\end{aligned} \tag{1.17}$$

For (1.14) we observe that

$$0 \le \|f-g\|^2 = \|f\|^2 - \langle f, g\rangle - \langle g, f\rangle + \|g\|^2,$$

hence

$$\langle f, g\rangle + \langle g, f\rangle \le \|f\|^2 + \|g\|^2.$$

[6] In (1.12), $\|f\| \cdot \|g\|$ denotes the product of the norms of $f$ and $g$. In the remainder of the text a product of two norms will be written simply as $\|\cdot\|\|\cdot\|$. With this convention, (1.12) will read $|\langle f, g\rangle| \le \|f\|\|g\|$.

Upon insertion of this inequality into (1.17) one obtains (1.14). Finally, to obtain (1.15), suppose for example that $\|g\| \leq \|f\|$; then, by using (1.13):

$$\|f\| - \|g\| = \|f - g + g\| - \|g\| \leq \big(\|f - g\| + \|g\|\big) - \|g\| = \|f - g\|. \qquad \square$$

**1.1.2.** There are two types of convergence in a Hilbert space, called strong convergence and weak convergence, defined in terms of the norm and the scalar product respectively:

***Strong convergence***: A sequence of vectors $\{f_n\}_{n\in\mathbb{N}}$ in a Hilbert space $\mathcal{H}$ is said to converge strongly to a vector $f \in \mathcal{H}$ if $\|f_n - f\|$ converges to 0 when $n \to \infty$. One then writes $\text{s-lim}_{n\to\infty} f_n = f$.

***Weak convergence***: A sequence of vectors $\{f_n\}_{n\in\mathbb{N}}$ of $\mathcal{H}$ is said to converge weakly to a vector $f \in \mathcal{H}$ if, for each vector $g \in \mathcal{H}$, the sequence of complex numbers $\{\langle f_n, g\rangle\}_{n\in\mathbb{N}}$ converges to $\langle f, g\rangle$ when $n \to \infty$. One then writes $\text{w-lim}_{n\to\infty} f_n = f$.

The postulate (H3) is a Cauchy criterion for strong convergence: a strong Cauchy sequence has a limit (in $\mathcal{H}$). The Cauchy criterion is also true for weak convergence: if $\{f_n\}_{n\in\mathbb{N}}$ is weakly Cauchy (i.e. such that, for each $g \in \mathcal{H}$, the sequence $\{\langle f_n, g\rangle\}_{n\in\mathbb{N}}$ is a numerical Cauchy sequence), then there exists a vector $f$ in $\mathcal{H}$ such that $\{f_n\}$ converges weakly to $f$. By using (1.7) it is easy to prove the uniqueness of the limit of Cauchy sequences (see Proposition 1.3).

**Proposition 1.1.** *(a) Let $\{f_n\}_{n\in\mathbb{N}}$ be a sequence of vectors in $\mathcal{H}$. Then*

$$\underset{n\to\infty}{\text{s-lim}}\, f_n = f \quad \Longleftrightarrow \quad \underset{n\to\infty}{\text{w-lim}}\, f_n = f \quad \textit{and} \quad \lim_{n\to\infty} \|f_n\| = \|f\|.$$

*(b) If* $\text{s-lim}_{n\to\infty} f_n = f$ *and* $\text{s-lim}_{n\to\infty} g_n = g$, *then* $\lim_{n\to\infty}\langle f_n, g_n\rangle = \langle f, g\rangle$.

PROOF. (a) Assume that $\text{s-lim}_{n\to\infty} f_n = f$. By using (1.12) one obtains that

$$|\langle f_n, g\rangle - \langle f, g\rangle| = |\langle f_n - f, g\rangle| \leq \|f_n - f\|\|g\| \to 0 \quad \text{as } n \to \infty,$$

so $\text{w-lim}_{n\to\infty} f_n = f$. On the other hand, by using (1.15) one obtains $\big|\|f_n\| - \|f\|\big| \leq \|f_n - f\| \to 0$ as $n \to \infty$, hence $\lim_{n\to\infty}\|f_n\| = \|f\|$.

To obtain the implication $\Leftarrow$ we observe that $\|f_n - f\|^2 = \|f_n\|^2 + \|f\|^2 - \langle f_n, f\rangle - \langle f, f_n\rangle$. If $\text{w-lim}_{n\to\infty} f_n = f$ and $\lim_{n\to\infty}\|f_n\| = \|f\|$, the expression on the right-hand side converges to $\|f\|^2 + \|f\|^2 - \|f\|^2 - \|f\|^2 = 0$, so that $\text{s-lim}_{n\to\infty} f_n = f$.

(b) This follows from (1.12) and the result of (a):

$$\begin{aligned}|\langle f_n, g_n\rangle - \langle f, g\rangle| &= |\langle f_n, g_n - g\rangle + \langle f_n - f, g\rangle| \leq |\langle f_n, g_n - g\rangle| + |\langle f_n - f, g\rangle| \\ &\leq \|f_n\|\|g_n - g\| + \|f_n - f\|\|g\| \to \|f\| \cdot 0 + 0 \cdot \|g\| = 0. \qquad \square\end{aligned}$$

**Example 1.2.** Let $\{e_n\}_{n\in\mathbb{N}}$ be an infinite orthonormal sequence in a Hilbert space $\mathcal{H}$ of infinite dimension, i.e. satisfying $\langle e_j, e_k\rangle = \delta_{jk}$. Then $\text{w-lim}_{n\to\infty} e_n = \mathbf{0}$, because for any vector $g$ in $\mathcal{H}$:

$$\|g\|^2 \geq \sum_{n=1}^{\infty} |\langle e_n, g\rangle|^2, \tag{1.18}$$

consequently $\lim_{n\to\infty} \langle e_n, g\rangle = 0$ [the sequence $\{e_n\}$ can be considered to be part of an orthonormal basis of $\mathcal{H}$, and for such a basis one has equality in (1.18) as explained after Eq. (1.10)]. Thus:

*Every infinite orthonormal sequence $\{e_n\}$ converges weakly to the zero vector* $\mathbf{0}$.

On the other hand such a sequence cannot be strongly convergent (if one assumes that $\text{s-}\lim_{n\to\infty} e_n = f$, then $\text{w-}\lim_{n\to\infty} e_n = f$ by Proposition 1.1(a), and by the uniqueness of a weak limit, one must then have $f = \mathbf{0}$, i.e. $\text{s-}\lim_{n\to\infty} e_n = \mathbf{0}$; this is impossible since $\|e_n - \mathbf{0}\| = \|e_n\| = 1$ for each $n$). This example shows in particular that in an infinite-dimensional Hilbert space the notion of weak convergence is really weaker than that of strong convergence.

**Proposition 1.3.** *The limit of a strong or weak Cauchy sequence is unique.*

PROOF. If $\{f_n\}$ is a strong Cauchy sequence, it is also weakly Cauchy [see Proposition 1.1(a)]; thus it is enough to consider weak convergence. So let us assume that $\text{w-}\lim_{n\to\infty} f_n = h_1$ and $\text{w-}\lim_{n\to\infty} f_n = h_2$. Then one has for each $g \in \mathcal{H}$:

$$\begin{aligned}\langle g, h_1 - h_2\rangle = \langle g, h_1\rangle - \langle g, h_2\rangle &= \lim_{n\to\infty} \langle g, f_n\rangle - \lim_{n\to\infty} \langle g, f_n\rangle \\ &= \lim_{n\to\infty} \left[\langle g, f_n\rangle - \langle g, f_n\rangle\right] = 0.\end{aligned}$$

Upon taking $g = h_1 - h_2$, one obtains that $\|h_1 - h_2\|^2 = 0$. By (1.7) one then has $h_1 - h_2 = \mathbf{0}$, i.e. $h_1 = h_2$. □

REMARK. If $\{f_n\}_{n\in\mathbb{N}}$ is a strong Cauchy sequence, then this sequence is bounded in $\mathcal{H}$, i.e. $\sup_{n\in\mathbb{N}}\|f_n\| < \infty$. This follows immediately from Proposition 1.1(a) (because $\|f_n\| \to \|f\|$). The same conclusion is true if one assumes only that $\{f_n\}$ is weakly Cauchy, but the proof is longer. This fact is often useful in the theory of Hilbert spaces, and we shall see in Chapter 2 other results of this type. All these results can be deduced from a general theorem (the Uniform Boundedness Principle) which we state below for completeness (the proof can be found in most textbooks, see for example [K]).

**Uniform Boundedness Principle**: Let $\Lambda$ be a set and, for each element $\lambda$ of $\Lambda$, let $\varphi_\lambda : \mathcal{H} \to [0, \infty)$ be a continuous mapping [7] such that $\varphi_\lambda(f + g) \leq \varphi_\lambda(f) + \varphi_\lambda(g)$ for all $f, g \in \mathcal{H}$. If for each fixed vector g the family $\{\varphi_\lambda(g)\}_{\lambda\in\Lambda}$ is bounded (i.e. $\varphi_\lambda(g) \leq M \;\forall\lambda \in \Lambda$, where $M < \infty$ is a constant depending only on $g$), then the family $\{\varphi_\lambda\}$ is uniformly bounded on the unit ball of $\mathcal{H}$, in other terms there exists a constant $c < \infty$ such that $\varphi_\lambda(h) \leq c$ for all $h \in \mathcal{H}$ satisfying $\|h\| \leq 1$ and all $\lambda \in \Lambda$.

To deduce for example the boundedness of a weakly convergent sequence $\{f_n\}_{n\in\mathbb{N}}$, take $\Lambda = \mathbb{N}$ and $\varphi_\lambda(g) \equiv \varphi_n(g) = |\langle f_n, g\rangle|$. For each $g \in \mathcal{H}$, the numerical sequence $\{|\langle f_n, g\rangle|\}_{n\in\mathbb{N}}$ is bounded because $\langle f_n, g\rangle$ converges to a finite limit; then one uses the fact that $\|f_n\| = \sup_{h\in\mathcal{H},\|h\|=1} |\langle f_n, h\rangle|$ [see Eq. (2.6)] to arrive at the inequality $\|f_n\| = \sup_{h\in\mathcal{H},\|h\|=1} \varphi_n(h) \leq c$ for all $n$.

---

[7] i.e. $\varphi_\lambda(f_n) \to \varphi_\lambda(f)$ when $f_n \to f$ strongly.

## 1.2 Vector-valued functions

Let $\Lambda$ be a set and $\mathcal{H}$ a Hilbert space. A ***vector-valued function*** on $\Lambda$ is a mapping $f\colon \Lambda \to \mathcal{H}$. It associates a vector in $\mathcal{H}$ with each element of $\Lambda$. Thus a sequence of vectors $\{f_n\}_{n\in\mathbb{N}}$ can be viewed as a function $f\colon \mathbb{N} \to \mathcal{H}$ by setting $f(n) = f_n$ for $n \in \mathbb{N}$. The most important case in our context is that where $\Lambda$ is a (finite or infinite) interval $J$; we write $f(s)$ for the value of a function $f\colon J \to \mathcal{H}$ at the point $s \in J$. So for each $s \in J$, $f(s)$ is a vector in $\mathcal{H}$. If $f(s)$ depends continuously on $s$, the family $\{f(s)\}_{s\in J}$ describes a curve in $\mathcal{H}$.

As for numerical functions, one can consider the notions of continuity or differentiability and define integrals of such functions. The derivative at a point of $J$ and the integral of a vector-valued function will be vectors in $\mathcal{H}$. Since derivatives and integrals are limits, these notions may be defined in the strong sense or in the weak sense (for example a function $f\colon J \to \mathcal{H}$ may be weakly differentiable but not strongly differentiable). Similarly one can define integrals in the sense of Riemann or in a more general sense, and one can integrate with respect to different measures on $J$ (the concept of a measure will be discussed in Section 1.4). We restrict ourselves here to definitions in the strong topology of $\mathcal{H}$, and we consider only integrals in the sense of Riemann (and with respect to Lebesgue measure).

A function $f\colon J \to \mathcal{H}$ is ***strongly continuous*** if one has for each $t \in J$:

$$\lim_{s\to t, s\in J} \|f(s) - f(t)\| = 0. \tag{1.19}$$

A vector-valued function $f$ on an open interval $J$ is ***strongly differentiable*** at the point $s \in J$ if there is a vector $g$ in $\mathcal{H}$ such that:

$$\lim_{\tau\to 0} \left\| \frac{1}{\tau}\left[f(s+\tau) - f(s)\right] - g \right\| = 0. \tag{1.20}$$

We shall write $g = f'(s)$. More generally, $f$ is ***strongly differentiable in J*** if $f$ is differentiable at each point $s \in J$, i.e. if there exists a function $f'\colon J \to \mathcal{H}$ such that

$$\lim_{\tau\to 0} \left\| \frac{1}{\tau}\left[f(s+\tau) - f(s)\right] - f'(s) \right\| = 0 \qquad \forall s \in J. \tag{1.21}$$

The function $f'$ is called the (strong) ***derivative*** of $f$, and one also writes

$$f'(s) = \frac{d}{ds} f(s) = \text{s-}\lim_{\tau\to 0} \tau^{-1}\big[f(s+\tau) - f(s)\big]. \tag{1.22}$$

To define the Riemann integral of a vector-valued function, one proceeds in analogy with the construction for numerical functions. Let $J = (a, b]$ be a finite interval. A collection of real numbers $\Pi = \{s_0, s_1, \ldots, s_N; u_1, \ldots, u_N\}$, with $a = s_0 < s_1 < \cdots < s_N = b$ and with $u_k \in (s_{k-1}, s_k]$, will be called a ***partition*** of $J$ (Figure 1.1). We set $|\Pi| = \max_{k=1,\ldots,N} |s_k - s_{k-1}|$.

If $f\colon J \to \mathcal{H}$, let

$$\Sigma(\Pi, f) = \sum_{k=1}^{N} (s_k - s_{k-1}) f(u_k). \tag{1.23}$$

This is a finite linear combination of vectors in $\mathcal{H}$, hence it defines a vector $\Sigma(\Pi, f)$ in $\mathcal{H}$. One chooses a sequence $\{\Pi_r\}_{r\in\mathbb{N}}$ of partitions of $J$ such that $\lim_{r\to\infty}|\Pi_r|=0$ and defines

$$\int_J f(s)\,ds \equiv \int_a^b f(s)\,ds = \underset{r\to\infty}{\text{s-lim}}\,\Sigma(\Pi_r, f) \tag{1.24}$$

if the limit exists and is independent of the sequence $\{\Pi_r\}$.

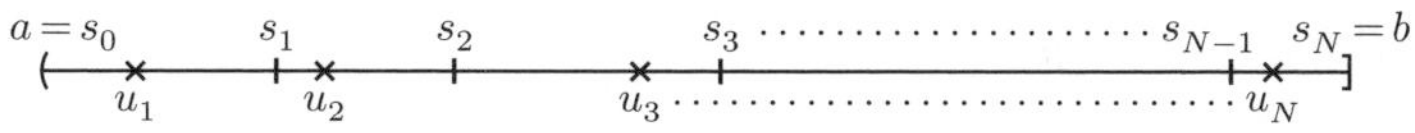


**Figure 1.1** Partition of an interval $(a, b]$.

The integral of a vector-valued function on a finite open interval $(a, b)$ or on a finite closed interval $[a, b]$ is defined similarly by suitably adjusting the definition of a partition of $J$: if $J = (a, b)$, then one must assume that $u_N \neq b$, and if $J = [a, b]$, then $u_1 = a$ is also admissible.

If $J$ is an infinite interval, one first defines the integral for each finite subinterval and then takes a sequence of finite subintervals converging to $J$ in an appropriate way. For example, if $J = (a, \infty)$ for some finite $a$, one defines (if the limit exists)

$$\int_a^\infty f(s)\,ds = \underset{b\to\infty}{\text{s-lim}} \int_a^b f(s)\,ds. \tag{1.25}$$

We state below some simple properties of these integrals and sufficient conditions for their existence. The proofs are analogous to those for numerical (real or complex) Riemann integrals; it suffices to replace in the latter proofs the absolute value $|f(s)|$ by the norm $\|f(s)\|$ (details can be found for example in Section 4.4 of [AJS]).

**Proposition 1.4.** *Let $(a, b]$ and $(b, c]$ be finite or infinite intervals and suppose that all integrals below exist. Then*

$$\int_a^b f(s)\,ds + \int_b^c f(s)\,ds = \int_a^c f(s)\,ds, \tag{1.26}$$

$$\int_a^b [\alpha f_1(s) + f_2(s)]\,ds = \alpha \int_a^b f_1(s)\,ds + \int_a^b f_2(s)\,ds, \tag{1.27}$$

$$\left\| \int_a^b f(s)\,ds \right\| \leq \int_a^b \|f(s)\|\,ds. \tag{1.28}$$

PROOF. (i) Assume that $a$ and $b$ are finite. Then (1.28) is obtained by using Proposition 1.1(a), the triangle inequality (1.13) and the definition of the numerical Riemann integral:

$$\left\| \int_a^b f(s)\,ds \right\| = \lim_{|\Pi_r|\to 0} \|\Sigma(\Pi_r, f)\| \leq \lim_{|\Pi_r|\to 0} \sum_{k=1}^{N_r} \|f(u_k)\|\,|s_k - s_{k-1}| = \int_a^b \|f(s)\|\,ds.$$

(ii) If for example $a$ is finite and $b = \infty$, one finds from Proposition 1.1 and (i) above that

$$\left\| \int_a^\infty f(s)ds \right\| = \lim_{b\to\infty} \left\| \int_a^b f(s)ds \right\| \leq \lim_{b\to\infty} \int_a^b \|f(s)\| ds = \int_a^\infty \|f(s)\| ds. \quad \square$$

**Proposition 1.5.** *(a) If $[a,b]$ is a finite closed interval and $f\colon [a,b] \to \mathcal{H}$ is strongly continuous, then $\int_a^b f(s)ds$ exists.*
*(b) If $a$ and $b$ are arbitrary ($a < b$), $f\colon (a,b) \to \mathcal{H}$ is strongly continuous and $\int_a^b \|f(s)\| ds < \infty$, then $\int_a^b f(s)ds$ exists.*
*(c) If $f$ is strongly differentiable on $(a,b)$ and its derivative $f'$ is strongly continuous and integrable on $[a,b]$, then*

$$\int_a^b f'(s)ds = f(b) - f(a). \tag{1.29}$$

Notice that, if $f$ is strongly continuous, then $\|f(s)\|$ depends continuously on $s$ by Proposition 1.1(a), so that $\int_a^b \|f(s)\| ds$ can be defined as a numerical Riemann integral.

## 1.3 Subsets and dual of a Hilbert space

**1.3.1.** We consider here certain types of subsets of a Hilbert space $\mathcal{H}$. A subset $\mathcal{D}$ of $\mathcal{H}$ is ***dense*** in $\mathcal{H}$ if, given any vector $f \in \mathcal{H}$ and $\varepsilon > 0$, there exists an element $g$ of $\mathcal{D}$ such that $\|f - g\| < \varepsilon$. Equivalently, $\mathcal{D}$ is dense in $\mathcal{H}$ if, given any vector $f \in \mathcal{H}$, there is a sequence $\{f_n\}_{n\in\mathbb{N}}$ in $\mathcal{D}$ such that $\lim_{n\to\infty} \|f - f_n\| = 0$. An example of a dense set is given by the collection of all finite linear combinations of vectors of an orthonormal basis $\{e_n\}$ of $\mathcal{H}$. In this case $\mathcal{D}$ is the set of all vectors of the form $\sum_{k=1}^N \alpha_k e_k$, with $\alpha_k \in \mathbb{C}$ and $N = 1, 2, 3, \ldots$ ($N < \infty$); if one assumes in addition that the real part and the imaginary part of each $\alpha_k$ are rational numbers, one obtains a dense subset of $\mathcal{H}$ that is *countable*.

A ***linear manifold*** $\mathcal{E}$ is a linear subset of $\mathcal{H}$, i.e. such that $f, g \in \mathcal{E}$ and $\alpha \in \mathbb{C} \Longrightarrow f + \alpha g \in \mathcal{E}$ (in other words each finite linear combination of elements of $\mathcal{E}$ also belongs to $\mathcal{E}$). An example is the already indicated set of all finite linear combinations of vectors of an orthonormal basis $\{e_n\}$ of $\mathcal{H}$; more generally, if $\mathcal{N}$ is an arbitrary subset of $\mathcal{H}$, the set of all finite linear combinations of vectors belonging to $\mathcal{N}$ is a linear manifold, called the ***linear manifold spanned by $\mathcal{N}$***. If the linear manifold spanned by a subset $\mathcal{N}$ is dense in $\mathcal{H}$, one says that $\mathcal{N}$ is a ***total set*** in $\mathcal{H}$ or a ***total subset of $\mathcal{H}$***. An example of a non-total subset is given by the vectors $e_2, e_4, e_6, \ldots$ of an orthonormal basis $\{e_n\}$.

A linear manifold $\mathcal{E}$ in $\mathcal{H}$ is clearly a complex linear vector space equipped with a scalar product (induced by the scalar product in $\mathcal{H}$, i.e. if $f, g \in \mathcal{E}$, then their scalar product $\langle f, g\rangle_{\mathcal{E}}$ in $\mathcal{E}$ is simply their scalar product $\langle f, g\rangle$ in $\mathcal{H}$). In general $\mathcal{E}$ will not be complete in the norm $\|\cdot\|$, i.e. the limit of a Cauchy sequence $\{f_n\}$, with $f_n \in \mathcal{E}$, will not belong to $\mathcal{E}$ but only to $\mathcal{H}$. If $\mathcal{E}$ is complete [i.e. if $\mathcal{E}$ satisfies *all* of the postulates (H1) - (H4)], then $\mathcal{E}$ will be called a ***subspace*** of $\mathcal{H}$. In what follows we use the symbol $\mathcal{M}$ for subspaces.

A linear manifold $\mathcal{E}$ determines uniquely a subspace $\overline{\mathcal{E}}$, called the ***closure*** of $\mathcal{E}$: a vector $f$ in $\mathcal{H}$ belongs to $\overline{\mathcal{E}}$ if and only if there exists a sequence $\{f_n\}_{n\in\mathbb{N}}$ in $\mathcal{E}$ that converges strongly to $f$ as $n \to \infty$. If $\mathcal{E}$ is dense in $\mathcal{H}$, then $\overline{\mathcal{E}} = \mathcal{H}$, otherwise $\overline{\mathcal{E}}$ is a subspace of $\mathcal{H}$ that is strictly smaller than $\mathcal{H}$.

More generally, an arbitrary subset $\mathcal{N}$ of $\mathcal{H}$ defines uniquely a subspace $\mathcal{M}_{\mathcal{N}}$, called the ***subspace spanned by*** $\boldsymbol{\mathcal{N}}$: $\mathcal{M}_{\mathcal{N}}$ is the smallest subspace containing $\mathcal{N}$, in other terms the closure $\overline{\mathcal{E}}$ of the linear manifold $\mathcal{E}$ spanned by $\mathcal{N}$.

EXAMPLES OF SUBSPACES
(a) A subspace of dimension 0 is $\{\mathbf{0}\}$, consisting of only the zero vector.
(b) Each vector $f \neq \mathbf{0}$ determines a one-dimensional subspace, the set $\{\alpha f \mid \alpha \in \mathbb{C}\}$, i.e. all multiples of the vector $f$.
(c) A subspace of infinite dimension of an infinite-dimensional Hilbert space $\mathcal{H}$ is obtained by taking all (finite and infinite) linear combinations belonging to $\mathcal{H}$ of the vectors $e_2, e_4, e_6, \ldots$ of an orthonormal basis $\{e_n\}_{n\in\mathbb{N}}$ of $\mathcal{H}$.

REMARK. The terminology concerning subspaces is not uniform; some authors say "subspace" for a linear manifold (they have in mind a subspace of the *vector space* $\mathcal{H}$) and "closed subspace" for what we call a subspace.

The scalar product leads to the notion of orthogonality in $\mathcal{H}$. Two vectors $f$ and $g$ of $\mathcal{H}$ are said to be ***orthogonal*** if $\langle f, g\rangle = 0$, and one then writes $f \perp g$. A sequence $\{f_1, f_2, f_3, \ldots\}$ of vectors of $\mathcal{H}$ is called ***orthonormal*** if $\langle f_j, f_k\rangle = \delta_{jk}$.

If $\mathcal{N}$ is a subset of $\mathcal{H}$, one defines its ***orthogonal complement*** $\mathcal{N}^\perp$ as the set of all vectors in $\mathcal{H}$ that are orthogonal to each vector in $\mathcal{N}$:

$$\mathcal{N}^\perp = \{f \in \mathcal{H} \mid \langle f, g\rangle = 0 \ \ \forall g \in \mathcal{N}\}. \tag{1.30}$$

We write $f \perp \mathcal{N}$ if $f \in \mathcal{N}^\perp$. The set $\mathcal{N}^\perp$ is a subspace of $\mathcal{H}$ (Problem 1.7). If $\mathcal{M}_{\mathcal{N}}$ is the subspace spanned by $\mathcal{N}$, one has

$$f \perp \mathcal{N} \iff f \perp \mathcal{M}_{\mathcal{N}}. \tag{1.31}$$

Indeed: (i) the implication $\Leftarrow$ is trivial. (ii) For the implication $\Rightarrow$ we first observe that $f \perp \mathcal{N}$ implies $f \perp \mathcal{E}$, where $\mathcal{E}$ is the linear manifold spanned by $\mathcal{N}$ (because each $g \in \mathcal{E}$ is of the form $g = \sum_{k=1}^{N} \alpha_k g_k$ with $g_k \in \mathcal{N}$ for some $N < \infty$, hence $\langle f, g\rangle = \sum_{k=1}^{N} \alpha_k \langle f, g_k\rangle = 0$). Finally $f \perp \mathcal{E}$ implies that $f \perp \overline{\mathcal{E}} \equiv \mathcal{M}_{\mathcal{N}}$ (because each element $g$ of the subspace $\overline{\mathcal{E}}$ is given as $g = \text{s-lim}_{n\to\infty} g_n$ with $g_n \in \mathcal{E}$, hence $\langle f, g\rangle = \lim_{n\to\infty} \langle f, g_n\rangle = \lim_{n\to\infty} 0 = 0$).

**Proposition 1.6.** *A subset $\mathcal{N}$ of $\mathcal{H}$ is total in $\mathcal{H}$ if and only if $f \perp \mathcal{N}$ implies that $f = \mathbf{0}$.*

PROOF. If $\mathcal{N}$ is total in $\mathcal{H}$ and $f \perp \mathcal{N}$, then $f$ is orthogonal to $\mathcal{H}$ by (1.31). In particular $f$ is then orthogonal to itself, i.e. $\langle f, f\rangle = \|f\|^2 = 0$, hence $f = \mathbf{0}$ by (1.7).

Conversely, suppose that $f \perp \mathcal{N}$ implies that $f = \mathbf{0}$. By virtue of (1.31) this means that the orthogonal complement of the subspace $\mathcal{M}_{\mathcal{N}}$ spanned by $\mathcal{N}$ is $\{\mathbf{0}\}$. Hence $\mathcal{M}_{\mathcal{N}} = \{\mathbf{0}\}^{\perp} = \mathcal{H}$ (here one applies the Projection Theorem given below). □

The next proposition generalises a well-known property of finite-dimensional vector spaces equipped with a scalar product. This is a fundamental result in the theory of Hilbert spaces, and it also holds in non-separable Hilbert spaces [the postulate (H4) is not used in its proof].

**Proposition 1.7 (Projection Theorem).** *Let $\mathcal{M}$ be a subspace of $\mathcal{H}$, $\mathcal{M}^{\perp}$ its orthogonal complement. Then each vector $f$ of $\mathcal{H}$ admits a unique decomposition into a component in $\mathcal{M}$ and a component in $\mathcal{M}^{\perp}$:*

$$f = f_1 + f_2 \qquad \text{with } f_1 \in \mathcal{M},\, f_2 \in \mathcal{M}^{\perp}. \tag{1.32}$$

PROOF. We fix a vector $f \in \mathcal{H}$, and we denote by $d$ the distance from $f$ to $\mathcal{M}$, i.e. $d = \inf_{g\in\mathcal{M}} \|f - g\|$. We shall use the following identity: if $h_1$ and $h_2$ are vectors in $\mathcal{H}$ then, by Eq. (1.17) with $f = h_1$ and $g = \mp h_2$:

$$\|h_1 - h_2\|^2 + \|h_1 + h_2\|^2 = 2\|h_1\|^2 + 2\|h_2\|^2. \tag{1.33}$$

(i) We first show that there is a unique vector $g_0 \in \mathcal{M}$ such that $d = \|f - g_0\|$.
Existence. Choose a sequence $\{g_k\}_{k\in\mathbb{N}}$ in $\mathcal{M}$ such that $\lim_{k\to\infty} \|f - g_k\| = d$. Let us take $h_1 = (f - g_j)/2$ and $h_2 = (f - g_k)/2$ in (1.33). We obtain

$$\Big\|\frac{1}{2}(g_j - g_k)\Big\|^2 = \frac{1}{2}\big\|f - g_j\big\|^2 + \frac{1}{2}\big\|f - g_k\big\|^2 - \Big\|f - \frac{1}{2}(g_j + g_k)\Big\|^2.$$

Since $g_j + g_k \in \mathcal{M}$, we have $\|f - (g_j + g_k)/2\|^2 \geq d^2$. Hence

$$\frac{1}{4}\big\|g_j - g_k\big\|^2 \leq \frac{1}{2}\big\|f - g_j\big\|^2 + \frac{1}{2}\big\|f - g_k\big\|^2 - d^2 \to 0 \quad \text{as } j, k \to \infty.$$

Thus $\{g_k\}$ is a Cauchy sequence in $\mathcal{H}$. We denote its limit by $g_0$. Since $\mathcal{M}$ is a subspace, we have $g_0 \in \mathcal{M}$. Consequently $\|f - g_0\| \geq d$. On the other hand

$$\|f - g_0\| \leq \|f - g_k\| + \|g_k - g_0\| \to d + 0 = d \quad \text{as } k \to \infty.$$

We have thus shown that $\|f - g_0\| = d$.
Uniqueness. Let us assume that there exist two vectors $g_0^{(1)}$ and $g_0^{(2)}$ in $\mathcal{M}$ such that $\|f - g_0^{(1)}\| = \|f - g_0^{(2)}\| = d$. By taking $h_1 = (f - g_0^{(1)})/2$ and $h_2 = (f - g_0^{(2)})/2$ in (1.33) one obtains as in (i) that

$$\frac{1}{4}\|g_0^{(1)} - g_0^{(2)}\|^2 \leq \frac{1}{2}\|f - g_0^{(1)}\|^2 + \frac{1}{2}\|f - g_0^{(2)}\|^2 - d^2 = 0.$$

Thus $\|g_0^{(1)} - g_0^{(2)}\| = 0$, i.e. $g_0^{(1)} = g_0^{(2)}$.

(ii) To prove (1.32) we set $f_1 = g_0$, where $g_0$ is the unique vector in $\mathcal{M}$ associated with $f$ according to (i). Then $f_2 = f - g_0$, and we have to show that $f_2 \in \mathcal{M}^\perp$, in other terms that $\langle h, f_2 \rangle = 0$ for each $h \in \mathcal{M}$. So let $h \neq \mathbf{0}$ be a fixed vector in $\mathcal{M}$ and set $a = \langle h, f_2 \rangle$. Let $g = f_1 + ah/\|h\|^2$. Clearly $g \in \mathcal{M}$, hence $\|f - g\| \geq d$ and consequently

$$\begin{aligned} d^2 \leq \|f-g\|^2 &= \langle f_2 - \frac{a}{\|h\|^2} h, f_2 - \frac{a}{\|h\|^2} h \rangle \\ &= \|f_2\|^2 - \frac{\overline{a}}{\|h\|^2}\langle h, f_2\rangle - \frac{a}{\|h\|^2}\langle f_2, h\rangle + \frac{|a|^2}{\|h\|^4}\|h\|^2 \\ &= \|f_2\|^2 - \frac{|a|^2}{\|h\|^2} - \frac{|a|^2}{\|h\|^2} + \frac{|a|^2}{\|h\|^2} = \|f_2\|^2 - \frac{|a|^2}{\|h\|^2} = d^2 - \frac{|a|^2}{\|h\|^2}. \end{aligned}$$

This implies that $|a|^2 \leq 0$, hence that $a = 0$.

(iii) Finally let us check the uniqueness of the decomposition (1.32). Assume that $g_1, g_2 \in \mathcal{M}$ and $h_1, h_2 \in \mathcal{M}^\perp$ are such that $g_1 + h_1 = g_2 + h_2$. We must show that $g_1 = g_2$ and $h_1 = h_2$. Now

$$\begin{aligned} 0 = \|\mathbf{0}\|^2 &= \|g_1 + h_1 - (g_2 + h_2)\|^2 \\ &= \|(g_1 - g_2) + (h_1 - h_2)\|^2 = \|g_1 - g_2\|^2 + \|h_1 - h_2\|^2, \end{aligned}$$

because $\langle g_1 - g_2, h_1 - h_2 \rangle = 0$. Thus $\|g_1 - g_2\| = \|h_1 - h_2\| = 0$, i.e. $g_1 = g_2$ and $h_1 = h_2$ by (1.7). □

The projection theorem shows that $\mathcal{H}$ may be viewed as the orthogonal (direct) sum of two mutually orthogonal subspaces $\mathcal{M}$ and $\mathcal{M}^\perp$. We recall that the ***orthogonal sum*** $\mathcal{H} \equiv \mathcal{H}_1 \oplus \mathcal{H}_2 \oplus \cdots \oplus \mathcal{H}_N$ of $N$ Hilbert spaces $\mathcal{H}_1, \ldots, \mathcal{H}_N$ is the Hilbert space formed by all $N$-tuples $\{f_1, \ldots, f_N\}$, where $f_k \in \mathcal{H}_k$, with the following scalar product:

$$\big\langle \{f_1, \ldots, f_N\}, \{g_1, \ldots, g_N\} \big\rangle_{\mathcal{H}} = \langle f_1, g_1 \rangle_{\mathcal{H}_1} + \cdots + \langle f_N, g_N \rangle_{\mathcal{H}_N}. \tag{1.34}$$

Each $\mathcal{H}_k$ is a subspace of $\mathcal{H}$, and if $j \neq k$ then the spaces $\mathcal{H}_j$ and $\mathcal{H}_k$, considered as subspaces of $\mathcal{H}$, are mutually orthogonal.

Occasionally we shall also use orthogonal sums of an infinite number of Hilbert spaces. The orthogonal sum $\mathcal{H} = \bigoplus_{k=1}^{\infty} \mathcal{H}_k$ of an infinite sequence $\{\mathcal{H}_1, \mathcal{H}_2, \ldots\}$ of Hilbert spaces is formed by all infinite sequences $\{f_1, f_2, \ldots\}$, with $f_k \in \mathcal{H}_k$ and such that $\sum_{k=1}^{\infty} \|f_k\|^2_{\mathcal{H}_k} < \infty$, the scalar product in $\mathcal{H}$ being defined as in (1.34) but with $N = \infty$. Again each $\mathcal{H}_k$ is a subspace of $\mathcal{H}$ (the set of all sequences $\{f_1, f_2, \ldots\}$ with $f_j = \mathbf{0}$ for all $j \neq k$), and the subspaces $\mathcal{H}_j$ and $\mathcal{H}_k$ are mutually orthogonal if $j \neq k$.

As an example, suppose that each $\mathcal{H}_k$ is a one-dimensional space, i.e. $\mathcal{H}_k = \mathbb{C}$ for each $k$. If $N$ is finite, the orthogonal sum $\mathcal{H} = \mathcal{H}_1 \oplus \mathcal{H}_2 \oplus \cdots \oplus \mathcal{H}_N$ is (isomorphic to) the space $\mathbb{C}^N$. If $N = \infty$ one obtains the space denoted by $\ell^2$, the Hilbert space of all infinite sequences $\{\alpha_1, \alpha_2, \ldots\}$ of complex numbers satisfying $\sum_{k=1}^{\infty} |\alpha_k|^2 < \infty$, with scalar product

$$\big\langle \{\alpha_1, \alpha_2, \ldots\}, \{\beta_1, \beta_2, \ldots\} \big\rangle = \sum_{k=1}^{\infty} \overline{\alpha_k} \beta_k. \tag{1.35}$$

Each (separable) infinite-dimensional Hilbert space $\mathcal{H}$ is isomorphic to $\ell^2$. Indeed $\mathcal{H}$ may be identified in a natural way with $\ell^2$ by choosing in $\mathcal{H}$ an orthonormal basis $\{e_n\}$ and by identifying the vector $f = \sum_k \alpha_k e_k$ (see Section 1.1) with the element of $\ell^2$ given by the sequence $\{\alpha_1, \alpha_2, \ldots\}$. In the terminology that will be introduced in §2.2.2, the correspondence $\mathcal{H} \ni f \mapsto \{\alpha_1, \alpha_2, \ldots\}$ defines a unitary operator from $\mathcal{H}$ to $\ell^2$.

**1.3.2.** We end this section with a characterisation of the dual of a Hilbert space. The ***dual*** $\mathcal{H}^*$ of $\mathcal{H}$ is by definition the set of all bounded linear functionals on $\mathcal{H}$, i.e. the set of all mappings $\varphi\colon \mathcal{H} \to \mathbb{C}$ having the following two properties:

$$\varphi(f + \alpha g) = \varphi(f) + \alpha\varphi(g) \qquad \text{(linearity)}$$
$$|\varphi(f)| \leq c\|f\| \qquad \forall f \in \mathcal{H}, \qquad \text{(boundedness)}$$

where $c$ is a constant (independent of $f$). The space $\mathcal{H}^*$ is a complex linear vector space (for example the sum of two bounded linear functionals is again a bounded linear functional), and $\mathcal{H}^*$ is also normed with the norm

$$\|\varphi\| \equiv \|\varphi\|_{\mathcal{H}^*} = \sup_{0 \neq f \in \mathcal{H}} \frac{|\varphi(f)|}{\|f\|}. \tag{1.36}$$

Each vector $g$ of $\mathcal{H}$ determines an element $\varphi_g$ of $\mathcal{H}^*$ by setting

$$\varphi_g(f) = \langle g, f \rangle \qquad \text{(where } f \text{ varies over } \mathcal{H}). \tag{1.37}$$

Indeed, the scalar product (for fixed $g$) is linear in $f$, and [use (1.12)]

$$\|\varphi_g\|_{\mathcal{H}^*} = \sup_{0 \neq f \in \mathcal{H}} \frac{1}{\|f\|} |\langle g, f \rangle| \leq \sup_{0 \neq f \in \mathcal{H}} \frac{\|g\|\|f\|}{\|f\|} = \|g\|.$$

In fact one has

$$\|\varphi_g\|_{\mathcal{H}^*} = \|g\|, \tag{1.38}$$

because the supremum is attained for $f = g$: $\langle g, g \rangle / \|g\| = \|g\|^2/\|g\| = \|g\|$.

We have thus shown that each vector of $\mathcal{H}$ determines an element of the dual space $\mathcal{H}^*$. It is an important fact that the converse is also true (so that one may identify $\mathcal{H}^*$ and $\mathcal{H}$):

**Proposition 1.8 (Riesz Lemma).** *Let $\mathcal{H}$ be a Hilbert space and $\varphi$ a bounded linear functional on $\mathcal{H}$. Then there exists a unique vector $g$ in $\mathcal{H}$ such that*

$$\varphi(f) = \langle g, f \rangle \qquad \forall f \in \mathcal{H}. \tag{1.39}$$

*In particular:*

$$\|\varphi\|_{\mathcal{H}^*} = \|g\|_{\mathcal{H}} \equiv \|g\|. \tag{1.40}$$

PROOF. (i) Define $\mathcal{M}$ to be the set of all vectors $f \in \mathcal{H}$ for which $\varphi(f) = 0$. Let us show that $\mathcal{M}$ is a subspace:

(1) $f_1, f_2 \in \mathcal{M}$ and $\alpha \in \mathbb{C} \Longrightarrow \varphi(f_1 + \alpha f_2) = \varphi(f_1) + \alpha\varphi(f_2) = 0 + 0 = 0$,

(2) if $\{f_n\}$ is a sequence of vectors in $\mathcal{M}$ such that $\lim_{m,n\to\infty} \|f_m - f_n\| = 0$, then the sequence $\{f_n\}$ has a limit $f$ in $\mathcal{H}$, and we must show that this vector $f$ belongs to $\mathcal{M}$ (completeness of $\mathcal{M}$), i.e. that $\varphi(f) = 0$. Now

$$|\varphi(f)| = |\varphi(f - f_n + f_n)| \le |\varphi(f - f_n)| + |\varphi(f_n)| \le \|\varphi\|_{\mathcal{H}^*} \|f - f_n\|$$

because $\varphi(f_n) = 0$. As $\|f - f_n\| \to 0$ when $n \to \infty$ and $\varphi$ is bounded, one must have $\varphi(f) = 0$.

(ii) If $\varphi \equiv 0$ we take $g = \mathbf{0}$; indeed one then has $0 = \varphi(f) = \langle \mathbf{0}, f\rangle$ for each $f \in \mathcal{H}$. If $\varphi \not\equiv 0$ there is a vector $h$ in $\mathcal{H}$ such that $\varphi(h) \neq 0$. By the Projection Theorem we can write $h = h_1 + h_2$ with $h_1 \in \mathcal{M}$ and $h_2 \in \mathcal{M}^\perp$. Then

$$\varphi(h_2) = \varphi(h - h_1) = \varphi(h) - \underbrace{\varphi(h_1)}_{0} = \varphi(h) \neq 0,$$

in particular $h_2 \neq \mathbf{0}$ (because $\varphi(\mathbf{0}) = 0$).

If $f$ is an arbitrary vector in $\mathcal{H}$, let us consider the vector $f - \big[\varphi(f)/\varphi(h_2)\big] h_2$. We then have

$$\varphi\Big(f - \frac{\varphi(f)}{\varphi(h_2)} h_2\Big) = \varphi(f) - \frac{\varphi(f)}{\varphi(h_2)} \varphi(h_2) = 0,$$

hence the vector $f - \big[\varphi(f)/\varphi(h_2)\big] h_2$ belongs to $\mathcal{M}$. Since $h_2 \perp \mathcal{M}$ we obtain

$$\Big\langle h_2, f - \frac{\varphi(f)}{\varphi(h_2)} h_2 \Big\rangle = 0,$$

i.e. $\varphi(h_2)\langle h_2, f\rangle = \varphi(f)\|h_2\|^2$, and consequently

$$\varphi(f) = \frac{\varphi(h_2)}{\|h_2\|^2}\langle h_2, f\rangle \qquad \forall f \in \mathcal{H}.$$

Upon setting $g = \big[\overline{\varphi(h_2)}/\|h_2\|^2\big] h_2$ (which defines a vector in $\mathcal{H}$ because $\|h_2\| \neq 0$), one obtains (1.39).

(iii) We finally show that $g$ is unique. If $g_1, g_2$ are two vectors such that $\varphi(f) = \langle g_1, f\rangle = \langle g_2, f\rangle$ for all $f \in \mathcal{H}$, then $\langle g_1 - g_2, f\rangle = 0$ for all $f \in \mathcal{H}$. Hence $g_1 - g_2 = \mathbf{0}$ by Proposition 1.6. □

Let us add that the identification of $\mathcal{H}^*$ with $\mathcal{H}$ is not linear but *anti-linear*: if $\varphi_1$ and $\varphi_2$ are elements of $\mathcal{H}^*$ such that $\varphi_1(f) = \langle g_1, f\rangle$ and $\varphi_2(f) = \langle g_2, f\rangle$ for all $f \in \mathcal{H}$, then

$$(\varphi_1 + \alpha\varphi_2)(f) = \varphi_1(f) + \alpha\varphi_2(f) = \langle g_1, f\rangle + \alpha\langle g_2, f\rangle = \langle g_1 + \overline{\alpha} g_2, f\rangle,$$

hence the vector in $\mathcal{H}$ representing the linear functional $\varphi_1 + \alpha\varphi_2$ is the vector $g_1 + \overline{\alpha} g_2$.

## 1.4 Measures, integrals and $L^p$ spaces

A rigourous treatment of the theory of $L^p$ spaces requires familiarity with measure theory and integration in the sense of Lebesgue. We shall here present a description of these spaces and of some aspects of measure and integration theory, without giving the proofs. The complete theory is treated in numerous books.

**1.4.1.** Consider a set $\mathcal{O}$ and a collection $\mathcal{A}$ of subsets of $\mathcal{O}$ having the structure of a ***σ-algebra***, i.e. satisfying: (1) $\mathcal{A}$ contains the empty set $\varnothing$ and $\mathcal{O}$ itself, (2) if $\mathcal{V} \in \mathcal{A}$, then its complementary set $\mathcal{O}\backslash\mathcal{V}$ also belongs to $\mathcal{A}$, (3) $\mathcal{A}$ is stable under the operation of taking countable unions in the sense of set theory: if $\mathcal{V}_k \in \mathcal{A}$ for $k = 1, 2, 3, \ldots$, then $\bigcup_k \mathcal{V}_k \in \mathcal{A}$. If these conditions are satisfied, then $\mathcal{A}$ is also stable under the operation of taking countable intersections, and the couple $(\mathcal{O}, \mathcal{A})$ is called a ***measurable space***. We use the letter $x$ for the elements of $\mathcal{O}$ (the "points"). A ***measure*** $m$ on $(\mathcal{O}, \mathcal{A})$ is a mapping from $\mathcal{A}$ to $[0, \infty]$ that is $\sigma$-additive, i.e. such that $m(\bigcup_k \mathcal{V}_k) = \sum_k m(\mathcal{V}_k)$ for each countable family $\{\mathcal{V}_k\}$ of disjoint elements of $\mathcal{A}$ (i.e. satisfying $\mathcal{V}_j \cap \mathcal{V}_k = \varnothing$ if $j \neq k$), and with $m(\varnothing) = 0$. A measurable space $(\mathcal{O}, \mathcal{A})$ together with a measure $m$ defined on it is called a ***measure space***, denoted by $(\mathcal{O}, \mathcal{A}, m)$.

We see that a measure $m$ on $(\mathcal{O}, \mathcal{A})$ assigns a "weight" (which may be infinite) to each subset $\mathcal{V}$ of $\mathcal{O}$ belonging to $\mathcal{A}$, and this assignment is additive (the weight of a finite or countable union of disjoint subsets of $\mathcal{O}$ is the sum of the weights of these subsets). A set $\mathcal{V}$ of measure zero will be called a ***null set*** with respect to $m$. If $m(\mathcal{O}) < \infty$, then $m$ is called a ***finite measure***. If $\mathcal{O}$ can be expressed as the union of a countable collection of elements of $\mathcal{A}$ each of which is of finite measure, then $m$ is called a ***σ-finite measure***. In particular every finite measure is $\sigma$-finite. An example of a $\sigma$-finite measure that is not finite is given by the Lebesgue measure on $\mathbb{R}$ (see §1.4.2). A finite measure with $m(\mathcal{O}) = 1$ is a ***probability measure***; in applications the set $\mathcal{O}$ then represents the possible outcomes of an experiment, the elements of the $\sigma$-algebra $\mathcal{A}$ correspond to the events of interest in the experiment, and the measure $m(\mathcal{V})$ is the probability that the event $\mathcal{V}$ will occur (an example in quantum mechanics will be discussed on page 168).

A function $\varphi\colon \mathcal{O} \to \mathbb{R}$ assigns a real number $\varphi(x)$ to each point $x$ of $\mathcal{O}$. Given a $\sigma$-algebra $\mathcal{A}$ of subsets of $\mathcal{O}$, such a function $\varphi$ is said to be a ***measurable function*** if, for each interval $J$ in $\mathbb{R}$, the set $\varphi^{-1}(J) := \{x \in \mathcal{O} \mid \varphi(x) \in J\}$ belongs to the $\sigma$-algebra $\mathcal{A}$. It is seen that measurability is not simply a property of the function $\varphi$ but depends very much on the $\sigma$-algebra $\mathcal{A}$. $\varphi$ may be measurable with respect to certain $\sigma$-algebras and non-measurable with respect to other $\sigma$-algebras of subsets of $\mathcal{O}$.

The important fact for integration theory in the sense of Lebesgue is that each measurable function is the limit of a sequence of simple measurable functions (one could take this characterisation as the definition of a measurable function). A ***simple function*** is a function of the form $\varphi = \sum_{k=1}^{N} \alpha_k \chi_{\mathcal{V}_k}$ with $\alpha_k \in \mathbb{R}$ and $N < \infty$, where $\chi_{\mathcal{V}_k}$ is the characteristic function of the subset $\mathcal{V}_k$ of $\mathcal{O}$; a simple function is measurable if $\mathcal{V}_k \in \mathcal{A}$ for $k = 1, \ldots, N$. We recall that the ***characteristic function*** $\chi_{\mathcal{V}}$ of a set $\mathcal{V}$ is defined as

$$\chi_{\mathcal{V}}(x) = 1 \quad \text{if } x \in \mathcal{V}, \quad \chi_{\mathcal{V}}(x) = 0 \quad \text{if } x \notin \mathcal{V}. \tag{1.41}$$

Thus, if $\varphi$ is a measurable function, there exists a sequence $\{\varphi_n\}$ of measurable simple functions such that $\varphi(x) = \lim_{n\to\infty} \varphi_n(x)$ for each $x \in \mathcal{O}$. Furthermore, if $\varphi \geq 0$, the sequence $\{\varphi_n\}$ can be chosen non-decreasing, i.e. such that $0 \leq \varphi_n(x) \leq \varphi_m(x) \leq \varphi(x)$ for each $x \in \mathcal{O}$ and $m \geq n$.

If $\varphi$ is a measurable simple function, say $\varphi = \sum_{k=1}^N \alpha_k \chi_{\mathcal{V}_k}$, then the ***integral*** of $\varphi$ with respect to the measure $m$ on $(\mathcal{O}, \mathcal{A})$ is defined to be the real number

$$\int_{\mathcal{O}} \varphi(x) m(dx) = \sum_{k=1}^{N} \alpha_k m(\mathcal{V}_k). \tag{1.42}$$

This has a sense provided that the sets $\mathcal{V}_k$ are such that $m(\mathcal{V}_k) < \infty$. If this is the case, the function $\varphi = \sum_{k=1}^N \alpha_k \chi_{\mathcal{V}_k}$ is called a ***simple m-integrable function***.

If $\varphi$ is a general measurable function (not necessarily simple), one can define its integral by approximating $\varphi$ by a sequence of simple functions $\{\varphi_n\}$. One considers first measurable functions $\varphi \geq 0$ and defines

$$\int_{\mathcal{O}} \varphi(x) m(dx) = \lim_{n\to\infty} \int_{\mathcal{O}} \varphi_n(x) m(dx), \tag{1.43}$$

where $\{\varphi_n\}$ is such that $0 \leq \varphi_n(x) \leq \varphi(x)$ and $\lim_{n\to\infty} \varphi_n(x) = \varphi(x)$ for all $x \in \mathcal{O}$. This makes sense if each of the simple approximating functions $\varphi_n$ is $m$-integrable and if the limit in (1.43) is finite. If $\varphi$ is not $\geq 0$, one applies the preceding definition to its positive and its negative part, i.e. one decomposes $\varphi$ into $\varphi = \varphi_+ - \varphi_-$ with $\varphi_+ \geq 0$ and $\varphi_- \geq 0$; the integral of $\varphi_+$ and that of $\varphi_-$ are defined as in (1.43), and $\varphi$ is called ***m-integrable*** if each of these two integrals is finite. Then one sets

$$\int_{\mathcal{O}} \varphi(x) m(dx) = \int_{\mathcal{O}} \varphi_+(x) m(dx) - \int_{\mathcal{O}} \varphi_-(x) m(dx). \tag{1.44}$$

If $\varphi : \mathcal{O} \to \mathbb{C}$ is a complex function, it is $m$-integrable if its real part and its imaginary part are $m$-integrable in the sense of the preceding definition. The integral of $\varphi$ is obtained by integrating separately its real part and its imaginary part.

The Lebesgue type integral introduced above generalises the (proper) Riemann integral on the real line. As an illustration we shall compare in §1.4.2 below the Lebesgue integral on $\mathbb{R}$ with the Riemann integral for continuous functions. For measurable functions without continuity properties, only the Lebesgue integral has a meaning.

In the situations considered in this text, the underlying set $\mathcal{O}$ will be a subset of $\mathbb{R}^n$ $(n = 1, 2, 3, \ldots)$, often a finite open interval $(a, b)$ or a semi-infinite interval $(a, \infty)$, or $\mathbb{R}^n$ itself. In each of these situations the $\sigma$-algebra $\mathcal{A}$ will be the Borel $\sigma$-algebra or the Lebesgue $\sigma$-algebra of the considered set $\mathcal{O}$. Mostly the measure $m$ will be the

Lebesgue measure $m(dx) = dx$ [$m(dx) = d^n x$ if $n \geq 2$]. In the next subsection we explain these terms in somewhat more detail.

**1.4.2.** If $\mathcal{B}$ is an arbitrary collection of subsets of a set $\mathcal{O}$, there is a smallest $\sigma$-algebra $\mathcal{A}$ containing $\mathcal{B}$ (i.e. such that, if $\mathcal{A}'$ is a $\sigma$-algebra with $\mathcal{B} \subseteq \mathcal{A}'$, then $\mathcal{A} \subseteq \mathcal{A}'$). This minimal $\sigma$-algebra $\mathcal{A}$, called ***the $\sigma$-algebra generated by $\mathcal{B}$***, is simply the intersection of all $\sigma$-algebras containing $\mathcal{B}$ (this intersection is not empty because the collection of *all* subsets of $\mathcal{O}$ is a $\sigma$-algebra). In certain situations a measure $m$ on $\mathcal{A}$ is completely determined if it is given on $\mathcal{B}$. An important example is the Borel-Lebesgue measure on the real line which we shall now describe.

So let us consider the case $\mathcal{O} = \mathbb{R}$. We take for $\mathcal{B}$ the set of all half-open intervals $(a, b]$ with $-\infty < a \leq b < +\infty$. The $\sigma$-algebra generated by the collection of all these intervals is called the ***Borel $\sigma$-algebra*** of $\mathbb{R}$; we denote it by $\mathcal{A}_{\mathrm{B}}$, and its elements will be called ***Borel sets***.[8] Let us take for the measure of an interval $(a, b]$ its length $\ell$, i.e. $m((a, b]) = \ell((a, b]) \equiv b - a$. The thus defined mapping $m\colon \mathcal{B} \to [0, \infty)$ has an extension to a measure on $(\mathbb{R}, \mathcal{A}_{\mathrm{B}})$, often called the ***Borel measure*** on $\mathbb{R}$. This measure associates with each Borel set of $\mathbb{R}$ a non-negative number (which can be infinite) in a $\sigma$-additive way, and the number associated with an interval is its length. The length of a general Borel set $\mathcal{V}$ is obtained in the following manner: one considers covers of $\mathcal{V}$ of the form $\mathcal{V} \subseteq \bigcup_k J_k$, where $\{J_k\}_{k \in \mathbb{N}}$ is a countable collection of half-open intervals [$J_k = (a_k, b_k]$, $a_k < b_k$], and one then defines

$$\ell(\mathcal{V}) = \inf \sum_k \ell(J_k) \equiv \inf \sum_k (b_k - a_k), \tag{1.45}$$

where the infimum is over all covers of $\mathcal{V}$ of the form specified above. This length function has the properties of a measure on $(\mathbb{R}, \mathcal{A}_{\mathrm{B}})$, viz. the Borel measure $m_{\mathrm{B}}$ on the real line: $m_{\mathrm{B}}(\mathcal{V}) = \ell(\mathcal{V})$ for $\mathcal{V} \in \mathcal{A}_{\mathrm{B}}$.

We mention some examples of null sets with respect to Borel measure (Borel sets of length zero). The simplest example is a single point $\{x_0\}$. This set is the intersection of all intervals $(x_0 - k^{-1}, x_0]$, $k = 1, 2, 3, \ldots$ . The length of such an interval is $m_{\mathrm{B}}\big((x_0 - k^{-1}, x_0]\big) = 1/k$; since $\{x_0\}$ is contained in each of these intervals, its measure must be less than $1/k$ for each $k = 1, 2, 3, \ldots$, hence equal to zero.[9] A second example is that of a countable set $\{x_1, x_2, x_3, \ldots\}$; as $m_{\mathrm{B}}$ is $\sigma$-additive and the measure of each $\{x_k\}$ is zero, one has $m_{\mathrm{B}}\big(\{x_1, x_2, x_3, \ldots\}\big) = \sum_{k=1}^{\infty} m_{\mathrm{B}}(\{x_k\}) = 0$. There also exist Borel null sets that are not countable, for example the ***Cantor set*** $\mathfrak{C}$ which is a subset of the interval $[0, 1]$ obtained by removing from $[0, 1]$ a countable union

[8] The same $\sigma$-algebra is obtained by starting with the set of all open intervals or with that of all closed intervals. Thus $\mathcal{A}_{\mathrm{B}}$ contains all open, closed and half-open intervals. In fact each open and each closed subset of $\mathbb{R}$ is a Borel set (this is a consequence of the fact that each open subset of $\mathbb{R}$ is the union of countably many disjoint open intervals), but there is no simple characterisation of individual Borel sets directly in terms of unions and intersections of open and closed sets. The use of intervals of the form $(a, b]$ may be justified by the fact that such intervals fit together in a nice way, and this will be crucial for the definition of the more general Stieltjes measures (see Remark 4.6).

[9] A basic property of a measure $m$ is that, if $\mathcal{V}, \mathcal{W} \in \mathcal{A}$ are such that $\mathcal{V} \subseteq \mathcal{W}$, then $m(\mathcal{V}) \leq m(\mathcal{W})$, which is a straightforward consequence of the fact that a measure is additive.

of disjoint open subintervals of total length 1. This is done recursively in the following manner (Figure 1.2): first one removes the open middle third $(1/3, 2/3)$ from the interval $[0, 1]$, next one removes the middle thirds $(1/9, 2/9)$ and $(7/9, 8/9)$ of the remaining intervals, and so on. The Cantor set is a closed Borel set, it has Lebesgue measure zero and is uncountable. The reader may find more details for example in Vol. I of [RS] or in [GO].

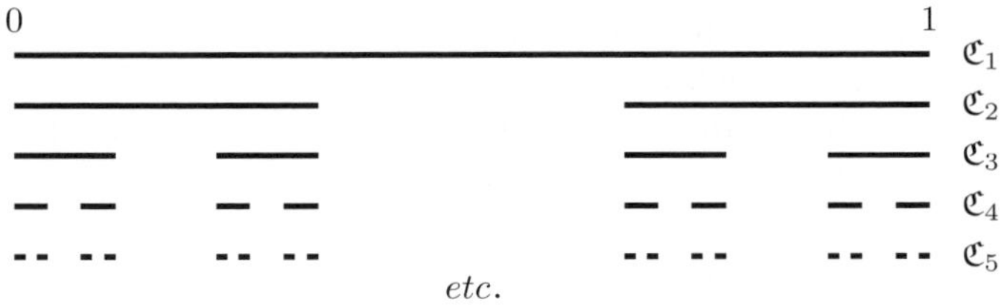


**Figure 1.2** Construction of the Cantor set.

As another illustration, let us compare the construction of the Riemann integral and of the Lebesgue integral for a continuous non-negative function $f$ on some finite interval $[a, b]$. Both are defined as limits of sequences of finite sums. The difference resides in the fact that the Riemann integral is formulated in terms of partitions of the domain $[a, b]$ of $f$, whereas the Lebesgue integral uses partitions of the range of $f$. For the Riemann integral, let $\Pi$ be a partition of $[a, b]$ as on page 6. The contribution to the Riemann sum $\Sigma(\Pi, f)$ of an individual subinterval $J$ of this partition is just $f(u')\ell(J)$, as indicated in Figure 1.3(a) [$f(u')$ is the value of $f$ at some point $u' \in J$]. The integral is of course given by the area under the graph of $f$. For the Lebesgue integral one partitions the ordinate axis into disjoint intervals $\{\Delta_\ell\}$. The contribution to the integral of a fixed one of these intervals, say of $\Delta_k$, is $f_k\ell(\mathcal{V}_k)$, where $f_k$ is some number in $\Delta_k$ and $\mathcal{V}_k = f^{-1}(\Delta_k)$ is the Borel set on the $x$-axis formed by all points $x \in [a, b]$ at which $f$ assumes a value in $\Delta_k$ [Figure 1.3(b)]. It is clear that, upon adding up the contributions from all intervals $\Delta_\ell$ and passing to the limit where their length tends to zero, the value of the integral is again equal to the area under the graph of $f$. Observe that, for a continuous function $f$, the set $\mathcal{V}_k \equiv f^{-1}(\Delta_k)$ is simply a finite union of intervals, whereas for an arbitrary measurable function on $(\mathbb{R}, \mathcal{A}_{\mathrm{B}})$ it could be any Borel subset of $\mathbb{R}$.

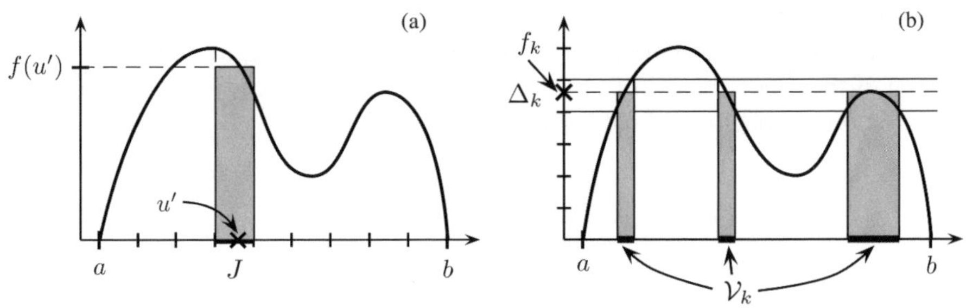


**Figure 1.3** (a) Riemann integration; (b) Lebesgue integration.

By the prescription (1.45) one can assign a length to any subset $\mathcal{V}$ of $\mathbb{R}$ (not only to Borel sets). It seems that, by using this length function on the $\sigma$-algebra $\mathcal{A}_{\max}$ consisting of *all* subsets of $\mathbb{R}$, one might obtain a measure on this largest possible $\sigma$-algebra, with the property that the measure of any interval coincides with the length of that interval. As a matter of fact this maximal extension of the Borel measure in terms of the length function is not a measure (it is not $\sigma$-additive on $\mathcal{A}_{\max}$). It is impossible to extend the Borel measure to the class of all subsets of $\mathbb{R}$ in such a way that the extension is still a measure which is invariant under translations.[10]

Of course there exist extensions of the Borel measure to $\sigma$-algebras larger than $\mathcal{A}_{\mathrm{B}}$. One such extension is frequently used in mathematics, namely the extension to the ***Lebesgue σ-algebra*** $\mathcal{A}_{\mathrm{L}}$, called Lebesgue measure. The elements $\mathcal{V}$ of $\mathcal{A}_{\mathrm{L}}$ have the form $\mathcal{V} = \mathcal{W} \cup \mathcal{V}_0$, where $\mathcal{W}$ is a Borel set ($\mathcal{W} \in \mathcal{A}_{\mathrm{B}}$) and $\mathcal{V}_0$ can be any subset of some Borel set of measure zero.[11] The Lebesgue measure of $\mathcal{V}$ is just the Borel measure of $\mathcal{W}$, the Lebesgue measure of $\mathcal{V}_0$ is zero: $m_{\mathrm{L}}(\mathcal{V}) = m_{\mathrm{B}}(\mathcal{W}) = \ell(\mathcal{W})$. So $\mathcal{A}_{\mathrm{L}}$ contains, in addition to the Borel sets, all subsets of Borel sets of measure zero, in fact all subsets of $\mathbb{R}$ of length zero. This is not trivial, as shown for example by the fact that there exist subsets of the Cantor set which are not Borel sets (we recall that the Cantor set is a Borel set, and it is of measure zero). The measure space $(\mathbb{R}, \mathcal{A}_{\mathrm{L}}, m_{\mathrm{L}})$ is ***complete*** in the sense that the $\sigma$-algebra $\mathcal{A}_{\mathrm{L}}$ contains all subsets of sets of measure zero, whereas the measure space $(\mathbb{R}, \mathcal{A}_{\mathrm{B}}, m_{\mathrm{B}})$ is not complete. Some statements in measure theory depend on the completeness of the underlying measure space.

From now on we shall use the standard notation $dx$ for the measure determined by the length function $\ell$ of Eq. (1.45) and call this measure simply ***Lebesgue measure***: $m_{\mathrm{B}}(dx) = m_{\mathrm{L}}(dx) = dx$. The underlying $\sigma$-algebra will be either the Borel $\sigma$-algebra $\mathcal{A}_{\mathrm{B}}$ or the Lebesgue $\sigma$-algebra $\mathcal{A}_{\mathrm{L}}$. In contexts where it is necessary, we shall specify which of these two $\sigma$-algebras is involved.

Above we have introduced the Lebesgue measure on the real line. Its analogue on $\mathbb{R}^n$ is defined in much the same way by taking for $\mathcal{B}$ the collection of all subsets of the form $\{\vec{x} \in \mathbb{R}^n \mid -\infty < a_k < x_k \leq b_k < +\infty, k = 1, \ldots, n\}$ and for the measure of such a subset its n-dimensional volume [here we use the notation $\vec{x} = (x_1, x_2, \ldots, x_n)$ for the individual points of $\mathbb{R}^n$]. A Borel set in $\mathbb{R}^n$ is an element of the $\sigma$-algebra generated by this collection $\mathcal{B}$, and the Lebesgue measurable sets are obtained by adding to the Borel $\sigma$-algebra all subsets of Borel sets of measure zero.

Other important measures in this text are Stieltjes measures on $(\mathbb{R}, \mathcal{A}_{\mathrm{B}})$, obtained by considering weight functions different from the length function on the set $\mathcal{B}$ of half-open intervals. In quantum mechanics, measures of this type occur in the characterisation of the spectral parts of observables; they are introduced in Section 4.1 and play an important role throughout Chapter 4.

---

[10] One uses the term ***outer measure*** for the extension of a measure given on some $\sigma$-algebra $\mathcal{A}$ to the maximal $\sigma$-algebra consisting of all subsets of $\mathcal{O}$. In general an outer measure does not satisfy all conditions imposed on a measure, i.e. an outer measure is not a measure.

[11] So $\mathcal{V}_0$ is a subset of $\mathbb{R}$ such that there exists a Borel set $\mathcal{U}$ satisfying $\ell(\mathcal{U}) = 0$ and $\mathcal{V}_0 \subset \mathcal{U}$.

**1.4.3.** We next consider $L^p$ spaces on a general measure space $(\mathcal{O}, \mathcal{A}, m)$. Two measurable functions (real or complex) $\varphi_1$ and $\varphi_2$ on $\mathcal{O}$ are said to be ***equivalent*** if they are equal $m$-almost everywhere, i.e. if they differ at most on a null set with respect to the measure $m$. Let $\mathcal{W} = \{x \in \mathcal{O} \mid \varphi_1(x) \neq \varphi_2(x)\}$ be the set of all points $x \in \mathcal{O}$ at which these functions have different values; this set belongs to $\mathcal{A}$, hence $m(\mathcal{W})$ is well defined, and $\varphi_1$ and $\varphi_2$ are equivalent if and only if $m(\mathcal{W}) = 0$.[12]

We now consider equivalence classes of measurable functions. The important fact is that, for any $t > 0$, the integral of $|\varphi_1 - \varphi_2|^t$ is zero if $\varphi_1$ and $\varphi_2$ are equivalent and that, if one of these two functions is $m$-integrable, then

$$\int_{\mathcal{O}} \varphi_1(x) m(dx) = \int_{\mathcal{O}} \varphi_2(x) m(dx).$$

If $p \in [1, +\infty)$, one defines the space $L^p(\mathcal{O}, m) \equiv L^p(\mathcal{O}, \mathcal{A}, m)$ to be the set of equivalence classes of measurable functions $f \colon \mathcal{O} \to \mathbb{C}$ satisfying

$$\|f\|_p := \left[ \int_{\mathcal{O}} |f(x)|^p m(dx) \right]^{1/p} < \infty. \tag{1.46}$$

$L^p(\mathcal{O}, m)$ is a *normed* vector space, the norm being given by (1.46), and this space is complete with respect to this norm (in other terms it is a *Banach space*). By considering equivalence classes of functions rather than individual functions as elements of $L^p(\mathcal{O}, m)$, one ensures that the only element of $L^p(\mathcal{O}, m)$ with norm zero is the equivalence class of the function $f \equiv 0$ (so the zero vector is unique, as in a Hilbert space).

When considering functions $f \colon \mathcal{O} \to \mathbb{C}$, one should each time specify whether $f$ is to be interpreted as an individual function or as an element of an $L^p$ space, i.e. as an equivalence class of functions. In most situations encountered in this text the precise meaning of $f$ will be clear from the context and there will be no need to mention it in detail.

An important result in the context of $L^p$ spaces is the Dominated Convergence Theorem (also called Lebesgue Convergence Theorem) which gives a sufficient condition for interchanging limits and integrals. This criterion is frequently used in our text.

**Proposition 1.9 (Dominated Convergence Theorem).** *Assume that*
*(i)* $f_n, h \in L^1(\mathcal{O}, m)$,
*(ii) for each* $n \in \mathbb{N}$, *one has* $|f_n(x)| \le h(x)$ *m-a.e.,*
*(iii)* $\lim_{n\to\infty} f_n(x) = f(x)$ *m-a.e.*
*Then the limit function* $f$ *is* $m$*-integrable and one has*

$$\lim_{n\to\infty} \int_{\mathcal{O}} f_n(x) m(dx) = \int_{\mathcal{O}} f(x) m(dx). \tag{1.47}$$

[12] A statement involving the points of $\mathcal{O}$ is said to hold ***m-almost everywhere*** (abbreviated $m$-a.e.) if the set of points $x \in \mathcal{O}$ at which the statement is not satisfied is a null set with respect to the measure $m$. One then also says that the statement holds at ***almost all*** points of $\mathcal{O}$.

The following inclusion relations between $L^p$ spaces are useful:

$$L^p(\mathcal{O},m)\cap L^q(\mathcal{O},m)\subseteq L^r(\mathcal{O},m) \quad \text{if } 1\le p\le r\le q\le\infty, \text{ [13]} \tag{1.48}$$

$$f\in L^q(\mathcal{O},m) \text{ and } m(\mathcal{V})<\infty \implies \chi_{\mathcal{V}}f\in L^r(\mathcal{O},m) \quad \forall r\in[1,q]. \tag{1.49}$$

PROOF. Let us write $f=f_>+f_<$ with $f_>(x)=f(x)$ if $|f(x)|>1$ and $f_>(x)=0$ otherwise (hence $f_<(x)=f(x)$ if $|f(x)|\le 1$ and $f_<(x)=0$ otherwise). Writing $L^p$ for $L^p(\mathcal{O},m)$ and assuming that $s<\infty$, we have the following implications:

$$f\in L^s \implies f_>\in L^t \qquad \forall t\in[1,s] \qquad (\text{since } |f_>(x)|^t\le|f(x)|^s) \tag{1.50}$$

$$f\in L^s \implies f_<\in L^u \qquad \forall u\in[s,\infty] \qquad (\text{since } |f_<(x)|^u\le|f(x)|^s). \tag{1.51}$$

To prove (1.48), let $f\in L^p\cap L^q$. If $r\in[p,q]$, then $f_>\in L^r$ [by (1.50) with $s=q$ and $t=r$] and $f_<\in L^r$ [by (1.51) with $s=p$ and $u=r$]. Hence $f\equiv f_>+f_<\in L^r$. To verify (1.49) we observe that, under the conditions of (1.49), one has $f_>\in L^r$ for $r<q$ [apply (1.50) as above], hence $\chi_{\mathcal{V}}f_>\in L^r$. On the other hand $|(\chi_{\mathcal{V}}f_<)(x)|\le\chi_{\mathcal{V}}(x)$, and $\chi_{\mathcal{V}}$ belongs to each $L^p$ if $m(\mathcal{V})<\infty$ (one has $\|\chi_{\mathcal{V}}\|_p=[m(\mathcal{V})]^{1/p}$). □

Among the spaces $L^p(\mathcal{O},m)$, only that for $p=2$ is a Hilbert space (Problem 1.14). The scalar product in $L^2(\mathcal{O},m)$ is given as follows:

$$\langle f,g\rangle=\int_{\mathcal{O}}\overline{f(x)}g(x)m(dx). \tag{1.52}$$

By the Schwarz inequality this number is finite if $f$ and $g$ belong to $L^2(\mathcal{O},m)$ (and the integral in (1.52) depends only on the equivalence classes of $f$ and $g$). We use the simplified notation $\|\cdot\|$ for the norm $\|\cdot\|_2$. In quantum mechanics, when $\mathcal{O}$ is the configuration space of a physical system [typically $\mathcal{O}=\mathbb{R}$ or $\mathbb{R}^n$ or an interval $(a,b)$], the elements of $L^2(\mathcal{O},m)$ (or the representative functions of an equivalence class) are usually called ***wave functions***; we refer to Remark 2.32 for further considerations on this terminology.

Before considering more explicit cases, let us mention the following generalisation (we restrict ourselves to $p=2$): instead of looking at functions defined on $\mathcal{O}$ with values in $\mathbb{C}$, one may consider functions with values in a Hilbert space $\mathcal{K}$ (vector-valued functions in the sense of Section 1.2). In analogy with the discussion above, one can introduce a Hilbert space $L^2(\mathcal{O};\mathcal{K},m)$; its elements are equivalence classes of such functions,[14] with scalar product and norm given by

$$\langle f,g\rangle=\int_{\mathcal{O}}\langle f(x),g(x)\rangle_{\mathcal{K}}\,m(dx) \qquad \text{and} \qquad \Big[\int_{\mathcal{O}}\|f(x)\|_{\mathcal{K}}^2\,m(dx)\Big]^{1/2}, \tag{1.53}$$

where $\langle\cdot,\cdot\rangle_{\mathcal{K}}$ and $\|\cdot\|_{\mathcal{K}}$ denote the scalar product and the norm in $\mathcal{K}$ respectively. For $\mathcal{K}=\mathbb{C}$ one regains the space $L^2(\mathcal{O},m)$ considered before.

[13] The space $L^\infty(\mathcal{O},m)$ will be defined in Section 2.5. The proof of (1.48) is given here for $q<\infty$.

[14] A function $f\colon\mathcal{O}\to\mathcal{K}$ is measurable if for each $g\in\mathcal{K}$ the numerical function $x\mapsto\langle g,f(x)\rangle_{\mathcal{K}}$ is measurable in the sense specified before.

**1.4.4.** One may introduce $L^p$ spaces on the real line $\mathbb{R}$ in terms of Lebesgue measure, by taking either $\mathcal{A}_{\rm B}$ or $\mathcal{A}_{\rm L}$ for the $\sigma$-algebra. Both definitions lead to the same space: $L^p(\mathbb{R}, \mathcal{A}_{\rm B}, m_{\rm B}) = L^p(\mathbb{R}, \mathcal{A}_{\rm L}, m_{\rm L}) \equiv L^p(\mathbb{R}, dx)$. This follows from the fact that null sets are irrelevant in an integral, by remembering that the elements of an $L^p$ space are *equivalence classes* of measurable functions [of course an equivalence class with respect to Lebesgue measure on $(\mathbb{R}, \mathcal{A}_{\rm L})$ contains more functions than the corresponding equivalence class with respect to Lebesgue measure on $(\mathbb{R}, \mathcal{A}_{\rm B})$, since $\mathcal{A}_{\rm B}$ is strictly smaller than $\mathcal{A}_{\rm L}$]. Another way of understanding the relation $L^p(\mathbb{R}, \mathcal{A}_{\rm B}, m_{\rm B}) = L^p(\mathbb{R}, \mathcal{A}_{\rm L}, m_{\rm L})$ is to use the fact that these spaces can be defined in terms of Riemann integrals without having recourse to measure theory, as explained towards the end of the present subsection.

When the measure defining an $L^p$ space on a Borel subset $\mathcal{V}$ of $\mathbb{R}$ or of $\mathbb{R}^n$ is the Lebesgue measure, we use the simplified notation $L^p(\mathcal{V})$ for the space $L^p(\mathcal{V}, m)$; it is understood that the measure in $L^p(\mathcal{V})$ is the Lebesgue measure. Also we then say "almost everywhere" instead of "$m$-almost everywhere" or of "Lebesgue almost everywhere".

If $\mathcal{V}$ is a Borel set in $\mathbb{R}$, for example an interval, the space $L^2(\mathcal{V})$ is a Hilbert space, consisting of all equivalence classes of measurable functions $f\colon \mathcal{V} \to \mathbb{C}$ that are square-integrable over $\mathcal{V}$ (with respect to the Lebesgue measure on $\mathcal{V}$, which is the restriction of the Lebesgue measure on $\mathbb{R}$ to the collection of Borel sets contained in $\mathcal{V}$). The space $L^2(\mathcal{V})$ can also be considered in a natural way as a subspace of $L^2(\mathbb{R})$, namely the subspace of all (equivalence classes of) functions $f$ in $L^2(\mathbb{R})$ that are zero outside $\mathcal{V}$, i.e. which satisfy $f(x) = 0$ if $x \notin \mathcal{V}$ (except possibly on a Borel set $\mathcal{W}$ of points $x \notin \mathcal{V}$ with $|\mathcal{W}| = 0$, where $|\mathcal{W}|$ denotes the Lebesgue measure of $\mathcal{W}$). In what follows we do not distinguish between these two meanings of $L^2(\mathcal{V})$; it will each time be clear from the context whether the space $L^2(\mathcal{V})$ is considered as a Hilbert space without reference to $L^2(\mathbb{R})$ or rather as a subspace of $L^2(\mathbb{R})$.

If $\mathcal{V}$ and $\mathcal{V}'$ are two disjoint Borel subsets of $\mathbb{R}$, then $L^2(\mathcal{V})$ and $L^2(\mathcal{V}')$ are mutually orthogonal as subspaces of $L^2(\mathbb{R})$. Also $L^2(\mathbb{R}\setminus\mathcal{V})$ is the orthogonal complement [in $L^2(\mathbb{R})$] of the subspace $L^2(\mathcal{V})$.

Examples of dense linear manifolds in the Hilbert space $L^2(\mathbb{R})$ are (modulo equivalence) the Schwartz space $\mathcal{S}(\mathbb{R})$, consisting of infinitely differentiable functions of rapid decay, and the set $C_0^\infty(\mathbb{R})$ of all infinitely differentiable functions of compact support. A function $f\colon \mathbb{R} \to \mathbb{C}$ belongs to $\mathcal{S}(\mathbb{R})$ if it is infinitely differentiable and if $f$ and all its derivatives converge to zero at infinity more rapidly than any inverse power of $|x|$, i.e. if

$$\sup_{x\in\mathbb{R}} |x|^k \left|\frac{d^\ell f(x)}{dx^\ell}\right| < \infty \qquad \forall k, \ell = 0, 1, 2, \dots \tag{1.54}$$

A function $f\colon \mathbb{R} \to \mathbb{C}$ belongs to $C_0^\infty(\mathbb{R})$ if it is infinitely differentiable and if there is a *finite* interval $J$ (depending on $f$) such that $f(x) = 0$ for all $x \notin J$. Clearly $C_0^\infty(\mathbb{R}) \subset \mathcal{S}(\mathbb{R})$, so that the density of $\mathcal{S}(\mathbb{R})$ in $L^2(\mathbb{R})$ follows from that of $C_0^\infty(\mathbb{R})$. The density of the latter set is discussed in §1.4.5.

More generally, if $-\infty \leq a < b \leq +\infty$, a function $f\colon (a,b) \to \mathbb{C}$ belongs to $C_0^\infty((a,b))$ if $f$ is infinitely differentiable and $f(x) = 0$ in some neighbourhood of $x = a$ and in some neighbourhood of $x = b$. The set $C_0^\infty((a,b))$ is dense in $L^p((a,b))$ for each $p \in [1,\infty)$; more precisely the set of equivalence classes of the functions in $C_0^\infty(\mathbb{R})$ is a dense set in $L^p((a,b))$.

Similar definitions and results apply in $L^p(\mathbb{R}^n)$. $C_0^\infty(\mathbb{R}^n)$ is the set of all infinitely differentiable functions $f\colon \mathbb{R}^n \to \mathbb{C}$ such that $f(\vec{x}) = 0$ for $|\vec{x}| > M$ for some number $M < \infty$ (depending on $f$); it is dense in $L^p(\mathbb{R}^n)$ for each $p \in [1,\infty)$. The Schwartz space $\mathcal{S}(\mathbb{R}^n)$ consists of infinitely differentiable functions satisfying

$$\sup_{\vec{x}\in\mathbb{R}^n} |\vec{x}|^k \left| \frac{\partial^{\ell_1+\cdots+\ell_n} f(\vec{x})}{\partial x_1^{\ell_1}\cdots\partial x_n^{\ell_n}} \right| < \infty \qquad \forall k, \ell_1, \ldots, \ell_n = 0, 1, 2, \ldots \tag{1.55}$$

It is clear that $C_0^\infty(\mathbb{R}^n) \subset \mathcal{S}(\mathbb{R}^n)$. In certain considerations the Schwartz space $\mathcal{S}(\mathbb{R}^n)$ is more suitable because it is invariant under Fourier transformation: if $f$ belongs to $\mathcal{S}(\mathbb{R}^n)$, then so does its Fourier transform $\tilde{f}$. The Fourier transformation will be introduced in Section 2.5.

The fact that $L^p(\mathbb{R}^n)$ or $L^p((a,b))$ contains a dense set consisting of continuous functions (modulo equivalence) allows one to characterise these spaces in terms of Riemann integrals. For example each continuous function $f$ defines an element of $L^p((a,b))$ if $\int_a^b |f(x)|^p\,dx < \infty$, where the integral can be interpreted as a Riemann integral. One can then define $L^p((a,b))$ as the norm closure of $C_0^\infty((a,b))$, i.e. one takes as elements of $L^p((a,b))$ the (equivalence classes of) sequences $\{f_n\}$ such that each $f_n$ belongs to $C_0^\infty((a,b))$ and $\lim_{m,n\to\infty}\int_a^b |f_n(x) - f_m(x)|^p\,dx = 0$ (Riemann integral !); two such sequences $\{f_n\}$ and $\{g_n\}$ are equivalent if their difference converges to zero in $L^p$ norm.

For completeness we recall that the ***support*** of a function $f\colon \mathcal{O} \to \mathbb{C}$, where $\mathcal{O}$ is an open subset of $\mathbb{R}^n$ ($n = 1, 2, 3, \ldots$), is the closure of the set of points $\vec{x} \in \mathcal{O}$ at which $f(\vec{x}) \neq 0$. So the support of $f$ is a closed subset of $\mathbb{R}^n$, which need not be contained in $\mathcal{O}$ because $\mathcal{O}$ is assumed to be open.[15] If the support of $f$ is contained in $\mathcal{O}$ and bounded,[16] then $f$ is said to have ***compact support in*** $\boldsymbol{\mathcal{O}}$ (see the example of the functions in $C_0^\infty((a,b))$ mentioned above).

**1.4.5.** We end with some remarks concerning the fact that $C_0^\infty(\mathbb{R})$ is dense in $L^p(\mathbb{R})$ for $1 \leq p < \infty$. This can be proved for example as follows: one shows (1) that each measurable function $f$ satisfying $\int_{\mathbb{R}} |f(x)|^p\,dx < \infty$ is a limit (in $L^p$ norm) of a sequence of bounded functions of compact support; (2) that a bounded function of compact support is a limit (in $L^p$ norm) of a sequence of continuous functions of compact support; and (3) that a continuous function of compact support is a limit (in $L^p$ norm) of a sequence of functions belonging to $C_0^\infty(\mathbb{R})$. Some details on each of these three steps are given below. Then one concludes the argument as follows. Consider an

[15] If for example $f(\vec{x}) \neq 0$ at all points of $\mathcal{O}$, then its support is the closure of $\mathcal{O}$ in $\mathbb{R}^n$.

[16] A subset of $\mathbb{R}^n$ is ***compact*** if it is closed and bounded.

element of $L^p(\mathbb{R})$ and let $\varepsilon > 0$. Choose a representative $f$ of the considered equivalence class. By (1) there is a bounded function of compact support $g_1$ in $L^p(\mathbb{R})$ such that $\|f - g_1\|_p < \varepsilon/3$. By (2) there is a continuous function of compact support $g_2$ in $L^p(\mathbb{R})$ such that $\|g_1 - g_2\|_p < \varepsilon/3$. By (3) there exists $g_3$ in $C_0^\infty(\mathbb{R})$ such that $\|g_2 - g_3\|_p < \varepsilon/3$. Then $\|f - g_3\|_p < \varepsilon$, which proves the denseness of $C_0^\infty(\mathbb{R})$ in $L^p(\mathbb{R})$ (it is easily seen that the result is independent of the choice of a representative in the equivalence class under consideration).

(1) Let $f\colon \mathbb{R} \to \mathbb{C}$ satisfy $\int_{\mathbb{R}} |f(x)|^p dx < \infty$. For $n \in \mathbb{N}$ define $f_n$ as follows: $f_n(x) = f(x)$ if $|f(x)| \leq n$ and $|x| < n$, and $f_n(x) = 0$ otherwise. Clearly $f_n$ is bounded ($|f_n(x)| \leq n$ for each $x$) and of compact support (i.e. $f_n$ is zero outside some finite interval). Furthermore the sequence $\{f_n\}$ converges to $f$ in $L^p$ norm: one has $f_n \in L^p(\mathbb{R})$ (because $|f_n(x)| \leq |f(x)|$ for all $x$) and

$$\|f - f_n\|_p = \Big[\int_{\mathbb{R}} |f(x) - f_n(x)|^p \, dx\Big]^{1/p}.$$

As $n \to \infty$ the integrand converges to zero almost everywhere (for each $x \in \mathbb{R}$ for which $|f(x)| \neq \infty$), and for each $n$ it is bounded by the function $h(x) = |f(x)|^p$ (because $f(x) - f_n(x)$ is equal to $f(x)$ or to 0). Since $h \in L^1(\mathbb{R})$, the result of (1) follows from the Dominated Convergence Theorem.

(2) Here one uses the fact that the measurable space $(\mathcal{O}, \mathcal{A})$ is special: one has $\mathcal{O} = \mathbb{R}$ and $\mathcal{A} = \mathcal{A}_{\mathrm{B}}$. In this situation one can approximate a measurable function by a sequence of ***step functions***; these are particular simple functions, namely functions of the form $\varphi = \sum_{k=1}^{N} \alpha_k \chi_{\mathcal{V}_k}$, where $\alpha_k$ are constants and $\mathcal{V}_k$ are *intervals* of the form $(a_k, b_k]$, see Figure 1.4(a). If $g$ is a measurable function with $|g(x)| \leq M$ for all $x$, there exists a sequence $\{\varphi_n\}$ of step functions such that $\varphi_n(x) \to g(x)$ and $|\varphi_n(x)| \leq M$ for all $x$.

So let $g$ be a bounded function ($|g(x)| \leq M$) with support in an interval $[-N, +N]$ for some $N \in (0, \infty)$. Then there exists a sequence $\{g_n\}$ of step functions, with support in the interval $[-N, N]$, such that $|g_n(x)| \leq M$ and $\lim_{n\to\infty} g_n(x) = g(x)$ for all $x \in \mathbb{R}$. By applying the Dominated Convergence Theorem as in (1) (observe that $|g(x) - g_n(x)|^p \leq (2M)^p$ and that the domain of integration $[-N, N]$ is bounded), one sees that $\|g - g_n\|_p \to 0$. Hence, given $\varepsilon > 0$, one can choose a step function $h$ with support in $[-N, N]$ such that $\|g - h\|_p < \varepsilon/2$. It is clear that one can approximate $h$ in $L^p$ norm by a continuous function $h_0$ with support in the interval $[-N - 1, N + 1]$ and with $\|h - h_0\|_p < \varepsilon/2$ [slightly round off the real part and the imaginary part of the function $h$ at its points of discontinuity, as illustrated in Figure 1.4(b) for a real-valued step function, or else use a regularisation as in (3) below]. By the triangle inequality one then has $\|g - h_0\|_p \leq \|g - h\|_p + \|h - h_0\|_p < \varepsilon$.

**Figure 1.4** (a) Step function; (b) approximate step function.

(3) One now applies a method of ***regularisation***.[17] Choose a non-negative function $\theta \in C_0^\infty(\mathbb{R})$ such that $\theta(x) = 0$ if $|x| \geq 1$ and $\int_{-1}^{1} \theta(x)dx = 1$, for example $\theta(x) = 0$ if $|x| \geq 1$ and $\theta(x) = \gamma \exp[-1/(1-x^2)]$ if $|x| < 1$, where the constant $\gamma$ is such that $\int_{-1}^{1} \theta(x)dx = 1$. For $0 < \varepsilon < 1$, set $\theta_\varepsilon(x) = \varepsilon^{-1}\theta(x/\varepsilon)$. Clearly $\theta_\varepsilon \in C_0^\infty(\mathbb{R})$, $\theta_\varepsilon(x) \geq 0$, $\theta_\varepsilon(x) = 0$ if $|x| \geq \varepsilon$ and $\int_{-\infty}^{\infty} \theta_\varepsilon(x)dx = 1$. A function $f \in L^p(\mathbb{R})$ is approximated by a family $\{f_\varepsilon\}$ of smooth functions defined by

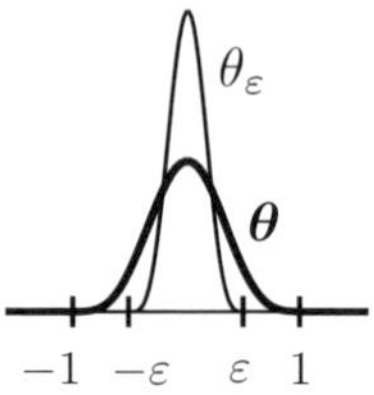


$$f_\varepsilon(x) = \int_{-\infty}^{\infty} \theta_\varepsilon(x-y)f(y)dy.$$

$f_\varepsilon(x)$ is some sort of average of $f$ over a small neighbourhood of $x$ (over the interval $[x-\varepsilon, x+\varepsilon]$). For each fixed $\varepsilon$ the function $f_\varepsilon$ is infinitely differentiable (one may interchange the derivatives with the integral, and the integrand is infinitely differentiable with respect to $x$). It can be shown that $\|f - f_\varepsilon\|_p \to 0$ as $\varepsilon \to 0$.

We consider the situation in which $f$ is continuous and has compact support, say $f(x) = 0$ if $|x| \geq N$. Then the support of $f_\varepsilon$ is contained in $[N-\varepsilon, N+\varepsilon]$, because f(y) is zero if $|y| > N$ and $\theta_\varepsilon(x-y)$ is zero if $|x-y| > \varepsilon$. Let us show that $\|f - f_\varepsilon\|_p \to 0$ as $\varepsilon \to 0$. Since $\int_{-\infty}^{\infty} \theta_\varepsilon(x)dx = 1$, we have

$$\begin{aligned} f_\varepsilon(x) - f(x) &= \int_{-\infty}^{\infty} \theta_\varepsilon(x-y)f(y)dy - \int_{-\infty}^{\infty} \theta_\varepsilon(z)f(x)dz \\ &= \int_{-\infty}^{\infty} \theta_\varepsilon(z)\left[f(x-z) - f(x)\right]dz. \end{aligned}$$

One may consider $z \mapsto f(x-z) - f(x)$ as a vector-valued function of $z$ with values in $L^p(\mathbb{R}, dx)$. By using the triangle inequality [for the continuous sum, e.g. as in (1.28)], one finds that

$$\begin{aligned} \|f - f_\varepsilon\|_p &\leq \int_{-\infty}^{\infty} \theta_\varepsilon(z)\|f(\cdot - z) - f(\cdot)\|_p dz \\ &= \int_{-\infty}^{\infty} dz\, \theta_\varepsilon(z)\left[\int_{-\infty}^{\infty} dx\, |f(x-z) - f(x)|^p\right]^{1/p} \\ &= \int_{-1}^{+1} dw\, \theta(w)\left[\int_{-N-1}^{N+1} |f(x-\varepsilon w) - f(x)|^p dx\right]^{1/p}. \end{aligned} \tag{1.56}$$

To obtain the last expression we have made the change of variables $z \mapsto w = z/\varepsilon$ and taken into account the hypotheses that $\theta(x) = 0$ if $|x| > 1$ and $f(x) = 0$ if $|x| \geq N$. To see that the expression (1.56) converges to zero when $\varepsilon \to 0$, one applies twice the Dominated Convergence Theorem:

(i) Let us first consider the integral with respect to $dx$ (for fixed $w$). For each $x$ the integrand $|f(x-\varepsilon w) - f(x)|^p$ tends to zero as $\varepsilon \to 0$, because $f$ is continuous.

[17] See for example Chapter II of [Ad].

Furthermore the integrand is bounded: since $|f(x)| \leq M < \infty$ for all $x \in \mathbb{R}$, one has $|f(x - \varepsilon w) - f(x)|^p \leq (2M)^p$, which is integrable on $[-N-1, N+1]$. By Proposition 1.9 with $h(x) = (2M)^p$ we obtain

$$\lim_{\varepsilon \to 0} \left[ \int_{-N-1}^{N+1} |f(x - \varepsilon w) - f(x)|^p \, dx \right]^{1/p} = 0.$$

We notice that

$$\left[ \int_{-N-1}^{N+1} |f(x - \varepsilon w) - f(x)|^p \, dx \right]^{1/p} \leq \left[ \int_{-N-1}^{N+1} (2M)^p \, dx \right]^{1/p} = 2M(2N+2)^{1/p}.$$

(ii) One can now apply Proposition 1.9 to the integral with respect to $dw$ in (1.56). It suffices to observe that, for each fixed $w$, the integrand converges to zero as $\varepsilon \to 0$ [by the result of (i)] and that it is majorised by the function $h(w) = 2M(2N+2)^{1/p}\theta(w)$.

## Bibliographical Notes

The books by Akhiezer and Glazman [AG], Riesz and Sz.-Nagy [RN] and Weidmann [W1] provide excellent presentations of the theory of Hilbert spaces and linear operators. There are numerous books on measure theory and integration; the following is a selective list: [B], [CK], [D], [H], [Na] [R], [Ra], [ST], Volume I of [RS]. $L^p$ spaces are also discussed in many texts, see for example [CV] or [R].

## Problems

**1.1.** Let $\mathcal{H}$ be a Hilbert space. Verify the parallelogram identity

$$\|f + g\|^2 + \|f - g\|^2 = 2\|f\|^2 + 2\|g\|^2 \tag{1.57}$$

and the inequality

$$\Big\| \sum_{k=1}^{N} f_k \Big\|^2 \leq N \sum_{k=1}^{N} \|f_k\|^2. \tag{1.58}$$

**1.2.** Show that, in a finite-dimensional Hilbert space, weak convergence and strong convergence of sequences of vectors are equivalent.

**1.3.** Prove that, if $\{f_n\}$ converges weakly to $f$ and $\|f_n\| \leq \|f\|$ for each $n$, then $\{f_n\}$ converges strongly to $f$.

**1.4.** Let $\{f_n\}$ be an infinite sequence of mutually orthogonal vectors, i.e. satisfying $\langle f_j, f_k \rangle = 0$ if $j \neq k$, and let $F_N = \sum_{k=1}^{N} f_k$. Show that the following three statements are equivalent: (i) $\text{s-}\lim_{N\to\infty} F_N$ exists, (ii) $\text{w-}\lim_{N\to\infty} F_N$ exists, (iii) $\sum_{k=1}^{\infty} \|f_k\|^2 < \infty$.

**1.5.** In Proposition 1.5(c), prove that the values of $f$ at the endpoints $a$ and $b$ of the open interval under consideration are well defined.

**1.6.** Let $\mathcal{N}$ be a total subset of a Hilbert space $\mathcal{H}$. Let $\{f_n\}$ be a bounded sequence in $\mathcal{H}$ ($\|f_n\| \leq M < \infty$) such that $\lim_{n\to\infty}\langle f_n, g\rangle = \langle f, g\rangle$ for some $f \in \mathcal{H}$ and each $g \in \mathcal{N}$. Prove that $\{f_n\}$ converges weakly to $f$. Show by means of an example that the condition $\|f_n\| \leq M < \infty$ cannot be dropped.

**1.7.** Show that, if $\mathcal{N}$ is any subset of a Hilbert space $\mathcal{H}$, then $\mathcal{N}^\perp$ is a subspace of $\mathcal{H}$ and $(\mathcal{N}^\perp)^\perp$ is equal to the subspace spanned by $\mathcal{N}$.

**1.8.** (a) Prove that $\ell^2$ is a Hilbert space.
(b) Find an orthonormal basis in $\ell^2$.

**1.9.** Let $BC(\mathbb{R}^n)$ be the space of all bounded continuous functions $f\colon \mathbb{R}^n \to \mathbb{C}$. This space is a normed (even complete) space for the norm $\|f\|_\infty = \sup_{\vec{x}\in\mathbb{R}^n}|f(\vec{x})|$. Show that $C_0^\infty(\mathbb{R}^n)$ is not dense in $BC(\mathbb{R}^n)$.

**1.10.** Give an example of a square-integrable continuous function $f$ on $\mathbb{R}$ which does not satisfy $\lim_{|x|\to\infty} f(x) = 0$.

**1.11.** Give an example of a sequence $\{f_n\}$ of integrable functions on $\mathbb{R}$ (with respect to Lebesgue measure) which converges almost everywhere to a function $f$ but such that $\int_{\mathbb{R}} f_n(x)\,dx$ does not converge to $\int_{\mathbb{R}} f(x)\,dx$.

**1.12.** Decide which of the following functions belong to $L^p(\mathbb{R}^n)$:
$f(\vec{x}) = |\vec{x}|^{-\alpha}$, $f(\vec{x}) = (1+|\vec{x}|)^{-\alpha}$, $f(\vec{x}) = e^{-x_1}$, $f(\vec{x}) = e^{-|\vec{x}|}$, $f(\vec{x}) = 1$
($1 \leq p \leq \infty$, $n \in \mathbb{N}$, $\alpha \in \mathbb{R}$).
Note: See page 57 for the definition of $L^\infty(\mathbb{R}^n)$.

**1.13.** Let $f\colon \mathbb{R} \to \mathbb{C}$ be a continuous function vanishing outside some finite interval. For $n = 1, 2, 3, \ldots$ define $f_n$ by $f_n(x) = f(x-n)$. By using the fact that the set of continuous functions of compact support is dense in $L^2(\mathbb{R})$, show that the sequence $\{f_n\}$ converges weakly but not strongly to **0** in $L^2(\mathbb{R})$.

**1.14.** Show that if $p \neq 2$, then $L^p$ is not a Hilbert space for the norm $\|\cdot\|_p$.
[Hint: use the polarisation identity.]

**1.15.** Consider the collection $\mathcal{A}_{\mathrm{L}}$ of all subsets of $\mathbb{R}$ of the form $\mathcal{V} = \mathcal{W} \cup \mathcal{V}_0$, with $\mathcal{W}$ a Borel set and $\mathcal{V}_0$ a subset of some Borel set of measure zero. Verify that $\mathcal{A}_{\mathrm{L}}$ is a $\sigma$-algebra.

## CHAPTER 2

# Linear Operators

Throughout this text we are concerned with linear mappings from a Hilbert space into itself or into another Hilbert space, called linear operators. In the present chapter we introduce and discuss various types of operators that frequently occur in quantum mechanics. In Section 2.2 we consider projections, isometric and unitary operators. Section 2.3 is devoted to compact and Hilbert-Schmidt operators and Section 2.5 to multiplication operators in $L^2$ spaces. Often quantum-mechanical operators are unbounded. Section 2.4 is concerned with some of the subtleties that one meets in such situations, especially with the notion of closedness of an operator. In Section 2.6 we define the resolvent and the spectrum of closed operators.

A very important class of linear operators are the self-adjoint operators, used to describe observables in quantum mechanics. For unbounded operators the property of being self-adjoint is rather subtle. Some preliminary explanations, pointing out the difference between symmetric and self-adjoint operators, are given in Section 2.4, and Chapter 3 will be entirely devoted to a detailed study of the relations between symmetric and self-adjoint operators. The present chapter ends with some criteria for the preservation of self-adjointness under perturbations (Section 2.7); in particular we prove the self-adjointness of some Hamiltonians of non-relativistic quantum mechanics as perturbations of the self-adjoint free Hamiltonian $\vec{P}^2/2m$.

## 2.1 The algebra $\mathcal{B}(\mathcal{H})$

**2.1.1.** Let $\mathcal{H}$ be a Hilbert space. We denote by $\mathcal{B}(\mathcal{H})$ the set of all bounded linear operators on $\mathcal{H}$. An element of $\mathcal{B}(\mathcal{H})$ is a mapping $A$ that associates with each vector $f \in \mathcal{H}$ another vector $Af$ belonging to $\mathcal{H}$, and this mapping is linear and bounded, i.e. $A(\alpha f + g) = \alpha Af + Ag$ and there is a finite constant $M$ (depending on $A$) such that $\|Af\| \leq M\|f\|$ for each vector $f$ in $\mathcal{H}$. The infimum of all possible numbers M is called the ***norm*** of the operator $A$ and denoted by $\|A\|$:

$$\begin{aligned}\|A\| &= \inf\{M \in \mathbb{R} \mid \|Af\| \le M\|f\| \ \forall f \in \mathcal{H}\} \\ &= \sup_{0 \ne f \in \mathcal{H}} \frac{\|Af\|}{\|f\|} = \sup_{g \in \mathcal{H}, \|g\|=1} \|Ag\|. \end{aligned} \tag{2.1}$$

The elements of $\mathcal{B}(\mathcal{H})$ will usually just be called ***bounded operators***.

Observe that (2.1) implies that

$$\|Af\| \le \|A\| \|f\| \qquad \forall f \in \mathcal{H}. \tag{2.2}$$

If $A \in \mathcal{B}(\mathcal{H})$, then the image $\{Af_n\}$ of each strong Cauchy sequence $\{f_n\}$ is again a strong Cauchy sequence:

$$\operatorname*{s-lim}_{n\to\infty} f_n = f \quad \Longrightarrow \quad \operatorname*{s-lim}_{n\to\infty} Af_n = Af. \tag{2.3}$$

This follows from (2.2) which implies that $\|Af - Af_n\| \le \|A\| \|f - f_n\|$.

The following expression for the operator norm is sometimes useful: If $\mathcal{D}_1$ and $\mathcal{D}_2$ are arbitrary dense linear manifolds in $\mathcal{H}$ (in particular if $\mathcal{D}_1 = \mathcal{D}_2 = \mathcal{H}$), then

$$\|A\| = \sup_{f \in \mathcal{D}_1, g \in \mathcal{D}_2, \|f\|=\|g\|=1} |\langle f, Ag\rangle|. \tag{2.4}$$

PROOF. Let $g$ be a vector in $\mathcal{H}$ with $\|g\| = 1$. Let $\{g_n\}$ be a sequence in $\mathcal{D}_2$ converging strongly to $g$. One may assume that $\|g_n\| = 1$ [if $\operatorname{s-lim}_{n\to\infty} h_n = h$, with $h \ne \mathbf{0}$, then $\operatorname{s-lim}_{n\to\infty}(h_n/\|h_n\|) = h/\|h\|$, which is easy to see because $\|h_n\| \to \|h\|$ by Proposition 1.1(a)]. Then, by (2.3) and Proposition 1.1, $\|Ag\| = \lim_{n\to\infty} \|Ag_n\|$. Hence, if we restrict the vectors $g$ in the last expression in (2.1) to the set $\{g \in \mathcal{D}_2 \mid \|g\| = 1\}$, we obtain the same supremum:

$$\|A\| = \sup_{g \in \mathcal{D}_2, \|g\|=1} \|Ag\|. \tag{2.5}$$

Next let $h \in \mathcal{H}$, $h \ne \mathbf{0}$. Then, if $\|f\| = 1$, one has $|\langle f, h\rangle| \le \|f\| \|h\| = \|h\|$, with equality for $f = h/\|h\|$. So $\|h\| = \sup_{f \in \mathcal{H}, \|f\|=1} |\langle f, h\rangle|$. By using an argument as above (approximating a vector $f \in \mathcal{H}$, $\|f\| = 1$, by a sequence $\{f_n\}$ in $\mathcal{D}_1$, with $\|f_n\| = 1$), one sees that

$$\|h\| = \sup_{f \in \mathcal{D}_1, \|f\|=1} |\langle f, h\rangle|. \tag{2.6}$$

By taking $h = Ag$ in (2.6) and inserting this expression into (2.5), one obtains the formula (2.4). □

Let us briefly look at the structure of the set $\mathcal{B}(\mathcal{H})$. It is clear that $\mathcal{B}(\mathcal{H})$ is a *linear vector space* over the field $\mathbb{C}$: if $A, B \in \mathcal{B}(\mathcal{H})$ and $\alpha \in \mathbb{C}$, then $\alpha A + B$ is again a bounded linear operator, given by $(\alpha A + B)f = \alpha Af + Bf$, and it is bounded:

$$\|\alpha A + B\| \le |\alpha| \|A\| + \|B\|, \tag{2.7}$$

because $\|(\alpha A + B)f\| \leq \|\alpha Af\| + \|Bf\| \leq |\alpha|\|A\|\|f\| + \|B\|\|f\|$ [we have used (1.13)]. $\mathcal{B}(\mathcal{H})$ is an *algebra*, i.e. a vector space equipped with an operation of multiplication. The product of two bounded operators $A$ and $B$ is simply denoted by $AB$ and is just the composition of the two mappings: $(AB)f = A(Bf)$ is the vector obtained from $f$ by acting first with $B$ on $f$ and then with $A$ on the image vector $Bf$. The multiplication in $\mathcal{B}(\mathcal{H})$ has the following properties:

$$(\alpha A + B)C = \alpha AC + BC \qquad \text{and} \qquad C(\alpha A + B) = \alpha CA + CB.$$

To see that $AB$ is bounded, observe that $\|ABf\| \leq \|A\|\,\|Bf\| \leq \|A\|\,\|B\|\,\|f\|$ as a consequence of (2.2), so that

$$\|AB\| \leq \|A\|\|B\|. \tag{2.8}$$

**2.1.2.** The algebra $\mathcal{B}(\mathcal{H})$ is involutive (called a ***-algebra**). The involution, denoted by an asterisk, associates with each operator $A \in \mathcal{B}(\mathcal{H})$ another element $A^*$ of $\mathcal{B}(\mathcal{H})$, called the ***adjoint*** of $A$, such that $(A^*)^* = A$ [one usually writes $A^{**}$ for $(A^*)^*$]. Given $A$, its adjoint $A^*$ is defined by the relation

$$\langle A^*f, g\rangle = \langle f, Ag\rangle \qquad \forall f, g \in \mathcal{H}. \tag{2.9}$$

This relation determines uniquely an element $A^*$ of $\mathcal{B}(\mathcal{H})$: for fixed $f$, the mapping $g \mapsto \langle f, Ag\rangle$ is a bounded linear functional on $\mathcal{H}$ (it is linear in $g$, and $|\langle f, Ag\rangle| \leq \|f\|\|A\|\|g\|$). By virtue of the Riesz Lemma (Proposition 1.8) there exists a unique vector $f^* \in \mathcal{H}$ such that $\langle f, Ag\rangle = \langle f^*, g\rangle$ for each $g \in \mathcal{H}$. One sets $A^*f = f^*$, which defines $A^*$ on each vector $f$ of $\mathcal{H}$. We must check that the thus-defined mapping $A^*$ is linear and bounded:

(i) One has

$$\begin{aligned}\langle A^*(\alpha f + h), g\rangle &= \langle \alpha f + h, Ag\rangle = \overline{\alpha}\langle f, Ag\rangle + \langle h, Ag\rangle \\ &= \overline{\alpha}\langle f^*, g\rangle + \langle h^*, g\rangle = \overline{\alpha}\langle A^*f, g\rangle + \langle A^*h, g\rangle = \langle \alpha A^*f + A^*h, g\rangle.\end{aligned}$$

So $A^*(\alpha f + h) - \alpha A^*f - A^*h$ is orthogonal to each $g \in \mathcal{H}$, hence equal to the zero vector (Proposition 1.6), i.e. $A^*(\alpha f + h) = \alpha A^*f + A^*h$.

(ii) By using (2.4) one finds that

$$\|A^*\| = \sup_{f,g\in\mathcal{H},\|f\|=\|g\|=1} |\langle g, A^*f\rangle| = \sup_{f,g\in\mathcal{H},\|f\|=\|g\|=1} |\langle Ag, f\rangle| = \|A\|.$$

Hence $A^*$ is bounded and

$$\|A^*\| = \|A\|. \tag{2.10}$$

The definition (2.9) also implies that, if $A, B \in \mathcal{B}(\mathcal{H})$, then [1]

$$(AB)^* = B^*A^* \qquad \text{and} \qquad A^{**} = A. \tag{2.11}$$

[1] The relation $A^{**} = A$ is obtained from Proposition 1.6 by observing that $\langle A^{**}f, g\rangle = \langle f, A^*g\rangle = \langle Af, g\rangle$ for all $f, g \in \mathcal{H}$.

Another useful relation is the following:

$$\|A^*A\| = \|A\|^2. \tag{2.12}$$

Indeed one has on the one hand $\|A^*A\| \leq \|A^*\|\|A\| = \|A\|^2$ and, on the other hand,

$$\begin{aligned}\|A\|^2 = \sup_{0\neq f\in\mathcal{H}} \frac{\|Af\|^2}{\|f\|^2} &= \sup_{0\neq f\in\mathcal{H}} \frac{1}{\|f\|^2}\langle Af, Af\rangle = \sup_{0\neq f\in\mathcal{H}} \frac{1}{\|f\|^2}\langle f, A^*Af\rangle \\ &\leq \sup_{0\neq f\in\mathcal{H}} \frac{1}{\|f\|^2}\|f\|\|A^*A\|\|f\| = \|A^*A\|.\end{aligned}$$

An operator $A\in\mathcal{B}(\mathcal{H})$ is ***self-adjoint*** if $A^* = A$, i.e. if

$$\langle Af, g\rangle = \langle f, Ag\rangle \qquad \forall f, g\in\mathcal{H}. \tag{2.13}$$

Together with (1.5), Eq. (2.13) implies that $\langle f, Af\rangle$ is *real* for each $f\in\mathcal{H}$ if $A$ is self-adjoint. The norm of a bounded self-adjoint operator $A$ can be expressed as follows:

$$\|A\| = \sup_{f\in\mathcal{D},\|f\|=1} |\langle f, Af\rangle|, \tag{2.14}$$

where $\mathcal{D}$ is an arbitrary dense linear manifold in $\mathcal{H}$.

PROOF. We set $a = \sup_{f\in\mathcal{D},\|f\|=1}|\langle f, Af\rangle|$. Observe that $|\langle h, Ah\rangle| \leq a\|h\|^2$ for each $h\in\mathcal{D}$ (because $h/\|h\|$ is a vector of norm 1 if $h\neq \mathbf{0}$).

The inequality $a\leq\|A\|$ follows from (2.4). To obtain the converse inequality we must show that $a\geq|\langle f, Ag\rangle|$ for all $f, g\in\mathcal{D}$ with $\|f\| = \|g\| = 1$. Now $|\langle f, Ag\rangle| = |\langle e^{i\alpha}f, Ag\rangle|$ for each real $\alpha$, and for fixed $f$ and $g$ one can find a real number $\alpha$ such that $\langle e^{i\alpha}f, Ag\rangle$ is real and non-negative. Hence it suffices to show that $a\geq\langle f, Ag\rangle$ for all $f, g\in\mathcal{D}$ with $\|f\| = \|g\| = 1$ and such that $\langle f, Ag\rangle\in\mathbb{R}$.

For this we use the following generalisation of the polarisation identity (1.11):

$$\begin{aligned}4\langle f, Ag\rangle &= \langle f+g, A(f+g)\rangle - \langle f-g, A(f-g)\rangle \\ &\quad - i\langle f+ig, A(f+ig)\rangle + i\langle f-ig, A(f-ig)\rangle.\end{aligned} \tag{2.15}$$

Since $\langle h, Ah\rangle$ is real for each $h\in\mathcal{H}$, the real part of $4\langle f, Ag\rangle$ is given by the sum of the first two terms on the right-hand side of the preceding equation. So, if $f$ and $g$ are such that $\langle f, Ag\rangle\in\mathbb{R}$, one has [by also using (1.33)]

$$\begin{aligned}\langle f, Ag\rangle &= \frac{1}{4}\left[\langle f+g, A(f+g)\rangle - \langle f-g, A(f-g)\rangle\right] \\ &\leq \frac{1}{4}\left[|\langle f+g, A(f+g)\rangle| + |\langle f-g, A(f-g)\rangle|\right] \\ &\leq \frac{1}{4}\left[a\|f+g\|^2 + a\|f-g\|^2\right] \\ &= \frac{a}{4}\left[2\|f\|^2 + 2\|g\|^2\right] = a \qquad \text{if } \|f\| = \|g\| = 1. \qquad \square\end{aligned}$$

**2.1.3.** One may consider various types of convergence of sequences of operators $\{A_n\}$ in $\mathcal{B}(\mathcal{H})$, in particular convergence in norm (also called uniform convergence), strong convergence and weak convergence. The last two notions are defined in terms of the corresponding notions for sequences of vectors introduced in Section 1.2. More precisely:

***Strong convergence***: A sequence $\{A_n\}_{n\in\mathbb{N}}$ of operators in $\mathcal{B}(\mathcal{H})$ converges strongly to an operator $A \in \mathcal{B}(\mathcal{H})$ if, for each vector $f \in \mathcal{H}$, the sequence of vectors $\{A_n f\}$ converges strongly to $Af$, i.e. if $\|Af - A_n f\| \to 0$ as $n \to \infty$ for each $f \in \mathcal{H}$. One then writes $\text{s-}\lim_{n\to\infty} A_n = A$.

***Weak convergence***: A sequence $\{A_n\}_{n\in\mathbb{N}}$ of operators in $\mathcal{B}(\mathcal{H})$ converges weakly to an operator $A \in \mathcal{B}(\mathcal{H})$ if, for each vector $f \in \mathcal{H}$, the sequence of vectors $\{A_n f\}$ converges weakly to $Af$, i.e. if $\langle g, Af\rangle - \langle g, A_n f\rangle \to 0$ as $n \to \infty$ for all $f, g \in \mathcal{H}$. One then writes $\text{w-}\lim_{n\to\infty} A_n = A$.

***Convergence in norm*** (***uniform convergence***): A sequence $\{A_n\}_{n\in\mathbb{N}}$ of operators in $\mathcal{B}(\mathcal{H})$ converges in norm to an operator $A \in \mathcal{B}(\mathcal{H})$ if $\|A - A_n\| \to 0$ as $n \to \infty$. One then writes $\text{u-}\lim_{n\to\infty} A_n = A$.

Some useful properties of these notions are as follows:

(1) It is clear that strong convergence of $\{A_n\}$ to $A$ implies weak convergence of $\{A_n\}$ to $A$ (as a consequence of Proposition 1.1).

(2) Convergence in norm implies strong convergence: if $\|A - A_n\| \to 0$, then one has for each $f \in \mathcal{H}$: $\|Af - A_n f\| \leq \|A - A_n\|\|f\| \to 0$.

(3) If $\{A_n\}$ converges weakly to $A$, then the adjoint sequence $\{A_n^*\}$ converges weakly to $A^*$, because $\langle A_n^* f, g\rangle = \langle f, A_n g\rangle \to \langle f, Ag\rangle = \langle A^* f, g\rangle$. However strong convergence of a sequence $\{A_n\}$ does not imply strong convergence of the adjoint sequence $\{A_n^*\}$ (only its weak convergence).

(4) If $\{A_n\}$ converges weakly to $A$ and each $A_n$ is self-adjoint, then the limit $A$ is also self-adjoint. Indeed $\langle f, Ag\rangle - \langle Af, g\rangle = \lim_{n\to\infty}[\langle f, A_n g\rangle - \langle A_n f, g\rangle] = 0$ if $A_n^* = A_n$.

If a sequence $\{A_n\}_{n\in\mathbb{N}}$ of operators belonging to $\mathcal{B}(\mathcal{H})$ is weakly Cauchy, i.e. if $\lim_{m,n\to\infty}[\langle f, A_m g\rangle - \langle f, A_n g\rangle] = 0$ for all $f, g \in \mathcal{H}$, then there exists an operator $A \in \mathcal{B}(\mathcal{H})$ such that $\text{w-}\lim_{n\to\infty} A_n = A$. Analogous statements hold for sequences that are strongly Cauchy and for sequences that are Cauchy in norm. We omit the proof of these statements (it involves the Uniform Boundedness Principle cited at the end of Section 1.1).

**Proposition 2.1.** *Let* $A, B, A_n, B_n \in \mathcal{B}(\mathcal{H})$ *be such that* $A = \text{s-}\lim_{n\to\infty} A_n$ *and* $B = \text{s-}\lim_{n\to\infty} B_n$. *Then the sequence* $\{A_n B_n\}_{n\in\mathbb{N}}$ *is strongly convergent and satisfies* $\text{s-}\lim_{n\to\infty} A_n B_n = AB$.

PROOF. We use the fact that, as a consequence of the Uniform Boundedness Principle (Section 1.1), a strongly (or only weakly) convergent sequence of operators is bounded: there is a constant $M < \infty$ such that $\|A_n\| \leq M$ for all $n \in \mathbb{N}$ (in our applications of Proposition 2.1 the validity of this inequality will always be evident). So let $f \in \mathcal{H}$. Then

$$\begin{aligned}\|ABf - A_nB_nf\| &= \|(A - A_n)Bf + A_n(B - B_n)f\| \\ &\leq \|(A - A_n)Bf\| + \|A_n(B - B_n)f\| \\ &\leq \|(A - A_n)Bf\| + M\|(B - B_n)f\|, \end{aligned} \tag{2.16}$$

which converges to 0 when $n \to \infty$. □

Proposition 2.1 remains true if one replaces everywhere strong convergence by norm convergence. However it has no analogue for weak convergence (i.e. if $\{A_n\}$ and $\{B_n\}$ are weakly convergent sequences, then the product sequence $\{A_nB_n\}$ may not be weakly Cauchy).

**Proposition 2.2.** *Let $\{A_n\}_{n\in\mathbb{N}}$ be a sequence in $\mathcal{B}(\mathcal{H})$ such that $\|A_n\| \leq M < \infty$ for all $n \in \mathbb{N}$. Let $\mathcal{M}$ be a subspace of $\mathcal{H}$ and suppose that there is a subset $\mathcal{N}$ of $\mathcal{M}$ which is total in $\mathcal{M}$ and such that, for each $f \in \mathcal{N}$, the sequence $\{A_nf\}$ is strongly Cauchy. Then the limit* $\text{s-lim}_{n\to\infty} A_ng$ *exists for each $g \in \mathcal{M}$. If furthermore $A$ is an operator in $\mathcal{B}(\mathcal{H})$ such that* $\text{s-lim}_{n\to\infty} A_nf = Af$ *for each $f \in \mathcal{N}$, then one has* $\text{s-lim}_{n\to\infty} A_ng = Ag$ *for each $g \in \mathcal{M}$.*

PROOF. Let $\mathcal{E}$ be the linear manifold spanned by $\mathcal{N}$ (so that $\mathcal{M} = \overline{\mathcal{E}}$). It is clear that $\{A_nf\}$ will be strongly Cauchy for each $f \in \mathcal{E}$. So let $g \in \mathcal{M}$ and $\varepsilon > 0$; choose a vector $h \in \mathcal{E}$ such that $\|g - h\| \leq \varepsilon/(3M)$. Then by the triangle inequality:

$$\begin{aligned}\|A_mg - A_ng\| &\leq \|A_m(g - h)\| + \|A_mh - A_nh\| + \|A_n(h - g)\| \\ &\leq 2M\|g - h\| + \|A_mh - A_nh\| \leq 2\varepsilon/3 + \|A_mh - A_nh\|.\end{aligned}$$

As $\{A_nh\}$ is strongly Cauchy, one has $\|A_mh - A_nh\| \leq \varepsilon/3$ provided that $m, n \geq N(\varepsilon)$ for some $N(\varepsilon) \in (0, \infty)$. So $\|A_mg - A_ng\| < \varepsilon$ provided that $m, n > N(\varepsilon)$, i.e. $\{A_ng\}$ is strongly Cauchy. The proof of the second part of the Proposition is similar, it suffices to replace $A_m$ by $A$ in the arguments given above. □

**2.1.4.** We continue with our remarks on the structure of $\mathcal{B}(\mathcal{H})$. We have seen that $\mathcal{B}(\mathcal{H})$ is a $*$-algebra. We have also seen that $\mathcal{B}(\mathcal{H})$ is a *normed algebra*, the norm being given by (2.1), and that $\mathcal{B}(\mathcal{H})$ is *complete* with respect to this norm [a norm convergent sequence of elements of $\mathcal{B}(\mathcal{H})$ has a limit in $\mathcal{B}(\mathcal{H})$]. A complete normed algebra is called a ***Banach algebra***. Here we even have a $*$-algebra, in which case one uses the term ***B*-algebra*** (a complete normed algebra equipped with an involution). A B*-algebra in which the identity (2.12) (i.e. $\|A^*A\| = \|A\|^2$) is satisfied is called a ***C*-algebra***. The notion of a C*-algebra is used for abstract algebras; in our context the algebra $\mathcal{B}(\mathcal{H})$ is realised by operators acting in a Hilbert space $\mathcal{H}$, and there are other topologies on $\mathcal{B}(\mathcal{H})$ than just the norm topology (in particular the strong topology and the weak topology). A C*-algebra of operators on a Hilbert space $\mathcal{H}$ (i.e. a $*$-subalgebra of $\mathcal{B}(\mathcal{H})$ that is complete in the norm $\|\cdot\|$) and which is closed in the weak topology is called a ***W*-algebra*** or a ***von Neumann algebra***. In particular $\mathcal{B}(\mathcal{H})$ itself is a *von Neumann algebra*.

**2.1.5.** In analogy with vector-valued functions discussed in Section 1.2, one may consider functions defined on a set $\Lambda$ with values in $\mathcal{B}(\mathcal{H})$. An important case in what follows (see Section 2.6) is that in which $\Lambda$ is an open subset of the complex plane $\mathbb{C}$. Here we consider the case $\Lambda = J$, where $J$ is an interval in $\mathbb{R}$. We write $A(s)$ for the value of a function $A\colon J \to \mathcal{B}(\mathcal{H})$ at the point $s \in J$. Again there are various notions of continuity and differentiability of such functions (uniform, strong or weak) and different possibilities for defining integrals of such functions. We recall below a few definitions. For the strong and the weak topologies one refers to the case of vector-valued functions already considered in Section 1.2 by acting with $A(s)$ on the vectors of $\mathcal{H}$, i.e. by considering for each vector $f \in \mathcal{H}$ the vector-valued function $s \mapsto A(s)f$.

A function $A\colon J \to \mathcal{B}(\mathcal{H})$ is ***strongly continuous*** if, for each $f \in \mathcal{H}$ and for each $t \in J$, one has

$$\lim_{s\to t, s\in J} \|A(s)f - A(t)f\| = 0. \tag{2.17}$$

It is ***continuous in norm*** if one has for each $t \in J$:

$$\lim_{s\to t, s\in J} \|A(s) - A(t)\| = 0. \tag{2.18}$$

A function $A\colon J \to \mathcal{B}(\mathcal{H})$ on an open interval $J$ is ***strongly differentiable on J*** if there exists a function $A'\colon J \to \mathcal{B}(\mathcal{H})$ (called the ***strong derivative*** of $A$) such that for each $f \in \mathcal{H}$ and for each $s \in J$:

$$\lim_{\tau\to 0} \Big\|\frac{1}{\tau}[A(s+\tau)f - A(s)f] - A'(s)f\Big\| = 0. \tag{2.19}$$

It is ***differentiable in norm*** on $J$ if there exists a function $A'\colon J \to \mathcal{B}(\mathcal{H})$ such that for each $s \in J$:

$$\lim_{\tau\to 0} \Big\|\frac{1}{\tau}[A(s+\tau) - A(s)] - A'(s)\Big\| = 0. \tag{2.20}$$

As regards integrals, we shall use only Riemann integrals in the strong sense. If $J$ is a finite interval, $\Pi = \{s_0, s_1, \ldots, s_N; u_1, \ldots, u_N\}$ a partition of $J$ (as in Section 1.2) and $A\colon J \to \mathcal{B}(\mathcal{H})$ an operator-valued function, we define

$$\Sigma(\Pi, A) = \sum_{k=1}^{N} (s_k - s_{k-1})A(u_k). \tag{2.21}$$

$\Sigma(\Pi, A)$ is a finite sum of operators in $\mathcal{B}(\mathcal{H})$, so (2.21) defines an operator in $\mathcal{B}(\mathcal{H})$. One then takes a sequence $\{\Pi_r\}_{r\in\mathbb{N}}$ of partitions of $J$ with $\lim_{r\to\infty} |\Pi_r| = 0$ and sets for each $f \in \mathcal{H}$:

$$\Big[\int_J A(s)ds\Big] f = \operatorname*{s-lim}_{r\to\infty} \Sigma(\Pi_r, A)f = \operatorname*{s-lim}_{r\to\infty} \sum_{k=1}^{N_r} (s_k - s_{k-1})A(u_k)f, \tag{2.22}$$

if the limit exists and is independent of the sequence $\{\Pi_r\}$. This defines the operator $\int_J A(s)ds$ [Eq. (2.22) gives its action on each vector $f \in \mathcal{H}$].

If $J$ is an infinite interval, one first defines the integral for each finite subinterval as in (2.22) and then takes a sequence of finite subintervals of $J$ converging to $J$. For example, if $J = (a, \infty)$ for some finite $a$, one defines (if the limit exists)

$$\int_a^\infty A(s)\,ds = \operatorname*{s-lim}_{b\to\infty} \int_a^b A(s)\,ds. \tag{2.23}$$

The properties of these integrals are easily obtained by applying Propositions 1.4 and 1.5. In the following proposition we state some results that will be useful further on in this text.

**Proposition 2.3.** *Let $J$ be a finite or infinite interval, and let $A\colon J \to \mathcal{B}(\mathcal{H})$ and $f\colon J \to \mathcal{H}$.*
*(a) If the integrals below exist and $B \in \mathcal{B}(\mathcal{H})$, then*

$$B\int_J f(s)\,ds = \int_J Bf(s)\,ds \qquad \text{and} \qquad B\int_J A(s)\,ds = \int_J BA(s)\,ds \tag{2.24}$$

*(i.e. one may commute bounded operators with the operation of integration).*
*(b) One has*

$$\Big\|\int_J A(s)\,ds\Big\| \le \int_J \|A(s)\|\,ds. \tag{2.25}$$

*(c) If $A(\cdot)$ is strongly continuous on $J$ and $\int_J \|A(s)\|\,ds < \infty$, then the integral $\int_J A(s)\,ds$ exists and belongs to $\mathcal{B}(\mathcal{H})$.*
*(d) If $[a,b]$ is finite and $A\colon [a,b] \to \mathcal{B}(\mathcal{H})$ is strongly continuous, then $\int_a^b A(s)\,ds$ exists and belongs to $\mathcal{B}(\mathcal{H})$.*

**Remark 2.4.** If one assumes the function $A$ to be continuous in the norm of $\mathcal{B}(\mathcal{H})$ or in some other norm on $\mathcal{B}(\mathcal{H})$, and if the interval $[a,b]$ is finite, then the approximating Riemann sums $\Sigma(\Pi_r, A)$ are Cauchy in the norm under consideration. If all of the operators $A(s)$, $s \in [a,b]$ belong to a sub-algebra $\mathcal{B}_0$ of $\mathcal{B}(\mathcal{H})$ which is complete in this norm, then the integral $\int_a^b A(s)\,ds$ will also belong to $\mathcal{B}_0$. This observation will be useful later, and two important examples are as follows (compact and Hilbert-Schmidt operators will be introduced in Section 2.3):
(a) $\mathcal{B}_0 = \mathcal{K}(\mathcal{H})$, the set of all compact operators, equipped with the norm (2.1),
(b) $\mathcal{B}_0 = \mathcal{B}_2(\mathcal{H})$, the set of all Hilbert-Schmidt operators, equipped with the Hilbert-Schmidt norm $\|\cdot\|_{\mathrm{HS}}$.

**2.1.6.** We denote by $\mathcal{R}(A)$ the ***range*** of the operator $A$, i.e. the linear manifold formed by the set of vectors $Ag$ as $g$ varies over $\mathcal{H}$:

$$\mathcal{R}(A) = A\mathcal{H} = \big\{f \in \mathcal{H} \,\big|\, \exists\, g \in \mathcal{H} \text{ with } f = Ag\big\}. \tag{2.26}$$

An operator A is ***invertible*** if $f \neq g$ implies $Af \neq Ag$, or equivalently if $Af = \mathbf{0}$ implies that $f = \mathbf{0}$ (the only vector mapped by $A$ onto the zero vector $\mathbf{0}$ is the vector $\mathbf{0}$, i.e. the number $0$ is not an eigenvalue of $A$). If A is invertible, there exists a linear mapping $A^{-1}\colon \mathcal{R}(A) \to \mathcal{H}$ satisfying $A^{-1}Af = f$ for all $f \in \mathcal{H}$ and $AA^{-1}g = g$ for all $g \in \mathcal{R}(A)$. In our context an interesting situation is that in which $A^{-1}$ also belongs to $\mathcal{B}(\mathcal{H})$; we then say that $A$ is ***invertible in $\mathcal{B}(\mathcal{H})$***. For this to be the case it is necessary that $\mathcal{R}(A) = \mathcal{H}$ and that $A^{-1}$ is bounded (in fact the latter condition is superfluous, see the footnote on page 68).

We denote by $I$ the ***identity operator*** defined by $If = f$ for all $f \in \mathcal{H}$ and by $O$ the ***zero operator*** ($Of = \mathbf{0}$ for all $f \in \mathcal{H}$). Clearly $I$ is invertible in $\mathcal{B}(\mathcal{H})$, with $I^{-1} = I$. It is useful to know that operators close to the identity operator are also invertible in $\mathcal{B}(\mathcal{H})$. More precisely, if $A \in \mathcal{B}(\mathcal{H})$ is such that $\|A\| < 1$, then $I - A$ is invertible in $\mathcal{B}(\mathcal{H})$:

**Proposition 2.5.** *Let $A \in \mathcal{B}(\mathcal{H})$ with $\|A\| < 1$. Then $I - A$ is invertible in $\mathcal{B}(\mathcal{H})$, in particular $\mathcal{R}(I - A) = \mathcal{H}$ and $(I - A)^{-1} \in \mathcal{B}(\mathcal{H})$. Furthermore $\|(I - A)^{-1}\| \leq [1 - \|A\|]^{-1}$, and $(I - A)^{-1}$ is given by the following series which converges in the uniform topology (in operator norm), called the* ***Neumann series****:*

$$(I - A)^{-1} = I + A + A^2 + \cdots = \sum_{k=0}^{\infty} A^k. \tag{2.27}$$

PROOF. The series (2.27) is convergent in norm, because by (2.8) and (2.7) we have for $n > m$

$$\Big\|\sum_{k=m}^{n} A^k\Big\| \leq \|A^m\| \sum_{k=0}^{n-m} \|A^k\| \leq \|A\|^m \sum_{k=0}^{\infty} \|A\|^k = \|A\|^m \big[1 - \|A\|\big]^{-1},$$

which converges to zero as $m \to \infty$ since $\|A\| < 1$. Thus the series (2.27) defines an operator $B$ in $\mathcal{B}(\mathcal{H})$, with

$$\|B\| = \Big\|\sum_{k=0}^{\infty} A^k\Big\| \leq \big[1 - \|A\|\big]^{-1} < \infty.$$

We now observe that

$$(I - A)(I + A + A^2 + \cdots + A^n) = I - A^{n+1}. \tag{2.28}$$

As $\|A^{n+1}\| \leq \|A\|^{n+1} \to 0$ as $n \to \infty$, Eq. (2.28) implies that

$$(I - A)B \equiv (I - A)\Big(\sum_{k=0}^{\infty} A^k\Big) = I.$$

Consequently $\mathcal{R}(I - A) = \mathcal{H}$. A similar calculation shows that $(\sum_{k=0}^{\infty} A^k)(I - A) \equiv B(I - A) = I$. Hence $I - A$ is invertible and $(I - A)^{-1} = B$. In particular $(I - A)^{-1}$ is given by the series (2.27). □

**2.1.7.** So far we have considered operators mapping $\mathcal{H}$ into $\mathcal{H}$. Occasionally we deal with operators between different Hilbert spaces. Let $\mathcal{H}$ and $\mathcal{K}$ be two Hilbert spaces. We denote by $\mathcal{B}(\mathcal{H},\mathcal{K})$ the set of all bounded linear operators from $\mathcal{H}$ to $\mathcal{K}$. An element of $\mathcal{B}(\mathcal{H},\mathcal{K})$ is a linear mapping that associates with each vector $f\in\mathcal{H}$ a vector $Af\in\mathcal{K}$, with

$$\|A\| = \sup_{0\neq f\in\mathcal{H}} \frac{\|Af\|_{\mathcal{K}}}{\|f\|_{\mathcal{H}}} < \infty. \tag{2.29}$$

Here $\|\cdot\|_{\mathcal{H}}$ denotes the norm in $\mathcal{H}$ and $\|\cdot\|_{\mathcal{K}}$ that in $\mathcal{K}$. The properties of $\mathcal{B}(\mathcal{H},\mathcal{K})$ are analogous to those of $\mathcal{B}(\mathcal{H})$; the adjoint is defined by

$$\langle A^*f, g\rangle_{\mathcal{H}} = \langle f, Ag\rangle_{\mathcal{K}} \qquad \forall f\in\mathcal{K} \text{ and } \forall g\in\mathcal{H}. \tag{2.30}$$

$A^*$ is an operator that maps $\mathcal{K}$ into $\mathcal{H}$, i.e. $A^*\in\mathcal{B}(\mathcal{K},\mathcal{H})$. Also, if $A\in\mathcal{B}(\mathcal{H},\mathcal{K})$ is invertible, then $A^{-1}$ is a mapping $\mathcal{K}\to\mathcal{H}$ [defined on $\mathcal{R}(A)\subseteq\mathcal{K}$].

**2.1.8.** When the Hilbert space $\mathcal{H}$ is an $L^2$ space, some care may be required to make sure that an operator, often given by its action on individual functions, is indeed a mapping between equivalence classes of functions. We mention here a few frequently encountered situations, assuming that $\mathcal{H} = L^2(\mathcal{O}, m)$.

An operator $A$ in $L^2(\mathcal{O}, m)$ is said to be an ***integral operator*** if there exists a measurable function $a\colon \mathcal{O}\times\mathcal{O}\to\mathbb{C}$ such that for all $f$ in the domain of definition of $A$ (this latter concept will be discussed in §2.4.1):

$$(Af)(x) = \int_{\mathcal{O}} a(x,y)f(y)m(dy). \tag{2.31}$$

The function $a$ is called the ***kernel*** of $A$. Observe that the value of the integral in (2.31) is the same for all functions $f$ belonging to the same equivalence class, so that $A$ maps all representatives of an equivalence class onto the same function $(Af)(x)$.

A ***multiplication operator*** $A$ in $L^2(\mathcal{O}, m)$ is an operator given as follows:

$$(Af)(x) = a(x)f(x). \tag{2.32}$$

The value of the function $Af$ at the point $x\in\mathcal{O}$ is obtained by multiplying the value of $f$ at the point $x$ by the number $a(x)$, where $a\colon\mathcal{O}\to\mathbb{C}$ is a fixed measurable function. In other terms the function $Af$ is the product of the two functions $a$ and $f$. It is clear from (2.32) that equivalent functions get mapped onto equivalent functions. Properties of multiplication operators are presented in Section 2.5.

Somewhat more delicate are ***differential operators***, i.e. operators involving derivatives of the functions $f$ on which they act. Their definition on equivalence classes may involve subtleties of differential calculus which will not be discussed here. Usually such operators will be defined only on some dense subset of the Hilbert space (on equivalence classes containing a sufficiently differentiable representative $f$). We shall be more explicit when treating specific classes of differential operators.

## 2.2 Projections and isometries

**2.2.1.** The Projection Theorem (Proposition 1.7) shows that, if $\mathcal{M}$ is a subspace of a Hilbert space $\mathcal{H}$ and $\mathcal{M}^\perp$ its orthogonal complement, then each vector $f$ in $\mathcal{H}$ can be written in a unique way as $f = f_1 + f_2$ with $f_1 \in \mathcal{M}$ and $f_2 \in \mathcal{M}^\perp$. The correspondence $f \mapsto f_1$ defines a linear mapping $P_\mathcal{M}$ on $\mathcal{H}$ with range $\mathcal{R}(P_\mathcal{M}) = \mathcal{M}$, called the projection onto $\mathcal{M}$ (it would be more precise to use the term *orthogonal projection*, because $P_\mathcal{M}$ annihilates the orthogonal complement of $\mathcal{M}$; as we consider only orthogonal projections, we usually omit the specification "orthogonal" and simply use the term projection). Since $\|f_1\|^2 \leq \|f_1\|^2 + \|f_2\|^2 = \|f\|^2$, we see that $P_\mathcal{M}$ is bounded, with $\|P_\mathcal{M}\| = 1$ (if dim $\mathcal{M} \geq 1$). Consequently $P_\mathcal{M} \in \mathcal{B}(\mathcal{H})$. It is also clear that $P_\mathcal{M}^2 = P_\mathcal{M}$. In addition one has $P_\mathcal{M}^* = P_\mathcal{M}$. Indeed, if $f$ and $g$ are decomposed as $f = f_1 + f_2$ and $g = g_1 + g_2$ respectively, then $\langle P_\mathcal{M} f, g\rangle = \langle f_1, g\rangle = \langle f_1, g_1\rangle = \langle f, g_1\rangle = \langle f, P_\mathcal{M} g\rangle$ (using the relation $\mathcal{M} \perp \mathcal{M}^\perp$).

It is interesting to know that the two properties $P_\mathcal{M}^2 = P_\mathcal{M}$ and $P_\mathcal{M}^* = P_\mathcal{M}$ characterise an orthogonal projection. In our context an operator $E \in \mathcal{B}(\mathcal{H})$ is called a ***projection*** if $E^* = E = E^2$. Given a projection $E$ there exists a (unique) subspace $\mathcal{M}$ of $\mathcal{H}$ such that $E = P_\mathcal{M}$, i.e. such that $E$ is the orthogonal projection onto $\mathcal{M}$. Indeed, set

$$\mathcal{M} \equiv \mathcal{M}(E) = \{f \in \mathcal{H} \mid Ef = f\}. \tag{2.33}$$

It is easy to see that $\mathcal{M}$ is a subspace (this also follows from Proposition 2.38 proved in Section 2.6). If $g \perp \mathcal{M}$, then one has for each $h \in \mathcal{H}$:

$$\langle Eg, h\rangle = \langle g, E^* h\rangle = \langle g, Eh\rangle. \tag{2.34}$$

Since $E^2 h = Eh$, one has $Eh \in \mathcal{M}$ [set $f = Eh$ in (2.33)]. Consequently (2.34) implies that $\langle Eg, h\rangle = 0$, hence $Eg = \mathbf{0}$ by Proposition 1.6. So, if a vector $f \in \mathcal{H}$ is decomposed into $f = f_1 + f_2$ with $f_1 \in \mathcal{M}$ and $f_2 \in \mathcal{M}^\perp$, one has $Ef = Ef_1 + Ef_2 = f_1 + \mathbf{0}$, i.e. $Ef = P_\mathcal{M} f$.

RÉSUMÉ. With each subspace $\mathcal{M}$ there is associated a projection $P_\mathcal{M}$ describing the orthogonal projection of the vectors of $\mathcal{H}$ onto $\mathcal{M}$, and with each projection $E$ is associated a subspace $\mathcal{M}(E)$ of $\mathcal{H}$ such that $P_{\mathcal{M}(E)} = E$. This subspace is just the range of the operator $E$: $\mathcal{M}(E) = \mathcal{R}(E)$. It is evident that $P_{\mathcal{M}^\perp} = I - P_\mathcal{M}$ and $P_\mathcal{H} = I$, $P_{\{\mathbf{0}\}} = O$.

It is clear that geometric relations between subspaces correspond to algebraic relations between the associated projections. In the next proposition we specify some important examples, without giving the details of the proof.

**Proposition 2.6.** *Let $\mathcal{M}$ and $\mathcal{N}$ be subspaces of $\mathcal{H}$, and let $P_\mathcal{M}$ and $P_\mathcal{N}$ be the associated projections.*
*(a) If $P_\mathcal{M} P_\mathcal{N} = P_\mathcal{N} P_\mathcal{M}$, then $P_\mathcal{M} P_\mathcal{N}$ is a projection, the associated subspace being $\mathcal{M} \cap \mathcal{N}$, i.e. $P_\mathcal{M} P_\mathcal{N} = P_{\mathcal{M}\cap\mathcal{N}}$.*
*(b) If $\mathcal{M} \subseteq \mathcal{N}$, then $P_\mathcal{M} P_\mathcal{N} = P_\mathcal{N} P_\mathcal{M} = P_\mathcal{M}$.*

*(c) If $\mathcal{M} \perp \mathcal{N}$, then $P_{\mathcal{M}} P_{\mathcal{N}} = P_{\mathcal{N}} P_{\mathcal{M}} = O$ and $P_{\mathcal{M} \oplus \mathcal{N}} = P_{\mathcal{M}} + P_{\mathcal{N}}$.*
*(d) If $P_{\mathcal{M}} P_{\mathcal{N}} = O$, then $\mathcal{M} \perp \mathcal{N}$.*

If $\{E_n\}$ is a strongly convergent sequence of projections and $E$ denotes their limit (i.e. $E = \text{s-lim}_{n\to\infty} E_n$), then $E$ is also a projection. Indeed $E$ is self-adjoint [see point (4) on page 31] and $E^2 = \text{s-lim}_{n\to\infty} E_n^2 = \text{s-lim}_{n\to\infty} E_n = E$, where the first equality follows from Proposition 2.1.

**2.2.2.** An operator $\Omega$ belonging to $\mathcal{B}(\mathcal{H})$ is called an ***isometry*** (or an ***isometric operator***) if

$$\Omega^*\Omega = I. \tag{2.35}$$

**Proposition 2.7.** *Let $\Omega$ be an isometry. Then*
*(a) $\Omega$ preserves all scalar products:*

$$\langle \Omega f, \Omega g\rangle = \langle f, g\rangle \qquad \forall f, g \in \mathcal{H}. \tag{2.36}$$

*In particular:*

$$\|\Omega f\| = \|f\| \qquad \forall f \in \mathcal{H}. \tag{2.37}$$

*(b) If $\dim \mathcal{H} \neq 0$, then $\|\Omega\| = 1$.*
*(c) $\Omega\Omega^*$ is a projection, its corresponding subspace is $\mathcal{M}(\Omega\Omega^*) = \mathcal{R}(\Omega)$.*
*(d) $\Omega$ is invertible.*
*(e) One has $\Omega^* g = \Omega^{-1} g$ if $g \in \mathcal{R}(\Omega)$ and $\Omega^* g = \mathbf{0}$ if $g \perp \mathcal{R}(\Omega)$.*
*(f) If an operator $W \in \mathcal{B}(\mathcal{H})$ satisfies $\|Wf\| = \|f\|$ for all vectors $f \in \mathcal{H}$, then $W$ is an isometry.*

PROOF. (a)

$$\langle \Omega f, \Omega g\rangle = \langle f, \Omega^*\Omega g\rangle = \langle f, g\rangle.$$

(b)

$$\|\Omega\| = \sup_{0\neq f\in\mathcal{H}} \frac{\|\Omega f\|}{\|f\|} = \sup_{0\neq f\in\mathcal{H}} \frac{\|f\|}{\|f\|} = 1.$$

(c) Setting $F = \Omega\Omega^*$ and using (2.35) and (2.11), we obtain

$$F^2 = \Omega(\Omega^*\Omega)\Omega^* = \Omega I \Omega^* = \Omega\Omega^* = F \text{ and } F^* = (\Omega\Omega^*)^* = \Omega^{**}\Omega^* = \Omega\Omega^* = F.$$

Thus $F$ is a projection. By (2.33) we have $\mathcal{M}(F) = \{f \in \mathcal{H} \mid \Omega\Omega^* f = f\}$. Consequently, if $f \in \mathcal{M}(F)$, then $f \in \mathcal{R}(\Omega)$, i.e. $\mathcal{M}(F) \subseteq \mathcal{R}(\Omega)$. On the other hand, if $g = \Omega f \in \mathcal{R}(\Omega)$, then $Fg = \Omega\Omega^* g = \Omega\Omega^*\Omega f = \Omega I f = \Omega f = g$, which shows that $g \in \mathcal{M}(F)$. Thus $\mathcal{R}(\Omega) \subseteq \mathcal{M}(F)$.

(d) If $\Omega f = \mathbf{0}$, then $\|f\| = \|\Omega f\| = 0$ by (2.37), hence $f = \mathbf{0}$. This shows that $\Omega$ is invertible [but not necessarily invertible in $\mathcal{B}(\mathcal{H})$, because $\Omega^{-1}$ is defined only on $\mathcal{M}(F)$].

(e) Let $g = \Omega f \in \mathcal{R}(\Omega)$. Then $\Omega^{-1} g = f = \Omega^*\Omega f = \Omega^* g$. On the other hand, if $g \perp \mathcal{R}(\Omega)$, then $\langle \Omega^* g, f\rangle = \langle g, \Omega f\rangle = 0$ for all $f \in \mathcal{H}$, hence $\Omega^* g$ is orthogonal to $\mathcal{H}$, so that $\Omega^* g = \mathbf{0}$ by Proposition 1.6.

(f) The relation $W^*W = I$ can be obtained by using the polarisation identity (1.11):

$$\begin{aligned} 4\langle f, W^*Wg\rangle &= 4\langle Wf, Wg\rangle \\ &= \|W(f+g)\|^2 - \|W(f-g)\|^2 - i\|W(f+ig)\|^2 + i\|W(f-ig)\|^2 \\ &= \|f+g\|^2 - \|f-g\|^2 - i\|f+ig\|^2 + i\|f-ig\|^2 \\ &= 4\langle f, g\rangle = 4\langle f, Ig\rangle. \end{aligned}$$

□

The preceding proposition shows that an isometric operator maps $\mathcal{H}$ onto a subspace $\mathcal{M}(\Omega\Omega^*)$ by preserving the length (i.e. the norm) of all vectors as well as the angles between vectors (i.e. the scalar products). This subspace has the same dimension as $\mathcal{H}$: if $\{e_k\}$ is an orthonormal basis of $\mathcal{H}$, then $\{\Omega e_k\}$ is an orthonormal basis of $\mathcal{M}(\Omega\Omega^*) \equiv \mathcal{R}(\Omega)$. An important special case of an isometry is that of a ***unitary operator*** $U$ for which $\mathcal{M}(UU^*) = \mathcal{H}$. Thus $U$ is unitary if it is isometric and $F \equiv UU^* = I$; in other terms U is unitary if

$$U^*U = I \qquad \underline{\text{and}} \qquad UU^* = I. \tag{2.38}$$

In this case $U^*g = U^{-1}g$ for each $g \in \mathcal{H}$, and $U$ is invertible in $\mathcal{B}(\mathcal{H})$, i.e. one has $U^{-1} \in \mathcal{B}(\mathcal{H})$.

In a finite-dimensional Hilbert space, each isometry is unitary [the first equation in (2.38) implies the second one]. This is so because in a finite-dimensional Hilbert space $\mathcal{H}$, a subspace having the same dimension as $\mathcal{H}$ must be equal to $\mathcal{H}$. If $\mathcal{H}$ is infinite-dimensional, it can be mapped isometrically onto a proper subspace (also infinite-dimensional). An example is given by the shift operator: let $\{e_k\}_{k\in\mathbb{N}}$ be an orthonormal basis of $\mathcal{H}$ and $\Omega e_k = e_{k+1}$, hence $\Omega f = \sum_k \alpha_k e_{k+1}$ if $f = \sum_k \alpha_k e_k$. This operator is isometric but not unitary, its range is the subspace spanned by the vectors $e_2, e_3, e_4, \ldots$, i.e. $\mathcal{R}(\Omega)$ is the orthogonal complement of the vector $e_1$ (see also Problem 2.29 at the end of this chapter).

**Proposition 2.8.** *Let $U$ be unitary and $\mathcal{N}$ a subset of $\mathcal{H}$ which is invariant under $U$ and under $U^*$ (so $Uf \in \mathcal{N}$ and $U^*f \in \mathcal{N}$ for each $f \in \mathcal{N}$). Let $\mathcal{M}$ be the subspace spanned by $\mathcal{N}$. Then*
*(a) $\mathcal{M}$ and $\mathcal{M}^\perp$ are invariant under $U$ and under $U^*$, and the restriction $U_0$ of $U$ to $\mathcal{M}$ (considered as an operator in the Hilbert space $\mathcal{M}$) is unitary.*
*(b) One has $(U\mathcal{M})^\perp = (U\mathcal{N})^\perp = U(\mathcal{N}^\perp)$.*

PROOF. (a) We observe that $\mathcal{M}^\perp = \mathcal{N}^\perp$ [see (1.31)]. If $f \in \mathcal{M}^\perp$ and $g \in \mathcal{N}$, then $\langle Uf, g\rangle = \langle f, U^*g\rangle = 0$ because $U^*g \in \mathcal{N}$. Hence $Uf \in \mathcal{N}^\perp \equiv \mathcal{M}^\perp$, which shows that $U$ leaves $\mathcal{M}^\perp$ invariant. Similarly one verifies that $U^*$ leaves $\mathcal{M}^\perp$ invariant.

Suppose now that $f \in \mathcal{M}^\perp$ and $h \in \mathcal{M}$. Then $\langle f, Uh\rangle = \langle U^*f, h\rangle = 0$ because $U^*f \in \mathcal{M}^\perp$. This shows that $Uh \in (\mathcal{M}^\perp)^\perp \equiv \mathcal{M}$, hence that $U$ leaves $\mathcal{M}$ invariant. Similarly one sees that $\mathcal{M}$ is also invariant under $U^*$.

We denote by $(U^*)_0$ the restriction of $U^*$ to $\mathcal{M}$, considered as an operator in $\mathcal{M}$, and by $U_0^*$ the adjoint (in $\mathcal{M}$) of $U_0$. Then, for $f, g \in \mathcal{M}$:

$$\langle f, U_0 g\rangle = \langle f, Ug\rangle = \langle U^*f, g\rangle = \langle (U^*)_0 f, g\rangle,$$

consequently $U_0^* = (U^*)_0$. This implies that $U_0^* U_0 = (U^*)_0 U_0$. But if $f \in \mathcal{M}$, one has $(U^*)_0 U_0 f = U^* U f = f$, so that $U_0^* U_0 = I_0 \equiv$ the identity operator in the Hilbert space $\mathcal{M}$. Similarly one shows that $U_0 U_0^* = I_0$. This establishes the unitarity of $U_0$ (as an operator in $\mathcal{M}$).

(b) This is easily obtained by taking into account (1.31). □

If $\mathcal{H}$ and $\mathcal{K}$ are two Hilbert spaces, an operator $V \in \mathcal{B}(\mathcal{H}, \mathcal{K})$ is called ***unitary*** if it is isometric (i.e. $\|Vf\|_{\mathcal{K}} = \|f\|_{\mathcal{H}}$ for each $f \in \mathcal{H}$) and surjective (i.e. $\mathcal{R}(V) = \mathcal{K}$).

**2.2.3.** A generalisation of the notion of isometry is that of a partial isometry. An operator $\Omega$ belonging to $\mathcal{B}(\mathcal{H})$ is called a ***partial isometry*** if

$$\Omega^* \Omega = E, \tag{2.39}$$

where $E$ is a projection. In the next proposition we collect some properties of this type of operators.

**Proposition 2.9.** *Let $\Omega$ be a partial isometry. Then*

*(a)*
$$\Omega E = \Omega. \tag{2.40}$$

*(b)*
$$\langle \Omega f, \Omega g \rangle = \langle Ef, Eg \rangle \qquad \forall f, g \in \mathcal{H}. \tag{2.41}$$

*(c) One has $\|\Omega\| = 1$ except when $E = O$.*

*(d) $\Omega\Omega^*$ is a projection, and the associated subspace is $\mathcal{M}(\Omega\Omega^*) = \mathcal{R}(\Omega)$.*

*(e) Let $\mathcal{M}$ be a subspace and $W : \mathcal{M} \to \mathcal{H}$ a linear operator satisfying $\|Wf\| = \|f\|$ for all $f \in \mathcal{M}$. If one defines an operator $\Omega$ by $\Omega f = Wf$ if $f \in \mathcal{M}$ and $\Omega f = \mathbf{0}$ if $f \in \mathcal{M}^\perp$, then $\Omega$ is a partial isometry with $\Omega^*\Omega = P_{\mathcal{M}}$.*

PROOF. (a) Let $f \in \mathcal{H}$. Then we have with (2.39):

$$\begin{aligned} \|\Omega E f - \Omega f\|^2 &= \langle \Omega(Ef - f), \Omega(Ef - f) \rangle = \langle Ef - f, \Omega^*\Omega(Ef - f) \rangle \\ &= \langle Ef - f, E(Ef - f) \rangle = 0, \end{aligned}$$

since $E(Ef - f) = E^2 f - Ef = Ef - Ef = \mathbf{0}$. Thus $\Omega E f - \Omega f = \mathbf{0}$, hence $\Omega E f = \Omega f$.

The proof of (b) - (e) is similar to that of the corresponding statements in Proposition 2.7 and is left as an exercise. □

One sees that a partial isometry $\Omega$ maps the subspace $\mathcal{M}(E)$ isometrically onto the subspace $\mathcal{M}(F)$, where $F = \Omega\Omega^*$, whereas $\Omega f = \mathbf{0}$ for $f \perp \mathcal{M}(E)$ [if $f \perp \mathcal{M}(E)$, then by (2.40): $\Omega f = \Omega E f = \Omega \mathbf{0} = \mathbf{0}$]. In the language of operators between two Hilbert spaces, $\Omega$ would be called a unitary operator from $\mathcal{M}(E)$ to $\mathcal{M}(F)$. The subspaces $\mathcal{M}(E)$ and $\mathcal{M}(F)$ are called the ***initial set*** and the ***final set*** of $\Omega$ respectively.

## 2.3 Compact operators

Compact operators in an infinite-dimensional Hilbert space have properties that are very similar to those of operators in a finite-dimensional Hilbert space (for example they have no continuous spectrum, see Example 4.24). These operators are very useful in numerous theoretical developments and appear in various situations in this text.

**2.3.1.** We first consider operators $A$ defined by equations of the form

$$Af = \sum_{k=1}^{N} \langle g_k, f\rangle h_k, \tag{2.42}$$

where $\{g_k, h_k\}$ are $2N$ fixed vectors in $\mathcal{H}$ and $N < \infty$. In the notations used by physicists, $A$ is written as $A = \sum_{k=1}^{N} |h_k\rangle\langle g_k|$. The range $\mathcal{R}(A)$ of $A$ is contained in the finite-dimensional subspace spanned by $h_1, \ldots, h_N$, and $Af = \mathbf{0}$ if $f$ is orthogonal to the subspace spanned by $g_1, \ldots, g_N$. Thus $A$ may be considered to be an operator acting in the finite-dimensional subspace $\mathcal{M}$ spanned by $\{g_1, \ldots, g_N, h_1, \ldots, h_N\}$ in the sense that it is zero on $\mathcal{M}^\perp$ and its range is also contained in $\mathcal{M}$. $A$ can be described by a matrix acting on the vectors in $\mathcal{M}$. An operator of the form (2.42) with $N < \infty$ is called a ***finite rank operator***. One can show that each operator $A$ with $\dim \mathcal{R}(A) < \infty$ is a finite rank operator. Some simple properties of this type of operators are given in the following proposition:

**Proposition 2.10.** *Let $A$ be a finite rank operator. Then:*
*(a) $A$ belongs to $\mathcal{B}(\mathcal{H})$.*
*(b) $A^*$ is a finite rank operator.*
*(c) If $C \in \mathcal{B}(\mathcal{H})$, then $CA$ and $AC$ are finite rank operators.*
*(d) If $B$ is a second finite rank operator and $\alpha \in \mathbb{C}$, then $A + \alpha B$ is also a finite rank operator.*

PROOF. (a) This follows from the estimate $\|Af\| \leq \sum_{k=1}^{N} \|g_k\|\|f\|\|h_k\|$, which implies that $\|A\| \leq \sum_{k=1}^{N}\|g_k\|\|h_k\|$. For (b) we observe that for $e, f \in \mathcal{H}$:

$$\langle A^*e, f\rangle = \langle e, Af\rangle = \sum_{k=1}^{N} \langle g_k, f\rangle\langle e, h_k\rangle = \langle\big(\sum_{k=1}^{N} \langle h_k, e\rangle g_k\big), f\rangle,$$

which shows that $A^*e = \sum_{k=1}^{N} \langle h_k, e\rangle g_k$. Thus $A^*$ has the same form as $A$ but with $g_k$ and $h_k$ interchanged; one has $A^* = \sum_{k=1}^{N} |g_k\rangle\langle h_k|$. For (c) it suffices to observe that $CAf = \sum_{k=1}^{N} \langle g_k, f\rangle Ch_k$, which shows that $CA = \sum_{k=1}^{N} |Ch_k\rangle\langle g_k|$. Also $ACf = \sum_{k=1}^{N} \langle C^*g_k, f\rangle h_k$, i.e. $AC = \sum_{k=1}^{N} |h_k\rangle\langle C^*g_k|$. Finally (d) is evident. □

An operator $A \in \mathcal{B}(\mathcal{H})$ is ***compact*** if there exists a sequence $\{A_N\}$ of finite rank operators such that $\|A - A_N\| \to 0$ as $N \to \infty$, i.e. if $A$ is the norm limit of a sequence of finite rank operators. In particular each finite rank operator is compact, and in a finite-dimensional Hilbert space each operator is compact. We denote by $\mathcal{K}(\mathcal{H})$ the set of all compact operators in the Hilbert space $\mathcal{H}$. Some elementary properties of compact operators are given in the next proposition.

**Proposition 2.11.** *(a) $A$ is compact if and only if $A^*$ is compact.*
*(b) If $A \in \mathcal{K}(\mathcal{H})$ and $B \in \mathcal{B}(\mathcal{H})$, then $AB \in \mathcal{K}(\mathcal{H})$ and $BA \in \mathcal{K}(\mathcal{H})$.*
*(c) If $A_1$ and $A_2$ are compact, then so is $A_1 + \alpha A_2$ ($\alpha \in \mathbb{C}$).*
*(d) The uniform limit of a sequence of compact operators is compact: if $A_n \in \mathcal{K}(\mathcal{H})$ for $n = 1, 2, \ldots$, $A \in \mathcal{B}(\mathcal{H})$ and $\|A - A_n\| \to 0$ as $n \to \infty$, then $A \in \mathcal{K}(\mathcal{H})$.*

PROOF. (i) Let $A \in \mathcal{K}(\mathcal{H})$ and let $\{A_N\}$ be a sequence of finite rank operators such that $\|A - A_N\| \to 0$ as $N \to \infty$. We use the results of Proposition 2.10. Each $A_N^*$ is a finite rank operator and $\|A^* - A_N^*\| = \|A - A_N\| \to 0$ [using (2.10)]. Hence $A^*$ is compact. This proves (a). Next we observe that $A_N B$ and $BA_N$ are finite rank operators, and we have $\|AB - A_N B\| \leq \|A - A_N\|\|B\| \to 0$ and $\|BA - BA_N\| \leq \|B\|\|A - A_N\| \to 0$. This proves (b). The result of (c) is easily obtained from the definitions by using (2.7).

(ii) We now prove (d). Let $\varepsilon > 0$. For each $n$ choose a finite rank operator $A_{N(n)}$ such that $\|A_n - A_{N(n)}\| \leq \varepsilon/2$. Next take $n_0$ so large that $\|A - A_n\| < \varepsilon/2$ for $n \geq n_0$. Then (2.7) implies that $\|A - A_{N(n)}\| \leq \|A - A_n\| + \|A_n - A_{N(n)}\| < \varepsilon$ for $n \geq n_0$. Consequently $\|A - A_{N(n)}\| \to 0$ as $n \to \infty$, hence $A$ is compact. □

Compact operators have the property of improving the type of convergence of a sequence of vectors or a sequence of operators (they transform weak convergence into strong convergence and strong convergence into uniform convergence):

**Proposition 2.12.** *Let $A$ be a compact operator.*
*(a) If $\{f_n\}_{n\in\mathbb{N}}$ is a sequence of vectors that converges weakly to a vector $f$, then the sequence $\{Af_n\}$ converges strongly to $Af$.*
*(b) Suppose that $B_n$ ($n \in \mathbb{N}$) and $B$ belong to $\mathcal{B}(\mathcal{H})$ and that $\text{s-lim}_{n\to\infty} B_n = B$. Then $\|B_n A - BA\| \to 0$ and $\|AB_n^* - AB^*\| \to 0$ as $n \to \infty$; in other terms $\text{u-lim}_{n\to\infty} B_n A = BA$ and $\text{u-lim}_{n\to\infty} AB_n^* = AB^*$.*

PROOF. We use the Uniform Boundedness Principle (see the end of Section 1.1). Again, in our applications of Proposition 2.12, it will each time be evident that the considered sequence of vectors or operators is bounded.

(a) We set $\varphi_n = f_n - f$, so that $\text{w-lim}_{n\to\infty} \varphi_n = \mathbf{0}$. By the Uniform Boundedness Principle there exists a real number $M$ such that $\|\varphi_n\| \leq M$ for each $n$. Given $\varepsilon > 0$, we choose a finite rank operator $T$ [of the form (2.42)] such that $\|A - T\| < \varepsilon/(2M)$. Then

$$\|A\varphi_n\| \leq \|(A - T)\varphi_n\| + \|T\varphi_n\| \leq \varepsilon/2 + \sum_{k=1}^{N} |\langle g_k, \varphi_n\rangle|\|h_k\|.$$

Since $\text{w-lim}_{n\to\infty} \varphi_n = \mathbf{0}$, there exists $n_0 \in \mathbb{N}$ such that $|\langle g_k, \varphi_n\rangle|\|h_k\| < \varepsilon/(2N)$ for each $k = 1, \ldots, N$ and for all $n > n_0$. Thus $\|A\varphi_n\| < \varepsilon$ for all $n > n_0$, i.e. $\text{s-lim}_{n\to\infty} A\varphi_n = \mathbf{0}$.

(b) Set $C_n = B_n - B$, so that $\text{s-lim}_{n\to\infty} C_n = O$. There exists a real number $M$ such that $\|C_n\| \le M$ for each $n$. Let $\{A_N\}$ be a sequence of finite rank operators satisfying $\|A - A_N\| \to 0$ as $N \to \infty$. Then [assuming $A_N$ to be of the form (2.42)]:

$$\begin{aligned}\|C_n A f\| &\le M\|(A - A_N)f\| + \|C_n A_N f\| \\ &\le M\|A - A_N\|\|f\| + \sum_{k=1}^{N} |\langle g_k, f\rangle| \|C_n h_k\| \\ &\le \Big[M\|A - A_N\| + \sum_{k=1}^{N} \|g_k\|\|C_n h_k\|\Big] \|f\|.\end{aligned}$$

Given $\varepsilon > 0$ we first choose $N$ such that $\|A - A_N\| < \varepsilon/(2M)$ and then a number $n_0$ such that $\|g_k\|\|C_n h_k\| < \varepsilon/(2N)$ for all $n > n_0$ and each $k = 1, \dots, N$. One obtains $\|C_n A\| < \varepsilon$ if $n > n_0$, which proves the first assertion in (b). The second one follows from the first one by observing that $\|AB_n^* - AB^*\| = \|B_n A^* - BA^*\|$ and by remembering that $A^*$ is also compact. □

**Example 2.13.** A projection $P = P_{\mathcal{M}}$ is compact if and only if $\dim \mathcal{M} < \infty$. Indeed: (i) if $\dim \mathcal{M} < \infty$, $P_{\mathcal{M}}$ is a finite rank operator, given by $P_{\mathcal{M}} = \sum_{k=1}^{\dim \mathcal{M}} |e_k\rangle\langle e_k|$ for any orthonormal basis $\{e_k\}$ of $\mathcal{M}$. (ii) If $\dim \mathcal{M} = \infty$, one can choose an infinite orthonormal sequence $\{e_n\}$ in $\mathcal{M}$ (for example an orthonormal basis of $\mathcal{M}$). This sequence converges weakly to $\mathbf{0}$ but is not strongly convergent (see Example 1.2); so, by Proposition 2.12(a), $P_{\mathcal{M}}$ cannot be compact.

**2.3.2.** There are various special classes of compact operators that are of considerable use in the mathematical study of quantum-mechanical problems. We present here the most important one for our purposes, namely the class of Hilbert-Schmidt operators. These operators are useful for proving compactness of certain operators and play a role in scattering theory for the description of scattering cross sections.

The ***Hilbert-Schmidt norm*** $\|A\|_{\text{HS}}$ of an operator $A \in \mathcal{B}(\mathcal{H})$ is defined as follows:

$$\|A\|_{\text{HS}} = \Big[\sum_k \|Ae_k\|^2\Big]^{1/2}, \tag{2.43}$$

where $\{e_k\}$ is an orthonormal basis of $\mathcal{H}$. A priori this number depends on the chosen orthonormal basis, and it may be infinite. Let us first show that $\|A\|_{\text{HS}}$ depends only on $A$, i.e. that the sum in (2.43) is the same in each orthonormal basis of $\mathcal{H}$. So let $\{g_k\}$ be an arbitrary orthonormal basis of $\mathcal{H}$. Then, by developing $Ag_k$ in the basis $\{e_k\}$ (see Section 1.1), one has

$$\sum_k \|Ag_k\|^2 = \sum_k \sum_j |\langle e_j, Ag_k\rangle|^2 = \sum_k \sum_j |\langle A^* e_j, g_k\rangle|^2 = \sum_j \|A^* e_j\|^2 \tag{2.44}$$

(the interchange of the order of summation is permitted because all terms are non-negative). By considering the basis $\{e_k\}$ as being fixed, one sees that $\sum_k \|Ag_k\|^2$ is identical to the same number $\sum_j \|A^* e_j\|^2$ for each orthonormal basis $\{g_k\}$ and hence independent of this basis. The equation (2.44) also shows that

$$\|A\|_{\text{HS}} = \|A^*\|_{\text{HS}}. \tag{2.45}$$

An operator $A \in \mathcal{B}(\mathcal{H})$ is called a ***Hilbert-Schmidt operator*** if $\|A\|_{\text{HS}} < \infty$. The set of all Hilbert-Schmidt operators in $\mathcal{H}$ will be denoted by $\mathcal{B}_2(\mathcal{H})$ or simply by $\mathcal{B}_2$.

**Proposition 2.14.** *(a) One has $A \in \mathcal{B}_2(\mathcal{H})$ if and only if $A^* \in \mathcal{B}_2(\mathcal{H})$.*
*(b) Each Hilbert-Schmidt operator is compact, i.e. $\mathcal{B}_2(\mathcal{H}) \subseteq \mathcal{K}(\mathcal{H})$.*
*(c) If $A_1 \in \mathcal{B}_2(\mathcal{H})$ and $A_2 \in \mathcal{B}_2(\mathcal{H})$, then $A_1 + \alpha A_2 \in \mathcal{B}_2(\mathcal{H})$ $(\alpha \in \mathbb{C})$.*
*(d) One has*

$$\|A\| \leq \|A\|_{\text{HS}}. \tag{2.46}$$

*(e) If $A \in \mathcal{B}_2(\mathcal{H})$ and $B \in \mathcal{B}(\mathcal{H})$, then $AB \in \mathcal{B}_2(\mathcal{H})$ and $BA \in \mathcal{B}_2(\mathcal{H})$, with*

$$\|AB\|_{\text{HS}} \leq \|B\|\|A\|_{\text{HS}} \qquad \textit{and} \qquad \|BA\|_{\text{HS}} \leq \|B\|\|A\|_{\text{HS}}. \tag{2.47}$$

PROOF. The result of (a) follows from (2.45) and that of (c) from the definition (2.43) together with the inequality (1.14):

$$\begin{aligned}\|A_1 + \alpha A_2\|_{\text{HS}}^2 &= \sum_k \|(A_1 + \alpha A_2)e_k\|^2 \leq 2\sum_k \|A_1 e_k\|^2 + 2\sum_k |\alpha|^2 \|A_2 e_k\|^2 \\ &= 2\|A_1\|_{\text{HS}}^2 + 2|\alpha|^2 \|A_2\|_{\text{HS}}^2.\end{aligned}$$

For (d), let $f \neq \mathbf{0}$. Choose an orthonormal basis $\{e_k\}$ of $\mathcal{H}$ such that $e_1 = f/\|f\|$. Then

$$\frac{\|Af\|^2}{\|f\|^2} = \|Ae_1\|^2 \leq \sum_k \|Ae_k\|^2 = \|A\|_{\text{HS}}^2.$$

Since this inequality holds for each $f \neq \mathbf{0}$, one obtains (2.46). For (e) one has

$$\|BA\|_{\text{HS}}^2 = \sum_k \|BAe_k\|^2 \leq \|B\|^2 \sum_k \|Ae_k\|^2 = \|B\|^2 \|A\|_{\text{HS}}^2,$$

which gives the second inequality in (2.47). The first one then also follows: $\|AB\|_{\text{HS}} = \|B^*A^*\|_{\text{HS}} \leq \|B^*\|\|A^*\|_{\text{HS}} = \|B\|\|A\|_{\text{HS}}$ [by (2.45) and (2.10)].

It remains to prove (b). If $\dim \mathcal{H} < \infty$, this is trivial. So assume that $\dim \mathcal{H} = \infty$ and fix an orthonormal basis $\{e_k\}$ of $\mathcal{H}$. Let $A$ be a Hilbert-Schmidt operator. Define $A_N$ as follows by its action on $f = \sum_{k=1}^{\infty} \alpha_k e_k$: $A_N\left[\sum_{k=1}^{\infty} \alpha_k e_k\right] = \sum_{k=1}^{N} \alpha_k A e_k$; so $A_N e_n = Ae_n$ if $n \leq N$ whereas $A_N e_n = \mathbf{0}$ if $n > N$. $A_N$ is a finite rank operator: $A_N f = \sum_{k=1}^{N} \langle e_k, f\rangle A e_k$, and $\|A - A_N\|_{\text{HS}}^2 = \sum_{k=N+1}^{\infty} \|Ae_k\|^2 \to 0$ as $N \to \infty$ because $A \in \mathcal{B}_2$. This implies with (2.46) that $\|A - A_N\| \to 0$ as $N \to \infty$, hence $A \in \mathcal{K}(\mathcal{H})$. □

In an $L^2$ space the Hilbert-Schmidt operators have a simple and very useful characterisation as integral operators. Let $\mathcal{H} = L^2(\mathcal{O}, m)$. We recall that an operator $A$ is said to be an ***integral operator*** if there exists a measurable function $a\colon \mathcal{O} \times \mathcal{O} \to \mathbb{C}$, called the ***kernel*** of $A$, such that for all $f$ in the domain of definition of $A$ (see §2.4.1 for this concept):

$$(Af)(x) = \int_{\mathcal{O}} a(x,y) f(y) m(dy). \tag{2.48}$$

A measurable function $a\colon \mathcal{O} \times \mathcal{O} \to \mathbb{C}$ is called a ***Hilbert-Schmidt kernel*** if

$$M_a := \int_{\mathcal{O}} \int_{\mathcal{O}} |a(x,y)|^2 \, m(dx) m(dy) < \infty, \tag{2.49}$$

in other terms if the function $a$ belongs to $L^2(\mathcal{O}\times\mathcal{O}, m\times m)$.

**Proposition 2.15.** *Let $\mathcal{H} = L^2(\mathcal{O}, m)$. Then $A \in \mathcal{B}_2(\mathcal{H})$ if and only if $A$ is an integral operator with Hilbert-Schmidt kernel. Furthermore*

$$\|A\|_{\mathrm{HS}}^2 = M_a = \int_{\mathcal{O}} \int_{\mathcal{O}} |a(x,y)|^2 \, m(dx) m(dy). \tag{2.50}$$

REMARK. The preceding proposition, combined with (2.46), illustrates the usefulness of Hilbert-Schmidt operators: in certain cases it is possible to estimate the norm of an operator by calculating or estimating an integral. Another result of this type is given in §2.5.5.

PROOF OF PROPOSITION 2.15. To simplify the notations we do not specify the measures in the occurring $L^2$ spaces (this concerns either the measure $m$ or the product measure $m\times m$). We use the fact that $L^2(\mathcal{O}\times\mathcal{O}) = L^2(\mathcal{O}) \otimes L^2(\mathcal{O})$, the tensor product of $L^2(\mathcal{O})$ with itself. More precisely we use the following fact: if the functions $\{e_k(x)\}_{k\in\mathbb{N}}$ form an orthonormal basis of $L^2(\mathcal{O})$ and $\overline{e_k}$ denotes the complex conjugate of $e_k$, then the collection $\{\overline{e_k}(x)\}_{k\in\mathbb{N}}$ is also an orthonormal basis of $L^2(\mathcal{O})$, and the functions $\{e_j(x)\overline{e_k}(y)\}_{j,k\in\mathbb{N}}$ give an orthonormal basis of $L^2(\mathcal{O}\times\mathcal{O})$.

(i) Let $A$ be an integral operator with Hilbert-Schmidt kernel $a(x,y)$, and let $\mathcal{V} = \{x \in \mathcal{O} \mid \int_{\mathcal{O}} |a(x,y)|^2 m(dy) = \infty\}$. Then $m(\mathcal{V}) = 0$ because $a(\cdot,\cdot) \in L^2(\mathcal{O}\times\mathcal{O})$. Now let $f \in L^2(\mathcal{O})$ and $x \notin \mathcal{V}$. If the integral in Eq. (2.48) is interpreted as a scalar product, one obtains by applying the Schwarz inequality that

$$|(Af)(x)|^2 \leq \int_{\mathcal{O}} |a(x,y)|^2 \, m(dy) \cdot \int_{\mathcal{O}} |f(y')|^2 \, m(dy').$$

Integration of this inequality with respect to $x$ leads to $\|Af\|^2 \leq M_a \|f\|^2$. This shows that $A$ associates with each vector $f$ of $\mathcal{H}$ an element $Af$ of $\mathcal{H}$ and that $A \in \mathcal{B}(\mathcal{H})$ with $\|A\| \leq M_a$.

Let $\{e_k\}$ be an orthonormal basis of $L^2(\mathcal{O})$. By setting $\alpha_{jk} = \langle e_j, Ae_k\rangle$ we have $Ae_k = \sum_j \alpha_{jk} e_j$ and

$$\|A\|_{\mathrm{HS}}^2 = \sum_k \|Ae_k\|^2 = \sum_{j,k} |\alpha_{jk}|^2. \tag{2.51}$$

Let us also consider the orthonormal basis $\{e_j\overline{e_k}\}$ of $L^2(\mathcal{O} \times \mathcal{O})$. One can develop $a(\cdot,\cdot)$ in this basis: $a = \sum_{j,k} \beta_{jk} e_j \overline{e_k}$, and one then has $M_a = \sum_{j,k} |\beta_{jk}|^2$. Now observe that

$$\begin{aligned}\alpha_{jk} = \langle e_j, Ae_k\rangle &= \int_{\mathcal{O}}\int_{\mathcal{O}} \overline{e_j(x)}a(x,y)e_k(y)m(dx)m(dy)\\ &= \langle e_j\overline{e_k}, a\rangle_{L^2(\mathcal{O}\times\mathcal{O})} = \beta_{jk}.\end{aligned} \tag{2.52}$$

Eqs. (2.51) and (2.52) imply that $\|A\|_{\mathrm{HS}}^2 = \sum_{j,k}|\alpha_{jk}|^2 = \sum_{j,k}|\beta_{jk}|^2 = M_a$. In particular $A$ belongs to $\mathcal{B}_2$.

(ii) For $A \in \mathcal{B}_2$, we define $\alpha_{jk}$ as above and set $a_N(x,y) = \sum_{j,k\le N}\alpha_{jk}e_j(x)\overline{e_k(y)}$, where $N = 1, 2, \ldots$. Notice that $a_N$ is a finite linear combination of functions belonging to $L^2(\mathcal{O}\times\mathcal{O})$, hence $a_N \in L^2(\mathcal{O}\times\mathcal{O})$. If $M > N$ we have

$$\begin{aligned}\int_{\mathcal{O}}\int_{\mathcal{O}} |a_M(x,y) - a_N(x,y)|^2 m(dx)m(dy) &= \sum_{j=1}^{N}\sum_{k=N+1}^{M}|\alpha_{jk}|^2 + \sum_{j=N+1}^{M}\sum_{k=1}^{M}|\alpha_{jk}|^2\\ &\le \sum_{j=1}^{\infty}\sum_{k=N+1}^{\infty}|\alpha_{jk}|^2 + \sum_{j=N+1}^{\infty}\sum_{k=1}^{\infty}|\alpha_{jk}|^2.\end{aligned}$$

Since $A \in \mathcal{B}_2$, the double sum in (2.51) is convergent. Thus the preceding inequality implies that $\{a_N\}$ is a strong Cauchy sequence in $L^2(\mathcal{O}\times\mathcal{O})$. We denote its limit by $a_\infty$. By (i) above this function $a_\infty$ is the kernel of a Hilbert-Schmidt operator which we denote by $A_\infty$. To know that $A$ is an integral operator with Hilbert-Schmidt kernel, we still must prove that $A_\infty = A$. Since each of these two operators belongs to $\mathcal{B}(\mathcal{H})$, it suffices to verify that $A_\infty e_k = Ae_k$ for each vector $e_k$ of the basis $\{e_j\}$. We know already that $Ae_k = \sum_j \alpha_{jk}e_j$. If we write $A_\infty e_k = \sum_j \gamma_{jk}e_j$, we have by (2.52):

$$\gamma_{jk} = \langle e_j, A_\infty e_k\rangle = \langle e_j\overline{e_k}, a_\infty\rangle_{L^2(\mathcal{O}\times\mathcal{O})} = \lim_{N\to\infty}\langle e_j\overline{e_k}, a_N\rangle_{L^2(\mathcal{O}\times\mathcal{O})} = \alpha_{jk}. \qquad \square$$

**Remark 2.16.** We mention some further interesting properties of the set $\mathcal{B}_2(\mathcal{H})$ (for more details and proofs, see e.g. Chapter 3 of [A] or Section VI.6 of [RS]). For $A, B \in \mathcal{B}_2(\mathcal{H})$, set

$$\langle A,B\rangle = \sum_k \langle Ae_k, Be_k\rangle \equiv \mathrm{Tr}A^*B, \tag{2.53}$$

where $\{e_k\}$ is an orthonormal basis of $\mathcal{H}$ and Tr stands for *trace*. From the polarisation identity (1.11) one finds that

$$\langle A,B\rangle = \frac{1}{4}\left(\|A+B\|_{\mathrm{HS}}^2 - \|A-B\|_{\mathrm{HS}}^2 + i\|A-iB\|_{\mathrm{HS}}^2 - i\|A+iB\|_{\mathrm{HS}}^2\right), \tag{2.54}$$

which shows that the sum in (2.53) is finite and independent of the chosen orthonormal basis.

(a) Eq. (2.53) defines a scalar product on the set $\mathcal{B}_2(\mathcal{H})$ of all Hilbert-Schmidt operators in $\mathcal{H}$, and $\mathcal{B}_2(\mathcal{H})$ is a Hilbert space with this scalar product (and clearly the norm in this Hilbert space is the Hilbert-Schmidt norm $\|\cdot\|_{\mathrm{HS}}$). It should be noted that the vectors (i.e. the elements) of this Hilbert space $\mathcal{B}_2(\mathcal{H})$ are linear operators acting in another Hilbert space, namely in $\mathcal{H}$. As a Hilbert space $\mathcal{B}_2(\mathcal{H})$ is isomorphic to the tensor product $\mathcal{H}\otimes\mathcal{H}$. The fact that $\mathcal{B}_2$ is complete with respect to the norm $\|\cdot\|_{\mathrm{HS}}$

has the following consequence (see Remark 2.4): Let $[a,b]$ be a finite interval and $s \mapsto A(s)$ a function defined on $[a,b]$ with values in $\mathcal{B}_2(\mathcal{H})$ which is continuous in Hilbert-Schmidt norm (i.e. $\|A(s)-A(t)\|_{\rm HS} \to 0$ as $s \to t$). Then $\int_a^b A(s)ds$ belongs to $\mathcal{B}_2(\mathcal{H})$.

(b) Eqs. (2.53) and (2.54) show that the product of two Hilbert-Schmidt operators has a finite trace. Such products define a special class of compact operators, the ***trace class*** $\mathcal{B}_1(\mathcal{H})$; an operator $A \in \mathcal{B}(\mathcal{H})$ belongs to $\mathcal{B}_1(\mathcal{H})$ if and only if it can be expressed as $A = BC$ with $B, C \in \mathcal{B}_2(\mathcal{H})$. A trace class operator admits a finite trace which has the same value in each orthonormal basis of $\mathcal{H}$. It is clear from Proposition 2.14(e) that, if $A \in \mathcal{B}_1(\mathcal{H})$ and $B \in \mathcal{B}(\mathcal{H})$, then $AB$ and $BA$ belong to $\mathcal{B}_1(\mathcal{H})$. Trace class operators play an important role in quantum mechanics. For example the states of quantum-mechanical systems are described by self-adjoint positive trace class operators $\rho$ with $\mathrm{Tr}\,\rho = 1$, called *density matrices* or *density operators*. If $A$ is a bounded self-adjoint operator describing some observable quantity, then $\rho A$ belongs to $\mathcal{B}_1(\mathcal{H})$, so that its trace $\mathrm{Tr}\,\rho A$ is well defined. The number $\mathrm{Tr}\,\rho A$ represents the expectation value of the observable $A$ in the state $\rho$.

## 2.4 Unbounded operators

**2.4.1.** In quantum mechanics one often meets operators that are not bounded. Such operators cannot be defined on all vectors of the underlying Hilbert space $\mathcal{H}$ but only on a suitable subset of $\mathcal{H}$. To assure that linearity makes sense, it is necessary for this subset to be stable under the operation of taking finite linear combinations, so this subset should be a linear manifold in $\mathcal{H}$. One is thus led to the following definition.

A ***linear operator*** in a Hilbert space $\mathcal{H}$ is a couple $\{\mathcal{D}(A), A\}$, where $\mathcal{D}(A)$ is a linear manifold in $\mathcal{H}$, and $A$ is a linear mapping from $\mathcal{D}(A)$ to $\mathcal{H}$. Thus with each vector $f \in \mathcal{D}(A)$ there is associated a vector $Af \in \mathcal{H}$ in such a way that

$$A(\alpha f + g) = \alpha Af + Ag \qquad \forall f, g \in \mathcal{D}(A), \forall \alpha \in \mathbb{C}.$$

The linear manifold $\mathcal{D}(A)$ is called the domain of definition or simply the ***domain*** of the operator. It is customary to speak simply of an ***operator*** $A$ instead of $\{\mathcal{D}(A), A\}$ (it is understood that the mapping $A$ is defined on a certain linear manifold $\mathcal{D}(A)$ and that it is linear [2]). We denote by $\mathcal{R}(A)$ the ***range*** of $A$ and by $\mathcal{N}(A)$ its ***null space***:

$$\mathcal{R}(A) = A\mathcal{D}(A) = \{f \in \mathcal{H} \mid \exists g \in \mathcal{D}(A) \text{ with } f = Ag\} \tag{2.55}$$

$$\mathcal{N}(A) = \{f \in \mathcal{D}(A) \mid Af = \mathbf{0}\}. \tag{2.56}$$

Two operators $A$ and $B$ are ***equal*** if $\mathcal{D}(A) = \mathcal{D}(B)$ and $Af = Bf$ for each vector $f$ in $\mathcal{D}(A)$. If $\mathcal{D}(A) \subseteq \mathcal{D}(B)$ and $Bf = Af$ for each $f \in \mathcal{D}(A)$ (the smaller of the two domains), one says that $B$ is an ***extension*** of $A$ or that $A$ is the ***restriction*** of $B$ to the linear manifold $\mathcal{D}(A)$, and one writes $A \subseteq B$. The ***sum*** $A+B$ of two operators $A$ and

[2] If $\mathcal{D}(A) = \mathcal{H}$, one says that $A$ is an ***everywhere defined*** operator.

$B$ is defined as follows: $\mathcal{D}(A+B) = \mathcal{D}(A) \cap \mathcal{D}(B)$ and $(A+B)f = Af + Bf$ for $f \in \mathcal{D}(A+B)$. For the ***product*** $AB$ one has: $\mathcal{D}(AB) = \{f \in \mathcal{D}(B) \mid Bf \in \mathcal{D}(A)\}$ and $(AB)f = A(Bf)$ for $f \in \mathcal{D}(AB)$. $\mathcal{D}(A+B)$ and $\mathcal{D}(AB)$ are linear manifolds, but it may happen that they consist of only the zero vector, even if $\mathcal{D}(A)$ and $\mathcal{D}(B)$ are dense in $\mathcal{H}$ (see e.g. Problem 2.26). If $A \in \mathcal{B}(\mathcal{H})$, then clearly $\mathcal{D}(A+B) = \mathcal{D}(B)$ and $\mathcal{D}(AB) = \mathcal{D}(B)$.

In quantum mechanics one often uses commutation relations. When unbounded operators are involved some caution is required. We recall that if $A,B \in \mathcal{B}(\mathcal{H})$, then their ***commutator*** $[A,B]$ is defined as $[A,B] = AB - BA$. If at least one of the operators $A$ and $B$ is unbounded, the products in the preceding definition are not defined on all vectors (possibly only on the zero vector **0**). The commutator is a priori defined on the domain $\mathcal{D}([A,B]) = \{f \in \mathcal{D}(A) \cap \mathcal{D}(B) \mid Bf \in \mathcal{D}(A) \text{ and } Af \in \mathcal{D}(B)\}$. Nevertheless, in certain situations it is possible by different methods to give a meaning to $[A,B]$ on a larger domain. For example, if $A$ and $B$ are self-adjoint, the number $\langle f, [A,B]g\rangle$ makes sense for $f, g \in \mathcal{D}(A) \cap \mathcal{D}(B)$ if it is interpreted as $\langle Af, Bg\rangle - \langle Bf, Ag\rangle$ (the self-adjointness of unbounded operators will be discussed in §2.4.4).

If the operator $B$ belongs to $\mathcal{B}(\mathcal{H})$, then $\mathcal{D}([A,B]) = \{f \in \mathcal{D}(A) \mid Bf \in \mathcal{D}(A)\}$. In this case $A$ and $B$ commute if $ABf = BAf$ for each $f$ in this domain; in operator language the commutativity of $A$ and $B$ would be written as $[A,B] \subseteq O$. If $\mathcal{D}(A) \neq \mathcal{H}$, then the zero operator $O$ is a strict extension of $[A,B]$ because $\mathcal{D}(O) = \mathcal{H}$ whereas $\mathcal{D}([A,B]) \neq \mathcal{H}$.

An operator $A$ is ***bounded*** if there exists a real number $M < \infty$ such that $\|Af\| \leq M\|f\|$ for all $f \in \mathcal{D}(A)$. In this case one can define the (finite) norm of $A$ in analogy with (2.1):

$$\|A\|_{\mathcal{D}(A)} = \inf\{M \in \mathbb{R} \mid \|Af\| \leq M\|f\| \ \forall f \in \mathcal{D}(A)\} = \sup_{0 \neq f \in \mathcal{D}(A)} \frac{\|Af\|}{\|f\|}. \tag{2.57}$$

A bounded operator with domain $\mathcal{D}(A)$ has a natural bounded extension to the subspace $\overline{\mathcal{D}(A)}$ spanned by $\mathcal{D}(A)$. This extension is denoted $\overline{A}$ and called the ***closure of A***. We explain this in the particular case in which the domain $\mathcal{D}(A)$ is dense in $\mathcal{H}$ (hence $\overline{\mathcal{D}(A)} = \mathcal{H}$); the operator $\overline{A}$ then belongs to $\mathcal{B}(\mathcal{H})$.

So let us assume that $A$ is bounded and that $\mathcal{D}(A)$ is dense in $\mathcal{H}$. Given a vector $f \in \mathcal{H}$, choose a sequence $\{f_n\}$ in $\mathcal{D}(A)$ such that $f = \text{s-lim}_{n\to\infty} f_n$. Since the operator $A$ is bounded, the sequence $\{Af_n\}$ is also strongly Cauchy: $\|Af_m - Af_n)\| \leq \|A\|_{\mathcal{D}(A)}\|f_m - f_n\| \to 0$ as $m, n \to \infty$. Let $h = \text{s-lim}_{n\to\infty} Af_n$ and set $\overline{A}f = h$. Notice that $\overline{A}$ is well defined, i.e. the vector $h$ is independent of the sequence $\{f_n\}$ converging to $f$: if $\{g_n\}$ is another sequence in $\mathcal{D}(A)$ satisfying $\text{s-lim}_{n\to\infty} g_n = f$, then the sequence $\{Ag_n\}$ also converges to h:

$$\|h - Ag_n\| = \|(h - Af_n) + (Af_n - Ag_n)\| \leq \|h - Af_n\| + \|A\|_{\mathcal{D}(A)}\|f_n - g_n\| \to 0,$$

because $\|f_n - g_n\| \leq \|f_n - f\| + \|f - g_n\|$. So $\overline{A}$ is well defined on each vector $f \in \mathcal{H}$, and it is an extension of $A$: if $f \in \mathcal{D}(A)$, one may take $f_n = f$ for each $n$, so that $\overline{A}f = Af$. Finally let us show that the norm of $\overline{A}$ [as an operator in $\mathcal{B}(\mathcal{H})$] is identical with the norm of $A$ [as an operator defined on $\mathcal{D}(A)$], i.e. that

$$\|\overline{A}\| \equiv \sup_{0\neq f\in\mathcal{H}} \frac{\|\overline{A}f\|}{\|f\|} = \|A\|_{\mathcal{D}(A)} \equiv \sup_{0\neq g\in\mathcal{D}(A)} \frac{\|Ag\|}{\|g\|}.$$

For this we observe that, if $f$ and $\{f_n\}$ are as above, one has (apply twice Proposition 1.1)

$$\|\overline{A}f\| = \lim_{n\to\infty} \|Af_n\| \leq \|A\|_{\mathcal{D}(A)} \lim_{n\to\infty} \|f_n\| = \|A\|_{\mathcal{D}(A)}\|f\|,$$

hence $\|\overline{A}\| \leq \|A\|_{\mathcal{D}(A)}$. The opposite inequality is evident.

One can try to apply the preceding construction also to an unbounded operator $A$ in order to obtain a natural extension of $A$. Assume again that $\mathcal{D}(A)$ is dense in $\mathcal{H}$. Consider a vector $f$ in $\mathcal{H}$ and a sequence $\{f_n\}$ of vectors in $\mathcal{D}(A)$ converging strongly to $f$. Here the sequence $\{Af_n\}$ will not be Cauchy in general. But assume that $\{Af_n\}$ is (strongly) Cauchy and denote its limit again by $h$. Then it is natural to set $\overline{A}f = h$ as before. However, in order for this to make sense, one has to know that each sequence $\{f_n\}$ with the indicated properties leads to the same vector $h$, in other words: if $\{g_n\}$ is any other sequence in $\mathcal{D}(A)$ converging strongly to $f$ and such that $\{Ag_n\}$ is Cauchy, then the limit of $\{Ag_n\}$ must be the same as that of $\{Af_n\}$ (i.e. the vector $h$). If this condition is satisfied, then $\overline{A}f$ is well defined.

These considerations lead to the introduction of two important classes of operators, each of them containing $\mathcal{B}(\mathcal{H})$, namely the closable operators and the closed operators. We now describe these operators.

An operator $A$ is called closable if the condition on Cauchy sequences mentioned above is always satisfied: whenever $\{f_n\}$ and $\{g_n\}$ are Cauchy sequences in $\mathcal{D}(A)$ converging to the same limit and the sequences $\{Af_n\}$ and $\{Ag_n\}$ are also Cauchy, then $\text{s-lim}_{n\to\infty} Af_n = \text{s-lim}_{n\to\infty} Ag_n$. By observing that the sequence $\{f_n - g_n\}$ converges strongly to the zero vector $\mathbf{0}$, this condition then means that the limit of the sequence $\{Af_n - Ag_n\}$ – which is assumed to be Cauchy – must also be the vector $\mathbf{0}$. Thus the definition of a closable operator can be written as follows:

*An operator A is* ***closable*** *if the three properties*

(i) $f_n \in \mathcal{D}(A)$, (ii) $\text{s-lim}_{n\to\infty} f_n = \mathbf{0}$, and (iii) $\{Af_n\}$ *is strongly Cauchy together imply that* $\text{s-lim}_{n\to\infty} Af_n = \mathbf{0}$.

If $A$ is closable then, as illustrated above, one can associate with it a natural extension $\overline{A}$, called its ***closure***: the elements of $\mathcal{D}(\overline{A})$ are the vectors $f$ in $\mathcal{H}$ for which there exists a sequence $\{f_n\}$ in $\mathcal{D}(A)$ which converges strongly to $f$ and which is such that the sequence $\{Af_n\}$ is also strongly Cauchy, and then $\overline{A}f = \text{s-lim}_{n\to\infty} Af_n$. To see that $\overline{A}$ is an extension of $A$, it suffices to take $f_n = f$ for each $n$ if $f \in \mathcal{D}(A)$.

A special class of closable operators are the closed operators, defined as follows:

*An operator A is* ***closed*** *if the three conditions*

(i) $f_n \in \mathcal{D}(A)$, (ii) $\text{s-lim}_{n\to\infty} f_n = f$, and (iii) $\{Af_n\}$ *is strongly Cauchy together imply that* $f \in \mathcal{D}(A)$ *and* $\text{s-lim}_{n\to\infty} Af_n = Af$.

It is seen from the preceding two definitions that each closed operator is closable and equal to its closure. If $A$ is a closable operator, then its closure $\overline{A}$ is a closed operator (hence the closure of the operator $\overline{A}$ is identical with $\overline{A}$).[3]

If $A$ is closable and $B \in \mathcal{B}(\mathcal{H})$, then $A+B$ is also closable. If $A$ is closed and $B \in \mathcal{B}(\mathcal{H})$, then $A+B$ is closed. The null space of a closed operator $A$ is a subspace. It is strongly recommended to check these assertions as an exercise.

An unbounded closable or closed operator will have many closed extensions (see Chapter 3 and Problem 2.17 for examples); its closure is the minimal closed extension, as explained by the next proposition.

**Proposition 2.17.** *If an operator $A$ has a closed extension $B$, then $A$ is closable and $B$ is also an extension of the closure of $A$. In other terms: If $A \subseteq B$ and $B$ is closed, then $A$ is closable and $\overline{A} \subseteq B$.*

PROOF. Let $\{f_n\}$ be a sequence of vectors in $\mathcal{D}(A)$ such that $\text{s-lim}_{n\to\infty} f_n = \mathbf{0}$ and such that the sequence $\{Af_n\}$ is strongly Cauchy. Since $Af_n = Bf_n$, the sequence $\{f_n\}$ satisfies the three conditions (i) $f_n \in \mathcal{D}(B)$, (ii) $\text{s-lim}_{n\to\infty} f_n = \mathbf{0}$, and (iii) $\{Bf_n\}$ is strongly Cauchy. Since $B$ is closed, we must have $\mathbf{0} \in \mathcal{D}(B)$ (which is anyway the case) and $B\mathbf{0} \equiv \mathbf{0} = \text{s-lim}_{n\to\infty} Bf_n \equiv \text{s-lim}_{n\to\infty} Af_n$. This shows that $A$ is closable.

Now assume that $f \in \mathcal{D}(\overline{A})$. Then there exists a sequence $\{f_n\}$ in $\mathcal{D}(A)$ such that $\text{s-lim}_{n\to\infty} f_n = f$ and such that the sequence $\{Af_n\}$ is strongly Cauchy. But $Af_n = Bf_n$. Since $B$ is closed, we have $f \in \mathcal{D}(B)$ and $Bf = \text{s-lim}_{n\to\infty} Af_n = \overline{A}f$. This shows that $B$ is an extension of $\overline{A}$. □

The concept of invertibility of an operator is important. An operator $A$ is ***invertible*** if $\mathcal{N}(A) = \{\mathbf{0}\}$, in other terms if the conditions $f \in \mathcal{D}(A)$ and $Af = \mathbf{0}$ imply that $f = \mathbf{0}$. If $A$ is invertible, its inverse $A^{-1}$ gives a bijection from $\mathcal{R}(A)$ onto $\mathcal{D}(A)$: $\mathcal{D}(A^{-1}) = \mathcal{R}(A)$ and $\mathcal{R}(A^{-1}) = \mathcal{D}(A)$. If $A$ is invertible and closed, then its inverse $A^{-1}$ is also closed (exercise).

**2.4.2.** In relation to the concept of closure, the notion of the graph of an operator is frequently used. Consider the Hilbert space $\mathcal{H} \oplus \mathcal{H}$ (the orthogonal sum of two copies of $\mathcal{H}$). If $A$ is an operator in $\mathcal{H}$, then its ***graph*** $\Gamma(A)$ is the subset of $\mathcal{H} \oplus \mathcal{H}$ consisting of all pairs of vectors of $\mathcal{H}$ of the form $\{f, Af\}$ with $f \in \mathcal{D}(A)$.[4] It is clear that $\Gamma(A)$ is a linear manifold in $\mathcal{H} \oplus \mathcal{H}$. The norm in $\mathcal{H} \oplus \mathcal{H}$ of an element $\{f, Af\}$ of the graph $\Gamma(A)$ is given by $(\|f\|^2 + \|Af\|^2)^{1/2}$. A (strong) Cauchy sequence in $\Gamma(A)$ is a

---

[3] Assume that $A$ is closable, that $f_n \in \mathcal{D}(\overline{A})$, that $\text{s-lim}_{n\to\infty} f_n = f$ and that $\{\overline{A}f_n\}$ is Cauchy, say $\text{s-lim}_{n\to\infty} \overline{A}f_n = h$. To see that the operator $\overline{A}$ is closed, we must show that $f \in \mathcal{D}(\overline{A})$ and that $\overline{A}f = h$. Now, by the definition of the closure of $A$, there exists for each $n$ a sequence $\{g_k^{(n)}\}_{k\in\mathbb{N}}$ in $\mathcal{D}(A)$ such that $\|f_n - g_k^{(n)}\| \to 0$ and $\|Ag_k^{(n)} - \overline{A}f_n\| \to 0$ as $k \to \infty$. For each $n$, choose an integer $k(n)$ such that $\|f_n - g_{k(n)}^{(n)}\| < 1/n$ and $\|Ag_{k(n)}^{(n)} - \overline{A}f_n\| < 1/n$. Then $\{g_{k(n)}^{(n)}\}_{n\in\mathbb{N}}$ is a sequence in $\mathcal{D}(A)$ such that $\text{s-lim}_{n\to\infty} g_{k(n)}^{(n)} = f$ and such that $\{Ag_{k(n)}^{(n)}\}$ is Cauchy, namely $\text{s-lim}_{n\to\infty} Ag_{k(n)}^{(n)} = h$. Again by the definition of the closure of $A$, we have $f \in \mathcal{D}(\overline{A})$ and $\overline{A}f = \text{s-lim}_{n\to\infty} Ag_{k(n)}^{(n)} = h$.

[4] Observe the analogy with the graph of a function $\varphi\colon \mathbb{R} \to \mathbb{R}$ in the plane $\mathbb{R}^2$, given by the set of points $\{x, \varphi(x)\}$.

sequence $\{f_n, Af_n\}$ with $f_n \in \mathcal{D}(A)$ such that both $\{f_n\}$ and $\{Af_n\}$ are Cauchy (in $\mathcal{H}$). This shows that
(i) an operator $A$ is closed if and only if its graph $\Gamma(A)$ is closed in $\mathcal{H} \oplus \mathcal{H}$ (i.e. if and only if $\Gamma(A)$ is a subspace of $\mathcal{H} \oplus \mathcal{H}$);
(ii) if $A$ is closable, then the graph $\Gamma(\overline{A})$ of its closure coincides with the closure in $\mathcal{H} \oplus \mathcal{H}$ of the graph of $A$;
(iii) an operator $A$ is closable if and only if the closure of its graph contains no element of the form $\{\mathbf{0}, h\}$ with $h \neq \mathbf{0}$.
We shall use the properties (i) and (ii) in the proof of Proposition 2.22.

The expression given above for the norm of an element $\{f, Af\}$ of $\Gamma(A)$ [remember that $f \in \mathcal{D}(A)$] suggests that one may define a scalar product on $\mathcal{D}(A)$ by

$$\langle f, g\rangle_A = \langle f, g\rangle + \langle Af, Ag\rangle \qquad \text{if } f, g \in \mathcal{D}(A), \tag{2.58}$$

where $\langle\cdot,\cdot\rangle$ is the given scalar product in $\mathcal{H}$. If $A$ is closed, the linear manifold $\mathcal{D}(A)$, equipped with the scalar product (2.58), is a Hilbert space, and the norm associated with this scalar product, called the ***graph norm of A***, is given by

$$\|f\|_A = \left[\|f\|^2 + \|Af\|^2\right]^{1/2}. \tag{2.59}$$

**2.4.3.** We now turn to another very important question, the definition of the adjoint of an unbounded operator. We recall that, for an operator $A \in \mathcal{B}(\mathcal{H})$, the adjoint $A^*$ is another operator in $\mathcal{B}(\mathcal{H})$ defined by the condition that $\langle A^*f, g\rangle = \langle f, Ag\rangle$ for all $f, g \in \mathcal{H}$. If $A$ is not bounded, it is clearly necessary to restrict $g$ to $\mathcal{D}(A)$; because of this restriction one can no longer use the Riesz Lemma to conclude that $A^*$ exists. Also it seems clear that $A^*f$ will not be defined for all vectors $f$ in $\mathcal{H}$. So one has to specify not only the mapping $A^*$ but also its domain $\mathcal{D}(A^*)$. The maximal possible domain is obtained by applying the following definition in which one first defines the domain of $A^*$ in terms of the domain of $A$ and of the mapping $A$, and then one specifies the action of $A^*$ on the vectors in this domain:

**Definition 2.18.** *Let $A$ be an operator in $\mathcal{H}$ with domain $\mathcal{D}(A)$ that is dense in $\mathcal{H}$. Then the* ***adjoint*** *$A^*$ of $A$ is the following operator:*
*domain: a vector $f \in \mathcal{H}$ belongs to $\mathcal{D}(A^*)$ if there exists a vector $f^*$ in $\mathcal{H}$ such that*

$$\langle f^*, g\rangle = \langle f, Ag\rangle \qquad \forall g \in \mathcal{D}(A), \tag{2.60}$$

*mapping: if $f \in \mathcal{D}(A^*)$, one sets $A^*f = f^*$.*

COMMENTS. (i) The condition that $\mathcal{D}(A)$ be dense in $\mathcal{H}$ is necessary to ensure that $A^*$ is well defined, i.e. that the vector $f^*$ (if it exists) in (2.60) is unique. Indeed, if $f_1^*$ and $f_2^*$ are two vectors satisfying $\langle f_1^*, g\rangle = \langle f, Ag\rangle = \langle f_2^*, g\rangle$ for all $g \in \mathcal{D}(A)$, then $f_1^* - f_2^*$ is orthogonal to $\mathcal{D}(A)$, hence equal to the zero vector by Proposition 1.6 (provided that $\mathcal{D}(A)$ is dense in $\mathcal{H}$).

(ii) It is clear that, if $f, h \in \mathcal{D}(A^*)$ and $f^*, h^*$ are the vectors that correspond to them by (2.60), and if $\alpha \in \mathbb{C}$, then $\langle f^* + \alpha h^*, g\rangle = \langle f + \alpha h, Ag\rangle$ for all $g \in \mathcal{D}(A)$.

So $f+\alpha h \in \mathcal{D}(A^*)$, i.e. $\mathcal{D}(A^*)$ is a linear manifold, and $A^*(f+\alpha h) \equiv (f+\alpha h)^* = f^* + \alpha h^* = A^*f + \alpha A^*h$, i.e. the mapping $A^*$ is linear.

(iii) It may happen that $\mathcal{D}(A^*)$ is rather small, even that it consists of only the zero vector. One can show that, if $A$ is closable (and densely defined), then $\mathcal{D}(A^*)$ is dense in $\mathcal{H}$.

(iv) Finally we observe that, once $A^*$ has been defined, the relation (2.60) can be written as

$$\langle A^*f, g\rangle = \langle f, Ag\rangle \qquad \forall f \in \mathcal{D}(A^*),\ \forall g \in \mathcal{D}(A). \tag{2.61}$$

**Proposition 2.19.** *Let $A$ be an operator in $\mathcal{H}$ with dense domain $\mathcal{D}(A)$. Then*
*(a) $A^*$ is closed.*
*(b) One has $\mathcal{N}(A^*) = \mathcal{R}(A)^\perp$.*
*(c) If $B$ is an extension of $A$, then $A^*$ is an extension of $B^*$ : $A \subseteq B \Longrightarrow B^* \subseteq A^*$.*

PROOF. (a) Assume that $f_n \in \mathcal{D}(A^*)$, $\text{s-lim}_{n\to\infty} f_n = f$ and $\text{s-lim}_{n\to\infty} A^*f_n = h$. Then, for each $g \in \mathcal{D}(A)$:

$$\langle f, Ag\rangle = \lim_{n\to\infty} \langle f_n, Ag\rangle = \lim_{n\to\infty} \langle A^*f_n, g\rangle = \langle h, g\rangle.$$

Hence $f \in \mathcal{D}(A^*)$ and $A^*f = h$, which proves that $A^*$ is closed.

(b) Let $f \in \mathcal{N}(A^*)$, i.e. $f \in \mathcal{D}(A^*)$ and $A^*f = \mathbf{0}$. Then $0 = \langle A^*f, g\rangle = \langle f, Ag\rangle$ for all $g \in \mathcal{D}(A)$, hence $f \perp \mathcal{R}(A)$. For the opposite inclusion let $f \in \mathcal{R}(A)^\perp$. Then $\langle f, Ag\rangle = 0 = \langle \mathbf{0}, g\rangle$ for all $g \in \mathcal{D}(A)$, hence $f \in \mathcal{D}(A^*)$ and $A^*f = \mathbf{0}$.

(c) Let $f \in \mathcal{D}(B^*)$. Then $\langle B^*f, g\rangle = \langle f, Bg\rangle$ for all $g \in \mathcal{D}(B)$. In particular, since $B$ is an extension of $A$, one has $\langle B^*f, g\rangle = \langle f, Ag\rangle$ for all $g \in \mathcal{D}(A)$. This means that $f \in \mathcal{D}(A^*)$ and that $A^*f = B^*f$. Thus $A^*$ is defined on $\mathcal{D}(B^*)$ and coincides with $B^*$ on this domain, i.e. $A^*$ is an extension of $B^*$. □

**Proposition 2.20.** *Let $A$ be a closable operator with $\mathcal{D}(A)$ dense in $\mathcal{H}$. Then one has $\overline{A}^* \equiv (\overline{A})^* = A^*$, i.e. $A$ and its closure $\overline{A}$ have the same adjoint.*

PROOF. Since $A \subseteq \overline{A}$, one has $\overline{A}^* \subseteq A^*$ by Proposition 2.19(c). To verify the opposite inclusion, let $g \in \mathcal{D}(\overline{A})$ and choose a sequence $\{g_n\}$ in $\mathcal{D}(A)$ having the properties $\text{s-lim}_{n\to\infty} g_n = g$ and $\text{s-lim}_{n\to\infty} Ag_n = \overline{A}g$. Now let $f \in \mathcal{D}(A^*)$. Then

$$\langle A^*f, g\rangle = \lim_{n\to\infty} \langle A^*f, g_n\rangle = \lim_{n\to\infty} \langle f, Ag_n\rangle = \langle f, \overline{A}g\rangle.$$

Since this holds for each $g \in \mathcal{D}(\overline{A})$, we see that $f$ belongs to the domain of $\overline{A}^*$ and that $\overline{A}^*f = A^*f$. Hence $\overline{A}^*$ is an extension of $A^*$. □

The following properties of adjoint operators are obtained as simple applications of Definition 2.18; we leave the details as an exercise.

**Proposition 2.21.** *Let $A$ and $B$ be operators with dense domains in a Hilbert space $\mathcal{H}$.*
*(a) If $\mathcal{R}(B) \subseteq \mathcal{D}(A)$, then $B^*A^* \subseteq (AB)^*$.*
*(b) If $B \in \mathcal{B}(\mathcal{H})$ and $\alpha \neq 0$, then one has $\mathcal{D}((\alpha A + B)^*) = \mathcal{D}(A^*)$ and $(\alpha A + B)^* = \overline{\alpha}A^* + B^*$.*

If an operator $A$ is such that $\mathcal{D}(A)$ and $\mathcal{D}(A^*)$ are dense in the Hilbert space, then $A^*$ and its adjoint $(A^*)^* \equiv A^{**}$ are well defined. We know that $A^{**} = A$ if $A$ belongs to $\mathcal{B}(\mathcal{H})$. In the general case $A^{**}$ is an extension of $A$ [if $\mathcal{D}(A)$ and $\mathcal{D}(A^*)$ are dense]. We shall see that this extension is identical with the closure of $A$; hence, if $A$ is a closed operator, one again has $A^{**} = A$.

**Proposition 2.22.** *Let $A$ be an operator such that $\mathcal{D}(A)$ and $\mathcal{D}(A^*)$ are dense in $\mathcal{H}$. Then $A \subseteq A^{**}$. Furthermore $A$ is closable and $\overline{A} = A^{**}$. In particular, if $A$ is closed, one has $A^{**} = A$.*

PROOF. (i) Fix a vector $h \in \mathcal{D}(A)$. Then $\langle A^*f, h\rangle = \langle f, Ah\rangle$ for all $f \in \mathcal{D}(A^*)$. By virtue of Definition 2.18 (applied to $A^*$ instead of $A$) this implies that $h$ belongs to the domain of the adjoint of $A^*$, i.e. $h \in \mathcal{D}(A^{**})$, and that $A^{**}h = Ah$. This shows that $A^{**}$ is an extension of $A$.

(ii) Since $A^{**}$ is an adjoint, it is a closed operator (Proposition 2.19). So $A$ has a closed extension. By Proposition 2.17, $A$ is closable.

(iii) To see that $\overline{A} = A^{**}$ onc shows that the graph $\Gamma(\overline{A})$ of $\overline{A}$ is identical with the graph $\Gamma(A^{**})$ of $A^{**}$ (we know that each of these two graphs is a closed subset of $\mathcal{H} \oplus \mathcal{H}$). Let us introduce the following operator $U$ in $\mathcal{H} \oplus \mathcal{H}$: $U\{f,g\} = \{g, -f\}$. It is evident that $U$ is isometric: $\|U\{f,g\}\|^2_{\mathcal{H}\oplus\mathcal{H}} = \|g\|^2 + \|f\|^2 = \|\{f,g\}\|^2_{\mathcal{H}\oplus\mathcal{H}}$. Also $\mathcal{R}(U) = \mathcal{H} \oplus \mathcal{H}$. Consequently $U$ is a unitary operator in $\mathcal{H} \oplus \mathcal{H}$.

If $B$ is a linear operator in $\mathcal{H}$ with dense domain and if $\{g, Bg\} \in \Gamma(B)$, then $U\{g, Bg\} = \{Bg, -g\}$. Thus, if $f, f^* \in \mathcal{H}$:

$$\langle U\{g,Bg\}, \{f, f^*\}\rangle_{\mathcal{H}\oplus\mathcal{H}} = \langle Bg, f\rangle_{\mathcal{H}} - \langle g, f^*\rangle_{\mathcal{H}}. \tag{2.62}$$

Comparison with (2.60) shows that $f$ belongs to $\mathcal{D}(B^*)$ if and only if there exists a vector $f^* \in \mathcal{H}$ such that the expression (2.62) is zero for each element of $\Gamma(B)$, and then $f^* = B^*f$. Thus the graph $\Gamma(B^*)$ of $B^*$ is the orthogonal complement in $\mathcal{H} \oplus \mathcal{H}$ of the subspace $U\Gamma(B)$, i.e.

$$\Gamma(B^*) = [U\Gamma(B)]^\perp = U[\Gamma(B)^\perp] \tag{2.63}$$

[take into account Proposition 2.8(b) for the second equality].

By taking $B = A^*$ or $B = \overline{A}$, one obtains from (2.63) that

$$\Gamma(A^{**}) = [U\Gamma(A^*)]^\perp \qquad \text{and} \qquad \Gamma(\overline{A}^*) \equiv \Gamma(A^*) = U[\Gamma(\overline{A})]^\perp.$$

Since $U^2 = -I$, with $I$ the identity in $\mathcal{H} \oplus \mathcal{H}$, and since a subspace $\mathcal{M}$ is invariant if each of its elements is multiplied by $-1$ (i.e. $-\mathcal{M} = \mathcal{M}$), one obtains from the preceding two relations that

$$\Gamma(A^{**}) = U[\Gamma(A^*)^\perp] = U^2\Gamma(\overline{A}) = \Gamma(\overline{A}). \qquad \square$$

**2.4.4.** An operator $A$ is ***self-adjoint*** if $A^* = A$, i.e. if $\mathcal{D}(A^*) = \mathcal{D}(A)$ and $A^*f = Af$ for all $f \in \mathcal{D}(A)$. One must assume that $A$ has dense domain (so that $A^*$ exists). A self-adjoint operator is always closed (see Proposition 2.19).

For $A \in \mathcal{B}(\mathcal{H})$ the condition of self-adjointness is that

$$\langle Af, g \rangle = \langle f, Ag \rangle \qquad \forall f, g \in \mathcal{H}. \tag{2.64}$$

If $A$ is unbounded, this equation makes sense only if both $f$ and $g$ belong to $\mathcal{D}(A)$. An operator satisfying (2.64) for all $f$ and $g$ in its domain is called symmetric. So $A$ is a ***symmetric operator*** if $\mathcal{D}(A)$ is dense in $\mathcal{H}$ and

$$\langle Af, g \rangle = \langle f, Ag \rangle \qquad \forall f, g \in \mathcal{D}(A). \tag{2.65}$$

In particular, if $A$ is symmetric, one has $\langle f, Af \rangle \in \mathbb{R}$ for each $f \in \mathcal{D}(A)$.

In applications the validity of the condition (2.65) is often easy to verify, and sometimes physicists are not aware of the distinction between a symmetric and a self-adjoint operator. Eq (2.65) means only that $A \subseteq A^*$. Indeed, if $f \in \mathcal{D}(A)$ and (2.65) is satisfied, then there exists a vector $f^*$ (namely $f^* = Af$) such that $\langle f, Ag \rangle = \langle f^*, g \rangle$ for all $g$ in $\mathcal{D}(A)$, and since $A^*f = f^*$ by definition, one then has $A^*f = Af$. So (2.65) implies that $A^*$ is an extension of $A$. However the condition that $\mathcal{D}(A^*)$ be exactly the same as $\mathcal{D}(A)$ is very strong and is in general not satisfied for a symmetric operator. The relation between symmetric and self-adjoint operators is illustrated below and studied in detail in Chapter 3.

A symmetric operator $A$ is closable (because $A^*$ is a closed extension of $A$, see Proposition 2.17). Its closure $\overline{A}$ is also symmetric: If $f, g \in \mathcal{D}(\overline{A})$, then there are sequences $\{f_n\}$ and $\{g_n\}$ in $\mathcal{D}(A)$ such that $\text{s-lim}_{n\to\infty} f_n = f$, $\text{s-lim}_{n\to\infty} g_n = g$, $\text{s-lim}_{n\to\infty} Af_n = \overline{A}f$ and $\text{s-lim}_{n\to\infty} Ag_n = \overline{A}g$. Then [see Proposition 1.1(b)]

$$\langle \overline{A}f, g \rangle = \lim_{n\to\infty} \langle Af_n, g_n \rangle = \lim_{n\to\infty} \langle f_n, Ag_n \rangle = \langle f, \overline{A}g \rangle.$$

In general $\overline{A}$ will not be self-adjoint. If the closure $\overline{A}$ of a symmetric operator $A$ is self-adjoint (i.e. if $\overline{A}^* = \overline{A}$), one says that $A$ is ***essentially self-adjoint***, and $\mathcal{D}(A)$ is called a ***core*** for $\overline{A}$.[5] In this case the symmetric operator $A$ has exactly one self-adjoint extension, namely its closure $\overline{A}$. (Indeed, let $B = B^*$ be a self-adjoint extension of $A$; since $B$ is closed, it is also an extension of $\overline{A}$, see Proposition 2.17; thus $\overline{A} \subseteq B = B^*$. But $\overline{A} \subseteq B$ implies that $B^* \subseteq \overline{A}^*$, and $\overline{A}^* = \overline{A}$ by the assumption of essential self-adjointness of $A$. Thus one will have $\overline{A} \subseteq B = B^* \subseteq \overline{A}$, i.e. one must have $\overline{A} = B$.[6]) Evidently each self-adjoint operator is also essentially self-adjoint.

Thus, if $A$ is a symmetric operator that is essentially self-adjoint, it has exactly one symmetric closed extension, and this extension is self-adjoint.[7] In more general cases

---

[5] More generally, if $C$ is a closed operator with domain $\mathcal{D}(C)$, a linear submanifold $\mathcal{D}$ of $\mathcal{D}(C)$ is called a ***core*** for $C$ if the closure of the restriction of $C$ to $\mathcal{D}$ is equal to $C$. Thus, if $C_0$ is defined by $\mathcal{D}(C_0) = \mathcal{D} \subseteq \mathcal{D}(C)$ and $C_0 f = Cf$ for $f \in \mathcal{D}(C_0)$, then $\mathcal{D}$ is a core for the operator $C$ if $\overline{C_0} = C$. If $\mathcal{D}$ is chosen too small, then one will only have $\overline{C_0} \subsetneq C$ and $\mathcal{D}$ may not be a core for $C$. We refer to Chapter 3 for examples.

[6] The same reasoning allows one to show that the closure of an essentially self-adjoint operator $A$ is its only closed symmetric extension: one replaces the hypothesis that $B = B^*$ by the condition that $B$ be symmetric and closed, i.e. $B = \overline{B} \subseteq B^*$.

[7] This explains the terminology "essentially self-adjoint": except for passing to its closure, $A$ is self-adjoint.

the situation is considerably more complicated. There are symmetric operators having only one closed symmetric extension but this extension is not self-adjoint. In general a symmetric operator has many closed symmetric extensions; in certain cases none of these extensions is self-adjoint, in other cases an infinite number of them are self-adjoint (see Chapter 3 for details and examples).

As an illustration of these subtleties we consider the following situation. Let $A$ be a symmetric operator. If $B$ is an extension of $A$, then $B^*$ is the restriction of $A^*$ to a submanifold $\mathcal{D}(B^*)$ of $\mathcal{D}(A^*)$ [Proposition 2.19(c)]: if $A \subseteq B$, then $B^* \subseteq A^*$ (with strict inclusion in general). Thus in general $\mathcal{D}(B^*)$ will be smaller than $\mathcal{D}(A^*)$. Any symmetric extension $B$ of $A$ will satisfy $\mathcal{D}(A) \subseteq \mathcal{D}(B) \subseteq \mathcal{D}(B^*) \subseteq \mathcal{D}(A^*)$, and each of the four operators involved in this sequence acts as $A^*f$ on vectors $f$ in its domain. In particular $B$ will be the restriction of $A^*$ to $\mathcal{D}(B)$. So consider extensions $B$ of $A$ with $\mathcal{D}(A) \subseteq \mathcal{D}(B) \subseteq \mathcal{D}(A^*)$ and $Bf = A^*f$ for $f \in \mathcal{D}(B)$. Observe that the more one increases $\mathcal{D}(B)$, the more $\mathcal{D}(B^*)$ will shrink. [8] In order to obtain a self-adjoint extension of $A$, one has to find a domain $\mathcal{D}(B)$ such that $\mathcal{D}(B)$ and $\mathcal{D}(B^*)$ meet exactly, i.e. such that they become identical (which may be possible or not).

**Proposition 2.23.** *Let $A$ be a symmetric operator. Then $A$ is essentially self-adjoint if and only if $A^*$ is symmetric. More precisely:*
*(a) If $A^*$ is also symmetric, then $A$ is essentially self-adjoint.*
*(b) If $A$ is essentially self-adjoint, then $A^* = A^{**}$.*

PROOF. We recall that $A$ is essentially self-adjoint if and only if $\overline{A} = \overline{A}^*$. We use the facts that $A^{**} = \overline{A}$ and $\overline{A}^* = A^*$ if $A$ is symmetric (Propositions 2.22 and 2.20).
(a) Suppose that $A^*$ is symmetric, i.e. that $A^* \subseteq A^{**}$. This is equivalent to $\overline{A}^* \subseteq \overline{A}$. As seen before, $\overline{A}$ is also symmetric, i.e. $\overline{A} \subseteq \overline{A}^*$. One concludes that $\overline{A} = \overline{A}^*$.
(b) Now assume that $A$ is essentially self-adjoint, i.e. that $\overline{A} = \overline{A}^*$. As $\overline{A} = A^{**}$ and $\overline{A}^* = A^*$, it follows that $A^{**} = A^*$. □

EXAMPLE. As an example, consider in $L^2((0,1))$ the differential operator $-i(d/dx)$ [formally this is the momentum operator of a particle confined to the interval $(0,1)$]. Let us denote by $P_0$ the operator $-i(d/dx)$ with domain $\mathcal{D}_0 := C_0^\infty((0,1))$, and by $P_1$ the operator $-i(d/dx)$ with the domain $\mathcal{D}_1$ of all continuously differentiable functions on $[0,1]$. If $f$ and $g$ are continuously differentiable functions on $[0,1]$, then

$$\langle -if', g\rangle - \langle f, -ig'\rangle = i\big[\overline{f(1)}g(1) - \overline{f(0)}g(0)\big]. \tag{2.66}$$

Observe that the right-hand side is zero when $g$ belongs to the dense set $C_0^\infty((0,1))$. This shows that the operator $P_0$ is symmetric, and that $\mathcal{D}_1$ is contained in the domain of the adjoint of $P_0$, so that $P_1 \subseteq P_0^*$. Obviously one can find functions $f, g$ in $\mathcal{D}_1$ for which the right-hand side of (2.66) does not vanish, so that $P_1$ and hence also $P_0^*$ are not symmetric. In view of Proposition 2.23 we conclude that $P_0$ is symmetric but not essentially self-adjoint. Another way of arriving at this conclusion is as follows: the

[8] In order to have $f \in \mathcal{D}(B^*)$, there must exist a vector $f^*$ in $\mathcal{H}$ such that $\langle f, Bg\rangle = \langle f^*, g\rangle$ for all $g \in \mathcal{D}(B)$. Thus, when one increases $\mathcal{D}(B)$, the number of conditions imposed on $f$ increases, consequently $\mathcal{D}(B^*)$ will shrink.

assumption that $\overline{P_0} = P_0^*$ implies, together with the relation $\overline{P_0} \subseteq \overline{P_1} \subseteq P_0^*$ obtained above, that $\overline{P_0} = \overline{P_1}$, hence that $\overline{P_1}$ is also self-adjoint. This is not possible: the differential equation $-if' = if$ has the solution $f(x) = e^{-x}$ which is square-integrable on the interval $(0,1)$ and hence belongs to the domain of $P_1$. So the number $+i$ is an eigenvalue of the operator $P_1$, and since a self-adjoint operator can have only real eigenvalues, $\overline{P_1}$ is not self-adjoint. A detailed description of the self-adjoint extensions of the operator $P_0$ is given in §3.2.3.

We finally mention a result that will be discussed and proved at the beginning of the next chapter (Proposition 3.3). This result represents the basic self-adjointness criterion for symmetric operators and is used already in the next section to establish self-adjointness properties of multiplication operators. It involves knowledge of the range of $A+i$ and of $A-i$:

**Proposition 2.24.** *Let $A$ be a symmetric operator.*
*(a) $A$ is self-adjoint if and only if $\mathcal{R}(A+i) = \mathcal{H}$ and $\mathcal{R}(A-i) = \mathcal{H}$.*
*(b) $A$ is essentially self-adjoint if and only if $\mathcal{R}(A+i)$ and $\mathcal{R}(A-i)$ are dense in $\mathcal{H}$.*

**Remark 2.25.** If $A$ is an operator in $\mathcal{H}$ with domain $\mathcal{D}(A)$ and $U$ is a unitary operator in $\mathcal{H}$, one can define an operator $A_U$ by $\mathcal{D}(A_U) = U\mathcal{D}(A)$ [the linear manifold obtained by acting with $U$ on all vectors of $\mathcal{D}(A)$] and $A_U f = UAU^* f$ if $f \in \mathcal{D}(A_U)$. This makes sense because, if $f \in \mathcal{D}(A_U)$, then there exists a (unique) vector $g \in \mathcal{D}(A)$ such that $f = Ug$ (in fact $g = U^* f$), hence $A_U f = UAU^*(Ug) = UAIg = UAg$, and $Ag$ is well defined since $g \in \mathcal{D}(A)$. It is easy to check (exercise) that a unitary transformation preserves most properties of an operator; for example, if $A$ is closable or closed or symmetric or self-adjoint or essentially self-adjoint, then the operator $A_U$ also has the respective property. If $A$ is closable, then the closure $\overline{A_U}$ of $A_U$ is given by $\overline{A_U} = U\overline{A}U^* \equiv (\overline{A})_U$. The resolvent set, the spectrum (defined in Section 2.6), the eigenvalues and the norm of an operator are invariant under a unitary transformation.

## 2.5 Multiplication operators

**2.5.1.** A ***multiplication operator*** $A$ in $\mathcal{H} = L^2(\mathcal{O}, m)$ is an operator given as follows:

$$(Af)(x) = a(x)f(x), \tag{2.67}$$

i.e. the value of the function $Af$ at the point $x \in \mathcal{O}$ is obtained by multiplying the value of $f$ at the point $x$ by the number $a(x)$, where $a\colon \mathcal{O} \to \mathbb{C}$ is a fixed measurable function. In other terms the function $Af$ is the product of the two functions $a$ and $f$. The maximal domain on which this makes sense is clearly as follows:

$$\mathcal{D}(A) = \Big\{ f \in L^2(\mathcal{O}, m) \,\Big|\, \int_{\mathcal{O}} |a(x)|^2 |f(x)|^2 \, m(dx) < \infty \Big\}. \tag{2.68}$$

This domain is a linear manifold, and normally we consider a multiplication operator to be defined on its maximal domain.

The properties of such an operator depend on the function $a$. To characterise the multiplication operators belonging to $\mathcal{B}(\mathcal{H})$, it is useful to introduce the Banach space $L^p(\mathcal{O}, m)$ for $p = \infty$. A measurable function $f\colon \mathcal{O} \to \mathbb{C}$ is said to be ***essentially bounded*** if there exists a constant $M < \infty$ such that $|f(x)| \le M$ $m$-almost everywhere (in other terms if, for some $M < \infty$, the set $\mathcal{W} = \{x \in \mathcal{O} \mid |f(x)| > M\}$ belongs to the considered $\sigma$-algebra $\mathcal{A}$ and satisfies $m(\mathcal{W}) = 0$). The infimum of all numbers $M$ having this property is denoted by $\|f\|_\infty$:

$$\|f\|_\infty = \inf\{M \in \mathbb{R} \mid |f(x)| \le M \ \ m\text{-a.e.}\}. \tag{2.69}$$

$L^\infty(\mathcal{O}, m)$ is the set of equivalence classes (see page 19) of essentially bounded functions. A function $f$ representing an element of $L^\infty(\mathcal{O}, m)$ satisfies $|f(x)| \le \|f\|_\infty$ $m$-a.e.[9] For this reason one also writes $\|f\|_\infty = \operatorname{ess\,sup}_{x\in\mathcal{O}}|f(x)|$ (essential supremum, i.e. the supremum is taken modulo null sets). If for example $\mathcal{O} = (a, b)$ is an interval, $m$ is the Lebesgue measure and $f$ is continuous on $(a, b)$, then $\|f\|_\infty = \sup_{x\in(a,b)} |f(x)|$ [this supremum will be infinite if $f(x)$ diverges as $x \to a$ or $x \to b$, in which case $f$ will not define an element of $L^\infty((a, b))$].

**Proposition 2.26.** *Let $A$ be a multiplication operator in $\mathcal{H} = L^2(\mathcal{O}, m)$. Then $A$ belongs to $\mathcal{B}(\mathcal{H})$ if and only if $a \in L^\infty(\mathcal{O}, m)$, and in this case $\|A\| = \|a\|_\infty$.*

PROOF. (i) One has

$$\|Af\|^2 = \int_{\mathcal{O}} |a(x)|^2 |f(x)|^2 m(dx) \le \|a\|_\infty^2 \int_{\mathcal{O}} |f(x)|^2 m(dx) = \|a\|_\infty^2 \|f\|^2.$$

Thus, if $\|a\|_\infty < \infty$, one has $\mathcal{D}(A) = \mathcal{H}$ and $\|A\| \le \|a\|_\infty$.

(ii) Now assume that $a \notin L^\infty(\mathcal{O}, m)$. For $n \in \mathbb{N}$ set $\mathcal{W}_n = \{x \in \mathcal{O} \mid |a(x)| > n\}$. Then $\mathcal{W}_n$ belongs to $\mathcal{A}$ (because $a$ is a measurable function) and $m(\mathcal{W}_n) > 0$ because $a \notin L^\infty(\mathcal{O}, m)$. Let $\mathcal{V}_n$ be a subset of $\mathcal{W}_n$ with $\mathcal{V}_n \in \mathcal{A}$ and $0 < m(\mathcal{V}_n) < \infty$ [if $m(\mathcal{W}_n) < \infty$, one can take $\mathcal{V}_n = \mathcal{W}_n$]. Let $f_n = \chi_{\mathcal{V}_n}$, the characteristic function of $\mathcal{V}_n$ [see Eq. (1.41)]. One has $\|f_n\|^2 = m(\mathcal{V}_n) < \infty$, hence $f_n \in \mathcal{H} = L^2(\mathcal{O}, m)$, and

$$\|Af_n\|^2 = \int_{\mathcal{O}} |a(x)|^2 |\chi_{\mathcal{V}_n}(x)|^2 m(dx) = \int_{\mathcal{V}_n} |a(x)|^2 m(dx) \ge n^2 m(\mathcal{V}_n) = n^2 \|f_n\|^2.$$

This shows that $\|Af_n\|/\|f_n\| \ge n$. Since $n$ is arbitrary, $A$ cannot be bounded.

(iii) It remains to show that $\|A\| \ge \|a\|_\infty$ if $a \in L^\infty(\mathcal{O}, m)$. For this one proceeds as in (ii). If $\varepsilon > 0$ and $\mathcal{W}_\varepsilon = \{x \in \mathcal{O} \mid |a(x)| > \|a\|_\infty - \varepsilon\}$, then $\mathcal{W}_\varepsilon$ belongs to $\mathcal{A}$ and $m(\mathcal{W}_\varepsilon) > 0$. If $f \in L^2(\mathcal{O}, m)$ is such that $\|f\| \ne 0$ and $f(x) = 0$ for all $x \notin \mathcal{W}_\varepsilon$, then $\|Af\|/\|f| \ge \|a\|_\infty - \varepsilon$, which implies the desired inequality. □

**Example 2.27.** Let $A \in \mathcal{B}(\mathcal{H})$ be the operator of multiplication in $\mathcal{H} = L^2(\mathcal{O}, m)$ by a function $a \in L^\infty(\mathcal{O}, m)$. Observe that

$$\langle f, Ag\rangle = \int_{\mathcal{O}} \overline{f(x)} a(x) g(x) m(dx) = \int_{\mathcal{O}} \overline{\overline{a(x)} f(x)} g(x) m(dx). \tag{2.70}$$

[9] It can be shown that the infimum in (2.69) is attained, see e.g. [HS].

This shows that $A^*$ is the operator of multiplication by the complex conjugate function $\overline{a}$ of $a$: $(A^*f)(x) = \overline{a(x)}f(x)$. In particular:

(a) $A$ is *self-adjoint* if and only if $a$ is a real function.

(b) $A$ is a *projection* if and only if $a(x) = 0$ or $a(x) = 1$ $m$-a.e. If one defines $\mathcal{V} = \{x \in \mathcal{O} \mid a(x) = 1\}$, then $a$ is the characteristic function of the set $\mathcal{V}$ and $A$ is the projection onto the subspace $\mathcal{M}(A) = \{f \in L^2(\mathcal{O}, m) \mid f(x) = 0 \text{ if } x \notin \mathcal{V}\} = L^2(\mathcal{V}, m)$ [considered as a subspace of $\mathcal{H} = L^2(\mathcal{O}, m)$].

(c) $A$ is an *isometry* if and only if $|a(x)| = 1$ $m$-a.e., i.e. if $a(x) = e^{i\alpha(x)}$ $m$-a.e., with $\alpha\colon \mathcal{O} \to \mathbb{R}$. Then $A^*$ is the operator of multiplication by $e^{-i\alpha(x)}$, and one sees that $A$ is in fact unitary: *Each isometric multiplication operator is unitary.*

(d) $A$ is a *partial isometry* if $a(x) = e^{i\alpha(x)}\chi_{\mathcal{V}}(x)$, where $\alpha\colon \mathcal{O} \to \mathbb{R}$ and $\mathcal{V}$ is a subset of $\mathcal{O}$ belonging to $\mathcal{A}$. The initial set and the final set of $A$ are identical, given by $L^2(\mathcal{V}, m)$ [considered as a subspace of $L^2(\mathcal{O}, m)$].

As regards unbounded multiplication operators, we shall meet only a few simple examples. All the same we mention in Remark 2.31 some general results that are rather easy to obtain.

If $a$ is a real function, one sees from (2.70) that $\langle f, Ag\rangle = \langle Af, g\rangle$ for all $f$ and $g$ in the maximal domain $\mathcal{D}(A)$ given in (2.68). If this domain is dense in $L^2(\mathcal{O}, m)$, then it follows that the operator $A$ is symmetric. We first show by applying the self-adjointness criterion stated in Proposition 2.24 that $A$ is in fact self-adjoint.

**Proposition 2.28.** *Let $A$ be the multiplication operator in $\mathcal{H} = L^2(\mathcal{O}, m)$ by a real function, defined on its maximal domain* (2.68). *If this domain is dense in $\mathcal{H}$, then $A$ is self-adjoint.*

PROOF. We have seen above that $A$ is symmetric. By virtue of Proposition 2.24(a) it then suffices to show that $\mathcal{R}(A + i) = \mathcal{H}$ and $\mathcal{R}(A - i) = \mathcal{H}$. So let $g \in \mathcal{H}$. We must find a vector $f$ in $\mathcal{D}(A)$ such that $(A + i)f = g$. For this we define $f\colon \mathcal{O} \to \mathbb{C}$ by $f(x) = [a(x) + i]^{-1}g(x)$, where $x \in \mathcal{O}$. Since $\left|[a(x) + i]^{-1}\right| \leq 1$, we have $\|f\| \leq \|g\| < \infty$, so that the function $f$ defines an element of $L^2(\mathcal{O}, m)$. Furthermore, since $\left|a(x)[a(x) + i]^{-1}\right| \leq 1$, we have

$$\int_{\mathcal{O}} |a(x)|^2 |f(x)|^2 m(dx) = \int_{\mathcal{O}} \left|a(x)[a(x) + i]^{-1}\right|^2 \left|g(x)\right|^2 m(dx) \leq \|g\|^2.$$

Thus $f$ even belongs to $\mathcal{D}(A)$. Finally it is clear that $(A+i)f = g$. Hence $\mathcal{R}(A+i) = \mathcal{H}$. Similarly, taking $f(x) = [a(x) - i]^{-1}g(x)$, one sees that $\mathcal{R}(A - i) = \mathcal{H}$. □

**Proposition 2.29.** *(a) Let $a\colon \mathbb{R}^n \to \mathbb{R}$ be a continuous function. Then the multiplication operator by $a$ in $\mathcal{H} = L^2(\mathbb{R}^n)$ is essentially self-adjoint on $C_0^\infty(\mathbb{R}^n)$.*
*(b) Let $a$ be a real measurable function satisfying $|a(\vec{x})| \leq \kappa(1 + |\vec{x}|)^m$ for some constants $\kappa, m \in (0, \infty)$. Then the operator of multiplication by $a$ in $L^2(\mathbb{R}^n)$ is essentially self-adjoint on $C_0^\infty(\mathbb{R}^n)$ and on $\mathcal{S}(\mathbb{R}^n)$.*

PROOF. We consider the case $n = 1$. The proof for $n > 1$ is essentially the same. It is seen from the proof that the assumption of continuity of $a$ made in part (a) could be weakened to the condition that $a$ be locally bounded (i.e. bounded on each finite ball $\{\vec{x} \in \mathbb{R}^n \mid |\vec{x}| \leq R\}, 0 < R < \infty$).

(a) Let us denote by $A$ the multiplication operator by $a$ on the maximal domain $\mathcal{D}(A)$ given in (2.68) and by $A_0$ the restriction of $A$ to $\mathcal{D}(A_0) = C_0^\infty(\mathbb{R})$ [observe that $C_0^\infty(\mathbb{R})$ is contained in $\mathcal{D}(A)$ because the function $a$ is continuous, hence locally bounded]. We shall show that $\mathcal{R}(A_0 + i)$ is dense in $\mathcal{H}$. An analogous argument gives the denseness of $\mathcal{R}(A_0 - i)$, and the result of (a) then follows from Proposition 2.24.

Each vector $h \in L^2(\mathbb{R})$ can be approximated (in $L^2$ norm) by a sequence $\{h_j\}$ of functions of compact support. By taking $h_j(x) = h(x)$ if $|x| \leq j$ and $h_j(x) = 0$ if $|x| > j$, one has

$$\|h - h_j\|^2 = \int_{|x|>j} |h(x)|^2 dx,$$

which converges to zero as $j \to \infty$ because $h$ is square-integrable. Thus it suffices to show that each square-integrable function $g$ of compact support can be approximated in $L^2$ norm by a sequence of vectors $\{g_k\}$ of the form $g_k(x) = [a(x) + i]f_k(x)$, with $f_k \in \mathcal{D}(A_0) = C_0^\infty(\mathbb{R})$ [i.e. $g_k \in \mathcal{R}(A_0 + i)$].

So let $g \in L^2(\mathbb{R})$ be such that $g(x) = 0$ for $|x| > M$ ($M > 0$ is such that the support of $g$ is contained in the interval $[-M, M]$). We set $f(x) = [a(x) + i]^{-1}g(x)$. The function $f$ has compact support (identical with the support of $g$), and it is square-integrable (see the argument given in the preceding proof). One can choose a sequence $\{f_k\}$ in $C_0^\infty(\mathbb{R})$ such that $\operatorname{s-lim}_{k\to\infty} f_k = f$ and such that $f_k(x) = 0$ if $|x| > M + 1$ for each $k = 1, 2, 3, \ldots$ (see §1.4.5). The continuous function $a$ is bounded on the finite interval $[-M - 1, M + 1]$. If we set $C = \max_{|x|\leq M+1}|a(x) + i| < \infty$, then

$$\begin{aligned}\|(A_0 + i)f_k - g\|^2 &= \int_{|x|\leq M+1} \big|[a(x) + i][f_k(x) - f(x)]\big|^2 dx \\ &\leq C^2 \int_{|x|\leq M+1} \big|f_k(x) - f(x)\big|^2 dx = C^2\|f_k - f\|^2,\end{aligned}$$

which converges to zero as $k \to \infty$.

(b) The function $a$ considered in part (b) is locally bounded. By (a) and the observation made at the beginning of the proof, the operator of multiplication by $a$ is essentially self-adjoint on $C_0^\infty(\mathbb{R})$. Since $|a(x)| \leq \kappa(1 + |x|)^m$, we have $\mathcal{S}(\mathbb{R}) \subset \mathcal{D}(A)$, so $A$ is also essentially self-adjoint on the larger domain $\mathcal{S}(\mathbb{R})$. □

**Example 2.30.** We illustrate the preceding propositions by some simple operators from quantum mechanics.

(1) Let $(a, b)$ be an interval and $m \in \mathbb{N}$. Let $A$ be the multiplication operator by $a(x) = x^m$ in $\mathcal{H} = L^2((a, b))$. If the interval $(a, b)$ is finite, then $A \in \mathcal{B}(\mathcal{H})$ and $\|A\| = \max\{|a|^m, |b|^m\}$. If $b - a = \infty$, then $A$ is unbounded. For $(a, b) = \mathbb{R}$ and $m = 1$ one obtains the ***position operator*** $Q$ of quantum mechanics in $L^2(\mathbb{R})$:

$$(Qf)(x) = xf(x), \quad \mathcal{D}(Q) = \Big\{f\colon \mathbb{R} \to \mathbb{C} \,\Big|\, \int_{-\infty}^{\infty} (1 + x^2)|f(x)|^2 dx < \infty\Big\}. \quad (2.71)$$

The operator $Q$ is self-adjoint on this domain, and it is essentially self-adjoint on $C_0^\infty(\mathbb{R})$ and on $\mathcal{S}(\mathbb{R})$. The operator conjugate to $Q$ in $L^2(\mathbb{R})$ is the ***momentum operator*** $P$ which satisfies $[P, Q] \subseteq -iI$. $P$ is a *differential operator*, given formally by $(Pf)(x) = -if'(x) \equiv -i\, df(x)/dx$. More is said about this operator in the next subsection and in §3.2.3 (page 102). In the next chapter we also study differential operators on intervals $(a, b) \neq \mathbb{R}$.

(2) More generally, in dimension $n > 1$, let $a(\vec{x}) = x_j$, where $\vec{x} = (x_1, \ldots, x_n)$ and $j \in \{1, \ldots, n\}$. The associated multiplication operator $Q_j$ represents the $j$-th component of the n-dimensional position operator $\vec{Q} = (Q_1, \ldots, Q_n)$ in $L^2(\mathbb{R}^n)$, i.e. $(Q_j f)(\vec{x}) = x_j f(\vec{x})$. The operator $Q_j$ is self-adjoint on its maximal domain and essentially self-adjoint on $C_0^\infty(\mathbb{R}^n)$ and on $\mathcal{S}(\mathbb{R}^n)$. The operator conjugate to $Q_j$ is the $j$-th component $P_j$ of the n-dimensional momentum operator $\vec{P} = (P_1, \ldots, P_n)$, given formally by $(P_j f)(\vec{x}) = -i\partial f(\vec{x})/\partial x_j$.

To end this subsection we state some further properties of multiplication operators.

**Remark 2.31.** (1) If $a \in L^2(\mathcal{O}, m)$, then $\mathcal{D}(A)$ contains all functions in $L^2(\mathcal{O}, m)$ that also belong to $L^\infty(\mathcal{O}, m)$, i.e. $\mathcal{D}(A) \supset [L^2(\mathcal{O}, m) \cap L^\infty(\mathcal{O}, m)]$. For example in $\mathcal{H} = L^2(\mathbb{R}^n)$ one has $\mathcal{S}(\mathbb{R}^n) \subset \mathcal{D}(A)$ if $a \in L^2(\mathbb{R}^n)$, hence $\mathcal{D}(A)$ is dense in $\mathcal{H}$.

(2) If $A$ is the multiplication operator by a function $a$, let us denote by $\underline{A}$ the multiplication operator by $\overline{a}$ (the complex conjugate of $a$). Clearly the maximal domain of $\underline{A}$ is the same as that of $A$: $\mathcal{D}(\underline{A}) = \mathcal{D}(A)$. If $f$ and $g$ belong to this domain, then by Eq. (2.70): $\langle f, Ag\rangle = \langle \underline{A}f, g\rangle$. Thus, if $\mathcal{D}(A)$ is dense in $\mathcal{H} = L^2(\mathcal{O}, m)$, then the adjoint $A^*$ of $A$ exists and is an extension of $\underline{A}$ : $\underline{A} \subseteq A^*$. In this situation the domain of $A^*$ is dense in $\mathcal{H}$, hence $A^{**}$ is well defined. Since $A$ is closed by (3) below, we see from Proposition 2.22 that $A^{**} = A$.

(3) A multiplication operator is closed [on its maximal domain (2.68)]. This can be seen by using the following fact:

*If $f_n, f \in L^p(\mathcal{O}, m)$ and $\|f_n - f\|_p \to 0$ as $n \to \infty$, then there exists a subsequence $\{f_{n_k}\}$ of $\{f_n\}$ such that $f(x) = \lim_{k\to\infty} f_{n_k}(x)$ $m$-almost everywhere.*

So let us assume that $\{f_n\}$ is a sequence in $\mathcal{D}(A)$ such that $\|f_n - f\| \to 0$ (we take p = 2) and $\|Af_n - g\| \to 0$ as $n \to \infty$ for certain vectors $f, g \in L^2(\mathcal{O}, m)$. Let $\{f_{n_k}\}$ be a subsequence of $\{f_n\}$ converging $m$-almost everywhere to $f$. Then $\|Af_{n_k} - g\|$ converges to 0 as $k \to \infty$, hence there exists a subsequence of $\{f_{n_k}\}$, say $\{f_{n_{k_j}}\}$, such that $a(x) f_{n_{k_j}}(x) \to g(x)$ $m$-a.e. as $j \to \infty$. Since $f_{n_{k_j}} \to f(x)$ $m$-a.e., one then has $a(x)f(x) = g(x)$ $m$-a.e. Since $g \in L^2(\mathcal{O}, m)$, this shows that $f \in \mathcal{D}(A)$ and that $Af = g$. Thus $A$ is closed.

**2.5.2.** It is well known that a differential operator with constant coefficients becomes a multiplication operator under Fourier transformation. For example the momentum operator $P$ can be diagonalised by a Fourier transformation. We recall that the Fourier transform $\mathcal{F}f \equiv \tilde{f}$ of a function $f\colon \mathbb{R} \to \mathbb{C}$ is defined as

$$(\mathcal{F}f)(k) \equiv \tilde{f}(k) = \frac{1}{\sqrt{2\pi}} \int_{-\infty}^{\infty} e^{-ikx} f(x)\, dx. \tag{2.72}$$

The integral in (2.72) is an (improper) Riemann integral if $f$ is continuous and a Lebesgue integral in general, and $k$ varies on $\mathbb{R}$. For fixed $k$, the integral in (2.72) has a sense if $e^{-ikx}f(x)$ is an integrable function of $x$, i.e. if $f$ belongs to $L^1(\mathbb{R})$.[10] Under this condition one has

$$|\tilde{f}(k)| \leq \frac{1}{\sqrt{2\pi}} \int_{-\infty}^{\infty} |f(x)|\, dx = \frac{1}{\sqrt{2\pi}} \|f\|_1 \qquad \forall k \in \mathbb{R}, \tag{2.73}$$

hence $\tilde{f} \in L^\infty(\mathbb{R})$, and by applying the Dominated Convergence Theorem one finds that $\tilde{f}$ is a continuous function. If $f$ belongs to $\mathcal{S}(\mathbb{R})$, then so does $\tilde{f}$, and $\mathcal{F}$ is an isometric mapping on $\mathcal{S}(\mathbb{R})$, i.e.

$$\langle f, g \rangle \equiv \int_{-\infty}^{\infty} \overline{f(x)} g(x)\, dx = \int_{-\infty}^{\infty} \overline{\tilde{f}(k)} \tilde{g}(k)\, dk = \langle \tilde{f}, \tilde{g} \rangle \quad \text{if } f, g \in \mathcal{S}(\mathbb{R}). \tag{2.74}$$

If we set $f = g \in \mathcal{S}(\mathbb{R})$ we obtain $\|f\| = \|\tilde{f}\|$. This shows that $\mathcal{F}$ is bounded on $\mathcal{S}(\mathbb{R})$, consequently its closure belongs to $\mathcal{B}(\mathcal{H})$ and is an operator of norm 1. It is usual to denote this closure also by $\mathcal{F}$ (rather than by $\overline{\mathcal{F}}$) and to denote by $\tilde{f}$ the vector $\mathcal{F}f$ for each $f \in L^2(\mathbb{R})$. If $f \in L^2(\mathbb{R})$ also belongs to $L^1(\mathbb{R})$, then $\tilde{f}(k)$ can be defined by the formula (2.72) for each fixed $k$, and one has $\tilde{f} \in L^\infty(\mathbb{R}) \cap L^2(\mathbb{R})$. If $f \in L^2(\mathbb{R})$ but $f \notin L^1(\mathbb{R})$[11], one must apply the definition of the closure of an operator, in other terms one must approximate $f$ by a sequence $\{f_j\}$ with $f_j \in L^2(\mathbb{R}) \cap L^1(\mathbb{R})$, and one then obtains its Fourier transform $\tilde{f}$ as the strong limit of the sequence $\{\tilde{f}_j\}$. Usually one takes $f_j$ as follows: $f_j(x) = f(x)$ for $|x| \leq j$ and $f_j(x) = 0$ for $|x| > j$. Then, by (1.49), $f_j$ belongs to $L^1(\mathbb{R})$, hence $\tilde{f}_j(k)$ is given by the formula (2.72), and $\tilde{f}$ is the strong limit of the sequence $\{\tilde{f}_j\}$. One then writes

$$\tilde{f}(k) = \frac{1}{\sqrt{2\pi}} \operatorname*{l.i.m.}_{j \to \infty} \int_{-j}^{j} e^{-ikx} f(x)\, dx \tag{2.75}$$

[l.i.m. stands for "limit in the mean", i.e. the limit is not defined point-wise – for each $k$ – but in the $L^2$ sense, its existence is guaranteed by the fact that $L^2(\mathbb{R})$ is complete].

Equation (2.74), which is obtained first on $\mathcal{S}(\mathbb{R})$, holds for all $f, g \in \mathcal{H}$ (by a limiting argument):

$$\langle f, g \rangle = \langle \mathcal{F}f, \mathcal{F}g \rangle = \langle f, \mathcal{F}^*\mathcal{F}g \rangle \qquad \forall f, g \in \mathcal{H} = L^2(\mathbb{R}). \tag{2.76}$$

So $\mathcal{F}^*\mathcal{F} = I$, i.e. $\mathcal{F}$ is an isometric operator in the sense of Section 2.2. In fact $\mathcal{F}$ is unitary, since $\mathcal{F}^*$ is also isometric; $\mathcal{F}^*$ has properties analogous to those of $\mathcal{F}$ because, for $g \in \mathcal{S}(\mathbb{R})$, $\mathcal{F}^*g$ is given by a formula of the same type as (2.72):

$$(\mathcal{F}^*g)(z) = \frac{1}{\sqrt{2\pi}} \int_{-\infty}^{\infty} e^{+izx} g(x)\, dx \equiv \tilde{g}(-z). \tag{2.77}$$

[10] $\mathcal{F}$ is an integral operator as defined in §2.1.8, all representatives of an equivalence class in $L^1(\mathbb{R})$ get mapped onto the same function $(\mathcal{F}f)(\vec{k})$.

[11] There is no inclusion relation between $L^2(\mathbb{R})$ and $L^1(\mathbb{R})$, simply $L^2(\mathbb{R}) \cap L^1(\mathbb{R})$ is dense in $L^2(\mathbb{R})$. If $J = (a, b)$ is a finite interval, then one has $L^2(J) \subset L^1(J)$ because $\int_a^b |f(x)|\, dx \equiv \int_a^b 1|f(x)|\, dx \leq \|1\|_{L^2((a,b))} \|f\|_{L^2((a,b))} = (b-a)^{1/2} \|f\|_{L^2((a,b))}$ by the Schwarz inequality [see also (1.49)].

Thus $\mathcal{F}^* = \mathcal{F}^{-1}$ [see Proposition 2.7(e)], i.e. if $\tilde{f}(k)$ is given in terms of $f(x)$ by (2.72), then one can recover $f(x)$ by the formula

$$f(x) = \frac{1}{\sqrt{2\pi}} \int_{-\infty}^{\infty} e^{+ikx} \tilde{f}(k)\, dk \tag{2.78}$$

[provided that $f$ and $\tilde{f}$ belong to $L^2(\mathbb{R}) \cap L^1(\mathbb{R})$].

If $f \in \mathcal{S}(\mathbb{R})$ and $m \in \mathbb{N}$, then the Fourier transform of $(-id/dx)^m f(x)$ is $k^m \tilde{f}(k)$, which is easily checked by integrating $m$ times by parts. For example

$$\begin{aligned}[\mathcal{F}(-if')](k) &= \frac{-i}{\sqrt{2\pi}} \int_{-\infty}^{\infty} e^{-ikx} f'(x)\, dx = \frac{i}{\sqrt{2\pi}} \int_{-\infty}^{\infty} f(x) \frac{d}{dx} e^{-ikx}\, dx \\ &= \frac{k}{\sqrt{2\pi}} \int_{-\infty}^{\infty} f(x) e^{-ikx}\, dx = k\tilde{f}(k).\end{aligned}$$

This may be written as $\mathcal{F}P\mathcal{F}^* = Q$ on the domain $\mathcal{S}(\mathbb{R})$, i.e. after Fourier transformation the differential operator $P$ becomes the multiplication operator by the variable in $L^2(\mathbb{R})$. More generally one has $\mathcal{F}P^m\mathcal{F}^* = Q^m$ on $\mathcal{S}(\mathbb{R})$.

It is clear that $(-id/dx)^m$ is symmetric on $\mathcal{S}(\mathbb{R})$:

$$\int_{-\infty}^{\infty} \overline{f(x)} \left(-i\frac{d}{dx}\right)^m g(x)\, dx = \int_{-\infty}^{\infty} \overline{\left[\left(-i\frac{d}{dx}\right)^m f(x)\right]}\, g(x)\, dx \quad \text{if } f, g \in \mathcal{S}(\mathbb{R}).$$

Since the operator $Q^m$ is essentially self-adjoint on $\mathcal{S}(\mathbb{R})$ and since $\mathcal{F}$ maps $\mathcal{S}(\mathbb{R})$ onto $\mathcal{S}(\mathbb{R})$, one sees that the differential operator $(-id/dx)^m$ *is essentially self-adjoint on* $\mathcal{S}(\mathbb{R})$ (apply Remark 2.25 with $U = \mathcal{F}$).

Similar results hold in $n$ dimensions ($n \geq 2$). The n-dimensional Fourier transformation and its inverse are given by expressions that are analogous to (2.72) and (2.77):

$$(\mathcal{F}f)(\vec{k}) \equiv \tilde{f}(\vec{k}) = (2\pi)^{-n/2} \int_{\mathbb{R}^n} e^{-i\vec{k}\cdot\vec{x}} f(\vec{x})\, d^n x, \tag{2.79}$$

$$(\mathcal{F}^* g)(\vec{z}) = (2\pi)^{-n/2} \int_{\mathbb{R}^n} e^{+i\vec{z}\cdot\vec{x}} g(\vec{x})\, d^n x, \tag{2.80}$$

where for example $\vec{x} = (x_1, \ldots, x_n) \in \mathbb{R}^n$ and $\vec{k} \cdot \vec{x} = \sum_{j=1}^n k_j x_j$. After Fourier transformation, the differential operator $P_j = -i\partial/\partial x_j$ becomes multiplication by $k_j$: $[\mathcal{F}(-i\partial/\partial x_j f)](\vec{k}) = k_j \tilde{f}(\vec{k})$ if $f$ belongs to the Schwartz space $\mathcal{S}(\mathbb{R}^n)$ defined in (1.55). So, by Proposition 2.29, $P_j$ is essentially self-adjoint on $\mathcal{S}(\mathbb{R}^n)$, and we denote its closure also by $P_j$. Similar statements hold for other differential operators, in particular for the operator $-\Delta$ which is discussed in §2.5.3.

**Remark 2.32.** From the mathematical point of view the set of the (equivalence classes of) Fourier transforms $\tilde{f}$ of the elements $f$ of $L^2(\mathbb{R})$ is naturally identified with $L^2(\mathbb{R})$ [the Fourier transformation $\mathcal{F}$ is interpreted as a unitary operator in $L^2(\mathbb{R})$]. From the point of view of quantum mechanics, however, it is useful to distinguish between the two representations of this Hilbert space, because the variable $x$ in $f(x)$ is interpreted as *position* and the variable $k$ in $\tilde{f}(k)$ as *momentum* of a physical system (see Example 2.30). We refer to ***configuration space*** when the variable of the functions in $L^2(\mathbb{R})$ is interpreted as position, and to ***momentum space*** when this variable is momentum. In the latter case (in the momentum representation of the Hilbert space), the momentum operator $P$ is diagonalised, i.e. the vectors are identified with the Fourier transforms $\tilde{f}(k)$. We sometimes write $\tilde{L}^2(\mathbb{R})$ to specify that we consider the momentum representation of the Hilbert space $L^2(\mathbb{R})$ (and we adopt analogous conventions for n-dimensional systems). Note that, if one makes the distinction between the two mentioned representations of the space $L^2(\mathbb{R})$, then the Fourier transformation should be interpreted as a mapping between these two representations: $\mathcal{F}$ maps wave functions (viewed as defined on configuration space) to elements of $\tilde{L}^2(\mathbb{R})$, and $\mathcal{F}^* \equiv \mathcal{F}^{-1}$ maps back from the momentum representation to the position representation.

**2.5.3.** We consider here the differential operator $-\Delta \equiv -\sum_{j=1}^n \partial^2/\partial x_j^2$ in $L^2(\mathbb{R}^n)$. The ***free Hamiltonian*** $H_0$ of a non-relativistic quantum-mechanical particle in $\mathbb{R}^n$ is formally given by this differential operator:[12] one has $(H_0 f)(\vec{x}) = -\Delta f(\vec{x})$ if $f \in L^2(\mathbb{R}^n)$ is twice differentiable and $\Delta f$ is square-integrable, in particular if $f$ belongs to the Schwartz space $\mathcal{S}(\mathbb{R}^n)$. In momentum space (after Fourier transformation) this operator becomes multiplication by $\vec{k}^2 = \sum_{j=1}^n k_j^2$, i.e. $[\mathcal{F}(-\Delta f)](\vec{k}) = \vec{k}^2 \tilde{f}(\vec{k})$. Thus $H_0$ is essentially self-adjoint on $\mathcal{S}(\mathbb{R}^n)$, and in the sequel we use the notation $H_0$ for the unique self-adjoint extension (the closure) of the operator $-\Delta$ defined on $\mathcal{S}(\mathbb{R}^n)$; so $H_0 =$ closure of $-\Delta|_{\mathcal{S}(\mathbb{R}^n)}$. In fact, as shown in the Appendix to the present chapter (see page 81), the differential operator $-\Delta$ is even essentially self-adjoint on $C_0^\infty(\mathbb{R}^n)$.

The domain of $H_0$ is easy to specify after Fourier transformation of the wave functions. Since, in the representation $\tilde{L}^2(\mathbb{R}^n)$ of $\mathcal{H}$ (in momentum space), the operator $-\Delta$ becomes multiplication by the square of the variable, its domain is just the maximal domain of this multiplication operator [see (2.68)]:

$$
\begin{aligned}
D(H_0) &= \{f \in L^2(\mathbb{R}^n) \mid \vec{k}^2 \tilde{f}(\vec{k}) \in \tilde{L}^2(\mathbb{R}^n)\} \\
&= \{f \in L^2(\mathbb{R}^n) \mid \int_{\mathbb{R}^n} |\vec{k}^2 \tilde{f}(\vec{k})|^2 d^n k < \infty\}. \qquad (2.81)
\end{aligned}
$$

We see that $H_0 = \vec{P}^2 \equiv \sum_{j=1}^n P_j^2$.

Let us consider the operator of multiplication in $\tilde{L}^2(\mathbb{R}^n)$ by the function $\vec{k} \mapsto (\vec{k}^2 - z)^{-1}$, where $z$ is a fixed complex number. It is natural to denote this operator by $(H_0 - z)^{-1}$. If $z \notin [0, \infty)$, the preceding function belongs to $L^\infty(\mathbb{R}^n)$; then $(H_0 - z)^{-1} \in \mathcal{B}(\mathcal{H})$ and its norm is given as follows (see Proposition 2.26):

$$
\|(H_0 - z)^{-1}\| = \|(\vec{k}^2 - z)^{-1}\|_\infty = \sup_{\lambda \geq 0} |\lambda - z|^{-1} = [\text{dist}(z, [0, \infty))]^{-1}. \qquad (2.82)
$$

[12] We systematically set $\hbar = 1$ for Planck's constant and $m = 1/2$ for the mass.

Since the function $\vec{k} \mapsto \vec{k}^2(\vec{k}^2 - z)^{-1}$ is bounded if $z \notin [0, \infty)$, each vector in the range of $(H_0 - z)^{-1}$ belongs to $\mathcal{D}(H_0)$ and, if $f \in \mathcal{D}(H_0)$, then $f \in \mathcal{R}((H_0 - z)^{-1})$ [one has $f = (H_0 - z)^{-1}g$ for $g = (H_0 - z)f$]. The operator $(H_0 - z)^{-1}$ is called the *resolvent* of $H_0$, and we refer to Section 2.6 for a more general theory of resolvents.

In Section 2.7 it will be useful to know properties in configuration space of the functions $f$ belonging to the domain of $H_0$. In (2.81) this domain is given in the representation $\tilde{L}^2(\mathbb{R}^n)$, and to find properties of $f(\vec{x})$ one must make an inverse Fourier transformation. Since $\mathcal{D}(H_0) = \mathcal{R}((H_0 - z)^{-1})$, one must study properties of the inverse Fourier transform of functions of the form $\eta(\vec{k}) = (\vec{k}^2 - z)^{-1}\tilde{g}(\vec{k})$, with $\tilde{g}$ square-integrable [i.e. $\tilde{g} \in \tilde{L}^2(\mathbb{R}^n)$]. It turns out that the case of dimension $n = 1, 2$ and 3 is more elementary than the case $n \geq 4$. For $n = 1$, a detailed description of $\mathcal{D}(H_0)$ in configuration space is obtained in §3.2.4.

We first observe that, if $z \notin [0, \infty)$, then the function $\theta_z(\vec{k}) := (\vec{k}^2 - z)^{-1}$ belongs to $L^p(\mathbb{R}^n)$ for each $p > n/2$ but not for $p \leq n/2$; indeed $\theta_z$ is locally bounded, hence belongs locally to $L^p$ for each $p \in [1, \infty]$, and it suffices to consider its behaviour at infinity. By choosing spherical polar co-ordinates one sees that $\theta_z$ belongs to $L^p(\mathbb{R}^n)$ if and only if $p$ satisfies

$$\int_1^\infty \frac{k^{n-1}}{k^{2p}}\, dk < \infty,$$

i.e. if and only if $2p - (n - 1) > 1$ or $p > n/2$. In particular $\theta_z$ is square-integrable if $n = 1, 2$ or 3, but if $n \geq 4$ one has $\theta_z \in L^p(\mathbb{R}^n)$ only for $p > n/2$.

So, if $n = 1, 2$ or 3, the function $\eta(\vec{k}) = (\vec{k}^2 - z)^{-1}\tilde{g}(\vec{k})$ belongs to $L^1(\mathbb{R}^n)$, and one has

$$\begin{aligned}\big|[(H_0 - z)^{-1}g](\vec{x})\big| &= (2\pi)^{-n/2}\left|\int_{\mathbb{R}^n} e^{i\vec{k}\cdot\vec{x}}(\vec{k}^2 - z)^{-1}\tilde{g}(\vec{k})\, d^nk\right| \\ &\leq (2\pi)^{-n/2}\int_{\mathbb{R}^n} \big|(\vec{k}^2 - z)^{-1}\tilde{g}(\vec{k})\big|\, d^nk \leq \|\theta_z\|_2\|\tilde{g}\|_2,\end{aligned}$$

hence $(H_0 - z)^{-1}g$ is a bounded function of $\vec{x} \in \mathbb{R}^n$ (by a similar argument, using Proposition 1.9, one finds that this function is uniformly continuous). If $n \geq 4$, one can apply the Hölder inequality (2.121) to obtain that, as a function of $\vec{k}$, $\eta$ belongs to $L^r(\mathbb{R}^n)$ for $r$ satisfying $1/r = 1/p + 1/2$ with $p > n/2$ [hence for $r > 2n/(n + 4)$]. Then one can conclude by the Hausdorff-Young inequality (2.122) that, as a function of $\vec{x}$, $(H_0 - z)^{-1}g$ belongs to $L^{r'}(\mathbb{R}^n)$ with $1/r' = 1 - 1/r = 1/2 - 1/p$ (still with $p > n/2$). Summing up one has: [13]

$$\begin{aligned}&g \in \mathcal{D}(H_0) \text{ and } n = 1, 2 \text{ or } 3 \implies g \in L^\infty(\mathbb{R}^n) \\ &g \in \mathcal{D}(H_0) \text{ and } n \geq 4 \implies g \in L^s(\mathbb{R}^n)\ \forall s \in [2, 2n/(n-4)).\end{aligned} \tag{2.83}$$

**2.5.4.** Let us consider the behaviour of multiplication operators under Fourier transformation. Let $A$ be the operator of multiplication in $L^2(\mathbb{R})$ by a function $a$. If $a$ is a

[13] The restriction $s \geq 2$ on the second line is due to the fact that, in the argument involving the Hausdorff-Young inequality, one must require that $r \in [1, 2]$ (see the Appendix to Chapter 2), hence $r' \equiv s \geq 2$.

polynomial, then $\mathcal{F}A\mathcal{F}^*$ is a differential operator. Indeed the identity $\mathcal{F}P^m\mathcal{F}^* = Q^m$ implies that $\mathcal{F}^*Q^m\mathcal{F} = P^m$, and, since the passage from $\mathcal{F}$ to $\mathcal{F}^*$ simply amounts to changing $i$ to $-i$, one obtains $\mathcal{F}Q^m\mathcal{F}^* = (-P)^m = (+id/dx)^m$ on $\mathcal{S}(\mathbb{R})$. If $a$ is a bounded function, then formally $\mathcal{F}A\mathcal{F}^*$ is the operator of convolution by $\tilde{a}$, more precisely

$$(\mathcal{F}Af)(k) = \frac{1}{\sqrt{2\pi}}\int_{-\infty}^{\infty}\tilde{a}(k-k')\tilde{f}(k')dk', \tag{2.84}$$

where $\tilde{a}$ is the Fourier transform of the function $a$. This makes sense for example if one assumes that $a\in L^2(\mathbb{R})$. Indeed one then has

$$(\mathcal{F}Af)(k) = \frac{1}{\sqrt{2\pi}}\int_{-\infty}^{\infty}e^{-ikx}a(x)f(x)dx = \frac{1}{\sqrt{2\pi}}\langle e^{ik\cdot}\overline{a}(\cdot),f\rangle \equiv \frac{1}{\sqrt{2\pi}}\langle\alpha_k,f\rangle, \tag{2.85}$$

i.e. the integral can be interpreted as the scalar product in $L^2(\mathbb{R})$ of the two functions $\alpha_k(x) \equiv e^{ikx}\overline{a(x)}$ and $f(x)$. Now if $\tilde{a}$ denotes the Fourier transform of $a$, then the value of the Fourier transform $\widetilde{\alpha_k}$ of $\alpha_k$ at the point $k'$ is as follows:

$$\begin{aligned}\widetilde{\alpha_k}(k') &= \frac{1}{\sqrt{2\pi}}\int_{-\infty}^{\infty}e^{-ik'x}\left[e^{ikx}\overline{a(x)}\right]dx\\ &= \frac{1}{\sqrt{2\pi}}\overline{\int_{-\infty}^{\infty}e^{-i(k-k')x}a(x)dx} \equiv \overline{\tilde{a}(k-k')}.\end{aligned}$$

Consequently, by virtue of (2.76), Eq. (2.85) may be written as

$$(\mathcal{F}Af)(k) = \frac{1}{\sqrt{2\pi}}\langle\widetilde{\alpha_k},\tilde{f}\rangle = \frac{1}{\sqrt{2\pi}}\int_{-\infty}^{\infty}\tilde{a}(k-k')\tilde{f}(k')dk',$$

so that one obtains (2.84) in this case. In particular (2.84) shows that $\mathcal{F}A\mathcal{F}^*$ is an integral operator in $\tilde{L}^2(\mathbb{R})$.

Since the resolvent $(H_0 - z)^{-1}$ $[z\in\mathbb{C}\backslash[0,\infty)]$ is multiplication by $(\vec{k}^2 - z)^{-1}$ in momentum space, it will become an integral operator in the position variables, namely $[(H_0-z)^{-1}f](\vec{x}) = \int_{\mathbb{R}^n}G_z^{(0)}(\vec{x}-\vec{y})f(\vec{y})d^ny$. Its kernel $G_z^{(0)}$ is called the ***free Green's function***.

**2.5.5.** To end this section we present some compactness criteria for operators in the space $L^2(\mathbb{R}^n)$, $n\in\mathbb{N}$. These results will be important in later chapters. If $\varphi\colon\mathbb{R}^n\to\mathbb{C}$ is a measurable function, we denote by $\varphi(\vec{Q})$ the multiplication operator in $L^2(\mathbb{R}^n)$ by $\varphi(\vec{x})$, and we use the notation $\varphi(\vec{P})$ for the multiplication operator by $\varphi(\vec{k})$ acting on the Fourier transforms of the functions in $L^2(\mathbb{R}^n)$, i.e. the multiplication operator by $\varphi(\vec{k})$ in $\tilde{L}^2(\mathbb{R}^n)$; each of these operators is considered on its maximal domain as specified in (2.68). Thus

$$[\varphi(\vec{Q})f](\vec{x}) = \varphi(\vec{x})f(\vec{x})\ , \quad [\mathcal{F}\varphi(\vec{P})f](\vec{k}) = \varphi(\vec{k})\tilde{f}(\vec{k}). \tag{2.86}$$

The symbols $\vec{Q}$ and $\vec{P}$ stand for the n-dimensional position and momentum operator respectively, as defined in Example 2.30, with $(Q_jf)(\vec{x}) = x_jf(\vec{x})$ and $(\mathcal{F}P_jf)(\vec{k}) = k_j\tilde{f}(\vec{k})$.

The notations $\varphi(\vec{Q})$ and $\varphi(\vec{P})$ are in agreement with the functional calculus for the self-adjoint operators $\vec{Q}$ and $\vec{P}$ that is developed in Chapter 4. What is important in the present context is the fact that one often encounters operators of the form $\varphi(\vec{Q})\psi(\vec{P})$, and it is useful to know that such a product is a compact operator under suitable assumptions on the functions $\varphi$ et $\psi$. Before proving such a result one must be able to give a meaning to the product $\varphi(\vec{Q})\psi(\vec{P})$ as an operator in $\mathcal{B}(\mathcal{H})$. This is possible for example if the functions $\varphi$ et $\psi$ belong to the same space $L^p(\mathbb{R}^n)$ for some $p \geq 2$. The proof is simple if $p = 2$ or $p = \infty$; these two cases suffice for most applications in quantum mechanics where one considers physical systems in dimension $n = 1, 2$ or 3. In order to treat problems in configuration spaces of arbitrary dimension it is necessary to know the boundedness of $\varphi(\vec{Q})\psi(\vec{P})$ also for values of $p$ in $(2, \infty)$. We state this result in the following proposition and indicate the proof for $p \neq 2, \infty$ in the Appendix to this chapter.

**Proposition 2.33.** *Let $\mathcal{H} = L^2(\mathbb{R}^n)$, $p \in [2, \infty]$ and let $\varphi, \psi \colon \mathbb{R}^n \to \mathbb{C}$ be two functions belonging to $L^p(\mathbb{R}^n)$. Then the operators $\varphi(\vec{Q})\psi(\vec{P})$ and $\psi(\vec{P})\varphi(\vec{Q})$ are densely defined, their closures belong to $\mathcal{B}(\mathcal{H})$, and*

$$\|\varphi(\vec{Q})\psi(\vec{P})\| \leq (2\pi)^{-n/p}\|\varphi\|_p\|\psi\|_p\,, \quad \|\psi(\vec{P})\varphi(\vec{Q})\| \leq (2\pi)^{-n/p}\|\varphi\|_p\|\psi\|_p. \tag{2.87}$$

PROOF. We treat the case $\varphi(\vec{Q})\psi(\vec{P})$. The proof for $\psi(\vec{P})\varphi(\vec{Q})$ is analogous (or one can use that fact that $[\psi(\vec{P})\varphi(\vec{Q})]^* = \overline{\varphi}(\vec{Q})\overline{\psi}(\vec{P})$, where for example $\overline{\psi}$ denotes the complex conjugate function of $\psi$).

(i) If $p = \infty$, then both $\varphi(\vec{Q})$ and $\psi(\vec{P})$ are multiplication operators by essentially bounded functions and hence belong to $\mathcal{B}(\mathcal{H})$, with $\|\varphi(\vec{Q})\| = \|\varphi\|_\infty$ and $\|\psi(\vec{P})\| = \|\psi\|_\infty$ (see Proposition 2.26).

(ii) If $p = 2$, then the domain of $\psi(\vec{P})$ contains the Schwartz space $\mathcal{S}(\mathbb{R}^n)$ (see point (1) in Remark 2.31). If $f$ belongs to $\mathcal{S}(\mathbb{R}^n)$, there exists a finite constant $c$ such that $|[\psi(\vec{P})f](\vec{x})| \leq c$ for almost all $\vec{x} \in \mathbb{R}^n$ [see (2.88) below], hence one can again apply point (1) of Remark 2.31 to conclude that $\psi(\vec{P})f \in \mathcal{D}(\varphi(\vec{Q}))$. Consequently $\varphi(\vec{Q})\psi(\vec{P})$ is well defined on $\mathcal{S}(\mathbb{R}^n)$. To prove (2.87) for $p = 2$ it suffices to show that $\|\varphi(\vec{Q})\psi(\vec{P})f\| \leq (2\pi)^{-n/2}\|\varphi\|_2\|\psi\|_2\|f\|$ for $f \in \mathcal{S}(\mathbb{R}^n)$. For this we apply the Schwarz inequality as follows:

$$\left|[\psi(\vec{P})f](\vec{x})\right| = (2\pi)^{-n/2}\left|\int_{\mathbb{R}^n} e^{i\vec{k}\cdot\vec{x}}\psi(\vec{k})\tilde{f}(\vec{k})\,d^nk\right| \leq (2\pi)^{-n/2}\|\psi\|_2\|f\|. \tag{2.88}$$

The claimed inequality is easily obtained from (2.88):

$$\begin{aligned}\|\varphi(\vec{Q})\psi(\vec{P})f\|^2 &= \int_{\mathbb{R}^n}\left|\varphi(\vec{x})\right|^2\left|[\psi(\vec{P})f](\vec{x})\right|^2 d^nx \\ &\leq (2\pi)^{-n}\|\psi\|_2^2\|f\|^2\int_{\mathbb{R}^n}|\varphi(\vec{x})|^2 d^nx.\end{aligned}$$

(iii) The case $p \in (2, \infty)$ is treated in the Appendix to this chapter. □

**Proposition 2.34.** *Let $\mathcal{H} = L^2(\mathbb{R}^n)$ and $\varphi,\psi\colon \mathbb{R}^n \to \mathbb{C}$ measurable functions.*
*(a) $\varphi(\vec{Q})\psi(\vec{P})$ and $\psi(\vec{P})\varphi(\vec{Q})$ are Hilbert-Schmidt operators if and only if both functions $\varphi$ and $\psi$ belong to $L^2(\mathbb{R}^n)$ (or if one of these functions is zero almost everywhere), and*

$$\|\varphi(\vec{Q})\psi(\vec{P})\|_{\mathrm{HS}} = \|\psi(\vec{P})\varphi(\vec{Q})\|_{\mathrm{HS}} = (2\pi)^{-n/2}\|\varphi\|_2\|\psi\|_2. \tag{2.89}$$

*(b) If $\varphi$ and $\psi$ belong to $L^p(\mathbb{R}^n)$ for some $p \in [2,\infty)$, then $\varphi(\vec{Q})\psi(\vec{P})$ and $\psi(\vec{P})\varphi(\vec{Q})$ are compact operators.*[14]

PROOF. (a) Assume that $\varphi,\psi \in L^2(\mathbb{R}^n)$. Then, by the results of §2.5.4, $\varphi(\vec{Q})\psi(\vec{P})$ is an integral operator in $\tilde{L}^2(\mathbb{R}^n)$ with kernel $a(\vec{k},\vec{k}') = (2\pi)^{-n/2}\tilde{\varphi}(\vec{k}-\vec{k}')\psi(\vec{k}')$:

$$[\mathcal{F}\varphi(\vec{Q})\psi(\vec{P})f](\vec{k}) = (2\pi)^{-n/2}\int_{\mathbb{R}^n}\tilde{\varphi}(\vec{k}-\vec{k}')\psi(\vec{k}')\tilde{f}(\vec{k}')\,d^nk'.$$

The Hilbert-Schmidt norm of this operator is easily calculated by using Proposition 2.15 (setting $\vec{p} = \vec{k}-\vec{k}'$):

$$\begin{aligned}\|\varphi(\vec{Q})\psi(\vec{P})\|_{\mathrm{HS}}^2 &= (2\pi)^{-n}\int_{\mathbb{R}^n}\int_{\mathbb{R}^n}\left|\tilde{\varphi}(\vec{k}-\vec{k}')\psi(\vec{k}')\right|^2 d^nk\,d^nk' \\ &= (2\pi)^{-n}\int_{\mathbb{R}^n}\left|\tilde{\varphi}(\vec{p})\right|^2 d^np\int_{\mathbb{R}^n}\left|\psi(\vec{k}')\right|^2 d^nk',\end{aligned}$$

which gives (2.89) for $\varphi(\vec{Q})\psi(\vec{P})$. The operator $\psi(\vec{P})\varphi(\vec{Q})$ is treated by considering its adjoint (see the beginning of the preceding proof). The "only if" part of the proof of (a) is given in the Appendix to this chapter (page 81).

(b) We denote by $\chi_R$ the characteristic function of the ball $\{\vec{x}\in\mathbb{R}^n \mid |\vec{x}| \le R\}$, i.e. $\chi_R(\vec{x}) = 1$ if $|\vec{x}| \le R$ and $\chi_R(\vec{x}) = 0$ otherwise. By (1.49) the functions $\chi_R\varphi$ and $\chi_R\psi$ belong to $L^2(\mathbb{R}^n)$. Hence, if we set $C_R = \chi_R(\vec{Q})\varphi(\vec{Q})$ and $D_R = \chi_R(\vec{P})\psi(\vec{P})$, then $C_RD_R \in \mathcal{B}_2(\mathcal{H}) \subset \mathcal{K}(\mathcal{H})$ by the result of (a). So, to conclude that the operator $\varphi(\vec{Q})\psi(\vec{P})$ is compact, it suffices to show that $\|\varphi(\vec{Q})\psi(\vec{P})-C_RD_R\| \to 0$ as $R\to\infty$ and to apply Proposition 2.11(d). Now (2.87) implies that

$$\begin{aligned}\|\varphi(\vec{Q})\psi(\vec{P}) - C_RD_R\| &\le \|[\varphi(\vec{Q}) - C_R]\psi(\vec{P})\| + \|C_R[\psi(\vec{P}) - D_R]\| \\ &\le \|[1-\chi_R(\cdot)]\varphi(\cdot)\|_p\|\psi\|_p + \|\varphi\|_p\|[1-\chi_R(\cdot)]\psi(\cdot)\|_p,\end{aligned}$$

which converges to zero as $R\to\infty$ (for example for the first term one uses Proposition 1.9 with $h(\vec{x}) = |\varphi(\vec{x})|^p$). □

## 2.6 Resolvent and spectrum of an operator

Let $A$ be a *closed* operator. The ***resolvent set*** of $A$ is defined as the set of all complex numbers $z$ for which the operator $A-z \equiv A-zI$ is invertible with $(A-z)^{-1}\in\mathcal{B}(\mathcal{H})$.

[14] The value of $p$ must be the same for $\varphi$ and $\psi$, and $p=\infty$ is excluded.

This set is denoted by $\rho(A)$. Thus a complex number $z$ belongs to $\rho(A)$ if and only if (i) $\mathcal{N}(A-z)=\{\mathbf{0}\}$, (ii) $\mathcal{R}(A-z)=\mathcal{H}$ and (iii) $(A-z)^{-1}$ is bounded.[15] It is clear that, if $z\in\rho(A)$, then $(A-z)^{-1}$ maps $\mathcal{H}$ onto $\mathcal{D}(A)$ and gives a bijection between $\mathcal{H}$ and $\mathcal{D}(A)$. If $z\in\rho(A)$ one has

$$A(A-z)^{-1}=(A-z)(A-z)^{-1}+z(A-z)^{-1}=I+z(A-z)^{-1} \tag{2.90}$$

and, for $f\in\mathcal{D}(A)$:

$$(A-z)^{-1}Af=(A-z)^{-1}(A-z)f+z(A-z)^{-1}f=f+z(A-z)^{-1}f, \tag{2.91}$$

in particular

$$(A-z)^{-1}Af=A(A-z)^{-1}f \qquad \text{if } f\in\mathcal{D}(A). \tag{2.92}$$

For $z\in\rho(A)$ the operator $(A-z)^{-1}$ is called the ***resolvent*** of $A$ at the point $z$. The term "resolvent" is also used for the collection of operators $\{(A-z)^{-1}\mid z\in\rho(A)\}$, i.e. for the function $z\mapsto(A-z)^{-1}$ from $\rho(A)$ to $\mathcal{B}(\mathcal{H})$.

**Proposition 2.35.** *Let $A$ be a closed operator and $\rho(A)$ its resolvent set.*
*(a) One has the **first resolvent equation**: if $z_1,z_2\in\rho(A)$, then*

$$(A-z_1)^{-1}-(A-z_2)^{-1}=(z_1-z_2)(A-z_1)^{-1}(A-z_2)^{-1}. \tag{2.93}$$

*(b) If $z_1,z_2\in\rho(A)$, then $(A-z_1)^{-1}$ and $(A-z_2)^{-1}$ commute.*
*(c) The resolvent set $\rho(A)$ is an open subset of $\mathbb{C}$ (which may be empty).*
*(d) In each connected component of $\rho(A)$ the function $z\mapsto(A-z)^{-1}$ is holomorphic, i.e. at each point $z\in\rho(A)$ this function has a Taylor expansion converging in the norm of $\mathcal{B}(\mathcal{H})$.*

PROOF. (a,b) If $z_1,z_2\in\rho(A)$, then by (2.90):

$$z_1(A-z_1)^{-1}=-I+A(A-z_1)^{-1} \quad\text{and}\quad z_2(A-z_2)^{-1}=-I+A(A-z_2)^{-1},$$

hence

$$\begin{aligned}(z_1-z_2)(A-z_1)^{-1}(A-z_2)^{-1}&=\big[-(A-z_2)^{-1}+A(A-z_1)^{-1}(A-z_2)^{-1}\big]\\&\quad-\big[-(A-z_1)^{-1}+(A-z_1)^{-1}A(A-z_2)^{-1}\big]\\&=-(A-z_2)^{-1}+(A-z_1)^{-1}+\big[A(A-z_1)^{-1}-(A-z_1)^{-1}A\big](A-z_2)^{-1},\end{aligned}$$

and the last term on the bottom line is zero by (2.92). This proves (2.93). The result of (b) is a simple consequence of (2.93).

(c) Let $z\in\rho(A)$ and let $\zeta\in\mathbb{C}$ be such that $|z-\zeta|<\|(A-z)^{-1}\|^{-1}$. Then, by

[15] One can show that (iii) is a consequence of (i) and (ii) if $A$ is closed. In fact, by Problem 2.19(a), (i) and (ii) imply that $(A-z)^{-1}$ is closed and has domain $\mathcal{H}$, and it suffices to apply the Closed Graph Theorem (we omit its proof; the reader may consult a book, e.g. §51 of [AG] or §III.5.4 of [K]):
**Closed Graph Theorem**: *If $B$ is a closed operator with $\mathcal{D}(B)=\mathcal{H}$, then $B$ is bounded, i.e. $B\in\mathcal{B}(\mathcal{H})$.*

Proposition 2.5, the operator $I+(z-\zeta)(A-z)^{-1}$ has an inverse belonging to $\mathcal{B}(\mathcal{H})$, given by

$$\left[I+(z-\zeta)(A-z)^{-1}\right]^{-1}=\sum_{k=0}^{\infty}(z-\zeta)^k\left[(A-z)^{-1}\right]^k. \tag{2.94}$$

We claim that $\zeta$ belongs to $\rho(A)$. To see this we observe that

$$(A-\zeta)f=\left[I+(z-\zeta)(A-z)^{-1}\right](A-z)f \qquad \text{if } f\in\mathcal{D}(A). \tag{2.95}$$

Since each of the two operators on the right-hand side [i.e. $I+(z-\zeta)(A-z)^{-1}$ and $A-z$] is invertible, so is $A-\zeta$. To see that $\mathcal{R}(A-\zeta)=\mathcal{H}$, we fix a vector $g\in\mathcal{H}$ and set $f=(A-z)^{-1}[I+(z-\zeta)(A-z)^{-1}]^{-1}g$. Clearly $f\in\mathcal{D}(A)$ and $(A-\zeta)f=g$ by (2.95). This shows that $\mathcal{R}(A-\zeta)=\mathcal{H}$, and since $f=(A-\zeta)^{-1}g$, it also follows that

$$(A-\zeta)^{-1}=(A-z)^{-1}\left[I+(z-\zeta)(A-z)^{-1}\right]^{-1}\in\mathcal{B}(\mathcal{H}) \tag{2.96}$$

as a product of two operators each of which belongs to $\mathcal{B}(\mathcal{H})$. We have thus shown that each $z\in\rho(A)$ has a complex neighbourhood belonging also to $\rho(A)$; consequently $\rho(A)$ is an *open* subset of $\mathbb{C}$.

(d) The Taylor expansion in a neighbourhood of $z\in\rho(A)$ is obtained by inserting (2.94) into (2.96):

$$(A-\zeta)^{-1}=\sum_{k=0}^{\infty}(\zeta-z)^k\left[(A-z)^{-1}\right]^{k+1}. \tag{2.97}$$

□

The function $z\mapsto(A-z)^{-1}$ is infinitely differentiable, its derivatives exist in the uniform topology [in the norm of $\mathcal{B}(\mathcal{H})$]. They can be calculated from (2.97) [differentiate (2.97) $n$ times with respect to $\zeta$ and then set $\zeta=z$] or from (2.93):

$$\frac{d^n}{dz^n}(A-z)^{-1}=n!\,(A-z)^{-(n+1)} \qquad (n\geq 1) \tag{2.98}$$

with the convention that $(A-z)^{-k}=[(A-z)^{-1}]^k$. In particular one has

$$\frac{d}{dz}(A-z)^{-1}=(A-z)^{-2}. \tag{2.99}$$

The complementary set of $\rho(A)$ in the complex plane $\mathbb{C}$ is called the ***spectrum*** of $A$ and denoted by $\sigma(A)$:

$$\sigma(A)=\mathbb{C}\backslash\rho(A). \tag{2.100}$$

$\sigma(A)$ is a *closed* subset of $\mathbb{C}$. A complex number $z$ belongs to $\sigma(A)$ if $A-z$ is not invertible or if $\mathcal{R}(A-z)\neq\mathcal{H}$ (see the footnote on page 68). The spectrum of a closed operator $A$ contains in particular all eigenvalues of $A$. A complex number $z$ is an ***eigenvalue*** of $A$ if there exists a vector $f\neq\mathbf{0}$ in $\mathcal{D}(A)$ such that $Af=zf$, in other terms if $A-z$ is not invertible. Apart from the eigenvalues (if any) the spectrum of $A$ contains complex numbers $z$ for which $A-z$ is invertible but with unbounded or not densely defined inverse.

If $A$ is closed with dense domain in $\mathcal{H}$, then its adjoint $A^*$ exists and is closed (see Section 2.4). One then has a very simple relation between the spectrum of $A$ and that of $A^*$: the latter is obtained from the former by reflection at the real axis, i.e.

$$z \in \sigma(A^*) \Longleftrightarrow \overline{z} \in \sigma(A) \qquad \text{and} \qquad z \in \rho(A^*) \Longleftrightarrow \overline{z} \in \rho(A). \tag{2.101}$$

We do not give the proof in the general case (see Problem 2.20). In this text we are particularly interested in self-adjoint operators; as shown below, the spectrum of such an operator is a subset of $\mathbb{R}$, i.e. each $z \in \mathbb{C}$ with $\Im z \neq 0$ belongs to $\rho(A)$. The spectrum of symmetric operators will be discussed at the end of Section 3.1. A simple property of their eigenvalues is given in the next proposition.

**Proposition 2.36.** *(a) A symmetric operator has no non-real eigenvalues. In particular all eigenvalues of a self-adjoint operator are real.*
*(b) Eigenvectors associated with different eigenvalues of a symmetric operator are orthogonal to one another.*

PROOF. (a) Assume that $A \subseteq A^*$ and that $f \in \mathcal{D}(A)$ is such that $\|f\| \neq 0$ and $Af = zf$ for some $z \in \mathbb{C}$. Then

$$z\|f\|^2 = \langle f, zf\rangle = \langle f, Af\rangle = \langle Af, f\rangle = \langle zf, f\rangle = \overline{z}\|f\|^2.$$

Since $\|f\| \neq 0$, one must have $z = \overline{z}$.

(b) Assume that $Af = \lambda f$ and $Ag = \mu g$ with $f, g \in \mathcal{D}(A)$ and with $\lambda \neq \mu$. Then $\lambda\langle f, g\rangle = \langle Af, g\rangle = \langle f, Ag\rangle = \mu\langle f, g\rangle$ (remember that $\lambda$ and $\mu$ are real). Since $\lambda \neq \mu$, we must have $\langle f, g\rangle = 0$. □

The following identity will be important. Let $A$ be a symmetric operator and $z = \lambda + i\mu$ a complex number ($\lambda, \mu \in \mathbb{R}$). Then one has for each $f \in \mathcal{D}(A)$:

$$\begin{aligned}\|(A-z)f\|^2 &= \|(A-\lambda)f\|^2 + \mu^2\|f\|^2 - i\mu\langle (A-\lambda)f, f\rangle + i\mu\langle f, (A-\lambda)f\rangle \\ &= \|(A-\lambda)f\|^2 + \mu^2\|f\|^2.\end{aligned} \tag{2.102}$$

**Proposition 2.37.** *The spectrum of a self-adjoint operator A is real, i.e. $\sigma(A) \subseteq \mathbb{R}$. If $z = \lambda + i\mu$ with $\lambda, \mu \in \mathbb{R}$ and $\mu \neq 0$, then $z \in \rho(A)$ and $\|(A-z)^{-1}\| \leq 1/|\mu|$.*

PROOF. Assume that $z = \lambda + i\mu \in \mathbb{C}\setminus\mathbb{R}$. By (2.102) we have $\|(A-z)f\| \geq |\mu|\|f\|$ for each $f \in \mathcal{D}(A)$. Since $\mu \neq 0$, this implies that $A - z$ is invertible. Setting $f = (A-z)^{-1}g$ in (2.102), with $g \in \mathcal{R}(A-z)$, we get $\|(A-z)^{-1}g\| \leq |\mu|^{-1}\|g\|$. This shows that $(A-z)^{-1}$ is bounded, with norm $\leq 1/|\mu|$. Finally we see from Proposition 2.19(b) and Proposition 2.36(a) that $\mathcal{R}(A-z)^{\perp} = \mathcal{N}(A^* - \overline{z}) \equiv \mathcal{N}(A - \overline{z}) = \{\mathbf{0}\}$. So $\mathcal{R}(A-z)$ is dense in $\mathcal{H}$, and since $(A-z)^{-1}$ is bounded, we must have $\mathcal{R}(A-z) = \mathcal{H}$ (see the proof of Lemma 3.1(c) for more details about this last property). □

**Proposition 2.38.** *If $z$ is an eigenvalue of a closed operator $A$, denote by $\mathcal{M}(z) = \{f \in \mathcal{D}(A) \mid Af = zf\}$ the set of all eigenvectors of $A$ associated with this eigenvalue. Then $\mathcal{M}(z)$ is a subspace of $\mathcal{H}$. In particular the null space $\mathcal{N}(A)$ is a subspace.*

PROOF. Let $\{f_n\}$ be a (strong) Cauchy sequence in $\mathcal{M}(z)$. We denote its limit by $f$. We must show that $f \in \mathcal{M}(z)$, i.e. that $f \in \mathcal{D}(A)$ and $Af = zf$. For this we observe that $\|f_n - f_m\| \to 0$ as $m, n \to \infty$ and consequently $\|Af_n - Af_m\| = \|zf_n - zf_m\| = |z| \, \|f_n - f_m\| \to 0$ as $m, n \to \infty$. Thus the sequence $\{f_n\}$ and the sequence $\{Af_n\}$ are Cauchy. Since $A$ is closed, it follows that the limit of the sequence $\{f_n\}$ belongs to $\mathcal{D}(A)$, i.e. $f \in \mathcal{D}(A)$, and that $Af = \text{s-lim}_{n\to\infty} Af_n = \text{s-lim}_{n\to\infty} zf_n = z \cdot \text{s-lim}_{n\to\infty} f_n = zf$.

If $z = 0$ is an eigenvalue of $A$, then $\mathcal{N}(A) = \mathcal{M}(z = 0)$, so $\mathcal{N}(A)$ is a subspace by the preceding result. If $z = 0$ is not an eigenvalue of $A$, then $\mathcal{N}(A) = \{\mathbf{0}\}$, which is also a subspace. □

**Proposition 2.39.** *The spectrum of an operator $A \in \mathcal{B}(\mathcal{H})$ is contained in the disk $\{z \in \mathbb{C} \mid |z| \leq \|A\|\}$.*

PROOF. Let $z \in \mathbb{C}$ be such that $|z| > \|A\|$. One may write $A - z = -z(I - A/z)$. Since $\|A/z\| = \|A\|/|z| < 1$, Proposition 2.5 implies that the operator $A - z$ is invertible, with

$$(A - z)^{-1} = \big[-z(I - \frac{A}{z})\big]^{-1} = \frac{-1}{z}\big(I - \frac{A}{z}\big)^{-1} \in \mathcal{B}(\mathcal{H}).$$

Thus $z$ belongs to $\rho(A)$. □

Finally we point out that, if the operator $A$ is self-adjoint and $z \in \rho(A)$, then

$$[(A - z)^{-1}]^* = (A - \overline{z})^{-1} \qquad (A = A^*). \tag{2.103}$$

Indeed, if $f, g \in \mathcal{H}$, there exist vectors $f_0, g_0 \in \mathcal{D}(A)$ such that $f = (A - z)f_0$ and $g = (A - \overline{z})g_0$. Then

$$\begin{aligned}\langle f, [(A - z)^{-1}]^* g\rangle &= \langle (A - z)^{-1} f, g\rangle = \langle f_0, g\rangle = \langle f_0, (A - \overline{z})g_0\rangle \\ &= \langle (A - z)f_0, g_0\rangle = \langle f, (A - \overline{z})^{-1} g\rangle.\end{aligned}$$

## 2.7 Perturbations of self-adjoint operators

An important aspect of the theory of self-adjoint operators concerns the invariance of self-adjointness under perturbations. If $A$ is a self-adjoint operator and $B = B^*$ belongs to $\mathcal{B}(\mathcal{H})$, then it is easy to see that $A + B$ is also self-adjoint on the domain $\mathcal{D}(A + B) = \mathcal{D}(A)$. Thus the property of being self-adjoint is invariant under the addition of bounded self-adjoint operators. In many quantum-mechanical situations one considers the perturbation of a self-adjoint $A$ operator by an unbounded operator $B$, and it is important to know criteria implying the self-adjointness of $A + B$. The present section is devoted to this question.

**2.7.1.** If $A$ and $B$ are symmetric operators and if $\mathcal{D}(A) \cap \mathcal{D}(B)$ is dense in $\mathcal{H}$, then their sum $A + B$ is also symmetric: if $f, g \in \mathcal{D}(A + B) \equiv \mathcal{D}(A) \cap \mathcal{D}(B)$, then

$$\langle (A + B)f, g\rangle = \langle Af, g\rangle + \langle Bf, g\rangle = \langle f, Ag\rangle + \langle f, Bg\rangle = \langle f, (A + B)g\rangle.$$

However, even if both $A$ and $B$ are self-adjoint, the operator $A+B$ will not be self-adjoint in general (it will only be symmetric). In certain situations one can consider $B$ as a "perturbation" of $A$; roughly this means that, in a certain sense, $B$ is small compared to $A$, and under a condition of this type it is possible to prove that the sum $A+B$ is self-adjoint on a suitable domain, for example on $\mathcal{D}(A)$ for a certain type of perturbations $B$.

The simplest situation is that in which $B \in \mathcal{B}(\mathcal{H})$ ($B = B^*$). In this case the sum $A+B$ is self-adjoint on $\mathcal{D}(A)$. To see this, it suffices to show that $\mathcal{D}((A+B)^*) \subseteq \mathcal{D}(A+B) \equiv \mathcal{D}(A)$. So suppose that $f \in \mathcal{D}((A+B)^*)$, and let $f^* \in \mathcal{H}$ be such that $\langle f, (A+B)g\rangle = \langle f^*, g\rangle$ for all $g \in \mathcal{D}(A)$. Then

$$\langle f, (A+B)g\rangle = \langle f, Ag\rangle + \langle f, Bg\rangle = \langle f, Ag\rangle + \langle Bf, g\rangle = \langle f^*, g\rangle.$$

Thus one has $\langle f, Ag\rangle = \langle f^* - Bf, g\rangle$ for each $g \in \mathcal{D}(A)$. Since $f^* - Bf \in \mathcal{H}$, this means that $f \in \mathcal{D}(A^*) \equiv \mathcal{D}(A)$.

The preceding argument shows that the addition of a bounded self-adjoint operator $B$ to a self-adjoint operator $A$ does not change the self-adjointness. The same conclusion can be obtained when $B$ is unbounded but "small compared to $A$ modulo a bounded operator". Mathematically the concept "small compared to $A$" could for example be expressed as follows: $\|Bf\| < \varepsilon\|Af\|$ for some sufficiently small $\varepsilon > 0$ and for all $f \in \mathcal{D}(A) \cap \mathcal{D}(B)$. More precisely one says that $B$ is ***A-bounded*** (or ***relatively bounded with respect to A*** [16]) if $\mathcal{D}(A) \subseteq \mathcal{D}(B)$ and there exist constants $\alpha \geq 0$ and $\beta \geq 0$ such that

$$\|Bf\| \leq \alpha\|Af\| + \beta\|f\| \qquad \forall f \in \mathcal{D}(A). \tag{2.104}$$

The infimum of all numbers $\alpha$ for which such an inequality is true (for some $\beta$ which can depend on the value of $\alpha$) is called the ***A-bound*** of $B$ (or the ***relative bound*** of $B$ with respect to $A$). If this number is $< 1$, one may consider that $B$ is smaller than $A$ modulo a bounded term.

If $B \in \mathcal{B}(\mathcal{H})$, then (2.104) holds with $\alpha = 0$ and $\beta = \|B\|$, hence $B$ is $A$-bounded with relative bound 0. We shall see further on examples of *unbounded* operators $B$ having also $A$-bound 0, i.e. which are such that, for each $\varepsilon > 0$, there exists a number $\beta = \beta(\varepsilon) \in (0, \infty)$ with $\|Bf\| \leq \varepsilon\|Af\| + \beta(\varepsilon)\|f\|$ for all $f \in \mathcal{D}(A)$ [when $B$ is unbounded one must have $\beta(\varepsilon) \to \infty$ as $\varepsilon \to 0$]. The important fact is the following: if $A$ is self-adjoint and $B$ is symmetric with $A$-bound $< 1$, then $A+B$ is self-adjoint on $\mathcal{D}(A) \equiv \mathcal{D}(A) \cap \mathcal{D}(B)$. Before proving this we give a reformulation of the condition (2.104).

If $\mathcal{D}(A) \subseteq \mathcal{D}(B)$ and $z \in \rho(A)$ (the resolvent set of $A$), the operator $B(A-z)^{-1}$ is defined on each vector $g \in \mathcal{H}$ because $(A-z)^{-1}$ maps $\mathcal{H}$ onto $\mathcal{D}(A)$. We shall see, for a self-adjoint operator $A$, that the condition (2.104) is equivalent to the condition that $B(A-z)^{-1} \in \mathcal{B}(\mathcal{H})$. To obtain this result, it is useful to consider an inequality that is similar to (2.104) but contains the squares of the involved norms, viz.

$$\|Bf\|^2 \leq a^2\|Af\|^2 + b^2\|f\|^2 \qquad \forall f \in \mathcal{D}(A), \tag{2.105}$$

where $a \geq 0$ and $b \geq 0$ are constants.

[16] This definition holds for an arbitrary operator $A$ ($A$ does not have to be self-adjoint or closed).

**Lemma 2.40.** *Let $A$ be self-adjoint. For $a, b > 0$, the condition* (2.105) *is equivalent to* $\|B(A \pm i\frac{b}{a})^{-1}\| \leq a$. [17]

PROOF. By using (2.102) one sees that (2.105) is equivalent to

$$\|Bf\|^2 \leq a^2\big[\|Af\|^2 + \frac{b^2}{a^2}\|f\|^2\big] = a^2\|(A \pm i\frac{b}{a})f\|^2 \qquad \forall f \in \mathcal{D}(A). \tag{2.106}$$

Since $\pm ib/a$ belong to the resolvent set of $A$, each of the operators $(A \pm ib/a)^{-1}$ gives a bijection between $\mathcal{H}$ and $\mathcal{D}(A)$, so that the preceding inequality is equivalent to [18]

$$\|B(A \pm i\frac{b}{a})^{-1}g\| \leq a\|g\| \qquad \forall g \in \mathcal{H}. \tag{2.107}$$

□

**Lemma 2.41.** *Let $A$ be a self-adjoint operator and $\mathcal{D}(A) \subseteq \mathcal{D}(B)$.*
*(a) If* (2.104) *holds, then* (2.105) *is satisfied for each $a > \alpha$ (the value of $b$ depends on $a$).*
*(b) If* (2.104) *holds, there exists a number $\kappa > 0$ such that $B(A + i\kappa)^{-1} \in \mathcal{B}(\mathcal{H})$ and $B(A - i\kappa)^{-1} \in \mathcal{B}(\mathcal{H})$.*
*(c) If $B(A - z)^{-1} \in \mathcal{B}(\mathcal{H})$ for some $z \in \rho(A)$ [in particular if* (2.105) *is satisfied], then* (2.104) *holds with $\alpha = \|B(A - z)^{-1}\|$.*

PROOF. (a) If $\gamma, \eta \in \mathbb{R}$ and $\varepsilon > 0$ one has $0 \leq (\varepsilon^{1/2}\gamma - \varepsilon^{-1/2}\eta)^2 = \varepsilon\gamma^2 - 2\gamma\eta + \varepsilon^{-1}\eta^2$. This implies that

$$\begin{aligned}(\gamma + \eta)^2 &= \gamma^2 + 2\gamma\eta + \eta^2 \leq \gamma^2 + (\varepsilon\gamma^2 + \varepsilon^{-1}\eta^2) + \eta^2 \\ &= (1 + \varepsilon)\gamma^2 + (1 + \varepsilon^{-1})\eta^2.\end{aligned} \tag{2.108}$$

Assume now that (2.104) holds, and choose in (2.108) $\gamma = \alpha\|Af\|$ and $\eta = \beta\|f\|$. Then

$$\|Bf\|^2 \leq \big[\alpha\|Af\| + \beta\|f\|\big]^2 \leq (1 + \varepsilon)\alpha^2\|Af\|^2 + (1 + \varepsilon^{-1})\beta^2\|f\|^2.$$

By taking $\varepsilon$ sufficiently small one obtains the validity of (2.105) for any $a > \alpha$.

(b) This is immediate from (a) and Lemma 2.40.

(c) We set $\alpha = \|B(A - z)^{-1}\|$ and write for $f \in \mathcal{D}(A)$:

$$\begin{aligned}\|Bf\| = \|B(A - z)^{-1} \cdot (A - z)f\| &\leq \|B(A - z)^{-1}\|\|(A - z)f\| \\ &\leq \alpha\big[\|Af\| + |z|\|f\|\big].\end{aligned}$$

□

[17] If an equation or an expression contains double signs (for example $\pm$), it is understood that this means *two* equations or expressions, one by taking everywhere the upper sign, the other one by taking everywhere the lower sign.

[18] To see this equivalence, set $f = (A \pm i\frac{b}{a})^{-1}g$ in (2.106) or $g = (A \pm i\frac{b}{a})f$ in (2.107).

**Proposition 2.42.** *Let $A$ be a self-adjoint operator and $B$ an operator such that $\mathcal{D}(A) \subseteq \mathcal{D}(B)$.*
*(a) If $B(A-z)^{-1} \in \mathcal{B}(\mathcal{H})$ for some $z \in \rho(A)$, then $B$ is $A$-bounded.*
*(b) If $B$ is $A$-bounded, then $B(A-z)^{-1} \in \mathcal{B}(\mathcal{H})$ for each $z \in \rho(A)$ and, if $\mathfrak{b}$ denotes the $A$-bound of $B$, then*

$$\mathfrak{b} = \inf_{z \in \rho(A)} \|B(A-z)^{-1}\| = \inf_{\kappa > 0} \|B(A-i\kappa)^{-1}\| = \inf_{\kappa > 0} \|B(A+i\kappa)^{-1}\|. \quad (2.109)$$

PROOF. (a) This follows from Lemma 2.41(c).
(b) (i) By Lemma 2.41(b) there exists $\kappa > 0$ such that $B(A \pm i\kappa)^{-1} \in \mathcal{B}(\mathcal{H})$. If $z \in \rho(A)$ one has from the first resolvent equation (2.93):

$$B(A-z)^{-1} = [B(A-i\kappa)^{-1}] + (z-i\kappa)[B(A-i\kappa)^{-1}](A-z)^{-1}. \quad (2.110)$$

Clearly the right-hand side belongs to $\mathcal{B}(\mathcal{H})$, hence $B(A-z)^{-1} \in \mathcal{B}(\mathcal{H})$.
(ii) Now let $\mathfrak{b}$ be the $A$-bound of $B$. By Lemma 2.41(c), if $z \in \rho(A)$, then (2.104) is satisfied with $\alpha = \|B(A-z)^{-1}\|$, hence $\mathfrak{b} \le \inf_{z \in \rho(A)} \|B(A-z)^{-1}\|$. To obtain (2.109) it now suffices to show that, given any $\delta > 0$, there exists a real number $\kappa \neq 0$ such that $\|B(A \pm i\kappa)^{-1}\| \le \mathfrak{b} + \delta$. The inequality (2.104) is satisfied for any $\alpha > \mathfrak{b}$, hence for $\alpha = \mathfrak{b} + \delta/2$, and then Lemma 2.41(a) implies that (2.105) is satisfied with $a = \alpha + \delta/2 \equiv \mathfrak{b} + \delta$ and some $b > 0$. By virtue of Lemma 2.40 this means that $\|B(A \pm i\kappa)^{-1}\| \le \mathfrak{b} + \delta$ [with $\kappa = b/(\mathfrak{b} + \delta)$]. □

**Corollary 2.43.** *Let $A$ be self-adjoint. Then:*
*(a) $B$ is $A$-bounded with $A$-bound $\mathfrak{b} < 1 \iff$ there exists a number $z \in \rho(A)$ with $\|B(A-z)^{-1}\| < 1$.*
*(b) If $B$ is $A$-bounded with $A$-bound $\mathfrak{b} < 1$, there exists a number $\kappa > 0$ such that $\|B(A-i\kappa)^{-1}\| < 1$ and $\|B(A+i\kappa)^{-1}\| < 1$.*

PROOF. (a) is immediate from (2.109). (b) follows from the arguments in part (ii) of the preceding proof. □

In the next proposition we use the notion of semibounded self-adjoint operators. A self-adjoint operator $A$ is called ***lower semibounded*** or ***bounded from below*** if there exists a real number $\lambda$ such that the interval $(-\infty, \lambda)$ belongs to the resolvent set $\rho(A)$ of $A$. The operator $A$ is called ***upper semibounded*** if there exists $\lambda \in \mathbb{R}$ such that $(\lambda, +\infty) \subset \rho(A)$.

**Proposition 2.44 (Rellich-Kato Theorem).** *Let $A$ be a self-adjoint operator and $B$ an $A$-bounded symmetric operator with $A$-bound $\mathfrak{b} < 1$. Then:*
*(a) The operator $A + B$ is self-adjoint on $\mathcal{D}(A)$.*
*(b) $B$ is also $A + B$-bounded.*
*(c) If $A$ is semibounded, then $A + B$ is also semibounded.*

PROOF. (a) Since $\mathfrak{b} < 1$, there exists $\kappa > 0$ such that $\|B(A \pm i\kappa)^{-1}\| < 1$ [Corollary 2.43(b)]. The operator $A+B$ is symmetric on $\mathcal{D}(A)$. Let us show that $\mathcal{R}(A+B \pm i\kappa) = \mathcal{H}$. We write for $f \in \mathcal{D}(A) \equiv \mathcal{D}(A+B)$:

$$(A + B \pm i\kappa)f = \left[I + B(A \pm i\kappa)^{-1}\right](A \pm i\kappa)f. \quad (2.111)$$

As $\pm i\kappa \in \rho(A)$, the operators $(A \pm i\kappa)$ map $\mathcal{D}(A)$ onto $\mathcal{H}$. Since $\|B(A \pm i\kappa)^{-1}\| < 1$, the operators $I + B(A \pm i\kappa)^{-1}$ are invertible, with inverses in $\mathcal{B}(\mathcal{H})$ (by Proposition 2.5). Hence $I + B(A \pm i\kappa)^{-1}$ map $\mathcal{H}$ onto $\mathcal{H}$. Consequently Eq. (2.111) implies that $\mathcal{R}(A + B \pm i\kappa) = \mathcal{H}$. So $\mathcal{H} = \mathcal{R}(A + B \pm i\kappa) = \mathcal{R}(\kappa^{-1}(A + B \pm i\kappa)) = \mathcal{R}(\kappa^{-1}(A + B) \pm i)$. By Proposition 2.24, $\kappa^{-1}(A + B)$ (hence also $A + B$) is self-adjoint on $\mathcal{D}(A)$.

(b) For $f \in \mathcal{D}(A) \equiv \mathcal{D}(A + B)$ we may write $\|Af\| = \|(A + B)f - Bf\| \leq \|(A + B)f\| + \|Bf\|$. We insert this inequality into (2.104) to obtain

$$\|Bf\| \leq \alpha\|Af\| + \beta\|f\| \leq \alpha\|(A + B)f\| + \alpha\|Bf\| + \beta\|f\|. \tag{2.112}$$

Since $\mathfrak{b} < 1$, this inequality holds for some $\alpha < 1$. Assuming that $\alpha < 1$, (2.112) becomes

$$(1 - \alpha)\|Bf\| \leq \alpha\|(A + B)f\| + \beta\|f\|.$$

Hence $B$ is $(A+B)$-bounded with relative bound $\leq \alpha/(1-\alpha)$ [observe that in general the $(A + B)$-bound of $B$ will not be smaller than 1].

(c) We shall use the following fact that is proved in Chapter 4 [see (4.51)]: if $\varphi\colon \mathbb{R} \to \mathbb{C}$ is a continuous function [here $\varphi(\lambda) = (\lambda - z)^{-1}$ or $\varphi(\lambda) = \lambda(\lambda - z)^{-1}$ with $z \in \rho(A)$; see also the footnote on page 155], then the norm of the operator $\varphi(A)$ is equal to the supremum of $|\varphi(\lambda)|$ as $\lambda$ varies over the spectrum of $A$ (if $A$ is a multiplication operator, this result is contained in Proposition 2.26).

(i) Assume first that $A$ is a positive operator, i.e. such that $(-\infty, 0) \subset \rho(A)$. In this case the spectrum of $A$ is contained in $[0, +\infty)$, so that one has for $z < 0$:

$$\|(A-z)^{-1}\| \leq \sup_{\lambda \geq 0}(\lambda-z)^{-1} = (-z)^{-1} \quad \text{and} \quad \|A(A-z)^{-1}\| \leq \sup_{\lambda \geq 0} \lambda(\lambda-z)^{-1} = 1. \tag{2.113}$$

We must show that there exists $\lambda_0 \in \mathbb{R}$ such that the interval $(-\infty, \lambda_0)$ belongs to $\rho(A + B)$. For this we write $f = (A - z)^{-1}g$ in (2.104) (assuming that $z < 0$):

$$\|B(A - z)^{-1}g\| \leq \alpha\|A(A - z)^{-1}g\| + \beta\|(A - z)^{-1}g\| \qquad \forall g \in \mathcal{H}.$$

Combined with (2.113) this leads to

$$\|B(A - z)^{-1}\| \leq \alpha\|A(A - z)^{-1}\| + \beta\|(A - z)^{-1}\| \leq \alpha + \beta|z|^{-1}.$$

Since one can choose $\alpha < 1$, it is possible to find a number $\lambda_0 \in (-\infty, 0)$ such that $\|B(A-z)^{-1}\| < 1$ for each $z \leq \lambda_0$. For these values of $z$ the operator $I+B(A-z)^{-1}$ is invertible with inverse in $\mathcal{B}(\mathcal{H})$, i.e. $I+B(A-z)^{-1}$ maps $\mathcal{H}$ onto $\mathcal{H}$ if $z \in (-\infty, \lambda_0)$. Let us show that $(-\infty, \lambda_0) \subset \rho(A + B)$. For this let $z < \lambda_0$. One has on $\mathcal{D}(A)$:

$$A + B - z = \left[I + B(A - z)^{-1}\right](A - z). \tag{2.114}$$

(1) If $f \neq \mathbf{0}$ one has $(A - z)f \neq \mathbf{0}$ and $[I + B(A - z)^{-1}](A - z)f \neq \mathbf{0}$, because $A - z$ and $I + B(A - z)^{-1}$ are invertible. Hence $A + B - z$ is invertible.

(2) $A-z$ maps $\mathcal{D}(A)$ onto $\mathcal{H}$, and $I+B(A-z)^{-1}$ maps $\mathcal{H}$ onto $\mathcal{H}$. Hence $A+B-z$ maps $\mathcal{D}(A)$ onto $\mathcal{H}$, i.e. $\mathcal{R}(A + B - z) = \mathcal{H}$.

(3) One has $(A+B-z)^{-1} = (A-z)^{-1}[I + B(A-z)^{-1}]^{-1} \in \mathcal{B}(\mathcal{H})$.

(ii) The case of a general lower semibounded operator is easily reduced to the preceding one: if $(-\infty, \lambda) \subset \rho(A)$ for some $\lambda < 0$, then $(-\infty, 0) \subset \rho(A-\lambda)$, so that it suffices to replace the operator $A$ by $A-\lambda$ in the preceding considerations (and $A+B$ by $A+B-\lambda$). Upper semibounded operators are treated similarly [if $A \leq 0$, then $-A \geq 0$, hence there is a number $\lambda_0 < 0$ such that $(-\infty, \lambda_0) \subset \rho(-A-B)$, because $-B$ will be relatively bounded with respect to $-A$ with relative bound $< 1$. So $(|\lambda_0|, +\infty) \subset \rho(A+B)$]. □

**Corollary 2.45.** *Let $A$ be a self-adjoint operator and $B$ an $A$-bounded symmetric operator with $A$-bound $\mathfrak{b} < 1$. Then one has the* **second resolvent equation** *which holds for all $z \in \rho(A) \cap \rho(A+B)$:*

$$
\begin{aligned}
(A-z)^{-1} - (A+B-z)^{-1} &= (A-z)^{-1} B (A+B-z)^{-1} \\
&= (A+B-z)^{-1} B (A-z)^{-1}. \qquad (2.115)
\end{aligned}
$$

PROOF. If $z \in \rho(A) \cap \rho(A+B)$, then the operators $B(A+B-z)^{-1}$ and $B(A-z)^{-1}$ belong to $\mathcal{B}(\mathcal{H})$, hence each term in (2.115) is well defined in $\mathcal{B}(\mathcal{H})$. As $(A+B-z)^{-1}$ maps $\mathcal{H}$ onto $\mathcal{D}(A)$ one may write

$$
\begin{aligned}
B(A+B-z)^{-1} &= (A+B-z)(A+B-z)^{-1} - (A-z)(A+B-z)^{-1} \\
&= I - (A-z)(A+B-z)^{-1}.
\end{aligned}
$$

Upon multiplying this identity by $(A-z)^{-1}$, one obtains

$$
(A-z)^{-1} B (A+B-z)^{-1} = (A-z)^{-1} - (A+B-z)^{-1}.
$$

This proves the first equation in (2.115). The second one can be obtained in a similar way. □

**Remark 2.46.** (a) The condition that the $A$-bound of $B$ be strictly smaller than 1 is essential in the Rellich-Kato Theorem. In the case in which (2.104) is satisfied with $\alpha = 1$, it is possible to prove that $A+B$ is essentially self-adjoint on $\mathcal{D}(A)$ (Theorem of Wüst). The following simple example shows that one cannot expect more. Let $A$ be self-adjoint and unbounded, and take $B = -\gamma A$ with $\gamma \geq 0$. Then $\|Bf\| \leq \gamma \|Af\|$ for each $f \in \mathcal{D}(A)$, hence $B$ is $A$-bounded with relative bound $\gamma$. If $\gamma < 1$, then $A+B = (1-\gamma)A$, which is a self-adjoint operator on $\mathcal{D}(A)$. If $\gamma = 1$, then $A+B$ is the restriction of the zero operator $O$ to $\mathcal{D}(A)$; this operator [the restriction of $O$ to $\mathcal{D}(A)$] is essentially self-adjoint but not self-adjoint, since its adjoint is the zero operator with domain $\mathcal{H}$, and $\mathcal{D}(A) \neq \mathcal{H}$.

(b) A simple adjustment of the proof of Proposition 2.44 shows that, if $A$ is essentially self-adjoint and $B$ is symmetric and $A$-bounded with $A$-bound $\mathfrak{b} < 1$, then $A+B$ is essentially self-adjoint on $\mathcal{D}(A)$.

**2.7.2.** A particular class of relatively bounded operators are the relatively compact operators. Let $A$ be a closed operator with domain $\mathcal{D}(A)$. An operator $B$ is said to be ***relatively compact with respect to A*** (or ***A-compact***) if $\mathcal{D}(A) \subseteq \mathcal{D}(B)$ and if there exists a number $z \in \rho(A)$ such that $B(A-z)^{-1}$ is a compact operator. The $A$-compact operators are $A$-bounded with relative bound $0$, as shown in the following proposition.

**Proposition 2.47.** *Let $A$ be a self-adjoint operator and $B$ a symmetric operator that is $A$-compact. Then*
*(a) One has $B(A-z)^{-1} \in \mathcal{K}(\mathcal{H})$ for each $z \in \rho(A)$.*
*(b) $B$ is $A$-bounded with $A$-bound $\mathfrak{b} = 0$; in particular $A+B$ is self-adjoint on $\mathcal{D}(A)$.*
*(c) If $C$ is symmetric and $A$-bounded with $A$-bound $\mathfrak{b} < 1$, then $B$ is also $(A+C)$-compact.*

PROOF. (a) By hypothesis there is a number $z_0 \in \rho(A)$ with $B(A-z_0)^{-1} \in \mathcal{K}(\mathcal{H})$. For $z \in \rho(A)$ one has by the first resolvent equation (2.93):

$$B(A-z)^{-1} = \left[B(A-z_0)^{-1}\right]\left[I + (z-z_0)(A-z)^{-1}\right].$$

The first factor on the right-hand side is compact and the second one belongs to $\mathcal{B}(\mathcal{H})$. By Proposition 2.11(b) the operator $B(A-z)^{-1}$ is compact.

(b) We must show that, given $\varepsilon > 0$, there exists a number $z$ in the resolvent set of $A$ such that $\|B(A-z)^{-1}\| < \varepsilon$ [see (2.109)]. For this we take $z = i\mu$ with $\mu > 0$ and show that $\|B(A-i\mu)^{-1}\| \to 0$ as $\mu \to \infty$. We write

$$B(A-i\mu)^{-1} = B(A-i)^{-1}[(A-i)(A-i\mu)^{-1}] \tag{2.116}$$

and observe that

$$(A-i)(A-i\mu)^{-1} = \left[A-i\mu+i(\mu-1)\right](A-i\mu)^{-1} = I+i(\mu-1)(A-i\mu)^{-1}. \tag{2.117}$$

Thus $(A-i)(A-i\mu)^{-1}$ belongs to $\mathcal{B}(\mathcal{H})$ and one has (see Proposition 2.37):

$$\|(A-i)(A-i\mu)^{-1}\| \leq 1 + \frac{\mu-1}{\mu} \leq 2 \quad \forall \mu > 1. \tag{2.118}$$

We show below that $[(A-i)(A-i\mu)^{-1}]^*$ converges strongly to zero as $\mu \to \infty$. Since $B(A-i)^{-1} \in \mathcal{K}(\mathcal{H})$ by the result of (a), it then follows from (2.116), by using Proposition 2.12(b), that $\lim_{\mu\to\infty} \|B(A-i\mu)^{-1}\| = 0$.

Let us set $T_\mu = [(A-i)(A-i\mu)^{-1}]^* = (A+i)(A+i\mu)^{-1}$ [the second equality is formally evident; it can be established by taking the adjoint of the right-hand side of (2.117), using (2.103), and one sees that this amounts to replacing $i$ by $-i$]. If $f \in \mathcal{D}(A)$, one has for $\mu > 0$ [using (2.92) and then again Proposition 2.37]:

$$\|T_\mu f\| = \|(A+i\mu)^{-1}(A+i)f\| \leq \|(A+i\mu)^{-1}\|\|(A+i)f\| \leq \frac{1}{\mu}\|(A+i)f\|,$$

which converges to zero as $\mu \to \infty$. Since $\|T_\mu\| \leq 2$ by (2.118), it follows from Proposition 2.2 that $\text{s-lim}_{\mu\to\infty} T_\mu = O$.

(c) By the second resolvent equation (2.115) one has

$$B(A+C-i)^{-1} = B(A-i)^{-1} - B(A-i)^{-1}C(A+C-i)^{-1}$$
$$= B(A-i)^{-1}\big[I - C(A+C-i)^{-1}\big].$$

We have $B(A-i)^{-1} \in \mathcal{K}(\mathcal{H})$ by (a) and $C(A+C-i)^{-1} \in \mathcal{B}(\mathcal{H})$ by Propositions 2.44(b) and 2.42(b). As in (a), one obtains that $B(A+C-i)^{-1} \in \mathcal{K}(\mathcal{H})$. □

**2.7.3.** The results obtained above can be applied to prove the self-adjointness of quantum-mechanical Hamiltonians. We have seen in §2.5.3 that the free Hamiltonian $H_0$ of non-relativistic quantum mechanics is essentially self-adjoint on $C_0^\infty(\mathbb{R}^n)$, and we have specified its domain of self-adjointness in (2.81). We note that, in the sense explained in §2.5.5, $H_0$ is the operator $\varphi(\vec{P})$ for $\varphi(\vec{x}) = \vec{x}^2$.

Let us consider Hamiltonians of the form $H = H_0 + V$, where $V \equiv V(\vec{Q})$ denotes the operator of multiplication by a function $V(\vec{x})$ called a ***potential***. The following terminology will be used. We say that

(1) $V \in L^p_{\text{loc}}(\mathbb{R}^n)$ if $\int_{\mathcal{W}} |V(\vec{x})|^p d^n x < \infty$ for each *bounded* Borel subset $\mathcal{W}$ of $\mathbb{R}^n$,

(2) $V \in L^p(\mathbb{R}^n) + L^\infty(\mathbb{R}^n)$ if $V$ can be written in the form $V(\vec{x}) = V_1(\vec{x}) + V_2(\vec{x})$ with $V_1 \in L^p(\mathbb{R}^n)$ and $V_2 \in L^\infty(\mathbb{R}^n)$,

(3) $V \in L^p(\mathbb{R}^n) + L^\infty_\varepsilon(\mathbb{R}^n)$ if, given any $\delta > 0$, one can decompose $V$ into $V = V_1 + V_2$ with $V_1 \in L^p(\mathbb{R}^n)$, $V_2 \in L^\infty(\mathbb{R}^n)$ and $\|V_2\|_\infty < \delta$ (the decomposition may depend on the value of $\delta$).

**Example 2.48.** An important example in quantum mechanics is the ***Coulomb potential*** $V(\vec{x}) = \gamma|\vec{x}|^{-1}$ with $\gamma \in \mathbb{R}$ and $n = 3$. This potential is contained in none of the spaces $L^p(\mathbb{R}^3)$ (unless $\gamma = 0$). Indeed one has (for $p \neq \infty$)

$$\||\vec{x}|^{-1}\|_p = \left[4\pi \int_0^\infty r^{2-p}\,dr\right]^{1/p}. \tag{2.119}$$

If $p > 3$, the integral is convergent at infinity but divergent at zero, if $p = 3$ it diverges both at zero and at infinity, and if $p \in [1,3)$ it is convergent at zero but divergent at infinity. However the Coulomb potential (for $n = 3$) belongs to $L^p(\mathbb{R}^3) + L^\infty_\varepsilon(\mathbb{R}^3)$ for each $p \in [1,3)$: take $V_1(\vec{x}) = \gamma|\vec{x}|^{-1}$ for $|\vec{x}| \leq R$ and $V_1(\vec{x}) = 0$ for $|\vec{x}| > R$, hence $V_2(\vec{x}) = 0$ for $|\vec{x}| \leq R$ and $V_2(\vec{x}) = \gamma|\vec{x}|^{-1}$ for $|\vec{x}| > R$, where $R$ is a positive number. It is clear that, for any choice of $R$, one has $V_1 \in L^p(\mathbb{R}^3)$ if $1 \leq p < 3$ [apart from a factor $|\gamma|$, $\|V_1\|_p$ is given by (2.119) but with the upper limit of the integral replaced by $R$]. $V_2$ is a bounded function, and $\|V_2\|_\infty = \sup_{|\vec{x}|>R} \big|\gamma|\vec{x}|^{-1}\big| = |\gamma|R^{-1}$, which can be made arbitrarily small by choosing $R$ large enough.

Let us now discuss the properties that a potential should have so that it is a perturbation of $H_0$ in the sense considered here.

– In order for $V(\vec{Q})$ to be symmetric (and self-adjoint on its maximal domain), one must assume that $V$ is a *real* function.

– To study the condition $\mathcal{D}(H_0) \subseteq \mathcal{D}(V)$ we can use the characterisation of the functions in $\mathcal{D}(H_0)$ obtained in (2.83), since $V$ is a multiplication operator in $L^2(\mathbb{R}^n)$.

(i) Since $\mathcal{D}(H_0)$ contains $\mathcal{S}(\mathbb{R}^n)$, $V$ must be locally square-integrable, i.e. one must have $V \in L^2_{\text{loc}}(\mathbb{R}^n)$: if $\mathcal{W}$ is a bounded Borel set in $\mathbb{R}^n$, choose a function $f$ in $\mathcal{S}(\mathbb{R}^n)$ such that $f(\vec{x}) = 1$ for all $\vec{x} \in \mathcal{W}$. Since $\mathcal{S}(\mathbb{R}^n)$ belongs to $\mathcal{D}(H_0)$, one must have $Vf \in L^2(\mathbb{R}^n)$, hence

$$\int_{\mathcal{W}} |V(\vec{x})|^2 d^n x = \int_{\mathcal{W}} |V(\vec{x})f(\vec{x})|^2 d^n x \leq \int_{\mathbb{R}^n} |V(\vec{x})f(\vec{x})|^2 d^n x = \|Vf\|^2 < \infty.$$

(ii) If $n = 1, 2$ or $3$, the functions $f(\vec{x})$ in $\mathcal{D}(H_0)$ are bounded, so that $|V(\vec{x})f(\vec{x})| \leq |V(\vec{x})|\|f\|_\infty$. As a consequence one has $\mathcal{D}(H_0) \subseteq \mathcal{D}(V)$ if $V \in L^2(\mathbb{R}^n)$ and also if $V \in L^2(\mathbb{R}^n) + L^\infty(\mathbb{R}^n)$.

(iii) If $n \geq 4$ and $f \in \mathcal{D}(H_0)$, then $f \in L^s(\mathbb{R}^n)$ for each number $s$ satisfying $2 \leq s < 2n/(n-4)$. In order to have $Vf \in L^2(\mathbb{R}^n)$, assume that $V \in L^q(\mathbb{R}^n)$ for some $q$ to be determined. By the Hölder inequality (2.121) one has $\|Vf\|_2 \leq \|V\|_q\|f\|_s$ if $1/2 = 1/q + 1/s$. Since $s$ can be any number in $[2, 2n/(n-4))$, one concludes that $Vf \in L^2(\mathbb{R}^n)$ if $q > n/2$ [$s = 2 \Longrightarrow q = \infty$; $s = 2n/(n-4) \Longrightarrow q = n/2$].

**Proposition 2.49.** *Let $H = H_0 + V(\vec{Q})$ be a Schrödinger operator in $\mathcal{H} = L^2(\mathbb{R}^n)$, with $V$ a real-valued function.*
*(a) Let $n = 1, 2$ or $3$. If $V \in L^2(\mathbb{R}^n) + L^\infty(\mathbb{R}^n)$, then the operator $V(\vec{Q})$ is $H_0$-bounded with relative bound $\mathfrak{b} = 0$. In particular $H$ is self-adjoint on $\mathcal{D}(H_0)$. If $V \in L^2(\mathbb{R}^n) + L^\infty_\varepsilon(\mathbb{R}^n)$, then $V(\vec{Q})$ is $H_0$-compact.*
*(b) Let $n \geq 4$. If $V \in L^p(\mathbb{R}^n) + L^\infty(\mathbb{R}^n)$ for some $p > n/2$, then $V(\vec{Q})$ is $H_0$-bounded with relative bound $\mathfrak{b} = 0$. In particular $H$ is self-adjoint on $\mathcal{D}(H_0)$. If $V \in L^p(\mathbb{R}^n) + L^\infty_\varepsilon(\mathbb{R}^n)$ for some $p \in (n/2, \infty)$, then $V(\vec{Q})$ is $H_0$-compact.*

PROOF. We have already seen that $\mathcal{D}(H_0) \subseteq \mathcal{D}(V)$ (for all $n$). In (a) it is evident that $V \in L^2_{\text{loc}}(\mathbb{R}^n)$; in the situation of (b) this follows from (1.49) since $p > 2$.

(a) (i) Let $V = V_1 + V_2$ with $V_1 \in L^2(\mathbb{R}^n)$ and $V_2 \in L^\infty(\mathbb{R}^n)$. Let $\mu > 0$. Then $V_1(H_0 - i\mu)^{-1}$ is a Hilbert-Schmidt operator [see Proposition 2.34(a)]:

$$\begin{aligned}\|V_1(H_0 - i\mu)^{-1}\|_{\text{HS}} &= (2\pi)^{-n/2}\|V_1\|_2\|(\vec{k}^2 - i\mu)^{-1}\|_2 \\ &= (2\pi)^{-n/2}\|V_1\|_2\left[C_n\int_0^\infty \frac{k^{n-1}}{k^4 + \mu^2}\,dk\right]^{1/2},\end{aligned} \tag{2.120}$$

where $C_n$ is a constant (surface area of the unit sphere in $\mathbb{R}^n$). The integral in (2.120) is finite and converges to zero as $\mu \to \infty$ (if $n = 3$, write for example $k^2/(k^4 + \mu^2) = k^2/[(k^4 + \mu^2)^{1-\varepsilon}] \cdot (k^4 + \mu^2)^{-\varepsilon} \leq k^2/[(k^4 + \mu^2)^{1-\varepsilon}] \cdot \mu^{-2\varepsilon}$ and choose $\varepsilon < 1/4$). Thus $V_1$ is $H_0$-bounded with relative bound $0$ [see (2.109)]; since $V_2$ is a bounded operator, $V$ is $H_0$-bounded with relative bound $\mathfrak{b} = 0$.

(ii) If $V \in L^2(\mathbb{R}^n) + L^\infty_\varepsilon(\mathbb{R}^n)$ and $\delta > 0$ we choose a decomposition $V = V_{1,\delta} + V_{2,\delta}$ with $V_{1,\delta} \in L^2(\mathbb{R}^n)$, $V_{2,\delta} \in L^\infty(\mathbb{R}^n)$ and $\|V_{2,\delta}\|_\infty < \delta$. The operator $V_{1,\delta}(H_0 - i)^{-1}$ is compact (even Hilbert-Schmidt) and $\|V_{2,\delta}(H_0 - i)^{-1}\| \leq \|V_{2,\delta}\|_\infty\|(H_0 - i)^{-1}\| < \delta$. Consequently $V(H_0 - i)^{-1}$ is a compact operator as the uniform limit of the family of compact operators $\{V_{1,\delta}(H_0 - i)^{-1}\}$ as $\delta \to 0$ [Proposition 2.11(d)].

(b) The proof is similar. Let $V = V_1 + V_2$ with $V_1 \in L^p(\mathbb{R}^n)$ and $V_2 \in L^\infty(\mathbb{R}^n)$. In this case $V_1(H_0 - i\mu)^{-1}$ is compact [Proposition 2.34(b)] and the norm of this operator can be estimated by using (2.87):

$$\begin{aligned}\|V_1(H_0 - i\mu)^{-1}\| &\leq (2\pi)^{-n/p}\|V_1\|_p\|(\vec{k}^2 - i\mu)^{-1}\|_p \\ &= (2\pi)^{-n/p}\|V_1\|_p\Big[C_n\int_0^\infty \frac{k^{n-1}}{(k^4+\mu^2)^{p/2}}\,dk\Big]^{1/p}.\end{aligned}$$

Again, since $p > n/2$, the integral is finite and converges to zero as $\mu \to \infty$. We omit the remaining details of the proof. □

**Remark 2.50.** In view of the results of Example 2.48, part (a) of the preceding proposition implies in particular that the Coulomb Hamiltonian $H_C$ in $L^2(\mathbb{R}^3)$ is self-adjoint on $\mathcal{D}(H_0)$. Since $H_0$ is essentially self-adjoint on $C_0^\infty(\mathbb{R}^3)$ (see page 81), $H_C$ is also essentially self-adjoint on $C_0^\infty(\mathbb{R}^3)$ [see Remark 2.46(b)].

## Appendix to Chapter 2

**A.2.1. Proof of Proposition 2.33 for $2 < p < \infty$.** The proof involves the following two results:

(1) The ***Hölder inequality***: Let $(\mathcal{O}, \mathcal{A}, m)$ be a measure space. If $f \in L^p(\mathcal{O}, m)$, $g \in L^q(\mathcal{O}, m)$ and $1/r = 1/p + 1/q$, then the product function $fg$ belongs to $L^r(\mathcal{O}, m)$ and

$$\|fg\|_r \leq \|f\|_p\|g\|_q. \tag{2.121}$$

The proof is simple, see e.g. page 113 of [R].

(2) The Fourier transformation $\mathcal{F}$ is bounded from $L^q(R^n)$ to $L^{q'}(R^n)$ if $1 \leq q \leq 2$ and $q'$ is defined by $1/q' = 1 - 1/q$:

$$\begin{aligned}\|\tilde{f}\|_{q'} \equiv \Big[\int_{\mathbb{R}^n}|\tilde{f}(\vec{k})|^{q'}d^nk\Big]^{1/q'} &\leq (2\pi)^{n(1-2q^{-1})/2}\|f\|_q \\ &= (2\pi)^{n(1-2q^{-1})/2}\Big[\int_{\mathbb{R}^n}|f(\vec{x})|^q d^nx\Big]^{1/q}.\end{aligned} \tag{2.122}$$

This is called the ***Hausdorff-Young inequality***. For a proof see for example page 11 in Volume II of [RS]. If $q = 2$ one has $q' = 2$, and (2.122) expresses the unitarity of $\mathcal{F}$ as an operator in $L^2(\mathbb{R}^n)$. If $q = 1$, one has $q' = \infty$, and (2.122) coincides with the evident inequality $\|f\|_\infty \leq (2\pi)^{-n/2}\|f\|_1$ [see (2.73)]. If $q \in (1, 2)$, then $q' \in (2, \infty)$ and (2.122) is obtained by interpolating between the cases $q = 1$ and $q = 2$.

(i) We first show that $C_0^\infty(\mathbb{R}^n) \subset \mathcal{D}(\psi(\vec{P})\varphi(\vec{Q}))$. If $f \in C_0^\infty(\mathbb{R}^n)$ one can choose a bounded open set $\mathcal{V}$ in $\mathbb{R}^n$ such that $f(\vec{x}) = 0$ for all $\vec{x} \notin \mathcal{V}$. Clearly $\|\varphi(\vec{Q})f\|_p \leq \|\varphi\|_p\|f\|_\infty$, hence $\varphi(\vec{Q})f \in L^p(\mathbb{R}^n)$. Then, by (1.49), one has $\varphi(\vec{Q})f \in L^r(\mathbb{R}^n)$ for each $r \in [1, p]$. Since $p > 2$, this holds in particular for $r = 2$, which shows that $f$

belongs to $\mathcal{D}(\varphi(\vec{Q}))$. To see that $\varphi(\vec{Q})f \in \mathcal{D}(\psi(\vec{P}))$, we use the Hausdorff-Young inequality to find that $\mathcal{F}\varphi(\vec{Q})f \in L^s(\mathbb{R}^n)$ for each $s \in [2, \infty]$, in particular for $s = s_0 = [1/2 - 1/p]^{-1}$; since $1/2 = 1/p + 1/s_0$, one has $\psi(\cdot)[\mathcal{F}\varphi(\vec{Q})f](\cdot) \in \tilde{L}^2(\mathbb{R}^n)$ by the Hölder inequality.

(ii) To estimate the norm of $\psi(\vec{P})\varphi(\vec{Q})$, let $g \in L^2(\mathbb{R}^n)$ and define $q \in (1, 2)$ by $1/q = 1/p + 1/2$. Then (2.122) and the Hölder inequality imply that

$$\begin{aligned}\|\mathcal{F}\varphi(\vec{Q})g\|_{q'} \leq (2\pi)^{n(1-2q^{-1})/2}\|\varphi(\vec{Q})g\|_q &\leq (2\pi)^{n(1-2q^{-1})/2}\|\varphi\|_p\|g\|_2 \\ &\equiv (2\pi)^{-n/p}\|\varphi\|_p\|g\|_2.\end{aligned}$$

Now $1/p + 1/q' = 1/p + 1 - 1/q = 1/2$. So, by using again the Hölder inequality, one arrives at

$$\begin{aligned}\|\psi(\vec{P})\varphi(\vec{Q})g\| = \|\psi(\cdot)\left[\mathcal{F}\varphi(\vec{Q})g\right](\cdot)\|_2 &\leq \|\psi\|_p\|\mathcal{F}\varphi(\vec{Q})g\|_{q'} \\ &\leq (2\pi)^{-n/p}\|\psi\|_p\|\varphi\|_p\|g\|_2.\end{aligned}$$ □

**A.2.2. End of the proof of Proposition 2.34(a).** Let us assume that $\psi \in L^2(\mathbb{R}^n)$ but that $\varphi \notin L^2(\mathbb{R}^n)$. We must show that $\varphi(\vec{Q})\psi(\vec{P}) \notin \mathcal{B}_2(\mathcal{H})$ except when $\psi = 0$ a.e. [in which case $\varphi(\vec{Q})\psi(\vec{P}) = O$]. For $k \in \mathbb{N}$ we define $\varphi_k$ by

$$\varphi_k(\vec{x}) = \begin{cases} \varphi(\vec{x}) & \text{if } |\vec{x}| \leq k \text{ and } |\varphi(\vec{x})| \leq k \\ k & \text{if } |\vec{x}| \leq k \text{ and } |\varphi(\vec{x})| > k \\ 0 & \text{if } |\vec{x}| > k. \end{cases}$$

Clearly $\varphi_k \in L^2(\mathbb{R}^n)$ for each $k$, and $|\varphi_k(\vec{x})| \leq |\varphi(\vec{x})|$ for each $\vec{x} \in \mathbb{R}^n$. The last inequality implies that $\|\varphi(\vec{Q})g\| \geq \|\varphi_k(\vec{Q})g\|$ for each vector $g$. By taking into account the definition (2.43) of the Hilbert-Schmidt norm and Eq. (2.89), one then obtains $\|\varphi(\vec{Q})\psi(\vec{P})\|_{\text{HS}} \geq \|\varphi_k(\vec{Q})\psi(\vec{P})\|_{\text{HS}} = (2\pi)^{-n/2}\|\varphi_k\|_2\|\psi\|_2$. Since $\varphi \notin L^2(\mathbb{R}^n)$ and $\varphi_k(\vec{x}) \to \varphi(\vec{x})$ for almost all $\vec{x} \in \mathbb{R}^n$, one has $\|\varphi_k\|_2 \to \infty$ as $k \to \infty$ and hence $\|\varphi(\vec{Q})\psi(\vec{P})\|_{\text{HS}} = \infty$ (unless $\|\psi\|_2 = 0$). Thus $\varphi(\vec{Q})\psi(\vec{P}) \notin \mathcal{B}_2(\mathcal{H})$. The other situations in Proposition 2.34(a) are treated similarly. □

**A.2.3. $-\Delta$ is essentially self-adjoint on $C_0^\infty(\mathbb{R}^n)$.** Let $A_0 = \Delta|_{C_0^\infty(\mathbb{R}^n)}$ and $A_1 = \Delta|_{\mathcal{S}(\mathbb{R}^n)}$. We have seen that $A_1$ is essentially self-adjoint, so the essential self-adjointness of $A_0$ will follow if we can show that the closure of $A_0$ is an extension of $A_1$. To do this we fix a function $\theta \in C_0^\infty(\mathbb{R}^n)$ such that $\theta(\vec{x}) = 1$ for all $\vec{x}$ satisfying $|\vec{x}| \leq 1$, and we define $\theta_k$ by $\theta_k(\vec{x}) = \theta(\vec{x}/k)$, $k = 1, 2, \ldots$ (so $\theta_k(\vec{x}) = 1$ for all $|\vec{x}| \leq k$). For $f \in \mathcal{D}(A_1) \equiv \mathcal{S}(\mathbb{R}^n)$ we set $f_k = \theta_k f$. Then we have $f_k \in C_0^\infty(\mathbb{R}^n) \equiv \mathcal{D}(A_0)$ and $\|f_k - f\| \to 0$ as $k \to \infty$. Now

$$[\Delta f_k](\vec{x}) = \theta_k(\vec{x})\left[\Delta f\right](\vec{x}) + \frac{2}{k}[\vec{\nabla}\theta](\frac{\vec{x}}{k}) \cdot [\vec{\nabla}f](\vec{x}) + \frac{1}{k^2}[\Delta\theta](\frac{\vec{x}}{k})f(\vec{x})$$

and consequently

$$\|\Delta f - \Delta f_k\| \leq \|(1 - \theta_k)\Delta f\| + \frac{2}{k}\|\vec{\nabla}\theta\|_\infty \cdot \|\vec{\nabla}f\| + \frac{1}{k^2}\|\Delta\theta\|_\infty\|f\|.$$

This implies that $\lim_{k\to\infty} \|\Delta f - \Delta f_k\| = 0$. Since $A_0$ is closable, we conclude that $f$ belongs to $\mathcal{D}(\overline{A_0})$ and that $\overline{A_0} f = \Delta f \equiv A_1 f$, which shows that $A_1 \subseteq \overline{A_0}$. □

### Bibliographical Notes

For alternative reading we recommend the texts [AG], [RN] and [W1] already cited in the Bibliographical Notes to Chapter 1, also Volume I of [RS], [K] and Part I of [W2].

### Problems

**2.1.** Show that an operator $A \in \mathcal{B}(\mathcal{H})$ is completely (and uniquely) determined by its expectation values $A_{jk} = \langle e_j, Ae_k \rangle$ in any orthonormal basis $\{e_n\}$ of the underlying Hilbert space $\mathcal{H}$.

**2.2.** Let $\{e_n\}$ be an orthonormal basis of an infinite-dimensional Hilbert space. For $N = 1, 2, 3, \ldots$ let $A_N$ be the operator defined as follows:

$$A_N\Big(\sum_{k=1}^{\infty} \alpha_k e_k\Big) = \sum_{k=1}^{\infty} \alpha_{k+N} e_k.$$

(a) Show that $A_N \in \mathcal{B}(\mathcal{H})$ and determine its norm $\|A_N\|$.
(b) Prove that $\{A_N\}$ converges strongly to the zero operator $O$ as $N \to \infty$.
(c) Find $A_N^* \equiv (A_N)^*$.
(d) One knows that $A_N^* \to O$ weakly. Does $A_N^*$ converge strongly to $O$ ?

**2.3.** Let $\{e_n\}$ be an orthonormal basis of an infinite-dimensional Hilbert space. For $N = 1, 2, 3, \ldots$ let $B_N$ be the operator defined as follows: $B_N e_1 = e_N$ and $B_N e_k = \mathbf{0}$ for $k > 1$. Show that $\|B_N\| = 1$, that $\operatorname{w-lim}_{N\to\infty} B_N = O$ but that $\operatorname{s-lim}_{N\to\infty} B_N$ does not exist.

**2.4.** Find two sequences of bounded operators $\{A_n\}$ and $\{B_n\}$ in a Hilbert space $\mathcal{H}$ such that $\operatorname{w-lim}_{n\to\infty} A_n = O$ and $\operatorname{w-lim}_{n\to\infty} B_n = O$ but such that $\{A_n B_n\}$ does not converge weakly to $O$.

**2.5.** Let $\{A_n\}_{n\in\mathbb{N}}$ be a sequence of operators satisfying $\|A_n\| \le M < \infty$ for all $n \in \mathbb{N}$ and $\lim_{n\to\infty} \langle f, A_n g \rangle = 0$ for all $f$ and $g$ belonging to some dense set $\mathcal{D}$. Prove that $\operatorname{w-lim}_{n\to\infty} A_n = O$.

**2.6.** Let $A \in \mathcal{B}(\mathcal{H})$ be such that $\langle f, Af \rangle$ is real for each vector $f$ in $\mathcal{H}$. By using the polarisation identity, show that $A$ is self-adjoint.

**2.7.** Let $E$ be a projection and $\mathcal{M}(E) = \{f \in \mathcal{H} \mid Ef = f\}$. Prove that $\mathcal{M}(E)$ is a subspace.

**2.8.** Prove Proposition 2.6.

**2.9.** Let $\{e_n\}$ be an orthonormal basis of an infinite-dimensional Hilbert space. For $N = 1, 2, 3, \ldots$ let $P_N$ be the projection onto the subspace spanned by the vectors $e_N, e_{N+1}, e_{N+2}, \ldots$ Determine which of the following properties are satisfied by the sequence of operators $\{P_N\}$: (i) $\operatorname{u-lim}_{N\to\infty} P_N = O$, (ii) $\operatorname{s-lim}_{N\to\infty} P_N = O$, (iii) $\operatorname{w-lim}_{N\to\infty} P_N = O$.

**2.10.** Prove Proposition 2.8(b).

**2.11.** Prove Proposition 2.9(b) - (e).

**2.12.** Show that a bounded operator $A$ in a Hilbert space $\mathcal{H}$ is a finite rank operator if and only if its range $\mathcal{R}(A)$ is finite-dimensional.

**2.13.** Let $U$ be a unitary operator in $\mathcal{H}$ and $A \in \mathcal{B}(\mathcal{H})$. Show that $\|UA\| = \|AU\| = \|UAU^*\| = \|A\|$ and $\|UA\|_{\text{HS}} = \|AU\|_{\text{HS}} = \|UAU^*\|_{\text{HS}} = \|A\|_{\text{HS}}$.

**2.14.** Let $\{e_n\}$ be an orthonormal basis of an infinite-dimensional Hilbert space. Let $A$ be the operator defined as follows:

$$A\Big(\sum_{k=1}^{\infty} \alpha_k e_k\Big) = \sum_{k=1}^{\infty} \rho_k \alpha_k e_k,$$

where $\rho_k$ are complex numbers. Specify the conditions that have to be imposed on these numbers $\rho_k$ such that $A$ is: (i) bounded, (ii) compact, (iii) Hilbert-Schmidt, (iv) of finite rank, (v) an isometry, (vi) unitary, (vii) a partial isometry, (viii) a projection, (ix) self-adjoint.

**2.15.** Prove the following assertions.
(a) If $A$ is closed, then it is closable, and one has $\overline{A} = A$.
(b) The null space of a closed operator is a subspace.
(c) If $A$ is closed and $B \in \mathcal{B}(\mathcal{H})$, then $A + B$ is closed.
(d) If $A$ is closable and $B \in \mathcal{B}(\mathcal{H})$, then $A + B$ is closable.

**2.16.** Let $\{e_n\}$ be an orthonormal basis of an infinite-dimensional Hilbert space. Define a linear operator $A$ as follows: $\mathcal{D}(A)$ is the set of all finite linear combinations of vectors in $\{e_n\}$ and $Ae_k = ke_1$ for $k = 1, 2, 3, \ldots$.
(a) Show that $A$ is not closable.
(b) Show that $\mathcal{D}(A^*)$ is not dense in $\mathcal{H}$.
(c) Find a sequence $\{A_n\}$ of operators in $\mathcal{B}(\mathcal{H})$ converging strongly to $A$ on $\mathcal{D}(A)$.

**2.17.** Let $A$ be an unbounded closed operator, and let $g$ and $h$ be two vectors with $g \notin \mathcal{D}(A)$. Define an operator $B$ by $\mathcal{D}(B) = \{f + \alpha g \,|\, f \in \mathcal{D}(A), \alpha \in \mathbb{C}\}$ and $B(f + \alpha g) = Af + \alpha h$. Show that $B$ is closed. [By varying $h$, one obtains an uncountable number of closed extensions of $A$].
[Hint: Consider a Cauchy sequence $\{f_n + \alpha_n g\}$ in the domain of $B$ such that the sequence $\{B(f_n + \alpha_n g)\}$ is also Cauchy. Show that the sequence $\{|\alpha_n|\}$ must be bounded; hence there is a subsequence for which the numbers $\alpha_n$ converge to a limit $\alpha$.] If necessary, consult page 71 of [P] for the complete solution.

**2.18.** Prove Proposition 2.21.

**2.19.** Prove the following statements.
(a) If $A$ is invertible and closed, then $A^{-1}$ is closed.
(b) If in addition $A^{-1} \in \mathcal{B}(\mathcal{H})$ and $B$ is a closed operator, then $AB$ is closed.
(c) If $A$ is symmetric and invertible and its range $\mathcal{R}(A)$ is dense in $\mathcal{H}$, then $A^{-1}$ is symmetric.

**2.20.** If $A$ has dense domain in $\mathcal{H}$, is invertible and $A^{-1} \in \mathcal{B}(\mathcal{H})$, then $A^*$ is also invertible and $A^{*-1} = (A^{-1})^*$.

**2.21.** Let $\{A, \mathcal{D}(A)\}$ be a linear operator and $U$ a unitary operator in a Hilbert space $\mathcal{H}$. Define an operator $A_U$ by $\mathcal{D}(A_U) = U\mathcal{D}(A)$ and $A_U = UAU^*$ (see Remark 2.25).
(a) Show that if $A$ is closable, then so is $A_U$.
(b) Show that if $A$ is self-adjoint, then so is $A_U$.
(c) Assume that $A$ is closed. Show that $A_U$ and $A$ have the same spectrum and the same eigenvalues.

**2.22.** Show that a bounded symmetric operator is essentially self-adjoint.

**2.23.** (a) Let $A$ be essentially self-adjoint and let $B$ be a self-adjoint operator belonging to $\mathcal{B}(\mathcal{H})$. Show that $A + B$ is essentially self-adjoint on $\mathcal{D}(A)$.
(b) Let $V\colon \mathbb{R}^n \to \mathbb{R}$ be a bounded measurable function and $V(\vec{Q})$ the associated multiplication operator in $L^2(\mathbb{R}^n)$. Show that the operator $\vec{P}^2 + V(\vec{Q})$ is essentially self-adjoint on $C_0^\infty(\mathbb{R}^n)$.

**2.24.** Consider the differential operator $d^2/dx^2$ in $\mathcal{H} = L^2((0,1))$. Find dense linear manifolds $\mathcal{D}$ in $\mathcal{H}$ such that $d^2/dx^2$ is symmetric on $\mathcal{D}$.

**2.25.** Show that the canonical commutation relation $[A,B] = iI$ cannot be satisfied if both $A$ and $B$ belong to $\mathcal{B}(\mathcal{H})$.
[Hint: By assuming that $[A,B] = iI$ with $A,B \in \mathcal{B}(\mathcal{H})$ and by calculating $[A,B^n]$ for $n = 2,3,\ldots$, one can arrive at a contradiction.]

**2.26.** Find two dense linear manifolds $\mathcal{D}_1$ and $\mathcal{D}_2$ in $L^2(\mathbb{R})$ such that the restrictions of the position operator $Q$ to $\mathcal{D}_1$ and to $\mathcal{D}_2$ are essentially self-adjoint and such that $\mathcal{D}_1 \cap \mathcal{D}_2 = \{\mathbf{0}\}$.

**2.27.** Consider the infinite-dimensional matrix $A_{jk}$ given as follows:

$$A_{jk} = \begin{cases} 0 & \text{if } j = k \\ i/(j+k)^2 & \text{if } j > k \\ -i/(j+k)^2 & \text{if } j < k \end{cases}$$

with $j, k = 1, 2, 3, \ldots$ Let $\{e_n\}$ be an orthonormal basis of an infinite-dimensional Hilbert space $\mathcal{H}$ and let $A$ be the linear operator defined by setting $Ae_k = \sum_j A_{jk} e_j$. Show that $A$ belongs to $\mathcal{B}(\mathcal{H})$ and is self-adjoint.

**2.28.** Determine the spectrum of the operator $A$ in $L^2(\mathbb{R})$ defined by $(Af)(x) = f(x + x_0)$, where $x_0$ is a fixed real number.

**2.29.** Consider the shift operator $\Omega f = \sum_k \alpha_k e_{k+1}$ if $f = \sum_k \alpha_k e_k$, where $\{e_k\}_{k\in\mathbb{N}}$ is an orthonormal basis of an infinite-dimensional Hilbert space $\mathcal{H}$ (page 39).
(a) Show that $\Omega$ is isometric but not unitary and determine its adjoint $\Omega^*$.
(b) Find the eigenvalues (if any) of $\Omega$ and of $\Omega^*$.
(c) Determine the spectrum of $\Omega$.
[Hint for (c): Use Proposition 2.39 and (2.101).]

**2.30.** Let $A$ be a self-adjoint operator and $B$ an $A$-bounded operator with relative bound $\mathfrak{b}$. Furthermore let $\{B_k\}$ be a sequence of $A$-bounded operators such that $\|(B_k - B)(A + i)^{-1}\| \to 0$ as $k \to \infty$. If $\mathfrak{b}_k$ denotes the $A$-bound of $B_k$, show that $\mathfrak{b} = \lim_{k\to\infty} \mathfrak{b}_k$.

**2.31.** Let $H = H_0 + V(\vec{Q})$ be a Schrödinger operator in $L^2(\mathbb{R}^n)$.
(a) Show that, if $V$ is continuous and $V(\vec{x}) \to 0$ as $|\vec{x}| \to \infty$, then $V(\vec{Q})$ is a $H_0$-compact operator.
(b) Let $n = 1$ and let $V(x) = \sum_{k\in\mathbb{Z}} k^{0.999}\chi_{J_k}(x)$, where $\chi_{J_k}$ denotes the characteristic function of the interval $J_k = (k-k^{-3}, k+k^{-3})$. Show that $V(\vec{Q})$ is a $H_0$-compact operator.
**2.32.** Let $H = H_0 + 2\vec{A}(\vec{Q}) \cdot \vec{P} - i\,\mathrm{div}\vec{A}(\vec{Q}) + |\vec{A}(\vec{Q})|^2 + V(\vec{Q})$ be a Schrödinger operator in $L^2(\mathbb{R}^3)$, with $A_k, V : \mathbb{R}^3 \to \mathbb{R}$.
(a) If $A_k \in L^p(\mathbb{R}^3)$ for some $p > 3$, then $\vec{A}(\vec{Q}) \cdot \vec{P}$ is $H_0$-bounded with relative bound $\mathfrak{b} = 0$.
(b) If $A_k \in L^4(\mathbb{R}^3) + L^\infty(\mathbb{R}^3)$ and $V, \mathrm{div}\vec{A} \in L^2(\mathbb{R}^3) + L^\infty(\mathbb{R}^3)$, then the operator $2\vec{A}(\vec{Q}) \cdot \vec{P} - i\,\mathrm{div}\vec{A}(\vec{Q}) + |\vec{A}(\vec{Q})|^2 + V(\vec{Q})$ is $H_0$-bounded with relative bound 0.
REMARK. The operator $H$ is obtained by developing $H = [\vec{P} + \vec{A}(\vec{Q})]^2 + V(\vec{Q})$, which (in convenient units) is the Hamiltonian of a non-relativistic particle in the potential V and the magnetic field $\vec{B} = \mathrm{rot}\,\vec{A}$.
**2.33.** Prove the result given in Remark 2.46(b).

# CHAPTER 3

# Symmetric Operators and their Extensions

When discussing unbounded operators in Section 2.4, we pointed out the difference between symmetric and self-adjoint operators. For the purposes of quantum mechanics it is very important to know the self-adjoint extensions, if any, of a given symmetric operator. In this chapter we present a beautiful method, due to J. von Neumann, for obtaining a complete characterisation of the closed symmetric extensions of general symmetric operators (Section 3.1). We then apply the results to first and second order differential operators, including Schrödinger operators with spherically symmetric potentials, and give criteria for such operators to be essentially self-adjoint.

We recall that $A$ is a symmetric operator if its domain $\mathcal{D}(A)$ is dense in $\mathcal{H}$ and $A^*$ is an extension of $A$, i.e. if

$$\langle Af, g\rangle = \langle f, Ag\rangle \qquad \forall f, g \in \mathcal{D}(A). \tag{3.1}$$

Since $A$ and its closure $\overline{A}$ have the same closed extensions (Proposition 2.17), it will be enough to consider symmetric operators $A$ that are closed.

## 3.1 The method of the Cayley transform

**3.1.1.** We begin with some preliminary results.

**Lemma 3.1.** *Let $B$ be a linear operator in a Hilbert space $\mathcal{H}$. Assume that there exists a constant $\kappa > 0$ such that $\|Bf\| \geq \kappa\|f\|$ for all $f \in \mathcal{D}(B)$. Then:*
*(a) $B$ is invertible,*
*(b) the inverse $B^{-1}$ of $B$ is bounded, with $\|B^{-1}\| \leq 1/\kappa$,*
*(c) the range $\mathcal{R}(B)$ of $B$ is closed (hence a subspace of $\mathcal{H}$) if and only if $B$ is a closed operator.*[1]

[1] It is important to note that even if $B$ is closed but $\mathcal{R}(B) \neq \mathcal{H}$, then $B^{-1}$ is not defined everywhere and so does not belong to $\mathcal{B}(\mathcal{H})$ but only to $\mathcal{B}(\mathcal{R}(B), \mathcal{H})$. The norm $\|B^{-1}\|$ appearing in (b) of the lemma means the norm $\|B^{-1}\|_{\mathcal{R}(B)}$ in the notation introduced in (2.57).

PROOF. (a) If $Bf = \mathbf{0}$, then $\|f\| \leq \kappa^{-1}\|Bf\| = 0$, hence $f = \mathbf{0}$. So $B$ is invertible.

(b) If $g \in \mathcal{R}(B)$, then $g = Bf$ for some $f \in \mathcal{D}(B)$. Then $\|B^{-1}g\| = \|f\| \leq \kappa^{-1}\|Bf\| = \kappa^{-1}\|g\|$. Thus $\|B\|_{\mathcal{R}(B)} \leq \kappa^{-1}$.

(c) (i) Assume $B$ to be closed. Let $\{g_n\}$ be a strong Cauchy sequence in $\mathcal{R}(B)$ and denote its limit by $g$. To see that $\mathcal{R}(B)$ is closed, we must find a vector $f \in \mathcal{D}(B)$ such that $g = Bf$. Since $g_n \in \mathcal{R}(B)$, there exists for each $n$ a vector $f_n \in \mathcal{D}(B)$ such that $g_n = Bf_n$. We then have $\|f_n - f_m\| \leq \kappa^{-1}\|Bf_n - Bf_m\| = \kappa^{-1}\|g_n - g_m\|$, hence the hypothesis that $\{g_n\}$ be strongly Cauchy implies that $\{f_n\}$ is strongly Cauchy. Set $f = \text{s-lim}_{n\to\infty} f_n$. We are now in the following situation: we have a sequence $\{f_n\}$ in $\mathcal{D}(B)$ which is strongly Cauchy and such that the sequence $\{Bf_n\}$ is also strongly Cauchy; since $B$ is closed, this implies that $f \in \mathcal{D}(B)$ and $Bf = g$ (see page 49).

(ii) Assume that $\mathcal{R}(B)$ is closed. Consider a sequence of vectors $\{f_n\}$ in $\mathcal{D}(B)$ such that $\text{s-lim}_{n\to\infty} f_n \equiv f$ and $\text{s-lim}_{n\to\infty} Bf_n \equiv g$ exist. To see that $B$ is closed, we must show that $f \in \mathcal{D}(B)$ and $Bf = g$. Since $\mathcal{R}(B)$ is a subspace, one has $g \in \mathcal{R}(B)$; if we set $h = B^{-1}g$, we must show that $f = h$, i.e. that $\text{s-lim}_{n\to\infty} f_n = h$. By writing $g_n = Bf_n$ as before, we have $\|f_n - h\| = \|B^{-1}(g_n - g)\| \leq \kappa^{-1}\|g_n - g\| \to 0$ as $n \to \infty$. □

We recall the identity (2.102): If $A$ is a symmetric operator and $z = \lambda + i\mu$ a complex number ($\lambda, \mu \in \mathbb{R}$), one has for each $f \in \mathcal{D}(A)$:

$$\|(A - z)f\|^2 = \|(A - \lambda)f\|^2 + \mu^2\|f\|^2. \tag{3.2}$$

This implies, together with Lemma 3.1, that if $\mu \neq 0$, then $(A - z)^{-1}$ exists and is bounded. However in general the closure of $(A - z)^{-1}$ does not belong to $\mathcal{B}(\mathcal{H})$ because $\mathcal{R}(A - z)$ need not be dense in $\mathcal{H}$. From Proposition 2.19(b) we see that $\mathcal{R}(A - z)^{\perp} = \mathcal{N}(A^* - \overline{z})$. So, if $A$ is a closed symmetric operator, we have:

$$(A - z)^{-1} \in \mathcal{B}(\mathcal{H}) \iff \mathcal{N}(A^* - \overline{z}) = \{\mathbf{0}\} \iff \overline{z} \text{ is not an eigenvalue of } A^*. \tag{3.3}$$

Thus we have:

**Proposition 3.2.** *If $A$ is a closed symmetric operator and $z$ a non-real complex number, then $z$ belongs to the spectrum of $A$ if and only if $\overline{z}$ is an eigenvalue of $A^*$.*

If $A$ is self-adjoint and $z = \lambda + i\mu$ with $\mu \neq 0$, then $\overline{z}$ cannot be an eigenvalue of $A^* \equiv A$ (Proposition 2.36), hence $(A - z)^{-1}$ exists, is everywhere defined and bounded. Consequently (as already seen in Proposition 2.37), the complex number $z$ belongs to the resolvent set of $A$, i.e. $\mathbb{C}\setminus\mathbb{R} \subseteq \rho(A)$.

The preceding statement has a converse: if $A$ is a symmetric operator such that $(A - z)^{-1}$ belongs to $\mathcal{B}(\mathcal{H})$ for each non-real $z$, then $A$ is self-adjoint. In fact it suffices to know that $(A - z)^{-1}$ and $(A - \overline{z})^{-1}$ belong to $\mathcal{B}(\mathcal{H})$ for a single non-real $z$, which we may take to be for example the number $i$ (see Remark 3.10 for further comments on this point). This is the **fundamental self-adjointness criterion**:

**Proposition 3.3.** *Let $A$ be a symmetric operator in a Hilbert space $\mathcal{H}$.*
*(a) $A$ is self-adjoint if and only if $\mathcal{R}(A + i) = \mathcal{H}$* ***and*** *$\mathcal{R}(A - i) = \mathcal{H}$.*
*(b) $A$ is essentially self-adjoint if and only if $\mathcal{R}(A + i)$* ***and*** *$\mathcal{R}(A - i)$ are dense in $\mathcal{H}$.*

PROOF. (a) The implication $\Rightarrow$ has been noted above. For the implication $\Leftarrow$ suppose that $\mathcal{R}(A \pm i) = \mathcal{H}$. To conclude that $A$ is self-adjoint, it suffices to show that $\mathcal{D}(A^*) \subseteq \mathcal{D}(A)$ (because we have $A \subseteq A^*$ for $A$ symmetric). So let $g \in \mathcal{D}(A^*)$. Since $\mathcal{R}(A - i) = \mathcal{H}$, there is a vector $h \in \mathcal{D}(A)$ such that $(A^* - i)g = (A - i)h$. But $(A - i)h = (A^* - i)h$ because $A$ is symmetric, hence $(A^* - i)(g - h) = \mathbf{0}$. Thus $g - h \in \mathcal{N}(A^* - i) = \mathcal{R}(A + i)^\perp$ [see Proposition 2.19(b)]. Since $\mathcal{R}(A + i) = \mathcal{H}$, this implies that $g - h = \mathbf{0}$. Consequently the vector $g$ belongs to $\mathcal{D}(A)$ (as it is equal to $h$).

(b) This proof is left as an exercise [use the result of (a) and the fact that $\mathcal{R}(\overline{B})^\perp = \mathcal{R}(B)^\perp$ if $B$ is a closable operator]. □

**Proposition 3.4.** *Let $A$ be a symmetric operator in $\mathcal{H}$ such that $\mathcal{R}(A + \lambda) = \mathcal{H}$ for some $\lambda \in \mathbb{R}$. Then $A$ is self-adjoint.*

PROOF. The proof is a simple adaptation of the preceding one [for $g \in \mathcal{D}(A^*)$, choose $h \in \mathcal{D}(A)$ such that $(A^* + \lambda)g = (A + \lambda)h$ to obtain that $(A^* + \lambda)(g - h) = \mathbf{0}$]. □

**3.1.2.** If $A$ is a closed symmetric operator that is not self-adjoint (or a symmetric operator that is not essentially self-adjoint) then, by (3.3) and Proposition 3.3, at least one of the numbers $-i$ and $+i$ is an eigenvalue of its adjoint $A^*$. The multiplicities of these eigenvalues, i.e. the dimensions of the two subspaces $\mathcal{R}(A-i)^\perp$ and $\mathcal{R}(A+i)^\perp$, should be a measure for the degree of non-self-adjointness of $A$. These multiplicities are called the ***deficiency indices*** $\nu_+$ and $\nu_-$ of $A$:

$$\nu_+ = \dim \mathcal{N}(A^* + i) \equiv \dim \mathcal{R}(A - i)^\perp, \quad \nu_- = \dim \mathcal{N}(A^* - i) \equiv \dim \mathcal{R}(A + i)^\perp. \tag{3.4}$$

In terms of the deficiency indices, Proposition 3.3 can be rephrased as follows.

**Corollary 3.5.** *(a) A closed symmetric operator is self-adjoint if and only if $\nu_+ = \nu_- = 0$.*
*(b) A symmetric operator is essentially self-adjoint if and only if $\nu_+ = \nu_- = 0$.*

To simplify the notations we set $\mathcal{M}_+ = \mathcal{R}(A-i)$ and $\mathcal{M}_- = \mathcal{R}(A+i)$. We assume $A$ to be closed so that $\mathcal{M}_+$ and $\mathcal{M}_-$ are subspaces. The operator $A-i$ gives a bijection between $\mathcal{D}(A)$ and $\mathcal{M}_+$, whereas $A + i$ is a bijection between $\mathcal{D}(A)$ and $\mathcal{M}_-$. If $f \in \mathcal{D}(A)$, one has $\|(A + i)f\| = \|(A - i)f\|$ as a consequence of (3.2). It follows that the mapping $W\colon (A + i)f \mapsto (A - i)f$ $[f \in \mathcal{D}(A)]$ defines a bijection between $\mathcal{M}_-$ and $\mathcal{M}_+$ and that one has $\|Wg\| = \|g\|$ for each $g \in \mathcal{M}_-$. The mapping $W$ is given by

$$W = (A - i)(A + i)^{-1} \tag{3.5}$$

and is called the ***Cayley transform*** of the symmetric operator $A$. $W$ is a unitary operator when viewed as an operator from $\mathcal{M}_-$ to $\mathcal{M}_+$ or, if one defines $\Omega \in \mathcal{B}(\mathcal{H})$ by $\Omega g = Wg$ if $g \in \mathcal{M}_-$ and $\Omega g = \mathbf{0}$ if $g \perp \mathcal{M}_-$, then $\Omega$ is a partial isometry with initial set $\mathcal{M}_-$ and final set $\mathcal{M}_+$ [see Proposition 2.9(e)]. By virtue of Proposition 3.3, $\Omega$ is unitary if and only if $A$ is self-adjoint.

To recover $A$ from its Cayley transform $W$, observe that for $f \in \mathcal{D}(A)$:

$$(I+W)(A+i)f = (A+i)f + (A-i)f = 2Af \tag{3.6}$$

and

$$(I-W)(A+i)f = (A+i)f - (A-i)f = 2if. \tag{3.7}$$

Eq. (3.7) implies that $\mathcal{D}(A) = (I-W)\mathcal{M}_-$. Let us show that the operator $(I-W)$ is invertible on $\mathcal{M}_-$: If $g \in \mathcal{M}_-$ there exists $f \in \mathcal{D}(A)$ such that $g = (A+i)f$. If $(I-W)g = \mathbf{0}$, then (3.7) implies that $f = \mathbf{0}$ and consequently $g = (A+i)f = \mathbf{0}$. This proves that $(I-W)$ is invertible on $\mathcal{M}_-$. Its inverse $(I-W)^{-1}$ is defined on $\mathcal{D}(A)$ and maps $\mathcal{D}(A)$ onto $\mathcal{M}_-$, and by (3.7) one has

$$(A+i)f = 2i(I-W)^{-1}f \qquad \forall f \in \mathcal{D}(A). \tag{3.8}$$

By inserting this relation into (3.6) one obtains $A$ in terms of $W$:

$$Af = \frac{1}{2}(I+W)(A+i)f = i(I+W)(I-W)^{-1}f \qquad \text{if } f \in \mathcal{D}(A). \tag{3.9}$$

For each closed symmetric extension of $A$ one can repeat the preceding considerations. Let us denote by $\mathbb{A}$ such an extension and by $\mathbb{W} = (\mathbb{A}-i)(\mathbb{A}+i)^{-1}$ its Cayley transform, and let us write $\mathbb{M}_+ = \mathcal{R}(\mathbb{A}-i)$ and $\mathbb{M}_- = \mathcal{R}(\mathbb{A}+i)$. Since $\mathbb{A}$ is an extension of $A$ [hence $\mathcal{D}(A) \subseteq \mathcal{D}(\mathbb{A})$ and $\mathbb{A}f = Af$ for $f \in \mathcal{D}(A)$], it is clear that $\mathcal{R}(A-i) \subseteq \mathcal{R}(\mathbb{A}-i)$, i.e. $\mathcal{M}_+ \subseteq \mathbb{M}_+$. Thus $\mathcal{M}_+$ is a subspace of $\mathcal{H}$ contained in $\mathbb{M}_+$, and we can write $\mathbb{M}_+ = \mathcal{M}_+ \oplus \mathcal{N}_+$, where $\mathcal{N}_+$ is the orthogonal complement of $\mathcal{M}_+$ in $\mathbb{M}_+$ (i.e. $\mathcal{N}_+ = \mathbb{M}_+ \cap [\mathcal{M}_+]^\perp$). Then each $f \in \mathbb{M}_+$ has a unique decomposition into $f = f_1 + f_2$ with $f_1 \in \mathcal{M}_+$ and $f_2 \in \mathcal{N}_+$. In the same way one finds that $\mathcal{M}_- \subseteq \mathbb{M}_-$, and one may write $\mathbb{M}_- = \mathcal{M}_- \oplus \mathcal{N}_-$.

Let us have a closer look at the structure of $\mathbb{W}$ relative to the above orthogonal sum decompositions. The operator $\mathbb{W}\colon \mathbb{M}_- \to \mathbb{M}_+$ is unitary. If $f \in \mathcal{M}_-$, then $(\mathbb{A}+i)^{-1}f = (A+i)^{-1}f \in \mathcal{D}(A)$, so that $\mathbb{W}f = (\mathbb{A}-i)(\mathbb{A}+i)^{-1}f = (A-i)(A+i)^{-1}f = Wf$: On $\mathcal{M}_-$ the operators $\mathbb{W}$ and $W$ coincide. The difference between them resides in the fact that $\mathbb{W}$ is defined on a larger subspace than $W$. The action of $\mathbb{W}$ on $\mathcal{N}_-$ is as follows: the restriction of $\mathbb{W}$ to $\mathcal{N}_-$ is a unitary operator $U$ from $\mathcal{N}_-$ onto $\mathcal{N}_+$; indeed, if $f_1 \in \mathcal{M}_-$ and $f_2 \in \mathcal{N}_-$, then $\langle \mathbb{W}f_1, \mathbb{W}f_2\rangle = \langle f_1, f_2\rangle = 0$, hence $\mathbb{W}f_2$ must be orthogonal to the subspace $\mathbb{W}\mathcal{M}_- \equiv W\mathcal{M}_- = \mathcal{M}_+$, i.e. $\mathbb{W}$ maps $\mathcal{N}_-$ into $\mathcal{N}_+$. As $\mathcal{R}(\mathbb{W}) = \mathbb{M}_+ = \mathcal{M}_+ \oplus \mathcal{N}_+$, $\mathbb{W}$ must map $\mathcal{N}_-$ *onto* $\mathcal{N}_+$.

We summarise the situation (see also Figure 3.1). The property that $\mathbb{A}$ is a closed symmetric extension of $A$ is reflected in the Cayley transforms $\mathbb{W}$ of $\mathbb{A}$ and $W$ of $A$ by the fact that $\mathcal{D}(\mathbb{W}) = \mathcal{D}(W) \oplus \mathcal{N}_- = \mathcal{M}_- \oplus \mathcal{N}_-$, where $\mathcal{N}_-$ is a subspace orthogonal to $\mathcal{M}_-$, and the restriction of $\mathbb{W}$ to $\mathcal{N}_-$ gives a unitary mapping $U$ from $\mathcal{N}_-$ onto a subspace $\mathcal{N}_+$ orthogonal to $\mathcal{M}_+$ (in particular dim $\mathcal{N}_-$ = dim $\mathcal{N}_+$):

$$\mathbb{W} = W \oplus U \qquad \text{on} \quad \mathcal{M}_- \oplus \mathcal{N}_-. \tag{3.10}$$

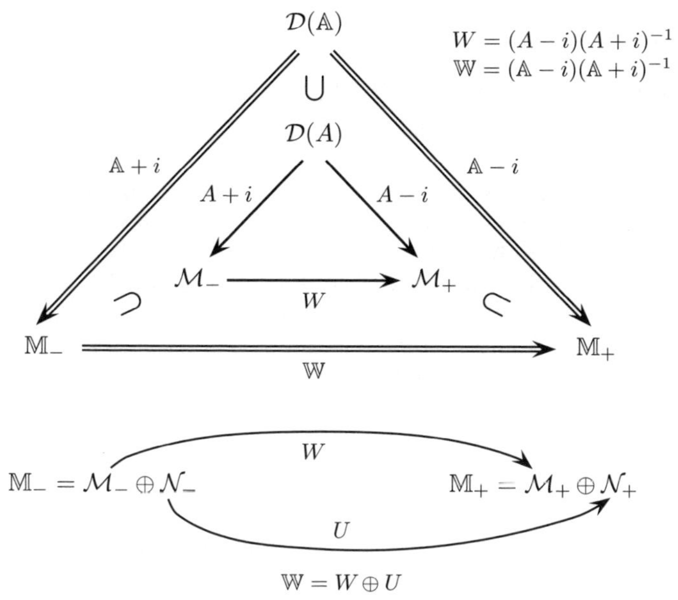


**Figure 3.1** Cayley transform of a symmetric operator $A$ and of a symmetric extension $\mathbb{A}$ of $A$.

Evidently, once a closed symmetric extension $\mathbb{A}$ of $A$ has been chosen, the three quantities $\mathcal{N}_-$, $\mathcal{N}_+$, and $U$ are uniquely determined. What is important is the fact that given $A$, then by an arbitrary choice of (i) a subspace $\mathcal{N}_-$ in $[\mathcal{M}_-]^\perp$, (ii) a subspace $\mathcal{N}_+$ in $[\mathcal{M}_+]^\perp$ with $\dim \mathcal{N}_+ = \dim \mathcal{N}_-$, and (iii) a unitary mapping $U$ from $\mathcal{N}_-$ onto $\mathcal{N}_+$, one determines a unique closed symmetric extension $\mathbb{A}$ of $A$ having the property that its Cayley transform is $\mathbb{W} = W \oplus U$ as in (3.10). The set of all closed symmetric extensions of $A$ can thus be parametrised by all possible choices of the three quantities $\mathcal{N}_-$, $\mathcal{N}_+$ and $U$, as indicated. We now verify this claim and then point out some important consequences. Examples are considered in Sections 3.2 and 3.3.

Verification. Let $A$ be a closed symmetric operator and W its Cayley transform. Let $\mathcal{N}_-$ and $\mathcal{N}_+$ be subspaces of $\mathcal{H}$ with $\mathcal{N}_- \perp \mathcal{M}_-$, $\mathcal{N}_+ \perp \mathcal{M}_+$, $\dim \mathcal{N}_- = \dim \mathcal{N}_+$, and let $U: \mathcal{N}_- \to \mathcal{N}_+$ be a unitary operator. Define $\mathbb{M}_- = \mathcal{M}_- \oplus \mathcal{N}_-$, $\mathbb{M}_+ = \mathcal{M}_+ \oplus \mathcal{N}_+$ and let $\mathbb{W}: \mathbb{M}_- \to \mathbb{M}_+$ be given by (3.10).

(i) Let us show that $I - \mathbb{W}$ is invertible on $\mathbb{M}_-$. Suppose that $g \in \mathbb{M}_-$ is such that $(I - \mathbb{W})g = \mathbf{0}$. If $f \in \mathbb{M}_-$, one obtains by taking into account the isometry of $\mathbb{W}$ that $\langle g, (I - \mathbb{W})f \rangle = \langle g, f \rangle - \langle g, \mathbb{W}f \rangle = \langle \mathbb{W}g, \mathbb{W}f \rangle - \langle g, \mathbb{W}f \rangle = -\langle (I - \mathbb{W})g, \mathbb{W}f \rangle = -\langle \mathbf{0}, \mathbb{W}f \rangle = 0$. Thus $g$ is orthogonal to the set $\{(I - \mathbb{W})f \mid f \in \mathbb{M}_-\}$, hence in particular to the set $\{(I - W)f \mid f \in \mathcal{M}_-\} \equiv \mathcal{D}(A)$ [see (3.7)]. Since $\mathcal{D}(A)$ is dense in $\mathcal{H}$, one obtains from Proposition 1.6 that $g = \mathbf{0}$.

(ii) Define $\mathbb{A} = i(I + \mathbb{W})(I - \mathbb{W})^{-1}$, with domain $\mathcal{D}(\mathbb{A}) = (I - \mathbb{W})\mathbb{M}_-$. It is evident that $\mathbb{A}$ is an extension of $A$. Let us then check that $\mathbb{A}$ is a closed symmetric operator. If $f, g$ belong to $\mathcal{D}(\mathbb{A})$, there exist $f_0$ and $g_0$ in $\mathbb{M}_-$ such that $f = (I - \mathbb{W})f_0$

and $g = (I - \mathbb{W})g_0$. Clearly one has $\mathbb{A}f = i(I + \mathbb{W})f_0$ and $\mathbb{A}g = i(I + \mathbb{W})g_0$, and therefore (by using the isometry of $\mathbb{W}$):

$$\begin{aligned}\langle f, \mathbb{A}g\rangle &= i\langle (I - \mathbb{W})f_0, (I + \mathbb{W})g_0\rangle = i\big[\langle f_0, \mathbb{W}g_0\rangle - \langle \mathbb{W}f_0, g_0\rangle\big]\\ &= -i\langle (I + \mathbb{W})f_0, (I - \mathbb{W})g_0\rangle = \langle \mathbb{A}f, g\rangle.\end{aligned}$$

This shows that $\mathbb{A}$ is symmetric. To see that it is closed, we apply Lemma 3.1(c) with $B = \mathbb{A} + i$; it suffices to show that $\mathcal{R}(\mathbb{A} + i)$ is closed (i.e. a subspace). Now if $f$ and $f_0$ are as above, one has $(\mathbb{A} + i)f = i(I + \mathbb{W})f_0 + i(I - \mathbb{W})f_0 = 2if_0$. Since $f_0$ may be any vector in $\mathbb{M}_-$, this shows that $\mathcal{R}(\mathbb{A} + i) = \mathbb{M}_-$, which is a subspace. Let us also observe that $(\mathbb{A} - i)f = i(I + \mathbb{W})f_0 - i(I - \mathbb{W})f_0 = 2i\mathbb{W}f_0$, and since $\mathbb{W}$ is unitary as an operator from $\mathbb{M}_-$ to $\mathbb{M}_+$, it follows that $\mathcal{R}(\mathbb{A} - i) = \mathbb{M}_+$.

(iii) Let again $\mathbb{A} = i(I + \mathbb{W})(I - \mathbb{W})^{-1}$. Denote by $\mathbb{W}' = (\mathbb{A} - i)(\mathbb{A} + i)^{-1}$ its Cayley transform. Let us show that $\mathbb{W}' = \mathbb{W}$. Remember that $\mathbb{W}'$ is the mapping $(\mathbb{A}+i)f \mapsto (\mathbb{A}-i)f$ for $f \in \mathcal{D}(\mathbb{A})$. With the expressions obtained above for $(\mathbb{A}+i)f$ and $(\mathbb{A} - i)f$, it follows that $\mathbb{W}'$ is the mapping $2if_0 \mapsto 2i\mathbb{W}f_0$, i.e. $f_0 \mapsto \mathbb{W}f_0$ for $f_0 \in \mathbb{M}_-$. This shows that $\mathbb{W}'$ is identical with $\mathbb{W}$. □

We now give some important consequences of the preceding results, namely (1) the von Neumann Formulas characterising the adjoint and the symmetric extensions of a symmetric operator $A$, and (2) a description of the different types of symmetric extensions of $A$ in terms of its deficiency indices.

**Proposition 3.6 (First von Neumann Formula).** *Let $A$ be a closed symmetric operator. Then $f \in \mathcal{D}(A^*)$ if and only if*

$$f = g + h_+ + h_- \tag{3.11}$$

*with $g \in \mathcal{D}(A)$, $h_+ \in \mathcal{N}(A^* + i)$ and $h_- \in \mathcal{N}(A^* - i)$, and*

$$A^*f = Ag - ih_+ + ih_-. \tag{3.12}$$

*Furthermore the decomposition* (3.11) *is unique* ($g$, $h_+$ *and* $h_-$ *are uniquely determined by* $f$).

PROOF. (i) It is clear that each vector of the form (3.11) belongs to $\mathcal{D}(A^*)$, because $\mathcal{D}(A)$, $\mathcal{N}(A^* + i)$ and $\mathcal{N}(A^* - i)$ are all contained in $\mathcal{D}(A^*)$.

(ii) Let $f \in \mathcal{D}(A^*)$. By the Projection Theorem one can write $(A^* + i)f = f_1 + f_2$ with $f_1 \in \mathcal{M}_-$ and $f_2 \perp \mathcal{M}_-$. Then there exists a vector $g \in \mathcal{D}(A)$ such that $f_1 = (A + i)g$. We set $h_- = -if_2/2$. One has $A^*g = Ag$ and $A^*f_2 = if_2$ [since $(\mathcal{M}_-)^\perp \equiv \mathcal{R}(A + i)^\perp = \mathcal{N}(A^* - i)$, see Proposition 2.19(b)], hence $A^*h_- = f_2/2$. So

$$\begin{aligned}A^*(f - g - h_-) &= \big[f_1 + f_2 - if\big] - Ag - f_2/2\\ &= \big[(A + i)g + f_2 - if\big] - Ag - f_2/2\\ &= -i(f - g) + f_2/2 = -i(f - g - h_-).\end{aligned}$$

Thus $(f-g-h_-) \in \mathcal{N}(A^*+i)$ and we have obtained (3.11) with $h_+ = f-g-h_-$. Eq. (3.12) is now evident.

(iii) To obtain the uniqueness of the decomposition (3.11) we must show that, if $g+h_++h_- = \mathbf{0}$ with $g \in \mathcal{D}(A)$, $h_+ \in \mathcal{N}(A^*+i)$ and $h_- \in \mathcal{N}(A^*-i)$, then $g = h_+ = h_- = \mathbf{0}$. Now if $g+h_++h_- = \mathbf{0}$, then $A^*(g+h_++h_-) \equiv Ag-ih_++ih_- = \mathbf{0}$, and by combining these two equations one obtains

$$(A+i)g = (ih_+ - ih_-) - i(h_+ + h_-) = -2ih_- \quad \text{and} \quad (A-i)g = 2ih_+. \tag{3.13}$$

The first equation in (3.13) shows that $h_- \in \mathcal{R}(A+i) = \mathcal{N}(A^*-i)^\perp$. Since $h_- \in \mathcal{N}(A^*-i)$ by assumption, one has $h_- \in \mathcal{N}(A^*-i) \cap \mathcal{N}(A^*-i)^\perp = \{\mathbf{0}\}$. So $h_- = \mathbf{0}$. In the same way one obtains that $h_+ = \mathbf{0}$ by using the second equation in (3.13). Finally $g = -h_+ - h_- = \mathbf{0}$. □

**Proposition 3.7 (Second von Neumann Formula).** *Let $A$ be a closed symmetric operator and let $\mathbb{A}$ be the closed symmetric extension of $A$ determined by three quantities $\mathcal{N}_-$, $\mathcal{N}_+$ and $U$ as before. Then $f \in \mathcal{D}(\mathbb{A})$ if and only if*

$$f = g + h - Uh \tag{3.14}$$

*with $g \in \mathcal{D}(A)$ and $h \in \mathcal{N}_-$, and then*

$$\mathbb{A}f = Ag + ih + iUh. \tag{3.15}$$

*Furthermore the decomposition* (3.14) *is unique.*

PROOF. If $f \in \mathcal{D}(\mathbb{A})$, there is a unique $f_0 \in \mathbb{M}_-$ such that $f = (I - \mathbb{W})f_0$, and $f_0$ has a unique decomposition into $f_0 = f_1 + f_2$ with $f_1 \in \mathcal{M}_-$ and $f_2 \in \mathcal{N}_-$. So $f = (I-W)f_1 + (I-U)f_2$. We take $g = (I-W)f_1$, which belongs to $\mathcal{D}(A)$, and $h = f_2$. Then $\mathbb{A}f = i(I+\mathbb{W})f_0 = i(I+W)f_1 + i(I+U)f_2 = Ag + ih + iUh$. The uniqueness of the decomposition (3.14) follows from Proposition 3.6 since $A^*$ is an extension of $\mathbb{A}$. □

**Deficiency indices and maximal symmetric extensions**

(a) If $A$ is a closed symmetric operator with $\mathcal{M}_+ = \mathcal{H}$ or $\mathcal{M}_- = \mathcal{H}$, i.e. if one of the deficiency indices of $A$ is zero, then $A$ has no proper symmetric extension (if $\mathbb{A}$ is a symmetric extension of $A$, then $\mathbb{A}$ is equal to $A$); indeed one then necessarily has $\mathcal{N}_+ = \{\mathbf{0}\}$ or $\mathcal{N}_- = \{\mathbf{0}\}$. Such an operator is called a ***maximal symmetric operator***. The following two possibilities may occur:
(i) $\nu_+ = \nu_- = 0$. In this case $A$ is *self-adjoint*,
(ii) $\nu_+ = 0, \nu_- \neq 0$ or $\nu_+ \neq 0, \nu_- = 0$. The operator $A$ is symmetric but not self-adjoint and *has no self-adjoint extension*.

(b) If $\mathcal{M}_+ \neq \mathcal{H}$ and $\mathcal{M}_- \neq \mathcal{H}$ but $\dim [\mathcal{M}_+]^\perp = \dim [\mathcal{M}_-]^\perp$, the deficiency indices of $A$ are equal: $\nu_+ = \nu_- \neq 0$. Again one must consider two possibilities:

(i) The deficiency indices of $A$ are finite ($0 \neq \nu_+ = \nu_- < \infty$). In this case *all closed maximal symmetric extensions of $A$ are self-adjoint*. The number of self-adjoint extensions of $A$ is infinite (and uncountable). Indeed, a maximal symmetric extension $\mathbb{A}$ is obtained by taking $\mathcal{N}_- = [\mathcal{M}_-]^\perp$, $\mathcal{N}_+ = [\mathcal{M}_+]^\perp$ [so that $\mathcal{R}(\mathbb{A}+i) = \mathcal{R}(\mathbb{A}-i) = \mathcal{H}$] and by choosing an arbitrary isometric operator $U$ from $[\mathcal{M}_-]^\perp$ to $[\mathcal{M}_+]^\perp$ ($U$ is automatically unitary because $\dim[\mathcal{M}_+]^\perp = \dim[\mathcal{M}_-]^\perp < \infty$). Already if $\nu_+ = \nu_- = 1$ the number of such operators $U$ is uncountable: Fix a vector $e_-$ in $[\mathcal{M}_-]^\perp$ and a vector $e_+$ in $[\mathcal{M}_+]^\perp$ with $\|e_-\| = \|e_+\| = 1$ and set $Ue_- = e^{i\alpha}e_+$, with $\alpha \in [0, 2\pi)$. For each value of $\alpha$ one obtains a self-adjoint extension of $A$, and these extensions are all different [because their Cayley transforms are different, see for example (3.10)]. If $\nu_+ = \nu_- \equiv \nu > 1$, the self-adjoint extensions of $A$ are in one-to-one correspondence with the elements $U$ of the group of all unitary $\nu \times \nu$ matrices; this group is described by $\nu^2$ continuous parameters.

(ii) The deficiency indices of $A$ are infinite: $\nu_+ = \nu_- = \infty$. In this case *$A$ has self-adjoint extensions as well as non-self-adjoint maximal symmetric extensions* (the number of self-adjoint extensions is again uncountably infinite). To obtain a self-adjoint extension one takes $\mathcal{N}_- = [\mathcal{M}_-]^\perp$, $\mathcal{N}_+ = [\mathcal{M}_+]^\perp$ and chooses for $U$ an arbitrary unitary operator from $\mathcal{N}_-$ onto $\mathcal{N}_+$ (remember that $\dim[\mathcal{M}_+]^\perp = \dim[\mathcal{M}_-]^\perp = \infty$ by assumption). On the other hand one may take $\mathcal{N}_- = [\mathcal{M}_-]^\perp$, choose for $\mathcal{N}_+$ an arbitrary infinite-dimensional strict subspace of $[\mathcal{M}_+]^\perp$ (thus $\mathcal{N}_+ \neq [\mathcal{M}_+]^\perp$, $\dim \mathcal{N}_+ = \infty$) and take for $U$ an arbitrary unitary operator from $\mathcal{N}_-$ onto $\mathcal{N}_+$. The associated extension is symmetric but not self-adjoint; if one denotes its deficiency indices by $n_+$ and $n_-$, then one has $n_- = 0$, $n_+ \neq 0$ ($n_+ = \dim[\mathcal{M}_+ \oplus \mathcal{N}_+]^\perp$, and by choosing $\mathcal{N}_+$ appropriately one may obtain any value $n_+ \in \{1, 2, 3, \ldots, \infty\}$). It is clear that one can in a similar way obtain maximal symmetric extensions with $n_+ = 0$ and arbitrary $n_- > 0$.

(c) If the deficiency indices of $A$ are different ($\nu_+ \neq \nu_-$), then at least one of them is finite. In this case *$A$ has no self-adjoint extension at all.* A special case has already been discussed in (a,ii) (one deficiency index $= 0$, the other one $\neq 0$). If $\nu_- \neq 0$, $\nu_+ \neq 0$ but $\nu_- \neq \nu_+$, consider for example the case in which $\nu_+ < \nu_-$ (one then necessarily has $\nu_+ < \infty$). The maximal symmetric extensions of $A$ are obtained by taking $\mathcal{N}_+ = [\mathcal{M}_+]^\perp$ and $\mathcal{N}_-$ a subspace of $[\mathcal{M}_-]^\perp$ with $\dim \mathcal{N}_- = \dim \mathcal{N}_+ = \nu_+$. It is clear that $\mathcal{N}_-$ is strictly smaller than $[\mathcal{M}_-]^\perp$, hence the deficiency indices $n_+$ and $n_-$ of the obtained extension $\mathbb{A}$ are $n_+ = 0$, $n_- = \nu_- - \nu_+$. Since $\nu_- - \nu_+ > 0$, the operator $\mathbb{A}$ is not self-adjoint.

**3.1.3.** We now describe a situation in which a symmetric operator always has self-adjoint extensions. A mapping $K\colon \mathcal{H} \to \mathcal{H}$ is called a ***conjugation*** if it has the following properties:
(1) $K$ is anti-linear, i.e. $K(\alpha f + g) = \overline{\alpha}Kf + Kg$ if $f, g \in \mathcal{H}$ and $\alpha \in \mathbb{C}$,
(2) $K^2 = I$,
(3) $\langle Kf, Kg\rangle = \langle g, f\rangle \equiv \overline{\langle f, g\rangle} \quad \forall f, g \in \mathcal{H}$.

**Example 3.8.** Let $\mathcal{H} = L^2(\mathcal{O}, m)$. Then the mapping $f \mapsto \overline{f}$, i.e. $f(x) \mapsto \overline{f(x)}$ for each $x \in \mathcal{O}$, is a conjugation.

**Proposition 3.9.** *Let $A$ be a symmetric operator. Assume that there exists a conjugation $K$ that maps $\mathcal{D}(A)$ into $\mathcal{D}(A)$ and commutes with $A$, i.e. such that $K\mathcal{D}(A) \subseteq \mathcal{D}(A)$ and $AKf = KAf$ for each $f \in \mathcal{D}(A)$. Then the deficiency indices of $A$ are equal; in particular $A$ has self-adjoint extensions.*

PROOF. We set $\mathcal{N}_+ = \mathcal{R}(A-i)^\perp$ and $\mathcal{N}_- = \mathcal{R}(A+i)^\perp$.

(i) Let $h \in \mathcal{D}(A)$. By the relations $K^2 = I$ and $K\mathcal{D}(A) \subseteq \mathcal{D}(A)$ we have $h = K^2h = K(Kh) \in K\mathcal{D}(A)$. This shows that $\mathcal{D}(A) \subseteq K\mathcal{D}(A)$. Hence we have $K\mathcal{D}(A) = \mathcal{D}(A)$.

(ii) Let $f \in \mathcal{N}_-$ and $g \in \mathcal{D}(A)$. Then [using $KA = AK$ and (1), then (3)]:

$$\langle Kf, (A-i)Kg\rangle = \langle Kf, K(A+i)g\rangle = \langle (A+i)g, f\rangle = 0.$$

Since $\{Kg \mid g \in \mathcal{D}(A)\} = \mathcal{D}(A)$ by (i), this shows that $Kf$ is orthogonal to $\mathcal{M}_+$, i.e. $Kf \in \mathcal{N}_+$. Thus $K$ maps $\mathcal{N}_-$ into $\mathcal{N}_+$. Similarly one shows that $K$ maps $\mathcal{N}_+$ into $\mathcal{N}_-$.

(iii) If $f \in \mathcal{N}_-$, then $f = K^2f = K(Kf)$. Since $Kf \in \mathcal{N}_+$ by (ii), this shows that $K$ is surjective as a mapping from $\mathcal{N}_+$ to $\mathcal{N}_-$ (K maps $\mathcal{N}_+$ *onto* $\mathcal{N}_-$). By (3) $K$ preserves the orthogonality of vectors and their norm; so if $\{e_n\}$ is an orthonormal basis of $\mathcal{N}_+$, then$\{Ke_n\}$ is an orthonormal basis of $\mathcal{N}_-$. This shows that $\dim \mathcal{N}_- = \dim \mathcal{N}_+$. □

**Remark 3.10.** If $A$ is a linear operator, one defines its ***domain of regularity*** $G(A)$ as the set of all complex numbers $z$ for which there exists a number $\kappa(z) > 0$ such that $\|(A-z)f\| \geq \kappa(z)\|f\|$ for all $f \in \mathcal{D}(A)$. One shows that $G(A)$ is an open subset of $\mathbb{C}$. Then one associates with each $z \in G(A)$ a ***deficiency subspace*** of $A$ at the point $z$, namely $\mathcal{R}(A-z)^\perp$, and shows that the dimension of this subspace is constant (i.e. independent of $z$) in each connected subset of $G(A)$ (see for example Section 8.1 of [W1] for the proof). If $A$ is symmetric, one sees from (3.2) that $G(A)$ contains all complex numbers $z = \lambda + i\mu$ with $\mu \neq 0$; hence the open upper half-plane ($\mu > 0$) and the open lower half-plane ($\mu < 0$) belong to $G(A)$, and in each of these half-planes the dimension of $\mathcal{R}(A-z)^\perp$ is independent of $z$ and equal to the deficiency index $\nu_+$ and $\nu_-$ respectively. This has the following consequences:

(a) Our choice of $z = \pm i$ for defining the deficiency indices has no particular significance, one could take any pair of complex numbers $\{z_+, z_-\}$ with $\Im z_+ > 0$ and $\Im z_- < 0$.

(b) In the self-adjointness criterion of Proposition 3.3 one may replace $\mathcal{R}(A-i)$ by $\mathcal{R}(A-z)$, where $z$ is any complex number with $\Im z \neq 0$, provided that $\mathcal{R}(A+i)$ is simultaneously replaced by $\mathcal{R}(A-\overline{z})$.

(c) Assume that the domain of regularity $G(A)$ of a symmetric operator $A$ contains a *real* point. Then $G(A)$ is connected, hence $\nu_+ = \nu_-$. Such an operator does have self-adjoint extensions [see point (b) in the discussion at the end of §3.1.2].

(d) Let $A$ be a closed symmetric operator. If $\nu_+ = 0$, then $\mathcal{R}(A-z) = \mathcal{H}$ for all $z$ with $\Im z > 0$, hence the upper half-plane belongs to the resolvent set $\rho(A)$. On the other hand, if $\nu_+ > 0$, the entire upper half-plane belongs to the spectrum of $A$. Similar results hold for $\nu_-$ and points $z$ in the lower half-plane. This leads to the

following characterisation of the spectrum of a closed symmetric operator $A$ (recalling that the spectrum is closed as a subset of the complex plane $\mathbb{C}$):
(1) if $\nu_+ = \nu_- = 0$: $\sigma(A)$ is contained in $\mathbb{R}$ ($A$ is self-adjoint),
(2) if $\nu_+ = 0$, $\nu_- > 0$: $\sigma(A)$ is the closed lower half-plane $\{z \in \mathbb{C} \mid \Im z \leq 0\}$,
(3) if $\nu_+ > 0$, $\nu_- = 0$: $\sigma(A)$ is the closed upper half-plane,
(4) if $\nu_+ > 0$ and $\nu_- > 0$: $\rho(A)$ is empty and $\sigma(A)$ is the whole plane.

## 3.2 Differential operators with constant coefficients

**3.2.1.** We begin with some preliminary results. Let us consider a ***differential expression*** $\mathcal{L}$ of the form

$$\mathcal{L} = p_N(x)\Big(\frac{d}{dx}\Big)^N + p_{N-1}(x)\Big(\frac{d}{dx}\Big)^{N-1} + \cdots + p_1(x)\frac{d}{dx} + p_0(x), \tag{3.16}$$

where $p_0, \ldots, p_N$ are functions of the real variable $x$, called the ***coefficients*** of the differential expression $\mathcal{L}$. By applying this expression to functions $f(x)$ one can associate to it ***linear differential operators*** in function spaces, in particular in $L^2(J)$, where $J \subseteq \mathbb{R}$ is an interval. One can choose various domains for these differential operators, but the elements of such domains should contain in their equivalence class a function that is at least $N$ times differentiable. One can similarly consider differential expressions involving partial derivatives ($n$ variables, $n \geq 2$) and associate with them differential operators in $L^2(\mathcal{W})$, where $\mathcal{W}$ is an open set in $\mathbb{R}^n$.

Suppose that one has associated with the expression (3.16) a differential operator $A$ in $L^2(J)$, with domain $\mathcal{D}(A)$, on some interval $J = (a, b)$. By partial integration one obtains for $f, g \in \mathcal{D}(A)$ (assuming that the coefficients $p_k$ are sufficiently differentiable):

$$\begin{aligned}\langle f, Ag\rangle &= \sum_{k=0}^{N} \int_a^b \overline{f(x)} p_k(x) g^{(k)}(x)\,dx \\ &= \sum_{k=0}^{N} (-1)^k \int_a^b \Big[\frac{d^k}{dx^k}\{p_k(x)\overline{f(x)}\}\Big] g(x)\,dx + P_N(b) - P_N(a),\end{aligned}$$

where the expression $P_N(b) - P_N(a)$ contains the boundary terms at $x = b$ and at $x = a$ that appear in the course of the integrations by parts. In order for $A$ to be a symmetric operator, this expression must vanish and one must have

$$\sum_{k=0}^{N} (-1)^k \frac{d^k}{dx^k}\Big[\overline{p_k(x)} f(x)\Big] = \sum_{k=0}^{N} p_k(x) \frac{d^k}{dx^k} f(x) \qquad \forall f \in \mathcal{D}(A). \tag{3.17}$$

It is easy to avoid the boundary terms by taking for the domain of $A$ only functions $f\colon (a, b) \to \mathbb{C}$ such that $f, f', f'', \ldots, f^{(N-1)}$ are zero at $x = a$ and at $x = b$, or even more simply $\mathcal{D}(A) = C_0^\infty((a, b))$ [then each $f \in \mathcal{D}(A)$ vanishes in some neighbourhood of $a$ and in some neighbourhood of $b$]. The condition (3.17) is a condition on the functions $p_0, \ldots, p_N$ that we shall not specify here in more detail. If this condition is satisfied, one speaks of a ***self-adjoint differential expression***; this does *not* imply that $A$ is a self-adjoint or essentially self-adjoint operator if $\mathcal{D}(A)$ is chosen such that $P_N(b) - P_N(a) = 0$. The operator $A$ will then be symmetric and may possibly have self-adjoint extensions, as we shall see. In order to be cautious we use the terminology ***formally self-adjoint differential expression*** rather than "self-adjoint differential expression".

In the case of a differential expression in several variables, one proceeds analogously; the expression will be formally self-adjoint if the coefficients satisfy certain conditions (example: an expression of the form $-\Delta + V(\vec{x})$ if $V$ is a real function), and if one chooses as domain of $A$ for example $\mathcal{D}(A) = C_0^\infty(\mathcal{W})$ if $\mathcal{H} = L^2(\mathcal{W})$ [then each function $f$ in $\mathcal{D}(A)$ vanishes in some neighbourhood of the boundary of the set $\mathcal{W} \subseteq \mathbb{R}^n$].

If $\mathcal{L}$ is a formally self-adjoint differential expression, defined for $x$ in an open subset $\mathcal{W}$ of $\mathbb{R}$ or $\mathbb{R}^n$, the differential operator in $L^2(\mathcal{W})$ with domain $C_0^\infty(\mathcal{W})$ is often called the ***minimal operator*** associated with $\mathcal{L}$. We shall denote it by $A_0$.[2] Sometimes one uses the term ***maximal operator*** associated with $\mathcal{L}$ for the adjoint of the minimal operator, and we denote this operator by $A_+$, i.e. $A_+ = A_0^*$. If $\mathcal{L} = \sum_{k=0}^N p_k(x)(d/dx)^k$ as in (3.16), the maximal operator can be interpreted as follows: an element $f$ of $L^2(\mathcal{W})$ belongs to the domain of $A_+$ if and only if the distribution $\sum_{k=0}^N (d/dx)^k [\overline{p_k(x)} f(x)]$ is a function and this function belongs to $L^2(\mathcal{W})$, and then $A_+ f = \sum_{k=0}^N (d/dx)^k [\overline{p_k(x)} f(x)]$, where the derivatives are taken in the sense of distributions. We recall that a ***distribution*** is a continuous linear mapping from a function space [here $C_0^\infty(\mathcal{W})$] to $\mathbb{C}$. If $T$ is a distribution, one denotes by $T(f)$ its value for the function $f$. The derivative $T'$ of $T$ is again a distribution defined by $T'(f) = -T(f')$, and by iteration one has $T^{(k)}(f) = (-1)^k T(f^{(k)})$. The condition that $f$ belongs to the domain of $A_0^*$ is $\langle f, A_0 g\rangle = \langle f^*, g\rangle$ for some $f^* \in L^2(\mathcal{W})$ and all $g \in C_0^\infty(\mathcal{W})$, and one has $A_0^* f = f^*$; so by definition $f^*$ is the distribution associated with $\sum_{k=0}^N (d/dx)^k [\overline{p_k(x)} f(x)]$ [interpreted in the sense of distributions, i.e. the derivatives are taken in the sense of distributions; the signs $(-1)^k$ appearing in (3.17) are incorporated into the definition of the distributional derivatives].

In what follows we restrict ourselves to the cases $N = 1$ and $N = 2$ which are by far the most important ones in quantum mechanics (in particular the momentum operator for $N = 1$, non-relativistic Hamiltonians $-\Delta + V(\vec{x})$ or $[-i\vec{\nabla} - \vec{A}(\vec{x})]^2 + V(\vec{x})$ for $N = 2$, called ***Schrödinger operators***). The case $N > 2$ can be treated similarly. Let us write the condition of formal self-adjointness for $N = 1$ and $N = 2$:

$\underline{N = 1}$: The differential expression $\mathcal{L} = p_1(x)(d/dx) + p_0(x)$ is formally self-adjoint if $p_1$ is purely imaginary, i.e. $p_1(x) = i\phi(x)$ with $\phi(x)$ real, and if $\overline{p_0(x)} + i\phi'(x) = p_0(x)$, i.e. $\Im p_0(x) = \phi'(x)/2$. A special case is obtained by taking for $\phi$ a constant function, i.e. $\phi(x) = \gamma \in \mathbb{R}$ for each $x$; then $p_0(x)$ is real. Thus: if $\mathcal{L}$ has the form $\mathcal{L} = i\gamma(d/dx) + p(x)$, where $p$ is a real function and $\gamma$ a real constant, then $\mathcal{L}$ is a

[2] Some authors use the term "minimal operator" for the closure of $A_0$ or for the restriction of the maximal operator to functions having compact support in $\mathcal{W}$.

formally self-adjoint differential expression.

$\underline{N=2}$: Here it is convenient to write $\mathcal{L}$ in a more symmetrical form:

$$\mathcal{L} = (d/dx)q_2(x)(d/dx) + q_1(x)(d/dx) + q_0(x).$$

It is easy to check (integration by parts) that this differential expression is formally self-adjoint if $q_2(x)$ is real, $q_1(x) = i\phi(x)$ for a real function $\phi$ and $\Im q_0(x) = \phi'(x)/2$. In particular, if we consider again a constant function $\phi$, then $\mathcal{L}$ is a formally self-adjoint differential expression if it is of the form $\mathcal{L} = (d/dx)q(x)(d/dx) + i\gamma(d/dx) + V(x)$, where $q$ and $V$ are real functions and $\gamma$ is a real constant.

Below we shall study the above two types of operators, restricting ourselves to $p(x) = 0$ (in the case $N = 1$) and to $q(x) = 1$ (in the case $N = 2$). The latter class ($N = 2$) contains non-relativistic Schrödinger Hamiltonians in one dimension and Hamiltonians occurring in the radial Schrödinger equation for three-dimensional quantum systems with spherically symmetric potentials. We treat here the simplest case in which $\gamma = 0$ and $V = 0$. The case in which $V \neq 0$ and Hamiltonians in dimension $n > 1$ with spherically symmetric potentials will be discussed in Section 3.3. Sufficient conditions for self-adjointness with potentials without spherical symmetry in dimension $n > 1$ have been obtained in §2.7.3.

**3.2.2.** We begin with some preliminary considerations on differentiability properties of functions defined on an open interval $J = (a, b)$. A function $g\colon J \to \mathbb{C}$ is said to be ***absolutely continuous*** if it is an indefinite integral (in the sense of Lebesgue). Thus $g$ is absolutely continuous if it can be expressed in the form $g(x) = \int_{x_0}^{x} h(x)dx + \text{const.}$, with fixed $x_0 \in J$ and with $h\colon J \to \mathbb{C}$ a locally integrable function [which means that $h \in L^1(\mathfrak{J})$ for each bounded closed subinterval $\mathfrak{J}$ of $J$]. It is clear that absolutely continuous functions $g$ are continuous on $J$. Moreover it can be shown that they are differentiable almost everywhere in $J$, with $g' = h$ (almost everywhere). Of course each function of class $C^1$ on $J$ is absolutely continuous on this interval. On the other hand there exist continuous functions that are differentiable almost everywhere but which are not absolutely continuous, i.e. not equal to the indefinite integral of their derivative (an example is given by the Cantor function described in Example 4.5).

An absolutely continuous function $g$ may be diverging at the endpoints of the interval $J$, because its derivative $g'$ is assumed to be only locally integrable on $J$. Such divergence cannot occur if $g'$ is globally integrable on $J$, i.e. if $g$ is an absolutely continuous function with $g' \equiv h \in L^1(J)$. Indeed, if $|h|$ is integrable over $J$, then $g$ has boundary values at the end points of this interval: $g(a) \equiv \lim_{x \searrow a} g(x)$ and $g(b) \equiv \lim_{x \nearrow b} g(x)$ exist.[3] We shall use the fact that, for functions of this type, the integration by parts formula is valid: If $f$ and $g$ are absolutely continuous on $J = (a, b)$, with $f'$ and $g'$ in $L^1(J)$, then

$$\int_a^b f(x)g'(x)dx + \int_a^b f'(x)g(x)dx = f(b)g(b) - f(a)g(a). \tag{3.18}$$

[3] For example, if $b$ is finite, then $g(b-\varepsilon) - g(b-\delta) = \int_{b-\delta}^{b-\varepsilon} h(y)dy \to 0$ as $\delta, \varepsilon \to +0$, so that $\lim_{x \nearrow b} g(x)$ exists.

Details on the preceding results are given in numerous textbooks, see for example §23 - §26 of [RN] or Section 18 of [HS].

In the theory of ordinary differential operators one often encounters continuous functions $g$ with square-integrable derivative $g' \equiv h$, i.e. satisfying $h \in L^2(J)$. Recalling from (1.49) that $h \in L^2(J)$ implies that $h \in L^1(\mathfrak{J})$ for each finite subinterval $\mathfrak{J}$ of $J$, it follows that such a function $g$ is absolutely continuous on $J$. In this context the following fact will be important:

**Proposition 3.11.** *Let $g$ be an absolutely continuous function on $J = (a, b)$ such that $g' \in L^2(J)$.*
*(a) If $a$ is finite, then the boundary value $g(a)$ exists, and similarly for the endpoint $b$.*
*(b) If $J$ is an infinite interval and $g$ itself is also square-integrable on $J$, then $g(x) \to 0$ as $x$ tends to an infinite endpoint of $J$ (e.g. as $x \to +\infty$ if $b = +\infty$).*

PROOF. (a) If $a$ is finite and $c \in (a, b)$, then the assumption made on $g'$ implies that $g' \in L^2((a, c)) \subseteq L^1((a, c))$, see (1.49). Thus $g(a)$ exists.

(b) We treat the case $b = +\infty$. One has $(d/dx)|g(x)|^2 = \overline{g(x)}g'(x) + \overline{g'(x)}g(x)$. Hence, for any $c \in (a, b)$:

$$|g(x)|^2 - |g(c)|^2 = \int_c^x \frac{d}{dy}|g(y)|^2 dy = \int_c^x \overline{g(y)}g'(y)\,dy + \int_c^x \overline{g'(y)}g(y)\,dy.$$

The integrals on the right-hand side have limits as $x \to \infty$, because $g$ and $g'$ belong to $L^2([c, \infty))$ and these integrals can be interpreted as scalar products. This shows that the limit of $|g(x)|^2$ as $x \to \infty$ exists. Since $|g(x)|^2$ is integrable at infinity, this limit must be zero. □

In our discussion of ordinary differential operators we use the following terminology and convention: An element $f$ of $L^2(J)$ is said to be absolutely continuous if its equivalence class contains an absolutely continuous function, and in our equations involving $f$ it is understood that we work with this continuous representative. For example, if $f$ and $g$ are absolutely continuous elements of $L^2(J)$ such that $f'$ and $g'$ belong to $L^2(J)$, then the integrals in (3.18) may be interpreted as scalar products between elements of $L^2(J)$ (they are independent of the chosen representatives in the equivalence classes of $f$ and $g$), but for a large class of representatives the right-hand side will not make sense.

With the preceding convention in mind we introduce an operator $B$ in $L^2(J)$ as follows:

$$\mathcal{D}(B) = \{g \in L^2(J) \mid \exists\, h \in L^2(J) \text{ such that } g(x) = \int_0^x h(y)dy + g(0)\}, \quad (3.19)$$

and $Bg = -ih = -ig'$ [we assume that the point $x = 0$ belongs to $J$ or is an endpoint of $J$, so that $g(0)$ is defined]. The elements of $\mathcal{D}(B)$ are the equivalence classes of the (absolutely continuous) functions specified on the right-hand side of (3.19), and by Proposition 3.11 their boundary values $g(a)$ and $g(b)$ are well defined (with $g(a) = 0$ if $a = -\infty$, $g(b) = 0$ if $b = +\infty$).

**3.2.3.** We now discuss the differential expression $\mathcal{L} = -i(d/dx)$ which formally corresponds to the momentum operator. We consider this expression on $\mathbb{R}$, on a semi-infinite interval [which we take to be $\mathbb{R}_+ = (0, \infty)$] and on a finite interval $(a, b)$ (we shall set $a = 0$ and $b = 1$). So let $J = \mathbb{R}$, $\mathbb{R}_+$ or $(0, 1)$ and denote by $C_0^\infty(J)$ the set of all complex functions of class $C^\infty$ on $J$ each of which vanishes in some neighbourhood of each endpoint of $J$. The minimal operator $P_0$ is given by $\mathcal{D}(P_0) = C_0^\infty(J)$ and $P_0 f = -if'$ for $f \in C_0^\infty(J)$. We first show that the adjoint of $P_0$ (i.e. the maximal operator) coincides with the operator $B$ introduced above, which gives a useful description of the domain of $P_0^*$ [namely the domain introduced in (3.19)].

**Proposition 3.12.** *The operator $B$ is the adjoint of the minimal operator: $B = P_0^*$. Hence $B$ coincides with the maximal operator $P_+$, and the closure of the minimal operator is the adjoint of $B$, i.e. $\overline{P_0} = B^*$.*

PROOF. The last assertion follows from the first one because the closure of an operator $C$ is equal to $C^{**}$ (Proposition 2.22). So let us determine the adjoint $P_0^*$ of $P_0$. An element $g$ of $L^2(J)$ belongs to $\mathcal{D}(P_0^*)$ if and only if there exists a vector $g^* \in L^2(J)$ such that $\langle g, -if' \rangle = \langle g^*, f \rangle$ for all $f \in C_0^\infty(J)$. By interpreting $g$ as a distribution on the space of test functions $C_0^\infty(J)$, this means that the derivative (in the distributional sense) of $g$ is equal to $ig^*$, hence that this derivative is a function.[4] One can then define an absolutely continuous function $G$ on $J$ by $G(x) = \int_0^x g^*(y)dy$. The function $g^*$ is locally of class $L^1$, hence $G$ is bounded on each finite subinterval $\mathfrak{J}$ of $J$ and thus belongs to $L^2(\mathfrak{J})$. For $f \in C_0^\infty(\mathfrak{J})$ one obtains upon integrating by parts that $\langle g^*, f \rangle = \langle G', f \rangle = -\langle G, f' \rangle$ (the boundary terms are zero because $f$ vanishes in some neighbourhood of each endpoint of $\mathfrak{J}$). It follows that $ig + G$ is orthogonal to the set of functions $\{f' \mid f \in C_0^\infty(\mathfrak{J})\}$. We shall see below that this implies that $ig + G$ is (equivalent to) a constant function, i.e. $g(x) = iG(x) + \text{const.}$ a.e. in $\mathfrak{J}$. Since $\mathfrak{J}$ is arbitrary, this shows that the elements of $\mathcal{D}(P_0^*)$ are absolutely continuous functions with square-integrable derivative, and a square-integrable absolutely continuous function on $J$ belongs to $\mathcal{D}(P_0^*)$ if its derivative (which will be equal to $g^*$ in the preceding notations) belongs to $L^2(J)$. Then one will have $P_0^* g = g^* = -ig'$. Thus $P_0^* = B$.

To see that $ig + G$ is constant on each finite subinterval $\mathfrak{J} = (c, d)$ of $J$, we denote by $g_0$ the restriction of $ig + G$ to $\mathfrak{J}$, more precisely $g_0(x) = ig(x) + G(x)$ if $x \in \mathfrak{J}$ and $g_0(x) = 0$ otherwise. $g_0$ belongs to $L^2(\mathfrak{J})$ and we may assume that $g_0 \neq \mathbf{0}$ (otherwise there is nothing to prove). Observe now that if $\varphi \in C_0^\infty(\mathfrak{J})$ is such that $\int_c^d \varphi(x)dx = 0$, then the function $\vartheta$ defined as $\vartheta(x) \equiv \int_c^x \varphi(y)dy$ belongs to $C_0^\infty(\mathfrak{J})$ and satisfies $\vartheta'(x) = \varphi(x)$. Hence we have $\langle \varphi, g_0 \rangle = 0$. Thus, by setting $\mathcal{D}_1 = \{h \in C_0^\infty(\mathfrak{J}) \mid \int_c^d h(x)dx = 0\}$, we have $\langle h, g_0 \rangle = 0 \ \forall h \in \mathcal{D}_1$. Let us fix a function $e \in C_0^\infty(\mathfrak{J})$ such that $\int_c^d e(x) = 1$ and observe that, for any $f \in C_0^\infty(\mathfrak{J})$, the function $f - \big[\int_c^d f(x)dx\big] e$ belongs to $\mathcal{D}_1$. Hence

$$\langle f, g_0 \rangle - \langle \Big[\int_c^d f(x)dx\Big] e, g_0 \rangle = 0 \qquad \forall f \in C_0^\infty(\mathfrak{J}).$$

[4] The function $g$ determines a distribution $T$ by $T(f) = \langle g, f \rangle$; its distributional derivative $T'$ is given by $T'(f) = -\langle g, f' \rangle \equiv \langle ig, -if' \rangle \equiv \langle ig^*, f \rangle$.

This equation may be rewritten as

$$\int_c^d \overline{f(x)}\,\big[g_0(x) - \langle e, g_0\rangle\big]\,dx = 0 \qquad \forall f \in C_0^\infty(\mathfrak{J}).$$

Since $C_0^\infty(\mathfrak{J})$ is dense in $L^2(\mathfrak{J})$, one must have $g_0 - \langle e, g_0\rangle = \mathbf{0}$ (see Proposition 1.6). Thus $g_0(x) = \langle e, g_0\rangle$ for almost all $x \in \mathfrak{J}$. □

Knowing the domain of the maximal operator, we can now determine the deficiency indices of the minimal operator $P_0$. If $g \in \mathcal{R}(P_0 \mp i)^\perp = \mathcal{N}(B \pm i) \subset \mathcal{D}(B)$, then $g$ is absolutely continuous and satisfies $-ig' \pm ig = \mathbf{0}$, i.e. $g' = \pm g$. Since $g$ is absolutely continuous, this implies that $g'$ has the same property. If $g''$ denotes the derivative of $g'$, we have $g'' = \pm g'$. Thus $g''$ is absolutely continuous, and by iterating this argument one finds that $g$ is of class $C^\infty$. It is easy to calculate $g$: the general solution of $g' = g$ is given by $g_+(x) = c_+ e^x$ and that of $g' = -g$ by $g_-(x) = c_- e^{-x}$, where $c_+$ and $c_-$ are arbitrary complex constants. One sees that the deficiency indices can assume only the values 1 or 0 [depending on whether $g_+$ and $g_-$ belong to $L^2(J)$ or not]. Let us consider separately the cases $J = \mathbb{R}$, $J = \mathbb{R}_+$ and $J = (0,1)$.

(1) $\underline{J = \mathbb{R}}$: One has $g_+ \notin L^2(\mathbb{R})$ and $g_- \notin L^2(\mathbb{R})$ (assuming that $c_\pm \neq 0$). Thus the deficiency indices are $\nu_+ = \nu_- = 0$. The minimal operator [the operator $-i(d/dx)$ with domain $C_0^\infty(\mathbb{R})$] is *essentially self-adjoint* [see Corollary 3.5(b)]. Recall that we have already established the essential self-adjointness of $-i(d/dx)$ on $\mathcal{S}(\mathbb{R})$ in §2.5.2 by a different method.

(2) $\underline{J = \mathbb{R}_+ = (0,\infty)}$: One has $g_+ \notin L^2(J)$ but $g_- \in L^2(J)$, and the deficiency indices are $\nu_+ = 0$, $\nu_- = 1$. The closure of the minimal operator is a maximal symmetric operator that is not self-adjoint. The differential expression $-i(d/dx)$ determines *no self-adjoint operator* in $L^2(\mathbb{R}_+)$.

(3) $\underline{J = (0,1)}$: Here $g_+ \in L^2(J)$ and $g_- \in L^2(J)$. The deficiency indices of $P_0$ are $\nu_+ = \nu_- = 1$. Every closed maximal symmetric extension of $P_0$ is self-adjoint. The differential expression $-i(d/dx)$ determines *an uncountable family of self-adjoint operators* in $L^2((0,1))$.

To characterise the self-adjoint extensions of $P_0$ that exist in case (1) and in case (3), it is useful to know the domain of the closure $\overline{P_0} = B^*$ of the minimal operator $P_0$.

**Proposition 3.13.** *The domain of the closure $\overline{P_0}$ of the minimal operator is as follows. An element $f$ of $L^2(J)$ belongs to $\mathcal{D}(\overline{P_0})$ if and only if it has the following properties:*
*(1) $f$ is absolutely continuous,*
*(2) $f' \in L^2(J)$,*
*(3) if $J \neq \mathbb{R}$, then:* *(3a) if $J = (0,\infty)$: $f(0) = 0$*
*(3b) if $J = (0,1)$: $f(0) = f(1) = 0$.*
*If $f \in \mathcal{D}(\overline{P_0})$, then $\overline{P_0} f = -if'$.*

PROOF. (i) If $J = \mathbb{R}$, the closure of $P_0$ is equal to $B$ because $P_0$ is essentially self-adjoint. Thus the elements of $\mathcal{D}(\overline{P_0})$ are absolutely continuous functions with derivative in $L^2(\mathbb{R})$ and converging to zero as $x \to \pm\infty$ [see Proposition 3.11(b)].

(ii) In the other two cases let us denote by $B_0$ the restriction of the operator $B$ to the domain $\mathcal{D}(B_0)$ consisting of all $f \in L^2(J)$ having the properties (1), (2) and (3) of the proposition [remember that the elements of $\mathcal{D}(B)$ are characterised by (1) and (2) and that we use the convention made at the end of §3.2.2]. We must show that $B_0 = \overline{P_0}$. We use the following identity which is a special case of (3.18) (integration by parts): If $f, g \in \mathcal{D}(B)$ and $J = (a, b)$, then

$$\langle Bf, g\rangle - \langle f, Bg\rangle = \langle -if', g\rangle - \langle f, -ig'\rangle = i\big[\overline{f(b)}g(b) - \overline{f(a)}g(a)\big]. \qquad (3.20)$$

If $J = (0, \infty)$ or $J = (0, 1)$ we have $a = 0$ and $b = \infty$ or $b = 1$. If $f \in \mathcal{D}(B_0)$, the condition (3) ensures that $f(a) \equiv f(0) = 0$; if $b = \infty$ we have $f(b) \equiv f(\infty) = 0$ by Proposition 3.11, and if $b = 1$ we have $f(b) = 0$ by the assumption (3). Thus (3.20) implies that

$$\langle B_0 f, g\rangle - \langle f, P_0^* g\rangle = 0 \qquad \forall f \in \mathcal{D}(B_0),\ \forall g \in \mathcal{D}(P_0^*) \equiv \mathcal{D}(B).$$

This shows that the adjoint $P_0^{**} \equiv \overline{P_0}$ of $P_0^*$ is an extension of $B_0$: $B_0 \subseteq \overline{P_0}$.

To obtain the inclusion $\overline{P_0} \subseteq B_0$, let us fix $f \in \mathcal{D}(\overline{P_0}) \equiv \mathcal{D}(P_0^{**})$. Since $B$ is an extension of $P_0^{**}$ one has

$$\langle Bf, g\rangle - \langle f, Bg\rangle \equiv \langle P_0^{**} f, g\rangle - \langle f, P_0^* g\rangle = 0 \qquad \forall g \in \mathcal{D}(P_0^*). \qquad (3.21)$$

Let us choose a particular $g \in \mathcal{D}(P_0^*) \equiv \mathcal{D}(B)$: we take an infinitely differentiable function $g\colon [0, \infty) \to \mathbb{R}_+$ such that $g(0) = 1$ and $g(x) = 0$ for $x \geq 1$. As $g(b) = 0$ (for $b = 1$ as well as for $b = \infty$), Eq. (3.20) becomes [for this particular $g$ and for $f \in \mathcal{D}(\overline{P_0})$]:

$$\langle Bf, g\rangle - \langle f, Bg\rangle = -i\overline{f(0)}g(0) = -i\overline{f(0)}.$$

Comparison with (3.21) shows that one must have $f(0) = 0$. Thus the elements of $\mathcal{D}(\overline{P_0})$ satisfy $f(0) = 0$. This proves the inclusion $\overline{P_0} \subseteq B_0$ if $J = (0, \infty)$.

In the case $J = (0, 1)$ we still have to show that $f(1) = 0$ if $f \in \mathcal{D}(\overline{P_0})$. This is done by repeating the argument given above, by choosing the particular function $g$ such that $g(0) = 0$ and $g(1) = 1$. □

We can now specify the self-adjoint extensions of the minimal operator.

$J = \mathbb{R}$: We have already seen that the minimal operator is essentially self-adjoint. Its unique self-adjoint extension is the *momentum operator* $P$. In configuration space, i.e. in $L^2(\mathbb{R})$, one has $(Pf)(x) = -if'(x)$ and the domain of $P$ is the set of absolutely continuous $f\colon \mathbb{R} \to \mathbb{C}$ such that $f$ and $f'$ belong to $L^2(\mathbb{R})$. In momentum space, i.e. in $\tilde{L}^2(\mathbb{R})$, P is the operator of multiplication by the variable: $(\mathcal{F}Pf)(k) = k\tilde{f}(k)$, and its domain in this representation is the set of (equivalence classes of) measurable functions $g \equiv \tilde{f}\colon \mathbb{R} \to \mathbb{C}$ satisfying $\int_{-\infty}^{\infty}(1 + k^2)|g(k)|^2 dk < \infty$.

$J = (0, 1)$: Let $\mathbb{P}$ be a self-adjoint extension of $P_0$ (hence of $\overline{P_0}$). The operator $\mathbb{P}$ is the restriction of $P_0^*$ to a linear submanifold of $\mathcal{D}(P_0^*) \equiv \mathcal{D}(B)$. If we take in (3.20) $f = g \in \mathcal{D}(\mathbb{P})$, the expression on the right-hand side (which is $i[|f(1)|^2 - |f(0)|^2]$) must be zero because $\mathbb{P}$ is assumed to be self-adjoint. If $f \in \mathcal{D}(\overline{P_0})$, one has $f(0) = f(1) = 0$, hence the condition $|f(1)|^2 - |f(0)|^2 = 0$ is automatically satisfied. If $f \in \mathcal{D}(\mathbb{P})$ but $f \notin \mathcal{D}(\overline{P_0})$, one will have $f(0) \neq 0$ or $f(1) \neq 0$; since $|f(1)|^2 - |f(0)|^2 = 0$, one then has in fact $f(0) \neq 0$ *and* $f(1) \neq 0$, and there exists a number $\alpha \in [0, 2\pi)$ such that $f(1) = e^{i\alpha} f(0)$. A priori $\alpha$ depends on $f$: $\alpha = \alpha(f)$. By taking in (3.20) two vectors $f \neq g$ belonging to $\mathcal{D}(\mathbb{P}) \backslash \mathcal{D}(\overline{P_0})$, one must have

$$\overline{f(1)}g(1) - \overline{f(0)}g(0) = \left[e^{-i\alpha(f)}e^{i\alpha(g)} - 1\right]\overline{f(0)}g(0) = 0,$$

consequently $\alpha(f) = \alpha(g)$. Thus the number $\alpha$ is a constant depending only on the self-adjoint extension $\mathbb{P}$. If we denote by $\mathbb{P}_\alpha$ the restriction of $P_0^*$ to the domain

$$\mathcal{D}(\mathbb{P}_\alpha) = \{f \in \mathcal{D}(P_0^*) \big| f(1) = e^{i\alpha} f(0)\}, \tag{3.22}$$

then we have shown that $\mathbb{P}_\alpha$ is an extension of $\mathbb{P}$: $\mathbb{P} \subseteq \mathbb{P}_\alpha$. Now $\mathbb{P}_\alpha$ is symmetric [the last expression in (3.20) is zero if $f, g \in \mathcal{D}(\mathbb{P}_\alpha)$], hence $\mathbb{P}_\alpha$ is a symmetric extension of the self-adjoint operator $\mathbb{P}$. This is possible only if $\mathbb{P}_\alpha = \mathbb{P}$.

These considerations show that any self-adjoint extension of the minimal operator $P_0$ must be equal to one of the operators $\mathbb{P}_\alpha$, $\alpha \in [0, 2\pi)$. In fact *each* $\mathbb{P}_\alpha$ *is a self-adjoint extension of* $P_0$ because each $\mathbb{P}_\alpha$ is a self-adjoint operator: if $g \in \mathcal{D}(\mathbb{P}_\alpha^*)$, then (3.20) implies that one must have

$$\overline{f(1)}g(1) - \overline{f(0)}g(0) = \left[e^{-i\alpha}g(1) - g(0)\right]\overline{f(0)} = 0 \qquad \forall f \in \mathcal{D}(\mathbb{P}_\alpha).$$

By taking $f$ such that $f(0) \neq 0$, one sees that one must have $g(1) = e^{i\alpha} g(0)$, hence $g \in \mathcal{D}(\mathbb{P}_\alpha)$. An interpretation of the operators $\mathbb{P}_\alpha$ is given below, and some additional properties of these operators can be found in §3.2.5.

We have thus found all self-adjoint extensions of the operator $P_0$ in $L^2((0,1))$. Let us indicate how the same result follows from von Neumann's formulas (3.14) and (3.15). If $\mathbb{P}$ is a self-adjoint extension of $P_0$, one has $f \in \mathcal{D}(\mathbb{P})$ if and only if $f$ is of the form $f = g + h - Uh$ with $g \in \mathcal{D}(\overline{P_0})$, $h \in \mathcal{N}(P_0^* - i)$ and $U : \mathcal{N}(P_0^* - i) \to \mathcal{N}(P_0^* + i)$ unitary. Now $\mathcal{N}(P_0^* - i)$ is the one-dimensional subspace spanned by $g_-$, hence $h(x) = \kappa e^{-x}$ with $\kappa$ varying over $\mathbb{C}$. Similarly $\mathcal{N}(P_0^* + i) = \{\kappa e^x \mid \kappa \in \mathbb{C}\}$; so $Uh \equiv U(\kappa e^{-x})$ is a multiple of $e^x$, say $U(\kappa e^{-x}) = \kappa_+ e^x$, and, since $U$ must be unitary, one must have $\|\kappa e^{-x}\| = \|\kappa_+ e^x\|$. By calculating these norms [in $L^2((0,1))$] one sees that this condition is equivalent to $|\kappa_+| = |\kappa|/e$. So, if we set $\kappa_+ = \beta\kappa/e$ with $|\beta| = 1$, we have

$$f \in \mathcal{D}(\mathbb{P}) \iff f(x) = g(x) + \kappa e^{-x} - \beta\kappa e^{x-1} \text{ a.e.,} \quad \text{with } g \in \mathcal{D}(\overline{P_0}). \tag{3.23}$$

Eq. (3.15) shows that $[\mathbb{P}f](x) = [\overline{P_0}g + ih + iUh](x) = -ig'(x) + i\kappa e^{-x} + i\beta\kappa e^{x-1}$, so that, by (3.23), $\mathbb{P}f = -if'$. Hence $\mathbb{P}$ is given by $-i(d/dx)$. Finally let us consider the relation between $f(0)$ and $f(1)$ if $f$ is of the form (3.23). By using the fact that $g(0) = g(1) = 0$ one obtains

$$f(1) = \frac{\kappa}{e} - \beta\kappa = \frac{\kappa}{e}(1 - \beta e)\,, \quad f(0) = \kappa - \frac{\beta\kappa}{e} = \frac{\kappa}{e}(e - \beta).$$

Hence $f(1) = (1 - \beta e)(e - \beta)^{-1} f(0) = -\beta(\overline{\beta} - e)(\beta - e)^{-1} f(0) \equiv e^{i\alpha} f(0)$, because $|-\beta(\overline{\beta} - e)(\beta - e)^{-1}| = 1$ (the value of $\alpha$ can be determined in terms of that of $\beta$).

The spectrum of the operators $\mathbb{P}_\alpha$ is quite easy to determine (see Problem 3.3). It is an interesting fact that each real number $\lambda$ is an eigenvalue of some $\mathbb{P}_\alpha$. Let us also point out an interpretation of these operators. Since each $\mathbb{P}_\alpha$ acts in the same way as $\overline{P_0}$ on functions in $L^2((0,1))$ having support away from 0 and 1, it is clear that one must consider what happens at the endpoints of the interval $J = (0,1)$. Let us recall from quantum mechanics that, by exponentiation, the momentum operator generates translations in configuration space: in $L^2(\mathbb{R})$, the operators $\{\exp(-iP_0a)\}_{a\in\mathbb{R}}$ give a unitary representation of the translation group, i.e. $[\exp(-iP_0a)f](x) = f(x-a)$ (this will be discussed in detail in Section 5.1; see in particular Example 5.4). So one expects that, if $f$ has compact support in $(0,1)$ and if $a$ is sufficiently small [so that the set $\{x-a\}_{x\in\text{supp}f}$ is still contained in $(0,1)$], one will have $[\exp(-i\mathbb{P}_\alpha a)f](x) = f(x-a)$. To understand the situation when these conditions are not satisfied, let us take for $f$ an eigenvector of $\mathbb{P}_\alpha$, i.e. $f(x) = \exp(i\lambda_n x)$ with $\lambda_n = 2\pi n + \alpha$ $(n \in \mathbb{Z})$. Then $[\exp(-i\mathbb{P}_\alpha a)f](x) = [\exp(-i\lambda_n a)f](x) = \exp(i\lambda_n(x-a))$, with $x \in [0,1]$. It follows that, for $0 < a < 1$:

$$[\exp(-i\mathbb{P}_\alpha a)f](x) = \begin{cases} f(x-a) & \text{if } x-a \geq 0 \\ e^{-i\alpha}f(x-a+1) & \text{if } x-a \leq 0 \end{cases}$$

Thus, for $0 < a < 1$, the action of $\exp(-i\mathbb{P}_\alpha a)$ corresponds to a translation by $a$ to the right in such a way that the part of the translated function falling above the endpoint 1 gets re-injected into $J$ at the other endpoint 0, by acquiring an additional phase factor $e^{-i\alpha}$. Similarly, for $-1 < a < 0$, $\exp(-i\mathbb{P}_\alpha a)$ acts as translation to the left with re-injection at the right endpoint 1 with an additional phase factor $e^{+i\alpha}$. The action of $\exp(-i\mathbb{P}_\alpha a)$ for $|a| \geq 1$ is then obtained by using the group property of the family $\{\exp(-i\mathbb{P}_\alpha a)\}_{a\in\mathbb{R}}$.

The above considerations also shed some light on the fact that on the interval $(0,\infty)$ the symmetric minimal operator has no self-adjoint extension. If we consider the case where $a < 0$ (translation to the left), then on a finite interval the part of a function $f$ that gets translated outside $J$ on the left is re-injected (with a change of phase) at the right endpoint, thus preserving the norm of the function (unitarity); on the interval $(0,\infty)$ re-injection at the right endpoint, i.e. at $x = \infty$, is clearly not possible.

**3.2.4.** We now consider differential operators of the second order and begin with the simplest case $\mathcal{L} = -d^2/dx^2$. Formally this differential expression corresponds to the Hamiltonian of a free particle (the kinetic energy operator); as for the momentum operator one encounters problems when one tries to associate with it a self-adjoint operator in $L^2(J)$ when $J$ is a finite or semi-infinite interval. We denote the minimal operator by $K_0$: $\mathcal{D}(K_0) = C_0^\infty(J)$, $(K_0f)(x) = -f''(x)$, and we shall again take $J = (0,1)$, $J = (0,\infty)$ or $J = \mathbb{R}$. The minus sign in $\mathcal{L}$ is clearly irrelevant for the question of self-adjointness, it has been inserted so as to work with the usual expression of quantum mechanics.

The determination of the extensions of $K_0$ is carried out in a very similar way to that employed for the momentum operator. We shall first characterise the maximal operator (the adjoint of the minimal operator $K_0$) by proceeding in analogy with

Proposition 3.12. Then we indicate the self-adjoint extensions of $K_0$ and some of their properties.

**Proposition 3.14.** *The domain $\mathcal{D}(K_0^*)$ of the maximal operator $K_0^*$ is given as follows (modulo equivalence of functions):*
$g \in \mathcal{D}(K_0^*) \iff g$ *is continuously differentiable, its derivative $g'$ is an absolutely continuous function and $g''$ belongs to $L^2(J)$. If $g \in \mathcal{D}(K_0^*)$, then $(K_0^* g) = -g''$.*

PROOF. The proof is similar to that of Proposition 3.12. An element $g$ of $L^2(J)$ belongs to $\mathcal{D}(K_0^*)$ if and only if there exists $g^* \in L^2(J)$ such that $\langle g, -f''\rangle = \langle g^*, f\rangle$ for all $f \in C_0^\infty(J)$. Thus the second derivative (in the sense of distributions) of $g$ is equal to $-g^*$, so this second derivative is a function (which is locally of class $L^1$). One can then define a function $G$ of class $C^1$ with absolutely continuous derivative $G'$ by $G(x) = \int_0^x dy \int_0^y g^*(z)dz$. This function $G$ is bounded on each finite subinterval $\mathfrak{J}$ of $J$ and thus belongs to $L^2(\mathfrak{J})$. For $f \in C_0^\infty(\mathfrak{J})$ one obtains after integration by parts (observe that the boundary terms are zero): $\langle g^*, f\rangle = \langle G'', f\rangle = \langle G, f''\rangle$. Hence $g + G$ is orthogonal to the set of functions $\{f'' \mid f \in C_0^\infty(\mathfrak{J})\}$. We shall show below that this implies that $(g + G)(x) = c_1 + c_2 x$ for certain constants $c_1$ and $c_2$ and for $x \in \mathfrak{J}$, so that $g(x) = -G(x) + c_1 + c_2 x$ a.e. in $\mathfrak{J}$. Since $\mathfrak{J}$ is arbitrary, this shows that the elements of $\mathcal{D}(K_0^*)$ are functions of class $C^1$ with absolutely continuous derivatives, and a function on $J$ with these properties belongs to $\mathcal{D}(K_0^*)$ if its second derivative belongs to $L^2(J)$. It is clear that one will have (using the above notations): $K_0^* g = g^* = G'' = -g''$.

To see that $g + G$ is a linear function on each finite subinterval $\mathfrak{J} = (c, d)$ of $J$, we denote by $g_0$ the restriction of $g + G$ to $\mathfrak{J}$, more precisely: $g_0(x) = g(x) + G(x)$ if $x \in \mathfrak{J}$ and $g_0(x) = 0$ otherwise. $g_0$ belongs to $L^2(\mathfrak{J})$, and we know that $\langle g_0, f''\rangle = 0$ for all $f \in C_0^\infty(\mathfrak{J})$. Let us define $\mathcal{D}_2$ by

$$\mathcal{D}_2 = \{h \in C_0^\infty(\mathfrak{J}) \Bigm| \int_c^d h(x)dx = 0 \text{ and } \int_c^d xh(x)dx = 0\}.$$

For $\varphi \in \mathcal{D}_2$ and $x \in (c, d)$ we set $\phi(x) = \int_c^x dy \int_c^y \varphi(z)dz$. Clearly $\phi'' = \varphi$, and $\phi$ belongs to $C_0^\infty(\mathfrak{J})$: (i) $\phi(x) = 0$ in some neighbourhood of $x = c$ because $\varphi$ has this property, (ii) upon integrating by parts (with respect to $y$) one obtains

$$\phi(x) \equiv \int_c^x dy \int_c^y \varphi(z)dz = x\int_c^x \varphi(z)dz - \int_c^x y\varphi(y)dy,$$

and for $x$ so close to $d$ that $\varphi(y) = 0$ for $y > x$, each term on the right-hand side is zero because $\varphi \in \mathcal{D}_2$. Thus we have $\langle g_0, \varphi\rangle \equiv \langle g_0, \phi''\rangle = 0$. So $\langle g_0, h\rangle = 0$ for all $h \in \mathcal{D}_2$. We now fix two functions $e_1, e_2 \in C_0^\infty(\mathfrak{J})$ such that (see the example given at the end of the proof)

$$\int_c^d e_1(x)dx = 1\,, \quad \int_c^d xe_1(x)dx = 0\,, \quad \int_c^d e_2(x)dx = 0 \text{ and } \int_c^d xe_2(x)dx = 1.$$

Observe that, for any $f \in C_0^\infty(\mathfrak{J})$, $f - [\int_c^d f(x)dx]e_1 - [\int_c^d xf(x)dx]e_2$ belongs to $\mathcal{D}_2$. Hence

$$\langle g_0, f\rangle - \Big[\int_c^d f(x)dx\Big]\langle g_0, e_1\rangle - \Big[\int_c^d xf(x)dx\Big]\langle g_0, e_2\rangle = 0 \qquad \forall f \in C_0^\infty(\mathfrak{J}).$$

This equation can be rewritten as follows:

$$\int_c^d \big[\overline{g_0(x)} - \langle g_0, e_1\rangle - \langle g_0, e_2\rangle x\big] f(x)dx = 0 \qquad \forall f \in C_0^\infty(\mathfrak{J}).$$

Since $C_0^\infty(\mathfrak{J})$ is dense in $L^2(\mathfrak{J})$, one must have $g_0(x) = \langle e_1, g_0\rangle + \langle e_2, g_0\rangle x$ for almost all $x \in \mathfrak{J}$.

We finally indicate how to obtain $e_1$ and $e_2$ if $0 < c < d$. Choose a function $\theta$ as indicated in the figure, i.e. such that $\theta > 0$ on $(c, \kappa)$, $\theta < 0$ on $(\kappa, d)$ for some $\kappa \in (c, d)$ and $\int_c^d \theta(x)dx = 0$. Clearly $\alpha := \int_c^d x\theta(x)dx < 0$ and $\beta := \int_c^d x^{-1}\theta(x)dx > 0$. One can then take $e_2(x) = \alpha^{-1}\theta(x)$ and $e_1(x) = \beta^{-1}x^{-1}\theta(x)$. □

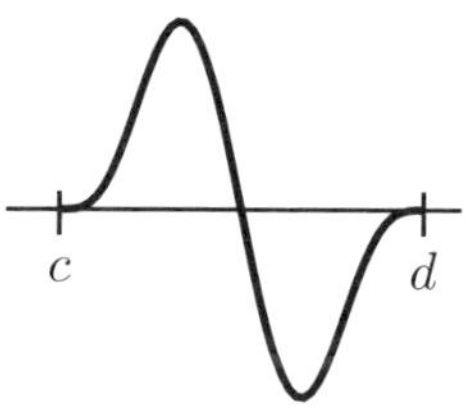


To characterise the self-adjoint extensions of $K_0$, the following remark is useful: If $g \in \mathcal{D}(K_0^*)$ then, in addition to $g''$, its first derivative $g'$ also belongs to $L^2(J)$. This fact will be discussed in relation with Proposition 3.16. By taking into account Proposition 3.11, one then finds that, if $g \in \mathcal{D}(K_0^*)$, then the boundary values of $g$ and of $g'$ at the endpoints of $J$ exist, and that these boundary values are zero at $-\infty$ and at $+\infty$ if $J$ extends to $-\infty$ or to $+\infty$.

Let us now determine the deficiency indices of the operator $K_0$. The functions in $\mathcal{N}(K_0^* \pm i)$ satisfy the differential equation $-f'' \pm if = 0$, i.e. $f'' = \pm if$. Two linearly independent solutions are as follows:

for $f'' = if$: $\quad f_1(x) = \exp[(1+i)x/\sqrt{2}]$, $f_2(x) = \exp[-(1+i)x/\sqrt{2}]$

for $f'' = -if$: $\quad f_3(x) = \exp[(1-i)x/\sqrt{2}]$, $f_4(x) = \exp[-(1-i)x/\sqrt{2}]$.

Each of these four functions is square-integrable on $(0, 1)$, and none of them belongs to $L^2(\mathbb{R})$. For $J = (0, \infty)$ one has $f_2, f_4 \in L^2((0, \infty))$ and $f_1, f_3 \notin L^2((0, \infty))$. So the deficiency indices of the minimal operator are as follows:

(1) $J = \mathbb{R}$: $\nu_+ = \nu_- = 0$, i.e. $K_0$ is essentially self-adjoint on $C_0^\infty(\mathbb{R})$.

(2) $J = (0, \infty)$: $\nu_+ = \nu_- = 1$. $K_0$ has a one-parameter family of self-adjoint extensions.

(3) $J = (0, 1)$: $\nu_+ = \nu_- = 2$. $K_0$ has a four-parameter family of self-adjoint extensions.

In case (1) the domain of the unique self-adjoint extension of $K_0$, i.e. the domain of the closure $\overline{K_0}$ of $K_0$, is equal to $\mathcal{D}(K_0^*)$ given by Proposition 3.14 with $J = \mathbb{R}$. In the other two cases one obtains the domains of different self-adjoint extensions by restricting the functions in $\mathcal{D}(K_0^*)$ to suitable linear submanifolds. The following identity is easily verified by twice integrating by parts: if $f, g \in \mathcal{D}(K_0^*)$ and $J = (a, b)$, then

$$\begin{aligned}\langle K_0^* f, g\rangle - \langle f, K_0^* g\rangle &\equiv -\langle f'', g\rangle + \langle f, g''\rangle \\ &= \overline{f(b)}g'(b) - \overline{f'(b)}g(b) - \overline{f(a)}g'(a) + \overline{f'(a)}g(a). \quad (3.24)\end{aligned}$$

To obtain a self-adjoint extension of $K_0$, one must choose a linear submanifold of $\mathcal{D}(K_0^*)$ in such a way that the right-hand side of (3.24) is zero for $f$ and $g$ in this submanifold; this expression is given in terms of the values of $f, f', g$ and $g'$ at the endpoints $a$ and $b$ of $J$, so the choice of a suitable submanifold is reduced to imposing *boundary conditions* at $x = a$ and at $x = b$. We have mentioned that, if $a = -\infty$, then each function in $\mathcal{D}(K_0^*)$ vanishes at $x = a$; more precisely $f(x) \to 0$ as $x \to -\infty$. Similarly, if $b = +\infty$ and $f \in \mathcal{D}(K_0^*)$, then $f(x) \to 0$ as $x \to +\infty$. This explains the fact that, if $J = \mathbb{R}$, then the right-hand side of (3.24) is always zero, i.e. that $K_0^*$ is self-adjoint. If $J = (0, \infty)$, then the right-hand side of (3.24) is $-\overline{f(0)}g'(0) + \overline{f'(0)}g(0)$, and this expression is zero if one imposes the boundary condition $h'(0) = \beta h(0)$ for some $\beta \in \mathbb{R}$ [i.e. the right-hand side of (3.24) is zero if $f$ and $g$ satisfy this boundary condition at $x = 0$]. The case $\beta = \infty$ is also possible [rewriting the preceding boundary condition in the form $h(0) = \beta^{-1}h'(0)$]. Upon identifying $\beta$ with $\cot\alpha$ we obtain:

<u>If $J = (0, \infty)$</u>, then, for each $\alpha \in [0, \pi)$, the operator $K(\alpha)$ defined by $K(\alpha)g = -g''$, <u>with domain</u>

$$\begin{aligned}\mathcal{D}(K(\alpha)) = \{g \in L^2((0,\infty)) \mid\ & g \text{ is continuously differentiable, } g' \text{ is absolutely} \\ & \text{continuous, } g'' \in L^2((0,\infty)) \text{ and } g'(0)\sin\alpha = g(0)\cos\alpha\}\end{aligned}$$

is a self-adjoint extension of $K_0$, and $K_0$ has no other self-adjoint extensions. The self-adjoint extension for $\alpha = 0$, characterised by the boundary condition $g(0) = 0$, is the most important one in quantum mechanics (see §3.3.3).

<u>If $J = (0, 1)$</u>, one must impose two boundary conditions, involving the values of the functions in $\mathcal{D}(K_0^*)$ at $x = 0$ and at $x = 1$. One can impose conditions analogous to those discussed above at each of the points $x = 0$ and $x = 1$ (one then speaks of ***separated boundary conditions***), or one can impose ***mixed boundary conditions*** involving at the same time the values of these functions at $x = 0$ and at $x = 1$. This leads to the following self-adjoint extensions of $K_0$:

<u>Wtih separated boundary conditions:</u> $K(\beta, \gamma)g = -g''$ with domain

$$\begin{aligned}\mathcal{D}(K(\beta,\gamma)) = \{g \in L^2((0,1)) \mid\ & g \text{ is continuously differentiable, } g' \text{ is absolutely continuous, } \\ & g'' \in L^2((0,1)),\ g'(0)\sin\beta = g(0)\cos\beta \text{ and } g'(1)\sin\gamma = g(1)\cos\gamma\}.\end{aligned}$$

Here $\beta, \gamma \in [0, \pi)$. One speaks of ***Dirichlet conditions*** at $x = 0$ if $\beta = 0$ [this is the condition $g(0) = 0$] and of ***Neumann conditions*** at $x = 0$ if $\beta = \pi/2$ [this is the condition $g'(0) = 0$]. The same terminology applies at $x = 1$. If $\beta = \gamma = 0$ one speaks of the ***Dirichlet Hamiltonian***.

<u>With mixed boundary conditions:</u> $K^{(\rho)}g = -g''$ with domain

$$\begin{aligned}\mathcal{D}(K^{(\rho)}) = \{g \in L^2((0,1)) \mid\ & g \text{ is continuously differentiable, } g' \text{ is absolutely} \\ & \text{continuous, } g'' \in L^2((0,1)),\ g(1) = e^{i\rho}g(0) \text{ and } g'(1) = e^{i\rho}g'(0)\},\end{aligned}$$

where $\rho \in [0, 2\pi)$. Of special interest is the case of ***periodic boundary conditions*** obtained by taking $\rho = 0$ : $g(1) = g(0)$ and $g'(1) = g'(0)$.

The general self-adjoint extension of $K_0$ in $L^2((0,1))$ can be characterised as follows.[5] To each vector $f \in \mathcal{D}(K_0^*)$ we associate two elements $\mathfrak{f}$ and $\mathfrak{f}'$ of $\mathbb{C}^2$ by setting $\mathfrak{f} = \{f(1), f(0)\}$ and $\mathfrak{f}' = \{f'(1), -f'(0)\}$. Then the right-hand side of (3.24), for the case $J = (0,1)$, is just

$$\langle \mathfrak{f}, \mathfrak{g}' \rangle_{\mathbb{C}^2} - \langle \mathfrak{f}', \mathfrak{g} \rangle_{\mathbb{C}^2}. \tag{3.25}$$

Each self-adjoint extension of $K_0$ is characterised by the following restriction on the boundary values $\mathfrak{f}$ and $\mathfrak{f}'$ of the elements $f \in \mathcal{D}(K_0^*)$:

$$(\cos T)\mathfrak{f} - (\sin T)\mathfrak{f}' = 0, \tag{3.26}$$

where $T$ is a self-adjoint operator in $\mathbb{C}^2$ (given by a Hermitian $2 \times 2$ matrix; the matrices $\cos T$ and $\sin T$ are defined by the usual power series for these functions). In other terms, if the self-adjoint extension of $K_0$ associated with such an operator $T$ is denoted by $K_T$, then $\mathcal{D}(K_T) = \{f \in \mathcal{D}(K_0^*) \mid \text{the couple } (\mathfrak{f}, \mathfrak{f}') \text{ satisfies (3.26)}\}$. It is easy to see that the expression (3.25) does vanish if $f, g \in \mathcal{D}(K_T)$. Indeed, using the relation $I_{\mathbb{C}^2} = (\cos T)^2 + (\sin T)^2$ along with the self-adjointness of $\cos T$ and of $\sin T$ and with the fact that these operators commute, one finds that (all scalar products being in $\mathbb{C}^2$)

$$\begin{aligned}
\langle \mathfrak{f}, \mathfrak{g}' \rangle &= \langle (\cos T)\mathfrak{f}, (\cos T)\mathfrak{g}' \rangle + \langle (\sin T)\mathfrak{f}, (\sin T)\mathfrak{g}' \rangle \\
&= \langle (\sin T)\mathfrak{f}', (\cos T)\mathfrak{g}' \rangle + \langle (\sin T)\mathfrak{f}, (\cos T)\mathfrak{g} \rangle \\
&= \langle (\cos T)\mathfrak{f}', (\sin T)\mathfrak{g}' \rangle + \langle (\cos T)\mathfrak{f}, (\sin T)\mathfrak{g} \rangle.
\end{aligned}$$

Similarly one obtains

$$\langle \mathfrak{f}', \mathfrak{g} \rangle = \langle (\cos T)\mathfrak{f}', (\sin T)\mathfrak{g}' \rangle + \langle (\cos T)\mathfrak{f}, (\sin T)\mathfrak{g} \rangle.$$

The operator $K(\beta, \gamma)$ defined above by separated boundary conditions corresponds to the operator $T$ given by the matrix $\begin{pmatrix} \gamma & 0 \\ 0 & -\beta \end{pmatrix}$.

**Remark 3.15.** In order to arrive at an interpretation of a boundary condition of the type $g'(a)\sin\beta = g(a)\cos\beta$, let us consider a time-dependent wave function in $L^2((0,1))$ and write its development in the basis $\{exp(ik_n x)\}_{n\in\mathbb{Z}}$, where $k_n = 2\pi n$:

$$f(x,t) = \sum_{n\in\mathbb{Z}} a_n e^{i(k_n x - k_n^2 t)}.$$

Let us assume that the time evolution of $f$ is governed by the operator $K(\beta, \gamma)$. Then, if the derivative $f_{,t} \equiv \partial f/\partial t$ of $f$ with respect to $t$ is square-integrable, $f_{,t}$ will belong to the domain of $K(\beta, \gamma)$ for each $t$ (see Proposition 5.1). Taking $t = 0$, we have

$$\frac{\partial}{\partial t} f(x,t)\big|_{t=0} \equiv f_{,t}(x,t)\big|_{t=0} = -i \sum_{n\in\mathbb{Z}\setminus\{0\}} k_n^2 a_n e^{ik_n x},$$

[5] We refer to Appendix A.1 and Appendix A.2 of [RK] or to §125 of [AG] for a detailed derivation.

and the boundary condition $\partial f_{,t}(x,t)/\partial x\,|_{t=0} = (\cot\beta) f_{,t}(x,t)$ at $x = 0$ reads

$$\sum_{n\in\mathbb{Z}\setminus\{0\}} (ik_n)k_n^2 a_n = \cot\beta \sum_{n\in\mathbb{Z}\setminus\{0\}} k_n^2 a_n. \tag{3.27}$$

A single mode ($a_n = 0$ except for one value of $n$) cannot satisfy this equation. However, since $k_{-n} = -k_n$, (3.27) can be satisfied if $a_n$ and $a_{-n}$ are related as follows:

$$a_n = \frac{ik_n + \cot\beta}{ik_n - \cot\beta}\, a_{-n}$$

[then one has $ik_n a_n + ik_{-n}a_{-n} = \cot\beta(a_n + a_{-n})$, and of course $k_{-n}^2 = k_n^2$]. So the boundary condition $g'(0)\sin\beta = g(0)\cos\beta$ selects wave functions for which the amplitude of each incoming mode at $x = 0$ ($n < 0$) differs from that of the corresponding reflected mode ($n > 0$) by a phase factor $(ik_n + \cot\beta)(ik_n - \cot\beta)^{-1}$.

**3.2.5.** The symmetric operator $K_0 = -d^2/dx^2$ is the square of $P_0 = -i(d/dx)$, both operators having the domain $C_0^\infty(J)$. Consequently there must exist certain relations between self-adjoint extensions of $P_0$ and those of $K_0$. We specify some of these relations below.

If $J = \mathbb{R}$, both $P_0$ and $K_0$ are essentially self-adjoint, and the self-adjoint closure $\overline{K_0} \equiv H_0$ is the square of the momentum operator $P \equiv \overline{P_0}$.

Let us discuss the case $J = (0,1)$. We first consider the operator $\mathbb{P}_\alpha^* \mathbb{P}_\alpha \equiv \mathbb{P}_\alpha^2$, where $\mathbb{P}_\alpha$ is the self-adjoint extension of $P_0$ defined in (3.22). A function $f$ belongs to $\mathcal{D}(\mathbb{P}_\alpha^2)$ if and only if $f \in \mathcal{D}(\mathbb{P}_\alpha)$ and $f' \in \mathcal{D}(\mathbb{P}_\alpha)$, i.e. if and only if $f$ and $f'$ are absolutely continuous, $f'$ and $f''$ belong to $L^2((0,1))$ and satisfy $f(1) = e^{i\alpha} f(0)$ and $f'(1) = e^{i\alpha} f'(0)$. Then clearly $f$ is of class $C^1$, and it is seen that $\mathcal{D}(\mathbb{P}_\alpha^2) = \mathcal{D}(K^{(\alpha)})$, where $K^{(\alpha)}$ is the operator $K^{(\rho)}$ with $\rho = \alpha$ introduced in §3.2.4. This shows that

$$\mathbb{P}_\alpha^* \mathbb{P}_\alpha \equiv \mathbb{P}_\alpha^2 = K^{(\alpha)}. \tag{3.28}$$

Next let us consider the operator $P_0^* \overline{P_0}$ in $L^2((0,1))$. As above, a function $f$ belongs to $\mathcal{D}(P_0^*\overline{P_0})$ if and only if we have $f \in \mathcal{D}(\overline{P_0})$ and $f' \in \mathcal{D}(P_0^*)$, i.e. if and only if $f$ and $f'$ are absolutely continuous, $f'$ and $f''$ belong to $L^2((0,1))$ and $f(0) = f(1) = 0$. This domain coincides with the domain of the operator $K(0,0)$ [the operator $K(\beta,\gamma)$ of page 107 with $\beta = \gamma = 0$]. Hence [6]

$$P_0^* \overline{P_0} = K(0,0). \tag{3.29}$$

It follows from Eqs. (3.28) and (3.29) that the operators $K^{(\alpha)}$ and $K(0,0)$ are positive. Other self-adjoint extensions of $K_0$ need not have this property. However, a self-adjoint extension of $K_0$ can have at most two negative eigenvalues. We refer

[6] The operator $P_0^*\overline{P_0}$ is self-adjoint. This is a particular case of a very general theorem stating that if $A$ is closed and densely defined, then the operator $A^*A$ is self-adjoint (see for example §51 in [AG] or §V.3.7 in [K]). For the self-adjointness of the square of a self-adjoint operator (e.g. of $\mathbb{P}_\alpha$), see also Problem 3.12 or Chapter 4.

to Example 4.25 for a proof of this fact and for other results on the relation between the spectra of different self-adjoint extensions of symmetric operators with finite deficiency indices.

We should point out that the familiar commutation relations between the operators $P$, $H_0$ and $Q$ in $L^2(\mathbb{R})$ have no analogue in $L^2((0,1))$. Of course the operator $\mathbb{P}_\alpha$ commutes with its square $K^{(\alpha)}$ but not with other self-adjoint extensions of $K_0$. Also none of the operators $\mathbb{P}_\alpha$ is canonically conjugate to the position operator $Q$ (multiplication by the variable $x$) in $L^2((0,1))$. Even though $[\mathbb{P}_\alpha, Q]g = [P_0, Q]g = -ig'$ if $g \in C_0^\infty((0,1))$, there are functions $f$ and $g$ in $\mathcal{D}(\mathbb{P}_\alpha)$ such that $\langle \mathbb{P}_\alpha f, Qg\rangle - \langle Qf, \mathbb{P}_\alpha g\rangle$ is different from $-i\langle f, g\rangle$ (Problem 3.11).

**3.2.6.** To end these considerations, let us show that the functions in $\mathcal{D}(K_0^*)$, and also their first derivatives, vanish at infinity if $J$ is an infinite interval. The argument is based on the following result which is a particular case of the fact that, if $f$ is a function of class $C^m$ ($m$ times continuously differentiable), then the intermediate derivatives $f^{(k)}$, $1 \le k \le m-1$, can be estimated in terms of $f$ and $f^{(m)}$.

**Proposition 3.16.** *Let $J = (a,b)$ be an interval. Let $\mathcal{E}(J)$ be the set of functions $f \colon J \to \mathbb{C}$ that are continuously differentiable and such that $f'$ is absolutely continuous. Then there exist constants $C_1 > 0$ and $C_2 > 0$ such that for all $f \in \mathcal{E}(J)$:*

$$\int_a^b |f'(x)|^2\,dx \le C_1 \int_a^b |f(x)|^2\,dx + C_2 \int_a^b |f''(x)|^2\,dx, \tag{3.30}$$

*where the value $+\infty$ is admitted for these integrals (if the function in the integrand is not square-integrable on $J$). In particular: if $f \in \mathcal{E}(J)$ is such that $f$ and $f''$ belong to $L^2(J)$, then one also has $f' \in L^2(J)$.*

We give the proof of this proposition in the Appendix to this chapter. As an application, let $K_0$ be the minimal operator associated with $-d^2/dx^2$ in $L^2(J)$ and observe that $\mathcal{D}(K_0^*) \subseteq \mathcal{E}(J)$. If $g \in \mathcal{D}(K_0^*)$, then $g$ and $g''$ belong to $L^2(J)$, hence so does $g'$ by the preceding proposition. So $g$ and $g'$ are absolutely continuous, belong to $L^2(J)$, and their derivatives also belong to $L^2(J)$; the vanishing of their boundary values at infinity then follows from Proposition 3.11(b).

**Remark 3.17.** If one equips the linear manifolds $\mathcal{D}(P_0^*)$ and $\mathcal{D}(K_0^*)$ with the graph norm of the respective operator [see (2.59)], they become Hilbert spaces. These spaces are particular cases of Sobolev spaces, denoted $\mathcal{H}^1(J)$ and $\mathcal{H}^2(J)$ respectively:

$\mathcal{H}^1(J) = \{f \colon J \to \mathbb{C} \mid f$ is absolutely continuous, $f \in L^2(J)$ and $f' \in L^2(J)\}$,
with norm $\|f\|_{\mathcal{H}^1} = [\|f\|^2 + \|f'\|^2]^{1/2}$.

$\mathcal{H}^2(J) = \{f \colon J \to \mathbb{C} \mid f$ is continuously differentiable, $f'$ is absolutely continuous, $f \in L^2(J)$ and $f'' \in L^2(J)\}$,
with norm $\|f\|_{\mathcal{H}^2} = [\|f\|^2 + \|f'\|^2 + \|f''\|^2]^{1/2}$.

More generally one considers the following ***Sobolev spaces*** $\mathcal{H}^{m,p}$ for $m = 0, 1, 2, \dots$ and $p \in [1, \infty)$:

$\mathcal{H}^{m,p}$ is the set of (equivalence classes of) functions $f\colon J \to \mathbb{C}$ such that $f$ is $m-1$ times continuously differentiable, $f^{(m-1)}$ is absolutely continuous and $f, f', \ldots, f^{(m)}$ belong to $L^p(J)$, with norm $\|f\|_{\mathcal{H}^{m,p}} = [\sum_{k=0}^{m} \int_J |f^{(k)}(x)|^p dx]^{1/p}$.
These spaces are Banach spaces (complete normed linear vector spaces), even Hilbert spaces if $p = 2$; the notation $\mathcal{H}^m(J)$ is frequently used for $\mathcal{H}^{m,2}(J)$. The theory of Sobolev spaces is developed for example in [Ad].

## 3.3 Schrödinger operators

**3.3.1.** We now consider differential operators corresponding to Hamiltonians of non-relativistic quantum mechanics, i.e. operators given formally by $\mathcal{L} = -d^2/dx^2 + V(x)$ in $L^2(J)$, where $J = (0,1)$, $(0,\infty)$ or $\mathbb{R}$ and $V$ is a *real* function assumed to be continuous on the open interval $J$ [so $V$ could diverge at the endpoints of $J$, for example at $x = 0$ and/or at $x = \infty$ if $J = (0,\infty)$]. We denote by $A_0$ the minimal operator associated with $\mathcal{L}$ : $\mathcal{D}(A_0) = C_0^\infty(J)$, $[A_0 f](x) = -f''(x) + V(x)f(x)$.

If $f \in C_0^\infty(J)$, $f$ is zero in some neighbourhood of each endpoint of $J$ and $V$ is continuous on the support of $f$. Hence $Vf$ is a bounded function which is zero outside some bounded subinterval of $J$ (outside the support of $f$ which is contained in some finite interval), in particular $Vf \in L^2(J)$. This shows that $A_0$ is well defined on $C_0^\infty(J)$. Since $V$ is assumed to be real, $A_0$ is a symmetric operator. Complex conjugation (Example 3.8) maps $C_0^\infty(J)$ into $C_0^\infty(J)$ and commutes with $A_0$; by Proposition 3.9 the deficiency indices of $A_0$ are equal: $\nu_+ = \nu_-$. We shall see that their possible values are 0, 1 and 2; thus each closed maximal symmetric extension of $A_0$ is self-adjoint. We begin by specifying the domain of the maximal operator $A_0^*$:

**Proposition 3.18.** *The domain $\mathcal{D}(A_0^*)$ of the maximal operator $A_0^*$ is as follows (modulo equivalence of functions):*
*(a) $g \in \mathcal{D}(A_0^*) \iff g$ is continuously differentiable, the derivative $g'$ is an absolutely continuous function, and the function $-g'' + Vg$ belongs to $L^2(J)$.*
*(b) If $g \in \mathcal{D}(A_0^*)$, then $g'' \in L^2(\mathfrak{J})$ for each subinterval $\mathfrak{J}$ of $J$ on which the function $V$ is bounded, in particular for each subinterval of the form $\mathfrak{J} = (a,b)$ with $a$ and $b$ in the interior of $J$ (hence different from the endpoints of $J$). For $g \in \mathcal{D}(A_0^*)$ one has $A_0^* g = -g'' + Vg$.*

PROOF. Let $g \in \mathcal{D}(A_0^*)$ and let $\mathfrak{J}$ be an open subinterval of $J$ such that $|V(x)| \leq M < \infty$ for all $x \in \mathfrak{J}$. Denote by $g_0$ the restriction of $g$ to $\mathfrak{J}$, more precisely $g_0(x) = g(x)$ if $x \in \mathfrak{J}$ and $g_0(x) = 0$ otherwise. The operator of multiplication by $V$ in $L^2(\mathfrak{J})$ is bounded, hence $Vg_0 \in L^2(\mathfrak{J})$. In particular one has for each $f \in C_0^\infty(\mathfrak{J})$:

$$\langle f, A_0^* g\rangle = \langle A_0 f, g\rangle = \langle -f'' + Vf, g\rangle = \langle -f'', g_0\rangle + \langle f, Vg_0\rangle,$$

i.e.

$$\langle -f'', g_0\rangle \equiv \langle K_0 f, g_0\rangle = \langle f, A_0^* g - Vg_0\rangle \equiv \int_{\mathfrak{J}} \overline{f(x)}\left[(A_0^* g)(x) - V(x)g(x)\right] dx. \tag{3.31}$$

Since $(A_0^* g)(x) - V(x)g(x)$ is square-integrable on $\mathfrak{J}$, (3.31) shows that $g_0$ belongs to the domain of the adjoint $K_0^*$ of the minimal operator $K_0$ associated with $-d^2/dx^2$ in $L^2(\mathfrak{J})$. By Proposition 3.14 (applied in $\mathfrak{J}$) this implies that $g$ is continuously differentiable on $\mathfrak{J}$, $g'$ is absolutely continuous on $\mathfrak{J}$, $g''$ belongs to $L^2(\mathfrak{J})$ and $(A_0^* g)(x) - V(x)g(x) = -g''(x)$ for $x \in \mathfrak{J}$. In particular the restriction to $\mathfrak{J}$ of $-g'' + Vg$ belongs to $L^2(\mathfrak{J})$.

Since these considerations apply in each subinterval $\mathfrak{J} = (a, b)$ of $J$ with $a$ and $b$ different from the endpoints of $J$, we have $(A_0^* g)(x) = -g''(x) + V(x)g(x)$ almost everywhere in $J$. Since $A_0^* g$ belongs by assumption to $L^2(J)$, we conclude that $-g'' + Vg \in L^2(J)$. □

We have seen that, if $g \in \mathcal{D}(A_0^*)$, then $-g'' + Vg$ belongs to $L^2(J)$ and both $-g''$ and $Vg$ belong to $L^2(\mathfrak{J})$ if $\mathfrak{J}$ lies in the interior of $J$ (so $-g''$ and $Vg$ are square-integrable except possibly in a neighbourhood of the endpoints of $J$). If $V$ is bounded on the entire interval $J$, then $Vg \in L^2(J)$ and consequently also $-g'' \in L^2(J)$. If $V$ is unbounded near at least one endpoint of $J$, then in general $-g''$ and $Vg$ will not belong separately to $L^2(J)$, only their sum will be in $L^2(J)$. If $V$ is bounded in a neighbourhood of one endpoint of $J$, then $Vg \in L^2(\mathfrak{J})$ for each subinterval $\mathfrak{J}$ of $J$ not containing the other endpoint. Some further local properties of the functions in the domain of $A_0^*$ are described in the next proposition; the term "local" means that one considers these functions in a subinterval of $J$ or that one localises them by multiplying them by a smooth function $\theta$ which vanishes outside some subinterval of $J$ (a result of this type is given in part (c); this result will be used later, and we refrain from presenting a general discussion of the method of localisation).

**Proposition 3.19.** *Let $\mathcal{L} = -d^2/dx^2 + V(x)$ be a formally self-adjoint differential expression on an open interval $J$. Let $A_0$ be the minimal operator associated with $\mathcal{L}$ in $L^2(J)$, hence $\mathcal{D}(A_0) = C_0^\infty(J)$, and let $A_0^*$ be the adjoint of $A_0$ in $L^2(J)$.*
*(a) Let $\mathfrak{J}$ be an open subinterval of $J$. Denote by $B_0$ the minimal operator associated with $\mathcal{L}$ in $L^2(\mathfrak{J})$, $\mathcal{D}(B_0) = C_0^\infty(\mathfrak{J})$, and denote by $B_0^*$ the adjoint of $B_0$ in $L^2(\mathfrak{J})$. If $g \in \mathcal{D}(A_0^*)$, then the restriction of $g$ to $\mathfrak{J}$ belongs to $\mathcal{D}(B_0^*)$.*
*(b) If $\mathfrak{J}$ is an open subinterval of $J$ on which the function $V$ is bounded and if $g \in \mathcal{D}(A_0^*)$, then the restriction of $g$ to $\mathfrak{J}$ belongs to $\mathcal{D}(K_0^*)$, where $K_0^*$ is the adjoint in $L^2(\mathfrak{J})$ of the minimal operator $K_0$ associated with $-d^2/dx^2$ in $L^2(\mathfrak{J})$ [hence $\mathcal{D}(K_0) = C_0^\infty(\mathfrak{J})$ and $K_0 f = -f''$].*
*(c) Let $\theta\colon J \to \mathbb{C}$ be a function of class $C^2$ such that $\theta(x) = 1$ in some neighbourhood of one of the endpoints of $J$, and $\theta(x) = 0$ in some neighbourhood of the other endpoint. If $g \in \mathcal{D}(A_0^*)$, then $\theta g$ also belongs to $\mathcal{D}(A_0^*)$.*

PROOF. (a) is an immediate consequence of Proposition 3.18(a) (applied in $J$ and in $\mathfrak{J}$). (b) has been obtained in the proof of Proposition 3.18.

(c) Suppose for example that $\theta(x) = 0$ in some neighbourhood of $b$ and $\theta(x) = 1$ near $a$. One has $(\theta g)' = \theta g' + \theta' g$ and $(\theta g)'' = \theta'' g + 2\theta' g' + \theta g''$. Since $\theta$ is of class $C^2$ and $g$ belongs to $\mathcal{D}(A_0^*)$, one sees that $\theta g$ is continuously differentiable and that $(\theta g)'$ is absolutely continuous. Also $-(\theta g)'' + V\theta g = \theta[-g'' + Vg] - \theta'' g - 2\theta' g'$. Clearly $\theta[-g'' + Vg]$ and $\theta'' g$ belong to $L^2(J)$ (because $g \in \mathcal{D}(A_0^*)$, see Proposition

3.18(a), and because $\theta$ and $\theta''$ are bounded). $\theta'$ is bounded, $\theta'(x) = 0$ near $a$ and near $b$, hence $\theta'$ is non-zero only on a finite subinterval in the interior of $J$. Since $g'$ is absolutely continuous, it is bounded on the support of $\theta'$, hence $\theta' g'$ is a bounded function with support in a finite interval, so that $\theta' g' \in L^2(J)$. Summing up, one has $-(\theta g)'' + V\theta g \in L^2(J)$, hence $\theta g \in \mathcal{D}(A_0^*)$ by Proposition 3.18(a). □

For $f, g \in \mathcal{D}(A_0^*)$ and $x \in J$ we set

$$[f,g]_x = \overline{f'(x)}g(x) - \overline{f(x)}g'(x). \tag{3.32}$$

By Proposition 3.18(a) this expression is well defined for each $x \in J$, and $[f,g]_x$ is continuous in $x$. So if $J = (a,b)$ with $-\infty \leq a < b \leq +\infty$, then $[f,g]_x$ is a continuous function on $(a,b)$. Let us show that $[f,g]_a := \lim_{x \searrow a}[f,g]_x$ and $[f,g]_b := \lim_{x \nearrow b}[f,g]_x$ exist. For this observe that, if $a < \lambda < \mu < b$, then (using Proposition 3.18(b) and the hypothesis that $V$ is real, and integrating by parts):

$$\int_\lambda^\mu \left[\overline{f(x)}(A_0^* g)(x) - \overline{(A_0^* f)(x)}g(x)\right] dx = \int_\lambda^\mu \left[-\overline{f(x)}g''(x) + \overline{f''(x)}g(x)\right] dx$$
$$= \left[-\overline{f(x)}g'(x) + \overline{f'(x)}g(x)\right]\Big|_\lambda^\mu = [f,g]_\mu - [f,g]_\lambda.$$

The integrand of the first integral belongs to $L^1((a,b))$ [scalar product between functions of $L^2((a,b))$]. In fact, as $\lambda \to a$ and $\mu \to b$, the left-hand side converges to $\langle f, A_0^* g\rangle - \langle A_0^* f, g\rangle$ [scalar product in $L^2(J)$!]. Thus the right-hand side must have limits as $\lambda \to a$ and/or $\mu \to b$, and one has

$$\langle f, A_0^* g\rangle - \langle A_0^* f, g\rangle = [f,g]_b - [f,g]_a \qquad \forall f, g \in \mathcal{D}(A_0^*). \tag{3.33}$$

We observe that (3.24) is a special case of this identity.

**Remark 3.20.** Below we shall use the following fact concerning solutions of differential equations (see e.g. Section 2.1 of [Br] or page 150 of Volume II of [RS]): Let $(\lambda, \mu)$ be an interval and $W\colon (\lambda,\mu) \to \mathbb{C}$ a continuous function. Denote by $\mathfrak{X}$ the set of solutions $h$ of the differential equation $-h''(x) + W(x)h(x) = 0$, $x \in (\lambda, \mu)$. Then $\mathfrak{X}$ is a vector space of dimension 2 (each solution is a linear combination of two fixed linearly independent solutions) and the functions $h \in \mathfrak{X}$ are of class $C^2$ [twice continuously differentiable] on $(\lambda, \mu)$.

We return to the situation considered in this section: $J = (a,b)$ is a (finite or infinite) interval and $V\colon J \to \mathbb{R}$ a continuous function which could diverge at $a$ and/or at $b$. Let $c \in (a,b)$ be an interior point of $J$, and consider the subinterval $(a,c]$ [or similarly $[c,b)$] of $J$. $V$ is continuous on $(a,c)$ and bounded on $(a',c]$ for each $a' \in (a,c]$. Let $\mathfrak{X}$ be the two-dimensional vector space formed by the solutions of the differential equation $-h''(x) + V(x)h(x) = ih(x)$ on $(a,c)$ [take $W(x) = V(x) - i$ in Remark 3.20]. Each element of $\mathfrak{X}$ is continuous on $(a,c)$ and has a finite limit as $x \to c$ [in fact $h$ is of class $C^2$ on $(a,c]$ and even has a continuation of class $C^2$ to $(a, c+\delta)$ for sufficiently small $\delta > 0$]. Thus the elements of $\mathfrak{X}$ are square-integrable on $(a',c)$ for each $a' \in (a, c)$. As $a' \to a$, the norm in $L^2((a', c))$ of an element $h$ of $\mathfrak{X}$ will either remain bounded or diverge; in the first case $h$ belongs to $L^2((a, c))$, in the second case one has $h \notin L^2((a, c))$. It is a very important fact that there exists at least one function $h \not\equiv 0$ in $\mathfrak{X}$ that is square-integrable on $(a, c)$:

**Proposition 3.21 (Weyl's Alternative).** *Let $(a, b)$ be an interval, $V\colon (a, b) \to \mathbb{R}$ continuous and $c \in (a, b)$. Then either all solutions of $-h''(x) + V(x)h(x) = ih(x)$ belong to $L^2((a, c))$ or there exists exactly one (up to a numerical factor) solution $h \neq \mathbf{0}$ in $L^2((a, c))$.*

Following H. Weyl, one says that $V$ is in the ***limit circle case*** at $a$ if all solutions of $-h''(x) + V(x)h(x) = ih(x)$ are in $L^2((a, c))$ and that $V$ is in the ***limit point case*** at $a$ if only one (up to a numerical factor) solution of this differential equation belongs to $L^2((a, c))$. One applies the same terminology at the other endpoint $b$, replacing $L^2((a, c))$ by $L^2((c, b))$ [Weyl's alternative also holds in $L^2((c, b))$, the proof is completely analogous]. We also observe that any solution $h$ will have the property that $h''$ is a continuous function on $(a, b)$ because we have assumed $V$ to be continuous on this interval. A comment on the origin of the limit point-limit circle terminology can be found in Problem 3.10.

PROOF OF PROPOSITION 3.21. It suffices to show that there exists at least one $h \in L^2((a, c))$ (not identically zero) satisfying $-h''(x) + V(x)h(x) = ih(x)$. For this we denote by $B_0$ the minimal operator associated with the differential expression $\mathcal{L} = -d^2/dx^2 + V(x)$ in $L^2((a, c))$ and by $B_0^*$ its adjoint in $L^2((a, c))$. We first show that $B_0^*$ is not symmetric. For this we choose two functions $f, g\colon (a, c] \to \mathbb{C}$ of class $C^2$ that vanish in some neighbourhood of $x = a$ and satisfy $f(c) = g'(c) = 0$ and $f'(c) = g(c) = 1$. Clearly $f$ and $g$ belong to $\mathcal{D}(B_0^*)$ [see Proposition 3.18(a)]. Hence (3.33), applied in $L^2((a, c))$, gives

$$\langle f, B_0^* g\rangle - \langle B_0^* f, g\rangle = [f, g]_c - [f, g]_a = 1,$$

since $[f, g]_a = 0$. This shows that $B_0^*$ is not symmetric.

Since $B_0^*$ is not symmetric, $B_0$ is not essentially self-adjoint (Proposition 2.23). Hence, by Proposition 3.3, either $\mathcal{R}(B_0 + i)$ or $\mathcal{R}(B_0 - i)$ is not dense in $\mathcal{H}$, or equivalently either $\mathcal{N}(B_0^* - i) \neq \{\mathbf{0}\}$ or $\mathcal{N}(B_0^* + i) \neq \{\mathbf{0}\}$. If $\mathcal{N}(B_0^* - i) \neq \{\mathbf{0}\}$, then any $h \not\equiv 0$ in $\mathcal{N}(B_0^* - i)$ satisfies $-h''(x) + V(x)h(x) = ih(x)$. If $\mathcal{N}(B_0^* + i) \neq \{\mathbf{0}\}$, let $g \neq \mathbf{0}$ be any element of $\mathcal{N}(B_0^* + i)$; then $g$ satisfies $-g''(x) + V(x)g(x) = -ig(x)$, hence its complex conjugate $h \equiv \overline{g}$ is a non-zero solution of $-h''(x) + V(x)h(x) = ih(x)$. □

**Proposition 3.22.** *Let $A_0$ be the minimal operator associated with $\mathcal{L} = -d^2/dx^2 + V(x)$ on an interval $J = (a, b)$, with $V\colon J \to \mathbb{R}$ continuous. Denote the pair of deficiency indices of $A_0$ by $(\nu_+, \nu_-)$. Then*

*(a) The deficiency indices of $A_0$ are $(2, 2)$ if and only if $V$ is in the limit circle case at $a$ and at $b$.*

*(b) The deficiency indices of $A_0$ are $(1, 1)$ if and only if $V$ is in the limit circle case at one endpoint of $J$ (at $a$ or at $b$) and in the limit point case at the other endpoint.*

*(c) $A_0$ is essentially self-adjoint if and only if $V$ is in the limit point case at $a$ and at $b$.*

PROOF. We use the fact that $A_0$ has equal deficiency indices. In (i) we prove the "if" part for all three situations and in (ii) the "only if" part.

(i) If $V$ is in the limit circle case at $a$ and at $b$, then each solution of $A_0^*h = ih$ belongs to $L^2(J)$. Thus $\mathcal{N}(A_0^* - i)$ is two-dimensional, i.e. the deficiency indices of $A_0$ are $\nu_+ = \nu_- = 2$. If we assume for example that $V$ is in the limit point case at $a$ and in the limit circle case at $b$, then there exists exactly one (up to a numerical factor) non-zero solution of $A_0^*h = ih$ satisfying $h \in L^2((a,c))$ if $c < b$. Since $V$ is in the limit circle case at $b$, this solution belongs to $L^2((c,b))$, hence also to $L^2((a,b))$. Thus $\dim \mathcal{N}(A_0^* - i) = 1$, i.e. $\nu_+ = \nu_- = 1$. Finally assume that $V$ is in the limit point case at both endpoints of $J$. This case is rather subtle. At each endpoint of $J$ there exists exactly one (up to a numerical factor) solution $h \not\equiv 0$ of $A_0^*h = ih$ that is square-integrable near the respective endpoint. If one denotes by $h_a$ the solution which is square-integrable near $a$ [so $h_a \not\equiv 0$ and $h_a \in L^2((a,c))$ for each $c \in (a,b)$] and if one continues this solution to all of $J$, it could a priori coincide with a multiple of the solution which is square-integrable in a neighbourhood of $b$. Part (c) of the proposition asserts that this is never the case (under the assumption that $V$ is continuous on $J$). To see this one uses the fact (proved in the following lemma) that $[f,g]_a = [f,g]_b = 0$ for all $f,g \in \mathcal{D}(A_0^*)$ if $V$ is in the limit point case at $a$ and at $b$. Eq. (3.33) then implies that the maximal operator $A_0^*$ is symmetric, and the essential self-adjointness of $A_0$ follows from Proposition 2.23.

(ii) If $A_0$ has deficiency indices $(2,2)$, then each solution of $-h''(x)+V(x)h(x) = ih(x)$ belongs to $L^2((a,b))$, hence to $L^2((a,c))$ as well as to $L^2((c,b))$ for any $c \in (a,b)$. Thus $V$ is in the limit circle case at both endpoints of $J$, and (a) is completely proved. It follows that if the deficiency indices of $A_0$ are $(1,1)$ or $(0,0)$, then $V$ must be in the limit point case at least at one endpoint of $J$, say at $a$. If the deficiency indices are $(0,0)$ then we must have the limit point case also at $b$ [otherwise, by the "if" part of (b), $A_0$ would have deficiency indices $(1,1)$, i.e. a contradiction]. Similarly, if the deficiency indices are $(1,1)$, one must have the limit circle case at $b$ [otherwise, by the "if" part of (c), $A_0$ would have deficiency indices $(0,0)$, i.e. a contradiction]. □

**Lemma 3.23.** *Let $J = (a,b)$. Let $\mathcal{L} = -d^2/dx^2 + V(x)$ be a formally self-adjoint differential expression on $J$, and let $A_0$ be the minimal operator associated with $\mathcal{L}$ in $L^2(J)$.*
*(a) Let $z \in \mathbb{C}$ and let $f,g$ be two solutions of $\mathcal{L}\varphi = z\varphi$, and let $W[f,g] = f(x)g'(x) - f'(x)g(x)$ be their Wronskian at $x \in J$. Then $W[f,g]$ is a constant [i.e. independent of $x \in (a,b)$]. Furthermore $W[f,g] \neq 0$ if and only if $f$ and $g$ are linearly independent.*
*(b) If $f \in \mathcal{D}(\overline{A_0})$ and $g \in \mathcal{D}(A_0^*)$, then $[f,g]_a = [f,g]_b = 0$.*
*(c) If $V$ is in the limit point case at $a$ (or at $b$), then $[f,g]_a = 0$ (or $[f,g]_b = 0$) for all $f,g \in \mathcal{D}(A_0^*)$.*

PROOF. (a) We first verify that $f(x)g'(x) - f'(x)g(x)$ does not depend on $x$:

$$\frac{d}{dx}\big[f(x)g'(x) - f'(x)g(x)\big] = f(x)g''(x) - f''(x)g(x)$$
$$= f(x)\big[(V(x)-z)g(x)\big] - \big[(V(x)-z)f(x)\big]g(x) = 0.$$

Next assume that $W[f,g]=0$ and for example that $g$ is not identically zero. Choose a number $c\in(a,b)$ such that $g(c)\neq 0$ and set $\rho=f(c)/g(c)$ so that $f(c)=\rho g(c)$. Since $W[f,g]=f(c)g'(c)-f'(c)g(c)=0$ one also has $f'(c)=\rho g'(c)$. Since the solution $\varphi$ of an ordinary differential equation of second order is uniquely determined by prescribing the values of $\varphi$ and $\varphi'$ at one point, and since $\rho g$ is a solution of $\mathcal{L}\varphi=z\varphi$ with initial conditions $\varphi(c)=f(c)=\rho g(c)$, $\varphi'(c)=f'(c)=\rho g'(c)$, one must have $f(x)=\rho g(x)$ on $(a,b)$. Thus, if $W[f,g]=0$, then $f$ and $g$ are linearly dependent. The converse is easy to obtain.

(b) Let $\theta\colon J\to\mathbb{R}$ be of class $C^2$ and such that $\theta(x)=1$ in some neighbourhood of $a$ and $\theta(x)=0$ near $b$, and set $g_0=\theta g$. Then $g_0\in\mathcal{D}(A_0^*)$ by Proposition 3.19(c). Since $g_0$ vanishes in some neighbourhood of $b$, we have $[f,g_0]_b=0$. On the other hand $g_0(x)=g(x)$ near $a$, hence $[f,g]_a=[f,g_0]_a$. Thus, by using also the hypothesis that $f\in\mathcal{D}(\overline{A_0})$, the identity (3.33) (applied to the couple $\{f,g_0\}$) becomes $[f,g]_a=[f,g_0]_a=\langle A_0^*f,g_0\rangle-\langle f,A_0^*g\rangle=\langle\overline{A_0}f,g_0\rangle-\langle f,A_0^*g\rangle=0$, because $A_0^*$ is the adjoint of $\overline{A_0}$ (Proposition 2.20). Hence $[f,g]_a=0$. By interchanging the roles of $a$ and $b$ in this argument one finds that $[f,g]_b=0$.

(c) Let us treat the situation in which $V$ is in the limit point case at $a$. Fix a number $c\in(a,b)$, set $\mathfrak{J}=(a,c)$ and let $B_0$ be the minimal operator associated with the differential expression $\mathcal{L}=-d^2/dx^2+V(x)$ in $L^2(\mathfrak{J})$, as in Proposition 3.19(a). The deficiency indices of $B_0$ are $(1,1)$; so, by the First von Neumann Formula (3.11), the domain $\mathcal{D}(B_0^*)$ of the adjoint of $B_0$ is the linear manifold spanned by $\mathcal{D}(\overline{B_0})$ and two linearly independent functions belonging to $\mathcal{D}(B_0^*)\backslash\mathcal{D}(\overline{B_0})$ [in (3.11) these functions are the solutions of $B_0^*g=\pm ig$, but clearly one obtains the same linear manifold if one chooses any two linearly independent functions in $\mathcal{D}(B_0^*)\backslash\mathcal{D}(\overline{B_0})$].

Let $\varphi_1,\varphi_2$ be linearly independent solutions of $\mathcal{L}\varphi=i\varphi$ on $(a,c)$. $\varphi_1$ and $\varphi_2$ are of class $C^2$, in particular square-integrable, on each interval of the form $(a',c)$ with $a'>a$. Let $\theta$ be a function of class $C^\infty$ on $(a,c)$ such that $\theta(x)=0$ in some neighbourhood of $a$ and $\theta(x)=1$ near $c$, and set $h_1=\theta\varphi_1$ and $h_2=\theta\varphi_2$. Then $h_1$ and $h_2$ are linearly independent and belong to $\mathcal{D}(B_0^*)$ [use Proposition 3.18(a) in $L^2((a,c))$ by observing that $h_k''\in L^2((a,c))$ and that $V$ is bounded on the support of $\theta$]. We show below that $h_1$ and $h_2$ do not belong to $\mathcal{D}(\overline{B_0})$, which implies that $\mathcal{D}(B_0^*)$ is the linear manifold spanned by $\mathcal{D}(\overline{B_0})$ and the two functions $h_1$ and $h_2$. Thus, if $f\in\mathcal{D}(B_0^*)$, there exist a vector $h\in\mathcal{D}(\overline{B_0})$ and constants $\alpha$, $\beta$ such that $f=h+\alpha h_1+\beta h_2$, in particular $f(x)=h(x)$ in some neighbourhood of $a$. If $g$ is another element of $\mathcal{D}(B_0^*)$, one will have $[f,g]_a=[h,g]_a$, and the last expression is 0 by the result of (b). Hence $[f,g]_a=0$ for all $f,g\in\mathcal{D}(B_0^*)$. Since the restriction to $(a,c)$ of a function in $\mathcal{D}(A_0^*)$ belongs to $\mathcal{D}(B_0^*)$ by Proposition 3.19(a), we have $[f,g]_a=0$ for all $f,g\in\mathcal{D}(A_0^*)$.

It remains to show that $h_1,h_2\notin\mathcal{D}(\overline{B_0})$. Let $\varepsilon>0$ be such that $\theta(x)=1$ for all $x\in(c-\varepsilon,c)$. Then $\mathcal{L}h_1=ih_1$ and $\mathcal{L}h_2=ih_2$ on $(c-\varepsilon,c)$. Assume for example that $h_2\in\mathcal{D}(\overline{B_0})$, and denote by $\overline{h_1}$ the complex conjugate of the function $h_1$. It is clear that $\overline{h_1}\in\mathcal{D}(B_0^*)$, hence one must have $[\overline{h_1},h_2]_c=0$ by the result of (b). But $[\overline{h_1},h_2]_c=h_1'(c)h_2(c)-h_1(c)h_2'(c)$, hence $W[h_1,h_2]=0$. This contradicts the fact that $W[h_1,h_2]\neq 0$ since $h_1$ and $h_2$ are linearly independent [cf. (a)]. □

We briefly indicate how one obtains the self-adjoint extensions of $A_0$ in the situation in which this operator is not essentially self-adjoint. Recall Eq. (3.33):

$$\langle f, A_0^* g\rangle - \langle A_0^* f, g\rangle = [f,g]_b - [f,g]_a \qquad \forall f,g \in \mathcal{D}(A_0^*). \tag{3.34}$$

As in the case of the operator $-d^2/dx^2$ discussed in §3.2.4, one has to restrict $A_0^*$ to a linear submanifold of its domain, by imposing boundary conditions at $x = a$ and/or at $x = b$, so that the right-hand side of (3.34) is zero.

We first consider the case in which the deficiency indices of $A_0$ are $(1,1)$. Assume for example that $V$ is in the limit circle case at $a$ and in the limit point case at $b$. Then $[f,g]_b = 0$ for all $f, g \in \mathcal{D}(A_0^*)$, hence it suffices to impose a boundary condition at $x = a$. In §3.2.4 the number $a$ was finite, $g(a)$ and $g'(a)$ were well defined for each $g \in \mathcal{D}(K_0^*)$, and one imposed a condition of the type $g'(a) = \alpha g(a)$. Here it is more convenient to formulate a condition of the form $[f_0, g]_a = 0$, where $f_0$ is a *fixed* function belonging to $\mathcal{D}(A_0^*)$ but not to $\mathcal{D}(\overline{A_0})$, because $[f,g]_a$ is well defined for any $f, g \in \mathcal{D}(A_0^*)$ [we recall that, by the Second von Neumann Formula (3.14), the domain of a self-adjoint extension of $A_0$ is determined by exactly one function $f_0$ of the above type if the deficiency indices of $A_0$ are $(1,1)$, because $\dim \mathcal{N}_- = 1$ in this case].[7]

If the deficiency indices of $A_0$ are $(2,2)$, one can impose boundary conditions of the preceding type at $a$ and at $b$, i.e. $[f_0, g]_a = 0$ and $[f_1, g]_b = 0$ with $f_0$, $f_1$ two functions in $\mathcal{D}(A_0^*)\backslash\mathcal{D}(\overline{A_0})$, or one can impose mixed boundary conditions. We omit the details and refer to Proposition 8.29 of [W1] for a precise formulation. The following result is interesting: If the deficiency indices of $A_0$ are $(2,2)$, then the spectrum of each self-adjoint extension of $A_0$ consists of only eigenvalues (no continuous spectrum !); each eigenvalue is of multiplicity 1 (the subspace spanned by the eigenvectors associated with each fixed eigenvalue is one-dimensional), and the eigenvalues form a discrete set, i.e. their only possible accumulation points are $\pm\infty$.

**3.3.2.** It remains to see under which conditions a potential $V$ is in the limit point case or in the limit circle case at the endpoints of an interval $J$. The answer depends on the behaviour of $V$ near these endpoints, and one must distinguish between the case of a finite endpoint and that of an infinite one. To simplify the notations we consider the situation of an endpoint at $x = 0$ and that of an endpoint at $x = \infty$ (the other situations are entirely analogous). We give for each of these two situations a simple criterion permitting one to decide, for a class of relatively well behaved potentials, whether they are in the limit point case or in the limit circle case. A considerable number of finer criteria are presented in Chapter XIII of [DS]. We begin with an auxiliary result which is useful in that it leads to a simplification when determining the deficiency indices, in the sense that it will suffice to study the equation $A_0^* f = \mathbf{0}$ rather than $A_0^* f = if$.

**Lemma 3.24.** *Let $(a,b)$ be an interval, $V\colon (a,b) \to \mathbb{R}$ continuous and $c \in (a,b)$. Assume that there exists a number $z_0 \in \mathbb{C}$ such that all solutions of the differential*

---

[7] For example the condition $g'(0) = \alpha g(0)$ used in §3.2.4 is equivalent to $[f_0, g]_0 = 0$ for $f_0(x) = \exp[(1+i)x/\sqrt{2}] + \kappa \exp[(1-i)x/\sqrt{2}]$ with $\kappa = (1-i+i\sqrt{2}\alpha)/(1+i-i\sqrt{2}\alpha)$, or for $f_0(x) = \overline{\alpha}x + 1$.

*equation* $-h''(x) + V(x)h(x) = z_0 h(x)$ *belong to* $L^2((c,b))$. *Then for each* $z \in \mathbb{C}$, *all solutions of* $-h''(x)+V(x)h(x) = zh(x)$ *belong to* $L^2((c,b))$ [*and similarly with* $L^2((c,b))$ *replaced by* $L^2((a,c))$].

The proof of this lemma is given in the Appendix to this chapter. We point out the especially useful particular case mentioned before the lemma:

**Corollary 3.25.** *(a)* $V$ *is in the limit circle case at* $x = a$ *if and only if all solutions of* $-h''(x) + V(x)h(x) = 0$ *belong to* $L^2((a,c))$ [*with* $c \in (a,b)$].
*(b)* $V$ *is in the limit point case at* $x = a$ *if and only if there exists a solution of* $-h''(x) + V(x)h(x) = 0$ *that does not belong to* $L^2((a,c))$.

We first study the deficiency indices at infinity. The next result shows that in practically all cases of physical interest one has the limit point case at infinity, hence that it is not necessary to impose boundary conditions at infinity to define a self-adjoint Hamiltonian.

**Proposition 3.26.** *If* $V(x) \geq -\kappa x^2$ *in some neighbourhood of* $x = \infty$, *where* $\kappa$ *is a real constant, then* $V$ *is in the limit point case at* $x = \infty$. *Similarly, if* $V(x) \geq -\kappa x^2$ *in some neighbourhood of* $x = -\infty$, *then* $V$ *is in the limit point case at* $x = -\infty$.

PROOF. Without loss of generality we can assume that $\kappa > 0$. Let $c > 0$ be such that $V(x) \geq -\kappa x^2$ for $x \geq c$.

(i) Let us first prove the following result: If $g$ is a real solution of $\mathcal{L}g = 0$ with $g \in L^2((c,\infty))$, then $x^{-1}g'$ belongs to $L^2((c,\infty))$. For this, let us denote by $\|g\|$ the norm of $g$ in $L^2((c,\infty))$. Since $g'' = Vg$, we have for each $y > c$:

$$\int_c^y \frac{g''(x)g(x)}{x^2}\,dx = \int_c^y \frac{V(x)g(x)^2}{x^2}\,dx \geq -\kappa \int_c^y g(x)^2\,dx \geq -\kappa\|g\|^2. \tag{3.35}$$

On the other hand, upon integrating by parts, one obtains the following expression for the left-hand side of this inequality:

$$\int_c^y \frac{g''(x)g(x)}{x^2}\,dx = \frac{g'(y)g(y)}{y^2} - \frac{g'(c)g(c)}{c^2} - \int_c^y \frac{g'(x)^2}{x^2}\,dx + 2\int_c^y \frac{g'(x)g(x)}{x^3}\,dx. \tag{3.36}$$

We set $\Psi(y) = \left[\int_c^y x^{-2}g'(x)^2\,dx\right]^{1/2}$. By the Schwarz inequality one has (by using $1/x^2 \leq 1/c^2$):

$$\Big|\int_c^y \frac{g'(x)g(x)}{x^3}\,dx\Big| \leq \frac{1}{c^2}\|g\|\Psi(y).$$

Together with (3.36) and (3.35) this implies that

$$\frac{g'(y)g(y)}{y^2} \geq \frac{g'(c)g(c)}{c^2} + \Psi(y)^2 - \frac{2}{c^2}\|g\|\Psi(y) - \kappa\|g\|^2. \tag{3.37}$$

Suppose that $x^{-1}g' \notin L^2((c,\infty))$. Then $\Psi(y) \to \infty$ as $y \to \infty$, so that there exists $y_0 \in (c,\infty)$ such that the right-hand side of (3.37) is positive for $y \geq y_0$. Then

$g'(y)g(y) > 0$ for $y \geq y_0$, i.e. $g'$ and $g$ have the same sign on $(y_0, \infty)$. This implies that $|g(y)| > |g(y_0)|$ for $y > y_0$, which contradicts the hypothesis that $g \in L^2((c, \infty))$. Therefore one must have $x^{-1}g' \in L^2((c, \infty))$.

(ii) Let us now consider two linearly independent solutions $f_1, f_2$ of $\mathcal{L}f = 0$. We may assume $f_1$ and $f_2$ to be real (the differential equation $\mathcal{L}f = 0$ has two linearly independent real solutions). Let $W[f_1, f_2]$ be their Wronskian [see Lemma 3.23(a)]. One has $W[f_1, f_2] \neq 0$ and $f_1(x)f_2'(x) - f_1'(x)f_2(x) = W[f_1, f_2]$ for each $x \in (c, \infty)$. Observe that $W[f_1, f_2]x^{-1} \notin L^1((c, \infty))$ because $\int_c^\infty x^{-1}dx = \infty$. Now

$$W[f_1, f_2]x^{-1} = f_1(x)\left[x^{-1}f_2'(x)\right] - \left[x^{-1}f_1'(x)\right]f_2(x).$$

If both $f_1$ and $f_2$ belong to $L^2((c, \infty))$, then the right-hand side of the preceding equation belongs to $L^1((c, \infty))$ [by the result of (i) and the Schwarz inequality], a contradiction. Thus at least one of these two functions is not square-integrable at infinity, i.e. $V$ is in the limit point case at $x = \infty$. □

**Example 3.27.** Consider the Hamiltonian of a harmonic oscillator which corresponds to the differential expression $\mathcal{L} = -d^2/dx^2 + \gamma x^2$ on $J = \mathbb{R}$, with $\gamma > 0$. The potential $V(x) = \gamma x^2$ is in the limit point case at $x = -\infty$ and at $x = +\infty$ (for any value of $\gamma \in \mathbb{R}$). So the Hamiltonian of a harmonic oscillator (in one dimension) is essentially self-adjoint on $C_0^\infty(\mathbb{R})$!

**Remark 3.28.** The result of the above proposition shows that a potential $V$ is in the limit point case at infinity if it is bounded below near infinity (in particular if it converges to zero at infinity) and even if it tends to $-\infty$ but not too rapidly. One can show that $V$ is in the limit circle case at $x = +\infty$ if $V(x) \to -\infty$ and $\int_c^\infty |V(x)|^{-1/2}\,dx < \infty$ for some $c < \infty$, for example if $V(x) = -\gamma x^\alpha$ with $\gamma > 0$ and $\alpha > 2$.[8] Observe that in classical mechanics a potential of the last type [$V(x) = -\gamma x^\alpha$ with $\gamma > 0$ and $\alpha > 2$] corresponds to a repulsive force increasing in strength as $x$ approaches $\pm\infty$; a classical particle submitted to such a force ($\alpha > 2$) would reach $x = \infty$ or $x = -\infty$ after a finite time. For not too bizarre potentials there is a correspondence between the classical behaviour and Weyl's Alternative: $V$ is in the limit point case at infinity if and only if a classical particle, with arbitrary initial condition, would require an infinite time to arrive at infinity, and $V$ is in the limit circle case at infinity if and only if there exist initial conditions under which a classical particle reaches infinity after a finite time.[9]

Let us now discuss the deficiency indices at $x = 0$. We consider the case $J = (0, a)$ for some $a \in (0, \infty]$. The next proposition concerns potentials that are positive in a neighbourhood of $x = 0$. This result can be applied to purely repulsive potentials, i.e. satisfying $V'(x) < 0$ for $x \in (0, b)$ for some $b > 0$; the proposition shows that strongly repulsive potentials are in the limit point case. On the other hand potentials that are only moderately repulsive (for example Coulomb potentials) are in the limit circle case (Proposition 3.30). The case of attractive potentials [for example $V'(x) > 0$ for $x \in (0, b)$, $V(x) \to -\infty$ as $x \to 0$] is considered in Proposition 3.30(b).

---

[8] For a proof see [DS], Theorem XIII.6.20.

[9] For a more detailed discussion see [RS], Volume II, Appendix to Section X.1.

**Proposition 3.29.** *Assume that* $V(x) \geq \frac{3}{4}x^{-2}$ *in a neighbourhood of* $x = 0$ [*i.e. for each* $x \in (0, b)$, *for some* $b > 0$]. *Then* $V$ *is in the limit point case at* $x = 0$.

PROOF. Consider first the differential equation $-f''(x) + (3/4)x^{-2}f(x) = 0$. Two linearly independent solutions are $f_1(x) = x^{-1/2}$ and $f_2(x) = x^{3/2}$. Let us fix a number $\gamma > 0$ and denote by $f_0$ the unique solution satisfying the initial conditions $f_0(b) = 0$, $f_0'(b) = -\gamma$. $f_0$ is a linear combination of $f_1$ and $f_2$, given by $f_0(x) = (\gamma/2)(b^{3/2}x^{-1/2} - b^{-1/2}x^{3/2})$. Observe that $f_0 \notin L^2((0, b))$. Furthermore $f_0(x) > 0$ and $f_0' < 0$ on the interval $(0, b)$.

To prove the proposition it suffices to exhibit a solution $g$ of $-g''(x)+V(x)g(x) = 0$ satisfying $g(x) \geq f_0(x)$ on $(0, b)$ [then $g$ will not belong to $L^2((0, b))$]. We fix $g$ by the initial conditions $g(b) = 0$, $g'(b) = -2\gamma$. Then $g$ will be real and we have

$$g(x) - f_0(x) = -\int_x^b bg'(z)dz + \int_x^b f_0'(z)dz = \int_x^b [f_0'(z) - g'(z)]dz, \qquad (3.38)$$

so that it suffices to know that $f_0'(x) > g'(x)$ for all $x \in (0, b)$. This inequality is satisfied in a neighbourhood of $x = b$ as a consequence of the initial conditions at $x = b$ imposed on the functions $f_0$ and $g$. So let $y$ be the largest number in $(0, b)$ such that $f_0'(y) = g'(y)$ (assuming that such a number exists). Then $g(x) > f_0(x)$ on $(y, b]$ by (3.38). By using also the hypothesis that $V(x) \geq \frac{3}{4}x^{-2}$, we obtain:

$$\begin{aligned} f_0'(y) &= f_0'(b) - \int_y^b f_0''(x)dx = -\gamma - \int_y^b \frac{3}{4}x^{-2}f_0(x)dx \\ &> -\gamma - \int_y^b \frac{3}{4}x^{-2}g(x)dx \geq -\gamma - \int_y^b V(x)g(x)dx \\ &= -\gamma - \int_y^b g''(x)dx = -\gamma - g'(b) + g'(y) = \gamma + g'(y) > g'(y), \end{aligned}$$

in contradiction with the hypothesis that $f_0'(y) = g'(y)$. So there is no number $y$ having the stated properties, hence $f_0'(x) > g'(x)$ on $(0, b)$. □

**Proposition 3.30.** *(a) Assume that* $|x^2V(x)| \leq \beta < 3/4$ *in some neighbourhood of* $x = 0$ [*i.e. for each* $x \in (0, b)$, *for some* $b > 0$]. *Then* $V$ *is in the limit circle case at* $x = 0$.
*(b) If* $V$ *is monotone increasing in an interval* $(0, b)$ *with* $b > 0$ [*i.e.* $V(y) > V(x)$ *if* $0 < x < y < b$], *then* $V$ *is in the limit circle case at* $x = 0$.

PROOF. We give the proof of (a) and refer to Theorem XIII.6.24 of [DS] for that of (b). The arguments for proving (a) are similar to those of the preceding proof.

(i) One considers first the differential equation $-f''(x) + \beta x^{-2}f(x) = 0$ (we can assume that $\beta \neq 0$). Making the ansatz $f(x) = x^\rho$ one obtains the following pair of linearly independent solutions: $f_1(x) = x^{\rho_1}$ and $f_2(x) = x^{\rho_2}$, where $\rho_1$ and $\rho_2$ are given by $\rho_1 = [1 + (1 + 4\beta)^{1/2}]/2$ and $\rho_2 = [1 - (1 + 4\beta)^{1/2}]/2$. One has $\rho_1 > 1$ and $\rho_2 > -1/2$, because $\beta < 3/4$. Thus $f_1$ and $f_2$, and consequently all solutions $f$

of $-f''(x) + \beta x^{-2} f(x) = 0$, are square-integrable on $(0, b)$.

(ii) Now consider a solution $g$ of $-g''(x) + V(x)g(x) = 0$. Let $f_0$ be the solution of $-f''(x) + \beta x^{-2} f(x) = 0$ satisfying the initial conditions $f_0(b) = |g(b)| + 1$, $f_0'(b) = -|g'(b)| - 1$. The function $f_0$ is real, and $f_0 \in L^2((0, b))$ as seen in (i). Clearly there is a neighbourhood of $x = b$ in which $f_0(x) > |g(x)|$ and $f_0'(x) < -|g'(x)|$. Let $y$ be the largest number in $(0, b)$ for which one of the preceding two inequalities is violated, i.e. the largest number in $(0, b)$ such that either $|g(y)| = f_0(y)$ or $f_0'(y) = -|g'(y)|$. We shall show that such a $y$ cannot exist. Hence one will have $|g(x)| < f_0(x)$ for all $x \in (0, b)$, which shows that $g \in L^2((0, b))$. Since $g$ is an arbitrary solution of $-g''(x) + V(x)g(x) = 0$, $V$ is in the limit circle case at $x = 0$.

($\alpha$) Suppose that $y \in (0, b)$, $f_0(y) = |g(y)|$ and $f_0'(x) \leq -|g'(x)|$ for $x \in [y, b]$. Then

$$\begin{aligned} f_0(y) &= f_0(b) - \int_y^b f_0'(x)\,dx = |g(b)| + 1 - \int_y^b f_0'(x)\,dx \\ &> |g(b)| + 1 + \int_y^b |g'(x)|\,dx \geq |g(b)| + 1 + \left| \int_y^b g'(x)\,dx \right| \\ &= |g(b)| + 1 + |g(b) - g(y)| \geq |g(b)| + 1 + \big[|g(y)| - |g(b)|\big] > |g(y)|, \end{aligned}$$

where the second inequality from the end is obtained by applying (1.15) in the Hilbert space $\mathbb{C}$. We have obtained a contradiction with the hypothesis that $f_0(y) = |g(y)|$.

($\beta$) Suppose that $y \in (0, b)$, $f_0'(y) = -|g'(y)|$ and $f_0(x) \geq |g(x)|$ for $x \in [y, b]$. Then

$$\begin{aligned} f_0'(y) &= f_0'(b) - \int_y^b f_0''(x)\,dx = -|g'(b)| - 1 - \int_y^b f_0''(x)\,dx \\ &= -|g'(b)| - 1 - \int_y^b \beta x^{-2} f_0(x)\,dx \leq -|g'(b)| - 1 - \int_y^b |V(x)||g(x)|\,dx \\ &= -|g'(b)| - 1 - \int_y^b |g''(x)|\,dx \leq -|g'(b)| - 1 - \left| \int_y^b g''(x)\,dx \right| \\ &= -|g'(b)| - 1 - \big|g'(b) - g'(y)\big| < -|g'(y)|, \end{aligned}$$

where the last inequality is obtained as in ($\alpha$). We have obtained a contradiction with the hypothesis that $f_0'(y) = -|g'(y)|$. □

**3.3.3.** The above considerations have implications for Schrödinger operators with spherically symmetric potentials in $n$ dimensions ($n \geq 2$). We discuss the most important case $n = 3$. The Hilbert space is $\mathcal{H} = L^2(\mathbb{R}^3)$. We first recall the ***partial wave decomposition*** of $L^2(\mathbb{R}^3)$. The wave functions $f(\vec{x})$ in $L^2(\mathbb{R}^3)$ can be considered, in spherical polar co-ordinates, as functions of $r$, $\theta$ and $\phi$, where

$$r = |\vec{x}|, \quad \cos\theta = \frac{x_3}{r} \quad \text{and} \quad \tan\phi = \frac{x_2}{x_1},$$

so

$$x_1 = r\sin\theta\cos\phi, \quad x_2 = r\sin\theta\sin\phi \quad \text{and} \quad x_3 = r\cos\theta.$$

Let us write $f(r, \theta, \phi)$ for the wave functions in spherical polar co-ordinates. One has

$$\|f\|^2 = \int_0^\infty r^2 dr \int_0^\pi \sin\theta \, d\theta \int_0^{2\pi} d\phi \, |f(r, \theta, \phi)|^2. \tag{3.39}$$

So, for almost all $r$, $f(r, \theta, \phi)$ must be square-integrable over the angular variables $\theta$ and $\phi$ (otherwise the integral with respect to $dr$ would be infinite, i.e. one would have $\|f\| = \infty$). Thus, for almost all $r$, $f(r, \theta, \phi)$ belongs to the Hilbert space $L^2(S^2)$, the set of (equivalence classes of) functions $g\colon S^2 \to \mathbb{C}$ that are square-integrable with respect to the measure $\sin\theta \, d\theta \, d\phi$. Here $S^2 = \{\vec{x} \in \mathbb{R}^3 \mid |\vec{x}| = 1\}$ denotes the unit sphere in $\mathbb{R}^3$, and the scalar product in $L^2(S^2)$ is given by

$$\langle g, h \rangle_{L^2(S^2)} = \int_0^\pi \sin\theta \, d\theta \int_0^{2\pi} d\phi \, \overline{g(\theta, \phi)} h(\theta, \phi). \tag{3.40}$$

We see that $L^2(\mathbb{R}^3)$ can be identified with $L^2((0, \infty); L^2(S^2), r^2 dr)$, the Hilbert space of square-integrable functions defined on $(0, \infty)$ with values in $L^2(S^2)$, with measure $r^2 dr$ on $(0, \infty)$. [10]

An orthonormal basis of $L^2(S^2)$ is given by the surface spherical harmonics $Y_\ell^m$, where $\ell$ assumes the values $0, 1, 2, 3, \ldots$ and $m = \ell, \ell - 1, \ell - 2, \ldots, -\ell$. [11] If $f$ is a wave function, one can develop $f(r, \theta, \phi)$, for each fixed $r$, in this basis $\{Y_\ell^m\}$:

$$f(r, \theta, \phi) = \sum_{\ell=0}^{\infty} \sum_{m=-\ell}^{\ell} f_{\ell m}(r) Y_\ell^m(\theta, \phi) \tag{3.41}$$

[more precisely, if the function $f$ is given, this development makes sense for almost all $r > 0$, namely for those values of $r$ for which $f(r, \theta, \phi)$ belongs to $L^2(S^2)$ as a function of $\theta$ and $\phi$]. The coefficient of $Y_\ell^m$ depends on $r$; for this reason it has been denoted $f_{\ell m}(r)$ (for fixed $\ell$ and $m$ it is a function of the radial variable $r$). Since the functions $\{Y_\ell^m\}$ form an orthonormal basis of $L^2(S^2)$, one has for fixed $r$:

$$\|f(r, \theta, \phi)\|^2_{L^2(S^2)} \equiv \int_0^\pi \sin\theta \, d\theta \int_0^{2\pi} d\phi |f(r, \theta, \phi)|^2 = \sum_{\ell=0}^{\infty} \sum_{m=-\ell}^{\ell} |f_{\ell m}(r)|^2, \tag{3.42}$$

and consequently

$$\|f\|^2 = \sum_{\ell=0}^{\infty} \sum_{m=-\ell}^{\ell} \int_0^\infty r^2 dr |f_{\ell m}(r)|^2. \tag{3.43}$$

The wave functions of the form $f(r, \theta, \phi) = f_{\ell m}(r) Y_\ell^m(\theta, \phi)$, for fixed $\ell$ and $m$, form a subspace of $L^2(\mathbb{R}^3)$ which we denote by $\mathcal{H}_{\ell m}$. One has $\mathcal{H}_{\ell m} \perp \mathcal{H}_{\ell' m'}$ if $\ell \neq \ell'$ or $m \neq m'$. We set $\mathcal{H}_\ell = \bigoplus_{m=-\ell}^{\ell} \mathcal{H}_{\ell m}$. Thus the wave functions in $\mathcal{H}_\ell$ are of the form

[10] This is a special case of the situation considered in relation with (1.53): $\mathcal{O} = (0, \infty)$, $\mathcal{K} = L^2(S^2)$ and $m(dr) = r^2 dr$.

[11] See for example Section 2.1 of [Ne].

$f(r,\theta,\phi) = \sum_{m=-\ell}^{\ell} f_{\ell m}(r) Y_\ell^m(\theta,\phi)$, with $f_{\ell m} \in L^2((0,\infty), r^2 dr)$ [so each $f_{\ell m}$ is a complex-valued function defined on $(0,\infty)$ which is square-integrable with respect to the measure $r^2 dr$]. The wave functions in $\mathcal{H}_\ell$ are called wave functions in the ***partial wave of angular momentum $\ell$***, $\mathcal{H}_\ell$ is the associated ***partial wave subspace***. Since $L^2(\mathbb{R}^3) = \bigoplus_{\ell=0}^{\infty} \mathcal{H}_\ell$, each wave function $g$ admits a unique decomposition $g = \sum_{\ell=0}^{\infty} g_\ell$, with $g_\ell \in \mathcal{H}_\ell$ (this represents the partial wave decomposition of $g$).

Let us now consider the free Hamiltonian $H_0 = \vec{P}^2$ introduced in §2.5.3. We use the fact that each $\mathcal{H}_{\ell m}$ (hence also each $\mathcal{H}_\ell$) reduces the operator $H_0$, i.e. that the restriction of $H_0$ to $\mathcal{H}_{\ell m}$ defines a self-adjoint operator $H_{0,\ell m}$ in $\mathcal{H}_{\ell m}$, the domain of $H_{0,\ell m}$ being $\mathcal{H}_{\ell m} \cap \mathcal{D}(H_0)$ (some details on this are given in the Appendix to this chapter). $H_{0,\ell m}$ is an ordinary differential operator. To see this we recall that, on smooth functions, $H_0$ acts as $-\Delta$. In spherical polar co-ordinates the Laplacian has the following form:

$$\Delta = \frac{\partial^2}{\partial r^2} + \frac{2}{r}\frac{\partial}{\partial r} + \frac{1}{r^2}\Delta_S, \tag{3.44}$$

where $\Delta_S$ is the spherical Laplacian (acting on functions of $\theta$ and $\phi$):

$$\Delta_S = \frac{\partial^2}{\partial \theta^2} + \cot\theta \frac{\partial}{\partial \theta} + \frac{1}{\sin^2\theta}\frac{\partial^2}{\partial \phi^2}. \tag{3.45}$$

The spherical harmonics are the eigenfunctions of $\Delta_S$, more precisely: $\Delta_S Y_\ell^m(\theta,\phi) = -\ell(\ell+1) Y_\ell^m(\theta,\phi)$. Thus, if $f \in \mathcal{H}_{\ell m}$ is twice differentiable, then

$$\begin{aligned} -\Delta f(r,\theta,\phi) &= -\Delta\big[f_{\ell m}(r) Y_\ell^m(\theta,\phi)\big] \\ &= \Big[-\big(\frac{d^2}{dr^2} + \frac{2}{r}\frac{d}{dr}\big) f_{\ell m}(r) + \frac{\ell(\ell+1)}{r^2} f_{\ell m}(r)\Big] Y_\ell^m(\theta,\phi), \end{aligned}$$

i.e. $H_{0,\ell m}$ is a self-adjoint extension in $L^2((0,\infty), r^2 dr)$ of the ordinary differential operator

$$-\big(\frac{d^2}{dr^2} + \frac{2}{r}\frac{d}{dr}\big) + \frac{\ell(\ell+1)}{r^2}.$$

In particular $H_{0,\ell m}$ is independent of $m$, it depends only on the value of $\ell$. For this reason we simply write $H_{0,\ell}$ for these operators. In quantum mechanics $H_{0,\ell}$ is called the ***free radial Hamiltonian of angular momentum $\ell$***.

In order to be able to apply the results obtained earlier in this section, it is useful to map $L^2((0,\infty), r^2 dr)$ in a unitary way onto $L^2((0,\infty), dr) \equiv L^2(\mathbb{R}_+)$. The unitary mapping $U$ is simple: if $f \in L^2((0,\infty), r^2 dr)$, then $Uf$ defined by $(Uf)(r) = rf(r)$ belongs to $L^2(\mathbb{R}_+)$ and

$$\|Uf\|^2_{L^2(\mathbb{R}_+)} = \int_0^\infty |f(r)|^2 r^2 dr = \|f\|^2_{L^2((0,\infty), r^2 dr)}.$$

Also, if $g \in L^2(\mathbb{R}_+)$ is twice differentiable and $f \equiv U^{-1} g$ is the associated element of $L^2((0,\infty), r^2 dr)$, i.e. $f(r) = r^{-1} g(r)$, then a short calculation (using the Leibniz rule) shows that

$$-\big(\frac{d^2}{dr^2} + \frac{2}{r}\frac{d}{dr}\big) f(r) = -\big(\frac{d^2}{dr^2} + \frac{2}{r}\frac{d}{dr}\big) r^{-1} g(r) = -r^{-1} g''(r).$$

This implies that

$$[UH_{0,\ell}f](r) = r[-r^{-1}g''(r) + \ell(\ell+1)r^{-2}f(r)] = -g''(r) + \ell(\ell+1)r^{-2}g(r).$$

This shows that the operator $H_{0,\ell}$ is unitarily equivalent to the differential operator $-d^2/dr^2 + \ell(\ell+1)r^{-2}$ in $L^2((0,\infty))$. More precisely:

*$UH_{0,\ell}U^{-1}$ is a self-adjoint extension of the minimal operator associated with the differential expression $-d^2/dr^2 + \ell(\ell+1)r^{-2}$ in $L^2((0,\infty))$.*

If $\ell \neq 0$ this minimal operator is essentially self-adjoint [the deficiency indices are $(0,0)$ by Propositions 3.26 and 3.29], so that in this case $UH_{0,\ell}U^{-1}$ is the closure of the minimal operator. If $\ell = 0$, however, the minimal operator has deficiency indices $(1,1)$ because the potential $V \equiv 0$ is in the limit circle case at $r = 0$. Hence $UH_{0,0}U^{-1}$ coincides with one of the self-adjoint extensions of the minimal operator associated with $-d^2/dr^2$ in $L^2((0,\infty))$, i.e. with one of the operators $K(\alpha)$ described on page 107. To know the value of $\alpha$, we recall from (2.83) that the wave functions in $\mathcal{D}(H_0)$ are bounded on $\mathbb{R}^3$. Since the spherical harmonic $Y_0^0$ is a constant (i.e. independent of $\theta$ and $\phi$), the elements in $\mathcal{H}_{0,0}$ are just functions $f(r)$, and if $f \in \mathcal{D}(H_{0,0})$, then $f(r)$ must be bounded and continuous, in particular $f(0) < \infty$. Hence the function $g = Uf$ which is associated with $f$ in $L^2(\mathbb{R}_+)$, i.e. $g(r) = rf(r)$, must satisfy $g(0) = 0$. By comparing with the definition of $K(\alpha)$ given on page 107, one sees that $UH_{0,0}U^{-1}$ is the operator $K(\alpha)$ for $\alpha = 0$.

Note that the non-self-adjointness in the partial wave of angular momentum $\ell = 0$ appears in a somewhat artificial way through the introduction of spherical polar co-ordinates and the separation of variables (the point $\vec{x} = 0$ is singular for spherical polar co-ordinates). The operator $-\Delta$ is essentially self-adjoint on $C_0^\infty(\mathbb{R}^3)$ but not on $C_0^\infty(\mathbb{R}^3 \setminus \{0\})$ [$C_0^\infty(\mathbb{R}^3 \setminus \{0\})$ is the set of functions in $C_0^\infty(\mathbb{R}^3)$ that vanish near $\vec{x} = 0$], and the fact of considering the minimal operators $UH_{0,\ell}U^{-1}$ on $C_0^\infty((0,\infty))$ amounts in $L^2(\mathbb{R}^3)$ to restricting $-\Delta$ to $C_0^\infty(\mathbb{R}^3 \setminus \{0\})$ [one avoids the singular point $\vec{x} = 0$ of the co-ordinate system].

Let us next consider Hamiltonians including a potential $V$, and assume that $V$ is spherically symmetric (and real, of course). In other words let us consider differential operators of the type $-\Delta + V(r)$ in three dimensions, assuming that $V$ is a real function that is continuous on $(0,\infty)$. If $V$ is square-integrable in a neighbourhood of the origin (as a function defined on $\mathbb{R}^3$), i.e. if $\int_0^1 |V(r)|^2 r^2 dr < \infty$, then $-\Delta + V(r)$ is defined on $C_0^\infty(\mathbb{R}^3)$. On the other hand, if $V(r)$ is too singular at $r = 0$, then the preceding integral will not be finite and one must consider the operator $-\Delta + V(r)$ on $C_0^\infty(\mathbb{R}^3 \setminus \{0\})$. In this situation one is naturally led to avoid the singular point $\vec{x} = 0$ of the spherical co-ordinate system (this point then coincides with the singular point of the potential).

The component in $\mathcal{H}_{\ell m}$ of a function $f$ belonging to $C_0^\infty(\mathbb{R}^3 \setminus \{0\})$ will have the form $f_{\ell m}(r) Y_\ell^m(\theta,\phi)$ with $f_{\ell m} \in C_0^\infty((0,\infty))$, and the function $g_{\ell m} \equiv Uf_{\ell m}$ corresponding in $L^2(\mathbb{R}_+)$ to $f_{\ell m}$ will also belong to $C_0^\infty((0,\infty))$. The deficiency indices of the operator $-\Delta + V(r)$ on $C_0^\infty(\mathbb{R}^3 \setminus \{0\})$ are equal to the sum of the deficiency indices of the minimal operators associated with $-d^2/dr^2 + \ell(\ell+1)r^{-2} + V(r)$ in

$L^2(\mathbb{R}_+)$, each of the partial deficiency indices being multiplied by $2\ell+1$ (because for fixed $\ell$, the parameter $m$ assumes $2\ell+1$ values). We give below a somewhat more precise description of this situation but leave the verification as an exercise. We denote by $\nu_\ell$ the value of the deficiency indices of $-d^2/dr^2+\ell(\ell+1)r^{-2}+V(r)$ in $L^2((0,\infty))$ (remember that the two deficiency indices are identical for each $\ell$).

Let us denote by $A$ the symmetric operator $[-\Delta+V]\big|_{C_0^\infty(\mathbb{R}^3\setminus\{0\})}$, i.e. the differential operator $-\Delta+V(r)$ in the Hilbert space $L^2(\mathbb{R}^3)$ with domain $C_0^\infty(\mathbb{R}^3\setminus\{0\})$. The deficiency indices of $A$ are equal (Proposition 3.9 applies with $K$ the complex conjugation). In the decomposition $\mathcal{H}=\bigoplus_{\ell m}\mathcal{H}_{\ell m}$ of the Hilbert space, the operator $A$ has the form $A=\bigoplus_{\ell m}A_{\ell m}$, where $A_{\ell m}$ is the following differential operator in the space $L^2((0,\infty),r^2dr)$:

$$A_{\ell m}=-\frac{d^2}{dr^2}+\frac{2}{r}\frac{d}{dr}+\frac{\ell(\ell+1)}{r^2}+V(r),\qquad \mathcal{D}(A_{\ell m})=C_0^\infty((0,\infty)).$$

Let $B_\ell=UA_{\ell m}U^{-1}$ be the differential operator associated with $A_{\ell m}$ in $L^2(\mathbb{R}_+)$, also with domain $\mathcal{D}(B_\ell)=C_0^\infty((0,\infty))$ [$U\colon L^2((0,\infty),r^2dr)\to L^2(\mathbb{R}_+)$ is given as above by $(Uf)(r)=rf(r)$, and $UA_{\ell m}U^{-1}$ is independent of $m$]. One has

$$\mathcal{N}(A^*_{\ell m}-i)=U^{-1}\mathcal{N}(B^*_\ell-i)\equiv U^{-1}[\mathcal{R}(B_\ell+i)^\perp].$$

Thus the deficiency indices of $A_{\ell m}$ are equal to those of $B_\ell$, i.e. to $\nu_\ell$ (for each $m\in\{\ell,\ell-1,\dots,-\ell\}$). Moreover

$$\mathcal{N}(A^*-i)=\bigoplus_{\ell,m}\mathcal{N}(A^*_{\ell m}-i)$$

and consequently

$$\nu\equiv\dim\mathcal{N}(A^*-i)=\sum_{\ell=0}^{\infty}\sum_{m=-\ell}^{\ell}\dim\mathcal{N}(A^*_{\ell m}-i)=\sum_{\ell=0}^{\infty}(2\ell+1)\nu_\ell.$$

If $\nu<\infty$, each closed maximal symmetric extension of $A$ is self-adjoint. If $\nu=\infty$, $A$ has self-adjoint extensions and also non-self-adjoint maximal symmetric extensions. To determine these extensions one applies the theory of the Cayley transform in the subspace $\mathcal{H}_{\text{def}}:=\bigoplus_{\ell,\nu_\ell\neq 0}\mathcal{H}_\ell$ (the subspace spanned by those partial wave subspaces $\mathcal{H}_\ell$ for which $\nu_\ell\neq 0$). Let us mention some possibilities for self-adjoint extensions.

Spherically symmetric self-adjoint extensions

In the representation $\mathcal{H}=\bigoplus_{\ell m}\mathcal{H}_{\ell m}$ of the Hilbert space, one considers self-adjoint extensions of $A$ of the form $H=\bigoplus_{\ell m}H_{\ell m}$ (i.e. the restriction of $H$ to the subspace $\mathcal{H}_{\ell m}$ is $H_{\ell m}$). The operators $H_{\ell m}$ are chosen as follows [after identification of $\mathcal{H}_{\ell m}$ with $L^2((0,\infty),r^2dr)$]: (i) if $\nu_\ell=0$, $H_{\ell m}$ is the unique self-adjoint extension of $A_{\ell m}$, (ii) for the values of $\ell$ for which $\nu_\ell\neq 0$, one takes $H_{\ell m}=U^{-1}\mathbb{B}_{\ell m}U$, where $\mathbb{B}_{\ell m}$ is one of the self-adjoint extensions of $B_\ell$ (which may depend on $m$). The operator $H$ leaves each $\mathcal{H}_{\ell m}$ invariant ($H$ commutes with the usual representation of the rotation group in $L^2(\mathbb{R}^3)$; it is clear that each maximal symmetric extension of $A$ that commutes with this representation is self-adjoint, even if the deficiency indices of $A$ are $\infty$, because such an extension must leave each $\mathcal{H}_{\ell m}$ invariant and is therefore of the form indicated here).

Self-adjoint extensions without spherical symmetry

Suppose for example that there are at least two values of $\ell$ for which $\nu_\ell \neq 0$, say $\ell_1$ and $\ell_2$. Consider the subspace $\mathcal{H}_{\ell_1} \oplus \mathcal{H}_{\ell_2}$ of $L^2(\mathbb{R}^3)$, and denote by $A_{(o)}$ the restriction of $A$ to this subspace. The dimension of $[\mathcal{R}(A_{(o)} \pm i)]^\perp \cap \mathcal{H}_{\ell_j}$ is non-zero for $j = 1$ and $j = 2$. Let $\mathbb{A}_0$ be a self-adjoint extension of $A_{(o)}$ (in $\mathcal{H}_{\ell_1} \oplus \mathcal{H}_{\ell_2}$). If the operator $U$ in the Second von Neumann Formula (3.14) maps certain vectors belonging to the subspace $[\mathcal{R}(A_{(o)} + i)]^\perp \cap \mathcal{H}_{\ell_1}$ into $[\mathcal{R}(A_{(o)} - i)]^\perp \cap \mathcal{H}_{\ell_2}$, then $\mathbb{A}_0$ mixes $\mathcal{H}_{\ell_1}$ and $\mathcal{H}_{\ell_2}$; so any self-adjoint extension of $A$ coinciding on $\mathcal{H}_{\ell_1} \oplus \mathcal{H}_{\ell_2}$ with $\mathbb{A}_0$ will not commute with the representation of the rotation group.

EXAMPLES. (a) If $V(r) \geq \frac{3}{4} r^{-2}$ near $r = 0$ and $V(r) \geq -\kappa r^2$ in a neighbourhood of infinity, then $-\Delta + V$ is essentially self-adjoint on $C_0^\infty(\mathbb{R}^3 \setminus \{0\})$.

(b) If $r^2 V(r) \to 0$ as $r \to 0$ and $V(r) \geq -\kappa r^2$ in a neighbourhood of infinity, the minimal operator $B_\ell$ in $L^2((0,\infty))$ is essentially self-adjoint for $\ell > 0$, whereas for $\ell = 0$ its deficiency indices are $(1,1)$. For example for a Coulomb potential $V(r) = \gamma r^{-1}$ the deficiency indices of $-\Delta + \gamma r^{-1}$ on $C_0^\infty(\mathbb{R}^3 \setminus \{0\})$ are $(1,1)$. We have seen in Remark 2.50 that the Coulomb Hamiltonian $H_C \equiv -\Delta + \gamma r^{-1}$ is essentially self-adjoint on $C_0^\infty(\mathbb{R}^3)$: its restriction to $C_0^\infty(\mathbb{R}^3 \setminus \{0\})$ has deficiency indices $(1,1)$, it is non-self-adjoint in the partial wave of angular momentum $\ell = 0$. The restriction of $H_C$ [defined on its maximal domain, which is equal to $\mathcal{D}(H_0)$] to the partial wave subspace $\mathcal{H}_0$ of angular momentum $\ell = 0$ corresponds again to the boundary condition $g'(0)\sin\alpha = g(0)\cos\alpha$ with $\alpha = 0$ [hence to $g(0) = 0$] in $L^2((0,\infty))$, as for the free Hamiltonian $H_0$.

(c) If $V(r) = -\gamma r^{-\rho}$ near $r = 0$, with $\gamma > 0$ and $\rho > 2$, then for each $\ell$ the function $-\gamma r^{-\rho} + \ell(\ell+1) r^{-2}$ is monotone increasing near $r = 0$. Thus, by Proposition 3.30(b), the deficiency indices of $[-\Delta + V]|_{C_0^\infty(\mathbb{R}^3 \setminus \{0\})}$ are $(\infty, \infty)$.

## Appendix to Chapter 3

**A.3.1. Proof of Proposition 3.16**. One divides the interval $J = (a,b)$ into a finite (if $b - a < \infty$) or infinite number of disjoint subintervals of length $3\varepsilon$, where $\varepsilon > 0$ is fixed. It suffices to show that there exist constants $C_1$ and $C_2$ such that one has for each of these subintervals $\mathfrak{J}$ and for $f \in \mathcal{E}(J)$:

$$\int_{\mathfrak{J}} |f'(x)|^2 dx \leq C_1 \int_{\mathfrak{J}} |f(x)|^2 dx + C_2 \int_{\mathfrak{J}} |f''(x)|^2 dx$$

[then, by summing this inequality over all subintervals $\mathfrak{J}$ introduced above, one obtains (3.30)].

So let $\mathfrak{J} = (\alpha, \beta]$ be an interval of length $3\varepsilon$, and write $\mathfrak{J} = J_1 \cup J_2 \cup J_3$ with $J_1 = (\alpha, \alpha+\varepsilon]$, $J_2 = (\alpha+\varepsilon, \alpha+2\varepsilon]$, $J_3 = (\alpha+2\varepsilon, \beta]$; the length of $J_k$ $(k = 1,2,3)$ is thus $\varepsilon$. Let $x_1 \in J_1$ and $x_3 \in J_3$; since $f$ is of class $C^1$, there exist $y \in [x_1, x_3]$ such that $f(x_3) - f(x_1) = f'(y)(x_3 - x_1)$ (the point $y$ depends on $x_1$ and $x_3$). We thus have for each $x \in (\alpha, \beta]$

$$|f'(x)| = \left|f'(y) + \int_y^x f''(z)dz\right| \leq |f'(y)| + \int_\alpha^\beta |f''(z)|dz$$
$$\leq \frac{1}{\varepsilon}\Big[|f(x_1)| + |f(x_3)|\Big] + \int_\alpha^\beta |f''(z)|dz \tag{3.46}$$

(because $x_3 - x_1 > \varepsilon$). This inequality holds for all $x_1 \in J_1$, $x_3 \in J_3$ and $x \in (\alpha, \beta]$ ($y$ does not occur any more in the last expression). We integrate (3.46) with respect to $dx_1$ over $J_1$, which leads to

$$\varepsilon|f'(x)| \leq \frac{1}{\varepsilon}\Big[\int_{J_1} |f(x_1)|dx_1 + \varepsilon|f(x_3)|\Big] + \varepsilon\int_\alpha^\beta |f''(z)|dz.$$

We now integrate this last inequality with respect to $dx_3$ over $J_3$:

$$\varepsilon^2|f'(x)| \leq \int_{J_1} |f(x_1)|dx_1 + \int_{J_3} |f(x_3)|dx_3 + \varepsilon^2\int_\alpha^\beta |f''(z)|dz$$
$$\leq \int_\alpha^\beta |f(z)|dz + \varepsilon^2\int_\alpha^\beta |f''(z)|dz. \tag{3.47}$$

By applying the Schwarz inequality [with $g(z) \equiv 1$] to the integrals on the right-hand side of (3.47), one arrives at

$$\varepsilon^2|f'(x)| \leq \Big[3\varepsilon\int_\alpha^\beta |f(z)|^2dz\Big]^{1/2} + \varepsilon^2\Big[3\varepsilon\int_\alpha^\beta |f''(z)|^2dz\Big]^{1/2}.$$

Consequently [using the inequality $(a + b)^2 \leq 2(a^2 + b^2)$]:

$$|f'(x)|^2 \leq \frac{2}{\varepsilon^4}\,3\varepsilon\int_\alpha^\beta |f(z)|^2dz + 2\Big[3\varepsilon\int_\alpha^\beta |f''(z)|^2dz\Big]. \tag{3.48}$$

We now integrate (3.48) with respect to $dx$ over $(\alpha, \beta]$ (observe that the right-hand side does not depend on $x$):

$$\int_\alpha^\beta |f'(x)|^2dx \leq \frac{6}{\varepsilon^3}\,3\varepsilon\int_\alpha^\beta |f(z)|^2dz + 6\varepsilon\cdot 3\varepsilon\int_\alpha^\beta |f''(z)|^2dz$$
$$= \frac{18}{\varepsilon^2}\int_\alpha^\beta |f(z)|^2dz + 18\varepsilon^2\int_\alpha^\beta |f''(z)|^2dz.$$

Hence (3.30) is true [and it is possible to choose in (3.30) a constant $C_2$ that is arbitrarily small !]. □

**A.3.2. Proof of Lemma 3.24**. Let $f_1, f_2$ be two linearly independent solutions of the equation $\mathcal{L}f = z_0 f$ and $g$ a solution of $\mathcal{L}g = zg$. We have $f_1, f_2 \in L^2((c, b))$ by hypothesis. We assume that $W[f_1, f_2] = 1$ (if $W[f_1, f_2] \neq 1$ it suffices to replace $f_1$ by $f_1/W[f_1, f_2]$). For $x \in (c, b)$ we set

$$h(x) = f_1(x) \int_\kappa^x f_2(y) g(y)\, dy - f_2(x) \int_\kappa^x f_1(y) g(y)\, dy, \tag{3.49}$$

where $\kappa$ is a fixed number in $(c, b)$. One has

$$h'(x) = f_1'(x) \int_\kappa^x f_2(y) g(y)\, dy - f_2'(x) \int_\kappa^x f_1(y) g(y)\, dy,$$

because the terms coming from the derivatives of the integrals in (3.49) cancel. Since $f_j'' = (V - z_0) f_j$, we have

$$h''(x) = \big[V(x) - z_0\big] h(x) + \big[f_1'(x) f_2(x) - f_2'(x) f_1(x)\big] g(x),$$

i.e.

$$\mathcal{L} h = z_0 h + W[f_1, f_2] g = z_0 h + g.$$

It follows that

$$\begin{aligned}(\mathcal{L} - z_0)\big[g - (z - z_0) h\big] &= \big[(\mathcal{L} - z) g + (z - z_0) g\big] - (z - z_0)(\mathcal{L} - z_0) h \\ &= \big[(\mathcal{L} - z) g + (z - z_0) g\big] - (z - z_0) g = (\mathcal{L} - z) g = 0.\end{aligned}$$

Consequently $g - (z - z_0)h$ is a linear combination of $f_1$ and $f_2$: there exist constants $\alpha_1$ and $\alpha_2$ (depending on $\kappa$) such that

$$g(x) = \alpha_1 f_1(x) + \alpha_2 f_2(x) + (z - z_0) h(x).$$

Thus [by using the inequality (1.58)]:

$$|g(x)|^2 \leq 3|\alpha_1|^2 |f_1(x)|^2 + 3|\alpha_2|^2 |f_2(x)|^2 + 3|z - z_0|^2 |h(x)|^2. \tag{3.50}$$

Next, applying the Schwarz inequality to the integrals in (3.49), one gets for $x \in (\kappa, b)$:

$$\begin{aligned}|h(x)|^2 &\leq 2\Big\{|f_1(x)|^2 \int_\kappa^x |f_2(y)|^2 dy + |f_2(x)|^2 \int_\kappa^x |f_1(y)|^2 dy\Big\} \int_\kappa^x |g(z)|^2 dz \\ &\leq 2\Big\{|f_1(x)|^2 \int_\kappa^b |f_2(y)|^2 dy + |f_2(x)|^2 \int_\kappa^b |f_1(y)|^2 dy\Big\} \int_\kappa^x |g(z)|^2 dz.\end{aligned}$$

Upon inserting this inequality into (3.50) and then integrating over an interval $(\kappa, \nu)$ with $\kappa < \nu < b$, one obtains

$$\begin{aligned}\int_\kappa^\nu |g(x)|^2 dx &\leq 3|\alpha_1|^2 \int_\kappa^b |f_1(x)|^2 dx + 3|\alpha_2|^2 \int_\kappa^b |f_2(x)|^2 dx \\ &\quad + 6|z - z_0|^2 \Big\{\int_\kappa^b |f_1(x)|^2 dx \int_\kappa^b |f_2(y)|^2 dy + \int_\kappa^b |f_2(x)|^2 dx \int_\kappa^b |f_1(y)|^2 dy\Big\} \\ &\quad \times \int_\kappa^\nu |g(z)|^2 dz.\end{aligned} \tag{3.51}$$

As $f_1, f_2 \in L^2((c,b))$, one can choose $\kappa$ sufficiently close to $b$ so that

$$12|z-z_0|^2 \int_\kappa^b |f_1(x)|^2 dx \int_\kappa^b |f_2(y)|^2 dy < \frac{1}{2}.$$

For such a choice of $\kappa$ and for $\nu \in (\kappa, b)$, (3.51) gives

$$\frac{1}{2}\int_\kappa^\nu |g(x)|^2 dx \leq 3|\alpha_1|^2 \int_\kappa^b |f_1(x)|^2 dx + 3|\alpha_2|^2 \int_\kappa^b |f_2(x)|^2 dx.$$

Since the right-hand side of this inequality is finite and independent of $\nu$, one obtains by letting $\nu \to b$ that $\int_\kappa^b |g(x)|^2 dx < \infty$, i.e. that $g$ is square-integrable in some neighbourhood of $b$ and hence on $(c,b)$.

The preceding argument applies to each solution $g$ of $\mathcal{L}g = zg$. Thus the lemma is proved. □

**A.3.3. The restriction of $H_0$ to a partial wave subspace is self-adjoint.**

(i) We first show that each of the subspaces $\mathcal{H}_{\ell m}$ is invariant under Fourier transformation. For this we introduce the analogue of the partial wave decomposition (3.41) in momentum space. One can introduce spherical polar co-ordinates for the variable $\vec{k}$ in $\tilde{L}(\mathbb{R}^3)$ and decompose a function $\tilde{f}(\vec{k})$ into partial wave components in $\tilde{L}(\mathbb{R}^3)$: if we write $(k, \theta_k, \phi_k)$ for the vector $\vec{k}$ in spherical polar co-ordinates, then, as in (3.41), each $\tilde{f} \in \tilde{L}^2(\mathbb{R}^3)$ can be written as

$$\tilde{f}(k,\theta_k,\phi_k) = \sum_{\ell=0}^{\infty}\sum_{m=-\ell}^{\ell} \tilde{f}_{\ell m}(k) Y_\ell^m(\theta_k,\phi_k). \qquad (3.52)$$

The important fact is that if $f \in \mathcal{H}_{\ell m}$, i.e. $f(r,\theta,\phi) = f_{\ell m}(r) Y_\ell^m(\theta,\phi)$, then the Fourier transform $\tilde{f}$ of $f$ has the form $\tilde{f}(k,\theta_k,\phi_k) = \tilde{f}_{\ell m}(k) Y_\ell^m(\theta_k,\phi_k)$ *with the same values of* $\ell$ *and* $m$ [$\tilde{f}_{\ell m}(k)$ is obtained from $f_{\ell m}(r)$ by a certain formula which will become evident in the calculations below]. To see this one uses the formula for the development of a plane wave $\exp\{i\vec{k}\cdot\vec{x}\}$ in spherical harmonics (for fixed $\vec{k}$ and $r$, $\exp\{i\vec{k}\cdot\vec{x}\}$ is a function of the polar angles $\theta$ and $\phi$ of $\vec{x}$, hence it can be developed in spherical harmonics; see for example Chapter 2 of [Ne] for the details):

$$e^{i\vec{k}\cdot\vec{x}} = \frac{4\pi}{kr}\sum_{\ell=0}^{\infty}\sum_{m=-\ell}^{\ell} i^\ell \hat{j}_\ell(kr)\overline{Y_\ell^m(\theta_k,\phi_k)}Y_\ell^m(\theta,\phi), \qquad (3.53)$$

where $\hat{j}_\ell$ denotes the Riccati-Bessel functions:

$$\hat{j}_\ell(x) = (-1)^\ell x^{\ell+1}\Big(\frac{1}{x}\frac{d}{dx}\Big)^\ell \frac{\sin x}{x} \qquad (\text{if } x > 0).$$

By using the complex conjugate of (3.53) to express $e^{-i\vec{k}\cdot\vec{x}}$, one obtains for the Fourier transform of $f_{\ell m}(r)Y_\ell^m(\theta,\phi)$:

$$\begin{aligned}
\tilde{f}(k,\theta_k,\phi_k) &= (2\pi)^{-3/2}\int_{\mathbb{R}^3} e^{-i\vec{k}\cdot\vec{x}} f_{\ell m}(r) Y_\ell^m(\theta,\phi)\, d^3x \\
&= (2\pi)^{-3/2}\frac{4\pi}{k}\sum_{\ell'=0}^{\infty}\sum_{m'=-\ell'}^{\ell'} (-i)^{\ell'}\int_0^\infty \hat{\jmath}_{\ell'}(kr) f_{\ell m}(r) r dr \\
&\quad\times\int_0^\pi \sin\theta\, d\theta\int_0^{2\pi} d\phi\, \overline{Y_{\ell'}^{m'}(\theta,\phi)} Y_\ell^m(\theta,\phi) Y_{\ell'}^{m'}(\theta_k,\phi_k) \\
&= \sqrt{\frac{2}{\pi}}\frac{(-i)^\ell}{k}\int_0^\infty \hat{\jmath}_\ell(kr) f_{\ell m}(r) r dr\, Y_\ell^m(\theta_k,\phi_k) \equiv \tilde{f}_{\ell m}(k) Y_\ell^m(\theta_k,\phi_k).
\end{aligned} \tag{3.54}$$

(ii) We denote by $P_{\ell m}$ the projection in $L^2(\mathbb{R}^3)$ onto the subspace $\mathcal{H}_{\ell m}$ and show that $f \in \mathcal{D}(H_0) \Longrightarrow P_{\ell m} f \in \mathcal{D}(H_0)$ and $H_0 P_{\ell m} f \in \mathcal{H}_{\ell m}$. By using the decomposition (3.52) for the function $\tilde{f}$ and the hypothesis that $f \in \mathcal{D}(H_0)$, one finds

$$\|H_0 f\|^2 = \sum_{\ell=0}^{\infty}\sum_{m=-\ell}^{\ell}\int_0^\infty \left|k^2 \tilde{f}_{\ell m}(k)\right|^2 k^2 dk < \infty. \tag{3.55}$$

Since the Fourier transform of $P_{\ell m} f$ is $\tilde{f}_{\ell m}(k) Y_\ell^m(\theta_k,\phi_k)$ by (3.54), one sees from (3.55) that $P_{\ell m} f$ belongs to $\mathcal{D}(H_0)$ because $\|H_0 P_{\ell m} f\|^2 = \int_0^\infty \left|k^2 \tilde{f}_{\ell m}(k)\right|^2 k^2 dk$. Moreover the Fourier transform of $H_0 P_{\ell m} f$ is $k^2 \tilde{f}_{\ell m}(k) Y_\ell^m(\theta_k,\phi_k)$, consequently $H_0 P_{\ell m} f \in \mathcal{H}_{\ell m}$.

(iii) Let $H_{0,\ell m}$ be the restriction of $H_0$ to $\mathcal{D}_{\ell m} := P_{\ell m}\mathcal{D}(H_0)$ (considered as an operator in the Hilbert space $\mathcal{H}_{\ell m}$). $H_{0,\ell m}$ is a symmetric operator in $\mathcal{H}_{\ell m}$. To see that it is self-adjoint, it remains to be shown that $\mathcal{R}(H_{0,\ell m} \pm i) = \mathcal{H}_{\ell m}$ (Proposition 3.3). Now for $f \in \mathcal{D}(H_0)$:

$$f = \sum_{\ell=0}^{\infty}\sum_{m=-\ell}^{\ell} P_{\ell m} f \qquad \text{and} \qquad (H_0 \pm i) f = \sum_{\ell=0}^{\infty}\sum_{m=-\ell}^{\ell} (H_{0,\ell m} \pm i) P_{\ell m} f.$$

Consider a non-zero vector $g \in \mathcal{H}_{\ell_0 m_0}$. Since $H_0 \pm i$ are bijections between $\mathcal{D}(H_0)$ and $\mathcal{H}$, there exist unique vectors $f_\pm \in D(H_0)$ such that $g = (H_0 \pm i) f_\pm$:

$$g = \sum_{\ell=0}^{\infty}\sum_{m=-\ell}^{\ell} (H_{0,\ell m} \pm i) P_{\ell m} f_\pm.$$

Since $g \in \mathcal{H}_{\ell_0 m_0}$, only the term with $\ell = \ell_0$ and $m = m_0$ is non-zero in this sum. Consequently $g = (H_{0,\ell_0 m_0} \pm i) P_{\ell_0 m_0} f_\pm$, with $P_{\ell_0 m_0} f_\pm \in \mathcal{D}_{\ell_0 m_0}$. This shows that $\mathcal{R}(H_{0,\ell_0 m_0} \pm i) = \mathcal{H}_{\ell_0 m_0}$. □

REMARK. In the language of group theory, the situation can be described as follows. If $U(t) = \exp(-iH_0 t)$, the family $\{U(t)\}_{t\in\mathbb{R}}$ defines a strongly continuous unitary representation of the additive group $\mathbb{R}$ in $L^2(\mathbb{R}^3)$. The operators $U(t)$ leave each $\mathcal{H}_{\ell m}$

invariant, hence the restriction $\mathcal{U}_{\ell m}$ of this representation $U$ to $\mathcal{H}_{\ell m}$ defines a strongly continuous unitary representation of $\mathbb{R}$ in $\mathcal{H}_{\ell m}$. The infinitesimal generator $A_{\ell m}$ of this representation $\mathcal{U}_{\ell m}$ is a self-adjoint operator in $\mathcal{H}_{\ell m}$ (see Section 5.1), and this operator coincides with the restriction $H_{0,\ell m}$ of $H_0$ to $\mathcal{H}_{\ell m}$.

### Bibliographical Notes

The method of the Cayley transform is presented in many texts, e.g. in [AG], [W1]. For further results on ordinary differential operators the reader may consult [AG], [N] and also [DS]. In particular, second order operators (called ***Sturm-Liouville operators***) and Dirac operators are treated in Part II of [W2]. Details of the limit point-limit circle theory may be found in Section 6.2 of [P] or in Chapter 9 of [CL]. For distribution theory we recommend [FJ] or [GS].

### Problems

**3.1.** Prove Proposition 3.3(b).

**3.2.** Consider the operator $A$ in the space $L^2(\mathbb{R})\oplus L^2(\mathbb{R})$ defined as follows: $A\{f,g\} = \{g', -f'\}$, with $\{f,g\} \in \mathcal{D}(A) = C_0^\infty(\mathbb{R}) \oplus C_0^\infty(\mathbb{R})$.
(a) Show that $A$ is symmetric.
(b) Does $A$ have self-adjoint extensions?

**3.3.** Let $P_0$ be the minimal operator associated with $-i(d/dx)$ in $L^2((0,1))$.
(a) Determine the eigenvalues, the eigenvectors and the spectrum of the self-adjoint extensions $\mathbb{P}_\alpha$ of $P_0$, $\alpha \in [0, 2\pi)$. Show that each $\lambda \in \mathbb{R}$ belongs to the spectrum of some $\mathbb{P}_\alpha$.
(b) Determine the spectrum of the maximal operator $P_0^*$.
(c) Determine the spectrum of $\overline{P_0}$.

**3.4.** Let $P_0$ be the minimal operator associated with $-i(d/dx)$ in $L^2((0,\infty))$. Determine the spectrum and the eigenvalues of its closure $\overline{P_0}$.

**3.5.** Consider the differential expression $\mathcal{L} = -i(d/dx)$ on the following domains in $L^2([0,1])$: (i) $\mathcal{D}_1 = C^1([0,1])$, (ii) $\mathcal{D}_2 = \{f \in C^1([0,1]) \mid f(0) = f(1)\}$ and (iii) $\mathcal{D}_3 = \{f \in C^1([0,1]) \mid f(0) = f(1) = 0\}$. Determine in each case whether the corresponding differential operator is symmetric, essentially self-adjoint or self-adjoint.

**3.6.** In $L^2((0,1))$, let $P_0$ be the minimal operator associated with $-i(d/dx)$. Then one has $\overline{P_0}P_0^*|_{C_0^\infty((0,1))} = K_0$. Show that the operator $\overline{P_0}P_0^*$ (defined as an operator product on its maximal domain) coincides with one of the self-adjoint extensions of the operator $K_0$ given on page 107.

**3.7.** Let $K_0$ be the minimal operator associated with $-d^2/dx^2$ in $L^2((0,1))$. Determine the eigenvalues of the following self-adjoint extensions of $K_0$:
$K(0,0)$, $K(0,\pi/2)$, $K(\pi/2,\pi/2)$ and $K^{(\rho)}$, $\rho \in [0, 2\pi)$.

**3.8.** In $L^2((0,\infty))$, let $P_0$ be the minimal operator associated with $-i(d/dx)$. Find the domains of the self-adjoint operators $P_0^*\overline{P_0}$ and $\overline{P_0}P_0^*$ and relate these operators to one of the operators $K(\alpha)$ described on page 107.
See also Problem 4.7 for spectral properties of $K(\alpha)$.

**3.9.** (a) Show that the operator $K(0,0)$ in $L^2((0,1))$ is invertible and that its inverse is an integral operator the kernel of which is given by

$$G(x,y) = \begin{cases} (1-y)x & \text{if } x \leq y \\ (1-x)y & \text{if } x \geq y, \end{cases}$$

where $x, y \in (0,1)$.
(b) Show that $G$ is a Hilbert-Schmidt kernel.
(c) Show that, for each $z \in \rho(K(0,0))$, the resolvent $[K(0,0) - z]^{-1}$ is also a Hilbert-Schmidt operator.
[Hint: use the first resolvent equation].
REMARK. Since each Hilbert-Schmidt operator is an integral operator (see Proposition 2.15), $[K(0,0) - z]^{-1}$ is an integral operator. Its kernel is called the ***Green's function*** of $K(0,0)$ at the point $z$.

**3.10.** The limit point-limit circle terminology comes from attempts to study Schrödinger operators on an infinite interval $(0,\infty)$ as limits of such operators on finite intervals $(0,b)$ as $b \to \infty$. Let $f$ and $g$ be solutions of the differential equation $-\varphi''(x) + V(x)\varphi(x) = i\varphi(x)$ on $(0,\infty)$ satisfying $f(0) = g'(0) = 0$ and $-f'(0) = g(0) = 1$. For fixed $b > 0$, consider the set $C_b = \{z \in \mathbb{C} \mid h(x) := f(x) + zg(x)$ satisfies $h(b)\cos\gamma + h'(b)\sin\gamma = 0$ for some $\gamma \in [0,\pi)\}$. Each $\gamma \in [0,\pi)$ determines a point $z = z(\gamma)$ in $\mathbb{C}$, and as $\gamma$ varies, the points $z(\gamma)$ describe a closed curve $C_b$ in the complex plane. One can show that $C_b$ is a circle. As $b \to \infty$ these circles $C_b$ either converge to a limit circle $C_\infty$ or to a single point (a circle of radius 0). In the first case each solution of $-\varphi''(x) + V(x)\varphi(x) = i\varphi(x)$ belongs to $L^2((0,\infty))$, while in the second case only one solution (up to a numerical factor) has this property.
(a) Prove the statement made above that, for given $b > 0$, each $\gamma \in [0,\pi)$ determines a unique point $z = z(\gamma)$.
(b) For $V = 0$, calculate $f$ and $g$, show that $C_b$ is a circle, that $C_{b'}$ is contained in the interior of $C_b$ if $b' > b$, and that the circles $\{C_b\}$ shrink to a point when $b \to \infty$.

**3.11.** Show by means of examples that there are functions $f, g \in \mathcal{D}(\mathbb{P}_\alpha) \subset L^2((0,1))$ such that $\langle \mathbb{P}_\alpha f, Qg\rangle - \langle Qf, \mathbb{P}_\alpha g\rangle \neq -i\langle f, g\rangle$. Similarly show that there are functions $f, g \in \mathcal{D}(\mathbb{P}_\alpha) \cap \mathcal{D}(K(\beta,\gamma))$ such that $\langle \mathbb{P}_\alpha f, K(\beta,\gamma)g\rangle - \langle K(\beta,\gamma)f, \mathbb{P}_\alpha g\rangle \neq 0$.

**3.12.** By using results from Sections 2.4 and 2.6, verify that the square $A^2$ of an unbounded self-adjoint operator $A$ is self-adjoint, i.e. $(A^2)^* = A^2$.

# CHAPTER 4

# Spectral Theory of Self-Adjoint Operators

We have seen in Section 2.6 that the spectrum of a self-adjoint operator $A$ is a subset of the real axis, and it is well known from quantum mechanics that such a spectrum may involve different spectral types (eigenvalues, discrete spectrum, continuous spectrum). One of the topics of the present chapter concerns the definition and characterisation of these different spectral parts of $A$ (Sections 4.3 and 4.4). Before that, in Section 4.2, we study the concept of a spectral measure; this means a measure defined on $(\mathbb{R}, \mathcal{A}_{\mathrm{B}})$ the values of which are projection operators in a Hilbert space $\mathcal{H}$ rather than just real numbers. The fundamental result here will be the Spectral Theorem establishing a one-to-one correspondence between self-adjoint operators and spectral measures. This theorem somehow expresses the fact that each self-adjoint operator can be diagonalised, and it also furnishes a simple means for defining functions of self-adjoint operators.

Given a spectral measure, one can associate a numerical measure $m_f$ with each vector $f$ of the underlying Hilbert space $\mathcal{H}$. These measures $m_f$ are of the Stieltjes type (Section 4.1) and are the basis for obtaining the spectral parts of self-adjoint operators in Section 4.3.

## 4.1 Stieltjes measures

**4.1.1.** Let $F\colon \mathbb{R} \to \mathbb{R}$ be a bounded monotone function. Since each non-decreasing or non-increasing bounded sequence of real numbers has a limit, all of the following limits exist ($\lambda \in \mathbb{R}$):

$$F(\lambda - 0) := \lim_{\varepsilon \to +0} F(\lambda - \varepsilon), \qquad F(\lambda + 0) := \lim_{\varepsilon \to +0} F(\lambda + \varepsilon),$$

$$F(-\infty) := \lim_{\lambda \to -\infty} F(\lambda), \qquad F(+\infty) := \lim_{\lambda \to +\infty} F(\lambda).$$

If $F$ is continuous at the point $\lambda$, then $F(\lambda - 0) = F(\lambda + 0)$. If $F$ is discontinuous at the point $\lambda$, then $F(\lambda - 0) \neq F(\lambda + 0)$. We make the convention that, if $\lambda$ is a point of discontinuity of $F$, then the value of $F$ at the point $\lambda$ is chosen such that its jump at this point is already included in this value $F(\lambda)$, in other terms such that $F(\lambda) = F(\lambda + 0)$:

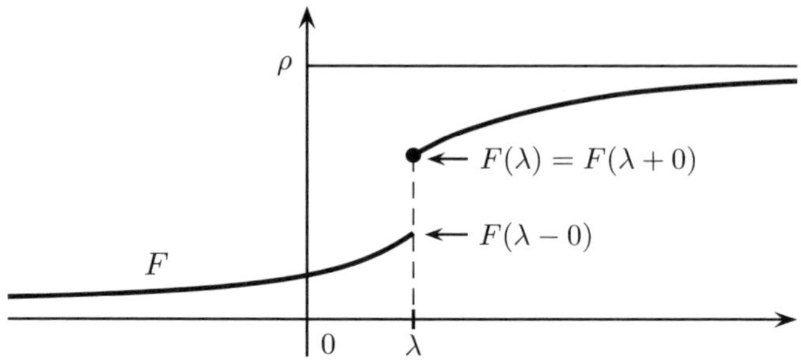


One could choose other conventions, for example $F(\lambda) = [F(\lambda - 0) + F(\lambda + 0)]/2$ or $F(\lambda) = F(\lambda - 0)$. If a monotone function has more than a finite number of discontinuities, then the set of all points of discontinuity is a *countable* subset of $\mathbb{R}$. The reader should verify this important simple fact (see Problem 4.1).

In this section we consider functions $F\colon \mathbb{R} \to \mathbb{R}$ having the following properties:
(1) $F$ is monotone non-decreasing, i.e. $\lambda \geq \mu \implies F(\lambda) \geq F(\mu)$,
(2) $F$ is right continuous, i.e. $F(\lambda) = F(\lambda + 0)$ for all $\lambda \in \mathbb{R}$,
(3) $F(-\infty) = 0$ and $\rho := F(+\infty) < \infty$.

A function $F$ with these properties generates a Borel measure $m^F$ on the real line, called the ***Borel-Stieltjes measure*** (or simply the ***Stieltjes measure***) associated with $F$. It is obtained in the same way as the Borel measure discussed in §1.4.2, by replacing the length of an interval $(a, b]$ by its Stieltjes measure defined as

$$m^F((a, b]) = F(b) - F(a) = \text{the variation of } F \text{ on } (a, b]. \tag{4.1}$$

The value of $m^F((a, b])$ is a number in $[0, \rho]$, and the condition of $\sigma$-additivity of $m^F$ on the collection of all half-open intervals is satisfied; for example if $(a, b]$ and $(b, c]$ are adjoining intervals, then $m^F((a, c]) = m^F((a, b]) + m^F((b, c])$. The measure of a general Borel set $\mathcal{V}$ is determined, in analogy with (1.45), by

$$m^F(\mathcal{V}) = \inf \sum_k m^F(J_k), \tag{4.2}$$

where again $\{J_k\}$ denotes a sequence of half-open intervals covering the set $\mathcal{V}$, and the infimum is taken over all such coverings of $\mathcal{V}$. Since $\rho < \infty$, the Stieltjes measure $m^F$ is a finite measure, i.e. $m^F(\mathbb{R}) < \infty$:

$$m^F(\mathbb{R}) = m^F\big(\textstyle\bigcup_{k=-\infty}^{\infty}(k, k+1]\big) = \displaystyle\sum_{k=-\infty}^{\infty} m^F\big((k, k+1]\big)$$

$$= \sum_{k=-\infty}^{\infty} \big[F(k+1) - F(k)\big] = F(+\infty) - F(-\infty) = \rho.$$

It is interesting to know that there is a converse of the preceding result: if $m$ is a finite measure on the Borel $\sigma$-algebra $\mathcal{A}_{\rm B}$ of $\mathbb{R}$, then there exists a (unique) function $F$ on $\mathbb{R}$ satisfying (1)-(3) such that the associated Stieltjes measure $m^F$ coincides with $m$, i.e. such that $m^F(\mathcal{V}) = m(\mathcal{V})$ for each $\mathcal{V} \in \mathcal{A}_{\rm B}$. Indeed, it is easy to verify that the following function $F$ has all the required properties:

$$F(\lambda) = m((-\infty, \lambda]). \tag{4.3}$$

We continue by discussing some examples that will be useful in the sequel. We say that a Stieltjes measure $m^F$ is *supported* on a Borel set $\mathcal{V}$ if $m^F(\mathbb{R}\backslash\mathcal{V}) = 0$.

**Example 4.1.** Let $F$ be a *piece-wise constant* function, for example

$$F(\lambda) = \begin{cases} 0 & \text{if } \lambda < +1 \\ k/(k+1) & \text{if } k \leq \lambda < k+1, \qquad k = 1, 2, \ldots \end{cases}$$

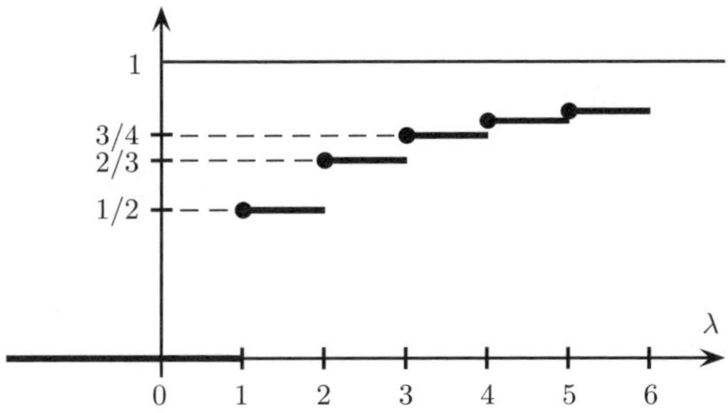


In this case one has $m^F(\{\lambda\}) = 0$ if $\lambda \notin \mathbb{N} \equiv \{1, 2, 3, \ldots\}$, whereas $m^F(\{n\}) = n/(n+1) - (n-1)/n = 1/[n(n+1)]$ if $n = 1, 2, 3, \ldots$, i.e. $m^F(\{n\})$ is equal to the jump of $F$ at the point $n$. For an arbitrary Borel set $\mathcal{V}$ one has:

$$m^F(\mathcal{V}) = \sum_{n \in \mathcal{V} \cap \mathbb{N}} \frac{1}{n(n+1)},$$

in particular $m^F(\mathcal{V}) = 0$ if $\mathcal{V} \cap \mathbb{N} = \varnothing$. A measure $m^F$ of this type is called a ***discrete measure***, it is supported on a discrete subset of $\mathbb{R}$ (a set consisting of isolated points).

**Example 4.2.** If the function $F$ assumes only the values 0 and 1, one obtains a Dirac measure. If the jump of $F$ is at the point $\lambda = 0$, the associated ***Dirac measure*** $\delta$ is as follows:

$$\delta(\mathcal{V}) = \begin{cases} 0 & \text{if } 0 \notin \mathcal{V} \\ 1 & \text{if } 0 \in \mathcal{V} \end{cases} \qquad (\mathcal{V} \in \mathcal{A}_{\rm B})$$

In this case $F$ is the ***Heaviside function***, viz.

$$F(\lambda) = \begin{cases} 0 & \text{if } \lambda < 0 \\ 1 & \text{if } \lambda \geq 0 \end{cases}$$

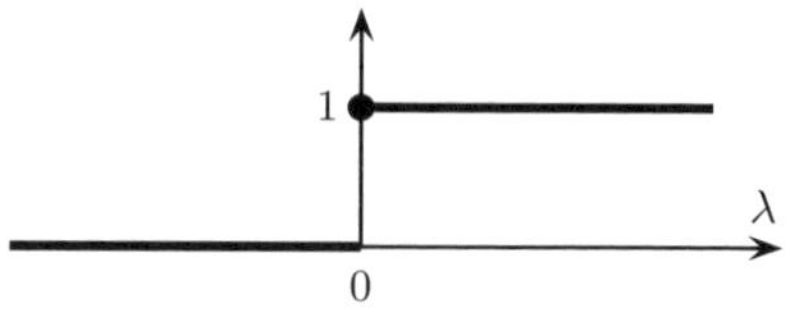


**Example 4.3.** A generalisation of the situation considered in Example 4.1 is that of a function $F$ which is continuous except for a countable number of jumps, say at the points of a sequence $\{\lambda_j\}$. These points need not form a discrete set (they need not be isolated from one another). For example, let $\{r_j\}$ be an enumeration of the rational numbers in the interval $(0, 1)$ and set

$$F(\lambda) = \sum_{r_j \leq \lambda} 2^{-j}.$$

This is a *jump function*: $F(\lambda) = 0$ for $\lambda \leq 0$, $F(\lambda) = 1$ for $\lambda \geq 1$, and on $(0, 1)$ the function $F$ is continuous except at the rational points, with a jump of $2^{-j}$ at $r_j$. The associated measure is given by

$$m^F(\mathcal{V}) = \sum_{r_j \in \mathcal{V}} m^F(\{r_j\}) = \sum_{r_j \in \mathcal{V}} 2^{-j}.$$

This measure is supported on the set of all rational numbers in $(0, 1)$, the measure of the rational point $r_j$ being $2^{-j}$. A measure of this type is called a ***pure point measure*** or an ***atomic measure***.[1]

**Example 4.4.** Let $F$ be a *continuous* function. Then each point of $\mathbb{R}$ has measure zero: if $\lambda \in \mathbb{R}$, then $\lambda \in (\lambda - \varepsilon, \lambda]$ for each $\varepsilon > 0$, hence $m^F(\{\lambda\}) \leq [F(\lambda) - F(\lambda - \varepsilon)]$ for each $\varepsilon > 0$; since $F$ is continuous, one must have $m^F(\{\lambda\}) = 0$. An example is given by the function $F(\lambda) = 1/2 + \pi^{-1}\arctan(\lambda)$:

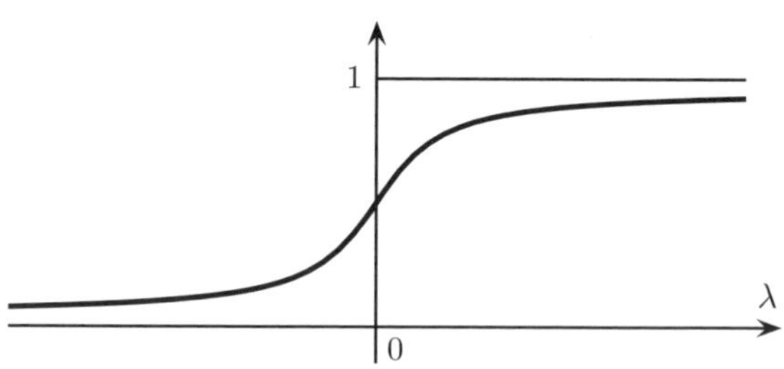


The explicit function $F(\lambda) = 1/2 + \pi^{-1}\arctan(\lambda)$ given in the preceding example is *absolutely continuous*, i.e. an indefinite integral: $F(\lambda) = \int_{-\infty}^{\lambda} [\pi(1 + \mu^2)]^{-1} d\mu$. In

[1] Some authors use the term "discrete measure" also in such a situation.

this case the measure $m^F(\mathcal{V})$ of a Borel set $\mathcal{V}$ is obtained by integrating the derivative of $F$ over $\mathcal{V}$:[2]

$$m^F(\mathcal{V}) = \int_{\mathcal{V}} F'(\mu)\, d\mu \equiv \int_{-\infty}^{\infty} F'(\mu)\chi_{\mathcal{V}}(\mu)\, d\mu = \frac{1}{\pi}\int_{\mathcal{V}} \frac{d\mu}{1+\mu^2}. \tag{4.4}$$

This is an example of an ***absolutely continuous measure*** on $\mathbb{R}$. There also exist continuous functions that cannot be expressed as an indefinite integral, as explained in the next example.

**Example 4.5.** Without entering into explicit mathematical derivations, we describe here the ***Cantor function***. This is a non-decreasing continuous function which is differentiable at each point in $\mathbb{R}$ not belonging to the Cantor set $\mathfrak{C}$ (see page 16), hence almost everywhere with respect to Lebesgue measure, and the derivative is zero at all these points: $F'(\lambda) = 0$ if $\lambda \notin \mathfrak{C}$. Thus $F$ is not the integral of its derivative, since $F(-\infty) = 0$ and $F(+\infty) = 1$. This function is illustrated in Figure 4.1:

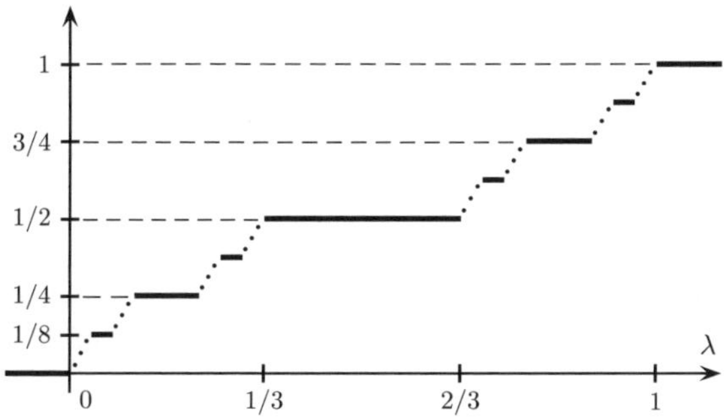


**Figure 4.1** The Cantor function.

The Cantor function is constant on each interval disjoint from $\mathfrak{C}$: $F(\lambda) = 0$ for $\lambda \leq 0$, $F(\lambda) = 1$ for $\lambda \geq 1$, $F(\lambda) = 1/2$ on $(1/3, 2/3)$; $F(\lambda) = 1/4$ on $(1/9, 2/9)$, $F(\lambda) = 3/4$ on $(7/9, 8/9)$: $F(\lambda) = 1/8$ on $(1/27, 2/27)$ etc., and its values at the points $\lambda$ of $\mathfrak{C}$ are defined by continuity. A more precise definition of the Cantor set and the Cantor function can be obtained in terms of the expression in base three of the numbers $\lambda \in (0, 1)$ (for details see for example [R]).

The measure $m^F$ associated with the Cantor function has the property that $m^F(\mathcal{V}) = 0$ if the intersection of $\mathcal{V}$ with the Cantor set is empty; thus this measure is entirely supported on the Cantor set, which is a Borel set of Lebesgue measure zero. A measure $m$ on $\mathbb{R}$ having these properties [i.e. a measure satisfying (i) $F(\lambda) \equiv m((-\infty, \lambda])$ is continuous in $\lambda$, (ii) there exists a Borel set $\mathcal{W}$ of Lebesgue measure zero such that $m(\mathcal{V}) = 0$ whenever $\mathcal{V} \cap \mathcal{W} = \varnothing$] is called a ***singular continuous measure***.

The Cantor function is an example of a ***singular continuous function***, i.e. a function $\Phi$ having the following properties: (i) $\Phi$ is continuous, (ii) $\Phi$ is differentiable

[2] $\chi_{\mathcal{V}}$ denotes the characteristic function of the Borel set $\mathcal{V}$, see (1.41).

except possibly on a null set $\mathcal{W}$ (with respect to Lebesgue measure), (iii) its derivative $\Phi'$ is zero at each point of $\mathbb{R}\backslash\mathcal{W}$.

**Remark 4.6.** To construct Stieltjes measures $m^F$ it is important to start by defining the measure for *half-open* intervals $(a, b]$. Indeed, contrarily to Lebesgue measure, the Stieltjes measure of an open interval $(a, b)$ can be different from that of $(a, b]$; for instance in Example 4.1 one has $m^F((0, 2)) = 1/2$ but $m^F((0, 2]) = 2/3$. It is easy to give expressions in terms of the function $F$ for the measure of intervals that are not of the type $(a, b]$:

$$m^F\big((a, b)\big) = F(b - 0) - F(a)\,, \quad m^F\big([a, b]\big) = F(b) - F(a - 0). \tag{4.5}$$

**4.1.2.** The preceding examples exhibit all typical features of Stieltjes measures on the real line: any Stieltjes measure $m$ on $\mathbb{R}$ admits a unique decomposition into a sum of three measures:

$$m = m_{\rm p} + m_{\rm sc} + m_{\rm ac}, \tag{4.6}$$

where $m_{\rm p}$ is a pure point measure, $m_{\rm sc}$ a singular continuous measure and $m_{\rm ac}$ an absolutely continuous measure. Each of the four measures occurring in this equation is associated with a unique function having the properties (1)-(3) stated at the beginning of the preceding subsection. Let us denote these functions by $F$, $F_{\rm p}$, $F_{\rm sc}$ and $F_{\rm ac}$, so that $m = m^F, m_{\rm p} = m^{F_{\rm p}}, m_{\rm sc} = m^{F_{\rm sc}}$ and $m_{\rm ac} = m^{F_{\rm ac}}$. In view of Eq. (4.3), one then has

$$F = F_{\rm p} + F_{\rm sc} + F_{\rm ac}, \tag{4.7}$$

in other terms a unique decomposition of the function $F$ associated with the measure $m$ into a sum of a jump function, a singular continuous function and an absolutely continuous function.

It is easy to construct examples of functions $F$ such that each of the three measures $m_{\rm p}^F$, $m_{\rm sc}^F$ and $m_{\rm ac}^F$ in the decomposition (4.6) of $m^F$ is non-trivial. One can take for example for $F_{\rm p}$ the Heaviside function (see Example 4.2), for $F_{\rm sc}$ the Cantor function and $F_{\rm ac}(\lambda) = 1/2 + \pi^{-1}\arctan(\lambda)$, and set $F = F_{\rm p} + F_{\rm sc} + F_{\rm ac}$.

The remainder of this subsection is devoted to some explanations concerning the above-mentioned decomposition of a general Stieltjes measure. This decomposition can be obtained either by decomposing the function $F$, as in Eq. (4.7), or by decomposing the measure $m$ itself, as in Eq. (4.6). The former method is based on special properties of monotone functions, the latter on general results from measure theory.

We first comment on the decomposition (4.7). Let $F$ be a function having the properties (1)-(3) of §4.1.1. One first collects all discontinuities of $F$ into a function $F_{\rm p}$. If $F$ has no discontinuities, take $F_{\rm p} = 0$. Otherwise let $\{\lambda_j\}$ be an enumeration of the points of discontinuity of $F$ (remember that this set is at most countable), and define

$$F_{\rm p}(\lambda) = \sum_{\lambda_j \leq \lambda} m(\{\lambda_j\}) \equiv \sum_{\lambda_j \leq \lambda} \big[F(\lambda_j) - F(\lambda_j - 0)\big]. \tag{4.8}$$

$F_{\rm p}$ is the the unique jump function having discontinuities of the same size and at the same points as the function $F$. Setting $F_{\rm c} = F - F_{\rm p}$, it is easy to see that $F_{\rm c}$ is a continuous function also satisfying (1) and (3) of §4.1.1 (Problem 4.1). Now each continuous monotone function $\phi$ is differentiable almost everywhere[3]: there exists a Borel set $\mathcal{V}$ of Lebesgue measure zero such that $\phi'(\lambda)$ exists and is finite at all points $\lambda \in \mathbb{R}\setminus\mathcal{V}$. Applying this result to the function $F_c$, one finds that its derivative $F'_c$ is equivalent to a non-negative integrable function on $\mathbb{R}$. One may thus set

$$F_{ac}(\lambda) = \int_{-\infty}^{\lambda} F'_c(\mu)\,d\mu. \tag{4.9}$$

It is then clear that $F_{sc} := F_c - F_{ac}$ is continuous and that $F'_{sc}(\lambda) = 0$ for $\lambda \in \mathbb{R}\setminus\mathcal{V}$. One can also check that $F_{sc}$ is monotone non-decreasing.

To obtain the decomposition (4.6) of a Stieltjes measure $m$, one can appeal to an abstract theorem on comparison of measures which will now be described. Let $(\mathcal{O}, \mathcal{A})$ be a measurable space and $m_0, m$ two measures on $(\mathcal{O}, \mathcal{A})$. One says that $m$ is ***absolutely continuous*** with respect to $m_0$ if $m(\mathcal{V}) = 0$ for each $\mathcal{V} \in \mathcal{A}$ such that $m_0(\mathcal{V}) = 0$, and that $m$ and $m_0$ are ***mutually singular*** if there exist $\mathcal{W}_1, \mathcal{W}_2 \in \mathcal{A}$ such that $\mathcal{W}_1 \cap \mathcal{W}_2 = \varnothing$, $\mathcal{W}_1 \cup \mathcal{W}_2 = \mathcal{O}$ and $m(\mathcal{W}_1) = m_0(\mathcal{W}_2) = 0$. Thus $m$ is absolutely continuous with respect to $m_0$ if all null sets with respect to $m_0$ are also null sets with respect to $m$, and $m$ and $m_0$ are mutually singular if $m$ is supported on an $m_0$-null set and $m_0$ is supported on an $m$-null set. The following theorem shows that, given two measures $m_0$ and $m$ on the same measurable space $(\mathcal{O}, \mathcal{A})$, one can express $m$ as a sum of two new measures one of which is absolutely continuous with respect to $m_0$ and the other one is singular with respect to $m_0$ (this result holds provided that both measures $m$ and $m_0$ are $\sigma$-finite, see page 14).

**Proposition 4.7 (Lebesgue Decomposition Theorem).** *Let $m, m_0$ be $\sigma$-finite measures on a measurable space $(\mathcal{O}, \mathcal{A})$. Then there exist two unique measures $m_1, m_2$ on $(\mathcal{O}, \mathcal{A})$ such that $m_1$ is absolutely continuous with respect to $m_0$, $m_2$ and $m_0$ are mutually singular and $m = m_1 + m_2$ [i.e. $m(\mathcal{V}) = m_1(\mathcal{V}) + m_2(\mathcal{V})$ for each $\mathcal{V} \in \mathcal{A}$].*

By taking for $m$ a Stieltjes measure and for $m_0$ the Lebesgue measure on $(\mathbb{R}, \mathcal{A}_B)$, one obtains from Proposition 4.7 a decomposition of $m$ into $m = m_{ac} + m_s$, where $m_{ac}$ is absolutely continuous with respect to Lebesgue measure and $m_s$ is singular with respect to Lebesgue measure[4]. The pure point part of the measure $m$ is of course contained in $m_s$, and it is easily isolated by setting for each Borel set $\mathcal{W}$:

$$m_p(\mathcal{W}) = \sum_{\lambda \in \mathcal{W}} m(\{\lambda\}) \equiv \sum_{\lambda \in \mathcal{W}} m_s(\{\lambda\}). \tag{4.10}$$

There are at most a countable number of contributions to the sum in Eq. (4.10), arising from the points of discontinuity of $F$ (the points having non-zero measure). The difference $m_s - m_p$ is a measure without atoms (i.e. such that $m(\{\lambda\}) = 0$ for each

[3] See the Bibliographical Notes at the end of this chapter.

[4] In this case [measurable space $(\mathbb{R}, \mathcal{A}_B)$] it is usual to speak simply of an ***absolutely continuous measure*** and of a ***singular measure***; it is understood that these notions are defined with respect to Lebesgue measure.

$\lambda \in \mathbb{R}$) and supported on a Borel set of Lebesgue measure zero, hence it is a singular continuous measure.

Let us finally mention the Radon-Nikodym Theorem which gives a characterisation of the absolutely continuous part of a measure with respect to some other measure. In the terminology of Proposition 4.7, the absolutely continuous part $m_1$ of the measure $m$ with respect to $m_0$ has the following property: there is a measurable function $f$ such that the measure $m_1(\mathcal{W})$ of any Borel set $\mathcal{W}$ can be obtained by integrating this function with respect to $m_0$ over $\mathcal{W}$. In the statement below we write $m$ in place of $m_1$ and assume that $m$ is absolutely continuous with respect to $m_0$.

**Proposition 4.8 (Radon-Nikodym Theorem).** *Let $m$ and $m_0$ be $\sigma$-finite measures on a measurable space $(\mathcal{O}, \mathcal{A})$ and assume that $m$ is absolutely continuous with respect to $m_0$. Then there exists a non-negative measurable function $f\colon \mathcal{O} \to [0, \infty]$ such that for each $\mathcal{V} \in \mathcal{A}$:*

$$m(\mathcal{V}) = \int_{\mathcal{V}} f(x) m_0(dx). \tag{4.11}$$

*$f$ is unique in the sense that if $g$ is another function having the above properties, then $g(x) = f(x)$ $m_0$-a.e.*

The function $f$ appearing in (4.11) is called the ***Radon-Nikodym derivative*** of $m$ with respect to $m_0$ and sometimes denoted by $f = dm/dm_0$.

For a Stieltjes measure $m^F$ on the measurable space $(\mathbb{R}, \mathcal{A}_{\mathrm{B}})$ we know from Eq. (4.9) that its absolutely continuous part $m^F_{\mathrm{ac}}$ can be written as an integral of the derivative of $F_{\mathrm{ac}}$ (or equivalently of $F_{\mathrm{c}}$) with respect to Lebesgue measure: $m^F_{\mathrm{ac}}(\mathcal{W}) = \int_{\mathcal{W}} F'_{\mathrm{ac}}(\mu)\, d\mu = \int_{\mathcal{W}} F'_{\mathrm{c}}(\mu)\, d\mu$. In this situation the derivative of $F_{\mathrm{ac}}$ coincides with the Radon-Nikodym derivative of the absolutely continuous part $m^F_{\mathrm{ac}}$ of $m^F$ with respect to Lebesgue measure.

**4.1.3.** Let us briefly consider integration with respect to a finite Stieltjes measure $m^F$. One can apply the general definition of an integral given in §1.4.1 by simply replacing in (1.44) the measure $m$ by the Stieltjes measure $m^F$ under consideration. This amounts to approximating the function $\varphi$ that is to be integrated by a sequence $\{\varphi_k\}$ of simple functions [observe that each measurable simple function is $m^F$-integrable because $m^F(\mathcal{V}) < \infty$ for each Borel set $\mathcal{V}$, since we assume that $m^F$ is a finite measure]. The integral obtained in this way is called the ***Lebesgue-Stieltjes integral***.

For a continuous function $\varphi$ one can also define the integral $\int_{-\infty}^{\infty} \varphi(\lambda) m^F(d\lambda)$ in the sense of Riemann (one then speaks of a ***Riemann-Stieltjes integral***). If $(a, b]$ is a finite half-open interval and $\Pi$ a partition of this interval (see Section 1.2), one sets

$$\Sigma(\Pi, \varphi) = \sum_{k=1}^{N} \varphi(u_k)\big[F(s_k) - F(s_{k-1})\big] = \sum_{k=1}^{N} \varphi(u_k)\, m^F((s_{k-1}, s_k]) \tag{4.12}$$

and

$$\int_a^b \varphi(\lambda)\, m^F(d\lambda) \equiv \int_a^b \varphi(\lambda)\, dF(\lambda) = \lim_{r \to \infty} \Sigma(\Pi_r, \varphi), \tag{4.13}$$

where $\{\Pi_r\}$ is a sequence of partitions of the interval $(a, b]$ with $\lim_{r\to\infty} |\Pi_r| = 0$. The second notation in (4.13) for this integral is the more usual one, and the limit in (4.13) exists for each function $\varphi$ that is continuous on $[a, b]$. The integral over $\mathbb{R}$ of a continuous function is defined as an improper integral (if this improper integral exists). If $\varphi\colon \mathbb{R} \to \mathbb{C}$ is *continuous and bounded*, then the integral $\int_{-\infty}^{\infty} \varphi(\lambda) m^F(d\lambda)$ exists (we assume that $m^F$ is a finite measure), and one has

$$\Big|\int_{-\infty}^{\infty} \varphi(\lambda) m^F(d\lambda)\Big| \leq \sup_{\mu\in\mathbb{R}} |\varphi(\mu)| \int_{-\infty}^{\infty} m^F(d\lambda) = \|\varphi\|_\infty m^F(\mathbb{R}). \tag{4.14}$$

Finally, if the function $F$ is absolutely continuous, the following identity is interesting: if $\varphi$ is a bounded measurable function, then

$$\int_a^b \varphi(\lambda)\, dF(\lambda) = \int_a^b \varphi(\lambda) F'(\lambda)\, d\lambda. \tag{4.15}$$

## 4.2 Spectral measures

**4.2.1.** A spectral measure is some kind of a generalisation of the Stieltjes measures discussed in the preceding section. Instead of considering non-decreasing functions defined on $\mathbb{R}$ with values in $\mathbb{R}$, one takes non-decreasing functions defined on $\mathbb{R}$ whose values are projections in a Hilbert space $\mathcal{H}$. Consider for each $\lambda \in \mathbb{R}$ a subspace $\mathcal{M}_\lambda$ and assume that this family of subspaces $\{\mathcal{M}_\lambda\}$ is *non-decreasing* in the sense that $\mathcal{M}_\lambda \subseteq \mathcal{M}_\mu$ if $\lambda \leq \mu$. Furthermore assume that the intersection of all subspaces $\mathcal{M}_\lambda$ consists of only the zero vector $\mathbf{0}$ and that the union of all $\mathcal{M}_\lambda$ is dense in $\mathcal{H}$. If we denote the projection onto $\mathcal{M}_\lambda$ by $E_\lambda$, i.e. $E_\lambda = P_{\mathcal{M}_\lambda}$, we can express the properties of the family $\{\mathcal{M}_\lambda\}$ in terms of the family of projections $\{E_\lambda\}$ (see in particular Proposition 2.6):

(a) the family $\{E_\lambda\}$ is non-decreasing, i.e. $E_\lambda E_\mu = E_{\min\{\lambda,\mu\}}$, (4.16)

(b) $\operatorname{s-lim}_{\lambda\to-\infty} E_\lambda = O \quad \text{and} \quad \operatorname{s-lim}_{\lambda\to+\infty} E_\lambda = I.$ (4.17)

REMARKS. (i) In view of (b) it is natural to set $E(-\infty) = O$ and $E(+\infty) = I$.
(ii) Equation (4.16) implies that

$$E_\lambda E_\mu = E_\mu E_\lambda \qquad \forall \lambda, \mu \in \mathbb{R}, \tag{4.18}$$

i.e. the family of projections $\{E_\lambda\}$ is a family of commuting operators. Equation (4.16) can also be written as $E_\lambda \leq E_\mu$ if $\lambda \leq \mu$, i.e. $E_\mu - E_\lambda$ is a positive operator if $\lambda \leq \mu$. We recall that $A$ is a ***positive operator*** if $\langle f, Af\rangle \geq 0$ for each vector $f$ in $\mathcal{D}(A)$.

A family of projections $\{E_\lambda\}_{\lambda\in\mathbb{R}}$ satisfying (4.16) and (4.17) is called a ***spectral family*** or a ***resolution of the identity***. It is easy to show by using (4.16) that the limits $\operatorname{s-lim}_{\varepsilon\to+0} E_{\lambda\pm\varepsilon}$ exist for each $\lambda \in \mathbb{R}$ (Problem 4.2). We use the notation $E_{\lambda\pm0}$ for these limits and observe that $E_{\lambda-0} \le E_\lambda \le E_{\lambda+0}$. As in the case of Stieltjes measures we consider only spectral families that are ***right continuous***, i.e. such that

$$E_{\lambda+0} \equiv \operatorname*{s-lim}_{\varepsilon\to+0} E_{\lambda+\varepsilon} = E_\lambda, \tag{4.19}$$

and we use the simplified terminology "spectral family" in place of "right continuous spectral family".

If $-\infty < a < b < +\infty$ one can define in analogy with (4.1)

$$E((a,b]) = E_b - E_a. \tag{4.20}$$

This associates with each interval of the form $(a,b]$ a positive operator $E((a,b])$. Clearly if $(a,b]$ and $(b,c]$ are two adjoining intervals, then $E((a,c]) = E((a,b]) + E((b,c])$. If $(a,b]$ and $(c,d]$ are two arbitrary intervals, then (4.16) implies that

$$\begin{aligned} E((a,b])E((c,d]) &= (E_b - E_a)(E_d - E_c) \\ &= E_{\min\{b,d\}} - E_{\min\{b,c\}} - E_{\min\{a,d\}} + E_{\min\{a,c\}}. \end{aligned}$$

- If $(a,b] \cap (c,d] = \varnothing$, for example if $a < b \le c < d$, then

$$E((a,b])E((c,d]) = E_b - E_b - E_a + E_a = O. \tag{4.21}$$

- If $(a,b] \subseteq (c,d]$, i.e. if $c \le a < b \le d$, then

$$E((a,b])E((c,d]) = E_b - E_c - E_a + E_c = E_b - E_a = E((a,b]). \tag{4.22}$$

- If $(a,b]$ and $(c,d]$ overlap without one of them being a subset of the other one, e.g. if $a < c < b < d$, then

$$E((a,b])E((c,d]) = E_b - E_c - E_a + E_a = E_b - E_c = E((c,b]). \tag{4.23}$$

These three formulas can be combined into the following one (with the convention that $E(\varnothing) = O$):

$$E((a,b])E((c,d]) = E((a,b] \cap (c,d]). \tag{4.24}$$

Consequences: (i) If $E((c,d]) = O$ and $(a,b] \subseteq (c,d]$, then $E((a,b]) = O$.
(ii) $E((a,b])$ is a *projection*: $E((a,b])$ is self-adjoint because $E_a$ and $E_b$ are self-adjoint, and the relation $E((a,b])^2 = E((a,b])$ is obtained by taking $a = c$, $b = d$ in (4.22).

As for numerical measures, one can extend the family of operators defined by (4.20) to the $\sigma$-algebra generated by the family of half-open intervals $(a,b]$, i.e. to the Borel $\sigma$-algebra $\mathcal{A}_{\mathrm{B}}$ of $\mathbb{R}$. In this way one associates a projection $E(\mathcal{V})$ with each Borel set $\mathcal{V}$ in $\mathcal{A}_{\mathrm{B}}$, and this association is $\sigma$-additive:

$$E(\textstyle\bigcup_k \mathcal{V}_k) = \displaystyle\sum_k E(\mathcal{V}_k) \tag{4.25}$$

if $\{\mathcal{V}_k\}$ is a countable family of disjoint elements of $\mathcal{A}_{\mathrm{B}}$. Furthermore

$$E(\varnothing) = O \qquad \text{and} \qquad E(\mathbb{R}) = I. \tag{4.26}$$

The collection of projections $\{E(\mathcal{V})\}$ is called the ***spectral measure*** associated with the spectral family $\{E_\lambda\}$ (the term "spectral" will become clear later). Thus a spectral measure is a $\sigma$-additive mapping from $\mathcal{A}_{\mathrm{B}}$ into the set of projections of a Hilbert space $\mathcal{H}$ satisfying also (4.26). The following property generalises (4.24):

$$E(\mathcal{V})E(\mathcal{W}) = E(\mathcal{V}\cap\mathcal{W}) \qquad \forall\,\mathcal{V},\mathcal{W}\in\mathcal{A}_{\mathrm{B}}. \tag{4.27}$$

Thus, given a spectral family $\{E_\lambda\}$, one can associate with each Borel set $\mathcal{V}$ of $\mathbb{R}$ a subspace $\mathcal{M}(\mathcal{V})$ of $\mathcal{H}$, viz. $\mathcal{M}(\mathcal{V}) = \{f\in\mathcal{H} \mid E(\mathcal{V})f = f\}$, and one has

$$\mathcal{M}(\mathcal{V})\subseteq\mathcal{M}(\mathcal{W}) \text{ if } \mathcal{V}\subseteq\mathcal{W} \quad\text{and}\quad \mathcal{M}(\mathcal{V})\perp\mathcal{M}(\mathcal{W}) \text{ if } \mathcal{V}\cap\mathcal{W}=\varnothing. \tag{4.28}$$

In analogy with (4.5) one has for example for the projection associated with an open or closed interval:

$$E((a,b)) = E_{b-0} - E_a, \qquad E([a,b]) = E_b - E_{a-0}. \tag{4.29}$$

The following consequence of (4.28) is useful. Let $\{\mathcal{V}_j\}_{j=1}^n$ be a collection of disjoint Borel subsets of $\mathbb{R}$ (i.e. $\mathcal{V}_j\cap\mathcal{V}_k=\varnothing$ if $j\neq k$), $\{\alpha_j\}_{j=1}^n$ a collection of complex numbers and $f\in\mathcal{H}$. Then, since $E(\mathcal{V}_j)f\perp E(\mathcal{V}_k)f$ if $j\neq k$:

$$\Big\|\sum_{j=1}^n \alpha_j E(\mathcal{V}_j)f\Big\|^2 = \sum_{j=1}^n |\alpha_j|^2\|E(\mathcal{V}_j)f\|^2. \tag{4.30}$$

One may take in (4.20) $a = b-\varepsilon$ and let $\varepsilon\to+0$, which defines the projection $E(\{b\})$ associated with the Borel set $\mathcal{V}=\{b\}$ consisting of the single point $b$:

$$E(\{b\}) = \operatorname*{s-lim}_{\varepsilon\to+0} E((b-\varepsilon,b]) = E_b - E_{b-0} \tag{4.31}$$

[equivalently: set $a=b$ in the second equation in (4.29)]. $E(\{b\})$ is a projection (since it is the strong limit of a sequence of projections, see the end of §2.2.1). One has:

(i) $E(\{b\}) = O$ (spectral measure of the point $b$ equal to zero) $\iff$ the spectral family $\{E_\lambda\}$ is (strongly) continuous at the point $\lambda=b$,

(ii) $E(\{b\}) \neq O$ (spectral measure of the point $b$ not equal to zero) $\iff$ the spectral family $\{E_\lambda\}$ is discontinuous at the point $\lambda=b$.

In case (ii) one can associate with the point $b$ a subspace $\mathcal{M}(\{b\})$ of non-zero dimension, namely

$$\mathcal{M}(\{b\}) = \{f\in\mathcal{H} \mid E(\{b\})f = f\} = \mathcal{R}\big(E(\{b\})\big). \tag{4.32}$$

The interpretation of this subspace will be discussed in §4.3.1. Let us mention here that the discontinuities of a spectral family form a *countable* subset of $\mathbb{R}$. Indeed, if $b_1\neq b_2$ are two points of discontinuity, then $E(\{b_1\})E(\{b_2\}) = O$ by (4.27), hence $\mathcal{M}(\{b_1\})$ is orthogonal to $\mathcal{M}(\{b_2\})$. If one chooses at each point of discontinuity $b$ a vector $f_b\neq\mathbf{0}$, one obtains a family $\{f_b\}$ of non-zero vectors that are pair-wise orthogonal; this family is countable since the Hilbert space is assumed to be separable [postulate (H4) of §1.1.1].

The ***support of a spectral family*** $\{E_\lambda\}$ is the following subset of $\mathbb{R}$:

$$\operatorname{supp}\{E_\lambda\} = \{\mu \in \mathbb{R} \mid E_{\mu+\varepsilon} - E_{\mu-\varepsilon} \neq O \ \forall \varepsilon > 0\}. \tag{4.33}$$

In other terms $\operatorname{supp}\{E_\lambda\}$ is the set of ***points of non-constancy*** (or ***points of increase***) of the spectral family $\{E_\lambda\}$. The complementary set in $\mathbb{R}$ is the set of ***points of constancy*** of the family $\{E_\lambda\}$, i.e. the set of points $\mu$ such that $E_{\mu+\varepsilon} = E_{\mu-\varepsilon}$ for some $\varepsilon > 0$; in each sufficiently small neighbourhood of such a point the spectral family does not vary. It is clear that *the set of points of constancy is an open set*, hence $\operatorname{supp}\{E_\lambda\}$ is closed.

An important case is that in which the support of $\{E_\lambda\}$ is *bounded*, hence contained in a finite interval $[M_-, M_+]$ for some real numbers $M_-$ and $M_+$ ($M_- < M_+$). In this case one has $E_\lambda = O$ for each $\lambda < M_-$, $E_\lambda = I$ for each $\lambda \geq M_+$ and $E(\mathcal{V}) = O$ if $\mathcal{V} \cap [M_-, M_+] = \varnothing$.

One often encounters spectral families that are *semibounded*, i.e. bounded from below or bounded from above. One says for example that a spectral family $\{E_\lambda\}$ is ***bounded from below*** if there is a number $M_- \in \mathbb{R}$ such that $E_\lambda = O$ for each $\lambda < M_-$. Such a number $M_-$ is a lower bound for $\{E_\lambda\}$, and the supremum of all numbers $M_-$ having this property is called the ***lower bound*** of $\{E_\lambda\}$.

If the support of $\{E_\lambda\}$ is bounded, then $E([M_-, M_+]) = I$, i.e. $E([M_-, M_+])f = f$ for each vector $f \in \mathcal{H}$. In the general case there exists no bounded interval $J$ such that $E(J) = I$. However $E(J)$ converges strongly to the identity operator $I$ when $J$ tends to $\mathbb{R}$. By taking for example $J = [-M, M]$, with $M > 0$, one obtains as a consequence of (4.17) that $\operatorname{s-lim}_{M\to\infty} E([-M, M]) = I$, i.e. $\|f - E([-M, M])f\| \to 0$ as $M \to \infty$ for each $f \in \mathcal{H}$. This shows that *the set of vectors $f$ of compact support with respect to the spectral family $\{E_\lambda\}$ is dense in $\mathcal{H}$* (a vector $f$ is said to have ***compact support with respect to $\{E_\lambda\}$*** if there exists a number $M \in (0, \infty)$ such that $E([-M, M])f = f$).

**Example 4.9.** Let $\mathcal{H} = L^2(\mathbb{R})$ and let $E_\lambda$ be the operator of multiplication by the characteristic function $\chi_{(-\infty,\lambda]}$ of the interval $(-\infty, \lambda]$:

$$[E_\lambda f](x) = \begin{cases} 0 & \text{if } x > \lambda \\ f(x) & \text{if } x \leq \lambda \end{cases} \tag{4.34}$$

[see (1.41)]. The associated spectral measure is easy to write: if $\mathcal{V}$ is a Borel set in $\mathbb{R}$, then $E(\mathcal{V})$ is the operator of multiplication by the characteristic function $\chi_{\mathcal{V}}$ of $\mathcal{V}$:

$$[E(\mathcal{V})f](x) = \begin{cases} 0 & \text{if } x \notin \mathcal{V} \\ f(x) & \text{if } x \in \mathcal{V}. \end{cases} \tag{4.35}$$

One has $E(\mathcal{V}) = O$ if the Lebesgue measure of $\mathcal{V}$ is zero and $E(\mathcal{V}) \neq O$ if the Lebesgue measure of $\mathcal{V}$ is non-zero [the range of $E(\mathcal{V})$ is the subspace $L^2(\mathcal{V})$ which is infinite-dimensional if the Lebesgue measure of $\mathcal{V}$ is non-zero]. In particular the support of this spectral family is the full real line $\mathbb{R}$, hence it is not bounded [if $\mu \in \mathbb{R}$

and $\varepsilon > 0$, then $E_{\mu+\varepsilon} - E_{\mu-\varepsilon}$ is the projection onto the subspace $L^2((\mu-\varepsilon, \mu+\varepsilon))$ which is infinite-dimensional]. The spectral family defined by (4.34) is continuous, one has $E(\{\lambda\}) = O$ for each $\lambda \in \mathbb{R}$.

**4.2.2.** Let $\{E_\lambda\}$ be a spectral family and $f$ a fixed vector in a Hilbert space $\mathcal{H}$. Consider the following function defined on $\mathbb{R}$:[5]

$$F_f(\lambda) = \|E_\lambda f\|^2 = \langle f, E_\lambda f\rangle. \tag{4.36}$$

This function satisfies the conditions (1) - (3) stated at the beginning of Section 4.1, with $\rho = \|f\|^2$. Indeed: (1) if $\mu \leq \lambda$, then $F_f(\lambda) = \langle f, E_\lambda f\rangle \geq \langle f, E_\mu f\rangle = F_f(\mu)$ because $E_\lambda - E_\mu$ is a positive operator, (2) $F_f$ is right continuous as a consequence of the right continuity of the spectral family $\{E_\lambda\}$, (3) $F_f(-\infty) = \lim_{\lambda\to-\infty}\langle f, E_\lambda f\rangle = \langle f, Of\rangle = 0$ and $F_f(+\infty) = \lim_{\lambda\to+\infty}\langle f, E_\lambda f\rangle = \langle f, If\rangle = \|f\|^2$ by (4.17). Thus each vector $f \in \mathcal{H}$ determines a finite Stieltjes measure on $\mathbb{R}$, viz. the Stieltjes measure associated with the non-decreasing function $F_f$. In what follows we denote this measure by $m_f$. One has $m_f((a,b]) = \|E((a,b])f\|^2 = \langle f, E((a,b])f\rangle$, and more generally

$$m_f(\mathcal{V}) = \|E(\mathcal{V})f\|^2 = \langle f, E(\mathcal{V})f\rangle \qquad \text{for } \mathcal{V} \in \mathcal{A}_{\mathrm{B}}. \tag{4.37}$$

For $\mu \in \mathbb{R}$ the Stieltjes measure $m_{E_\mu f}$ determined by the vector $E_\mu f$ is obtained from the Stieltjes measure $m_f$ determined by $f$ in the following way:

$$\begin{aligned} m_{E_\mu f}(\mathcal{V}) &= \|E(\mathcal{V})E_\mu f\|^2 = \|E(\mathcal{V})E((-\infty,\mu])f\|^2 \\ &= \|E(\mathcal{V} \cap (-\infty,\mu])f\|^2 = m_f(\mathcal{V} \cap (-\infty,\mu]). \end{aligned} \tag{4.38}$$

One can classify the vectors $f$ of the Hilbert space $\mathcal{H}$ with respect to the spectral family $\{E_\lambda\}$ by having recourse to properties of the associated Stieltjes measures $m_f$ (absolutely continuous or singular continuous or pure point measures). This will be discussed in detail in Section 4.3; for the moment we consider integrals with respect to these measures, and more generally with respect to a spectral measure (one then speaks of ***spectral integrals***).

**4.2.3.** Let $\lambda \mapsto A(\lambda)$, $\lambda \mapsto B(\lambda)$ be functions defined on $\mathbb{R}$ with values in $\mathcal{B}(\mathcal{H})$. One can consider integrals with respect to a spectral measure $\{E(\cdot)\}$ of the type $\int A(\lambda)E(d\lambda)$ or $\int E(d\lambda)B(\lambda)$ or $\int A(\lambda)E(d\lambda)B(\lambda)$ [observe that in general the operators $A(\lambda)$ and $B(\lambda)$ do not commute with the spectral projections $E(\mathcal{V})$]. If $A$ and $B$ are sufficiently smooth as functions of $\lambda$, these integrals will exist in the sense of Riemann-Stieltjes (as weak or strong or uniform limits of sequences of sums, by a construction that is analogous to that used in Section 4.1 to define numerical Riemann-Stieltjes integrals). Such integrals do occur in quantum mechanics, for example in scattering theory, as explained in Section 7.3. We refrain from giving sufficient conditions on the functions $A(\cdot)$ and $B(\cdot)$ that guarantee the existence of integrals of the type

[5] If $E$ is a projection, then $\|Ef\|^2 = \langle Ef, Ef\rangle = \langle f, E^*Ef\rangle = \langle f, E^2f\rangle = \langle f, Ef\rangle$.

$\int A(\lambda)E(d\lambda)B(\lambda)$. We rather discuss the simple situation in which these functions are multiples of the identity operator, hence of the form $\varphi(\cdot)I$, where $\varphi$ is a real or complex function defined on $\mathbb{R}$. So we consider operators of the form $\int \varphi(\lambda)E(d\lambda)$; we assume $\varphi$ to be continuous (the considerations are easily extended to the case in which $\varphi$ is piece-wise continuous, and this will be sufficient for our applications). Under these conditions one can define $\int \varphi(\lambda)E(d\lambda)$ in the sense of Riemann-Stieltjes (limit of sums if the domain of integration is bounded).

If $(a, b]$ is a finite interval and $\Pi$ a partition of this interval (as in Section 1.2), we set

$$\Sigma(\Pi; \varphi) = \sum_{k=1}^{N} \varphi(u_k)E((s_{k-1}, s_k]). \tag{4.39}$$

Clearly $\Sigma(\Pi; \varphi)$ is a well defined operator in $\mathcal{B}(\mathcal{H})$. The fundamental identity is as follows: let $\psi$ be another continuous function on $\mathbb{R}$ and $\varphi\psi$ the product of the functions $\varphi$ and $\psi$, i.e. $(\varphi\psi)(\lambda) = \varphi(\lambda)\psi(\lambda)$; then

$$\Sigma(\Pi; \varphi) \cdot \Sigma(\Pi; \psi) = \Sigma(\Pi; \varphi\psi). \tag{4.40}$$

Indeed one obtains by using (4.24):

$$\begin{aligned}\Sigma(\Pi; \varphi) \cdot \Sigma(\Pi; \psi) &= \sum_{j=1}^{N}\sum_{k=1}^{N} \varphi(u_j)\psi(u_k)E((s_{j-1}, s_j])E((s_{k-1}, s_k]) \\ &= \sum_{j=1}^{N}\sum_{k=1}^{N} \varphi(u_j)\psi(u_k)\delta_{jk}E((s_{k-1}, s_k]) \\ &= \sum_{k=1}^{N} \varphi(u_k)\psi(u_k)E((s_{k-1}, s_k]) = \Sigma(\Pi; \varphi\psi).\end{aligned}$$

**Proposition 4.10.** *Let $\{E_\lambda\}$ be a spectral family in a Hilbert space $\mathcal{H}$, $-\infty < a < b < +\infty$ and $\varphi\colon [a, b] \to \mathbb{C}$ a continuous function.*
*(a) Let $\{\Pi_r\}$ be a sequence of partitions of the interval $(a, b]$ with $|\Pi_r| \to 0$ as $r \to \infty$. Then the sequence of operators $\{\Sigma(\Pi_r; \varphi)\}_{r=1}^{\infty}$ is strongly convergent; its limit, denoted $\int_a^b \varphi(\lambda)E(d\lambda)$, belongs to $\mathcal{B}(\mathcal{H})$ and is independent of the sequence $\{\Pi_r\}$. Moreover the limit operator commutes with each $E_\lambda$ and satisfies*

$$\left\| \int_a^b \varphi(\lambda)E(d\lambda) \right\| = \sup_{\mu \in [a,b] \cap \operatorname{supp}\{E_\lambda\}} |\varphi(\mu)|. \tag{4.41}$$

*(b) If $\overline{\varphi}$ denotes the complex conjugate of the function $\varphi$, then $\int_a^b \overline{\varphi}(\lambda)E(d\lambda)$ is the adjoint of $\int_a^b \varphi(\lambda)E(d\lambda)$.*
*(c) One has for each vector $f \in \mathcal{H}$:*

$$\left\| \int_a^b \varphi(\lambda)E(d\lambda)f \right\|^2 = \int_a^b |\varphi(\lambda)|^2 m_f(d\lambda). \tag{4.42}$$

*(d) If $\psi\colon [a,b] \to \mathbb{C}$ is a second continuous function, then*

$$\int_a^b \varphi(\lambda)E(d\lambda) \cdot \int_a^b \psi(\mu)E(d\mu) = \int_a^b \varphi(\lambda)\psi(\lambda)E(d\lambda) \tag{4.43}$$

*(the product of two spectral integrals is equal to a single spectral integral, that of the product function).*

PROOF. (a) Let $\Pi = \{s_0, \dots, s_n; u_1, \dots, u_n\}$ and $\tilde{\Pi} = \{\tilde{s}_0, \dots, \tilde{s}_m; \tilde{u}_1, \dots, \tilde{u}_m\}$ be two partitions of the interval $(a,b]$. By arranging the numbers $s_0 = \tilde{s}_0 = a, s_1, \dots, s_{n-1}, s_n = \tilde{s}_m = b, \tilde{s}_1, \dots, \tilde{s}_{m-1}$ in increasing order, one obtains a subdivision of the interval $(a,b]$ into $N$ disjoint half-open subintervals $J_1, \dots, J_N$, with $N$ satisfying $\max\{m,n\} \le N \le m+n-1$ (see Figure 4.2). We then have $J_j \cap J_k = \varnothing$ if $j \neq k$, $(a,b] = \bigcup_{\ell=1}^N J_\ell$ and $E((a,b]) = \sum_{\ell=1}^N E(J_\ell)$. Hence

$$\Sigma(\Pi;\varphi) = \sum_{j=1}^n \varphi(u_j)E((s_{j-1},s_j]) = \sum_{\ell=1}^N \varphi(w_\ell)E(J_\ell),$$

$$\Sigma(\tilde{\Pi};\varphi) = \sum_{k=1}^m \varphi(\tilde{u}_k)E((\tilde{s}_{k-1},\tilde{s}_k]) = \sum_{\ell=1}^N \varphi(\tilde{w}_\ell)E(J_\ell),$$

where $w_\ell$ coincides with one of the numbers $u_1, \dots, u_n$ ($w_\ell = u_j$ where $j$ is such that $J_\ell \subseteq (s_{j-1}, s_j]$ ), $\tilde{w}_\ell$ coincides with one of the numbers $\tilde{u}_1, \dots, \tilde{u}_m$, and one has $|w_\ell - \tilde{w}_\ell| \le |\Pi| + |\tilde{\Pi}|$. Consequently, using (4.30) in the first and in the last step:

$$\begin{aligned}\|[\Sigma(\Pi;\varphi) - \Sigma(\tilde{\Pi};\varphi)]f\|^2 &= \sum_{\ell=1}^N |\varphi(w_\ell) - \varphi(\tilde{w}_\ell)|^2 \|E(J_\ell)f\|^2 \\ &\le \sup_{\ell=1,\dots,N} |\varphi(w_\ell) - \varphi(\tilde{w}_\ell)|^2 \sum_{\ell=1}^N \|E(J_\ell)f\|^2 \\ &= \sup_{\ell=1,\dots,N} |\varphi(w_\ell) - \varphi(\tilde{w}_\ell)|^2 \|E((a,b])f\|^2.\end{aligned}$$

Since $\varphi$ is continuous, it is uniformly continuous on the closed interval $[a,b]$, hence the supremum in the preceding expression can be made arbitrarily small by taking the

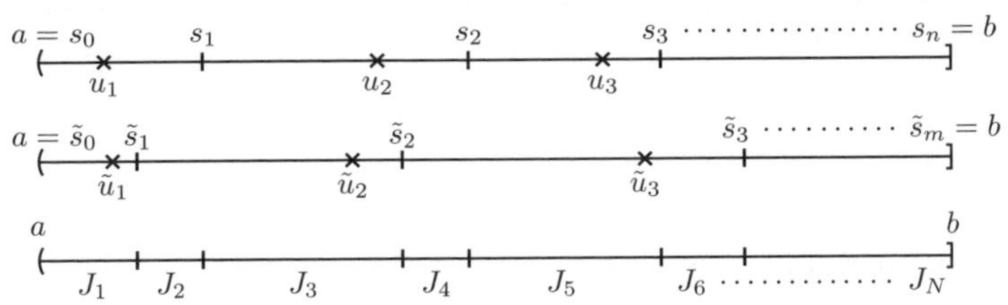


**Figure 4.2**

partitions $\Pi$ and $\tilde{\Pi}$ sufficiently fine (recall that $|w_\ell - \tilde{w}_\ell| \leq |\Pi| + |\tilde{\Pi}|$). In particular, if $\Pi$ and $\tilde{\Pi}$ belong to a sequence $\{\Pi_r\}$ satisfying $|\Pi_r| \to 0$ as $r \to \infty$, and if $\varepsilon > 0$, one will have $\|[\Sigma(\Pi_{r_1};\varphi) - \Sigma(\Pi_{r_2};\varphi)]f\|^2 < \varepsilon$ if $r_1, r_2 > R = R(\varepsilon)$. This implies the existence of the limit $\int_a^b \varphi(\lambda)E(d\lambda)$ and its independence on the sequence $\{\Pi_r\}$.

Since $E_\lambda$ commutes with each $\Sigma(\Pi_r;\varphi)$, it also commutes with the limit operator $\int_a^b \varphi(\lambda)E(d\lambda)$. The remaining statements of (a) will be established at the end of the proof.

(b) It is clear that $[\Sigma(\Pi;\varphi)]^* = \Sigma(\Pi;\overline{\varphi})$. Thus one has for all $f, g \in \mathcal{H}$:

$$\langle f, \Big[\int_a^b \varphi(\lambda)E(d\lambda)\Big]^* g\rangle = \langle \int_a^b \varphi(\lambda)E(d\lambda)f, g\rangle = \lim_{r\to\infty} \langle \Sigma(\Pi_r;\varphi)f, g\rangle$$
$$= \lim_{r\to\infty} \langle f, \Sigma(\Pi_r;\overline{\varphi})g\rangle = \langle f, \int_a^b \overline{\varphi}(\lambda)E(d\lambda)g\rangle.$$

(c) If $\Pi$ is a partition of $(a, b]$ as above, then [using (4.40)]:

$$\begin{aligned}\|\Sigma(\Pi;\varphi)f\|^2 &= \langle \Sigma(\Pi;\varphi)f, \Sigma(\Pi;\varphi)f\rangle = \langle f, \Sigma(\Pi;\overline{\varphi})\Sigma(\Pi;\varphi)f\rangle \\ &= \langle f, \Sigma(\Pi;|\varphi|^2)f\rangle = \sum_{k=1}^N |\varphi(u_k)|^2 \langle f, E((s_{k-1}, s_k])f\rangle \\ &= \sum_{k=1}^N |\varphi(u_k)|^2 m_f((s_{k-1}, s_k]).\end{aligned} \tag{4.44}$$

Consider a sequence $\{\Pi_r\}$ of partitions of $(a, b]$ with $\lim_{r\to\infty} |\Pi_r| = 0$. Write (4.44) for each $\Pi_r$ and let $r \to \infty$. The left-hand side converges to $\|\int_a^b \varphi(\lambda)E(d\lambda)f\|^2$ [use the implication $\Rightarrow$ of Proposition 1.1(a)], and the right-hand side converges to $\int_a^b |\varphi(\lambda)|^2 m_f(d\lambda)$ by (4.13). This proves (4.42).

(d) This result is obtained by using Proposition 2.1 and (4.40):

$$\int_a^b \varphi(\lambda)E(d\lambda) \cdot \int_a^b \psi(\lambda)E(d\lambda) = \operatorname*{s-lim}_{r\to\infty} \Sigma(\Pi_r;\varphi) \cdot \operatorname*{s-lim}_{r\to\infty} \Sigma(\Pi_r;\psi)$$
$$= \operatorname*{s-lim}_{r\to\infty} \Sigma(\Pi_r;\varphi)\Sigma(\Pi_r;\psi) = \operatorname*{s-lim}_{r\to\infty} \Sigma(\Pi_r;\varphi\psi) = \int_a^b (\varphi\psi)(\lambda)E(d\lambda).$$

We finally complete the proof of (a). We set $\omega = \sup_{\mu\in[a,b]\cap\operatorname{supp}\{E_\lambda\}} |\varphi(\mu)|$. Eq. (4.42) implies that

$$\Big\|\int_a^b \varphi(\lambda)E(d\lambda)f\Big\|^2 \leq \sup_{\mu\in[a,b]\cap\operatorname{supp}\{E_\lambda\}} |\varphi(\mu)|^2 \|f\|^2$$

(see also the Remark following this proof). It follows that $\|\int_a^b \varphi(\lambda)E(d\lambda)\| \leq \omega$. If $\omega \neq 0$, let $\delta \in (0, \omega)$. There exists $\lambda_0 \in (a, b) \cap \operatorname{supp}\{E_\lambda\}$ with $|\varphi(\lambda_0)| > \omega - \delta$. Since the function $\varphi$ is continuous, there exists $\varepsilon > 0$ such that $|\varphi(\mu)| > \omega - 2\delta$ for $\mu \in [\lambda_0 - \varepsilon, \lambda_0 + \varepsilon]$ and such that $[\lambda_0 - \varepsilon, \lambda_0 + \varepsilon] \subset (a, b)$. Clearly $E((\lambda_0 - \varepsilon, \lambda_0 + \varepsilon]) \neq O$.

For $f$ belonging to the subspace $\mathcal{M}((\lambda_0 - \varepsilon, \lambda_0 + \varepsilon])$, one obtains from Eq. (4.42) that $\|\int_a^b \varphi(\lambda)E(d\lambda)f\|^2 \geq (\omega - 2\delta)^2\|f\|^2$, i.e. $\|\int_a^b \varphi(\lambda)E(d\lambda)\| \geq (\omega - 2\delta)$. As $\delta$ is arbitrary, we have proved (4.41). □

REMARK. If $\lambda_0$ is a point in $(a, b]$ which does not belong to $\operatorname{supp}\{E_\lambda\}$, then there exists $\varepsilon > 0$ such that $E((\lambda_0 - \varepsilon, \lambda_0 + \varepsilon)) = O$, so that $(\lambda_0 - \varepsilon, \lambda_0 + \varepsilon) \cap \operatorname{supp}\{E_\lambda\} = \varnothing$. Then, if in a sequence of partitions $\{\Pi_r\}$, $r$ is sufficiently large so that $|\Pi_r| < \varepsilon$, and if $J_{\lambda_0}$ denotes the interval $(s_{k-1}, s_k]$ of this partition $\Pi_r$ that contains $\lambda_0$, one has $E(J_{\lambda_0}) = O$; thus the contribution of this interval $J_{\lambda_0}$ to the sum in (4.39) is zero, in other terms the value of $\varphi$ at the point $\lambda_0$ (even in some neighbourhood of $\lambda_0$) has no influence on the integral $\int_a^b \varphi(\lambda)E(d\lambda)$. This integral, in particular its norm as given by (4.41), involves only the values of $\varphi(\mu)$ for $\mu \in \operatorname{supp}\{E_\lambda\}$.

Proposition 4.10 gives the essential properties of spectral integrals on a finite interval. We next consider the case in which the domain of integration is $\mathbb{R}$. We define $\int_{-\infty}^{\infty} \varphi(\lambda)E(d\lambda)$ as the strong limit of $\int_{-M}^{M} \varphi(\lambda)E(d\lambda)$ as $M \to \infty$. One has to distinguish between different situations. We again assume $\varphi\colon \mathbb{R} \to \mathbb{C}$ to be continuous.

(1) Suppose that the support of the spectral family $\{E_\lambda\}$ is bounded. Then the limit $M \to \infty$ is not necessary, i.e. the operator $\int_{-M}^{M} \varphi(\lambda)E(d\lambda)$ is independent of $M$ for $M \geq M_0$ if $\operatorname{supp}\{E_\lambda\} \subset [-M_0, M_0]$. In this situation $\int_{-\infty}^{\infty} \varphi(\lambda)E(d\lambda)$ exists for each continuous $\varphi\colon \mathbb{R} \to \mathbb{C}$, belongs to $\mathcal{B}(\mathcal{H})$ and has the properties stated in Proposition 4.10.

(2) Suppose that the function $\varphi$ is bounded, i.e. $|\varphi(\lambda)| \leq \|\varphi\|_\infty < \infty$ for all $\lambda \in \mathbb{R}$. Again the strong limit of $\int_{-M}^{M} \varphi(\lambda)E(d\lambda)$ exists in $\mathcal{B}(\mathcal{H})$; to see this it suffices to show that for each vector $f \in \mathcal{H}$ one has $\|\int_a^b \varphi(\lambda)E(d\lambda)f\|^2 \to 0$ as $a, b \to -\infty$ and as $a, b \to +\infty$ (always with $a < b$). By (4.42) one has

$$\left\|\int_a^b \varphi(\lambda)E(d\lambda)f\right\|^2 \leq \|\varphi\|_\infty^2 \int_a^b m_f(d\lambda) = \|\varphi\|_\infty^2 m_f((a, b]),$$

and $m_f((a, b]) \to 0$ as $a, b \to -\infty$ and as $a, b \to +\infty$ since $m_f$ is a finite measure on $\mathbb{R}$. In this case it is easy to convince oneself that the statements (b) - (d) of Proposition 4.10 remain true for the integral over $\mathbb{R}$ [in other terms one may replace $\int_a^b$ by $\int_{-\infty}^{\infty}$ in Proposition 4.10(b) - (d), by assuming also that $\|\psi\|_\infty < \infty$ in (d)].

(3) Finally we consider the situation in which $\varphi$ is unbounded (a continuous function $\varphi$ on $\mathbb{R}$ is necessarily bounded on each finite interval $[-M, M]$ but could diverge at infinity, as does for example $\varphi(\lambda) = \lambda^2$). We shall see that in this case (if $\operatorname{supp}\{E_\lambda\}$ is not bounded) the limit of $\int_{-M}^{M} \varphi(\lambda)E(d\lambda)f$ exists only for a dense set ($\neq \mathcal{H}$) of vectors $f$. Thus $\int_{-\infty}^{\infty} \varphi(\lambda)E(d\lambda)$ will be an unbounded operator, with dense domain $\mathcal{D}_\varphi$. To determine this domain we first observe that each vector $f$ having compact support with respect to $\{E_\lambda\}$ belongs to $\mathcal{D}_\varphi$; indeed in this situation there exists a number $M_0 \in \mathbb{R}_+$ such that $E((-M_0, M_0])f = f$, hence the integral $\int_{-M}^{M} \varphi(\lambda)E(d\lambda)f$ is

independent of $M$ for $M > M_0$. This shows that $\mathcal{D}_\varphi$ is dense in $\mathcal{H}$. Next suppose that $f$ belongs to $\mathcal{D}_\varphi$. Then one must have [use $\Rightarrow$ of Proposition 1.1(a) and (4.42)]:

$$\infty > \Big\| \int_{-\infty}^{\infty} \varphi(\lambda) E(d\lambda) f \Big\|^2 = \lim_{M\to\infty} \Big\| \int_{-M}^{M} \varphi(\lambda) E(d\lambda) f \Big\|^2$$
$$= \lim_{M\to\infty} \int_{-M}^{M} |\varphi(\lambda)|^2 m_f(d\lambda) = \int_{-\infty}^{\infty} |\varphi(\lambda)|^2 m_f(d\lambda).$$

Consequently $f \in \mathcal{D}_\varphi \Longrightarrow \int_{-\infty}^{\infty} |\varphi(\lambda)|^2 m_f(d\lambda) < \infty$. On the other hand, if $f$ is such that this integral is finite, one will have $\| \int_a^b \varphi(\lambda) E(d\lambda) f \|^2 \to 0$ as $a, b \to -\infty$ and as $a, b \to +\infty$, hence $\int_{-M}^{M} \varphi(\lambda) E(d\lambda) f$ will be strongly Cauchy as $M \to \infty$. Summing up:

$$\boxed{\int_{-\infty}^{\infty} \varphi(\lambda) E(d\lambda) \text{ exists on } \mathcal{D}_\varphi = \{ f \in \mathcal{H} \,\Big|\, \int_{-\infty}^{\infty} |\varphi(\lambda)|^2 m_f(d\lambda) < \infty \}.} \tag{4.45}$$

**Lemma 4.11.** *Assume* $\varphi \colon \mathbb{R} \to \mathbb{C}$ *to be continuous. Then* $\int_{-\infty}^{\infty} \varphi(\lambda) E(d\lambda)$ *is a closed operator* [*on* $\mathcal{D}_\varphi$ *defined in* (4.45)] *and its adjoint is given by* $\int_{-\infty}^{\infty} \overline{\varphi}(\lambda) E(d\lambda)$ (*with* $\mathcal{D}_{\overline{\varphi}} = \mathcal{D}_\varphi$). *In particular* $\int_{-\infty}^{\infty} \varphi(\lambda) E(d\lambda)$ *is self-adjoint if* $\varphi$ *is a real function.*

PROOF. We set $\Phi = \int_{-\infty}^{\infty} \varphi(\lambda) E(d\lambda)$, with $\mathcal{D}(\Phi) = \mathcal{D}_\varphi$.

(i) Let $f$ and $g$ belong to $\mathcal{D}_\varphi \equiv \mathcal{D}_{\overline{\varphi}}$. Then [use Proposition 4.10(b)]:

$$\langle f, \Phi g \rangle = \langle f, \int_{-\infty}^{\infty} \varphi(\lambda) E(d\lambda) g \rangle = \lim_{M\to\infty} \langle f, \int_{-M}^{M} \varphi(\lambda) E(d\lambda) g \rangle$$
$$= \lim_{M\to\infty} \langle \int_{-M}^{M} \overline{\varphi}(\lambda) E(d\lambda) f, g \rangle = \langle \int_{-\infty}^{\infty} \overline{\varphi}(\lambda) E(d\lambda) f, g \rangle.$$

This shows that $\mathcal{D}_{\overline{\varphi}} \subseteq \mathcal{D}(\Phi^*)$ and that $\Phi^* f = \int_{-\infty}^{\infty} \overline{\varphi}(\lambda) E(d\lambda) f$ if $f \in \mathcal{D}(\overline{\varphi})$. Hence $\Phi^*$ is an extension of $\int_{-\infty}^{\infty} \overline{\varphi}(\lambda) E(d\lambda)$. To see that $\Phi^* = \int_{-\infty}^{\infty} \overline{\varphi}(\lambda) E(d\lambda)$, it now suffices to show that $\mathcal{D}(\Phi^*) \subseteq \mathcal{D}_{\overline{\varphi}}$. If $f \in \mathcal{D}(\Phi^*)$ there exists $f^* \in \mathcal{H}$ such that

$$\langle f, \int_{-\infty}^{\infty} \varphi(\lambda) E(d\lambda) g \rangle = \langle f^*, g \rangle \qquad \forall g \in \mathcal{D}_\varphi,$$

in particular for each $g$ having compact support with respect to $\{E_\lambda\}$. By taking into account (2.6), we then have for each $M \in (0, \infty)$:

$$\begin{aligned}\|E((-M,M])f^*\| &= \sup_{g\in\mathcal{H},\|g\|=1} |\langle E((-M,M])f^*,g\rangle| \\ &= \sup_{g\in\mathcal{H},\|g\|=1} \big|\langle f^*,E((-M,M])g\rangle\big| = \sup_{g\in\mathcal{H},\|g\|=1} \Big|\Big\langle f,\int_{-M}^{M}\varphi(\lambda)E(d\lambda)g\Big\rangle\Big| \\ &= \sup_{g\in\mathcal{H},\|g\|=1} \Big|\Big\langle \int_{-M}^{M}\overline{\varphi}(\lambda)E(d\lambda)f,g\Big\rangle\Big| = \Big\|\int_{-M}^{M}\overline{\varphi}(\lambda)E(d\lambda)f\Big\| \\ &= \sqrt{\int_{-M}^{M}|\varphi(\lambda)|^2 m_f(d\lambda)}.\end{aligned}$$

It follows that $\sup_{M>0}\int_{-M}^{M}|\varphi(\lambda)|^2 m_f(d\lambda) \le \|f^*\|^2 < \infty$. Consequently we have $\int_{-\infty}^{\infty}|\varphi(\lambda)|^2 m_f(d\lambda) < \infty$. This shows that $f\in\mathcal{D}(\Phi^*) \implies f\in\mathcal{D}_\varphi = \mathcal{D}_{\overline{\varphi}}$.

(ii) Since $\int_{-\infty}^{\infty}\varphi(\lambda)E(d\lambda)$ is the adjoint of $\int_{-\infty}^{\infty}\overline{\varphi}(\lambda)E(d\lambda)$, it is a closed operator [see Proposition 2.19(a)]. □

The following consequence of (4.24) will occasionally be used further on: if $f$ is a vector such that $E((b,c])f = f$ for some numbers $b < c$, then $\int_{-\infty}^{\infty}\varphi(\lambda)E(d\lambda)f = \int_b^c\varphi(\lambda)E(d\lambda)f$.

**4.2.4.** We proceed by discussing the significance of the preceding developments for the theory of self-adjoint operators. Let us consider the function $\varphi(\lambda) = \lambda$. If $\{E_\lambda\}$ is a spectral family, then $\int_{-\infty}^{\infty}\lambda E(d\lambda)$ defines a self-adjoint operator $A$, called ***the self-adjoint operator associated with $\{E_\lambda\}$***, with domain

$$\mathcal{D}(A) = \Big\{f\in\mathcal{H}\,\Big|\,\int_{-\infty}^{\infty}\lambda^2 m_f(d\lambda) \equiv \int_{-\infty}^{\infty}\lambda^2\langle f,E(d\lambda)f\rangle < \infty\Big\}, \tag{4.46}$$

and one has [6]

$$\langle g,Af\rangle = \int_{-\infty}^{\infty}\lambda\langle g,E(d\lambda)f\rangle \qquad \forall g\in\mathcal{H},\ \forall f\in\mathcal{D}(A). \tag{4.47}$$

The operator $A$ belongs to $\mathcal{B}(\mathcal{H})$ if and only if the support of $\{E_\lambda\}$ is bounded, in which case $\|A\| = \max\{|M_-|,|M_+|\}$, where $M_-$ and $M_+$ are the lower bound and the upper bound respectively of $\{E_\lambda\}$ [one has $\|A\| = \sup_{\mu\in\operatorname{supp}\{E_\lambda\}}|\mu|$ by (4.41)].

So each spectral family determines canonically a self-adjoint operator by the preceding description. The *Spectral Theorem* asserts the converse: given a self-adjoint operator in a Hilbert space $\mathcal{H}$, there exists a unique (right continuous) spectral family $\{E_\lambda\}$, called ***the spectral family of A***, such that $A = \int_{-\infty}^{\infty}\lambda E(d\lambda)$. Thus there is a one-to-one correspondence between self-adjoint operators and right continuous spectral families! We shall occasionally write $\{E_\lambda^A\}$ (rather than simply $\{E_\lambda\}$) for the spectral family of a self-adjoint operator $A$.

[6] If $g \neq f$, the integral in (4.47) is with respect to a complex measure $\langle g,E(\cdot)f\rangle$. One can interpret (4.47) as the scalar product between the vectors $g$ and $\int_{-\infty}^{\infty}\lambda E(d\lambda)f$, or one may decompose (4.47) into a sum of four Stieltjes integrals by taking into account the polarisation identity (2.15).

The Spectral Theorem is a fundamental result in the theory of operators in a complex Hilbert space (it generalises the fact that a Hermitian $n \times n$ matrix is diagonalisable). One finds in the literature various different proofs of this result; they are all rather long, and we shall discuss one of them in Section 4.4 and present another one in the Appendix of this chapter. For the moment we assume that this result is known and consider some aspects of the ***functional calculus*** of a self-adjoint operator. A comment on the quantum-mechanical meaning of the spectral family of a self-adjoint operator $A$ is made at the end of §4.3.4.

So let $A$ be a self-adjoint operator and $\{E_\lambda\}$ its spectral family. Then, as seen before, one can associate with each continuous function $\varphi$ defined on $\mathbb{R}$ an operator $\int_{-\infty}^{\infty} \varphi(\lambda) E(d\lambda)$. Henceforth this operator will be denoted $\varphi(A)$, and we now try to justify this notation. In order to avoid domain questions, we assume $A$ to be bounded, so that $\varphi(A) \in \mathcal{B}(\mathcal{H})$ for each $\varphi$. Since $A = \int_{-\infty}^{\infty} \lambda E(d\lambda)$, one has by Eq. (4.43): $A^2 = \left[\int_{-\infty}^{\infty} \lambda E(d\lambda)\right]^2 = \int_{-\infty}^{\infty} \lambda^2 E(d\lambda)$, which does coincide with $\varphi(A)$ for $\varphi(\lambda) = \lambda^2$. Similarly one will have $A^k = \int_{-\infty}^{\infty} \lambda^k E(d\lambda)$ for $\varphi(\lambda) = \lambda^k$ for $k = 1, 2, 3, \ldots$. We shall also discuss other functions $\varphi$ that justify the notation $\varphi(A)$ for $\int_{-\infty}^{\infty} \varphi(\lambda) E(d\lambda)$ [for example $\varphi(\lambda) = (\lambda - z)^{-1}$ for the resolvent of $A$: we shall see that $(A - z)^{-1} = \int_{-\infty}^{\infty} (\lambda - z)^{-1} E(d\lambda)$]; for more general functions $\varphi$ one should take this notation as a definition.

If the number $\mu$ belongs to the support of the spectral family of $A$ and if $\varepsilon > 0$, then $E((\mu - \varepsilon, \mu + \varepsilon]) \neq O$, and if $f \in \mathcal{M}((\mu - \varepsilon, \mu + \varepsilon])$, then

$$Af = \int_{\mu-\varepsilon}^{\mu+\varepsilon} \lambda E(d\lambda) f \approx \mu \int_{\mu-\varepsilon}^{\mu+\varepsilon} E(d\lambda) f = \mu f \quad \text{if } \varepsilon \text{ is sufficiently small.} \tag{4.48}$$

Thus $A$ is approximately multiplication by $\mu$ in $\mathcal{M}((\mu - \varepsilon, \mu + \varepsilon])$ if $\varepsilon$ is sufficiently small. A similar argument shows that $\int_{-\infty}^{\infty} \varphi(\lambda) E(d\lambda)$ is approximately multiplication by $\varphi(\mu)$ in $\mathcal{M}((\mu - \varepsilon, \mu + \varepsilon])$, thus justifying the notation $\varphi(A)$ for this operator. One should also note the analogy with multiplication operators discussed in Section 2.5; the Spectral Theorem is a more precise statement of the fact, often cited by physicists, that a self-adjoint operator can be diagonalised.[7]

[7] Since the operator $A$ is essentially multiplication by $\mu$ on subspaces of the form $\mathcal{M}((\mu - \varepsilon, \mu + \varepsilon])$, it can be approximately diagonalised by covering its spectrum by a disjoint union of intervals of length $2\varepsilon$, with very small $\varepsilon$, and writing $\mathcal{H}$ as the direct sum of the subspaces $\mathcal{M}(J)$ associated with the intervals $J$ in this partition. To obtain an exact diagonalisation one would have to consider the limit where $\varepsilon \to 0$. Now, if $\mu$ is a point in the continuous spectrum of $A$, then $\mathcal{M}(\{\mu\})$ will not be a subspace of $\mathcal{H}$, so that the mentioned limit has to be interpreted carefully. This can be done by using the notion of a direct integral of Hilbert spaces which generalises that of direct sums. Without entering into details we mention that each self-adjoint operator $A$ is diagonalisable in the sense that it is unitarily equivalent to the multiplication operator by the (real) variable in some direct integral, called a ***spectral representation*** of $A$. As a rather simple example we shall treat in Chapter 5 the diagonalisation of the free Hamiltonian $H_0$, see the discussion under point (2) on page 229 and the ensuing formulas for the spectral representation of $H_0$. In this example the direct integral reduces to a space of the form $L^2(\mathcal{O}; \mathcal{K}, m)$ [see the end of §1.4.3], with $\mathcal{O} = [0, \infty) =$ the spectrum of $H_0$. A formulation of the diagonalisation of a general self-adjoint operator is given for example in Volume I of [RS].

We return to functional calculus by also admitting unbounded operators $A$. We use the fact (which will be proved in Example 4.13) that the support of the spectral family of $A$ coincides with its spectrum: $\operatorname{supp}\{E_\lambda\} = \sigma(A)$, and we denote by $BC(\mathbb{R})$ the set of bounded continuous complex-valued functions defined on $\mathbb{R}$. The set $BC(\mathbb{R})$ has the structure of an algebra, i.e. it is a linear vector space over the field $\mathbb{C}$ equipped with an operation of multiplication. We have the following results on the ***bounded functional calculus*** of $A$:

(1) *With each bounded continuous function* $\varphi$ *is associated an operator* $\varphi(A)$ *in* $\mathcal{B}(\mathcal{H})$, *given by*

$$\varphi(A) = \int_{-\infty}^{\infty} \varphi(\lambda) E(d\lambda). \tag{4.49}$$

$\varphi(A)$ *commutes with each* $E_\lambda$, *its adjoint is given by*

$$\varphi(A)^* = \int_{-\infty}^{\infty} \overline{\varphi}(\lambda) E(d\lambda) \equiv \overline{\varphi}(A), \tag{4.50}$$

*and one has*

$$\|\varphi(A)\| = \sup_{\lambda \in \sigma(A)} |\varphi(\lambda)|. \tag{4.51}$$

(2) *The mapping* $\varphi \mapsto \varphi(A)$ *defines a homomorphism from the algebra* $BC(\mathbb{R})$ *into* $\mathcal{B}(\mathcal{H})$, *in other terms: if* $\varphi, \psi \in BC(\mathbb{R})$ *and* $\alpha \in \mathbb{C}$, *then*

$$(\alpha\varphi + \psi)(A) = \alpha\varphi(A) + \psi(A) \qquad \text{and} \qquad (\varphi\psi)(A) = \varphi(A)\psi(A) \tag{4.52}$$

[the operations on the left of the equality signs are in $BC(\mathbb{R})$, those on the right of these signs are in $\mathcal{B}(\mathcal{H})$]. *In particular one has*

$$\varphi(A)\psi(A) = \psi(A)\varphi(A) \qquad \forall \varphi, \psi \in BC(\mathbb{R}). \tag{4.53}$$

(3) $\varphi(A)$ *is self-adjoint if and only if* $\varphi(\lambda)$ *is real for* $\lambda \in \sigma(A)$.

(4) $\varphi(A)$ *is unitary if and only if* $|\varphi(\lambda)| = 1$ *for* $\lambda \in \sigma(A)$.

The result of (4) is obtained from (4.50) and (4.52): $\varphi(A)^*\varphi(A) = \varphi(A)\varphi(A)^* = (\overline{\varphi}\varphi)(A)$, which is equal to the identity operator $I$ if and only if $\overline{\varphi}\varphi \equiv 1$ on the spectrum of $A$.

*If* $\varphi$ *is continuous but unbounded, then Eq.* (4.49) *defines a closed operator* $\varphi(A)$ *with domain*

$$\mathcal{D}(\varphi(A)) = \Big\{ f \in \mathcal{H} \,\Big|\, \int_{-\infty}^{\infty} |\varphi(\lambda)|^2 m_f(d\lambda) < \infty \Big\}. \tag{4.54}$$

The relation (4.50) remains true for unbounded $\varphi$, however (4.52) is replaced by $\alpha\varphi(A) + \psi(A) \subseteq (\alpha\varphi + \psi)(A)$ and $\varphi(A)\psi(A) \subseteq (\varphi\psi)(A)$ (it may happen for example that $\psi$ is unbounded but $\varphi\psi$ is bounded).

**Example 4.12. Continuous unitary group associated with $A$**

We consider the function $\varphi_t(\lambda) = e^{-i\lambda t}$, with $t \in \mathbb{R}$, and introduce the notation $U_t$ for the operator $\varphi_t(A)$: $U_t = e^{-iAt}$. By (4) above, $U_t$ is a *unitary* operator. The family $\{U_t\}$ forms a *group*, more precisely a representation of the additive group $\mathbb{R}$, i.e.

$$U_s U_t = \int_{-\infty}^{\infty} e^{-i\lambda s} E(d\lambda) \int_{-\infty}^{\infty} e^{-i\mu t} E(d\mu) = \int_{-\infty}^{\infty} e^{-i\lambda s} e^{-i\lambda t} E(d\lambda)$$
$$= \int_{-\infty}^{\infty} e^{-i\lambda(s+t)} E(d\lambda) = U_{s+t}.$$

In particular $U_t^* = [U_t]^{-1} = U_{-t}$. Also the family $\{U_t\}$ is *strongly continuous* in $t$. To see this one can use the Dominated Convergence Theorem (Proposition 1.9): since $|e^{-i\lambda(t+\varepsilon)} - e^{-i\lambda t}|^2 \leq 4$ one obtains

$$\|U_{t+\varepsilon} f - U_t f\|^2 = \int_{-\infty}^{\infty} \left|e^{-i\lambda(t+\varepsilon)} - e^{-i\lambda t}\right|^2 m_f(d\lambda) \to 0 \quad \text{as } \varepsilon \to 0.$$

One can avoid the application of the Dominated Convergence Theorem by arguing as follows. For $M > 0$ set $E_M = E((-M, M])$ and write

$$\|U_{t+\varepsilon} f - U_t f\|^2 = \|[U_{t+\varepsilon} - U_t] E_M f\|^2 + \|[U_{t+\varepsilon} - U_t][I - E_M] f\|^2.$$

Let $\delta > 0$. Since $E_M$ converges strongly to $I$ as $M \to \infty$ and $\|U_\tau\| = 1$ for each $\tau$, one can choose $M$ sufficiently large so that $\|[U_{t+\varepsilon} - U_t][I - E_M] f\|^2 < \delta/2$ for all $t$ and all $\varepsilon$. With this fixed $M$ one has

$$\|[U_{t+\varepsilon} - U_t] E_M f\|^2 = \int_{-M}^{M} \left|e^{-i\lambda(t+\varepsilon)} - e^{-i\lambda t}\right|^2 m_f(d\lambda)$$
$$= \int_{-M}^{M} \left|e^{-i\lambda\varepsilon} - 1\right|^2 m_f(d\lambda) \leq \int_{-M}^{M} |\lambda\varepsilon|^2 m_f(d\lambda) \leq \varepsilon^2 M^2 \|f\|^2,$$

which is less than $\delta/2$ for all $\varepsilon$ satisfying $|\varepsilon| < \varepsilon_0$ for a suitable $\varepsilon_0 > 0$. It follows that $\|U_{t+\varepsilon} f - U_t f\|^2 < \delta$ for each $|\varepsilon| < \varepsilon_0$, thus proving the strong continuity of $\{U_s\}$ at $s = t$.

The function $\varphi_t(\lambda) = e^{-i\lambda t}$ can be written as a series $\varphi_t(\lambda) = \sum_{k=0}^{\infty} (-i\lambda t)^k / k!$ One can ask if $U_t$ is given by the corresponding series of operators, i.e. if

$$U_t = \sum_{k=0}^{\infty} \frac{(-it)^k}{k!} A^k.$$

When $A$ belongs to $\mathcal{B}(\mathcal{H})$, the preceding series converges in the sense of uniform convergence, and $U_t$ is given by this series. If $A$ is unbounded, the terms in the series make sense only on vectors $f$ belonging to $\mathcal{D}(A^k)$ for each $k = 1, 2, 3, \ldots$ The set of these vectors is dense in $\mathcal{H}$, as it contains all vectors having compact support with respect to the spectral family $\{E_\lambda\}$ of $A$. If $f$ is a vector of compact support with respect to $\{E_\lambda\}$, then the series $\sum_{k=0}^{\infty} [(-it)^k / k!] A^k f$ is strongly convergent and its sum equals $U_t f$. The verification of these statements is left as an exercise.

**Example 4.13. Resolvent of $A$**
If $z \in \mathbb{C}$ with $\Im z \neq 0$ or if $z \in \mathbb{R} \setminus \text{supp}\{E_\lambda\}$, then the function $\psi_z(\lambda) = (\lambda - z)^{-1}$ is continuous and bounded on the support of $\{E_\lambda\}$ (we recall that $\mathbb{R} \setminus \text{supp}\{E_\lambda\}$ is an open set). Thus $\psi_z(A) \in \mathcal{B}(\mathcal{H})$, with $\|\psi_z(A)\| = \sup_{\mu \in \text{supp}\{E_\lambda\}} |\mu - z|^{-1} = [\text{dist}(z, \text{supp}\{E_\lambda\})]^{-1}$.[8] One should use the notation $(A - z)^{-1}$ for $\psi_z(A)$. For this to be justified we show that $\psi_z(A)$ is the resolvent of $A$ at the point $z$.

(i) Let $f \in \mathcal{D}(A)$. Then

$$\|\psi_z(A)f\|^2 = \int_{-\infty}^{\infty} |(\mu - z)|^2 m_f(d\mu) \geq \left[\text{dist}(z, \text{supp}\{E_\lambda\})\right]^2 \int_{-\infty}^{\infty} m_f(d\mu)$$
$$= \left[\text{dist}(z, \text{supp}\{E_\lambda\})\right]^2 \|f\|^2.$$

Thus: if $\psi_z(A)f = \mathbf{0}$, then $\|f\| = 0$, hence $f = \mathbf{0}$, i.e. $\psi_z(A)$ is invertible.

(ii) It remains to be seen that $(A - z)\psi_z(A)g = g$ for each $g \in \mathcal{H}$ [which implies that $\mathcal{R}(A - z) = \mathcal{H}$] and that $\psi_z(A)(A - z)f = f$ for each $f \in \mathcal{D}(A)$. Let us check for example the second relation. Let $f$ be a fixed vector in $\mathcal{D}(A)$. One has $\psi_z(A) = \text{s-lim}_{M\to\infty} \int_{-M}^{M} (\lambda - z)^{-1} E(d\lambda)$ and $(A - z)f = \text{s-lim}_{M\to\infty} \int_{-M}^{M} (\lambda - z) E(d\lambda) f$. By using first Proposition 2.1 [or (2.16) if$A$ is unbounded] and then (4.43) one finds that

$$\psi_z(A)(A - z)f = \operatorname*{s-lim}_{M\to\infty} \Big[\int_{-M}^{M} (\lambda - z)^{-1} E(d\lambda) \cdot \int_{-M}^{M} (\mu - z) E(d\mu)\Big] f$$
$$= \operatorname*{s-lim}_{M\to\infty} \int_{-M}^{M} (\lambda - z)^{-1}(\lambda - z) E(d\lambda) f = \operatorname*{s-lim}_{M\to\infty} E((-M, M])f = f.$$

We have thus justified the identity $(A - z)^{-1} = \int_{-\infty}^{\infty} (\lambda - z)^{-1} E(d\lambda)$ if $\Im z \neq 0$ or if $z \in \mathbb{R} \setminus \text{supp}\{E_\lambda\}$. The set of these values of $z$ coincides with the resolvent set $\rho(A)$ of the operator $A$. Indeed, the preceding considerations show that $\mathbb{R} \setminus \text{supp}\{E_\lambda\} \subseteq \rho(A)$; on the other hand, if $\lambda_0$ is a real point in $\rho(A)$, then $(A - \lambda_0 - i\varepsilon)^{-1}$ has a limit in $\mathcal{B}(\mathcal{H})$ as $\varepsilon \to 0$; however, if $\lambda \in \text{supp}\{E_\lambda\}$, one has $\|(A - \lambda - i\varepsilon)^{-1}\| = 1/|\varepsilon|$, which shows that $\text{supp}\{E_\lambda\} \nsubseteq \rho(A)$. In conclusion, at each $z \in \rho(A)$ the resolvent of $A$ is given by

$$(A - z)^{-1} = \int_{-\infty}^{\infty} (\lambda - z)^{-1} E(d\lambda) \equiv \int_{\sigma(A)} (\lambda - z)^{-1} E(d\lambda) \tag{4.55}$$

and satisfies

$$\|(A - z)^{-1}\| = \left[\text{dist}(z, \sigma(A))\right]^{-1}. \tag{4.56}$$

If $A$ is bounded and if $M_-$ and $M_+$ denote the lower and upper bound respectively of $\{E_\lambda\}$, it is clear that $\sigma(A) \subseteq [M_-, M_+] \subseteq [-\|A\|, \|A\|]$. If $A$ is semibounded, for example bounded below, one has $\sigma(A) \subseteq [M_-, \infty)$, where $M_- \in \mathbb{R}$ is the supremum of all real numbers $M$ such that $E_\lambda = O$ for each $\lambda < M$.

[8] For a real $z$ the function $\psi_z$ is not continuous at $\lambda = z$. However, if $z \in \mathbb{R} \setminus \text{supp}\{E_\lambda\}$, then the point $z$ is at positive distance from $\text{supp}\{E_\lambda\}$, and one obtains the same operator $\psi_z(A)$ if one modifies the function $\psi_z$ in a sufficiently small neighbourhood of $z$ in such a way that the modified function is continuous and bounded.

**Remark 4.14.** (a) So far we have considered integrals $\int \varphi(\lambda)E(d\lambda)$ for continuous functions $\varphi$, which permits their definition in the sense of Riemann-Stieltjes. One can extend this functional calculus to functions that are not necessarily continuous by using (numerical) integrals of the Lebesgue-Stieltjes type. If for example $\varphi\colon \mathbb{R} \to \mathbb{C}$ is a *bounded* measurable function, then the integral $\int_{-\infty}^{\infty} \varphi(\lambda)m_f(d\lambda)$ exists for each vector $f \in \mathcal{H}$ and defines the quantity $\langle f, \varphi(A)f\rangle$. By applying the polarisation identity one obtains an expression for $\langle g, \varphi(A)f\rangle$ for all $f, g \in \mathcal{H}$, and in this way one defines a bounded linear operator $\varphi(A)$ with $\|\varphi(A)\| \leq \sup_{\lambda\in\sigma(A)} |\varphi(\lambda)|$. If $\varphi$ is the characteristic function $\chi_{\mathcal{V}}$ of a Borel set $\mathcal{V} \in \mathcal{A}_{\mathrm{B}}$, then $\varphi(A) \equiv \chi_{\mathcal{V}}(A) = E(\mathcal{V})$ is the spectral projection of $A$ associated with $\mathcal{V}$. For certain locally unbounded functions [e.g. $\varphi(\lambda) = (\lambda - \lambda_0)^{-1}$ for some $\lambda_0 \in \sigma(A)$], one can also give a meaning to $\varphi(A)$ as an unbounded operator by conveniently defining its domain.
(b) If the Fourier transform $\tilde{\varphi}$ of $\varphi$ belongs to $L^1(\mathbb{R})$, then $\varphi(A)$ admits the following representation in terms of the unitary group $\{e^{-iAt}\}$ associated with $A$:

$$\varphi(A) = \frac{1}{\sqrt{2\pi}} \int_{-\infty}^{\infty} \tilde{\varphi}(t)e^{iAt}\,dt. \tag{4.57}$$

Indeed

$$\langle f, \varphi(A)f\rangle = \int_{\mathbb{R}} \varphi(\lambda)m_f(d\lambda) = \int_{\mathbb{R}} m_f(d\lambda)\,\frac{1}{\sqrt{2\pi}} \int_{\mathbb{R}} e^{i\lambda t}\tilde{\varphi}(t)\,dt.$$

By interchanging the order of integration (Fubini's Theorem) one arrives at

$$\langle f, \varphi(A)f\rangle = \frac{1}{\sqrt{2\pi}} \int_{\mathbb{R}} dt\,\tilde{\varphi}(t) \int_{\mathbb{R}} e^{i\lambda t} m_f(d\lambda) = \frac{1}{\sqrt{2\pi}} \int_{\mathbb{R}} dt\,\tilde{\varphi}(t)\langle f, e^{iAt}f\rangle,$$

and (4.57) follows by applying the polarisation identity.

**Example 4.15. Spectral family of $Q$**
The spectral family of the position operator $Q$ in $L^2(\mathbb{R})$ has already been indicated in (4.34). In this case $\varphi(Q)$ is the operator of multiplication by $\varphi(x)$. To illustrate this fact, let us denote this multiplication operator by $\Phi$ and consider for example an element $f$ of $L^2(\mathbb{R})$ such that $f(x) = 0$ almost everywhere outside some finite interval $[a, b]$, and assume $\varphi$ to be continuous. If $\Pi$ is a partition of $(a, b]$ as in Section 1.2, one has $\big[\Phi f - \Sigma(\Pi; \varphi)f\big](x) = \big[\varphi(x) - \varphi(u_k)\big]f(x)$ if $x \in (s_{k-1}, s_k]$, and consequently

$$\|\Phi f - \Sigma(\Pi; \varphi)f\|^2 = \sum_{k=1}^{N} \int_{s_{k-1}}^{s_k} |\varphi(x) - \varphi(u_k)|^2 |f(x)|^2\,dx.$$

For given $\varepsilon > 0$, one can choose the partition $\Pi$ so fine that $|\varphi(x) - \varphi(u_k)| \leq \varepsilon$ for all $x \in (s_{k-1}, s_k]$ and all $k = 1, \ldots, N$, and then one will have $\|\Phi f - \Sigma(\Pi; \varphi)f\|^2 \leq \varepsilon^2\|f\|^2$. This implies that $\Phi f = \int_a^b \varphi(\lambda)E(d\lambda)f \equiv \varphi(Q)f$ [here $\{E(\cdot)\}$ is the spectral family of the operator $Q$ given in (4.35)].

**Example 4.16. Spectral family of $Q^2$**
The operator $Q^2$ in $L^2(\mathbb{R})$ is multiplication by $x^2$. The projection $E_\lambda$ will be a multiplication operator; one must have $(E_\lambda f)(x) = f(x)$ if $x^2 \leq \lambda$ and $(E_\lambda f)(x) = 0$

otherwise. Thus $E_\lambda = O$ if $\lambda \leq 0$, whereas for $\lambda > 0$, $E_\lambda$ is the operator of multiplication by the characteristic function of the interval $[-\sqrt{\lambda}, \sqrt{\lambda}]$:

$$[E_\lambda f](x) = \chi_{[-\sqrt{\lambda},\sqrt{\lambda}]}(x) f(x) \quad \text{if } \lambda > 0; \quad E_\lambda = O \quad \text{if } \lambda \leq 0. \tag{4.58}$$

One can also consider the analogue in $L^2(\mathbb{R}^n)$. As seen in Example 2.30, the position operator $\vec{Q}$ is then a multi-component operator: $\vec{Q} = (Q_1, \ldots, Q_n)$, where $Q_j$ is the multiplication operator by $x_j$:

$$[Q_j f](\vec{x}) = x_j f(\vec{x}) \qquad \text{with} \quad \vec{x} = (x_1, \ldots, x_n) \in \mathbb{R}^n. \tag{4.59}$$

$Q^2 \equiv \vec{Q}^2$ is the operator of multiplication by $\vec{x}^{\,2} \equiv x_1^2 + \cdots + x_n^2$. Again one has $E_\lambda = O$ if $\lambda \leq 0$ and, for $\lambda > 0$: $(E_\lambda f)(\vec{x}) = f(\vec{x})$ if $\vec{x}^{\,2} \leq \lambda$ and $(E_\lambda f)(\vec{x}) = 0$ if $\vec{x}^{\,2} > \lambda$. Thus, for $\lambda > 0$, $E_\lambda$ is multiplication by $\chi_{[0,\lambda^{1/2}]}(|\vec{x}|)$, in other terms: $E_\lambda = \chi_{[0,\lambda^{1/2}]}(|\vec{Q}|) = \chi_{[0,\lambda]}(\vec{Q}^2)$.

**Example 4.17. Spectral family of $P^2$**
We have seen in §2.5.2 that the momentum operator $P$ in $L^2(\mathbb{R})$ is given by $\mathcal{F}^* Q \mathcal{F}$, in other terms $P$ is the multiplication operator by the variable $k$ on the Fourier transforms $\tilde{f}$ of the functions $f \in L^2(\mathbb{R})$: $[\mathcal{F}Pf](k) = k\tilde{f}(k)$. Therefore the spectral family of $P^2$ has the same form as that of $Q^2$ given in Example 4.16, except for the replacement of the wave functions by their Fourier transforms:

$$[\mathcal{F}E_\lambda f](k) = \chi_{[-\sqrt{\lambda},\sqrt{\lambda}]}(k) \tilde{f}(k) \quad \text{if } \lambda > 0; \quad E_\lambda = O \quad \text{if } \lambda \leq 0. \tag{4.60}$$

In $L^2(\mathbb{R}^n)$ the momentum operator $\vec{P}$ is an $n$-component operator: $\vec{P} = (P_1, \ldots, P_n)$, where $P_j$ is given by

$$[\mathcal{F}P_j f](\vec{k}) = k_j \tilde{f}(\vec{k}) \qquad \text{with} \quad \vec{k} = (k_1, \ldots, k_n) \in \mathbb{R}^n, \tag{4.61}$$

and where $\tilde{f}(\vec{k})$ denotes the (n-dimensional) Fourier transform of $f$ [see (2.79)]. The operator $P^2 \equiv \vec{P}^2$ is multiplication by $\vec{k}^2 \equiv k_1^2 + \cdots + k_n^2$ on $\tilde{f}(\vec{k})$: $[\mathcal{F}\vec{P}^2 f](\vec{k}) = \vec{k}^2 \tilde{f}(\vec{k})$, and its spectral family is given as follows: $E_\lambda = O$ if $\lambda \leq 0$ and, for $\lambda > 0$: $[\mathcal{F}E_\lambda f](\vec{k}) = \tilde{f}(\vec{k})$ if $\vec{k}^{\,2} \leq \lambda$ and $[\mathcal{F}E_\lambda f](\vec{k}) = 0$ if $\vec{k}^2 > \lambda$. Thus, for $\lambda > 0$: $E_\lambda = \chi_{[0,\sqrt{\lambda}]}(|\vec{P}|) = \chi_{[0,\lambda]}(\vec{P}^2)$.

## 4.3 Spectral parts of a self-adjoint operator

Let $A$ be a self-adjoint operator, $\{E_\lambda\}$ its spectral family and $\sigma(A)$ its spectrum. In quantum mechanics one generally distinguishes between the point spectrum (often called "discrete spectrum"), composed essentially of the eigenvalues of $A$, and the continuous spectrum. We give here definitions and precise characterisations of the spectral parts of self-adjoint operators. We should mention that one can use different methods for subdividing the spectrum, and the spectral parts obtained by means of different definitions are not necessarily identical. We present here a subdivision of the spectrum based on a decomposition of the Hilbert space. Other definitions use for example properties of the operator $A - \lambda$ for $\lambda \in \sigma(A)$ [$A - \lambda$ not invertible, $A - \lambda$ invertible but with unbounded inverse and with $\mathcal{D}((A-\lambda)^{-1})$ dense or non-dense in $\mathcal{H}$].

We have seen in §4.2.2 that each vector $f \in \mathcal{H}$ determines a Stieltjes measure $m_f$ by the formula $m_f(\mathcal{V}) = \langle f, E(\mathcal{V})f \rangle = \|E(\mathcal{V})f\|^2$ for $\mathcal{V} \in \mathcal{A}_{\mathrm{B}}$, and by (4.6) this measure has a unique decomposition into a sum of three measures $m_{f,\mathrm{p}}$, $m_{f,\mathrm{sc}}$ and $m_{f,\mathrm{ac}}$. It is very interesting to know that each of these three measures coincides with the Stieltjes measure associated with some vector in $\mathcal{H}$, in other words that there exist three vectors $f_{\mathrm{p}}$, $f_{\mathrm{sc}}$ and $f_{\mathrm{ac}}$ in $\mathcal{H}$ such that $f = f_{\mathrm{p}} + f_{\mathrm{sc}} + f_{\mathrm{ac}}$ and such that the part $m_{f,\mathrm{p}}$ of the Stieltjes measure associated with the vector $f$ coincides with the Stieltjes measure associated with the vector $f_{\mathrm{p}}$ (i.e. $m_{f,\mathrm{p}} = m_{f_{\mathrm{p}}}$), etc.

We shall see that the decomposition of the vectors $f \in \mathcal{H}$ into $f = f_{\mathrm{p}} + f_{\mathrm{sc}} + f_{\mathrm{ac}}$ induces an orthogonal decomposition of the Hilbert space $\mathcal{H}$ into three subspaces that one denotes by $\mathcal{H}_{\mathrm{p}}(A)$, $\mathcal{H}_{\mathrm{sc}}(A)$ and $\mathcal{H}_{\mathrm{ac}}(A)$. The notation indicates that this decomposition depends on the considered spectral family $\{E_\lambda\}$, or equivalently on the self-adjoint operator $A$ associated with this spectral family. Thus each self-adjoint operator $A$ in $\mathcal{H}$ induces a decomposition of $\mathcal{H}$ into

$$\mathcal{H} = \mathcal{H}_{\mathrm{p}}(A) \oplus \mathcal{H}_{\mathrm{sc}}(A) \oplus \mathcal{H}_{\mathrm{ac}}(A) \tag{4.62}$$

in such a way that $m_g$ is a pure point measure (a measure that is supported on a countable set of points) if $g \in \mathcal{H}_{\mathrm{p}}(A)$, a singular continuous measure if $g \in \mathcal{H}_{\mathrm{sc}}(A)$ and an absolutely continuous measure if $g \in \mathcal{H}_{\mathrm{ac}}(A)$ (assuming $g \neq \mathbf{0}$).

**4.3.1.** We begin by the characterisation of $\mathcal{H}_{\mathrm{p}}(A)$. Let us first show that (with $\mu \in \mathbb{R}$)

$$\boxed{f \in \mathcal{M}(\{\mu\}) \iff f \in \mathcal{D}(A) \text{ and } Af = \mu f.} \tag{4.63}$$

To see this we use the following formula which is obtained from (4.42) with $(a, b) = (-\infty, \infty)$, $f \in \mathcal{D}(A)$ and $\varphi(\lambda) = \lambda - \mu$:

$$\|Af - \mu f\|^2 = \int_{-\infty}^{\infty} |\lambda - \mu|^2 m_f(d\lambda). \tag{4.64}$$

Since the integrand is strictly positive at each $\lambda \neq \mu$, one can have $\|Af - \mu f\|^2 = 0$ if and only if $m_f(\mathcal{V}) = 0$ for each Borel set $\mathcal{V}$ with $\mu \notin \mathcal{V}$, i.e. the measure $m_f$ is supported on $\{\mu\}$. [9]

The relation (4.63) expresses the fact that a number $\lambda \in \mathbb{R}$ is an eigenvalue of $A$ if and only if $\dim \mathcal{M}(\{\lambda\}) \neq 0$, and in this case $\mathcal{M}(\{\lambda\})$ is the subspace of $\mathcal{H}$ spanned by the set of eigenvectors of $A$ associated with the eigenvalue $\lambda$. Since the set of points of discontinuity of a spectral family $\{E_\lambda\}$ is countable [as observed after (4.32)], the

[9] If $f \in \mathcal{M}(\{\mu\})$, then the measure $m_f$ is supported on the set $\{\mu\}$ consisting of the single point $\mu$: $m_f$ is proportional to the Dirac measure at the point $\mu$, formally $m_f(d\lambda) = \|f\|^2 \delta(\lambda - \mu) d\lambda$.

eigenvalues of a self-adjoint operator form a countable subset of $\mathbb{R}$. [10] In the sequel we denote the eigenvalues by $\{\lambda_1, \lambda_2, \ldots\}$. The set of all eigenvalues of $A$ is called the ***point spectrum*** of $A$.

One then defines $\mathcal{H}_{\mathrm{p}}(A)$ as the subspace of $\mathcal{H}$ spanned by the set of all eigenvectors of $A$. In other terms

$$\mathcal{H}_{\mathrm{p}}(A) = \bigoplus \mathcal{M}(\{\lambda_j\}), \tag{4.65}$$

where the sum extends over all eigenvalues of $A$. Thus a vector $f$ belongs to $\mathcal{H}_{\mathrm{p}}(A)$ if and only if $f = \sum_j f_j$ with $Af_j = \lambda_j f_j$ and with $\sum_j \|f_j\|^2 < \infty$. It is clear that $f \in \mathcal{H}_{\mathrm{p}}(A)$ if and only if there exists a countable set $\mathcal{W}$ of real numbers such that $m_f(\mathbb{R}\backslash\mathcal{W}) = 0$, in other terms such that $E(\mathcal{W})f = f$. We write $E_{\mathrm{p}}$ for the projection onto the subspace $\mathcal{H}_{\mathrm{p}}(A)$:

$$E_{\mathrm{p}} = P_{\mathcal{H}_{\mathrm{p}}(A)} = \sum_j E(\{\lambda_j\}). \tag{4.66}$$

**4.3.2.** We continue by defining the other subspaces occurring in (4.62). We denote by $|\mathcal{V}|$ the Lebesgue measure of a Borel set $\mathcal{V}$. One sets

- $\mathcal{H}_{\mathrm{ac}}(A) = \{f \in \mathcal{H} \mid m_f \text{ is an absolutely continuous measure}\}$
  $\equiv \{f \in \mathcal{H} \mid \text{the function } \lambda \mapsto \|E_\lambda f\|^2 \text{ is absolutely continuous}\}$
- $\mathcal{H}_{\mathrm{sc}}(A) = \{f \in \mathcal{H} \mid m_f \text{ is a singular continuous measure}\}$
  $\equiv \{f \in \mathcal{H} \mid \text{the function } \lambda \mapsto \|E_\lambda f\|^2 \text{ is continuous and } \exists\, \mathcal{V} \in \mathcal{A}_{\mathrm{B}} \text{ with } |\mathcal{V}| = 0 \text{ and } E(\mathcal{V})f = f\}$
- $\mathcal{H}_{\mathrm{c}}(A) = \{f \in \mathcal{H} \mid m_f \text{ is a continuous measure}\}$
  $\equiv \{f \in \mathcal{H} \mid m_f(\{\lambda\}) = 0\ \forall \lambda \in \mathbb{R}\}$
  $\equiv \{f \in \mathcal{H} \mid \text{the function } \lambda \mapsto \|E_\lambda f\|^2 \text{ is continuous}\}$
  $\equiv \{f \in \mathcal{H} \mid \lambda \mapsto E_\lambda f \text{ is strongly continuous}\}$ [11]
- $\mathcal{H}_{\mathrm{s}}(A) = \{f \in \mathcal{H} \mid \text{the measure } m_f \text{ and the Lebesgue measure are mutually singular}\}$
  $\equiv \{f \in \mathcal{H} \mid \exists\, \mathcal{V} \in \mathcal{A}_{\mathrm{B}} \text{ with } |\mathcal{V}| = 0 \text{ and } E(\mathcal{V})f = f\}$.

We next discuss the properties of these subsets of $\mathcal{H}$. The following inclusions are evident:

$$\mathcal{H}_{\mathrm{ac}}(A) \subseteq \mathcal{H}_{\mathrm{c}}(A)\,,\ \mathcal{H}_{\mathrm{sc}}(A) \subseteq \mathcal{H}_{\mathrm{c}}(A)\,,\ \mathcal{H}_{\mathrm{p}}(A) \subseteq \mathcal{H}_{\mathrm{s}}(A)\,,\ \mathcal{H}_{\mathrm{sc}}(A) \subseteq \mathcal{H}_{\mathrm{s}}(A). \tag{4.67}$$

Let us verify first that $\mathcal{H}_{\mathrm{c}}(A) = \mathcal{H}_{\mathrm{p}}(A)^\perp$ [which also shows that $\mathcal{H}_{\mathrm{c}}(A)$ is a subspace]. If $f \in \mathcal{H}_{\mathrm{c}}(A)$, then clearly $E(\{\lambda\})f = \mathbf{0}$ for each $\lambda \in \mathbb{R}$, hence $E_{\mathrm{p}}f = \mathbf{0}$ and consequently $f \perp \mathcal{H}_{\mathrm{p}}(A)$. Conversely, if $f \in \mathcal{H}_{\mathrm{p}}(A)^\perp$, then $E_{\mathrm{p}}f = \mathbf{0}$, hence $E(\{\lambda\})f = \mathbf{0}$ for every eigenvalue $\lambda$ of $A$ and consequently for each $\lambda \in \mathbb{R}$ [by (4.63)]; thus $m_f(\{\lambda\}) = 0$ for each $\lambda \in \mathbb{R}$, i.e. $f \in \mathcal{H}_{\mathrm{c}}(A)$.

---

[10] This also follows from Proposition 2.36(b) (orthogonality of the eigenvectors associated with different eigenvalues) because we assume that each basis of $\mathcal{H}$ is countable [condition (H4) for separability of $\mathcal{H}$ in §1.1.1].

[11] See Problem 4.2(b) for the last identity.

Let us now show that $\mathcal{H}_{\rm sc}(A)$ is a subspace. If $f, g \in \mathcal{H}_{\rm sc}(A)$ and $\alpha \in \mathbb{C}$, then clearly $E_\lambda(f+\alpha g) = E_\lambda f + \alpha E_\lambda g$ is strongly continuous; also there exist $\mathcal{V}, \mathcal{W} \in \mathcal{A}_{\rm B}$ with $|\mathcal{V}| = |\mathcal{W}| = 0$ and $E(\mathcal{V})f = f$, $E(\mathcal{W})g = g$, and then one has $|\mathcal{V} \cup \mathcal{W}| = 0$ and $E(\mathcal{V} \cup \mathcal{W})(f+\alpha g) = f + \alpha g$. This shows that $f + \alpha g \in \mathcal{H}_{\rm sc}(A)$, hence that $\mathcal{H}_{\rm sc}(A)$ is a linear manifold. To see that it is a subspace, consider a strong Cauchy sequence $\{f_k\}$ belonging to $\mathcal{H}_{\rm sc}(A)$ and denote its limit by $f$; we have to show that $f \in \mathcal{H}_{\rm sc}(A)$. We know already that $f \in \mathcal{H}_{\rm c}(A)$ because $\mathcal{H}_{\rm c}(A)$ is a subspace. For each $k \in \mathbb{N}$ there exists a Borel set $\mathcal{V}_k$ with $|\mathcal{V}_k| = 0$ and $E(\mathcal{V}_k)f_k = f_k$. Then $\mathcal{V} := \bigcup_k \mathcal{V}_k$ is a Borel set with $|\mathcal{V}| \le \sum_k |\mathcal{V}_k| = 0$, and one has $E(\mathcal{V})f = \text{s-lim}_{k\to\infty} E(\mathcal{V})f_k = \text{s-lim}_{k\to\infty} f_k = f$. Thus we have shown that $f \in \mathcal{H}_{\rm sc}(A)$.

We next show that, if $f \in \mathcal{H}_{\rm c}(A)$ and $\mathcal{V} \in \mathcal{A}_{\rm B}$ with $|\mathcal{V}| = 0$, then one has $E(\mathcal{V})f \in \mathcal{H}_{\rm sc}(A)$. By the assumption made on $f$, we have $m_f(\{\lambda\}) \equiv \|E(\{\lambda\})f\|^2 = 0$ for all $\lambda \in \mathbb{R}$. Hence $m_{E(\mathcal{V})f}(\{\lambda\}) = \|E(\{\lambda\})E(\mathcal{V})f\|^2 = \|E(\mathcal{V})E(\{\lambda\})f\|^2 = 0$ for all $\lambda \in \mathbb{R}$, i.e. $E(\mathcal{V})f \in \mathcal{H}_{\rm c}(A)$. Now $E(\mathcal{V})[E(\mathcal{V})f] = [E(\mathcal{V})]^2 f = E(\mathcal{V})f$. Since $|\mathcal{V}| = 0$, the claim is proved.

We now assume that $f \in \mathcal{H}_{\rm c}(A)$ and $f \perp \mathcal{H}_{\rm sc}(A)$ and show that then $f \in \mathcal{H}_{\rm ac}(A)$ [which implies that $\mathcal{H}_{\rm ac}(A)$ is a subspace, the orthogonal complement in $\mathcal{H}_{\rm c}(A)$ of $\mathcal{H}_{\rm sc}(A)$, i.e. one has $\mathcal{H}_{\rm c}(A) = \mathcal{H}_{\rm sc}(A) \oplus \mathcal{H}_{\rm ac}(A)$]. We must prove that $E(\mathcal{V})f = \mathbf{0}$ for each Borel set $\mathcal{V}$ with $|\mathcal{V}| = 0$. We set $E(\mathcal{V})f = g$. One has $g \in \mathcal{H}_{\rm sc}(A)$ by the result of the preceding paragraph. So it is enough to prove that $g \perp \mathcal{H}_{\rm sc}(A)$. For this let $h \in \mathcal{H}_{\rm sc}(A)$. Then (again by the preceding paragraph) $E(\mathcal{V})h \in \mathcal{H}_{\rm sc}(A)$, hence $\langle h, g\rangle = \langle h, E(\mathcal{V})f\rangle = \langle E(\mathcal{V})h, f\rangle = 0$ because $E(\mathcal{V})h \in \mathcal{H}_{\rm sc}(A)$ and $f \perp \mathcal{H}_{\rm sc}(A)$.

The results obtained so far prove completely Eq. (4.62). We mention the identity $\mathcal{H}_{\rm s}(A) = \mathcal{H}_{\rm p}(A) \oplus \mathcal{H}_{\rm sc}(A)$ the proof of which we omit (Problem 4.3). These results will be collected in the first part of Proposition 4.19.

**Proposition 4.18.** *Let $\lambda \in \mathbb{R}$. Then:*

$$\begin{cases} f \in \mathcal{H}_{\rm p}(A) \implies E_\lambda f \in \mathcal{H}_{\rm p}(A) \\ f \in \mathcal{H}_{\rm ac}(A) \implies E_\lambda f \in \mathcal{H}_{\rm ac}(A) \\ f \in \mathcal{H}_{\rm sc}(A) \implies E_\lambda f \in \mathcal{H}_{\rm sc}(A). \end{cases} \tag{4.68}$$

PROOF. (i) For the first relation it is enough to see that, if $f$ is an eigenvector of $A$ (say $Af = \mu f$ for some $\mu \in \mathbb{R}$), then $E_\lambda f$ is also an eigenvector of $A$. Now the equation $Af = \mu f$ implies that $E(\{\mu\})f = f$ [see (4.63)]; thus $E(\{\mu\})[E_\lambda f] = E_\lambda E(\{\mu\})f = E_\lambda f$, so [again by (4.63)] the vector $E_\lambda f$ is indeed an eigenvector of $A$ (associated with the same eigenvalue $\mu$), or else $E_\lambda f = \mathbf{0}$. [12]

(ii) Let us show for example that, if $f \in \mathcal{H}_{\rm sc}(A)$, then $E_\lambda f \in \mathcal{H}_{\rm sc}(A)$. For this let $g \in \mathcal{H}_{\rm p}(A)$; then one has $\langle E_\lambda f, g\rangle = \langle f, E_\lambda g\rangle = 0$ because $E_\lambda g \in \mathcal{H}_{\rm p}(A)$ by (i) and $f \perp \mathcal{H}_{\rm p}(A)$. This shows that $E_\lambda f \in \mathcal{H}_{\rm p}(A)^\perp = \mathcal{H}_{\rm c}(A)$. Since $f \in \mathcal{H}_{\rm sc}(A)$ there exists a Borel set $\mathcal{V}$ with $|\mathcal{V}| = 0$ and $E(\mathcal{V})f = f$. As in (i) one obtains that $E(\mathcal{V})[E_\lambda f] = E_\lambda E(\mathcal{V})f = E_\lambda f$, hence $E_\lambda f$ belongs to $\mathcal{H}_{\rm sc}(A)$. □

[12] Evidently $E_\lambda f$ can be the zero vector; indeed this happens if and only if $\mu > \lambda$ because $E_\lambda E(\{\mu\}) = O$ if $\mu > \lambda$, whereas $E_\lambda E(\{\mu\}) = E(\{\mu\})$ if $\mu \le \lambda$.

The preceding proposition states that each of the subspaces $\mathcal{H}_{\mathrm{p}}(A)$, $\mathcal{H}_{\mathrm{sc}}(A)$ and $\mathcal{H}_{\mathrm{ac}}(A)$ is invariant under the spectral family (hence also under the spectral measure) of $A$. If $f$ is a vector in $\mathcal{H}$ and if one writes $f = f_{\mathrm{p}} + f_{\mathrm{sc}} + f_{\mathrm{ac}}$ for its decomposition in the representation (4.62), then the measure $m_f$ is the sum of three measures $m_{f_{\mathrm{p}}}$, $m_{f_{\mathrm{sc}}}$ and $m_{f_{\mathrm{ac}}}$ determined by the three components of $f$:

$$m_f(\mathcal{V}) = m_{f_{\mathrm{p}}}(\mathcal{V}) + m_{f_{\mathrm{sc}}}(\mathcal{V}) + m_{f_{\mathrm{ac}}}(\mathcal{V}) \qquad \forall\, \mathcal{V} \in \mathcal{A}_{\mathrm{B}}. \tag{4.69}$$

Indeed

$$\begin{aligned} m_f(\mathcal{V}) &= \langle f, E(\mathcal{V})f\rangle = \langle f_{\mathrm{p}} + f_{\mathrm{sc}} + f_{\mathrm{ac}}, E(\mathcal{V})f_{\mathrm{p}} + E(\mathcal{V})f_{\mathrm{sc}} + E(\mathcal{V})f_{\mathrm{ac}}\rangle \\ &= \langle f_{\mathrm{p}}, E(\mathcal{V})f_{\mathrm{p}}\rangle + \langle f_{\mathrm{sc}}, E(\mathcal{V})f_{\mathrm{sc}}\rangle + \langle f_{\mathrm{ac}}, E(\mathcal{V})f_{\mathrm{ac}}\rangle \\ &= m_{f_{\mathrm{p}}}(\mathcal{V}) + m_{f_{\mathrm{sc}}}(\mathcal{V}) + m_{f_{\mathrm{ac}}}(\mathcal{V}), \end{aligned}$$

because the cross terms in the last scalar product of the first line are zero [they involve vectors that are orthogonal to each other by Proposition 4.18: $E(\mathcal{V})f_{\mathrm{p}} \in \mathcal{H}_{\mathrm{p}}(A)$, $E(\mathcal{V})f_{\mathrm{sc}} \in \mathcal{H}_{\mathrm{sc}}(A)$ and $E(\mathcal{V})f_{\mathrm{ac}} \in \mathcal{H}_{\mathrm{ac}}(A)$].

**4.3.3.** We now study the action of the operator $A$ on the vectors belonging to the different subspaces introduced above. We shall see that $A$ leaves each of these subspaces invariant. If $\mathcal{M}$ is a subspace of $\mathcal{H}$ we denote by $A_{\mathcal{M}}$ the ***restriction*** of $A$ to $\mathcal{M}$:

$$\mathcal{D}(A_{\mathcal{M}}) = \mathcal{M} \cap \mathcal{D}(A) \qquad \text{and} \qquad A_{\mathcal{M}} f = Af \ \text{ for } f \in \mathcal{D}(A_{\mathcal{M}}). \tag{4.70}$$

**Proposition 4.19.** *Let $A$ be a self-adjoint operator in a Hilbert space $\mathcal{H}$.*
*(a) $\mathcal{H}$ can be decomposed as follows:*

$$\mathcal{H} = \mathcal{H}_{\mathrm{p}}(A) \oplus \mathcal{H}_{\mathrm{sc}}(A) \oplus \mathcal{H}_{\mathrm{ac}}(A) = \mathcal{H}_{\mathrm{p}}(A) \oplus \mathcal{H}_{\mathrm{c}}(A) = \mathcal{H}_{\mathrm{s}}(A) \oplus \mathcal{H}_{\mathrm{ac}}(A). \tag{4.71}$$

*The restriction of $A$ to one of the subspaces appearing in* (4.71) *defines a self-adjoint operator in this subspace. If these restrictions are denoted by $A_{\mathrm{p}}, A_{\mathrm{sc}}, A_{\mathrm{ac}}, A_{\mathrm{c}}$ and $A_{\mathrm{s}}$, then*

$$A = A_{\mathrm{p}} \oplus A_{\mathrm{sc}} \oplus A_{\mathrm{ac}} \equiv A_{\mathrm{p}} \oplus A_{\mathrm{c}} \equiv A_{\mathrm{s}} \oplus A_{\mathrm{ac}}. \tag{4.72}$$

*(b) If $\varphi$ is a bounded continuous function on $\mathbb{R}$, then each of the subspaces in* (4.71) *is invariant under $\varphi(A)$:*

$$\varphi(A) = \varphi(A_{\mathrm{p}}) \oplus \varphi(A_{\mathrm{sc}}) \oplus \varphi(A_{\mathrm{ac}}). \tag{4.73}$$

*Moreover one has*

$$\sigma(A) = \sigma(A_{\mathrm{p}}) \cup \sigma(A_{\mathrm{sc}}) \cup \sigma(A_{\mathrm{ac}}). \tag{4.74}$$

In what follows we denote the projections onto the subspaces appearing in (4.71) by $E_{\mathrm{p}}, E_{\mathrm{sc}}, E_{\mathrm{ac}}, E_{\mathrm{c}}$ and $E_{\mathrm{s}}$ respectively (all of these operators depend on $A$!). Each of the five operators $A_{\mathrm{p}}, A_{\mathrm{sc}}, A_{\mathrm{ac}}, A_{\mathrm{c}}$ and $A_{\mathrm{s}}$ introduced in (a) of the preceding proposition is self-adjoint as an operator in the respective subspace [for example $A_{\mathrm{sc}}$ is a self-adjoint operator in $\mathcal{H}_{\mathrm{sc}}(A)$, defined by $\mathcal{D}(A_{\mathrm{sc}}) = \mathcal{H}_{\mathrm{sc}}(A) \cap \mathcal{D}(A)$ and $A_{\mathrm{sc}} f = Af$ for $f \in \mathcal{D}(A_{\mathrm{sc}})$]. Eq. (4.74) expresses the spectrum of $A$ (a self-adjoint operator in $\mathcal{H}$) in terms of the spectra of $A_{\mathrm{p}}$, $A_{\mathrm{sc}}$ and $A_{\mathrm{ac}}$ (viewed as self-adjoint operators in the respective subspace).

PROOF OF PROPOSITION 4.19. Equation (4.71) has already been established.

(i) Let $f \in \mathcal{D}(A)$ and write $f = f_{\rm p} + f_{\rm sc} + f_{\rm ac} \equiv E_{\rm p} f + E_{\rm sc} f + E_{\rm ac} f$ in the first decomposition of $\mathcal{H}$ in (4.71). By (4.69) one has

$$\int_{-\infty}^{\infty} \lambda^2 m_{f_{\rm p}}(d\lambda) + \int_{-\infty}^{\infty} \lambda^2 m_{f_{\rm sc}}(d\lambda) + \int_{-\infty}^{\infty} \lambda^2 m_{f_{\rm ac}}(d\lambda) = \int_{-\infty}^{\infty} \lambda^2 m_f(d\lambda) < \infty. \tag{4.75}$$

This implies that each of the vectors $f_{\rm p}$, $f_{\rm sc}$ and $f_{\rm ac}$ belongs to $\mathcal{D}(A)$. We claim that $Af_{\rm p} \in \mathcal{H}_{\rm p}(A)$, $Af_{\rm sc} \in \mathcal{H}_{\rm sc}(A)$ and $Af_{\rm ac} \in \mathcal{H}_{\rm ac}(A)$. Let us verify for example the last inclusion: for each $M > 0$, let $\{\Pi_r^M\}$ be a sequence of partitions of the interval $(-M, M]$ with $|\Pi_r^M| \to 0$ as $r \to \infty$. Let $\Sigma_r^M$ be the operator associated with the partition $\{\Pi_r^M\}$ [defined as in (4.39) with $\Pi = \Pi_r^M$ and $\varphi(\lambda) = \lambda$] and observe that this operator maps the subspace $\mathcal{H}_{\rm ac}(A)$ into itself [because each $E(\mathcal{V})$ has this property by (4.68)]. Thus $\Sigma_r^M f_{\rm ac} \in \mathcal{H}_{\rm ac}(A)$. Since $\mathcal{H}_{\rm ac}(A)$ is a subspace, hence closed, one has

$$Af_{\rm ac} = \operatorname*{s-lim}_{M\to\infty} \left[ \operatorname*{s-lim}_{r\to\infty} \Sigma_r^M f_{\rm ac} \right] \in \mathcal{H}_{\rm ac}(A).$$

In a similar way one sees for example that $\varphi(A) f_{\rm ac} \in \mathcal{H}_{\rm ac}(A)$.

(ii) The preceding considerations show that each $f \in \mathcal{D}(A)$ has a unique decomposition $f = f_{\rm p} + f_{\rm sc} + f_{\rm ac}$ with $f_{\rm p} \in \mathcal{D}(A_{\rm p})$, $f_{\rm sc} \in \mathcal{D}(A_{\rm sc})$ and $f_{\rm ac} \in \mathcal{D}(A_{\rm ac})$, and that

$$Af = A_{\rm p} f_{\rm p} + A_{\rm sc} f_{\rm sc} + A_{\rm ac} f_{\rm ac}.$$

This confirms the decomposition (4.72) of $A$ and also implies that, for $z \in \mathbb{C}$:

$$(A - z)f = (A_{\rm p} - z) f_{\rm p} + (A_{\rm sc} - z) f_{\rm sc} + (A_{\rm ac} - z) f_{\rm ac}. \tag{4.76}$$

(iii) It is clear that the operators $A_{\rm p}$, $A_{\rm sc}$ and $A_{\rm ac}$ are symmetric [for example: if $f, g \in \mathcal{D}(A_{\rm p})$ then $\langle f, A_{\rm p} g\rangle = \langle f, Ag\rangle = \langle Af, g\rangle = \langle A_{\rm p} f, g\rangle$]. To see that they are self-adjoint, it suffices to show that $\mathcal{R}(A_{\rm p} - i) = \mathcal{R}(A_{\rm p} + i) = \mathcal{H}_{\rm p}(A)$, etc. (see Proposition 3.3). Now, since $A$ is self-adjoint, we have $\mathcal{R}(A \pm i) = \mathcal{H}$, i.e. $(A \pm i)\mathcal{D}(A) = \mathcal{H}$. By (4.76) this is possible only if $(A_{\rm p} \pm i)\mathcal{D}(A_{\rm p}) = \mathcal{H}_{\rm p}(A)$, $(A_{\rm sc} \pm i)\mathcal{D}(A_{\rm sc}) = \mathcal{H}_{\rm sc}(A)$ and $(A_{\rm ac} \pm i)\mathcal{D}(A_{\rm ac}) = \mathcal{H}_{\rm ac}(A)$, which establishes the self-adjointness of $A_{\rm p}$, $A_{\rm sc}$ and $A_{\rm ac}$.

(iv) Eq. (4.76) shows that $A - z$ is invertible if and only if each of the three operators on its right-hand side is invertible, and if this is the case then one has $(A - z)^{-1} = (A_{\rm p} - z)^{-1} \oplus (A_{\rm sc} - z)^{-1} \oplus (A_{\rm ac} - z)^{-1}$. This implies that $z \in \rho(A)$ if and only if $z \in \rho(A_{\rm p}) \cap \rho(A_{\rm sc}) \cap \rho(A_{\rm ac})$. It follows that

$$\begin{aligned}
\sigma(A) &= \mathbb{C}\backslash\rho(A) = \mathbb{C}\backslash\{\rho(A_{\rm p}) \cap \rho(A_{\rm sc}) \cap \rho(A_{\rm ac})\} \\
&= [\mathbb{C}\backslash\rho(A_{\rm p})] \cup [\mathbb{C}\backslash\rho(A_{\rm sc})] \cup [\mathbb{C}\backslash\rho(A_{\rm ac})] \\
&= \sigma(A_{\rm p}) \cup \sigma(A_{\rm sc}) \cup \sigma(A_{\rm ac}).
\end{aligned}$$

□

Let us now introduce the spectral parts of a self-adjoint operator $A$. Except for the point spectrum, they are defined in terms of the spectra of the restrictions of $A$ to the subspaces occurring in the decomposition (4.62) of the Hilbert space:

- ***point spectrum***: $\sigma_{\rm p}(A)$ = set of all eigenvalues of $A$,
- ***continuous spectrum***: $\sigma_{\rm c}(A) = \sigma(A_{\rm c}) \equiv$ spectrum of the self-adjoint operator $A_{\rm c}$ in $\mathcal{H}_{\rm c}$ [if $\dim \mathcal{H}_{\rm c}(A) \neq 0$],
- ***singular continuous spectrum***: $\sigma_{\rm sc}(A) = \sigma(A_{\rm sc})$ [if $\dim \mathcal{H}_{\rm sc}(A) \neq 0$],
- ***absolutely continuous spectrum***: $\sigma_{\rm ac}(A) = \sigma(A_{\rm ac})$ [if $\dim \mathcal{H}_{\rm ac}(A) \neq 0$],
- ***singular spectrum***: $\sigma_{\rm s}(A) = \sigma(A_{\rm s})$ [if $\dim \mathcal{H}_{\rm s}(A) \neq 0$].

In the case in which the dimension of the subspace involved is 0, one says that the respective part of the spectrum is empty. For example, if $\mathcal{H}_{\rm sc}(A) = \{\mathbf{0}\}$, one sets $\sigma_{\rm sc}(A) = \varnothing$.

The last four parts of the spectrum [$\sigma_{\rm c}(A), \sigma_{\rm sc}(A), \sigma_{\rm ac}(A)$ and $\sigma_{\rm s}(A)$] are *closed* subsets of $\mathbb{R}$ because each of them is the spectrum of a self-adjoint operator in a certain Hilbert space. The point spectrum however is not closed in general (see the examples given in §4.3.4); the spectrum $\sigma(A_{\rm p})$ of the self-adjoint operator $A_{\rm p}$ in $\mathcal{H}_{\rm p}(A)$ is equal to the closure of the set of eigenvalues of $A$:

$$\sigma(A_{\rm p}) = \overline{\sigma_{\rm p}(A)}. \tag{4.77}$$

Thus $\sigma(A_{\rm p})$ is composed of the eigenvalues of $A$ and of their accumulation points [if $\lambda$ is a real number such that $\mathrm{dist}(\lambda, \sigma_{\rm p}(A)) > 0$, then $\lambda \in \rho(A_{\rm p})$; the verification is left as an exercise, see Problem 4.12)].

By (4.74) the spectrum of $A$ is the union of the different spectral parts, more precisely

$$\sigma(A) = \overline{\sigma_{\rm p}(A)} \cup \sigma_{\rm sc}(A) \cup \sigma_{\rm ac}(A) = \overline{\sigma_{\rm p}(A)} \cup \sigma_{\rm c}(A) = \sigma_{\rm s}(A) \cup \sigma_{\rm ac}(A). \tag{4.78}$$

We say that $A$ has ***pure point spectrum*** if $\mathcal{H}_{\rm p}(A) = \mathcal{H}$, that it has ***purely continuous spectrum*** (or that its spectrum is purely continuous) if $\mathcal{H}_{\rm c}(A) = \mathcal{H}$ and that the spectrum of $A$ is ***purely absolutely continuous*** if $\mathcal{H}_{\rm ac}(A) = \mathcal{H}$. One says that *the continuous spectrum of A is absolutely continuous* if $\mathcal{H}_{\rm sc}(A) = \{\mathbf{0}\}$.

If $U$ is a unitary operator and $A_U = UAU^*$ as in Remark 2.25, then the spectral parts of $A_U$ and of $A$ (point spectrum, absolutely continuous spectrum, etc.) are identical [so $\sigma(A_U) = \sigma(A)$, $\sigma_{\rm p}(A_U) = \sigma_{\rm p}(A)$, etc.]. This follows from the fact that $m_{Uf}(\mathcal{V}) = m_f(\mathcal{V})$ for each vector $f$ and for each $\mathcal{V} \in \mathcal{A}_{\rm B}$ [observe that $\{UE_\lambda U^*\}$ is the spectral family of the operator $A_U$ and that $F_{Uf}^{A_U}(\lambda) \equiv \|UE_\lambda U^*(Uf)\|^2 = \|UE_\lambda f\|^2 = \|E_\lambda f\|^2 \equiv F_f^A(\lambda)$]. One has for example $\mathcal{H}_{\rm p}(A_U) = U\mathcal{H}_{\rm p}(A)$ (if $Af = \mu f$, then $A_U Uf = UAf = \mu Uf$; so if $f$ is an eigenvector of $A$, then $Uf$ is an eigenvector of $A_U$ associated with the same eigenvalue).

**4.3.4.** The spectral parts of a self-adjoint operator introduced above are defined in terms of a decomposition of the underlying Hilbert space. In general the different spectral parts will not be disjoint. For example one may have $\sigma_{\rm p}(A) \cap \sigma_{\rm c}(A) \neq \varnothing$ ($A$ may have eigenvalues embedded in its continuous spectrum). We now discuss a subdivision of the spectrum which is frequently used and which divides the spectrum into two disjoint parts.

The ***discrete spectrum*** $\sigma_{\mathrm{d}}(A)$ is the set of all eigenvalues of $A$ of that are *of finite multiplicity* and *isolated from the rest of the spectrum.* So $\lambda \in \sigma_{\mathrm{d}}(A)$ if and only if $0 < \dim \mathcal{M}(\{\lambda\}) < \infty$ and if there exists $\varepsilon > 0$ such that the interval $(\lambda-\varepsilon, \lambda+\varepsilon)$ is disjoint from the rest of the spectrum of $A$, i.e. such that $(\lambda-\varepsilon, \lambda+\varepsilon) \cap \sigma(A) = \{\lambda\}$. The ***essential spectrum*** $\sigma_{\mathrm{ess}}(A)$ is the complementary set of $\sigma_{\mathrm{d}}(A)$ in the spectrum of the operator $A$:

$$\sigma_{\mathrm{ess}}(A) = \sigma(A) \backslash \sigma_{\mathrm{d}}(A). \tag{4.79}$$

A point $\lambda \in \sigma(A)$ belongs to $\sigma_{\mathrm{ess}}(A)$ for example if $\lambda \in \sigma_{\mathrm{c}}(A)$, or if $\lambda$ is an eigenvalue of infinite multiplicity (isolated or non-isolated from the rest of the spectrum), or if $\lambda$ is an eigenvalue of finite multiplicity embedded in the continuous spectrum [in which case $\lambda$ belongs to both $\sigma_{\mathrm{p}}(A)$ and $\sigma_{\mathrm{c}}(A)$], or if $\lambda$ is an accumulation point of eigenvalues of $A$. We refer to the examples given at the end of this subsection.

One can characterise the points in the discrete spectrum and those in the essential spectrum as follows in terms of the family of subspaces $\{\mathcal{M}(\mathcal{V})\}$ associated with the spectral family of $A$:

$$\lambda \in \sigma_{\mathrm{d}}(A) \iff \exists\, \varepsilon_0 > 0 \text{ such that } 0 < \dim \mathcal{M}((\lambda-\varepsilon, \lambda+\varepsilon)) < \infty \;\forall \varepsilon \in (0, \varepsilon_0),$$
$$\lambda \in \sigma_{\mathrm{ess}}(A) \iff \dim \mathcal{M}((\lambda-\varepsilon, \lambda+\varepsilon)) = \infty \text{ for each } \varepsilon > 0.$$

The essential spectrum is closed: if $\{\lambda_n\}$ is a sequence of points in $\sigma_{\mathrm{ess}}(A)$ converging to a point $\lambda$, then $\mathcal{M}((\lambda-\varepsilon, \lambda+\varepsilon)) \supseteq \mathcal{M}((\lambda_n-\varepsilon/2, \lambda_n+\varepsilon/2))$ if $|\lambda-\lambda_n| < \varepsilon/2$, hence $\dim \mathcal{M}((\lambda-\varepsilon, \lambda+\varepsilon)) = \infty$ for each $\varepsilon > 0$.

Part (a) of the next proposition gives a useful characterisation of the essential spectrum; part (b) shows that the notion of the spectrum of a self-adjoint operator in an infinite-dimensional Hilbert space is a natural generalisation of the notion of an eigenvalue of Hermitian $n \times n$ matrices: $\lambda \in \sigma(A) \iff$ for each $\varepsilon > 0$ there exists a vector $f \in \mathcal{D}(A)$, with $\|f\| = 1$, such that $Af \approx \lambda f$ in the sense that $\|Af - \lambda f\| < \varepsilon$.

**Proposition 4.20 (Weyl's Criterion).** *Let $A$ be a self-adjoint operator in a Hilbert space $\mathcal{H}$.*

*(a) A real number $\lambda$ belongs to the essential spectrum $\sigma_{\mathrm{ess}}(A)$ of $A$ if and only if there exists a sequence $\{f_n\}$ of vectors in $\mathcal{D}(A)$ having the following properties: $\|f_n\| = 1$ for each $n$, $\mathrm{w}\text{-}\lim_{n\to\infty} f_n = \mathbf{0}$ and $\mathrm{s}\text{-}\lim_{n\to\infty}(A-\lambda)f_n = \mathbf{0}$.*

*(b) A real number $\lambda$ belongs to spectrum $\sigma(A)$ of $A$ if and only if there exists a sequence $\{f_n\}$ of vectors in $\mathcal{D}(A)$ satisfying $\|f_n\| = 1$ and $\mathrm{s}\text{-}\lim_{n\to\infty}(A-\lambda)f_n = \mathbf{0}$.*

PROOF. Let us fix $\lambda \in \mathbb{R}$ and set $J_n = [\lambda - 1/n, \lambda + 1/n]$ for $n = 1, 2, 3, \ldots$

(a) Suppose that $\lambda \in \sigma_{\mathrm{ess}}(A)$, so that $\dim \mathcal{M}(J_n) = \infty$ for each $n$. First take $f_1 \in \mathcal{M}(J_1)$ arbitrarily, with $\|f_1\| = 1$. Next choose a vector $f_2 \in \mathcal{M}(J_2)$ such that $f_2 \perp E(J_2)f_1$ and $\|f_2\| = 1$; then $\langle f_1, f_2\rangle = \langle f_1, E(J_2)f_2\rangle = \langle E(J_2)f_1, f_2\rangle = 0$. Continue recursively in this way: $f_n \in \mathcal{M}(J_n)$ is chosen such that $\|f_n\| = 1$ and such that $f_n$ is orthogonal to $E(J_n)f_1, \ldots, E(J_n)f_{n-1}$. One obtains in this way an orthonormal sequence $\{f_n\}$ [i.e. $\langle f_n, f_m\rangle = \delta_{nm}$] satisfying $E(J_n)f_n = f_n$. One has $\mathrm{w}\text{-}\lim_{n\to\infty} f_n = \mathbf{0}$ (see Example 1.2) and

$$\|(A-\lambda)f_n\|^2 = \int_{J_n} |\lambda-\mu|^2 m_{f_n}(d\mu) \le \frac{1}{n^2}\int_{-\infty}^{\infty} m_{f_n}(d\mu) = \frac{1}{n^2}\|f_n\|^2 = \frac{1}{n^2}.$$

Thus the sequence $\{f_n\}$ has the properties stated in the proposition.

Conversely, suppose that there exists a sequence $\{f_n\}$ having these three properties, and assume that $\lambda \notin \sigma_{\text{ess}}(A)$, i.e. that there exists a number $N \in \mathbb{N}$ such that $\dim \mathcal{M}(J_N) < \infty$. We must obtain a contradiction. For this let us consider an interval $(a, b)$ with $a < \lambda < b$. Then

$$\|(A-\lambda)f_n\|^2 = \int_{-\infty}^{a} |\lambda-\mu|^2 m_{f_n}(d\mu) + \int_{b}^{\infty} |\lambda-\mu|^2 m_{f_n}(d\mu)$$

$$+ \int_{a}^{b} |\lambda-\mu|^2 m_{f_n}(d\mu) \geq (\lambda-a)^2\|E((-\infty,a))f_n\|^2 + (b-\lambda)^2\|E((b,\infty))f_n\|^2.$$

Since $\|(A-\lambda)f_n\| \to 0$ as $n \to \infty$, this implies that $\|E((-\infty,a))f_n\| \to 0$ and $\|E((b,\infty))f_n\| \to 0$; consequently $\|E([a,b])f_n\| \to 1$ [because $1 = \|f_n\|^2 = \|E((-\infty,a))f_n\|^2 + \|E((b,\infty))f_n\|^2 + \|E([a,b])f_n\|^2$]. In particular, taking $a = \lambda - 1/N$ and $b = \lambda + 1/N$, we have $\|E(J_N)f_n\| \to 1$ as $n \to \infty$. If $\{e_1, \ldots, e_M\}$ is an orthonormal basis of $\mathcal{M}(J_N)$, we have $\|E(J_N)f_n\|^2 = \sum_{k=1}^{M} |\langle e_k, f_n\rangle|^2$. Now $\langle e_k, f_n\rangle \to 0$ as $n \to \infty$ for each fixed $k$ because $f_n$ converges weakly to zero, and the sum is finite by the assumption that $M \equiv \dim \mathcal{M}(J_N) < \infty$. Hence $\|E(J_N)f_n\| \to 0$ as $n \to \infty$, in contradiction with the relation $\|E(J_N)f_n\| \to 1$.

(b) $\Rightarrow$ In view of the result of (a), it suffices to consider the case $\lambda \in \sigma_{\text{d}}(A)$. In this case one can take $f_n = f$ for each $n$, where $f$ is a normalised eigenvector of $A$ associated with the eigenvalue $\lambda$: $Af = \lambda f$, $\|f\| = 1$.

$\Leftarrow$ Let $a < \lambda < b$. The argument given in the second part of the proof of (a) shows that $\|E([a,b])f_n\| \to 1$ as $n \to \infty$, which implies that $\dim \mathcal{M}([a,b]) \neq 0$. Hence $\lambda$ is a point of non-constancy of the spectral family $\{E_\lambda\}$, i.e. $\lambda \in \sigma(A)$. □

We can illustrate the different types of spectra by means of examples of self-adjoint operators. Most of the examples given below are of a purely mathematical nature, but we cite some operators of interest in physics without, however, proving the stated spectral properties (these proofs are often quite long, we shall return to this question and some further examples in Chapter 6). We also refer to Section 14.6 of [P] for a collection of examples of potentials in $\mathbb{R}$ or $\mathbb{R}^3$ some of which exhibit unusual spectral or scattering properties.

**Example 4.21. Operators with pure point spectrum**

(a) We begin with some well-known examples.

(i) *Projections*: if $P$ is a projection, then $\sigma(P) = \{0, 1\}$ if $P \neq O, I$. Indeed if $P$ is the projection onto the subspace $\mathcal{M}$ and if $\mathcal{M} \neq \{\mathbf{0}\}$ and $\mathcal{M} \neq \mathcal{H}$, then $\mathcal{M} = \mathcal{M}(\{1\})$ is the subspace of all eigenvectors of $P$ with eigenvalue $\lambda = 1$, whereas $\mathcal{M}^\perp = \mathcal{M}(\{0\})$ is the subspace of all eigenvectors of $P$ with eigenvalue $\lambda = 0$.

(ii) The Hamiltonian of the harmonic oscillator $H = P^2 + Q^2$ in $L^2(\mathbb{R})$. Its spectrum is discrete, the eigenvalues (each of them of multiplicity 1) are the numbers $\lambda_n = 2n + 1$ with $n = 0, 1, 2, \ldots$

(b) Let $\{e_1, e_2, \ldots\}$ be an orthonormal basis of an infinite-dimensional Hilbert space $\mathcal{H}$. An operator in $\mathcal{B}(\mathcal{H})$ is entirely determined by its action on each $e_k$.

(i) Let $A$ be defined by $Ae_k = k^{-1}e_k$. Each $e_k$ is an eigenvector of $A$, so $A$ has pure point spectrum. The point $\lambda = 0$ is not an eigenvalue of $A$, but it belongs to the spectrum of $A$ because it is an accumulation point of eigenvalues. Thus $\sigma(A) = \{0\} \cup \{1/k \mid k = 1, 2, 3 \ldots\}$, $\sigma_{\rm d}(A) = \sigma_{\rm p}(A) = \{1/k \mid k = 1, 2, 3 \ldots\}$ and $\sigma_{\rm ess}(A) = \{0\}$.

(ii) Let $A$ be determined by $Ae_1 = \mathbf{0}$ and $Ae_k = (k-1)^{-1}e_k$ for $k \geq 2$. Again each $e_k$ is an eigenvector of $A$, so $A$ has pure point spectrum. In contrast to the operator considered in (i), the point $\lambda = 0$ is now also an eigenvalue. Hence $\sigma(A) = \sigma_{\rm p}(A) = \{0\} \cup \{1/k \mid k = 1, 2, 3 \ldots\}$, $\sigma_{\rm d}(A) = \{1/k \mid k = 1, 2, 3 \ldots\}$ and $\sigma_{\rm ess}(A) = \{0\}$.

(iii) Let $\{r_1, r_2, \ldots\}$ be an enumeration of the rational numbers in $(0, 1)$ and $Ae_k = r_k e_k$. Again $A$ has pure point spectrum. Its point spectrum $\sigma_{\rm p}(A) = \{r_k\}$ is composed of all rational numbers in $(0, 1)$. This set is dense in $[0, 1]$. $A$ has no discrete spectrum, even though each of its eigenvalues is of multiplicity 1 (none of them is isolated from the rest of the spectrum). One has $\sigma(A) = \sigma_{\rm ess}(A) = [0, 1]$.

**Example 4.22. Spectrum of multiplication operators**

(a) The operators $Q, Q^2, P, P^2$ in $L^2(\mathbb{R})$ as well as the operators $\vec{Q}^2$ and $\vec{P}^2$ in $L^2(\mathbb{R}^n)$ $(n \geq 2)$ have purely absolutely continuous spectrum. Indeed the spectral families of these operators have been given in Examples 4.9 and 4.15 - 4.17, and in each case it is easy to see that the measures $m_f$ are absolutely continuous for all $f \in \mathcal{H}$. For example for $Q$ one has $m_f(\mathcal{V}) = \int_{\mathcal{V}} |f(x)|^2 dx$, its Radon-Nikodym derivative is therefore $|f(x)|^2$.

(b) More generally, if $A$ is the operator of multiplication in $L^2(\mathbb{R}^n)$ by a function $a \colon \mathbb{R}^n \to \mathbb{R}$, then the spectrum of $A$ is purely absolutely continuous if the function $a$ is continuously differentiable and $\text{grad}[a(\vec{x})] \neq 0$ almost everywhere. We omit the proof.

(c) Consider the multiplication operator $A$ by a real function $a$ in $\mathcal{H} = L^2([0, 1])$.

(i) For $\lambda \in \mathbb{R}$, let $\mathcal{W}_\lambda = \{x \in [0, 1] \mid a(x) = \lambda\}$. The number $\lambda$ is an eigenvalue of $A$ if and only if the Lebesgue measure of $\mathcal{W}_\lambda$ is non-zero. In this case the subspace $\mathcal{M}(\{\lambda\})$ of all eigenvectors associated with this eigenvalue is identical with $L^2(\mathcal{W}_\lambda)$, the set of functions $f$ in $L^2([0, 1])$ that vanish outside $\mathcal{W}_\lambda$. Each eigenvalue is of infinite multiplicity and thus belongs to the essential spectrum of $A$. By conveniently choosing the function $a$, one can again construct examples of operators with purely discrete spectrum or of operators with pure point spectrum that is dense in an interval or in $\mathbb{R}$.

(ii) If $a$ is the Cantor function, then the spectrum of $A$ is purely singular continuous: $\sigma(A) = \sigma_{\rm sc}(A) = \sigma_{\rm ess}(A) = \mathfrak{C}$ (the Cantor set).

(iii) It is easy to give multiplication operators with mixed spectrum. For example let $a(x) = 4x$ if $0 \leq x < 1/2$ and $a(x) = \mu$ for $1/2 \leq x \leq 1$, where $\mu$ is a fixed real number. Then $\sigma_{\rm ac}(A) = [0, 2]$, $\sigma_{\rm p}(A) = \{\mu\}$ and $\sigma_{\rm sc}(A) = \sigma_{\rm d}(A) = \varnothing$. If $\mu \in [0, 2]$, then $\sigma(A) = \sigma_{\rm ess}(A) = [0, 2]$. If $\mu \notin [0, 2]$, then $\sigma(A) = \sigma_{\rm ess}(A) = [0, 2] \cup \{\mu\}$.

**Example 4.23. Coulomb Hamiltonian**

An important operator in quantum mechanics is the Coulomb Hamiltonian $H_{\rm C} = \vec{P}^2 + \gamma/|\vec{Q}|$ in $L^2(\mathbb{R}^3)$, where $\gamma \neq 0$ is a real constant. Its spectrum contains an absolutely continuous part equal to $[0, \infty)$ and, if $\gamma < 0$, an infinite number of negative eigenvalues $\lambda_j = -\gamma^2/(4j^2)$, where $j = 1, 2, 3, \ldots$ These eigenvalues accumulate at $\lambda = 0$, and each of them is of finite multiplicity. Thus one has: (i) if $\gamma > 0$, then $\sigma(A) = \sigma_{\text{ac}}(A) = \sigma_{\text{ess}}(A) = [0, \infty)$ and $\sigma_{\text{p}}(A) = \sigma_{\text{sc}}(A) = \sigma_{\text{d}}(A) = \varnothing$, (ii) if $\gamma < 0$, then $\sigma(A) = [0, \infty) \cup \{\lambda_j \mid j = 1, 2, 3, \ldots\}$, $\sigma_{\text{ac}}(A) = \sigma_{\text{ess}}(A) = [0, \infty)$, $\sigma_{\text{p}}(A) = \sigma_{\text{d}}(A) = \{\lambda_j \mid j = 1, 2, 3, \ldots\}$ and $\sigma_{\text{sc}}(A) = \varnothing$.

**Example 4.24. Spectrum of compact operators**
Let $A$ be a compact self-adjoint operator in a Hilbert space $\mathcal{H}$. Let $\varepsilon > 0$. We claim that $\mathcal{M}([\varepsilon, \infty))$ and $\mathcal{M}((-\infty, -\varepsilon])$ are finite-dimensional subspaces. Indeed suppose for example that $\dim \mathcal{M}([\varepsilon, \infty)) = \infty$. Then there exists an infinite orthonormal sequence of vectors $\{e_n\}$ in $\mathcal{M}([\varepsilon, \infty))$. One has $\text{w-}\lim_{n\to\infty} e_n = \mathbf{0}$ (see Example 1.2), hence $\text{s-}\lim_{n\to\infty} Ae_n = \mathbf{0}$ by Proposition 2.12(a). But, as in the proof of Proposition 4.20(a), one has $\|Ae_n\|^2 = \int_\varepsilon^\infty \lambda^2 m_{e_n}(d\lambda) \geq \varepsilon^2 \|e_n\|^2 = \varepsilon^2 > 0$, in contradiction to the fact that $\|Ae_n\|^2 \to 0$.

Thus $A$ has pure point spectrum, each non-zero eigenvalue is of finite multiplicity, and for each $\varepsilon > 0$ there are at most a finite number of eigenvalues $\lambda$ with $|\lambda| \geq \varepsilon$. The point $\lambda = 0$ may or may not be an eigenvalue. However, if $\dim \mathcal{H} = \infty$, then $\lambda = 0$ must belong to the essential spectrum of $A$. The following three situations may occur:

(i) $\lambda = 0$ is an accumulation point of eigenvalues but not itself an eigenvalue,

(ii) $\lambda = 0$ is an accumulation point of eigenvalues and itself an eigenvalue (of finite or infinite multiplicity),

(iii) $\lambda = 0$ is an eigenvalue of infinite multiplicity and isolated from the other eigenvalues of $A$. Hence, in an infinite-dimensional Hilbert space one has $\sigma_{\text{ess}}(A) = \{0\}$ and $\sigma(A) = \sigma_{\text{ess}}(A) \cup \sigma_{\text{d}}(A)$, where $\sigma_{\text{d}}(A) \cap \big\{(-\infty, -\varepsilon] \cup [\varepsilon, \infty)\big\}$ is a finite set for each $\varepsilon > 0$.

**Example 4.25. Spectra of various self-adjoint extensions of a symmetric operator.** Let $A$ be a closed symmetric operator with equal and *finite* deficiency indices ($\nu_- = \nu_+ \equiv \nu < \infty$). Then:

(a) All self-adjoint extensions of $A$ have the same essential spectrum. In particular, if some self-adjoint extension of $A$ has purely discrete spectrum, then each self-adjoint extension of $A$ has this property.

(b) If $\sigma(\widehat{A}) \cap J = \varnothing$ for some self-adjoint extension $\widehat{A}$ of $A$ and some open interval $J$, then the spectrum in $J$ of any self-adjoint extension $\mathbb{A}$ of $A$, if not empty, consists of a finite number of eigenvalues of total multiplicity $\leq \nu$; in other terms the dimension of the range $\mathcal{M}^{\mathbb{A}}(J)$ of the projection $E^{\mathbb{A}}(J)$ is not greater than $\nu$.

(c) If $A$ is bounded from below, i.e. if $\langle f, Af \rangle \geq m\|f\|^2$ for some $m \in \mathbb{R}$ and all $f \in \mathcal{D}(A)$, then the part of the spectrum in $(-\infty, m)$ of any self-adjoint extension $\mathbb{A}$ of $A$ is discrete, and the total multiplicity of the eigenvalues of $\mathbb{A}$ in $(-\infty, m)$ is at most $\nu$, i.e. $\dim \mathcal{M}^{\mathbb{A}}((-\infty, m)) \leq \nu$.

The preceding results hold in particular for the self-adjoint extensions of the symmetric ordinary differential operators considered in Chapter 3. Below we give the proof of (b); the proof of (c) is similar, that of (a) is given for example in §8.3 of [W1].

PROOF OF (b). (i) We first consider the situation in which the interval $J$ is bounded, say $J = (a, b)$. Assume that the stated result is incorrect, i.e. that $\dim \mathcal{M}^{\mathbb{A}}(J) > \nu$. Let $f_1, f_2, \dots, f_{\nu+1}$ be $\nu+1$ linearly independent vectors in $\mathcal{M}^{\mathbb{A}}(J)$. By Proposition 3.7, each $f_k$ has the form $f_k = g_k + (I - U)h_k$, with $g_k \in \mathcal{D}(A)$ and $h_k \in \mathcal{N}(A^* - i)$ ($U$ is some partial isometry). Since $\dim \mathcal{N}(A^* - i) = \nu$, the $\nu + 1$ vectors $h_k$ must be linearly dependent. So there exist constants $\alpha_1, \dots, \alpha_{\nu+1}$, not all of them equal to zero, such that $\sum_{k=1}^{\nu+1} \alpha_k h_k = \mathbf{0}$. Then $g \equiv \sum_{k=1}^{\nu+1} \alpha_k g_k = \sum_{k=1}^{\nu+1} \alpha_k f_k$ is a vector in $\mathcal{M}^{\mathbb{A}}(J)$ belonging to $\mathcal{D}(A)$, and $g \neq \mathbf{0}$ because the vectors $f_k$ are linearly independent. Applying again Proposition 3.7, we see that $\|(A - c)g\| = \|(\widehat{A} - c)g\| = \|(\mathbb{A} - c)g\|$ for any constant $c$. To arrive at a contradiction, we now show that the second equality cannot be satisfied. We let $c = (a + b)/2$ be the midpoint of the interval $J$ and observe that [since $\sigma(\widehat{A}) \cap (a, b) = \varnothing$]:

$$\|(\widehat{A} - c)g\|^2 = \int_{|\lambda - c| \geq (b-a)/2} |\lambda - c|^2 m_g^{\widehat{A}}(d\lambda) \geq \left(\frac{b-a}{2}\right)^2 \|g\|^2,$$

whereas

$$\|(\mathbb{A} - c)g\|^2 = \int_{|\lambda - c| < (b-a)/2} |\lambda - c|^2 m_g^{\mathbb{A}}(d\lambda) < \left(\frac{b-a}{2}\right)^2 \|g\|^2.$$

(ii) If $J$ is unbounded, we have $\dim \mathcal{M}^{\mathbb{A}}(\mathcal{J}) \leq \nu$ for each bounded open subinterval $\mathcal{J}$ of $J$, which implies that $\dim \mathcal{M}^{\mathbb{A}}(J) \leq \nu$. □

Let us comment on the meaning of spectral projections in quantum mechanics. Observables are represented in standard quantum mechanics by self-adjoint operators. Let $A$ be a self-adjoint operator (for example the position operator or the Hamiltonian of a quantum-mechanical system) and let $\{E_\lambda\}$ be its spectral family. If $\mathcal{V}$ is a Borel set in $\mathbb{R}$, then the projection $E(\mathcal{V})$, which is also a self-adjoint operator, represents the observable "the value of $A$ lies in $\mathcal{V}$". More precisely, if $f$ is a vector in the underlying Hilbert space (a state vector for the system under consideration), the number $\|E(\mathcal{V})f\|^2/\|f\|^2$ gives the probability of obtaining a value in $\mathcal{V}$ if one measures the observable $A$ when the system is in the state described by $f$. If $\|f\| = 1$, then $m_f$ is a probability measure on $(\mathbb{R}, \mathcal{A}_{\mathrm{B}})$. For example if $A$ is the position operator $Q$ in $L^2(\mathbb{R})$ and $f$ is a vector with $\|f\| = 1$, then the number $\int_{\mathcal{V}} |f(x)|^2 dx$ represents the probability that the system is localised in $\mathcal{V}$ when it is in the state described by $f$. This interpretation says nothing about how the measurement is performed, it refers to some kind of ideal measurement, and the fact that $\|E(\mathcal{V})f\|^2/\|f\|^2$ is a probability implies that the measurement must be repeated many times (each time on a system in the same state $f$, i.e. prepared in the same way). Since $E(\mathcal{V}) = O$ if $\mathcal{V} \cap \sigma(A) = \varnothing$, i.e. if $\mathcal{V}$ is disjoint from the spectrum of $A$, one sees that the only possible values of a measurement of the observable $A$ (the only values that one would obtain with non-zero probability) are the values in the spectrum of $A$. More precisely: if $\lambda$ is an eigenvalue of $A$, then $E(\{\lambda\}) \neq O$, hence the probability of obtaining the value $\lambda$ in the course of a measurement is non-zero for certain states of the system. If $\lambda \in \sigma(A) \backslash \sigma_{\mathrm{p}}(A)$, then $E(\{\lambda\}) = O$, so the probability of obtaining exactly the value $\lambda$ in any measurement of $A$ is zero; however the probability of obtaining a value in an interval $(\lambda - \varepsilon, \lambda + \varepsilon)$ is non-zero for each $\varepsilon > 0$ and certain states of the system.

**4.3.5.** Finally we consider the behaviour of the spectrum under a perturbation. Let $A$ be a given self-adjoint operator and let $B$ be a second self-adjoint operator that can be considered in some sense as a perturbation of $A$, in particular such that $A+B$ is self-adjoint. What relations can be established between the spectral parts of $A$ and those of $A+B$? It is difficult to give answers in this generality, the spectrum of $A+B$ and its nature can be quite different from the corresponding quantities for $A$. Assume for example that $B$ is a multiple of the identity operator: $B = \alpha I$ with $\alpha \in \mathbb{R}$. Then $\sigma(A+B) = \sigma(A) + \alpha$, $\sigma_c(A+B) = \sigma_c(A) + \alpha$, etc. (so for example $\lambda \in \sigma(A+B) \iff (\lambda - \alpha) \in \sigma(A)$, i.e. $\sigma(A+B)$ is obtained by translating $\sigma(A)$ by $\alpha$ on the real line). In general, as subsets of $\mathbb{R}$, $\sigma(A+B)$ and $\sigma(A)$ can be entirely different, and similarly for the spectral parts of these operators, even if $B$ is very "gentle".

Let us mention another example that is well-known in quantum mechanics. Let $A$ be the operator $P^2$ in $L^2(\mathbb{R})$ and $B$ the potential of a square well, i.e. the multiplication operator by $V(x)$, with $V(x) = -V_0 < 0$ if $x \in [0,1]$ and $V(x) = 0$ otherwise ($V_0$ being a constant). The spectrum of $A$ is absolutely continuous and equal to $[0,\infty)$: $\sigma(A) = \sigma_{ac}(A) = \sigma_{ess}(A) = [0,\infty)$. The addition of $B$ does not change the essential spectrum but adds a finite number of negative eigenvalues (each of multiplicity 1): $\sigma_{ac}(A+B) = \sigma_{ess}(A+B) = [0,\infty)$, $\sigma(A+B) = [0,\infty) \cup \sigma_d(A+B)$, where $\sigma_d(A+B)$ is a finite set contained in $(-V_0, 0)$. This example shows that there must exist situations in which the spectrum or the spectral type of $A+B$ is identical with or very similar to the corresponding quantities for $A$.

The important general result is the invariance of the essential spectrum under a relatively compact perturbation (the preceding example is a special case of this situation). We have seen in Section 2.7 that, if $A$ is self-adjoint and $B$ is symmetric and relatively compact with respect to $A$, then $A+B$ is self-adjoint on $\mathcal{D}(A)$. We recall that the condition that $B$ be $A$-compact means that there exists a number $z \in \rho(A)$ such that $B(A-z)^{-1} \in \mathcal{K}(\mathcal{H})$.

**Proposition 4.26.** *Let $A$ be a self-adjoint operator in a Hilbert space $\mathcal{H}$. If $B$ is a symmetric operator which is $A$-compact, then the essential spectrum of the self-adjoint operator $A+B$ coincides with that of $A$: $\sigma_{ess}(A+B) = \sigma_{ess}(A)$.*

PROOF. (i) Let $\lambda \in \sigma_{ess}(A)$. We choose a sequence $\{f_n\}$ in $\mathcal{D}(A)$ satisfying $\|f_n\| = 1$ for each $n$, $\text{w-}\lim_{n\to\infty} f_n = \mathbf{0}$ and $\text{s-}\lim_{n\to\infty}(A-\lambda)f_n = \mathbf{0}$ [the existence of such a sequence is guaranteed by Proposition 4.20(a)]. By taking into account once more Proposition 4.20(a), we may conclude that $\lambda \in \sigma_{ess}(A+B)$ as soon as we have shown that $\text{s-}\lim_{n\to\infty}(A+B-\lambda)f_n = \mathbf{0}$, which here is equivalent to $\text{s-}\lim_{n\to\infty} Bf_n = \mathbf{0}$. For this we fix a number $z \in \rho(A)$ such that $B(A-z)^{-1} \in \mathcal{K}(\mathcal{H})$ and write

$$Bf_n = [B(A-z)^{-1}](A-z)f_n = [B(A-z)^{-1}](A-\lambda)f_n + (\lambda - z)[B(A-z)^{-1}]f_n.$$

Each term on the right-hand side converges strongly to zero: for the first term this follows from the hypothesis that $\text{s-}\lim_{n\to\infty}(A-\lambda)f_n = \mathbf{0}$, for the second one from the hypothesis that $\text{w-}\lim_{n\to\infty} f_n = \mathbf{0}$ combined with the fact that the compact operator $B(A-z)^{-1}$ maps each weakly convergent sequence onto a strongly convergent one [see Proposition 2.12(a)]. We have thus shown that $\sigma_{\text{ess}}(A) \subseteq \sigma_{\text{ess}}(A+B)$.

(ii) The converse inclusion is obtained in the same way by interchanging the roles of $A$ and $A+B$ in the preceding argument (more precisely by the substitutions $A \mapsto (A+B)$ and $B \mapsto -B$). The only additional input needed is to know that $B$ is also $(A+B)$-compact, i.e. that there exists a number $z' \in \rho(A+B)$ such that $B(A+B-z')^{-1} \in \mathcal{K}(\mathcal{H})$; this has been established in Proposition 2.47(c). □

**Example 4.27.** As an application of Proposition 4.26, let us consider Schrödinger operators of the type studied in Proposition 2.49. So let $H = H_0 + V(\vec{Q})$ in the Hilbert space $\mathcal{H} = L^2(\mathbb{R}^n)$, with $H_0 = \vec{P}^2$. Assume that $V \in L^2(\mathbb{R}^n) + L^\infty_\varepsilon(\mathbb{R}^n)$ if $n = 1, 2$ or 3 and that $V \in L^p(\mathbb{R}^n) + L^\infty_\varepsilon(\mathbb{R}^n)$ for some $p \in (n/2, \infty)$ if $n \geq 4$. Then $V(\vec{Q})$ is $H_0$-compact by Proposition 2.49, hence $H$ is lower semibounded and $\sigma_{\text{ess}}(H) = \sigma_{\text{ess}}(H_0) = [0, \infty)$. A particular case is that of the Coulomb Hamiltonian introduced in Example 4.23.

Proposition 4.26 expresses the invariance of the essential spectrum under a relatively compact perturbation. However nothing is said about the invariance of the spectral type. In fact the nature of the essential spectrum of $A+B$ can be entirely different from that of the essential spectrum of $A$. For example $A$ could have purely absolutely continuous spectrum and $A+B$ pure point spectrum [13], and this even if $B$ is a compact operator close to the zero operator. This statement is a special case of the following theorem (see §X.2 of [K] for the proof).

**Proposition 4.28 (Weyl-von Neumann).** *Let $A$ be an arbitrary self-adjoint operator and $\varepsilon > 0$. Then there exists a Hilbert-Schmidt operator $B$ with $\|B\|_{\text{HS}} < \varepsilon$ such that $A+B$ has pure point spectrum.*

In some situations that are important in quantum mechanics, the spectral type is at least partially preserved under a relatively compact perturbation. As an example, let $A = \vec{P}^2$ in $L^2(\mathbb{R}^n)$ and let $B = V(\vec{Q})$ be the multiplication operator by a real function $V$ converging to zero at infinity. The spectrum of $\vec{P}^2$ is purely absolutely continuous, equal to $[0, \infty)$, and for a large class of potentials $V$ of the indicated type (e.g. the potentials considered in Example 4.27) one can show that the essential spectrum of the operator $\vec{P}^2 + V(\vec{Q})$ is also purely absolutely continuous (the proof is quite technical, however; see Proposition 4.31 and Chapter 6 for some results in this direction).

## 4.4 The spectral theorem. The resolvent near the spectrum

Let $\{E_\lambda\}$ be a spectral family and $A = \int_{-\infty}^{\infty} \lambda E(d\lambda)$ the associated self-adjoint operator. We first prove an important formula which expresses the spectral projection

[13] In this case the eigenvalues of $A+B$ will be dense in $\sigma_{\text{ac}}(A)$; thus $A+B$ could be an operator of the type given in Example 4.21(b,iii).

$E((a,b]) \equiv E_b - E_a$ associated with a finite interval $(a,b]$ in terms of the resolvent of $A$. In order to present a somewhat formal explanation of this formula, recall that

$$\frac{1}{x-\lambda-i\varepsilon} - \frac{1}{x-\lambda+i\varepsilon} \to 2\pi i\delta(x-\lambda) \qquad \text{as } \varepsilon \to +0,$$

the convergence being in the sense of distributions. If we formally replace $x$ by a self-adjoint operator $A$, hence replacing $(x-\lambda \mp i\varepsilon)^{-1}$ by the resolvent $(A-z)^{-1}$ for $z = \lambda \pm i\varepsilon$, and then integrate with respect to $d\lambda$, we arrive at

$$\begin{aligned}\frac{1}{2\pi i}\int_a^b \left[(A-\lambda-i\varepsilon)^{-1} - (A-\lambda+i\varepsilon)^{-1}\right] d\lambda \to \int_a^b \delta(A-\lambda)\, d\lambda \\ = \chi_{(a,b)}(A) = E((a,b)).\end{aligned}$$

The following proposition shows that this formal argument is essentially correct and gives a precise meaning to the preceding formula (this formula is entirely correct if the endpoints $a$ and $b$ of the considered interval are not eigenvalues of $A$).

**Proposition 4.29 (Stone's Formula).** *Let $A = \int_{-\infty}^{\infty} \lambda E(d\lambda)$ be the self-adjoint operator associated with a spectral family $\{E_\lambda\}$. For $-\infty < a < b < \infty$ one has*

$$\frac{1}{2\pi i}\operatorname*{s-lim}_{\varepsilon\to+0}\int_a^b \left[(A-\lambda-i\varepsilon)^{-1}-(A-\lambda+i\varepsilon)^{-1}\right]d\lambda = E((a,b))+\frac{1}{2}E(\{a\})+\frac{1}{2}E(\{b\}) \tag{4.80}$$

*and*

$$E((a,b]) = \frac{1}{2\pi i}\operatorname*{s-lim}_{\delta\to+0}\operatorname*{s-lim}_{\varepsilon\to+0}\int_{a+\delta}^{b+\delta}\left[(A-\lambda-i\varepsilon)^{-1} - (A-\lambda+i\varepsilon)^{-1}\right] d\lambda \tag{4.81}$$

(*the order of the two limits in* (4.81) *is important*).

PROOF. (i) For $\varepsilon > 0$ and $\lambda \in \mathbb{R}$ set

$$\begin{aligned}\Psi_\varepsilon(\lambda) &= \frac{1}{2\pi i}\int_a^b \left[(x-\lambda-i\varepsilon)^{-1} - (x-\lambda+i\varepsilon)^{-1}\right] dx \\ &= \frac{1}{2\pi i}\int_a^b \frac{2i\varepsilon}{(x-\lambda)^2+\varepsilon^2}\, dx = \frac{1}{\pi}\left[\arctan\frac{b-\lambda}{\varepsilon} - \arctan\frac{a-\lambda}{\varepsilon}\right].\end{aligned}$$

Clearly $\Psi_\varepsilon$ is a continuous function of $\lambda$ and $|\Psi_\varepsilon(\lambda)| \le 1$ for all $\lambda \in \mathbb{R}$. If $J$ is an interval, we denote by $\chi_J$ its characteristic function [see (1.41)]. By using the fact that $\lim_{\varepsilon\to+0}\arctan[(x-\lambda)/\varepsilon] = +\pi/2$ if $\lambda < x$ and $= -\pi/2$ if $\lambda > x$, one easily sees that

$$\Psi_0(\lambda) := \lim_{\varepsilon\to+0}\Psi_\varepsilon(\lambda) = \frac{1}{2}\left[\chi_{(a,b)}(\lambda) + \chi_{[a,b]}(\lambda)\right]. \tag{4.82}$$

(ii) For each $\varepsilon > 0$, $\Psi_\varepsilon(A) \equiv \int_{-\infty}^{\infty}\Psi_\varepsilon(\lambda)E(d\lambda)$ is a well defined operator in $\mathcal{B}(\mathcal{H})$. Let us show that $\{\Psi_\varepsilon\}$ is strongly convergent as $\varepsilon \to 0$ and calculate its limit [the result is already suggested by (4.82)]. If $f \in \mathcal{H}$, we can write $f = E(\{a\})f + E(\{b\})f + f_0$, where $f_0 \in \big[\mathcal{M}(\{a\}) \oplus \mathcal{M}(\{b\})\big]^{\perp}$. One has

$$\Psi_\varepsilon(A)E(\{a\})f = \Psi_\varepsilon(a)E(\{a\})f = \frac{1}{\pi}\big[\arctan\frac{b-a}{\varepsilon}\big]E(\{a\})f,$$

which converges strongly to $(1/2)E(\{a\})f$ as $\varepsilon \to +0$. In the same way one finds that $\text{s-lim}_{\varepsilon\to+0}\,\Psi_\varepsilon(A)E(\{b\})f = (1/2)E(\{b\})f$. Moreover, since $E((a,b))f_0 = E((a,b])f_0 = \int_a^b E(\lambda)f_0$, one has

$$E((a,b))f_0 - \Psi_\varepsilon(A)f_0 = \int_a^b [1-\Psi_\varepsilon(\lambda)]E(d\lambda)f_0 - \int_{-\infty}^a \Psi_\varepsilon(\lambda)E(d\lambda)f_0 - \int_b^\infty \Psi_\varepsilon(\lambda)E(d\lambda)f_0.$$

By using the Dominated Convergence Theorem (Proposition 1.9) one finds that each term on the right-hand side converges strongly to zero; for example [see (4.42)]:

$$\Big\|\int_a^b [1-\Psi_\varepsilon(\lambda)]E(d\lambda)f_0\Big\|^2 = \int_a^b |1-\Psi_\varepsilon(\lambda)|^2 m_{f_0}(d\lambda),$$

and one has $|1-\Psi_\varepsilon(\lambda)|^2 \leq 4$ and $1-\Psi_\varepsilon(\lambda) \to 0$ $m_{f_0}$-a.e. on $(a,b]$ [the point $b$ is of measure zero: $m_{f_0}(\{b\}) = 0$]. We have thus shown that

$$\Psi_0(A) := \underset{\varepsilon\to+0}{\text{s-lim}}\ \Psi_\varepsilon(A) = E((a,b)) + \frac{1}{2}E(\{a\}) + \frac{1}{2}E(\{b\}).$$

(iii) To obtain the validity of (4.80) it now suffices to verify that, for $\varepsilon > 0$:

$$\Psi_\varepsilon(A) = \frac{1}{2\pi i}\int_a^b \big[(A-\lambda-i\varepsilon)^{-1} - (A-\lambda+i\varepsilon)^{-1}\big]\,d\lambda.$$

By virtue of the polarisation identity (2.15) this holds provided that one knows that

$$\langle f, \Psi_\varepsilon(A)f\rangle = \frac{1}{2\pi i}\int_a^b \langle f, \big[(A-\lambda-i\varepsilon)^{-1} - (A-\lambda+i\varepsilon)^{-1}\big]f\rangle\,d\lambda \quad \forall f \in \mathcal{H}. \tag{4.83}$$

To prove (4.83) we use the first resolvent equation (2.93) to write

$$\begin{aligned}(A-\lambda-i\varepsilon)^{-1} - (A-\lambda+i\varepsilon)^{-1} &= 2i\varepsilon(A-\lambda-i\varepsilon)^{-1}(A-\lambda+i\varepsilon)^{-1}\\ &= 2i\varepsilon\int_{-\infty}^\infty \frac{1}{(\mu-\lambda)^2+\varepsilon^2}\,E(d\mu).\end{aligned}$$

It follows that

$$\frac{1}{2\pi i}\int_a^b \big[(A-\lambda-i\varepsilon)^{-1} - (A-\lambda+i\varepsilon)^{-1}\big]\,d\lambda = \frac{\varepsilon}{\pi}\int_a^b d\lambda\int_{-\infty}^\infty \frac{1}{(\mu-\lambda)^2+\varepsilon^2}\,E(d\mu)$$

and consequently

$$\frac{1}{2\pi i}\int_a^b \langle f, \left[(A-\lambda-i\varepsilon)^{-1}-(A-\lambda+i\varepsilon)^{-1}\right] f\rangle\, d\lambda = \frac{\varepsilon}{\pi}\int_a^b d\lambda \int_{-\infty}^{\infty} \frac{m_f(d\mu)}{(\mu-\lambda)^2+\varepsilon^2}.$$

Since the integrand is positive, one may interchange the order of integration (Fubini's Theorem):

$$\frac{1}{2\pi i}\int_a^b \langle f, \left[(A-\lambda-i\varepsilon)^{-1}-(A-\lambda+i\varepsilon)^{-1}\right] f\rangle\, d\lambda$$

$$= \frac{\varepsilon}{\pi}\int_{-\infty}^{\infty} m_f(d\mu)\int_a^b d\lambda \frac{1}{(\mu-\lambda)^2+\varepsilon^2} = \int_{-\infty}^{\infty} m_f(d\mu)\Psi_\varepsilon(\mu) = \langle f, \Psi_\varepsilon(A) f\rangle,$$

which proves (4.83).

(iv) Finally we deduce (4.81) from (4.80). One has as a consequence of (4.80):

$$\frac{1}{2\pi i}\operatorname*{s\text{-}lim}_{\varepsilon\to+0}\int_{a+\delta}^{b+\delta}\left[(A-\lambda-i\varepsilon)^{-1}-(A-\lambda+i\varepsilon)^{-1}\right] d\lambda = E((a+\delta, b+\delta))$$
$$+\frac{1}{2}E(\{a+\delta\})+\frac{1}{2}E(\{b+\delta\}) = E_{b+\delta}-E_{a+\delta}+\frac{1}{2}E(\{a+\delta\})-\frac{1}{2}E(\{b+\delta\}). \tag{4.84}$$

Observe that if $\mu \in \mathbb{R}$ and $\delta > 0$, then $\|E(\{\mu+\delta\})f\|^2 \leq \|E((\mu, \mu+\delta])f\|^2 = \|(E_{\mu+\delta}-E_\mu)f\|^2 \to 0$ as $\delta \to 0$ (by the right continuity of the spectral family), i.e. $\operatorname{s-lim}_{\delta\to+0} E(\{\mu+\delta\}) = O$. Thus, by taking into account again the right continuity of $\{E_\lambda\}$, one sees that the strong limit as $\delta \to +0$ of the right-hand side of (4.84) is $E_b - E_a \equiv E((a,b])$, which proves (4.81). □

**Proposition 4.30 (Spectral Theorem).** *If $A$ is a self-adjoint operator in a Hilbert space $\mathcal{H}$, there exists a unique spectral family $\{E_\lambda\}$ such that $A = \int_{-\infty}^{\infty}\lambda E(d\lambda)$.*

REMARKS ON THE PROOF.

(a) Uniqueness: Let $\{E_\lambda\}$ and $\{\widehat{E}_\lambda\}$ be two spectral families defining the same self-adjoint operator $A$, i.e. such that $A = \int_{-\infty}^{\infty}\lambda E(d\lambda) = \int_{-\infty}^{\infty}\lambda \widehat{E}(d\lambda)$. Since the right-hand side of Stone's Formula (4.81) depends only on $A$ (it does not involve the spectral family in terms of which $A$ has been defined), it follows that $E((a,b]) = \widehat{E}((a,b])$ for all $a, b \in \mathbb{R}$ ($a < b$). Hence

$$E_\lambda = \operatorname*{s\text{-}lim}_{a\to-\infty} E((a,\lambda]) = \operatorname*{s\text{-}lim}_{a\to-\infty} \widehat{E}((a,\lambda]) = \widehat{E}_\lambda.$$

(b) Existence: There exist a fair number of different proofs of the Spectral Theorem. All of them are rather long. We give an essentially complete proof in the Appendix of this chapter, and we indicate here the strategy of another one of these proofs. If $\{E_\lambda\}$ exists, then $E((a,b])$ must be given by (4.81). The integral appearing in this formula is well defined for each $\delta > 0$ and each $\varepsilon > 0$, because $(A-\lambda \pm i\varepsilon)^{-1}$ are operators in $\mathcal{B}(\mathcal{H})$ if $A$ is self-adjoint [one has $\mathbb{C}\setminus\mathbb{R} \in \rho(A)$] and depend continuously on $\lambda$

[the functions $\lambda \mapsto (A-\lambda \pm i\varepsilon)^{-1}$ are even infinitely differentiable in norm, see Proposition 2.35(d)]. One must prove that the limits in (4.81) exist (at least weakly) and define for each $a < b$ a projection; that the collection of all these projections has the properties of a spectral measure; and that $A$ is the associated self-adjoint operator.

The limits (in the weak sense) in Stone's Formula are treated as follows. By taking into account the polarisation identity (2.15), one sees that it suffices to consider

$$\frac{1}{2\pi i}\lim_{\delta\to+0}\lim_{\varepsilon\to+0}\left[\int_{a+\delta}^{b+\delta}\langle f,(A-\lambda-i\varepsilon)^{-1}f\rangle\,d\lambda-\int_{a+\delta}^{b+\delta}\langle f,(A-\lambda+i\varepsilon)^{-1}f\rangle\,d\lambda\right]$$
$$=\lim_{\delta\to+0}\lim_{\varepsilon\to+0}\frac{1}{\pi}\int_{a+\delta}^{b+\delta}\Im\langle f,(A-\lambda-i\varepsilon)^{-1}f\rangle\,d\lambda \tag{4.85}$$

[because $\langle f,(A-\lambda+i\varepsilon)^{-1}f\rangle=\langle[A-\lambda+i\varepsilon)^{-1}]^{*}f,f\rangle=\langle(A-\lambda-i\varepsilon)^{-1}f,f\rangle$, since $[(A-z)^{-1}]^{*}=(A-\overline{z})^{-1}$ if $A$ is self-adjoint, see (2.103)]. Now $z\mapsto\Phi(z):=\langle f,(A-z)^{-1}f\rangle$ is a $\mathbb{C}$-valued holomorphic function in the upper half complex plane $\{z\in\mathbb{C}\mid\Im z>0\}$, and one has to control its behaviour at the boundary of the domain of analyticity (here for $\Im z\to+0$). For this one can use a theorem from the theory of analytic functions of a complex variable: If $\Phi$ is holomorphic in the upper half plane, has a non-negative imaginary part and satisfies the estimate $|\Phi(z)|\le C/\Im z$ for some constant $C$, then there exists a finite Stieltjes measure $m$ on $\mathbb{R}$ such that $\Phi(z)=\int_{-\infty}^{\infty}(\lambda-z)^{-1}m(d\lambda)$. Furthermore $m((a,b])$ is then given by the expression $\lim_{\delta\to+0}\lim_{\varepsilon\to+0}\pi^{-1}\int_{a+\delta}^{b+\delta}[\Im\Phi(\lambda+i\varepsilon)]d\lambda$. The application of this theorem gives for each $f\in\mathcal{H}$ the measure $m_f$; the remainder of the proof reduces essentially to direct verifications. The details of the complete proof are presented for example in §§65, 66 and 59 of [AG] or in §7.3 (Theorem 7.17) and Appendix B of [W1].

Stone's Formula (Proposition 4.29) suggests that the spectral properties of a self-adjoint operator should be determined by the behaviour of its resolvent $(A-z)^{-1}$ near the real axis. We illustrate this by some examples.

(a) If $\lambda\in\mathbb{R}$ belongs to the spectrum of $A$, then $\|(A-\lambda\pm i\varepsilon)^{-1}\|\to\infty$ as $\varepsilon\to0$. Indeed one then has $E((\lambda-\varepsilon,\lambda+\varepsilon))\neq O$ for each $\varepsilon>0$, so that for each $\varepsilon>0$ there exists a vector $f$ such that $\|f\|=1$ and $E((\lambda-\varepsilon,\lambda+\varepsilon))f=f$. Then

$$\|(A-\lambda\pm i\varepsilon)^{-1}f\|^2=\int_{-\infty}^{\infty}\left|\frac{1}{\mu-\lambda\pm i\varepsilon}\right|^2 m_f(d\mu)$$
$$=\int_{\lambda-\varepsilon}^{\lambda+\varepsilon}\left|\frac{1}{\mu-\lambda\pm i\varepsilon}\right|^2 m_f(d\mu)\ge\frac{1}{2\varepsilon^2}\int_{\lambda-\varepsilon}^{\lambda+\varepsilon}m_f(d\mu)=\frac{1}{2\varepsilon^2}\|f\|^2=\frac{1}{2\varepsilon^2}.$$

Hence $\|(A-\lambda\pm i\varepsilon)^{-1}\|\ge(1/\sqrt{2})\varepsilon^{-1}$.

In Stone's Formula one finds the difference $(A-\lambda-i\varepsilon)^{-1}-(A-\lambda+i\varepsilon)^{-1}$ which may have better behaviour as $\varepsilon\to0$ than each individual term, i.e. it could be less rapidly divergent than each term separately.

(b) If $\mu$ is a real number in the resolvent set $\rho(A)$ of $A$, then $(A-\mu)^{-1}\in\mathcal{B}(\mathcal{H})$ and $(A-\mu\pm i\varepsilon)^{-1}\to(A-\mu)^{-1}$ in norm as $\varepsilon\to0$. Since $\rho(A)$ is an open set, there exists $\kappa>0$ such that the interval $(\mu-2\kappa,\mu+2\kappa)$ also belongs to $\rho(A)$. If one considers

Stone's Formula (4.81) with $a = \mu - \kappa$ and $b = \mu + \kappa$, one may take the limit $\varepsilon \to +0$ under the integral sign, and one obtains that $E((a,b]) \equiv E((\mu-\kappa, \mu+\kappa)) = O$.

(c) If $\lambda \in \sigma(A)$, then $(A-\lambda-i\varepsilon)^{-1}f$ and $(A-\lambda+i\varepsilon)^{-1}f$ may be convergent as $\varepsilon \to 0$ for certain vectors $f$. In particular, this is the case if the measure $m_f$ is supported away from the point $\lambda$, more precisely if there exists $\kappa > 0$ such that $E([\lambda-\kappa, \lambda+\kappa])f = \mathbf{0}$. Indeed, if $E([\lambda-\kappa, \lambda+\kappa])f = \mathbf{0}$, then one has for example

$$\|(A-\lambda \pm i\varepsilon)^{-1}f\|^2 = \int_{-\infty}^{\lambda-\kappa} \left|\frac{1}{\mu-\lambda \pm i\varepsilon}\right|^2 m_f(d\mu)$$
$$+ \int_{\lambda+\kappa}^{\infty} \left|\frac{1}{\mu-\lambda \pm i\varepsilon}\right|^2 m_f(d\mu) \leq \frac{1}{\kappa^2}\int_{-\infty}^{\infty} m_f(d\mu) = \frac{1}{\kappa^2}\|f\|^2.$$

This inequality holds for each $\varepsilon > 0$ and even for $\varepsilon = 0$. In particular the integral $\int_{-\infty}^{\infty}(\mu-\lambda)^{-1}E(d\mu)f$ defines a vector in $\mathcal{H}$ [let us denote it by $(A-\lambda)^{-1}f$] and a calculation similar to the preceding one, using the Dominated Convergence Theorem (Proposition 1.9), shows that $\|(A-\lambda \pm i\varepsilon)^{-1}f - (A-\lambda)^{-1}f\|^2 \to 0$ as $\varepsilon \to 0$.

If $\lambda \notin \sigma_{\mathrm{p}}(A)$, the set of vectors $f$ of the type considered above ($m_f$ supported away from $\lambda$) is dense in $\mathcal{H}$, because $\operatorname{s-lim}_{\kappa \to 0} E([\lambda-\kappa, \lambda+\kappa]) = O$ in this case. On the other hand, if $\lambda$ is an eigenvalue of $A$, the set of these vectors is not dense in $\mathcal{H}$ because it contains only vectors that are orthogonal to the associated subspace $\mathcal{M}(\{\lambda\})$. To see this, let $g$ be an eigenvector of $A$ associated with the eigenvalue $\lambda$: $Ag = \lambda g$, $\|g\| = 1$. Then $(A-\lambda \pm i\varepsilon)^{-1}g = (\pm i\varepsilon)^{-1}g$ and $[(A-\lambda-i\varepsilon)^{-1} - (A-\lambda+i\varepsilon)^{-1}]g = 2i\varepsilon^{-1}g$, which are divergent as $\varepsilon \to 0$. We must show that, if $h \in \mathcal{H}$ is such that $\langle g, h\rangle \neq 0$, then $[(A-\lambda-i\varepsilon)^{-1} - (A-\lambda+i\varepsilon)^{-1}]h$ diverges as $\varepsilon \to 0$. Let us write $h = \langle g,h\rangle g + h_0$ with $\langle g, h_0\rangle = 0$. We observe that

$$\big[(A-\lambda-i\varepsilon)^{-1} - (A-\lambda+i\varepsilon)^{-1}\big]h = \frac{2i}{\varepsilon}\langle g,h\rangle g + \big[(A-\lambda-i\varepsilon)^{-1} - (A-\lambda+i\varepsilon)^{-1}\big]h_0.$$

Thus, since $\langle g, (A-\lambda \pm i\varepsilon)^{-1}h_0\rangle = \langle (A-\lambda \mp i\varepsilon)^{-1}g, h_0\rangle = (\pm i\varepsilon)^{-1}\langle g, h_0\rangle = 0$, we have

$$\|\big[(A-\lambda-i\varepsilon)^{-1} - (A-\lambda+i\varepsilon)^{-1}\big]h\|^2 = \frac{4}{\varepsilon^2}|\langle g,h\rangle|^2$$
$$+ \|\big[(A-\lambda-i\varepsilon)^{-1} - (A-\lambda+i\varepsilon)^{-1}\big]h_0\|^2 \to \infty \quad \text{as } \varepsilon \to 0.$$

(d) We have seen that if $\lambda \notin \sigma_{\mathrm{p}}(A)$, then $(A-\lambda \pm i\varepsilon)^{-1}f$ converges as $\varepsilon \to 0$ for a dense set $\mathcal{D}_\lambda$ of vectors $f$. This set $\mathcal{D}_\lambda$ depends on $\lambda$. A better situation would be that in which these limits exist for each $\lambda$ in some interval $J$ and for all $f$ in a dense set independent of $\lambda$. In such a situation one can conclude that the spectrum of $A$ in $J$ is absolutely continuous, i.e. that $\sigma_{\mathrm{s}}(A) \cap J = \varnothing$. In the next proposition we give a precise description of such a situation that is frequently met in applications. In order to obtain a somewhat better understanding of these results, we recall the following formula from the theory of distributions. If $\phi\colon \mathbb{R} \to \mathbb{C}$ is a test function, then

$$\lim_{\varepsilon \to +0} \int_{-\infty}^{\infty} \frac{\phi(\mu)}{\mu-\lambda-i\varepsilon}\, d\mu = i\pi\phi(\lambda) + \lim_{\delta \to +0}\Big[\int_{-\infty}^{\lambda-\delta} \frac{\phi(\mu)}{\mu-\lambda}\, d\mu + \int_{\lambda+\delta}^{\infty} \frac{\phi(\mu)}{\mu-\lambda}\, d\mu\Big]. \tag{4.86}$$

Let $f \in \mathcal{H}_{\mathrm{ac}}(A)$. Then there exists a non-negative function $\theta_f \in L^1(\mathbb{R})$ such that $m_f(\mathcal{V}) = \int_{\mathcal{V}} \theta_f(\mu)d\mu$ if $\mathcal{V}$ is a Borel set in $\mathbb{R}$. Assume that $\theta_f$ is smooth (for example that it is of class $C^1$). Then (4.86) implies that

$$\begin{aligned}\langle f, (A-\lambda-i\varepsilon)^{-1}f\rangle &= \int_{-\infty}^{\infty} \frac{\theta_f(\mu)}{\mu-\lambda-i\varepsilon}\, d\mu \\ &\to i\pi\theta_f(\lambda) + \lim_{\delta\to+0}\Big[\int_{-\infty}^{\lambda-\delta} \frac{\theta_f(\mu)}{\mu-\lambda}\, d\mu + \int_{\lambda+\delta}^{\infty} \frac{\theta_f(\mu)}{\mu-\lambda}\, d\mu\Big]\end{aligned}$$

as $\varepsilon \to +0$. The expression on the last line is well defined (the limit $\delta \to +0$ exists if $\theta_f$ is smooth and is called the principal value of the integral). Thus there exists a dense set of vectors of $\mathcal{H}_{\mathrm{ac}}(A)$ such that $\langle f, (A-\lambda-i\varepsilon)^{-1}f\rangle$ has a limit as $\varepsilon \to +0$ for each $\lambda \in \mathbb{R}$. Some sort of converse is as follows: if the limit of $\langle f, (A-\lambda-i\varepsilon)^{-1}f\rangle$ as $\varepsilon \to +0$ exists for each $\lambda$ in some interval $(a, b)$ and each $f$ in a dense subset of $\mathcal{H}$, uniformly in $\lambda$ for each fixed $f$, then the spectrum of $A$ in $(a, b)$ is purely absolutely continuous. This is the content of the next proposition.

**Proposition 4.31.** *Let $A$ be a self-adjoint operator and $J$ an open interval.*
*(a) Let $f \in \mathcal{H}$ be a vector such that, for each $\lambda \in J$, $\langle f, (A-\lambda-i\varepsilon)^{-1}f\rangle$ converges as $\varepsilon \to +0$. Assume in addition that the convergence is uniform in $\lambda$ on each finite closed subinterval $\mathfrak{J} = [a, b]$ of $J$ (i.e., if $\Phi(\lambda) := \lim_{\varepsilon\to+0}\langle f, (A-\lambda-i\varepsilon)^{-1}f\rangle$, then $\sup_{\lambda\in\mathfrak{J}}\big|\Phi(\lambda) - \langle f, (A-\lambda-i\varepsilon)^{-1}f\rangle\big| \to 0$ as $\varepsilon \to +0$). Then $E(J)f \in \mathcal{H}_{\mathrm{ac}}(A)$.*
*(b) If there exists a dense set $\mathcal{D}$ of vectors in $\mathcal{H}$ such that the hypotheses of (a) are satisfied for each $f \in \mathcal{D}$, then the spectrum of $A$ in $J$ is absolutely continuous, in other terms $E(J)\mathcal{H} \subseteq \mathcal{H}_{\mathrm{ac}}(A)$ and $\sigma_{\mathrm{s}}(A) \cap J = \varnothing$.*

The continuous spectrum of the Schrödinger operators that one usually encounters in quantum mechanics is absolutely continuous (this question will be treated in Chapter 6 where we present a method of verifying the hypotheses of Proposition 4.31). Nevertheless, many potentials also give rise to singular continuous spectrum [14]. Typically one can create singular continuous spectrum by an infinite sequence of potential barriers, and the spectrum of an absolutely continuous Hamiltonian can be transformed into a singular continuous spectrum by the addition of a (judiciously chosen) potential with arbitrarily small $L^\infty$ norm.

PROOF OF PROPOSITION 4.31. (a) (i) Consider a finite closed subinterval $\mathfrak{J}$ of $J$. Define $\mathfrak{D}$ by $\mathfrak{D} = \{z = \lambda + i\varepsilon \mid \lambda \in \mathfrak{J}, 0 < \varepsilon \le 1\}$, and let $\overline{\mathfrak{D}} = \{z = \lambda + i\varepsilon \mid \lambda \in \mathfrak{J}, 0 \le \varepsilon \le 1\}$ be its closure. Let $\phi\colon \overline{\mathfrak{D}} \to \mathbb{C}$ be as follows: $\phi(z) = \langle f, (A-z)^{-1}f\rangle$ for $z \in \mathfrak{D}$ and $\phi(z) \equiv \phi(\lambda) = \Phi(\lambda)$ if $z = \lambda \in \mathfrak{J}$, where $\Phi(\lambda)$ is defined in the statement of the proposition. $\phi$ is continuous on $\mathfrak{D}$ [continuity of the resolvent in $\rho(A)$], and $\phi(\lambda + i\varepsilon)$ converges [uniformly in $\lambda$] to $\Phi(\lambda) \equiv \phi(\lambda)$ as $\varepsilon \to 0$. Since the uniform limit of a sequence of continuous functions is continuous, it follows that $\Phi$ is a continuous function of $\lambda \in \mathfrak{J}$. Thus $\phi$ is a continuous function on $\overline{\mathfrak{D}}$. Since $\overline{\mathfrak{D}}$ is a compact set, it follows that $\phi$ is bounded and uniformly continuous.

[14] This question has been investigated for example by D. Pearson, see [13].

(ii) Assume that $J = (\alpha, \beta)$, where $-\infty \le \alpha < \beta \le +\infty$. Observe that $E(J) = \text{s-lim}_{a \searrow \alpha} \text{s-lim}_{b \nearrow \beta} E((a,b])$ (here $\alpha < a < b < \beta$ and $a \to \alpha$, $b \to \beta$). Hence, to conclude that $E(J)f \in \mathcal{H}_{\text{ac}}(A)$, it suffices to know that $E((a,b])f \in \mathcal{H}_{\text{ac}}(A)$ for all $a, b$ satisfying $\alpha < a < b < \beta$. In other terms it suffices to know [see (4.36)] that, for $\alpha < a < b < \beta$, the function $F_{E((a,b])f}(\lambda) := \big\langle E((a,b])f, E_\lambda E((a,b])f \big\rangle$ is absolutely continuous, i.e. of the form $\int_{-\infty}^{\lambda} g(\mu)d\mu$ with $g \in L^1(\mathbb{R})$. Now

$$\big\langle E((a,b])f, E_\lambda E((a,b])f \big\rangle = \begin{cases} O & \text{if } \lambda \le a \\ \langle f, E((a,\lambda])f \rangle & \text{if } a < \lambda \le b \\ \|E((a,b])f\|^2 & \text{if } \lambda \ge b. \end{cases}$$

Thus it suffices to show that, for $a < \lambda \le b$, $\langle f, E((a,\lambda])f \rangle$ is of the form $\int_a^\lambda g(\mu)d\mu$ with $g \in L^1([a,b])$. By virtue of Stone's Formula and (4.85) one has

$$\langle f, E((a,\lambda])f \rangle = \lim_{\delta \to +0} \lim_{\varepsilon \to +0} \frac{1}{\pi} \int_{a+\delta}^{\lambda+\delta} \big[\Im \langle f, (A - \mu - i\varepsilon)^{-1} f \rangle \big] d\mu. \tag{4.87}$$

If $\delta$ is sufficiently small, the interval $\mathfrak{J} \equiv [a+\delta, b+\delta]$ is contained in $J$. Then, by the result of (i), the integrand in (4.87) is bounded and uniformly continuous for $\mu \in \mathfrak{J}$ and $\varepsilon \in [0,1]$, and the Dominated Convergence Theorem (Proposition 1.9) allows one to take the limit $\varepsilon \to +0$ under the integral sign:

$$\langle f, E((a,\lambda])f \rangle = \lim_{\delta \to +0} \frac{1}{\pi} \int_{a+\delta}^{\lambda+\delta} \big[\Im \Phi(\mu)\big] d\mu = \frac{1}{\pi} \int_a^\lambda \big[\Im \Phi(\mu)\big] d\mu.$$

Since $\Phi$ is continuous on $[a,b]$, $\Im\Phi$ belongs to $L^1([a,b])$. Thus the proof of (a) is complete.

(b) One has $E(J)f \in \mathcal{H}_{\text{ac}}(A)$ for each $f \in \mathcal{D}$. Since $\mathcal{D}$ is assumed to be dense in $\mathcal{H}$, the set $\{E(J)f \mid f \in \mathcal{D}\}$ is dense in the subspace $\mathcal{M}(J) \equiv E(J)\mathcal{H}$ (as is easily verified [15]). Hence $\mathcal{M}(J) \subseteq \mathcal{H}_{\text{ac}}(A)$. □

The assumption of convergence of $\langle f, (A - \lambda - i\varepsilon)^{-1} f \rangle$ as $\varepsilon \to 0$ in Proposition 4.31 is somewhat too strong but sufficient for our applications. It would be enough to impose the boundedness of $\Im\langle f, (A - \lambda - i\varepsilon)^{-1} f \rangle$ near the real axis (see the proof of the next proposition). In this context it is very interesting to know that the boundedness of $\Im\langle f, (A - \lambda - i\varepsilon)^{-1} f \rangle$ (and hence absolute continuity) can be obtained as a consequence of the positivity of a commutator, as shown by the next proposition.

**Proposition 4.32 (Putnam's Theorem).** *Let $A$ and $H$ be bounded self-adjoint operators such that $[H, iA] \ge CC^*$ for some operator $C \in \mathcal{B}(\mathcal{H})$. Then one has for all $\lambda \in \mathbb{R}$, all $\varepsilon \ne 0$ and each $f \in \mathcal{H}$:*

$$\big|\Im\langle Cf, (H - \lambda - i\varepsilon)^{-1} Cf \rangle\big| \le 4\|A\|\|f\|^2, \tag{4.88}$$

*and $\mathcal{R}(C) \subseteq \mathcal{H}_{\text{ac}}(H)$. If furthermore $\mathcal{N}(C^*) = \{\mathbf{0}\}$, then the spectrum of $H$ is purely absolutely continuous.*

[15] If $\mathcal{D}$ is a dense subset of $\mathcal{H}$ and $\mathcal{M}$ a subspace of $\mathcal{H}$, then the projection of $\mathcal{D}$ into $\mathcal{M}$, i.e. the set $P_{\mathcal{M}}\mathcal{D}$, is a dense subset of $\mathcal{M}$.

Observe that $[H, iA]$ is self-adjoint and that $CC^* \geq O$. Thus the assumption of positivity of the commutator $[H, iA]$ entails absolute continuity of at least part of the spectrum of $H$. If $C^*$ is invertible, i.e. if $\mathcal{N}(C^*) = \{\mathbf{0}\}$, then $\mathcal{R}(C)$ is dense in the Hilbert space [see Proposition 2.19(b)], and then Proposition 4.32 leads to the conclusion that $\mathcal{H}_{\mathrm{ac}}(H) = \mathcal{H}$. Of course, since the hypotheses of the proposition are completely symmetric in $A$ and $H$, the same conclusions hold for the operator $A$. In concrete situations one is interested in knowing the spectral properties of a given operator $H$ (often a quantum-mechanical Hamiltonian, whence the notation), and the preceding proposition shows that a useful idea is to look for an auxiliary self-adjoint operator $A$ such that the commutator $[H, iA]$ is positive in some sense (the operator $A$ is then said to be *conjugate to* $H$). In Chapter 6 we present a method based on this idea that leads to absolute continuity also for unbounded operators $H$ and $A$.[16] We refer to the introduction of Chapter 6 for further comments.

PROOF OF PROPOSITION 4.32. We assume throughout the proof that $\varepsilon > 0$ and use the notation $R(z) = (H - z)^{-1}$ if $z \in \rho(H)$.

(i) For $f \in \mathcal{H}$ we have [use (2.103) and the first resolvent equation (2.93)]:

$$\begin{aligned} \Im\langle Cf, R(\lambda + i\varepsilon)Cf\rangle &= \frac{1}{2i}\langle Cf, [R(\lambda + i\varepsilon) - R(\lambda - i\varepsilon)]Cf\rangle \\ &= \varepsilon\big\langle Cf, R(\lambda + i\varepsilon)R(\lambda - i\varepsilon)Cf\big\rangle \\ &= \varepsilon\|R(\lambda - i\varepsilon)Cf\|^2 \leq \varepsilon\|R(\lambda - i\varepsilon)C\|^2\|f\|^2. \end{aligned} \tag{4.89}$$

Now (use $\|T\|^2 = \|TT^*\|$, Eqs. (2.14) and (2.103) and the hypothesis that $CC^* \leq [H, iA]$):

$$\begin{aligned} \|R(\lambda - i\varepsilon)C\|^2 &= \|R(\lambda - i\varepsilon)CC^*R(\lambda + i\varepsilon)\| \\ &= \sup_{f \in \mathcal{H}, \|f\|=1} \langle R(\lambda + i\varepsilon)f, CC^*R(\lambda + i\varepsilon)f\rangle \\ &\leq \sup_{f \in \mathcal{H}, \|f\|=1} \langle R(\lambda + i\varepsilon)f, [H, iA]R(\lambda + i\varepsilon)f\rangle \\ &= \|R(\lambda - i\varepsilon)[H, iA]R(\lambda + i\varepsilon)\| \\ &= \|R(\lambda - i\varepsilon)[H - \lambda + i\varepsilon, A]R(\lambda + i\varepsilon)\| \\ &\leq \|AR(\lambda + i\varepsilon)\| + \|R(\lambda - i\varepsilon)A\| + \|2i\varepsilon R(\lambda - i\varepsilon)AR(\lambda + i\varepsilon)\|. \end{aligned}$$

Since $\|R(\lambda \pm i\varepsilon)\| \leq \varepsilon^{-1}$, this implies that $\|R(\lambda - i\varepsilon)C\|^2 \leq 4\varepsilon^{-1}\|A\|$. Upon inserting this estimate for $\|R(\lambda - i\varepsilon)C\|^2$ into (4.89), one obtains the inequality (4.88).

(ii) Let $\{E_\lambda\}$ be the spectral family of $H$. If $J = (a, b]$ is a bounded half-open interval, one has by Stone's formula (4.81):

$$\langle Cf, E((a, b])Cf\rangle = \frac{1}{\pi} \lim_{\delta \to +0} \lim_{\varepsilon \to +0} \int_{a+\delta}^{b+\delta} \Im\langle Cf, (H - \lambda - i\varepsilon)^{-1}Cf\rangle \, d\lambda.$$

[16] It can be seen from the proof of Proposition 4.32 that the assumption of boundedness of $H$ in this proposition is not important, whereas the boundedness of $A$ is essential.

Now

$$\int_{a+\delta}^{b+\delta} \Im\langle Cf, (H-\lambda-i\varepsilon)^{-1}Cf\rangle \, d\lambda \leq 4\|A\|\|f\|^2(b-a)$$

hence

$$\langle Cf, E((a,b])f\rangle \equiv m_{Cf}(J) \leq 4\|A\|\|f\|^2|J|,$$

where $|\cdot|$ means Lebesgue measure. This implies that $m_{Cf}(\mathcal{V}) \leq 4\|A\|\|f\|^2|\mathcal{V}|$ for each Borel subset $\mathcal{V}$ of $\mathbb{R}$ (take into account the definitions (4.2) and (1.45) of these measures). Hence the measure $m_{Cf}$ is absolutely continuous with respect to Lebesgue measure, i.e. $Cf \in \mathcal{H}_{\text{ac}}(H)$. Finally, if $\mathcal{N}(C^*) = \{\mathbf{0}\}$, the set $\{Cf \mid f \in \mathcal{H}\}$ is dense in $\mathcal{H}$, as already pointed out before the proof. □

# Appendix to Chapter 4: Proof of the Spectral Theorem

**A.4.1. Bounded self-adjoint operators.** We construct the spectral family $\{E_\lambda\}$ for a bounded self-adjoint operator $A$. To motivate this construction, let us observe that the subspace $\mathcal{M}_\lambda$ (the range of the projection $E_\lambda$) is the subspace on which the operator $A_\lambda := A - \lambda I$ is negative. In more detail: one can write $A_\lambda = A_\lambda^+ + A_\lambda^-$ with $A_\lambda^+ = (A-\lambda)(I-E_\lambda)$ and $A_\lambda^- = (A-\lambda)E_\lambda$, which expresses $A - \lambda I$ as a sum of a positive operator and a negative operator (the positive and negative parts of $A - \lambda I$). The positive part $A_\lambda^+$ annihilates $\mathcal{M}_\lambda$, because one clearly has $A_\lambda^+ E_\lambda = O$. One can thus characterise the vectors $f$ in $\mathcal{M}_\lambda$ by the property that $A_\lambda^+ f = \mathbf{0}$. Consequently, in order to define $\mathcal{M}_\lambda$ in terms of the operator $A$, it suffices to find a means of exhibiting the positive part $T^+$ and the negative part $T^-$ of a self-adjoint operator $T$, i.e. to write $T = T^+ + T^-$ with $T^+ \geq O$ and $T^- \leq O$. Then one can set (by taking for $T$ the operator $A - \lambda I$): $\mathcal{M}_\lambda = \{f \in \mathcal{H} \mid (A-\lambda I)^+ f = \mathbf{0}\}$.

To define $T^+$ and $T^-$, we first construct an operator $|T| \geq O$ representing the absolute value of $T$, and then we set $T^+ = (T + |T|)/2$ and $T^- = (T - |T|)/2$. To obtain $|T|$, we proceed in analogy with the numerical case in which $|z|$ is the positive square root of $\overline{z}z$. Here one must define the positive square root of a self-adjoint positive operator $T^*T$. We begin with some preliminary results. We observe in particular that (4.90) below gives a generalisation of the Schwarz inequality (1.12).

**Lemma 4.33.** *Let $A$ be a bounded positive self-adjoint operator (hence $\langle h, Ah\rangle \geq 0$ for all $h \in \mathcal{H}$). Then*

*(a) for each $k = 1, 2, 3, \ldots$, the operator $A^k$ is positive,*

*(b) if $S \in \mathcal{B}(\mathcal{H})$, then $S^*AS$ is positive,*

*(c) one has for all $f, g \in \mathcal{H}$:*

$$|\langle g, Af\rangle|^2 \leq \langle f, Af\rangle\langle g, Ag\rangle, \tag{4.90}$$

*(d) if furthermore $A \leq I$, then $\|A\| \leq 1$.*

PROOF. (a) If $k$ is even, say $k = 2m$ with $m \in \mathbb{N}$, then $\langle f, A^k f\rangle = \|A^m f\|^2 \geq 0$. If $k$ is odd ($k = 2m+1, m \in \mathbb{N}$), then $\langle f, A^k f\rangle = \langle A^m f, AA^m f\rangle \geq O$ because $A$ is positive.

(b) One has $\langle f, S^*ASf\rangle = \langle Sf, ASf\rangle \geq O$.

(c) The inequality (4.90) is trivial if $\langle g, Af\rangle = 0$. So let us assume that $\langle g, Af\rangle \neq 0$. As $A$ is positive and self-adjoint, one has for each $\alpha \in \mathbb{C}$:

$$0 \leq \langle f + \alpha g, A(f + \alpha g)\rangle = \langle f, Af\rangle + |\alpha|^2\langle g, Ag\rangle + \overline{\alpha}\langle g, Af\rangle + \alpha\langle Af, g\rangle.$$

By setting $\alpha = -\lambda\langle g, Af\rangle$ with $\lambda \in \mathbb{R}$, this inequality becomes

$$0 \leq \langle f, Af\rangle + \lambda^2|\langle g, Af\rangle|^2\langle g, Ag\rangle - 2\lambda|\langle g, Af\rangle|^2 \qquad \forall \lambda \in \mathbb{R}. \tag{4.91}$$

This implies that $\langle g, Ag\rangle \neq 0$ [otherwise (4.91) would be violated for $\lambda \to +\infty$]. We may thus take $\lambda = 1/\langle g, Ag\rangle$; for this value of $\lambda$ and after multiplication by $\langle g, Ag\rangle$, the inequality (4.91) becomes:

$$0 \leq \langle f, Af\rangle\langle g, Ag\rangle - |\langle g, Af\rangle|^2.$$

(d) If $O \leq A \leq I$, one has by (2.14):

$$\|A\| = \sup_{f\in\mathcal{H}, \|f\|=1} \langle f, Af\rangle \leq \sup_{f\in\mathcal{H}, \|f\|=1} \langle f, If\rangle = \sup_{f\in\mathcal{H}, \|f\|=1} \|f\|^2 = 1. \qquad \square$$

**Proposition 4.34.** *Let A be a bounded positive self-adjoint operator. Then*
*(a) there exists a unique bounded positive self-adjoint operator B such that* $B^2 = A$ *(it is usual to denote this operator B by* $A^{1/2}$*),*
*(b) if* $C \in \mathcal{B}(\mathcal{H})$ *commutes with A, then it also commutes with B.*

PROOF. (i) We first assume that $A \leq I$ and set $T = I - A$. By writing $B = I - X$, the equation $B^2 = A$ becomes $X = (T + X^2)/2$. We try to solve this equation recursively, by setting

$$X_1 = \frac{1}{2}T \qquad \text{and} \qquad X_n = \frac{1}{2}\Big[T + (X_{n-1})^2\Big] \quad \text{for } n \geq 2. \tag{4.92}$$

Let us determine the properties of this family $\{X_n\}$.
(1) It is clear from (4.92) that each $X_n$ is a polynomial in $T$ of order $2^{n-1}$, and all coefficients of these polynomials are $\geq 0$.
(2) The operators $X_n$ commute with each other.
(3) As $O \leq A \leq I$, one has $O \leq T \leq I$, hence $\|T\| \leq 1$ [Lemma 4.33(d)]. It follows that $\|X_1\| \leq 1/2 < 1$ and by recurrence that $\|X_n\| \leq (\|T\| + \|X_{n-1}\|^2)/2 < 1$.
(4) Let us check that $O \leq X_n \leq I$. One has

$$\langle f, X_n f\rangle = \frac{1}{2}[\langle f, Tf\rangle + \|X_{n-1}f\|^2] \geq \frac{1}{2}\langle f, Tf\rangle \geq 0$$

and

$$\begin{aligned}\langle f, X_n f\rangle &= \frac{1}{2}[\langle f, Tf\rangle + \|X_{n-1}f\|^2] \leq \frac{1}{2}[\|f\|^2 + \|X_{n-1}\|^2\|f\|^2] \\ &\leq \frac{1}{2}[\|f\|^2 + \|f\|^2] = \|f\|^2 = \langle f, If\rangle.\end{aligned}$$

(5) Finally we show that $X_{n+1} - X_n \geq O$. We have the following recurrence relation [using (2) in the last step]:

$$X_{n+1} - X_n = \frac{1}{2}\big[(T + X_n^2) - (T + X_{n-1}^2)\big] = \frac{1}{2}(X_n^2 - X_{n-1}^2)$$
$$= \frac{1}{2}(X_n + X_{n-1})(X_n - X_{n-1}). \qquad (4.93)$$

As $X_2 - X_1 = T^2/8$, one deduces from Eq. (4.93) recursively that $X_{n+1} - X_n$ is a polynomial in $T$ (of order $2^n$), also with all coefficients $\geq 0$. Since $T$ is positive, so is $T^k$ for each $k \in \mathbb{N}$ [Lemma 4.33(a)]. Thus each term of this polynomial is positive, hence $X_{n+1} - X_n \geq O$.

(ii) Let us now show that the sequence $\{X_n\}$ is strongly convergent. We fix $f \in \mathcal{H}$. By (5) and (3) the numerical sequence $\{\langle f, X_n f\rangle\}$ is monotone non-decreasing and bounded above by $\|f\|^2$. Therefore this sequence is convergent. Now let $n > m$. Then $X_n - X_m \geq O$ by (5), and by using the inequality (4.90) (with $A = X_n - X_m$ and $g = X_n f - X_m f$) one sees that

$$\|X_n f - X_m f\|^4 = \big|\langle X_n f - X_m f, X_n f - X_m f\rangle\big|^2$$
$$\leq \big\langle f, X_n f - X_m f\big\rangle\big\langle X_n f - X_m f, (X_n - X_m)(X_n f - X_m f)\big\rangle.$$

The first scalar product on the last line converges to zero as $m, n \to \infty$ (see above). By (3) the second scalar product is majorised by $\|X_n - X_m\|^3\|f\|^2 \leq 8\|f\|^2$. This shows that $\{X_n f\}$ is strongly Cauchy for each $f \in \mathcal{H}$.

(iii) Let $X$ be the strong limit of the sequence of operators $\{X_n\}$. Since $\langle f, Xf\rangle = \lim_{n\to\infty}\langle f, X_n f\rangle$, one sees by taking into account (4) that $0 \leq \langle f, Xf\rangle \leq \|f\|^2$, which proves that $O \leq X \leq I$. Hence $B \equiv I - X$ is a positive operator. $B$ is self-adjoint, because $X$ is self-adjoint as the limit of a sequence of bounded self-adjoint operators. Finally, by letting $n \to \infty$ in (4.92), one obtains $X = (T + X^2)/2$ (observe that $X_n^2$ converges strongly to $X^2$ by Proposition 2.1). The equation $X = (T + X^2)/2$ is equivalent to $B^2 = A$.

As $X_n$ is a polynomial in $(I - A)$, $X_n$ commutes with each bounded operator $C$ satisfying $[C, A] = O$. By passing to the limit, one concludes that $X$ (and hence also $B$) commutes with these operators $C$. This proves part (b) of the proposition.

(iv) If the assumption $A \leq I$ made so far is not satisfied, let $A_0 = A/\|A\|$. Then $A_0$ is self-adjoint and satisfies $O \leq A_0 \leq I$, hence there exists a positive self-adjoint operator $B_0$ with $B_0^2 = A_0$. Then clearly $B := \|A\|^{1/2}B_0$ is self-adjoint and positive, and $B^2 = A$.

(v) Let us finally prove the uniqueness of $B$. Assume that $B'$ is a second positive self-adjoint operators satisfying ${B'}^2 = A$. Since $B'A = {B'}^3 = AB'$, the operator $B'$ commutes with $B$ [by part (b) of the proposition, applied with $C = B'$]. Thus

$$(B - B')B(B - B') + (B - B')B'(B - B') = (B - B')(B^2 - {B'}^2)$$
$$= (B - B')(A - A) = O. \qquad (4.94)$$

Each of the two terms on the left-hand side of (4.94) is a positive operator [Lemma 4.33(b)]. Hence each of these terms is zero, and consequently so is their difference:

$$O = (B - B')B(B - B') - (B - B')B'(B - B') = (B - B')^3.$$

It follows that $(B - B')^4 = O$, and by applying twice the identity (2.12):

$$O = \|(B - B')^4\| = \|(B - B')^2\|^2 = \|B - B'\|^4,$$

whence $B = B'$. □

**Corollary 4.35.** *Let $S$, $T$ and $Z$ be bounded self-adjoint operators satisfying $S \geq T$, $Z \geq O$ and $[Z,S] = [Z,T] = O$. Then $ZS = SZ \geq ZT = TZ$.*

PROOF. $Z^{1/2}$ commutes with $S$ and with $T$. By virtue of Lemma 4.33(b) one has $SZ - ZT = Z^{1/2}(S - T)Z^{1/2} \geq O$ if $S \geq T$. □

We can now define the positive and negative parts of a bounded self-adjoint operator $T$. One sets $|T| = (T^*T)^{1/2} \equiv (T^2)^{1/2}$, the positive square root of the self-adjoint operator $T^2$ (observe that $\langle f, T^2 f\rangle = \|Tf\|^2 \geq 0$). Let

$$T^+ = \frac{1}{2}\big(T + |T|\big) \qquad \text{and} \qquad T^- = \frac{1}{2}\big(T - |T|\big). \tag{4.95}$$

Let also $\mathcal{M}(T) = \{f \in \mathcal{H} \,|\, T^+ f = \mathbf{0}\}$. This is the closed subspace $\mathcal{N}(T^+)$ of all eigenvectors of the self-adjoint operator $T^+$ associated with the eigenvalue $0$. We write $E$ for the projection $P_{\mathcal{M}(T)}$ onto $\mathcal{M}(T)$, i.e. $E \equiv P_{\mathcal{M}(T)}$. In (A) - (J) below we specify some properties of these quantities.

- (A) *The operator $|T|$ (and hence also $T^+$ and $T^-$) commutes with each operator $C \in \mathcal{B}(\mathcal{H})$ satisfying $[C,T] = O$.*

This follows from Proposition 4.34(b).

- (B) *The operators $T, |T|, T^+$ and $T^-$ commute among themselves.*
- (C) *One has $|-T| = |T|$* [because $(-T)^2 = T^2$].
- (D) *(a) If $T \geq O$, then $|T| = T$, hence $T^+ = T$ and $T^- = O$.*
  *(b) If $T \leq O$, then $|T| = -T$, hence $T^+ = O$ and $T^- = T$.*
- (E) *One has*

$$T^+E = ET^+ = O \qquad \text{and} \qquad T^-E = ET^- = ET^-E = T^-. \tag{4.96}$$

Indeed: (i) The equation $T^+E = O$ is simply a reformulation of the definition of $\mathcal{M}(T)$ as the set of eigenvectors of $T^+$ with eigenvalue $0$. Then it follows that $ET^+ = (T^+E)^* = O^* = O$.

(ii) If $f = T^-g$, then (because $[T, |T|] = O$):

$$T^+f = T^+T^-g = \frac{1}{4}(T + |T|)(T - |T|)g = \frac{1}{4}(T^2 - |T|^2)g = \mathbf{0},$$

hence $f \in \mathcal{M}(T)$. In other terms $T^-g = ET^-g$ for each $g \in \mathcal{H}$, i.e. $ET^- = T^-$. It follows that $T^-E = (ET^-)^* = T^{-*} = T^-$ and then $ET^-E = ET^- = T^-$.

- (F) *The operator $E$ commutes with $T^+$ and with $T^-$, hence also with $T \equiv T^+ + T^-$ and with $|T| \equiv T^+ - T^-$. Furthermore*

$$ET = E(T^+ + T^-) = ET^- = T^- \tag{4.97}$$

*and*

$$(I - E)T = T - T^- = T^+. \tag{4.98}$$

These statements are simple consequences of (4.96).

- (G) *One has $T^+ \geq O$ and $T^- \leq O$.*

Indeed: (i) $-T^- = -ET^-E = E(T^+ - T^-)E = E|T|E \geq O$ by Lemma 4.33(b) (because $|T| \geq O$).

(ii) (4.96) implies that $(I - E)T^+(I - E) = T^+$ and $(I - E)T^-(I - E) = O$. Hence $T^+ = (I - E)(T^+ - T^-)(I - E) = (I - E)|T|(I - E) \geq O$.

- (H) *One has $T^+ \geq T^+ + T^- = T$* (because $T^- \leq O$).
- (I) *If $S$ is bounded and commutes with $T$, then $[S, E] = O$.*

Indeed: (i) Let $f \in \mathcal{M}(T)$. Then $ST^+f = \mathbf{0}$. But $S$ commutes with $|T|$, hence with $T^+$; thus $T^+Sf = \mathbf{0}$, i.e. $Sf \in \mathcal{M}(T)$ or $Sf = ESf$. In other terms one has $SE = ESE$. By taking the adjoint of this relation one gets $ES^* = ES^*E$. If we assume for the moment that $S$ is self-adjoint, this becomes $ES = ESE$. Since $ESE = SE$ as already seen, one then has $ES = SE$.

(ii) We have shown that $[S, T] = O$ implies that $[S, E] = O$ if $S$ is self-adjoint. For a general $S$, observe that $[S, T] = O$ implies that $[S^*, T] = O$. Since $S + S^*$ and $i(S - S^*)$ are self-adjoint, the result of (i) implies that $[S + S^*, E] = O$ and $i[S - S^*, E] = O$. By combining these two relations one obtains $[S, E] = O$.

- (J) *Let $S$ be a bounded self-adjoint operator satisfying $S \geq O, S \geq T$ and $[S, T] = O$. Then $S \geq T^+$.*

Indeed: since $I - E \geq O$ and $I - E$ commutes with $S$ and with $T$ by (I) and (F), Corollary 4.35 implies that $(I - E)S \geq (I - E)T$. On the other hand, since $ES = ESE$ and $S$ is positive, one has $ES \geq O$ [Lemma 4.33(b)]. Consequently $S = (I - E)S + ES \geq (I - E)T = T^+$.

Now let $A$ be a bounded self-adjoint operator. We set, for $\lambda \in \mathbb{R}$: $A_\lambda = A - \lambda I$, $A_\lambda^+ = (A_\lambda + |A_\lambda|)/2$, $A_\lambda^- = (A_\lambda - |A_\lambda|)/2$ and

$$\mathcal{M}_\lambda \equiv \mathcal{M}(A_\lambda) = \{f \in \mathcal{H} \mid A_\lambda^+ f = \mathbf{0}\}.$$

Furthermore let $E_\lambda$ be the projection onto the subspace $\mathcal{M}_\lambda$. In the next two lemmas we check that $\{E_\lambda\}$ is a spectral family (Lemma 4.36) and that $A$ is the self-adjoint operator associated with it (Lemma 4.37).

**Lemma 4.36.** *The family of projections $\{E_\lambda\}$ defined above is a spectral family, i.e. it satisfies* (4.16) *and* (4.17).

PROOF. (i) By the property (F), $A_\lambda$ commutes with $E_\lambda$. Since $A = A_\lambda + \lambda I$, it follows that $A$ commutes with each $E_\lambda$.

(ii) We now show that $\mathcal{M}_\lambda \subseteq \mathcal{M}_\mu$ if $\lambda \leq \mu$; this implies that $E_\lambda E_\mu = E_{\min\{\lambda,\mu\}}$ for all $\lambda, \mu \in \mathbb{R}$.

Since $[A_\lambda, A_\mu] = O$, one obtains from (A) [with $T = A - \lambda$, $C = A_\mu$] that $A_\mu$ commutes with $|A_\lambda|$ and with $A_\lambda^\pm$. By using again the property (A) [with $T = A_\mu$ and $C = |A_\lambda|$ or $A_\lambda^\pm$], one concludes that $|A_\mu|$ and $A_\mu^\pm$ commute with $|A_\lambda|$ and with $A_\lambda^\pm$. In particular we have $[A_\mu^+, A_\lambda^+] = O$.

Property (H) shows that, for $\mu \geq \lambda$:

$$A_\lambda^+ \geq A_\lambda = A - \lambda I \geq A - \mu I = A_\mu.$$

Since $A_\lambda^+ \geq O$, we can apply (J) ([with $S = A_\lambda^+, T = A_\mu$] to conclude that $A_\lambda^+ \geq A_\mu^+$. By virtue of Corollary 4.35 this implies that $A_\mu^+ A_\lambda^+ \geq (A_\mu^+)^2$.[17] Consequently we have $\langle f, A_\mu^+ A_\lambda^+ f\rangle \geq \|A_\mu^+ f\|^2$ for all $f \in \mathcal{H}$. If $f \in \mathcal{M}_\lambda$, i.e. if $A_\lambda^+ f = \mathbf{0}$, we obtain from this inequality that $A_\mu^+ f = \mathbf{0}$, i.e. $f \in \mathcal{M}_\mu$. This establishes the inclusion $\mathcal{M}_\lambda \subseteq \mathcal{M}_\mu$.

(iii) Next we show that $\mathcal{M}_\lambda = \{\mathbf{0}\}$ for all $\lambda < -\|A\|$. So let $\lambda < -\|A\|$ and set $\varepsilon = -\|A\| - \lambda$. We have $\varepsilon > 0$ and, by using (2.14):

$$\langle f, A_\lambda f\rangle = \langle f, Af\rangle - \lambda\|f\|^2 \geq \big(-\|A\| - \lambda\big)\|f\|^2 = \varepsilon\|f\|^2 = \langle f, \varepsilon I f\rangle,$$

i.e. $A_\lambda \geq \varepsilon I > O$. Consequently $A_\lambda^+ = A_\lambda$. If $f \in \mathcal{M}_\lambda$, i.e. if $A_\lambda^+ f = \mathbf{0}$, then

$$\varepsilon\|f\|^2 \leq \langle f, A_\lambda f\rangle = \langle f, A_\lambda^+ f\rangle = 0, \quad \text{hence } f = \mathbf{0}.$$

(iv) Finally we show that $\mathcal{M}_\lambda = \mathcal{H}$ if $\lambda > \|A\|$. Here we set $\varepsilon = \lambda - \|A\|$. Then

$$\langle f, A_\lambda f\rangle = \langle f, Af\rangle - \lambda\|f\|^2 \leq \big(\|A\| - \lambda\big)\|f\|^2 = -\varepsilon\|f\|^2.$$

Hence $A_\lambda \leq -\varepsilon I < O$ and consequently $A_\lambda^+ = O$. Thus $A_\lambda^+ f = \mathbf{0}$ for all $f \in \mathcal{H}$, i.e. $\mathcal{M}_\lambda = \mathcal{H}$. □

We have thus established that the projections $\{E_\lambda\}$ define a spectral family. We shall see in the following lemma that this spectral family is also right continuous. As a preliminary step we prove a useful auxiliary inequality: If $\lambda < \mu$, let $E((\lambda,\mu]) = E_\mu - E_\lambda$.[18] Then

$$\lambda E((\lambda,\mu]) \leq AE((\lambda,\mu]) \leq \mu E((\lambda,\mu]). \tag{4.99}$$

Indeed: by using the relations $[A, E((\lambda,\mu])] = O$ [see (i) of the preceding proof], $E((\lambda,\mu])^2 = E((\lambda,\mu]) = E((\lambda,\mu])E_\mu$ and as well as (4.97), one obtains

$$\begin{aligned}(\mu I - A)E((\lambda,\mu]) &= E((\lambda,\mu])(\mu I - A)E((\lambda,\mu]) \equiv -E((\lambda,\mu])A_\mu E((\lambda,\mu]) \\ &= -E((\lambda,\mu])E_\mu A_\mu E((\lambda,\mu]) = -E((\lambda,\mu])A_\mu^- E((\lambda,\mu]) \geq O\end{aligned}$$

[17] Take $S = A_\lambda^+$ and $T = Z = A_\mu^+$ in Corollary 4.35.

[18] Remember that $E((\lambda,\mu])$ is a positive operator, in fact a projection (see page 142).

(the final inequality is obtained from Lemma 4.33(b), because $-A_\mu^- \geq O$). This proves the second inequality in (4.99). The first one is obtained by a similar argument, by using the relation $E((\lambda,\mu]) = E((\lambda,\mu])(I - E_\lambda)$ and (4.98):

$$\begin{aligned}(A-\lambda)E((\lambda,\mu]) &= E((\lambda,\mu])(A-\lambda I)E((\lambda,\mu]) = E((\lambda,\mu])(I-E_\lambda)A_\lambda E((\lambda,\mu]) \\ &= E((\lambda,\mu])A_\lambda^+ E((\lambda,\mu]) \geq O.\end{aligned}$$

**Lemma 4.37.** *The spectral family $\{E_\lambda\}$ in Lemma* (4.36) *is right continuous, and $A$ is the self-adjoint operator associated with it.*

PROOF. (i) We first prove the right continuity of the family of projections $\{E_\lambda\}$. We fix $\lambda \in \mathbb{R}$. The number $\|E_{\lambda+\varepsilon}f\|^2$ is the square of the norm of the projection of $f$ onto the subspace $\mathcal{M}_{\lambda+\varepsilon}$. As $\mathcal{M}_{\lambda+\delta} \subseteq \mathcal{M}_{\lambda+\varepsilon}$ if $0<\delta<\varepsilon$, one has $\|E_{\lambda+\delta}f\|^2 \leq \|E_{\lambda+\varepsilon}f\|^2$. So the numerical sequence $\{\|E_{\lambda+\varepsilon}f\|^2\}$ is monotone non-increasing as $\varepsilon \to +0$, i.e. it has a limit. Hence (since $E_{\lambda+\varepsilon}E_{\lambda+\delta} = E_{\lambda+\delta}$ if $0<\delta<\varepsilon$):

$$\begin{aligned}\|E_{\lambda+\varepsilon}f - E_{\lambda+\delta}f\|^2 &= \|E_{\lambda+\varepsilon}f\|^2 + \|E_{\lambda+\delta}f\|^2 - \langle E_{\lambda+\varepsilon}f, E_{\lambda+\delta}f\rangle \\ &- \langle E_{\lambda+\delta}f, E_{\lambda+\varepsilon}f\rangle = \|E_{\lambda+\varepsilon}f\|^2 + \|E_{\lambda+\delta}f\|^2 - 2\|E_{\lambda+\delta}f\|^2 \\ &= \|E_{\lambda+\varepsilon}f\|^2 - \|E_{\lambda+\delta}f\|^2 \to 0 \quad \text{as } \delta,\varepsilon \to +0.\end{aligned}$$

This shows that $\{E_{\lambda+\varepsilon}\}$ is strongly Cauchy as $\varepsilon \to +0$. We define a projection $F_\lambda$ by $F_\lambda = \text{s-lim}_{\varepsilon\to+0}(E_{\lambda+\varepsilon} - E_\lambda) \equiv \text{s-lim}_{\varepsilon\to+0} E((\lambda,\lambda+\varepsilon])$. We must show that $F_\lambda = O$, in other terms that the equation $F_\lambda f = f$ implies that $f = \mathbf{0}$. For this we use (4.99) with $\mu = \lambda+\varepsilon$:

$$\lambda E((\lambda,\lambda+\varepsilon]) \leq AE((\lambda,\lambda+\varepsilon]) \leq (\lambda+\varepsilon)E((\lambda,\lambda+\varepsilon]).$$

Upon taking the limit $\varepsilon \to +0$, one obtains that for each $f \in \mathcal{H}$:

$$\langle f, \lambda F_\lambda f\rangle \leq \langle f, AF_\lambda f\rangle \leq \langle f, \lambda F_\lambda f\rangle.$$

So $\langle f, \lambda F_\lambda f\rangle = \langle f, AF_\lambda f\rangle$ for all $f \in \mathcal{H}$ and hence $\langle f, \lambda F_\lambda g\rangle = \langle f, AF_\lambda g\rangle$ for all $f, g \in \mathcal{H}$ [by polarisation, see (2.15)]. This shows that $AF_\lambda = \lambda F_\lambda$. Consequently, if $f$ satisfies $F_\lambda f = f$, one has $A_\lambda f = \mathbf{0}$. By taking into account Eq. (4.98) (with $T = A_\lambda$), one then obtains that $A_\lambda^+ f = (I - E_\lambda)A_\lambda f = \mathbf{0}$. Thus $f \in \mathcal{M}_\lambda \subseteq \mathcal{M}_{\lambda+\varepsilon}$ for each $\varepsilon > 0$, hence $E_\lambda f = E_{\lambda+\varepsilon} f = f$. It follows that $E_{\lambda+\varepsilon} f - E_\lambda f = \mathbf{0}$ for each $\varepsilon > 0$, hence $F_\lambda f = \mathbf{0}$. Since $F_\lambda f = f$, we have $f = \mathbf{0}$.

(ii) We now check that $A$ is the self-adjoint operator associated with $\{E_\lambda\}$. Let $(a,b]$ be an interval containing $[-\|A\|, \|A\|]$. Let us denote by $A'$ the self-adjoint operator associated with $\{E_\lambda\}$. Then, by (4.39):

$$A' = \underset{|\Pi|\to 0}{\text{s-lim}} \sum_{k=1}^{N} u_k E((s_{k-1}, s_k]), \tag{4.100}$$

where $\Pi = \{s_0, s_1, \ldots, s_N; u_1, \ldots, u_N\}$ is a partition of $(a,b]$. From (4.99) we deduce that

$$\sum_{k=1}^{N} s_{k-1}E((s_{k-1},s_k]) \leq A\sum_{k=1}^{N} E((s_{k-1},s_k]) \leq \sum_{k=1}^{N} s_k E((s_{k-1},s_k]). \tag{4.101}$$

We observe that $A\sum_{k=1}^{N} E((s_{k-1}, s_k]) = AE((s_0, s_N]) = A$. Thus the second inequality in (4.101) implies that $A \leq A'$ [we have taken $u_k = s_k$ in (4.100) to express $A'$]. Next, applying (4.100) with $u_k = s_{k-1} + |\Pi|/2$, we obtain from the first inequality in (4.101) that $A' \leq A$. Thus $\langle f, Af\rangle = \langle f, A'f\rangle$ for all $f \in \mathcal{H}$, and by polarisation $\langle f, Ag\rangle = \langle f, A'g\rangle$ for all $f, g \in \mathcal{H}$. So we have shown that $A = A'$. □

**Remark 4.38.** (a) One has $|A| = \int_{-\infty}^{\infty} |\lambda| E(d\lambda) = -\int_{-\infty}^{0} \lambda E(d\lambda) + \int_{0}^{\infty} \lambda E(d\lambda)$, hence

$$A^+ = \frac{1}{2}\Big[\int_{-\infty}^{0} \{(\lambda + (-\lambda)\} E(d\lambda) + \int_{0}^{\infty} (\lambda + \lambda) E(d\lambda)\Big] = \int_{0}^{\infty} \lambda E(d\lambda),$$

i.e. $A^+ = AE((0, \infty))$, and similarly

$$A^- = \int_{-\infty}^{0} \lambda E(d\lambda) = AE(-\infty, 0)) = AE((-\infty, 0]),$$

because $AE(\{0\}) = O$ [since $E(\{0\})\mathcal{H} = \mathcal{N}(A)$]. It is clear that $A^+ f = \mathbf{0}$ for each vector $f$ belonging to the subspace $E((-\infty, 0])\mathcal{H}$, and $A^+ f \neq \mathbf{0}$ for each non-zero $f$ in the subspace $E((0, \infty))\mathcal{H}$. Consequently, if $\mathcal{M}(A) = \{f \in \mathcal{H} \mid A^+ f = \mathbf{0}\}$, we have $\mathcal{M}(A) = E((-\infty, 0])\mathcal{H}$, i.e. the projection $E \equiv P_{\mathcal{M}(A)}$ associated with the self-adjoint operator $A$ is just the spectral projection $E((-\infty, 0])$. This observation will be useful in the proof of Proposition 4.39.
(b) If $A \geq O$ and $\lambda < 0$, then $A - \lambda I > O$, hence $A_\lambda^+ = A_\lambda > O$. It follows that $\mathcal{M}_\lambda \equiv \mathcal{N}(A_\lambda^+) = \{\mathbf{0}\}$. So $E_\lambda = O$ for all $\lambda < 0$. Similarly one finds that if $A \leq I$, then $\mathcal{M}_\lambda = \mathcal{H}$ and hence $E_\lambda = I$ for all $\lambda \geq 1$.

**A.4.2. Semibounded self-adjoint operators.** We have proved the Spectral Theorem under the assumption of boundedness of the operator $A$. The case of an unbounded self-adjoint operator can be reduced to that of a bounded operator. The simplest situation is that of a semibounded operator $A$. The case of a lower semibounded or upper semibounded operator is easily reduced to that of a positive operator; so we now discuss this latter situation.

Let $A = A^* \geq O$. Then one has for each $f \in \mathcal{D}(A)$:

$$\|(A + I)f\|^2 = \|Af\|^2 + 2\langle f, Af\rangle + \|f\|^2 \geq \|Af\|^2 + \|f\|^2. \tag{4.102}$$

Hence $A + I$ is invertible and $\|(A + I)^{-1}\| \leq 1$ (see Lemma 3.1). Also $\mathcal{R}(A + I)^\perp = \mathcal{N}(A + I) = \{\mathbf{0}\}$ [Proposition 2.19(b)]. Thus, again by Lemma 3.1, $Y := (A + I)^{-1}$ belongs to $\mathcal{B}(\mathcal{H})$, and $-1 \in \rho(A)$. The operator $Y$ is self-adjoint, and $O < Y \leq I$: the second inequality is satisfied because $\|Y\| \leq 1$ and the first one follows from the positivity of $A$: if $g \neq \mathbf{0}$, then

$$\begin{aligned}\langle g, Yg\rangle = \langle (A + I)(A + I)^{-1}g, (A + I)^{-1}g\rangle &\geq \langle (A + I)^{-1}g, (A + I)^{-1}g\rangle \\ &= \|(A + I)^{-1}g\|^2 > 0\end{aligned}$$

[for the last inequality observe that $(A+I)^{-1}g = \mathbf{0}$ only if $g = \mathbf{0}$, because $(A+I)^{-1}$ is a bijective mapping].

The correspondence $\lambda \mapsto (\lambda+1)^{-1}$ defines a bijection between the intervals $[0,\infty)$ and $(0,1]$, which will allow us to establish a one-to-one relation between the spectral projections $E_\lambda$ of $A$ and those of the bounded operator $Y$. We denote the spectral measure of $Y$ by $\{\mathbb{E}(\cdot)\}$ and observe that, by Remark 4.38(b), its support is contained in the interval $[0,1]$, so that $\mathbb{E}([0,1]) = I$. We define $E_\lambda$ as follows:

$$E_\lambda = O \quad \text{if } \lambda < 0 \quad , \quad E_\lambda = \mathbb{E}\Big(\big[\frac{1}{\lambda+1}, 1\big]\Big) \quad \text{if } \lambda \geq 0. \tag{4.103}$$

The projections $E_\lambda$ defined by (4.103) do form a spectral family, because
(i) If $0 \leq \lambda \leq \mu$ one has

$$\begin{aligned} E_\lambda E_\mu &= \mathbb{E}\Big(\big[\frac{1}{\lambda+1}, 1\big]\Big)\mathbb{E}\Big(\big[\frac{1}{\mu+1}, 1\big]\Big) = \mathbb{E}\Big(\big[\frac{1}{\lambda+1}, 1\big] \cap \big[\frac{1}{\mu+1}, 1\big]\Big) \\ &= \mathbb{E}\Big(\big[\frac{1}{\lambda+1}, 1\big]\Big) = E_\lambda \equiv E_{\min\{\lambda,\mu\}}. \end{aligned}$$

(ii) $\{E_\lambda\}$ is right continuous: if $\lambda \geq 0$, then

$$E_{\lambda+\varepsilon} - E_\lambda = \mathbb{E}\Big(\big[\frac{1}{\lambda+1+\varepsilon}, \frac{1}{\lambda+1}\big)\Big) \to 0 \qquad \text{as } \varepsilon \to +0.$$

(iii) One has $\text{s-}\lim_{\lambda\to+\infty} E_\lambda = \mathbb{E}((0,1])$, and $\mathbb{E}((0,1]) = \mathbb{E}([0,1]) = I$ (because $Y > O$, hence 0 is not an eigenvalue of $Y$).

Let us denote by $A' = \int_0^\infty \lambda E(d\lambda)$ the (positive) self-adjoint operator associated with the spectral family $\{E_\lambda\}$. By a simple calculation [consider partitions of an interval $(a,b]$ and make the change of variables $\lambda \mapsto \mu = (\lambda+1)^{-1}$] one finds that [19]

$$(A'+I)^{-1} = \int_0^\infty \frac{1}{\lambda+1}\, E(d\lambda) = \int_0^1 \mu\, \mathbb{E}(d\mu) = Y \equiv (A+I)^{-1}. \tag{4.104}$$

This relation implies that $\mathcal{D}(A') = \mathcal{D}(A)$ and that $(A+I)(A'+I)^{-1} = I$. Since $(A'+I)(A'+I)^{-1} = I$, we conclude that $Af = A'f$ for $f \in \mathcal{D}(A') = \mathcal{D}(A)$, whence $A' = A$: $\{E_\lambda\}$ is the spectral family of $A$!

REMARK. One can see that, if $A$ is assumed to be only symmetric, the preceding proof is not applicable (in fact the Spectral Theorem is not true without the assumption that $A$ be self-adjoint, or at least normal). Let us consider for example the symmetric (non-self-adjoint) operator $K_0 = -d^2/dx^2$ in $L^2((0,\infty))$, with domain $C_0^\infty((0,\infty))$. Its adjoint $K_0^*$ has $-1$ as eigenvalue: the equation $K_0^* g = -g$ has the square-integrable solution $g(x) = e^{-x}$ (see Proposition 3.14). Since $\mathcal{R}(K_0+I) \perp \mathcal{N}(K_0^*+I)$, this solution is orthogonal to $\mathcal{R}(K_0+I)$ and thus $(K_0+I)^{-1}$ cannot belong to $\mathcal{B}(\mathcal{H})$.

---

[19] The spectral integrals are defined by using partitions with subintervals $(s_{k-1}, s_k]$ that are open on the left and closed on the right. Now $E((c,d]) = \mathbb{E}([1/d, 1/c))$, which is equal to $\mathbb{E}((1/d, 1/c])$ if and only if $c$ and $d$ are not eigenvalues of $A$. Thus one should consider here only partitions of $(a,b]$ for which the endpoints $s_k$ of the involved subintervals are points of continuity of the spectral family $\{E_\lambda\}$. We recall that there are at most a countable number of points of discontinuity, see page 143.

**A.4.3. Non-semibounded self-adjoint operators.** If a self-adjoint operator $A$ has a spectral gap (i.e. if $\sigma(A) \neq \mathbb{R}$), then its resolvent set $\rho(A)$ contains a *real* number $\lambda_0$, and one can still apply the method used in §A.4.2 above to obtain the spectral family of $A$: the operator $Y := (A - \lambda_0)^{-1}$ belongs to $\mathcal{B}(\mathcal{H})$, and the correspondence $\lambda \mapsto (\lambda - \lambda_0)^{-1}$ defines a bijection between the spectrum of $A$ and that of $Y$. In the general case (without the hypothesis that $\sigma(A) \neq \mathbb{R}$), one relies on the following result which shows that it is possible to express $A$ as a sum of a positive and a negative operator. We refer to (4.70) for the concept of the restriction $A_{\mathcal{M}}$ of $A$ to a subspace $\mathcal{M}$.

**Proposition 4.39.** *If $A$ is a self-adjoint operator in a Hilbert space $\mathcal{H}$, there exist two mutually orthogonal subspaces $\mathcal{M}_-$ and $\mathcal{M}_+$ of $\mathcal{H}$ such that $\mathcal{H} = \mathcal{M}_- \oplus \mathcal{M}_+$ and such that the restrictions $\mathcal{A}_- := A_{\mathcal{M}_-}$ and $\mathcal{A}_+ := A_{\mathcal{M}_+}$ of $A$ to these subspaces are self-adjoint operators (in $\mathcal{M}_-$ and $\mathcal{M}_+$ respectively), with $\mathcal{A}_- \leq 0$ and $\mathcal{A}_+ > 0$.*

PROOF. Let us set $Z = (A+i)^{-1} + (A-i)^{-1}$. This operator is self-adjoint[20], with $\|Z\| \leq \|(A+i)^{-1}\| + \|(A-i)^{-1}\| \leq 2$ (see Proposition 2.37). Formally one has $Z = 2A/(A^2+1)$. Thus $A$ will be positive or negative on the same subspaces as $Z$. Consequently, if $\{\mathbb{E}_\mu\}$ denotes the spectral family of the bounded self-adjoint operator $Z$, we take $\mathcal{M}_- = \mathbb{E}((-\infty,0])\mathcal{H}$ and $\mathcal{M}_+ = \mathbb{E}((0,\infty))\mathcal{H}$. We then define $\mathcal{A}_- \equiv A_{\mathcal{M}_-}$ by

$$\mathcal{D}(\mathcal{A}_-) = \mathcal{M}_- \cap \mathcal{D}(A), \qquad \mathcal{A}_- f = Af \quad \text{for } f \in \mathcal{D}(\mathcal{A}_-). \tag{4.105}$$

(i) We first show that $\mathcal{D}(\mathcal{A}_-) = (A+i)^{-1}\mathcal{M}_- = (A-i)^{-1}\mathcal{M}_-$. We shall write $\mathbb{E}_- = \mathbb{E}((-\infty,0])$, so that $\mathcal{M}_- = \mathbb{E}_-\mathcal{H}$.

Since $(A \pm i)^{-1}$ commute with $Z$, property (I) of §A.4.1 (page 183) allows us to infer that $\left[(A \pm i)^{-1}, \mathbb{E}((-\infty,0])\right] = O$ (recall from Remark 4.38(a) that $\mathbb{E}_- \equiv \mathbb{E}((-\infty,0])$ is the projection $E \equiv P_{\mathcal{M}(Z)}$ associated with bounded the self-adjoint operator $Z$). Hence

$$[(A \pm i)^{-1}, \mathbb{E}_-] = O. \tag{4.106}$$

Now let $f \in \mathcal{D}(\mathcal{A}_-)$. Then, using (4.106):

$$f = \mathbb{E}_- f = \mathbb{E}_-(A \pm i)^{-1}(A \pm i) f = (A \pm i)^{-1}\mathbb{E}_-(A \pm i) f,$$

which shows that $f = (A \pm i)^{-1} g_\pm$ with $g_\pm \in \mathcal{M}_-$. Thus $\mathcal{D}(\mathcal{A}_-) \subseteq (A \pm i)^{-1}\mathcal{M}_-$. For the converse inclusion we observe that the ranges of $(A \pm i)^{-1}$ are contained in $\mathcal{D}(A)$ and that $(A \pm i)^{-1}$ map $\mathcal{M}_-$ into itself: if $h \in \mathcal{M}_-$, then $(A \pm i)^{-1}h = (A \pm i)^{-1}\mathbb{E}_- h = \mathbb{E}_-(A \pm i)^{-1} h \in \mathcal{M}_-$.

(ii) Next we show that $\mathcal{A}_-$ is a self-adjoint operator in the Hilbert space $\mathcal{M}_-$.

If $f \in \mathcal{D}(\mathcal{A}_-)$, then one has $f = (A+i)^{-1}g$ with $g \in \mathcal{M}_-$, as seen above. Thus $\mathcal{A}_- f = Af = g - if \in \mathcal{M}_-$. This shows that $\mathcal{A}_-$ is indeed an operator in $\mathcal{M}_-$. Its domain is dense in $\mathcal{M}_-$: If $h \in \mathcal{M}_-$ and $\varepsilon > 0$, there exists a vector $g_\varepsilon \in \mathcal{H}$ such that $\|h - (A+i)^{-1}g_\varepsilon\| < \varepsilon$. Then $\|h - (A+i)^{-1}\mathbb{E}_- g_\varepsilon\| = \|\mathbb{E}_-[h - (A+I)^{-1}g_\varepsilon]\| < \varepsilon$.

[20] Because $(A-i)^{-1} = (A+i)^{-1*}$, see (2.103).

Thus each vector $h \in \mathcal{M}_-$ is the strong limit of a sequence of vectors belonging to $(A+i)^{-1}\mathcal{M}_- = \mathcal{D}(\mathcal{A}_-)$.

Since $A = A^*$, it is clear that $\mathcal{A}_-$ is a symmetric operator in $\mathcal{M}_-$. Furthermore $(\mathcal{A}_- \pm i)\mathcal{D}(\mathcal{A}_-) = (A \pm i)\mathcal{D}(\mathcal{A}_-) = (A \pm i)(A \pm i)^{-1}\mathcal{M}_- = \mathcal{M}_-$. By Proposition 3.3 we conclude that $\mathcal{A}_-$ is self-adjoint.

(iii) Let us check that $\mathcal{A}_-$ is a negative operator. The identity $2A = (A-i)+(A+i)$, valid on $\mathcal{D}(A)$, implies that $2(A-i)^{-1}A(A+i)^{-1} = (A+i)^{-1} + (A-i)^{-1} = Z$. So let $f \in \mathcal{D}(\mathcal{A}_-)$. Writing $f = (A+i)^{-1}g$ with $g \in \mathcal{M}_-$, we then have

$$\langle f, \mathcal{A}_- f\rangle = \langle f, Af\rangle = \langle g, (A-i)^{-1}A(A+i)^{-1}g\rangle = \frac{1}{2}\langle g, Zg\rangle \leq 0.$$

(iv) By reasoning as above, one finds that the restriction $\mathcal{A}_+$ of $A$ to $\mathcal{M}_+$ is a positive self-adjoint operator in $\mathcal{M}_+$ [here one has $\langle f, \mathcal{A}_+ f\rangle > 0$ for each $f \neq \mathbf{0}$ in $\mathcal{M}_+$, with strict inequality, because $Z > O$ in $\mathcal{M}_+ = \mathbb{E}((0,\infty))\mathcal{H}$]. □

Since $\mathcal{A}_-$ and $\mathcal{A}_+$ are semibounded, one can obtain the spectral family $\{E_\lambda\}$ of $A$ from the spectral families of $\mathcal{A}_-$ and $\mathcal{A}_+$. If $\{\mathbb{E}_\mu^-\}$ denotes the spectral family of $\mathcal{A}_-$ in $\mathcal{M}_-$ and $\{\mathbb{E}_\mu^+\}$ that of $\mathcal{A}_+$ in $\mathcal{M}_+$, one has in the decomposition $\mathcal{H} = \mathcal{M}_- \oplus \mathcal{M}_+$ of the Hilbert space $\mathcal{H}$:

$$E_\lambda = \begin{cases} \mathbb{E}_\lambda^- \oplus O & \text{if } \lambda \leq 0 \\ I \oplus \mathbb{E}_\lambda^+ & \text{if } \lambda \geq 0 \end{cases}$$

Alternatively one could follow the proof of the Spectral Theorem given for bounded operators in §A.4.1: by applying Proposition 4.39 to the operator $A_\lambda \equiv A - \lambda I$, one obtains a decomposition of this operator into $A_\lambda = A_{\lambda -} \oplus A_{\lambda +}$ in $\mathcal{M}_{\lambda -} \oplus \mathcal{M}_{\lambda +}$, with $A_{\lambda -} \leq O$ and $A_{\lambda +} > O$. The spectral projection $E_\lambda$ of $A$ is the projection onto the subspace $\mathcal{M}_{\lambda -}$. By taking into account the arising domain questions, one can repeat the verifications presented in §A.4.1 (the details are given for example in §VI.5.3 of [K]).

REMARK. It is clear that the result of Proposition 4.39 is true in particular if $A$ is a bounded self-adjoint operator. For such an operator we have already obtained in §A.4.1 a decomposition into a sum of a positive operator $A^+$ and a negative operator $A^-$. The following difference should be noticed: the two operators $A^+$ and $A^-$ given in §A.4.1 are operators acting in $\mathcal{H}$, whereas the operators $\mathcal{A}_+$ and $\mathcal{A}_-$ constructed in Proposition 4.39 are operators acting in some subspace of $\mathcal{H}$ (in $\mathcal{M}_+$ and $\mathcal{M}_-$ respectively). The relation between these different operators is as follows: in the decomposition $\mathcal{H} = \mathcal{M}_- \oplus \mathcal{M}_+$, $\mathcal{A}_+$ acts in $\mathcal{M}_+$, $\mathcal{A}_-$ acts in $\mathcal{M}_-$, and one has $A^+ = O \oplus \mathcal{A}_+$ and $A^- = \mathcal{A}_- \oplus O$ (see also Remark 4.38).

## Bibliographical Notes

As regards measure theory, we refer to the Bibliographical Notes for Chapter 1. Stieltjes measures and integrals are also discussed in [CV]. Properties of monotone functions (in particular their differentiability almost everywhere) are treated in [RN] (§1-§3 of Chapter I and §1-§2 of Chapter VIII) and in [Na] (§6 of Chapter IX). The spectral theory of self-adjoint operators is presented for example in the texts cited in the Bibliographical Notes for Chapter 2.

## Problems

**4.1.** Let $F\colon \mathbb{R} \to [0, \rho]$ be a function satisfying the conditions (1), (2) and (3) given at the beginning of Section 4.1.
(a) Show that the set of points of discontinuity of $F$, if not empty, is at most countable.
(b) Let $F_{\mathrm{p}}$ be the unique jump function having the same discontinuities as $F$ [see Eq. (4.8)]. Show that $F - F_{\mathrm{p}}$ is continuous and monotone non-decreasing.
[Hint for (a): For $n \in \mathbb{N}$, consider the set of points $\lambda \in \mathbb{R}$ at which $F(\lambda) - F(\lambda - 0) > 1/n$.]
**4.2.** Let $\{E_\lambda\}_{\lambda\in\mathbb{R}}$ be a spectral family.
(a) Prove that the strong limits $\operatorname{s-lim}_{\varepsilon\to+0} E_{\mu\pm\varepsilon}$ exist for each $\mu \in \mathbb{R}$ and that these limits define projections $E_{\mu\pm0}$ satisfying $E_{\mu-0} \leq E_\mu \leq E_{\mu+0}$.
(b) Verify that for $\lambda, \mu \in \mathbb{R}$:

$$\|E_\lambda f - E_\mu f\|^2 = \big|\|E_\lambda f\|^2 - \|E_\mu f\|^2\big|. \tag{4.107}$$

Thus $\lambda \mapsto E_\lambda f$ is strongly continuous if and only if $\lambda \mapsto \|E_\lambda f\|^2$ is continuous.
**4.3.** Let $A$ be a self-adjoint operator. Show that $\mathcal{H}_{\mathrm{s}}(A) = \mathcal{H}_{\mathrm{p}}(A) \oplus \mathcal{H}_{\mathrm{sc}}(A)$.
**4.4.** For $\lambda \in \mathbb{R}$ denote by $C_\lambda(\mathbb{R})$ the set of continuous functions $\varphi\colon \mathbb{R} \to \mathbb{C}$ having compact support and satisfying $\varphi(\lambda) \neq 0$. Let $A$ be a self-adjoint operator. Show that the spectrum and the essential spectrum of $A$ are given by the following expressions:

$$\sigma(A) = \big\{\lambda\in\mathbb{R} \,\big|\, \varphi\in C_\lambda(\mathbb{R}) \implies \varphi(A) \neq O\big\}$$

$$\sigma_{\mathrm{ess}}(A) = \big\{\lambda\in\mathbb{R} \,\big|\, \varphi\in C_\lambda(\mathbb{R}) \implies \varphi(A) \text{ is not compact}\big\}.$$

**4.5.** Let $A$ be a bounded self-adjoint operator. Show that the continuous unitary one-parameter group associated with $A$ (Example 4.12) is given by the Taylor series for $e^{-iAt}$, with convergence in the operator norm. If $A$ is unbounded, this series converges strongly when applied to vectors of compact support with respect to the spectral family of $A$.
**4.6.** Let $A$ be the multiplication operator in $L^2(\mathcal{O}, m)$ by a real function $a \in L^\infty(\mathcal{O}, m)$. Determine the spectral projections of this operator and discuss its spectrum.

**4.7.** Consider the self-adjoint extensions $K(\alpha)$ of the minimal operator $-d^2/dx^2$ in $L^2((0,\infty))$, $\alpha \in [0,\pi)$ (see page 107).
(a) Determine the eigenvalues (if any) of $K(\alpha)$.
(b) Show that each $\lambda \geq 0$ belongs to the spectrum of $K(\alpha)$.
(c) What can be said about the nature of the spectrum of $K(\alpha)$ on the interval $[0,\infty)$ and on the interval $(-\infty, 0)$?
[Hints. For (b): Use Proposition 4.20 by considering functions of the form $f(x) = \vartheta(x)e^{i\sqrt{\lambda}x}$ with $\vartheta \in C_0^\infty((0,\infty))$. For (c): Example 4.25.]

**4.8.** Let $B$ be a convolution operator in $L^2(\mathbb{R})$: $(Bf)(x) = \int_{-\infty}^{\infty} b(x-y)f(y)dy$, where $b$ is a fixed function belonging to $L^1(\mathbb{R}) \cap L^2(\mathbb{R})$. Find the spectrum of $B$.

**4.9.** Consider a differential operator $H = -d^2/dx^2 + V(x)$ in $L^2(\mathbb{R})$, where $V$ is a bounded real-valued function of compact support with $\int_{\mathbb{R}} V(x) < 0$. Show that $\sigma_{\text{ess}}(H) = [0,\infty)$ and that $H$ has at least one negative eigenvalue.
[Hint: Show that one can find vectors $f \in L^2(\mathbb{R})$ for which $\langle f, Hf\rangle < 0$ by considering functions of the form $g_n(x) = g(x/n)$ with $g \in C_0^\infty(\mathbb{R})$ satisfying $g(x) = 1$ for $|x| \leq 1$.]

**4.10.** Let $\{E_\lambda\}$ be the spectral family of a self-adjoint operator $A$. Let $J$ be an interval and set $N = \dim \mathcal{R}(E(J)) \equiv \dim \mathcal{M}(J)$.
(a) If $N < \infty$, then the spectrum of $A$ in $J$ consists of only isolated eigenvalues of finite multiplicity. The sum of the multiplicities of these eigenvalues is equal to $N$.
(b) If $N = \infty$ and $J$ is bounded and closed, then $\sigma_{\text{ess}}(A) \cap J \neq \varnothing$.
(c) If $\dim \mathcal{R}(E_\mu) < \infty$ for some $\mu \in \mathbb{R}$, then $A$ is lower semibounded.

**4.11.** Let $A$ be a self-adjoint operator in a Hilbert space $\mathcal{H}$ and $\mathcal{M}(\mathcal{V}) = E(\mathcal{V})\mathcal{H}$, where $\{E(\cdot)\}$ denotes the spectral measure of $A$. Show that
(a) For each Borel set $\mathcal{V}$ the subspace $\mathcal{M}(\mathcal{V})$ is invariant under the operator $A$ [i.e. $f \in \mathcal{M}(\mathcal{V}) \cap \mathcal{D}(A) \Longrightarrow Af \in \mathcal{M}(\mathcal{V})$].
(b) If $\mathcal{V}$ is a bounded set, then the restriction $A_{\mathcal{M}(\mathcal{V})}$ of $A$ to the subspace $\mathcal{M}(\mathcal{V})$ [see (4.70)] is a bounded operator.
(c) If $\mathcal{V}$ is a closed set, then $\sigma(A_{\mathcal{M}(\mathcal{V})}) \subseteq \mathcal{V}$.

**4.12.** Let $A$ be a self-adjoint operator and let $\lambda \in \mathbb{R}$ be such that $\text{dist}(\lambda, \sigma_{\text{p}}(A)) > 0$. Show that $\lambda$ belongs to the resolvent set of the operator $A_{\text{p}}$.

**4.13.** Let $A$ be a bounded self-adjoint operator in an infinite-dimensional Hilbert space. Show that the following three statements are equivalent: (i) $\sigma_{\text{ess}}(A)$ is not empty and $A$ is compact, (ii) $\sigma_{\text{ess}}(A) = \{0\}$, (iii) $A$ is compact. What can be said if the condition of boundedness of $A$ is dropped?

**4.14.** Let $A$ be a compact self-adjoint operator.
(a) Show that $\|A\|$ or $-\|A\|$ must be an eigenvalue of $A$.
(b) Let $\{\lambda_n\}$ be the eigenvalues of $A$ and $m_n$ the multiplicity of $\lambda_n$. Prove that $A$ is a Hilbert-Schmidt operator if and only if $\sum_{n\in\mathbb{N}} m_n \lambda_n^2 < \infty$.

**4.15** Let $\{E_\lambda\}$ be a spectral family and $\mathcal{D}$ the set of vectors in $\mathcal{H}$ having compact support with respect to $\{E_\lambda\}$, i.e. $\mathcal{D} = \{f \in \mathcal{H} \mid E(J)f = f \text{ for some bounded interval } J\}$. Let $A$ be the self-adjoint operator associated with $\{E_\lambda\}$. Show that $A$ is essentially self-adjoint on $\mathcal{D}$.

**4.16.** An operator $A \in \mathcal{B}(\mathcal{H})$ is called a ***normal operator*** if it commutes with its adjoint, i.e. if $A^*A = AA^*$.
(a) $A$ is normal if and only if $\|A^*f\| = \|Af\|$ for each $f \in \mathcal{H}$.
(b) Denote by $\Re A = (A + A^*)/2$ the real part of $A$ and by $\Im A = (A - A^*)/(2i)$ its imaginary part. Then $A$ is normal if and only if $\Re A$ and $\Im A$ commute.
(c) A bounded self-adjoint operator is normal.
(d) Let $A$ be a self-adjoint operator and $\varphi\colon \mathbb{R} \to \mathbb{C}$ a bounded continuous function. Then $\varphi(A)$ is a normal operator.
(e) The resolvent $(A - z)^{-1}$ of a self-adjoint operator (for $z$ outside the spectrum of $A$) is a normal operator.
REMARK. The Spectral Theorem has an extension to normal operators.

**4.17.** ***Polar decomposition of a bounded operator***

Let $A \in \mathcal{B}(\mathcal{H})$ be a bounded operator in a Hilbert space $\mathcal{H}$.
(a) Show that the operator $A^*A$ is self-adjoint and positive. We denote by $|A|$ its positive square root: $|A| = (A^*A)^{1/2}$, i.e. $|A| = \varphi(A^*A)$ with $\varphi(\lambda) = +\lambda^{1/2}$ $(\lambda \geq 0)$.
(b) Show that there exists a partial isometry $U$ with initial set $\overline{\mathcal{R}(|A|)}$ and final set $\overline{\mathcal{R}(A)}$ such that $A = U|A|$ (see Section 2.2 for the notion of a partial isometry).
(c) Show that, if $A$ is invertible in $\mathcal{B}(\mathcal{H})$, then $U$ is unitary.
(d) Let $Q$ be the operator of multiplication by the variable in $L^2([-1,1])$: $(Qf)(x) = xf(x)$, $x \in [-1,1]$. Give the polar decomposition of $Q$.
(e) Let $A$ be the multiplication operator in $L^2(\mathcal{O}, m)$ by a function $a \in L^\infty(\mathcal{O}, m)$. Find the polar decomposition of $A$.

# CHAPTER 5

# Evolution Groups and Scattering Theory

In this chapter we develop the principal aspects of quantum scattering theory in a Hilbert space framework. If $H$ is the Hamiltonian of a quantum-mechanical system, and if $H$ does not depend on time, then the evolution of a state vector $f \in \mathcal{H}$ during a time interval of length $t$ is obtained by acting on $f$ with the unitary operator $e^{-iHt}$: if $f$ is the state vector of the considered system at time $t_0$, then the state vector at time $t_0 + t$ is given by $e^{-iHt}f$ (in quantum mechanics the time evolution of $f$ is given by $e^{-iHt/\hbar}f$, and in this text we set $\hbar = 1$). Thus, for time-independent Hamiltonians, the evolution of all possible states can be described in terms of the continuous unitary group $\{e^{-iHt}\}_{t\in\mathbb{R}}$ associated with the Hamiltonian $H$, as described in Example 4.12. Scattering theory deals with the behaviour of scattering states (normally the states associated with the continuous spectrum of $H$, as we shall see) at very large times, viz. for $t \to \pm\infty$, by comparing their evolution at very large times with a "simpler" evolution (called the free evolution, i.e. the evolution in the absence of interactions or external forces).

The family of operators $\{e^{-iHt}\}_{t\in\mathbb{R}}$ forms a strongly continuous one-parameter group of unitary operators. In Section 5.1 we derive some abstract mathematical properties of such groups, and in the following sections we present the basic aspects of scattering theory, in particular the definition of scattering states, wave operators, the scattering operator and scattering cross sections.

## 5.1 Evolution groups

**5.1.1.** If $A$ is a self-adjoint operator in a Hilbert space $\mathcal{H}$, then for each $t \in \mathbb{R}$ the operator $U_t := e^{-iAt}$ is unitary, and the collection $\{U_t\}_{t\in\mathbb{R}}$ is a unitary one-parameter group that is strongly continuous (see Example 4.12). To simplify the notations we shall say that it is an evolution group (we think of the situation that will be of interest in this chapter, in which $A$ will be the Hamiltonian of a quantum system). Thus by definition an ***evolution group*** is a family $\{U_t\}_{t\in\mathbb{R}}$ of unitary operators, indexed by a real parameter $t$, such that

$$U_t U_s = U_{t+s} \ \forall s, t \in \mathbb{R} \quad \text{and} \quad \operatorname*{s-lim}_{\varepsilon \to 0} \left[ U_{t+\varepsilon} - U_t \right] = O \ \forall t \in \mathbb{R}. \tag{5.1}$$

We observe the following consequences of this definition:

$$U_t U_s = U_s U_t \,, \qquad U_0 = I \qquad \text{and} \qquad U_t^* = U_{-t}. \tag{5.2}$$

The first identity in (5.2) follows from the first formula in (5.1); if one takes $s = t = 0$ in the first identity in (5.1) one obtains $U_0 U_0 = U_0$, and upon multiplying this relation by $U_0^*$ and using the isometry of $U_0$ (i.e. $U_0^* U_0 = I$), one arrives at $U_0 = I$. For the last identity in (5.2) we observe that the unitarity of $U_t$ implies that $U_t^* = [U_t]^{-1}$ (see Section 2.2) and that $U_t U_{-t} = U_{-t} U_t = U_0 = I$, hence $[U_t]^{-1} = U_{-t}$.

According to Example 4.12 each self-adjoint operator $A$ determines an evolution group by the formula $U_t = e^{-iAt}$ (the exponential being defined by the functional calculus or as a series, as explained in Example 4.12). It is an interesting fact that each evolution group is necessarily of this form, i.e. that, given an evolution group $\{U_t\}_{t \in \mathbb{R}}$, there exists a (unique) self-adjoint operator $A$ such that $U_t = e^{-iAt}$. This is the content of Stone's Theorem which we prove below. To motivate this theorem, consider the evolution group $e^{-iAt}$ associated with a self-adjoint operator $A$ and let us try to calculate $A$ from the group. Formally it is clear that $(d/dt)e^{-iAt} = -iAe^{-iAt}$, which suggests the relation $A = i(d/dt)e^{-iAt}\big|_{t=0}$. If $A$ is unbounded one has to consider this relation on the domain $\mathcal{D}(A)$ of $A$. Indeed, if $f \in \mathcal{D}(A)$, then the vector-valued function $t \mapsto e^{-iAt} f$ is strongly differentiable and its strong derivative, evaluated at $t = 0$, is equal to $-iAf$, i.e. one has $\lim_{t \to 0} \|t^{-1}[e^{-iAt} - I]f - (-iAf)\| = 0$ for each $f \in \mathcal{D}(A)$. This is easily obtained by applying the Dominated Convergence Theorem:

$$\left\| \frac{1}{t} \left[ e^{-iAt} - I \right] f + iAf \right\|^2 = \int_{-\infty}^{\infty} \left| \frac{e^{-i\lambda t} - 1}{t} + i\lambda \right|^2 m_f(d\lambda),$$

and $\left| t^{-1}[e^{-i\lambda t} - 1] + i\lambda \right|^2$ converges to zero for each fixed $\lambda$ as $t \to 0$ and can be majorised as follows:

$$\left| \frac{e^{-i\lambda t} - 1}{t} + i\lambda \right|^2 \le 2 \left| \frac{e^{-i\lambda t} - 1}{t} \right|^2 + 2|i\lambda|^2;$$

the right-hand side is integrable with respect to the measure $m_f$ if $f \in \mathcal{D}(A)$ since

$$\left| \frac{e^{-i\lambda t} - 1}{t} \right|^2 = \left| \frac{e^{-i\lambda t} - 1}{\lambda t} \right|^2 |\lambda|^2 \le |\lambda|^2 \quad \text{because} \quad \left| \frac{e^{-ix} - 1}{x} \right| \le 1 \ \forall x \in \mathbb{R}.$$

**Proposition 5.1 (Stone's Theorem).** *Let $\{U_t\}$ be an evolution group in a Hilbert space $\mathcal{H}$. Define a linear operator $A$ in $\mathcal{H}$ as follows:*

$$\mathcal{D}(A) = \{ f \in \mathcal{H} \mid \exists \operatorname*{s-lim}_{t \to 0} \frac{1}{t} \left[ U_t - I \right] f \}, \tag{5.3}$$

$$Af = \operatorname*{s-lim}_{t \to 0} \frac{i}{t} \left[ U_t - I \right] f \qquad \text{if } f \in \mathcal{D}(A). \tag{5.4}$$

*Then $\mathcal{D}(A)$ is dense in $\mathcal{H}$ and $A$ is a self-adjoint operator (called the* ***infinitesimal generator*** *of the group $\{U_t\}$). One has $U_t = e^{-iAt}$ (i.e. $\{U_t\}$ is the strongly continuous unitary group associated with $A$), the domain $\mathcal{D}(A)$ is invariant under each $U_t$ [i.e. $f \in \mathcal{D}(A) \Longrightarrow U_t f \in \mathcal{D}(A)$], and*

$$i\frac{d}{dt}U_t f = AU_t f = U_t Af \qquad \forall f \in \mathcal{D}(A). \tag{5.5}$$

PROOF. (i) Let $z = x + iy$ with $x, y \in \mathbb{R}$ and $y > 0$. We set $R_z = i\int_0^\infty e^{izs}U_s ds$. This integral exists by Proposition 2.3(c) because the integrand is strongly continuous and

$$\int_0^\infty \|e^{izs}U_s\|\, ds \le \int_0^\infty e^{-ys}\, ds = \frac{1}{y} < \infty.$$

By also taking into account (2.25) one sees that $R_z \in \mathcal{B}(\mathcal{H})$ and $\|R_z\| \le y^{-1}$. Let us also show that the range of $R_z$ is dense in $\mathcal{H}$. We must show that if $g \in \mathcal{H}$ is such that $\langle g, R_z f\rangle = 0$ for all $f \in \mathcal{H}$, i.e. if $g$ is orthogonal to $\mathcal{R}(R_z)$, then $g = \mathbf{0}$. For this we observe that $R_z U_t = U_t R_z$ for all $t \in \mathbb{R}$, which implies that $\langle g, U_t R_z f\rangle = \langle g, R_z U_t f\rangle = 0$ for all $f \in \mathcal{H}$ and all $t \in \mathbb{R}$. By inserting the definition of $R_z$ into this formula one obtains (setting $\tau = t + s$)

$$\langle g, U_t i\int_0^\infty e^{izs}U_s f\, ds\rangle \equiv i\int_0^\infty e^{izs}\langle g, U_{t+s}f\rangle\, ds \equiv ie^{-izt}\int_t^\infty e^{iz\tau}\langle g, U_\tau f\rangle\, d\tau = 0.$$

Hence $\int_t^\infty e^{iz\tau}\langle g, U_\tau f\rangle d\tau = 0$. Differentiating this identity with respect to $t$ leads to

$$\frac{d}{dt}\int_t^\infty e^{iz\tau}\langle g, U_\tau f\rangle\, d\tau = -e^{izt}\langle g, U_t f\rangle = 0.$$

Hence $\langle g, U_t f\rangle = 0$ for each $t \in \mathbb{R}$, in particular for $t = 0$. We have thus shown that $\langle g, f\rangle = 0$ for each $f \in \mathcal{H}$, which implies that $g = \mathbf{0}$.

(ii) We now show that $R_z f \in \mathcal{D}(A)$ for each $f \in \mathcal{H}$ and also calculate the action of $A$ on the vector $R_z f$. We have

$$\frac{i}{t}[U_t - I]R_z f = -\frac{1}{t}\int_0^\infty e^{izs}U_{t+s}f\, ds - \frac{i}{t}R_z f$$
$$= -\frac{1}{t}e^{-izt}\int_t^\infty e^{iz\tau}U_\tau f\, d\tau - \frac{i}{t}R_z f = \frac{i}{t}(e^{-izt} - 1)R_z f + \frac{1}{t}e^{-izt}\int_0^t e^{iz\tau}U_\tau f\, d\tau.$$

As $t \to 0$, the right-hand side converges strongly to $zR_z f + f$. Thus we have: (1) $R_z f \in \mathcal{D}(A)$ for all $f \in \mathcal{H}$ and (2) $AR_z f = zR_z f + f$, i.e.

$$(A - z)R_z f = f \qquad \forall f \in \mathcal{H}. \tag{5.6}$$

Since the set $R_z\mathcal{H}$ is dense in $\mathcal{H}$, (1) implies that $\mathcal{D}(A)$ is dense in $\mathcal{H}$.

(iii) By taking $z = i$ in (5.6) one sees that $\mathcal{R}(A - i) = \mathcal{H}$. One can repeat the arguments given in (i) and (ii) for $z = x + iy$ with $y < 0$, by setting in this case

$R_z = -i\int_{-\infty}^{0} e^{izs}U_s ds$. One finds that (5.6) holds also in this case, in particular (for $z=-i$) that $\mathcal{R}(A+i)=\mathcal{H}$. Thus, to conclude that $A$ is self-adjoint, it only remains to show that it is symmetric (see Proposition 3.3). So let $f,g\in\mathcal{D}(A)$. Then (using $U_t^* = U_{-t}$):

$$\begin{aligned}\langle Af,g\rangle &= \lim_{t\to 0}\langle it^{-1}(U_t-I)f,g\rangle = \lim_{t\to 0}\langle f,-it^{-1}(U_{-t}-I)g\rangle\\ &\equiv \lim_{s\to 0}\langle f,is^{-1}(U_s-I)g\rangle = \langle f,Ag\rangle.\end{aligned}$$

(iv) We now show that $\mathcal{D}(A)$ is invariant under the group. Assume that $f\in\mathcal{D}(A)$ and $s\in\mathbb{R}$. We must check that $U_sf\in\mathcal{D}(A)$. We observe that $U_s\big(it^{-1}[U_t-I]f\big) = it^{-1}[U_t-I]U_sf$. Since $f\in\mathcal{D}(A)$, the left-hand side converges strongly to $U_sAf$ as $t\to 0$. Consequently the right-hand side is strongly convergent as $t\to 0$, which means that $U_sf\in\mathcal{D}(A)$. In addition the strong limit of the right-hand side is equal to $AU_sf$, which establishes (5.5).

(v) We finally show that $U_t = e^{-iAt}$. Since $U_t$ and $e^{-iAt}$ belong to $\mathcal{B}(\mathcal{H})$, it suffices to prove that $\langle g,U_tf\rangle = \langle g,e^{-iAt}f\rangle$ for each $f$ in some dense subset $\mathcal{D}$ and all $g$ in $\mathcal{H}$. We take for $\mathcal{D}$ the set of vectors of compact support with respect to the spectral family of $A$ (see page 144). If $f$ has this property, then $f\in\mathcal{D}(A^k)$ for each $k\in\mathbb{N}$. Consequently, by (5.5), the mapping $t\mapsto U_tf$ is strongly of class $C^\infty$ and $(d/dt)^kU_tf = (-i)^kU_tA^kf$. From the argument given before the statement of Stone's Theorem, one sees that $t\mapsto e^{-iAt}f$ is also of class $C^\infty$, with derivatives $(d/dt)^ke^{-iAt}f = (-i)^ke^{-iAt}A^kf$. Thus one has for each $g\in\mathcal{H}$ and $f\in\mathcal{D}$:

$$\frac{d^k}{dt^k}\langle g,U_tf\rangle\big|_{t=0} = (-i)^k\langle g,A^kf\rangle = \frac{d^k}{dt^k}\langle g,e^{-iAt}f\rangle\big|_{t=0} \qquad \text{for } k=0,1,2,\dots$$

This shows that the functions $t\mapsto\langle g,U_tf\rangle$ and $t\mapsto\langle g,e^{-iAt}f\rangle$ have the same Taylor expansion at $t=0$, hence they are equal. Consequently $U_tf = e^{-iAt}f$ for $f\in\mathcal{D}$. □

**Corollary 5.2.** *Let $U_t$ and $A$ be as in Proposition 5.1. Then the resolvent of $A$ is given as follows in terms of the group $\{U_t\}$:*

$$\text{if } \Im z > 0: \qquad (A-z)^{-1} = i\int_0^\infty e^{izt}U_t\,dt \tag{5.7}$$

$$\text{if } \Im z < 0: \qquad (A-z)^{-1} = -i\int_{-\infty}^0 e^{izt}U_t\,dt. \tag{5.8}$$

PROOF. Since $A$ is self-adjoint, one has $z\in\rho(A)$ if $\Im z\neq 0$, hence $(A-z)^{-1}\in\mathcal{B}(\mathcal{H})$. Upon multiplying (5.6) by $(A-z)^{-1}$ one obtains $R_zf = (A-z)^{-1}f$ for each $f\in\mathcal{H}$, i.e. $R_z$ is identical with the resolvent of $A$ at the point $z$. By taking into account the definition of $R_z$ in the preceding proof one obtains (5.7) and (5.8). □

For explicitly given groups it is often easy to calculate the derivative of $U_tf$, hence the infinitesimal generator, for a set of particularly simple vectors $f$ [for example for $f\in\mathcal{S}(\mathbb{R}^n)$ in quantum mechanics of a particle]. If this set $\mathcal{D}$ of vectors is a core for the infinitesimal generator, i.e. if the restriction $A_0$ of the infinitesimal generator to this set of vectors is essentially self-adjoint, this determines the infinitesimal generator completely (it is simply the closure of $A_0$, both deficiency indices of $A_0$ being zero). The next proposition gives a criterion for this to hold.

**Proposition 5.3 (Nelson's Criterion).** *Let $\{U_t\}$ be an evolution group and $A$ its infinitesimal generator. Let $\mathcal{D}$ be a linear submanifold of $\mathcal{D}(A)$ having the following properties: (1) $\mathcal{D}$ is dense in $\mathcal{H}$, (2) $\mathcal{D}$ is invariant under the group, i.e. $U_t f \in \mathcal{D}$ for each $f \in \mathcal{D}$ and each $t \in \mathbb{R}$. Then the restriction of $A$ to $\mathcal{D}$ is essentially self-adjoint.*

PROOF. Let us denote the restriction of $A$ to $\mathcal{D}$ by $A_0$. Then $A_0$ is a symmetric operator. By Proposition 3.3(b) we must show that $\mathcal{R}(A_0 - i)^\perp \equiv \mathcal{N}(A_0^* + i) = \{\mathbf{0}\}$ and $\mathcal{R}(A_0 + i)^\perp \equiv \mathcal{N}(A_0^* - i) = \{\mathbf{0}\}$. Let us show for example that $\mathcal{N}(A_0^* - i) = \{\mathbf{0}\}$. Assume that $h \in \mathcal{N}(A_0^* - i)$, i.e. $h \in \mathcal{D}(A_0^*)$ and $A_0^* h = ih$. Let $f \in \mathcal{D}$. Then

$$\begin{aligned}\frac{d}{dt}\langle U_t f, h\rangle &= \langle -iAU_t f, h\rangle = i\langle A_0 U_t f, h\rangle \\ &= i\langle U_t f, A_0^* h\rangle = i\langle U_t f, ih\rangle = -\langle U_t f, h\rangle\end{aligned}$$

[using $U_t f \in \mathcal{D}(A)$ and $U_t f \in \mathcal{D}$ in the first and second equality respectively]. Thus the function $\phi$ defined as $\phi(t) = \langle U_t f, h\rangle$ satisfies the differential equation $\phi'(t) = -\phi(t)$; hence $\phi(t) = \phi(0)e^{-t}$. If $\phi(0) \neq 0$ then $|\phi(t)| \to \infty$ as $t \to -\infty$, which contradicts the fact that $|\langle U_t f, h\rangle| \leq \|f\|\|h\|$. Consequently $\phi(0) = 0$, i.e. $\langle f, h\rangle = 0$. This shows that $h$ is orthogonal to the dense set $\mathcal{D}$, hence $h = \mathbf{0}$. □

**Example 5.4.** Interesting examples of evolution groups are obtained by considering unitary representations of one-parameter transformation groups of a set $\mathcal{O}$. We consider here the ***translation group*** acting on $\mathbb{R}$: each value of the parameter $t$ induces a mapping $\mathbb{R} \to \mathbb{R}$ given by $\mathbb{R} \ni x \mapsto x + t$, and the collection of these mappings, as $t$ varies over $\mathbb{R}$, forms a one-parameter group of transformations of $\mathcal{O} \equiv \mathbb{R}$. If one defines operators $U_t$ in $\mathcal{H} = L^2(\mathbb{R})$ by $(U_t f)(x) = f(x-t)$, one obtains a continuous unitary one-parameter group in this Hilbert space. Clearly $U_s U_t = U(s+t)$, and by a simple change of variables one finds that

$$\langle g, U_t f\rangle = \int_{-\infty}^{\infty} \overline{g(x)} f(x-t)\,dx = \int_{-\infty}^{\infty} \overline{g(y+t)} f(y)\,dy = \langle U_{-t} g, f\rangle,$$

and similarly that $\|U_t f\|^2 = \|f\|^2$. These relations imply that $U_t \in \mathcal{B}(\mathcal{H})$ and that $U_t^* = U_{-t}$. Then $U_t^* U_t = U_{-t} U_t = U_0 = I$ and also $U_t U_t^* = I$, which gives the unitarity of $U_t$. To verify the strong continuity of $\{U_t\}$, we observe that

$$\|(U_{t+\varepsilon} - U_t)f\|^2 = \|(U_\varepsilon - I)f\|^2 = \int_{-\infty}^{\infty} |f(x-\varepsilon) - f(x)|^2\,dx. \tag{5.9}$$

If $f$ is continuous and bounded, the integrand converges to zero as $\varepsilon \to 0$ and is majorised by the constant $4\|f\|_\infty^2$. If furthermore $f$ vanishes outside some finite interval $[a, b]$, then for $\varepsilon \in [-1, 1]$ the domain of integration in (5.9) reduces to $[a-1, b+1]$, and an application of the Dominated Convergence Theorem (Proposition 1.9) shows that this integral converges to zero as $\varepsilon \to 0$. This gives in particular the strong continuity of $U_t f$ for each $f \in C_0^\infty(\mathbb{R})$. Since $C_0^\infty(\mathbb{R})$ is dense in $L^2(\mathbb{R})$, it follows by virtue of Proposition 2.2 that $U_t f$ is strongly continuous for each $f \in \mathcal{H}$.

Let us determine the infinitesimal generator of $\{U_t\}$. Differentiation with respect to $t$ of $(U_t f)(x)$, for fixed $x$, shows that $(\partial/\partial t)(U_t f)(x)|_{t=0} = -f'(x)$. This suggests that the infinitesimal generator should be $-i(d/dx)$. To make this precise one has to show for suitable vectors $f$ that $\|t^{-1}[U_t - I]f + f'\| \to 0$ as $t \to 0$. So consider

$$\|\frac{1}{t}[U_t f - f] + f'\|^2 = \int_{-\infty}^{\infty} \Big|\frac{1}{t}[f(x-t) - f(x)] + f'(x)\Big|^2 dx. \tag{5.10}$$

Let us take again $f \in C_0^\infty(\mathbb{R})$ and apply the Dominated Convergence Theorem as above. The integrand converges to zero as $t \to 0$, and is bounded by $4\|f'\|_\infty^2$ (since $|f(x-t) - f(x)| = |\int_x^{x-t} f'(y)dy| \le |t|\|f'\|_\infty$). Thus $C_0^\infty(\mathbb{R})$ belongs to the domain of the infinitesimal generator, and on $C_0^\infty(\mathbb{R})$ this operator is given by $-i(d/dx)$. We know that $-i(d/dx)$ is essentially self-adjoint on $C_0^\infty(\mathbb{R})$ and that its closure is the momentum operator $P$. So the infinitesimal generator of the translation group in $L^2(\mathbb{R})$ is the momentum operator $P$. One has $U_t = e^{-iPt}$.

A shorter way of arriving at the preceding result is to make a Fourier transformation and express the group $\{U_t\}$ in momentum space: $[\mathcal{F}U_t f](k) = e^{-ikt}\tilde{f}(k)$. In the momentum representation of the Hilbert space, $\{U_t\}$ is multiplication by $e^{-ikt}$ and thus coincides with the operator $\varphi(P)$ for $\varphi(\lambda) = e^{-i\lambda t}$, see Example 4.12. By the uniqueness of the infinitesimal generator we conclude that it must be the operator $P$.

In this context another useful example concerns the ***dilation group*** in $\mathbb{R}^n$, given by $\mathbb{R}^n \ni \vec{x} \mapsto e^{-t}\vec{x}$ (where again $t$ varies over $\mathbb{R}$). The unitary representation of this group in $L^2(\mathbb{R}^n)$ and its infinitesimal generator are discussed in Problem 5.4.

On various occasions we shall meet expressions involving products of evolution groups. To calculate the derivative of such an expression, one uses the Leibniz rule. However, some care is necessary when the infinitesimal generators of the occurring evolution groups are unbounded. The following result describes situations in which the formal application of the Leibniz rule is justified.

**Proposition 5.5.** *Let $A$ and $B$ be self-adjoint operators.*
*(a) Let $C \in \mathcal{B}(\mathcal{H})$ be such that $C\mathcal{D}(B) \subseteq \mathcal{D}(A)$. Then, for each $f \in \mathcal{D}(B)$, the vector-valued function $t \mapsto e^{iAt}Ce^{-iBt}f$ is strongly differentiable, and*

$$\frac{d}{dt}e^{iAt}Ce^{-iBt}f = ie^{iAt}(AC - CB)e^{-iBt}f. \tag{5.11}$$

*(b) Let $C$ be $B$-bounded and such that $C\mathcal{D}(B^2) \subseteq \mathcal{D}(A)$. Then, for $f \in \mathcal{D}(B^2)$, the vector-valued function $t \mapsto e^{iAt}Ce^{-iBt}f$ is strongly differentiable and its derivative is given by* (5.11).

PROOF. (a) Setting $U_t = e^{-iAt}$ and $W_t = e^{-iBt}$, we have for $f \in \mathcal{D}(B)$:

$$\frac{1}{\tau}[U_{t+\tau}^* CW_{t+\tau}f - U_t^* CW_t f] = \big[\frac{1}{\tau}(U_{t+\tau}^* - U_t^*]CW_t f + U_{t+\tau}^* C\big[\frac{1}{\tau}(W_{t+\tau} - W_t)f\big].$$

Since $CW_t f \in \mathcal{D}(A)$, the first term on the right-hand side is strongly convergent to $iU_t^* ACW_t f$ as $\tau \to 0$. For the second term we have $U_{t+\tau}^* C \to U_t^* C$ strongly and $\tau^{-1}(W_{t+\tau} - W_t)f \to -iBW_t f$ strongly as $\tau \to 0$. Then, by a straightforward adaptation of the argument in (2.16) (which requires the boundedness of $C$), one obtains that the second term on the right-hand side is strongly convergent to $-iU_t^* CBW_t f$.

(b) If $f \in \mathcal{D}(B^2)$, then $(B+i)f \in \mathcal{D}(B)$. Since the operator $C_0 := C(B+i)^{-1}$ belongs to $\mathcal{B}(\mathcal{H})$ and maps $\mathcal{D}(B)$ into $\mathcal{D}(A)$, we obtain by applying the result of (a) that $e^{iAt}Ce^{-iBt}f \equiv e^{iAt}C_0 e^{-iBt}(B+i)f$ is strongly differentiable, with derivative $ie^{iAt}(AC_0 - C_0 B)e^{-iBt}(B+i)f$. The last expression is identical with the right-hand side of (5.11). □

**5.1.2.** We continue by describing some results on the asymptotic behaviour ($t \to \pm\infty$) of $U_t f$. We begin with a classical result from Fourier analysis:

**Proposition 5.6 (Riemann-Lebesgue Lemma).** *If $f \in L^1(\mathbb{R}^n)$, then its Fourier transform $\tilde{f}$ is a continuous function converging to zero at infinity: $\tilde{f}(\vec{k}) \to 0$ as $|\vec{k}| \to \infty$.*

PROOF. To simplify the notations we consider the case $n = 1$. If $f$ belongs to $\mathcal{S}(\mathbb{R})$, then its Fourier transform $\tilde{f}$ also belongs to $\mathcal{S}(\mathbb{R})$; in particular the function $\tilde{f}$ is continuous and converges to zero at infinity. If $f \in L^1(\mathbb{R})$, there is a sequence $\{f_j\}$ of functions in $\mathcal{S}(\mathbb{R})$ such that $\|f - f_j\|_1 \to 0$ as $j \to \infty$, see §1.4.5. Then

$$\begin{aligned} |\tilde{f}(k) - \tilde{f}_j(k)| &= \frac{1}{\sqrt{2\pi}} \left| \int_{-\infty}^{\infty} e^{-ikx} [f(x) - f_j(x)]\, dx \right| \\ &\leq \frac{1}{\sqrt{2\pi}} \int_{-\infty}^{\infty} |f(x) - f_j(x)|\, dx = \frac{1}{\sqrt{2\pi}} \|f - f_j\|_1. \end{aligned} \tag{5.12}$$

Since the final expression is independent of $k$ and converges to zero as $j \to \infty$, one concludes that $\tilde{f}_j(k)$ converges to $\tilde{f}(k)$ as $j \to \infty$, uniformly in $k \in \mathbb{R}$. Hence $\tilde{f}$ is continuous (since each $\tilde{f}_j$ is continuous). To see that $\tilde{f}$ converges to zero at infinity, let us fix $\varepsilon > 0$. By (5.12) we can choose a number $j \in \mathbb{N}$ such that $|\tilde{f}(k) - \tilde{f}_j(k)| < \varepsilon/2$ for all $k \in \mathbb{R}$. Then, since $\tilde{f}_j(k)$ tends to zero at infinity, we can choose a real number $k_0 \in (0, \infty)$ such that $|\tilde{f}_j(k)| < \varepsilon/2$ for each $k \in \mathbb{R}$ with $|k| > k_0$. Thus one has for $|k| > k_0$:

$$|\tilde{f}(k)| \leq |\tilde{f}(k) - \tilde{f}_j(k)| + |\tilde{f}_j(k)| \leq \varepsilon/2 + \varepsilon/2 = \varepsilon,$$

which shows that $\tilde{f}$ converges to zero at infinity. □

In the next proposition we consider the asymptotic behaviour of vectors in the absolutely continuous subspace of the infinitesimal generator of an evolution group.

**Proposition 5.7.** *Let $U_t$ be an evolution group and $A$ its infinitesimal generator. Let $f \in \mathcal{H}_{ac}(A)$. Then:*

*(a) $U_t f$ converges weakly to zero as $t \to \pm\infty$.*

*(b) Let $B \in \mathcal{B}(\mathcal{H})$ be an $A$-compact operator. Then $\|BU_t f\| \to 0$ as $t \to \pm\infty$.*

*(c) Let $B \in \mathcal{B}(\mathcal{H})$ and assume that there exists an operator $C \in \mathcal{B}(\mathcal{H})$ with dense range (i.e. $\mathcal{R}(C)^\perp = \{\mathbf{0}\}$) and such that $BC \in \mathcal{K}(\mathcal{H})$ and $CU_t = U_t C$ for all $t \in \mathbb{R}$. Then $\|BU_t f\| \to 0$ as $t \to \pm\infty$.*

PROOF. (a) (i) If $h \in \mathcal{H}_{\text{ac}}(A)$, then $m_h$ [see the definition (4.37)] is an absolutely continuous measure, i.e. there is a non-negative function $\theta \in L^1_{\text{loc}}(\mathbb{R})$ such that $m_h(\mathcal{V}) = \int_{\mathcal{V}} \theta(\lambda) d\lambda$ for all $\mathcal{V} \in \mathcal{A}_{\text{B}}$. In particular $m_h(\mathbb{R}) = \int_{-\infty}^{\infty} \theta(\lambda) d\lambda = \|h\|^2 < \infty$, hence $\theta \equiv dm_h/d\lambda$ belongs to $L^1(\mathbb{R})$. Now observe that

$$\langle h, U_t h \rangle = \int_{-\infty}^{\infty} e^{-i\lambda t} m_h(d\lambda) = \int_{-\infty}^{\infty} e^{-i\lambda t} \theta(\lambda) d\lambda.$$

So $\varphi(t) := \langle h, U_t h \rangle$ is the Fourier transform of the function $\theta \in L^1(\mathbb{R})$, consequently $\varphi(t) \to 0$ as $|t| \to \infty$ by the Riemann-Lebesgue Lemma. Thus $\lim_{t\to\pm\infty} \langle h, U_t h \rangle = 0$ for each $h \in \mathcal{H}_{\text{ac}}(A)$.
(ii) We now show that $\langle g, U_t f \rangle \to 0$ as $t \to \pm\infty$ for each $g \in \mathcal{H}$ and each $f \in \mathcal{H}_{\text{ac}}(A)$. If $f \in \mathcal{H}_{\text{ac}}(A)$, then $U_t f \in \mathcal{H}_{\text{ac}}(A)$ for all $t \in \mathbb{R}$, hence $\langle g, U_t f \rangle = 0$ for all $t \in \mathbb{R}$ if $g \in \mathcal{H}_{\text{s}}(A) \equiv \mathcal{H}_{\text{ac}}(A)^\perp$. So it suffices to consider the case in which $g \in \mathcal{H}_{\text{ac}}(A)$. Then, by virtue of the polarisation identity (2.15), $\langle g, U_t f \rangle$ is a linear combination of four terms of the form $\langle g + \alpha f, U_t(g + \alpha f) \rangle$, and each of these terms converges to zero as $t \to \pm\infty$ by the result of (i), because $g + \alpha f \in \mathcal{H}_{\text{ac}}(A)$ for each $\alpha \in \mathbb{C}$.

(b) By Proposition 2.2 it is enough to show that $\|BU_t f\| \to 0$ as $|t| \to \infty$ for all vectors $f$ in some dense subset of $\mathcal{H}_{\text{ac}}(A)$. We take for this dense set the linear manifold $\mathcal{H}_{\text{ac}}(A) \cap \mathcal{D}(A)$. So let $f \in \mathcal{H}_{\text{ac}}(A) \cap \mathcal{D}(A)$; we set $g = (A+i)f$ and observe that $g$ belongs to $\mathcal{H}_{\text{ac}}(A)$ and that $f = (A+i)^{-1} g$. Thus $BU_t f = BU_t(A+i)^{-1} g = B(A+i)^{-1} U_t g$. Since $U_t g$ converges weakly to zero by (a) and $B(A+i)^{-1}$ is a compact operator by hypothesis, we obtain by applying Proposition 2.12(a) that $BU_t f$ converges strongly to zero: $\lim_{t\to\pm\infty} \|BU_t f\| = 0$.

(c) The argument here is similar to that in (b), replacing $(A+i)^{-1}$ by $C$ and $\mathcal{H}_{\text{ac}}(A) \cap \mathcal{D}(A)$ by $\mathcal{H}_{\text{ac}}(A) \cap \mathcal{R}(C)$. It suffices to know that $\mathcal{H}_{\text{ac}}(A) \cap \mathcal{R}(C)$ is dense in $\mathcal{H}_{\text{ac}}(A)$ and that $C$ maps the subspace $\mathcal{H}_{\text{ac}}(A)$ into itself. The last property is equivalent to $[C, E^A_{\text{ac}}] = O$, and this commutation relation can be deduced from the hypothesis that $C$ commutes with the group $\{U_t\}$. Then, since $\mathcal{R}(C)$ is dense in $\mathcal{H}$ by assumption, one sees that $C$ must map $\mathcal{H}_{\text{ac}}(A)$ onto a dense subset of $\mathcal{H}_{\text{ac}}(A)$. We omit the details since the result of (c) will not be used further on. □

Let us now consider vectors in $\mathcal{H}_{\text{c}}(A)$ that do not belong to $\mathcal{H}_{\text{ac}}(A)$. In general, when $f \in \mathcal{H}_{\text{c}}(A)$, then $U_t f$ is not weakly convergent to zero, as can be seen in explicit examples.[1] Nevertheless results similar to those of the preceding proposition are true if one introduces an average with respect to the parameter $t$. Since in many of our applications $t$ is interpreted as time, we shall then speak of a temporal mean. For example one can consider the average of $U_t f$ over an interval of length $T$, and this average converges weakly to zero as the length $T$ of the interval tends to infinity:

$$\frac{1}{T}\int_0^T U_t f\,dt \quad \text{and} \quad \frac{1}{T}\int_{-T}^0 U_t f\,dt \quad \text{and} \quad \frac{1}{2T}\int_{-T}^T U_t f\,dt \tag{5.13}$$

[1] See for example Problem 149 in Chapter XI, Volume III of [RS].

converge weakly to zero as $T \to \infty$, and for example $T^{-1}\int_0^T \|BU_t f\|^2 dt$ converges to zero as $T \to \infty$ if $f \in \mathcal{H}_c(A)$ and $B$ is $A$-compact. We give below a proof of this last result. As regards the first one, it will not be needed, but let us mention that there exists a much stronger result: if $f$ is orthogonal to the subspace $\mathcal{M}(\{0\}) \equiv \mathcal{N}(A)$ consisting of all eigenvectors of $A$ associated with the eigenvalue 0, then the expressions in (5.13) converge *strongly* to zero. This is a consequence of the following proposition (its proof is simple, see e.g. Problem 5.6 here or Proposition 4.20 in [A]):

**Proposition 5.8 (Mean Ergodic Theorem).** *Let $\{U_t\}$ be an evolution group, $A$ its infinitesimal generator and $E(\{0\})$ the projection onto $\mathcal{N}(A)$. Then one has for each $f \in \mathcal{H}$:*

$$\operatorname*{s\text{-}lim}_{T\to\pm\infty} \frac{1}{T}\int_0^T U_t f\, dt = E(\{0\})f.$$

*In particular, if $f \perp \mathcal{N}(A)$, then the preceding limits are* **0**.

Results of the type of those given in the next proposition and in Proposition 5.12 are often cited under the name of **RAGE Theorem** (e.g. Volume III of [RS] or Volume II of [W2]).

**Proposition 5.9.** *Let $\{U_t\}$ be an evolution group and $A$ its infinitesimal generator. Let $f \in \mathcal{H}_c(A)$. Then*
*(a) One has for each vector $h \in \mathcal{H}$:*

$$\lim_{T\to\infty} \frac{1}{T}\int_0^T \left|\langle h, U_t f\rangle\right|^2 dt = 0 \quad \textit{and} \quad \lim_{T\to\infty} \frac{1}{T}\int_{-T}^0 \left|\langle h, U_t f\rangle\right|^2 dt = 0. \tag{5.14}$$

*(b) If $B \in \mathcal{B}(\mathcal{H})$ is an $A$-compact operator, one has*

$$\lim_{T\to\infty} \frac{1}{T}\int_0^T \|BU_t f\|^2 dt = 0 \quad \textit{and} \quad \lim_{T\to\infty} \frac{1}{T}\int_{-T}^0 \|BU_t f\|^2 dt = 0. \tag{5.15}$$

*(c) More generally, let $B \in \mathcal{B}(\mathcal{H})$ and suppose that there exists an operator $C \in \mathcal{B}(\mathcal{H})$ with dense range (i.e. $\mathcal{R}(C)^\perp = \{\mathbf{0}\}$) and such that $CU_t = U_t C$ for all $t \in \mathbb{R}$ and $BC \in \mathcal{K}(\mathcal{H})$. Then (5.15) is true.*

PROOF. We shall prove the first relations in (5.14) and (5.15), and we shall omit the proof of (c). Before proving (a) we show that the result of (b) follows from that of (a).

(b) We may assume that $B \neq O$. Given $f \in \mathcal{H}_c(A)$ and $\varepsilon > 0$, we fix a vector $f_\varepsilon \neq \mathbf{0}$ in $\mathcal{H}_c(A) \cap \mathcal{D}(A)$ such that $\|f - f_\varepsilon\|^2 < \varepsilon/(9\|B\|^2)$, and we set $g = (A+i)f_\varepsilon$. Hence $BU_t f_\varepsilon = B(A+i)^{-1}U_t g$, and $g \in \mathcal{H}_c(A)$. We choose a finite rank operator $S$ such that $\|B(A+i)^{-1} - S\|^2 < \varepsilon/(9\|g\|^2)$, and we write $S$ in the form (2.42): $Sh = \sum_{k=1}^N \langle g_k, h\rangle h_k$ (where $h$ varies over $\mathcal{H}$ and $g_k, h_k$ are fixed vectors). Then, by using (1.58) with $N = 3$ and with general $N$, we may write:

$$\begin{aligned}\|BU_tf\|^2 &= \|BU_t(f-f_\varepsilon)+[B(A+i)^{-1}-S]U_tg+SU_tg\|^2\\ &\le 3\|B\|^2\|f-f_\varepsilon\|^2+3\|B(A+i)^{-1}-S\|^2\|g\|^2+3\Big\|\sum_{k=1}^N\langle g_k,U_tg\rangle h_k\Big\|^2\\ &< \frac{\varepsilon}{3}+\frac{\varepsilon}{3}+3N\sum_{k=1}^N\|h_k\|^2\big|\langle g_k,U_tg\rangle\big|^2.\end{aligned}$$

It follows that

$$\frac{1}{T}\int_0^T\|BU_tf\|^2dt \le \frac{2\varepsilon}{3}+3N\sum_{k=1}^N\|h_k\|^2\frac{1}{T}\int_0^T\big|\langle g_k,U_tg\rangle\big|^2dt.$$

Each term in the finite sum on the right-hand side converges to zero as $T\to\infty$, by (5.14). Thus there exists $T_0$ such that the right-hand side is less than $\varepsilon$ for each $T>T_0$, which proves (5.15).

(a) If $h\in\mathcal{H}_\mathrm{p}(A)$, then $\langle h,U_tf\rangle=0$ for all $t\in\mathbb{R}$ [assuming that $f\in\mathcal{H}_\mathrm{c}(A)$], hence (5.14) is trivially true. Thus it suffices to prove (5.14) under the assumption that $f$ and $h$ belong to $\mathcal{H}_\mathrm{c}(A)$. By using the polarisation identity (2.15) we can express $\langle h,U_tf\rangle$ as a linear combination of four terms of the form $\langle h+\alpha f,U_t(h+\alpha f)\rangle$. As $h+\alpha f\in\mathcal{H}_\mathrm{c}(A)$ if $f,h\in\mathcal{H}_\mathrm{c}(A)$, it then suffices to prove that

$$\lim_{T\to\infty}\frac{1}{T}\int_0^T\big|\langle g,U_tg\rangle\big|^2dt=0 \qquad \forall g\in\mathcal{H}_\mathrm{c}(A). \tag{5.16}$$

For this we denote by $\{E_\lambda\}$ the spectral family of $A$, and we fix $g\in\mathcal{H}_\mathrm{c}(A)$, $g\neq\mathbf{0}$. For $T>0$ we set

$$\Psi(T)=\frac{1}{T}\int_0^T\big|\langle g,U_tg\rangle\big|^2dt\equiv\frac{1}{T}\int_0^T dt\int_{-\infty}^\infty e^{i\lambda t}m_g(d\lambda)\int_{-\infty}^\infty e^{-i\mu t}m_g(d\mu). \tag{5.17}$$

It is clear that this multiple integral is absolutely convergent:

$$\Psi(T)\equiv\big|\Psi(T)\big|\le\frac{1}{T}\int_0^T dt\int_{-\infty}^\infty m_g(d\lambda)\int_{-\infty}^\infty m_g(d\mu)=\|g\|^4.$$

Thus we may change the order of integration in (5.17):

$$\begin{aligned}\Psi(T)&=\Big|\int_{-\infty}^\infty m_g(d\lambda)\int_{-\infty}^\infty m_g(d\mu)\,\frac{1}{T}\int_0^T e^{i\lambda t}e^{-i\mu t}dt\Big|\\ &\le\int_{-\infty}^\infty m_g(d\lambda)\int_{-\infty}^\infty m_g(d\mu)\left|\frac{e^{i(\lambda-\mu)T}-1}{i(\lambda-\mu)T}\right|.\end{aligned} \tag{5.18}$$

If $\lambda\neq\mu$ the integrand $\big|e^{i(\lambda-\mu)T}-1]/[i(\lambda-\mu)T]\big|$ converges to zero as $T\to\infty$, and it can be majorised by $\big|e^{i(\lambda-\mu)T}-1]/[i(\lambda-\mu)T]\big|\le 1$ for all $\lambda,\mu$ ($\lambda\neq\mu$) and all $T>0$. We show below that the subset of $\mathbb{R}\times\mathbb{R}$ defined as the set of points $\{(\lambda,\mu)\,\big|\,\lambda=\mu\}$

is a null set (with respect to the measure $m_g(d\lambda)m_g(d\mu)$ on $\mathbb{R}\times\mathbb{R}$). Assuming this result, we can apply the Dominated Convergence Theorem [Proposition 1.9] in (5.18) to conclude that $\Psi(T)\to 0$ as $T\to\infty$, which proves (5.16).

It remains to show that the diagonal $\{(\lambda,\mu)\mid\lambda=\mu\}$ of $\mathbb{R}\times\mathbb{R}$ is a null set for the measure $m_g(d\lambda)m_g(d\mu)$. The argument runs as follows. Since $g\in\mathcal{H}_c(A)$, the function $\mu\mapsto\langle g,E_\mu g\rangle$ is continuous on $\mathbb{R}$. It is even uniformly continuous: given $\varepsilon>0$, one can choose $M\in(0,\infty)$ such that $|\langle g,E_\mu g\rangle|<\varepsilon/2$ if $\mu<-M$ and $0\le\|g\|^2-\langle g,E_\mu g\rangle<\varepsilon/2$ if $\mu>M$ (hence $|\langle g,E_\mu g\rangle-\langle g,E_{\mu'}g\rangle|<\varepsilon$ if $\mu,\mu'\le -M$ and $|\langle g,E_\mu g\rangle-\langle g,E_{\mu'}g\rangle|<\varepsilon$ if $\mu,\mu'>M$). On the other hand, the function $\mu\mapsto\langle g,E_\mu g\rangle$, being continuous, is uniformly continuous on each finite interval, in particular on $[-M-1,M+1]$.

Now let $\epsilon>0$ be given. By the uniform continuity of $\mu\mapsto\langle g,E_\mu g\rangle$, there exists $\delta>0$ such that $|\langle g,E_{\lambda+\delta}g\rangle-\langle g,E_{\lambda-\delta}g\rangle|<\epsilon/\|g\|^2$ for all $\lambda\in\mathbb{R}$. Thus

$$\int_{-\infty}^{\infty}m_g(d\lambda)\int_{\lambda-\delta}^{\lambda+\delta}m_g(d\mu)=\int_{-\infty}^{\infty}m_g(d\lambda)\big[\langle g,E_{\lambda+\delta}g\rangle-\langle g,E_{\lambda-\delta}g\rangle\big]$$
$$<\frac{\epsilon}{\|g\|^2}\int_{-\infty}^{\infty}m_g(d\lambda)=\epsilon. \tag{5.19}$$

Let $\mathcal{O}_\delta=\{(\lambda,\mu)\,|\,|\lambda-\mu|<\delta\}\subset\mathbb{R}\times\mathbb{R}$. The set $\mathcal{O}_\delta$ is a neighbourhood of the diagonal $\{\lambda=\mu\}$ in $\mathbb{R}\times\mathbb{R}$. The inequality (5.19) shows that the measure of $\mathcal{O}_\delta$ is less than $\epsilon$; since $\epsilon$ is an arbitrary positive number, the measure of the diagonal $\{(\lambda,\mu)\mid\lambda=\mu\}$ must be zero. □

We have shown that if $f\in\mathcal{H}_{ac}(A)$, then $U_tf$ converges weakly to zero as $t\to\pm\infty$, whereas for $f\in\mathcal{H}_{sc}$ this is not necessarily the case. Nevertheless, if $f\in\mathcal{H}_{sc}(A)$, then $U_tf$ converges weakly (even strongly by Proposition 5.8) to zero in temporal mean. Also Proposition 5.9(a) asserts that then $\lim_{T\to\infty}T^{-1}\int_0^T|\langle g,U_tf\rangle|^2dt=0$ for each vector $g\in\mathcal{H}$. This implies that $|\langle g,U_tf\rangle|$ must be small for sufficiently many values of $t$, in particular that there must exist a sequence of real numbers $\{t_k\}$ tending to infinity such that $\langle g,U_{t_k}f\rangle\to 0$ as $k\to\infty$. This is a consequence of the following fact:

Let $\varphi\colon[0,\infty)\to[0,\infty)$ be a *bounded* non-negative function. Then:
(i) if $\varphi(t)\to 0$ as $t\to\infty$, one has $\lim_{T\to\infty}T^{-1}\int_0^T\varphi(t)\,dt=0$,
(ii) if $\lim_{T\to\infty}T^{-1}\int_0^T\varphi(t)\,dt=0$, there exists a sequence $\{t_k\}$ such that $t_k\to\infty$ and $\lim_{k\to\infty}\varphi(t_k)=0$.

PROOF. To check these properties, let $0<s<T$ and write

$$\frac{1}{T}\int_0^T\varphi(t)\,dt=\frac{1}{T}\int_0^s\varphi(t)\,dt+\frac{1}{T}\int_s^T\varphi(t)\,dt. \tag{5.20}$$

For each fixed $s$ the first term on the right-hand side converges to zero as $T\to\infty$. Thus one has for each fixed $s>0$:

$$\lim_{T\to\infty}\frac{1}{T}\int_0^T\varphi(t)\,dt=\lim_{T\to\infty}\frac{1}{T}\int_s^T\varphi(t)\,dt. \tag{5.21}$$

If $\varphi(t) \to 0$ as $t \to \infty$ and if $\varepsilon > 0$ is given, we choose $s > 0$ such that $\varphi(t) < \varepsilon$ for all $t > s$. Then $T^{-1}\int_s^T \varphi(t)dt < \varepsilon$ for each $T > s$, which by (5.21) proves (i).

The result of (ii) can be obtained by *reductio ad absurdum*. Let us assume that $\lim_{T\to\infty} T^{-1}\int_0^T \varphi(t)dt = 0$ and that for each $s > 0$ there is a number $\delta > 0$ such that $\varphi(t) \geq \delta$ for all $t > s$. Then $T^{-1}\int_s^T \varphi(t)dt > \delta(T-s)/T > \delta/2$ for each $T > 2s$. By taking into account (5.21) one obtains a contradiction with the assumption that $\lim_{T\to\infty} T^{-1}\int_0^T \varphi(t)dt = 0$. Hence, for each $k \in \mathbb{N}$ there exists a number $t_k > k$ such that $\varphi(t_k) < 1/k$ [if one sets $s = k$ above, one cannot have $\varphi(t) \geq 1/k \equiv \delta$ for all $t \geq k$]. □

In the following proposition we indicate a somewhat stronger result: it is possible to find a sequence $\{t_k\}$ that is independent of $f$ and $g$, i.e. a sequence $\{t_k\}$ converging to $+\infty$ such that $\langle g, U_{t_k} f\rangle \to 0$ for *all* $f \in \mathcal{H}_c(A)$ and for *all* $g \in \mathcal{H}$. We omit the proof, because the result will not be needed in this text (see the Corollary on page 343 of Volume III of [RS]).

**Proposition 5.10.** *Let $\{U_t\}$ be an evolution group and $A$ its infinitesimal generator. Then there exists a sequence $\{t_k\}$ of real numbers such that $t_k \to \infty$ and $U_{t_k} f$ converges weakly to zero as $k \to \infty$ for each $f \in \mathcal{H}_c(A)$. In particular, if $B \in \mathcal{B}(\mathcal{H})$ is an $A$-compact operator, one has $\lim_{k\to\infty} \|BU_{t_k} f\| = 0$ for each $f \in \mathcal{H}_c(A)$.*

## 5.2 Characterisation of the scattering states

In a scattering experiment particles are impinging from very far (from a great distance compared to the size of the scatterer) on a scatterer (a centre of force) and then observed again far away from the scatterer. The mathematical description of such situations involves states having the following property: under the evolution group characterising their temporal evolution they are localised very far from the scattering centre at very large times (positive or negative). State vectors having this property correspond to *scattering states*. One can also consider *bound states*; a particle is in a bound state if it remains localised near the scattering centre *at all times*. It is convenient (and useful if the interaction has spherical symmetry) to identify the scattering centre with the co-ordinate origin.

In quantum mechanics these notions of localisation can only be approximate as a consequence of the uncertainty relations. We present below a reasonable mathematical definition and study some of its consequences. Physicists often use other definitions which do not have a direct interpretation in terms of the temporal behaviour of states in configuration space: bound states are identified with eigenvectors of the Hamiltonian $H$ and scattering states with vectors in its continuous subspace $\mathcal{H}_c(H)$. This has its origin in the fact that, for most atomic systems, there is no difference between the two ways of defining bound states and scattering states (that in terms of the temporal behaviour in configuration space and that in terms of the spectral subspaces of the Hamiltonian, i.e. of the infinitesimal generator of the relevant evolution group).

Consider a quantum mechanical system that is described by the states in the Hilbert space $\mathcal{H} = L^2(\mathbb{R}^n)$. Let $H$ be its Hamiltonian (a self-adjoint operator in $\mathcal{H}$) and let $\{U_t = e^{-iHt}\}$ be the evolution group describing the time evolution of the system (here the parameter $t$ is interpreted as the ***time***). To describe the localisation in configuration space one uses the operators $\chi_{\mathcal{V}}(\vec{Q})$, where $\mathcal{V}$ is a Borel set in $\mathbb{R}^n$ (specifying the region of localisation that one considers) and $\chi_{\mathcal{V}}(\vec{Q})$ is the operator of multiplication by the characteristic function $\chi_{\mathcal{V}}$ of $\mathcal{V}$, i.e. the projection onto the subspace $L^2(\mathcal{V})$ of $L^2(\mathbb{R}^n)$ consisting of all wave functions $f(\vec{x})$ that vanish outside $\mathcal{V}$. We have in mind the situation in which the scattering centre is close to the origin of the co-ordinate system, and we shall use only very particular localisation regions $\mathcal{V}$, namely balls $B_R$ of radius $R$ centred at the origin: $B_R := \{\vec{x} \in \mathbb{R}^n \,\big|\, |\vec{x}| \leq R\}$. We use the notation $\chi_R$ for $\chi_{B_R}$, so that

$$[\chi_R(\vec{Q})f](\vec{x}) = f(\vec{x}) \text{ if } |\vec{x}| \leq R, \quad [\chi_R(\vec{Q})f](\vec{x}) = 0 \text{ if } |\vec{x}| > R. \tag{5.22}$$

A vector $f \in \mathcal{H}$ will be called a ***scattering state*** at $t = -\infty$ or at $t = +\infty$ respectively if

$$\lim_{t\to-\infty} \|\chi_R(\vec{Q})U_t f\| = 0 \text{ resp. } \lim_{t\to+\infty} \|\chi_R(\vec{Q})U_t f\| = 0 \quad \text{for each } R \in (0,\infty). \tag{5.23}$$

This means that at very large (negative or positive) times the probability of finding such a state in a finite part of configuration space is very small, or more precisely that such a state disappears from each given finite region of configuration space as time tends to infinity: for each bounded subset $\mathcal{V}$ of $\mathbb{R}^n$ one has

$$\lim_{t\to-\infty} \int_{\mathcal{V}} \big|[U_t f](\vec{x})\big|^2 d^n x = 0 \quad \text{resp.} \quad \lim_{t\to+\infty} \int_{\mathcal{V}} \big|[U_t f](\vec{x})\big|^2 d^n x = 0.$$

A vector $f \in \mathcal{H}$ will be called a ***bound state*** if

$$\lim_{R\to\infty} \sup_{t\in\mathbb{R}} \|[I - \chi_R(\vec{Q})]U_t f\| = 0 \tag{5.24}$$

or equivalently if, given $\varepsilon > 0$, there exists a finite ball $B_R$ such that at *all times* the system in the state $f$ has probability less than $\varepsilon$ to be localised outside this ball:

$$\frac{1}{\|f\|^2} \int_{|\vec{x}|\geq R} \big|[U_t f](\vec{x})\big|^2 d^n x < \varepsilon \quad \forall t \in \mathbb{R} \quad (\text{if } R \text{ is sufficiently large}).$$

More generally, one could consider states that are bound until some finite time (for example for all $t < 0$ but not necessarily as $t \to +\infty$) or states that are localised after some finite time (for all times $t > 0$ but not necessarily as $t \to -\infty$). This amounts to replacing in (5.24) $\sup_{t\in\mathbb{R}}$ by $\sup_{t\leq 0}$ or by $\sup_{t\geq 0}$ respectively.

It is interesting to know that, under a certain compactness condition [for example if the operator $\chi_R(\vec{Q})$ is $H$-compact for each $R \in (0,\infty)$], the set of bound states coincides with the subspace $\mathcal{H}_{\mathrm{p}}(H)$ spanned by the eigenvectors of the Hamiltonian $H$ and the set of scattering states with the subspace $\mathcal{H}_{\mathrm{c}}(H)$ if $H$ has no singular continuous spectrum. To treat Hamiltonians with singular continuous spectrum one has to slightly generalise the definition of the scattering states by introducing a time average, for example by requiring instead of the second condition in (5.23) that

$$\lim_{T\to\infty}\frac{1}{T}\int_0^T \|\chi_R(\vec{Q})U_t f\|^2\,dt = 0 \qquad \forall R\in(0,\infty). \tag{5.25}$$

This means that the state $f$ disappears from each finite region of configuration space in time average (roughly speaking: most of the time a system in such a state is localised far from the scattering centre, but from time to time it could return into a neighbourhood of the co-ordinate origin).

We give an abstract theorem that implies the above results. The only relevant properties of the localisation operators $\chi_R(\vec{Q})$ are the fact that they are projections and that $\text{s-lim}_{R\to\infty}\chi_R(\vec{Q}) = I$. Hence we now consider a general Hilbert space $\mathcal{H}$, an evolution group $\{U_t\}$ in $\mathcal{H}$ and a family $\{F_R\}_{R>0}$ of ***localising operators*** having the following properties[2]:

$$\text{(a) each } F_R \text{ is a projection:} \qquad F_R^2 = F_R = F_R^* \tag{5.26}$$

$$\text{(b)} \qquad \underset{R\to\infty}{\text{s-lim}}\, F_R = I. \tag{5.27}$$

For each sign of time we introduce subsets of $\mathcal{H}$ corresponding to the bound states, the scattering states and the scattering states in temporal mean. We denote these subsets by $\mathcal{M}_0^+(H)$, $\mathcal{M}_0^-(H)$, $\mathcal{M}_\infty^+(H)$, $\mathcal{M}_\infty^-(H)$, $\overline{\mathcal{M}}_\infty^+(H)$ and $\overline{\mathcal{M}}_\infty^-(H)$ respectively (the sign $+$ or $-$ corresponds to the sign of time, the subscript 0 indicates that it concerns states that are "close" to the origin, the subscript $\infty$ refers to states moving to infinity, and the over-bar on $\mathcal{M}$ indicates an average over time; it is clear that each of the preceding subsets depends on the Hamiltonian $H$ involved). Hence[3]

$$f\in\mathcal{M}_0^\pm(H) \iff \lim_{R\to\infty}\ \sup_{t\in[0,\pm\infty)} \|(I-F_R)U_t f\| = 0, \tag{5.28}$$

$$f\in\mathcal{M}_\infty^\pm(H) \iff \lim_{t\to\pm\infty}\|F_R U_t f\| = 0 \qquad \forall R<\infty, \tag{5.29}$$

$$f\in\overline{\mathcal{M}}_\infty^\pm(H) \iff \lim_{T\to\infty}\frac{1}{T}\int_0^{\pm T}\|F_R U_t f\|^2\,dt = 0 \quad \forall R<\infty. \tag{5.30}$$

Furthermore we define

$$\mathcal{M}_0(H) = \mathcal{M}_0^+(H)\cap\mathcal{M}_0^-(H) \tag{5.31}$$

$$\mathcal{M}_\infty(H) = \mathcal{M}_\infty^+(H)\cap\mathcal{M}_\infty^-(H) \tag{5.32}$$

$$\overline{\mathcal{M}}_\infty(H) = \overline{\mathcal{M}}_\infty^+(H)\cap\overline{\mathcal{M}}_\infty^-(H). \tag{5.33}$$

We begin by establishing some simple properties of these sets and some relations between them.

---

[2] In fact condition (a) could be weakened to: (a') $F_R^* = F_R\in\mathcal{B}(\mathcal{H})$.

[3] We recall from the footnote on page 73 that, if an equation or an expression contains double signs, it is understood that this means two equations or expressions, one by taking everywhere the upper sign, the other one by taking everywhere the lower sign. This convention will be frequently applied in the present chapter.

**Proposition 5.11.** *(a) Each of the subsets of $\mathcal{H}$ defined in* (5.28) - (5.33) *is a subspace of $\mathcal{H}$. Each of these subspaces is invariant under $U_t$, $t \in \mathbb{R}$.*
*(b) One has $\mathcal{M}_\infty^+(H) \subseteq \overline{\mathcal{M}}_\infty^+(H)$ and $\mathcal{M}_\infty^-(H) \subseteq \overline{\mathcal{M}}_\infty^-(H)$.*
*(c) One has $\mathcal{M}_0^+(H) \perp \overline{\mathcal{M}}_\infty^+(H)$, $\mathcal{M}_0^-(H) \perp \overline{\mathcal{M}}_\infty^-(H)$ and $\mathcal{M}_0(H) \perp \overline{\mathcal{M}}_\infty^\pm(H)$.*
*(d) One has $\mathcal{H}_{\rm p}(H) \subseteq \mathcal{M}_0(H)$ and $\overline{\mathcal{M}}_\infty^\pm(H) \subseteq \mathcal{H}_{\rm c}(H)$.*

PROOF. (a) These results are rather easy to obtain by using the triangle inequality, the inequality $\|f+g\|^2 \leq 2\|f\|^2 + 2\|g\|^2$ and the fact that $\|F_R\| = \|U_t\| = 1$. We discuss some typical situations.

(i) Let us show that $\mathcal{M}_\infty^+(H)$ is a *subspace*. If $f, g \in \mathcal{M}_\infty^+(H)$ and $\alpha \in \mathbb{C}$, then

$$\lim_{t\to\infty} \|F_R U_t(f+\alpha g)\| \leq \lim_{t\to\infty} \|F_R U_t f\| + |\alpha| \lim_{t\to\infty} \|F_R U_t g\|] = 0.$$

Thus $f + \alpha g \in \mathcal{M}_\infty^+(H)$, which implies that $\mathcal{M}_\infty^+(H)$ is a linear manifold. To see that this linear manifold is a subspace, let us consider a strong Cauchy sequence $\{f_k\}$ in $\mathcal{M}_\infty^+(H)$ and denote its limit by $f$: $\|f_k - f\| \to 0$ as $k \to \infty$. We must show that $f \in \mathcal{M}_\infty^+(H)$. One has

$$\begin{aligned}\|F_R U_t f\| &= \|F_R U_t(f - f_k) + F_R U_t f_k\| \\ &\leq \|F_R U_t(f-f_k)\| + \|F_R U_t f_k\| \leq \|f - f_k\| + \|F_R U_t f_k\|.\end{aligned}$$

Given $\varepsilon > 0$, first choose an integer $k$ such that $\|f - f_k\| < \varepsilon/2$, then a number $t_0$ such that $\|F_R U_t f_k\| < \varepsilon/2$ for $t > t_0$. One then has $\|F_R U_t f\| < \varepsilon$ for $t > t_0$, which shows that $f \in \mathcal{M}_\infty^+(H)$.

(ii) *Invariance under* $U_s$. It is evident that $\mathcal{M}_\infty^+(H)$ is invariant under each $U_s$. As a somewhat less trivial example, let us show that $f \in \mathcal{M}_0^+(H) \Longrightarrow U_s f \in \mathcal{M}_0^+(H)$. One has

$$\sup_{t\in[0,\infty)} \|(I - F_R)U_t U_s f\| = \sup_{t\in[0,\infty)} \|(I - F_R)U_{t+s} f\| = \sup_{\tau\in[s,\infty)} \|(I - F_R)U_\tau f\|.$$

If $s \geq 0$ one has $\sup_{\tau\in[s,\infty)} \|(I - F_R)U_\tau f\| \leq \sup_{\tau\in[0,\infty)} \|(I - F_R)U_\tau f\|$, which converges to zero as $R \to \infty$ by the assumption made on $f$. If $s < 0$ we write

$$\sup_{\tau\in[s,\infty)} \|(I - F_R)U_\tau f\| \leq \sup_{\tau\in[0,\infty)} \|(I - F_R)U_\tau f\| + \sup_{\tau\in[s,0]} \|(I - F_R)U_\tau f\|.$$

Again the first term on the right-hand side converges to zero as $R \to \infty$. For the second term, let $\varepsilon > 0$ be given; choose a number $b > 0$ such that $\|U_t f - f\| < \varepsilon/2$ for all $t \in [-b, b]$, which is possible by the strong continuity of the group $\{U_t\}$. Then one has $\|U_\kappa f - U_\sigma f\| < \varepsilon/2$ if $|\kappa - \sigma| \leq b$. Since $[s, 0]$ is a finite interval, we can choose a finite number of points $a_0, \ldots, a_N$ such that $a_0 = s < a_1 < \cdots < a_{N-1} < a_N = 0$ and $|a_k - a_{k-1}| < 2b$ for each $k = 1, \ldots, N$. Then choose for each $\tau \in [s, 0]$ one of these points, say $a_j$, such that $|\tau - a_j| < b$. With this choice one has

$$\begin{aligned}\|(I - F_R)U_\tau f\| &\leq \|(I - F_R)U_{a_j} f\| + \|(I - F_R)(U_\tau f - U_{a_j} f)\| \\ &\leq \|(I - F_R)U_{a_j} f\| + \varepsilon/2.\end{aligned}$$

Thus

$$\sup_{\tau\in[s,0]} \|(I-F_R)U_\tau f\| \leq \sup_{j=0,\dots,N} \|(I-F_R)U_{a_j} f\| + \varepsilon/2,$$

which is less than $\varepsilon$ if $R > R_0$ (for some $R_0$ depending on $\varepsilon$ and on $s$) by the hypothesis that $\text{s-lim}_{R\to\infty} F_R = I$.

(b) This is evident [take into account the observation (i) made before (5.20) with $\varphi(t) = \|F_R U_t f\|^2$].

(c) Since $\langle f, g\rangle = \langle U_t f, U_t g\rangle$ for all $f, g \in \mathcal{H}$ and all $t \in \mathbb{R}$, we get by using the inequality $|a+b|^2 \leq 2|a|^2 + 2|b|^2$ and the Schwarz inequality that

$$\begin{aligned}
|\langle f,g\rangle|^2 &= \frac{1}{T}\int_0^T |\langle U_t f, U_t g\rangle|^2\,dt \\
&= \frac{1}{T}\int_0^T \left|\langle U_t f, F_R U_t g\rangle + \langle (I-F_R)U_t f, U_t g\rangle\right|^2 dt \\
&\leq \frac{2}{T}\,\|f\|^2\int_0^T \|F_R U_t g\|^2\,dt + \frac{2}{T}\,\|g\|^2\int_0^T \|(I-F_R)U_t f\|^2\,dt. \qquad (5.34)
\end{aligned}$$

Now assume that $f \in \mathcal{M}_0^+(H)$ and $g \in \overline{\mathcal{M}}_\infty^+(H)$, and let $\delta > 0$ be given. Since $f \in \mathcal{M}_0^+(H)$, one can choose a number $R < \infty$ such that $\|(I-F_R)U_t f\|^2\|g\|^2 < \delta/4$ for each $t \geq 0$. Then the last term in the inequality (5.34) is $< \delta/2$ for each $T > 0$. Since $g \in \overline{\mathcal{M}}_\infty^+(H)$ the first term on the bottom line of (5.34) converges to zero as $T \to \infty$, hence it is $< \delta/2$ by a suitable choice of $T$. Thus $|\langle f,g\rangle|^2 < \delta$ for each $\delta > 0$, which shows that $f \perp g$.

(d) Let $f$ be an eigenvector of $H$: $Hf = \lambda f$ for some $\lambda \in \mathbb{R}$. Then $U_t f = e^{-i\lambda t} f$ and consequently $\|(I-F_R)U_t f\| = \|(I-F_R)f\|$, which does not depend on $t$. Since $\text{s-lim}_{R\to\infty}(I-F_R) = O$, one has

$$\lim_{R\to\infty}\sup_{t\in\mathbb{R}} \|(I-F_R)U_t f\| = \lim_{R\to\infty} \|(I-F_R)f\| = 0,$$

hence $f \in \mathcal{M}_0(H)$. Thus the set of all eigenvectors of $H$ belongs to $\mathcal{M}_0(H)$. Since $\mathcal{M}_0(H)$ and $\mathcal{H}_\mathrm{p}(H)$ are subspaces, it follows that $\mathcal{H}_\mathrm{p}(H) \subseteq \mathcal{M}_0(H)$. Then, since $\overline{\mathcal{M}}_\infty^\pm(H)$ are orthogonal to $\mathcal{M}_0(H)$, one has $\overline{\mathcal{M}}_\infty^\pm(H) \subseteq \mathcal{M}_0(H)^\perp \subseteq \mathcal{H}_\mathrm{p}(H)^\perp = \mathcal{H}_\mathrm{c}(H)$. □

In the next proposition we give sufficient conditions for having $\mathcal{M}_0(H) = \mathcal{H}_\mathrm{p}(H)$ and $\overline{\mathcal{M}}_\infty^\pm(H) = \mathcal{H}_\mathrm{c}(H)$.

**Proposition 5.12.** *Assume that for each $R < \infty$ the operator $F_R$ is $H$-compact, i.e. that $F_R(H-i)^{-1} \in \mathcal{K}(\mathcal{H})$ for each $R < \infty$. Then*
*(a) One has $\mathcal{M}_0(H) = \mathcal{H}_\mathrm{p}(H)$ and $\overline{\mathcal{M}}_\infty^\pm(H) = \overline{\mathcal{M}}_\infty(H) = \mathcal{H}_\mathrm{c}(H)$.*
*(b) If $H$ has no singular continuous spectrum, then $\mathcal{M}_\infty^\pm(H) = \mathcal{M}_\infty(H) = \mathcal{H}_\mathrm{ac}(H)$.*

REMARK. The conclusions of this proposition are true in various other situations. For example it suffices that one of the following two conditions is satisfied:
(i) $F_R E([-M,M]) \in \mathcal{K}(\mathcal{H})$ for each $R < \infty$ and each $0 < M < \infty$ [$\{E(\cdot)\}$ denotes the spectral measure of $H$],

(ii) there exists an operator $C \in \mathcal{B}(\mathcal{H})$ with dense range such that $CU_t = U_t C$ for all $t \in \mathbb{R}$ and such that $F_R C \in \mathcal{K}(\mathcal{H})$ for each $R < \infty$.

PROOF OF PROPOSITION 5.12. (a) By taking $B = F_R$ in Proposition 5.9(b) one obtains that for all vectors $f \in \mathcal{H}_c(H)$ and for all $R \in (0,\infty)$: $T^{-1}\int_0^{\pm T} \|F_R U_t f\|^2 dt \to 0$ as $T \to \infty$. Consequently one has $\mathcal{H}_c(H) \subseteq \overline{\mathcal{M}}_\infty^+(H) \cap \overline{\mathcal{M}}_\infty^-(H) \equiv \overline{\mathcal{M}}_\infty(H)$. The opposite inclusion $\overline{\mathcal{M}}_\infty^\pm(H) \subseteq \mathcal{H}_c(H)$ has already been obtained in Proposition 5.11(d). Hence one has $\overline{\mathcal{M}}_\infty^-(H) = \overline{\mathcal{M}}_\infty^+(H) = \overline{\mathcal{M}}_\infty(H) = \mathcal{H}_c(H)$. Then, since $\mathcal{M}_0(H) \perp \overline{\mathcal{M}}_\infty^\pm(H)$ [see Proposition 5.11(c)], it follows that $\mathcal{M}_0(H) \subseteq \mathcal{H}_p(H)$. The opposite inclusion is contained in Proposition 5.11(d); thus $\mathcal{M}_0(H) = \mathcal{H}_p(H)$.

(b) The argument is similar to that used in (a). By Proposition 5.7(b) we have $\lim_{t\to\pm\infty} \|F_R U_t f\| = 0$ for each $f \in \mathcal{H}_{ac}(H)$ and each $R < \infty$. Hence $\mathcal{H}_{ac}(H) \subseteq \mathcal{M}_\infty^+(H) \cap \mathcal{M}_\infty^-(H) \equiv \mathcal{M}_\infty(H)$. Thus $\mathcal{H}_{ac}(H) \subseteq \mathcal{M}_\infty(H) \subseteq \overline{\mathcal{M}}_\infty(H) = \mathcal{H}_c(H)$. Under the assumption that $\mathcal{H}_{ac}(H) = \mathcal{H}_c(H)$ all these inclusions must be equalities, hence one has in particular that $\mathcal{H}_{ac}(H) = \mathcal{M}_\infty(H)$. □

DISCUSSION. We return to the particularly important situation introduced at the beginning of this section in which $\mathcal{H} = L^2(\mathbb{R}^n)$ and $F_R = \chi_R(\vec{Q})$, with $\chi_R(\vec{Q})$ given by (5.22). To indicate the different situations without distinguishing too many cases we set $n = 3$; very similar results hold for the other values of $n$ (the conditions on the potential depend on $n$, as seen for example in Proposition 2.49).

**(A)** We first give an example showing that the conclusions of Proposition 5.12 are not automatically true. We take $H = \vec{Q}^2$ in $L^2(\mathbb{R}^3)$. Then $\chi_R(\vec{Q})$ commutes with $H$ and hence with $U_t = e^{-iHt}$. Consequently $\|\chi_R(\vec{Q})U_t f\| = \|U_t \chi_R(\vec{Q}) f\| = \|\chi_R(\vec{Q}) f\|$, which is independent of the parameter $t$ for each $f \in \mathcal{H}$ (the states do not propagate at all). Similarly $\|[I - \chi_R(\vec{Q})]U_t f\| = \|[I - \chi_R(\vec{Q})] f\|$. So each $f$ is a bound state in the sense of (5.24): $\mathcal{H} = \mathcal{M}_0(H)$. Since the spectrum of $\vec{Q}^2$ is purely absolutely continuous, one has $\mathcal{H} = \mathcal{H}_{ac}(H) = \mathcal{M}_0(H)$.

Evidently the operator $H$ considered in this example does not correspond to a physical Hamiltonian. A physical Hamiltonian contains a kinetic part, which is usually a function of the momentum operator $\vec{P}$. If the kinetic energy becomes infinite as the momentum tends to infinity, then, as shown below, the conclusions of Proposition 5.12 are true.

**(B)** Next let $H = \varphi(\vec{P})$, where $\varphi\colon \mathbb{R}^3 \to \mathbb{R}$ is a continuous function assumed to satisfy $\lim_{|\vec{k}|\to\infty} |\varphi(\vec{k})| = \infty$. In momentum space $\tilde{L}^2(\mathbb{R}^3)$, the resolvent $(H-i)^{-1}$ is the multiplication operator by $1/[\varphi(\vec{k}) - i]$, and the function $1/[\varphi(\vec{k}) - i]$ belongs to $L^2(\mathbb{R}^3) + L_\varepsilon^\infty(\mathbb{R}^3)$ (by the same reasoning as that applied in Example 2.48). Let us write

$$\chi_R(\vec{Q})(H-i)^{-1} = \chi_R(\vec{Q})\chi_R(\vec{P})(H-i)^{-1} + \chi_R(\vec{Q})[\varphi(\vec{P}) - i]^{-1}[I - \chi_R(\vec{P})].$$

The compactness of $\chi_R(\vec{Q})(H-i)^{-1}$ follows since $\chi_R(\vec{Q})\chi_R(\vec{P})$ is a Hilbert-Schmidt operator (by Proposition 2.34) and $\chi_R(\vec{Q})[\varphi(\vec{P}) - i]^{-1}$ is compact [this can be seen by following the arguments in part (ii) of the proof of Proposition 2.49(a), replacing $V$ by $[\varphi(\vec{P}) - i]^{-1}$ and $(H_0 - i)^{-1}$ by $\chi_R(\vec{Q})$].

As an example, let us take $\varphi(\vec{k}) = \vec{k}^2$, so that $H = H_0 = \vec{P}^2$ is the free Hamiltonian (the kinetic energy operator) of non-relativistic quantum mechanics. Since the functions $\vec{k} \mapsto 1/(|\vec{k}|^2 - i)$ and $\chi_R$ belong to $L^2(\mathbb{R}^3)$ one obtains directly from Proposition 2.34(a) that $\chi_R(\vec{Q})(H_0 - i)^{-1}$ is a compact operator. So $\|\chi_R(\vec{Q})e^{-iH_0t}f\| \to 0$ for each vector $f \in L^2(\mathbb{R}^3)$, i.e. *all state vectors are scattering states (without time average) for the free Hamiltonian* $H_0$. Evidently the case of a general $\varphi$ covers various other situations. Also one can easily adjust the preceding arguments to show for example that, for the free Dirac Hamiltonian, all vectors represent scattering states.

**(C)** Let us now consider Hamiltonians involving a potential, i.e. self-adjoint operators of the form $H = H_0 + V(\vec{Q})$, where $V$ is a real function defined on $\mathbb{R}^3$. In general $H$ will be a self-adjoint extension of the differential operator $-\Delta + V(\vec{x})$ defined on $C_0^\infty(\mathbb{R}^3)$, but let us consider the simple case in which $V(\vec{Q})$ is $H_0$-bounded with $H_0$-bound $< 1$, so that $H$ is self-adjoint on the domain $\mathcal{D}(H_0)$. In this case we can write [see also the second resolvent equation (2.115)]:

$$\begin{aligned}\chi_R(\vec{Q})(H - i)^{-1} &= \chi_R(\vec{Q})(H_0 - i)^{-1} \cdot (H_0 - i)(H - i)^{-1} \\ &= \big[\chi_R(\vec{Q})(H_0 - i)^{-1}\big]\big[I - V(\vec{Q})(H - i)^{-1}\big]. \qquad (5.35)\end{aligned}$$

The operator $\chi_R(\vec{Q})(H_0 - i)^{-1}$ is compact as seen in (B), and $[I - V(\vec{Q})(H - i)^{-1}] \in \mathcal{B}(\mathcal{H})$. Hence $\chi_R(\vec{Q})(H - i)^{-1}$ is compact, and the conclusions of Proposition 5.12 are true. This covers a fair number of Schrödinger Hamiltonians in $L^2(\mathbb{R}^3)$, in particular the Coulomb Hamiltonian: it suffices to have $V \in L^2(\mathbb{R}^3) + L^\infty(\mathbb{R}^3)$, see Proposition 2.49(a).

**(D)** Without giving the proofs we mention some further results that concern the case in which $V(\vec{Q})$ is not $H_0$-bounded. The details can be found for example in §2.5 and §5.1 of [A] (for the question of absorbed states see also §13.1 of [P]). When $V(\vec{Q})$ is not $H_0$-bounded, the domain of $H$ will in general be different from that of $H_0$. First of all one has to define what is meant by $H$. We assume that there exists a closed set $\mathcal{Z}$ in $\mathbb{R}^3$ of Lebesgue measure zero such that the potential is only moderately singular away from $\mathcal{Z}$ (more precisely such that $V$ is square-integrable on each bounded closed subset of $\mathbb{R}^3 \setminus \mathcal{Z}$). It is interesting to know that the results that follow are true for each self-adjoint extension $H$ of the minimal operator $-\Delta + V(\vec{x})$ defined on $C_0^\infty(\mathbb{R}^3 \setminus \mathcal{Z})$! This is due to the fact that in regions of configuration space where the potential is only moderately singular (bounded or square-integrable), $\mathcal{D}(H)$ and $\mathcal{D}(H_0)$ coincide [which is a local generalisation of Proposition 2.49(a) stating that the domains of $H$ and of $H_0$ are identical if the potential belongs to $L^2(\mathbb{R}^3) + L^\infty(\mathbb{R}^3)$]. More precisely:

*Let $H$ be a self-adjoint extension of the minimal operator. Let $\mathcal{W}$ be an open subset of $\mathbb{R}^3 \setminus \mathcal{Z}$ on which the potential $V$ is only moderately singular, i.e. such that $V\chi_{\mathcal{W}} \in L^2(\mathbb{R}^3) + L^\infty(\mathbb{R}^3)$. Then, for each $\theta \in C_0^\infty(\mathbb{R}^3)$ having support in $\mathcal{W}$, one has: $f \in \mathcal{D}(H_0) \Longrightarrow \theta(\vec{Q})f \in \mathcal{D}(H)$, and $f \in \mathcal{D}(H) \Longrightarrow \theta(\vec{Q})f \in \mathcal{D}(H_0)$.*

As $\mathcal{D}(H) \neq \mathcal{D}(H_0)$ in general, the operator $(H_0 - i)(H - i)^{-1}$ will not be defined on all vectors of $\mathcal{H}$ but only on those $f \in \mathcal{H}$ for which $(H - i)^{-1}f \in \mathcal{D}(H_0)$ [remember that $(H-i)^{-1}$ maps $\mathcal{H}$ onto the domain of $H$]. Thus the operator $(H_0-i)(H-i)^{-1}$ will not belong to $\mathcal{B}(\mathcal{H})$ in general, and the argument given in (5.35) to show that $\chi_R(\vec{Q})(H-i)^{-1}\in\mathcal{K}(\mathcal{H})$ cannot be applied any more. However a generalisation of this argument can be obtained by using the fact that $\mathcal{D}(H)$ and $\mathcal{D}(H_0)$ coincide locally in regions where the potential is only moderately singular. We just consider two particular situations, that in which the potential is moderately singular everywhere except at infinity (for example a potential of the type of the harmonic oscillator) and that where the potential is strongly singular only at the origin $\vec{x}=0$. In the first instance we have $\mathcal{Z}=\varnothing$, in the second one $\mathcal{Z}=\{0\}$.

(1) Suppose that $V\in L^2_{\text{loc}}(\mathbb{R}^3)$, i.e. that $\int_{|\vec{x}|<\rho}|V(\vec{x})|^2d^3x<\infty$ for each $\rho<\infty$; the behaviour of the potential at infinity can be entirely arbitrary. For fixed $R>0$ choose a function $\theta\in C_0^\infty(\mathbb{R}^3)$ such that $\theta(\vec{x})=1$ if $|\vec{x}|\leq R$ and $\theta(\vec{x})=0$ for $|\vec{x}|>R+1$. Since $\chi_R(\vec{x})\theta(\vec{x})=\chi_R(\vec{x})$, one has $\chi_R(\vec{Q})=\chi_R(\vec{Q})\theta(\vec{Q})$ and hence

$$\begin{aligned}\chi_R(\vec{Q})(H-i)^{-1}&=\chi_R(\vec{Q})\theta(\vec{Q})(H-i)^{-1}\\&=\chi_R(\vec{Q})(H_0-i)^{-1}\cdot(H_0-i)\theta(\vec{Q})(H-i)^{-1}.\end{aligned}\tag{5.36}$$

Since $\theta(\vec{Q})(H-i)^{-1}$ maps $\mathcal{H}$ into $\mathcal{D}(H_0)$, the operator $(H_0-i)\theta(\vec{Q})(H-i)^{-1}$ is defined on each vector of $\mathcal{H}$. This operator belongs to $\mathcal{B}(\mathcal{H})$ by virtue of the Closed Graph Theorem (page 68). Thus one can deduce from (5.36) that $\chi_R(\vec{Q})(H-i)^{-1}$ is a compact operator for each $R<\infty$ and hence that the conclusions of Proposition 5.12 are true.

(2) Suppose now that $V\in L^2_{\text{loc}}(\mathbb{R}^3\backslash\{0\})$, in other terms that $\int_{\rho<|\vec{x}|<\kappa}|V(\vec{x})|^2d^3x<\infty$ for all $0<\rho<\kappa<\infty$. Thus we admit that the potential could have bad behaviour at the origin (for example $V(\vec{x})=\gamma|\vec{x}|^{-405}$ for some real $\gamma\neq 0$) and also at infinity. In this case the argument of (1) does not apply any more. Here the range of the operator $\theta(\vec{Q})(H-i)^{-1}$ belongs to $\mathcal{D}(H_0)$ if $\theta\in C_0^\infty(\mathbb{R}^3\backslash\{0\})$, i.e. if $\theta$ vanishes not only in a neighbourhood of infinity but also near the origin $\vec{x}=0$, whereas in (1) the function $\theta$ had to be equal to 1 in the ball $B_R$. However one can still obtain interesting results by modifying in (5.36) the projection $F_R\equiv\chi_R(\vec{Q})$. We replace in the preceding definition of $F_R$ the function $\chi_R$ by the characteristic function of the domain $\mathcal{O}(R):=\{\vec{x}\in\mathbb{R}^3\mid R^{-1}<|\vec{x}|<R\}$. Then one may take for $\theta$ a function in $C_0^\infty(\mathbb{R}^3\backslash\{0\})$ which is equal to 1 on this domain [and for example $\theta(\vec{x})=0$ for $|\vec{x}|>R+1$ as well as for $|\vec{x}|<1/(2R)$]. With $F_R=\chi_{\mathcal{O}(R)}(\vec{Q})$ and $\theta$ as indicated, one has as in (5.36):

$$F_R(H-i)^{-1}=F_R\theta(\vec{Q})(H-i)^{-1}=F_R(H_0-i)^{-1}\cdot(H_0-i)\theta(\vec{Q})(H-i)^{-1},$$

which implies that $F_R(H-i)^{-1}\in\mathcal{K}(\mathcal{H})$.

To interpret the result of (2), assume (in order to avoid having to consider temporal means) that the continuous spectrum of $H$ is purely absolutely continuous (i.e. that $\sigma_{\text{sc}}(H)=\varnothing$). Then one has $\|F_RU_tf\|\to 0$ as $t\to\pm\infty$ for each $f\in\mathcal{H}_{\text{c}}(H)\equiv\mathcal{H}_{\text{ac}}(H)$, see Proposition 5.7(b). In physical terms, remembering the definition of $F_R$ in (2), a state in the continuous subspace of $\mathcal{H}$ will disappear at large times from each annular region $\{\vec{x}\in\mathbb{R}^3\mid\varepsilon<|\vec{x}|<R\}$ with $\varepsilon>0$ and $R<\infty$. For such states there are two possibilities: they may propagate to infinity, as do scattering states, or they may move closer and closer to the singularity of the potential (here the origin). In the second case the state will be absorbed at the singularity. It is therefore appropriate to introduce, besides the bound states and the scattering states, the ***absorbed states*** $\mathcal{M}^{\pm}_{\rm abs}(H)$:

$$f \in \mathcal{M}^{\pm}_{\rm abs}(H) \iff \lim_{t\to\pm\infty} \|[I - \chi_R(\vec{Q})]U_t f\| = 0 \qquad \forall R > 0, \tag{5.37}$$

in other terms

$$f \in \mathcal{M}^{\pm}_{\rm abs}(H) \iff \lim_{t\to\pm\infty} \int_{|\vec{x}|\geq\varepsilon} \left|(U_t f)(\vec{x})\right|^2 d^3x = 0 \qquad \forall \varepsilon > 0.$$

It can be shown again that $\mathcal{M}^{\pm}_{\rm abs}(H)$ are subspaces and invariant under each $U_t$. Their intersection $\mathcal{M}^{+}_{\rm abs}(H) \cap \mathcal{M}^{-}_{abs}(H)$ is contained in $\mathcal{M}_0(H)$: if $f$ is a vector belonging to $\mathcal{M}^{+}_{\rm abs}(H) \cap \mathcal{M}^{-}_{\rm abs}(H)$, then it describes a state that is localised very close to the origin if $|t| > t_0$ for a certain $t_0$, and during a finite time interval (here the interval $[-t_0, t_0]$) each vector remains essentially confined to a finite part of configuration space because of the continuity of the evolution [see the argument given in part (ii) of the proof of Proposition 5.11(a)]. Thus the states in $\mathcal{M}^{+}_{\rm abs}(H) \cap \mathcal{M}^{-}_{\rm abs}(H)$ are bound states. In such a situation, if $\mathcal{M}^{+}_{\rm abs}(H) \cap \mathcal{M}^{-}_{\rm abs}(H) \neq \{\mathbf{0}\}$, the set of bound states is not equal to the subspace spanned by the eigenvectors of the Hamiltonian, *there exist vectors in the absolutely continuous subspace $\mathcal{H}_{\rm ac}(H)$ that represent bound states!*

Under the assumptions that $V \in L^2_{\rm loc}(\mathbb{R}^3\backslash\{0\})$ and $\sigma_{\rm sc}(H) = \varnothing$ one can show that [4]

$$\mathcal{H}_{\rm c}(H) \equiv \mathcal{H}_{\rm ac}(H) = \mathcal{M}^{+}_{\infty}(H) \oplus \mathcal{M}^{+}_{\rm abs}(H) = \mathcal{M}^{-}_{\infty}(H) \oplus \mathcal{M}^{-}_{\rm abs}(H). \tag{5.38}$$

So, if for example $f \in \mathcal{M}^{-}_{\infty}(H)$ is a scattering state at $t = -\infty$ ($f$ describes a situation in which a particle arrives on the scattering centre from a large distance), then one can decompose $f$ into a part $f^{+}_{\infty}$ describing the scattered part after the interaction has taken place and a part $f^{+}_{\rm abs}$ that will be absorbed as $t \to +\infty$ [$f = f^{+}_{\infty} + f^{+}_{\rm abs}$ with $f^{+}_{\infty} \in \mathcal{M}^{+}_{\infty}(H)$ and $f^{+}_{\rm abs} \in \mathcal{M}^{+}_{\rm abs}(H)$]. A potential for which this situation is realised has been constructed by D. Pearson; in this example the potential is highly oscillatory and unbounded above and below near the point $\vec{x} = 0$ (see e.g. Section 14.6 of [P] or Volume III of [RS], pp. 70-71). It can be shown that no absorption occurs if the potential has only positive singularities, e.g. if $V$ is bounded below in some neighbourhood of the set $\mathcal{Z}$ containing its singularities (we refer to [1] which also contains various other applications of Proposition 5.12).

Since an absorbed state is essentially localised (for sufficiently large $t$) in a very small neighbourhood of the origin, it should have a very large kinetic energy at large times (as a consequence of the uncertainty relations). Indeed it is easy to show that, for each $0 < M < \infty$, one has $\lim_{t\to\pm\infty} \|E_0([0, M])U_t f\| = 0$ if $f \in \mathcal{M}^{\pm}_{\rm abs}(H)$; here $\{E_0(\cdot)\}$ denotes the spectral measure of the free Hamiltonian $H_0 = \vec{P}^2$. Thus, for each $M < \infty$, the probability that the kinetic energy of the state described by $f$ remains less than $M$ tends to zero as $t \to \pm\infty$ if $f \in \mathcal{M}^{\pm}_{\rm abs}(H)$.

[4] A more detailed decomposition of the Hilbert space $\mathcal{H}$ in terms of bound states, scattering states and absorbed states for a class of short range potentials is described in the Asymptotic Decomposition Theorem in Chapter 13 of [P].

## 5.3 Asymptotic condition. Wave operators

**5.3.1.** We consider again the situation in which $H$ is the Hamiltonian of a simple non-relativistic scattering system in $\mathcal{H} = L^2(\mathbb{R}^n)$, i.e. $H = H_0 + V(\vec{Q})$, where $H_0 = \vec{P}^2$ is the kinetic energy operator and the potential $V$ is small at infinity in some sense [for example $V(\vec{x}) \to 0$ as $|\vec{x}| \to \infty$, so that at large distances from the scattering centre there is essentially no force on the system, the system will be essentially free]. Here $H_0$ is the ***free Hamiltonian*** and $H$ is called the ***total Hamiltonian***. If $f$ is a scattering state of $H$, for example if $f \in \mathcal{M}_\infty^+(H)$, then for very large positive $t$ the particle described by $f$ is localised far from the scattering centre, i.e. in a region where the potential has very little effect, it should therefore behave essentially like a free particle. A mathematical theory of scattering phenomena must express this by means of a formula. One speaks of the ***asymptotic condition***. In physical terms this condition means that the scattering states of the total Hamiltonian $H$ become free as time tends to infinity. There are various possibilities for formulating this condition in mathematical terms; we give two examples, setting $U_t = e^{-iHt}$.

(1) One could require that, for each $f \in \mathcal{M}_\infty^+(H)$ and sufficiently many free observables $A$, the expectation values $\langle U_t f, AU_t f\rangle$ have limits as $t \to +\infty$ (and similarly for negative times). By a free observable, one understands a self-adjoint operator that is constant under the free evolution, i.e. such that $e^{iH_0 t}Ae^{-iH_0 t} = A$ for all $t$ (thus an operator $A$ that commutes with the free Hamiltonian); important examples are the components $P_j$ of the momentum operator $\vec{P}$. The idea here is that during a scattering experiment one measures observables of this type, and the measured values should be independent of the moment at which the measurement is done (provided that it is made sufficiently far from the scattering centre, i.e. at sufficiently large times). The effect of the scattering will then be given by the relations between the values of these observables at $t = -\infty$ and their values at $t = +\infty$. This type of theory (called ***algebraic scattering theory***, since one considers a collection of free observables having the structure of a C*-algebra or of a von Neumann algebra) requires rather refined mathematical techniques.

(2) More simply (but less generally) one can require that each scattering state of $H$ behaves asymptotically like a free state, in other terms that, given a vector $f \in \mathcal{M}_\infty^+(H)$ for example, there should exist a vector $f_+ \in \mathcal{M}_\infty(H_0) \equiv \mathcal{H}$ such that, as $t \to +\infty$, $U_t f$ approaches the vector $e^{-iH_0 t}f_+$, i.e. $\|U_t f - e^{-iH_0 t}f_+\| \to 0$ as $t \to +\infty$ (and similarly for negative times). Since $\|U_t f - e^{-iH_0 t}f_+\| = \|e^{iH_0 t}[U_t f - e^{-iH_0 t}f_+]\| = \|e^{iH_0 t}U_t f - f_+\|$, the preceding condition means that $\text{s-lim}_{t\to+\infty} e^{iH_0 t}U_t f$ exists (and is equal to $f_+$). Thus, by setting $U_t^0 = e^{-iH_0 t}$, one could formulate the asymptotic condition by requiring that, for each $f \in \mathcal{M}_\infty^\pm(H)$, the strong limits of $U_t^{0*}U_t f$ as $t \to \pm\infty$ exist. Observe that one also has

$$\|U_t f - e^{-iH_0 t}f_+\| = \|U_t^*[U_t f - e^{-iH_0 t}f_+]\| = \|f - U_t^* e^{-iH_0 t}f_+\|, \tag{5.39}$$

hence $f = \text{s-lim}_{t\to+\infty} U_t^* U_t^0 f_+$.

In what follows we study the second formulation of the asymptotic condition (it covers potentials $V$ that tend to zero at infinity more rapidly than $|\vec{x}|^{-1}$; on the other hand this condition is not satisfied for the Coulomb potential). We first consider an abstract setting: given two evolution groups in a Hilbert space $\mathcal{H}$, we study properties of limits of the type of those considered above (see §5.3.2). Then, in Section 5.4, we apply the obtained results to the case of Schrödinger Hamiltonians and we discuss scattering theory for such Hamiltonians.

**5.3.2.** Let us consider two self-adjoint operators $A$ and $B$ in a Hilbert space $\mathcal{H}$ and denote by $\{U_t\}$ and $\{W_t\}$ the associated evolution groups: $U_t = e^{-iAt}$, $W_t = e^{-iBt}$. The asymptotic condition can be formulated by requiring that the strong limits of $U_t^* W_t$ (or of $W_t^* U_t$, which is equivalent from the mathematical point of view) exist as $t \to \pm\infty$. Thus we shall here make the assumption that these limits exist on a certain set of vectors and establish their mathematical properties. By taking into account Proposition 2.2 we can assume without loss of generality that the set of vectors for which these limits exist form a subspace of $\mathcal{H}$. This subspace may be different for the limit $t \to +\infty$ and for the limit $t \to -\infty$. In this subsection we formulate our results only for the limit $t \to +\infty$; it is clear that analogous statements can be made for $t \to -\infty$.

So let $\mathcal{M}$ be a subspace of $\mathcal{H}$. We assume that $\mathcal{M}$ is invariant under the group $\{W_t\}$, i.e. that $W_t f \equiv e^{-iBt} f \in \mathcal{M}$ if $f \in \mathcal{M}$, for all $t \in \mathbb{R}$. We denote by $E$ the projection onto $\mathcal{M}$ (i.e. $E = P_{\mathcal{M}}$). We recall that the letter $E$ is also used for the projections of a spectral family; one should not confound these notations. The use of the letter $E$ for the projection onto $\mathcal{M}$ is justified by the fact that in many applications $E$ is in fact a spectral projection of the operator $B$. We denote the spectral measure of the operator $A$ by $\{E^A(\cdot)\}$ and that of $B$ by $\{E^B(\cdot)\}$.

The condition that $\mathcal{M}$ be invariant under $W_t$ can be expressed in purely operatorial form, it is equivalent to

$$EW_t = W_t E \qquad \forall t \in \mathbb{R}. \tag{5.40}$$

Verification. (i) We first observe that $f \in \mathcal{M}$ if and only if $Ef = f$. Thus the condition $W_t f \in \mathcal{M}$ if $f \in \mathcal{M}$ can be written in operatorial terms as $EW_t E = W_t E$. If this is satisfied, one finds by taking the adjoint that $EW_{-t} E = EW_{-t}$. Upon replacing $t$ by $-t$ (which is justified because we assume $\mathcal{M}$ to be invariant under *all* $W_t$) this leads to $EW_t E = EW_t$. Consequently (by comparing with the identity $EW_t E = W_t E$ given before), one has $EW_t = W_t E$.

(ii) Assume that $EW_t = W_t E$ for each $t \in \mathbb{R}$. Then, if $f \in \mathcal{M}$: $EW_t f = W_t E f = W_t f$, which shows that $W_t f \in \mathcal{M}$. □

For the remainder of this section we make the hypothesis that the strong limit of $U_t^* W_t$ as $t \to +\infty$ exists on $\mathcal{M}$. We denote this limit by $\Omega(A, B; E)$. So

$$\Omega(A, B; E) = \underset{t \to +\infty}{\text{s-lim}}\, U_t^* W_t E \equiv \underset{t \to +\infty}{\text{s-lim}}\, e^{iAt} e^{-iBt} E. \tag{5.41}$$

The operator $\Omega(A, B; E)$ is called a ***wave operator*** (at $t = +\infty$). Later, it will be denoted by $\Omega_+(A, B; E)$, since there will also be a wave operator at $t = -\infty$.

We refer to §2.2.2 and §2.2.3 for the definition of the notions used in the following proposition.

**Proposition 5.13.** *(a) $\Omega \equiv \Omega(A,B;E)$ is a partial isometry with initial set $\mathcal{M}$ satisfying $\Omega E = \Omega$ and $\Omega^*\Omega = E$. If $E = I$, then $\Omega$ is an isometry.*
*(b) The wave operator $\Omega$ intertwines $A$ and $B$, i.e. one has the following relations:*

$$U_t\Omega = \Omega W_t \qquad \forall t \in \mathbb{R}, \tag{5.42}$$

$$E^A(\mathcal{V})\Omega = \Omega E^B(\mathcal{V}) \qquad \forall \mathcal{V} \in \mathcal{A}_{\mathrm{B}}, \tag{5.43}$$

$$\varphi(A)\Omega = \Omega\varphi(B) \quad \text{if } \varphi\colon \mathbb{R} \to \mathbb{C} \text{ is continuous and bounded.} \tag{5.44}$$

*Moreover, if $f \in \mathcal{D}(B)$, then $\Omega f \in \mathcal{D}(A)$ and $A\Omega f = \Omega B f$.*

PROOF. (a) If $f \perp \mathcal{M}$ one has $\Omega f = \mathbf{0}$; if $f \in \mathcal{M}$, then $\|\Omega f\| = \lim_{t\to\infty} \|U_t^* W_t f\| = \lim_{t\to\infty} \|f\| = \|f\|$ [using Proposition 1.1(a)]. By Proposition 2.9(e) [or Proposition 2.7(f) if $E = I$], $\Omega$ is a partial isometry with initial set $\mathcal{M}$ (and an isometry if $E = I$). The fact that $\Omega E = \Omega$ is obvious in view of the definition (5.41), and $\Omega^*\Omega = E$ is contained in Proposition 2.9(e).

(b) One has

$$U_t\Omega = U_t\big[\operatorname*{s-lim}_{s\to+\infty} U_s^* W_s E\big] = \operatorname*{s-lim}_{s\to+\infty} U_t U_s^* W_s E = \operatorname*{s-lim}_{s\to+\infty} U_{s-t}^* W_s E.$$

By setting $\tau = s - t$ and using the hypothesis that $W_t$ commutes with $E$ this leads to

$$U_t\Omega = \operatorname*{s-lim}_{\tau\to+\infty} U_\tau^* W_{\tau+t} E = \operatorname*{s-lim}_{\tau\to+\infty} U_\tau^* W_\tau W_t E = \operatorname*{s-lim}_{\tau\to+\infty} U_\tau^* W_\tau E W_t = \Omega W_t.$$

This proves (5.42). Upon multiplying (5.42) by $ie^{izt}$ and then integrating with respect to $dt$ over $[0,\infty)$ or over $(-\infty,0]$ (if $z \in \mathbb{C}$ is such that $\Im z > 0$ or $\Im z < 0$), one obtains by taking into account (5.7) and (5.8):

$$(A - z)^{-1}\Omega = \Omega(B - z)^{-1} \qquad \forall z \in \mathbb{C} \text{ with } \Im z \neq 0. \tag{5.45}$$

Consequently, for $u, v \in \mathbb{R}$:

$$\int_u^v \big[(A-\lambda-i\varepsilon)^{-1}-(A-\lambda+i\varepsilon)^{-1}\big]\Omega d\lambda = \Omega\int_u^v \big[(B-\lambda-i\varepsilon)^{-1}-(B-\lambda+i\varepsilon)^{-1}\big]d\lambda.$$

By setting $u = a + \delta$, $v = b + \delta$ and then taking the limits $\varepsilon \to +0$ and $\delta \to +0$, one obtains (5.43) for $\mathcal{V} = (a, b]$ by taking into account Stone's Formula (4.81). By taking the limit $a \to -\infty$ one finds that $E_b^A\Omega = \Omega E_b^B$ for each $b \in \mathbb{R}$, which implies (5.43) for the spectral measures. Eq. (5.44) is a consequence of (5.43) and the definition of functions of an operator given in Section 4.2.

Finally, if $f \in \mathcal{D}(B)$, then $\Omega W_t f$ is strongly differentiable, its derivative at $t = 0$ is $-i\Omega B f$. Now $\Omega W_t f = U_t\Omega f$. Hence $U_t\Omega f$ is strongly differentiable; by Stone's Theorem (Proposition 5.1), $\Omega f$ belongs to the domain of $A$ and one has $A\Omega f = i(d/dt)U_t\Omega f|_{t=0} = i(d/dt)\Omega W_t f|_{t=0} = i[-i\Omega B f] = \Omega B f$. □

**Remark 5.14.** The argument used to obtain (5.43) from (5.42) (passage from the groups to the resolvents and then from the resolvents to the spectral projections) can also be applied to obtain the following result (one has twice the same evolution group and, in place of an intertwining operator $\Omega$, a bounded operator commuting with this group):

*Let $\{U_t\}$ be an evolution group and $\{E(\cdot)\}$ the spectral family of its infinitesimal generator. Let $C$ be an operator in $\mathcal{B}(\mathcal{H})$ such that $U_t C = C U_t$ for each $t \in \mathbb{R}$. Then $CE(\mathcal{V}) = E(\mathcal{V})C$ for each $\mathcal{V} \in \mathcal{A}_{\mathrm{B}}$.*

**Proposition 5.15.** *(a) Let $\mathcal{N} = \mathcal{R}(\Omega(A,B;E))$ be the final set of $\Omega \equiv \Omega(A,B;E)$. The projection $F \equiv P_{\mathcal{N}}$ onto $\mathcal{N}$ is given by*

$$F = \Omega\Omega^*. \tag{5.46}$$

*The limit*

$$\operatorname*{s-lim}_{t\to+\infty} W_t^* U_t F \equiv \operatorname*{s-lim}_{t\to+\infty} e^{iBt} e^{-iAt} F \equiv \Omega(B,A;F) \tag{5.47}$$

*exists, and one has*

$$\Omega(B,A;F) = [\Omega(A,B;E)]^*. \tag{5.48}$$

*(b) The subspace $\mathcal{N}$ is invariant under the group $\{U_t\}$:*

$$U_t F = F U_t \qquad \forall t \in \mathbb{R}. \tag{5.49}$$

*(c) If $\varphi\colon \mathbb{R} \to \mathbb{C}$ is continuous and bounded, then*

$$\varphi(A)F = F\varphi(A). \tag{5.50}$$

*(d) One has the following relations:*

$$\mathcal{M} \subseteq \mathcal{H}_{\mathrm{ac}}(B) \Longrightarrow \mathcal{N} \subseteq \mathcal{H}_{\mathrm{ac}}(A), \qquad \mathcal{M} \subseteq \mathcal{H}_{\mathrm{c}}(B) \Longrightarrow \mathcal{N} \subseteq \mathcal{H}_{\mathrm{c}}(A),$$
$$\mathcal{M} \subseteq \mathcal{M}_\infty^{\pm}(B) \Longrightarrow \mathcal{N} \subseteq \mathcal{M}_\infty^{\pm}(A), \quad \mathcal{M} \subseteq \overline{\mathcal{M}_\infty^{\pm}}(B) \Longrightarrow \mathcal{N} \subseteq \overline{\mathcal{M}_\infty^{\pm}}(A).$$

PROOF. (a) We refer to Propositions 2.7(c) and 2.9(d) for the fact that the projection onto $\mathcal{R}(\Omega)$ is $F = \Omega\Omega^*$. If $g \in \mathcal{R}(\Omega)$, there exists $f \in \mathcal{M}$ such that $g = \Omega f$. Then

$$\|W_t^* U_t g - f\| = \|U_t^* W_t (W_t^* U_t g - f)\| = \|g - U_t^* W_t f\| = \|\Omega f - U_t^* W_t f\|, \tag{5.51}$$

which converges to zero as $t \to +\infty$. So the limit $\operatorname{s-lim}_{t\to+\infty} W_t^* U_t g$ [hence the operator $\Omega(B,A;F)$] exists. Eq. (5.51) also shows that $\operatorname{s-lim}_{t\to+\infty} W_t^* U_t \Omega = E$, so that

$$\Omega(B,A;F) = \operatorname*{s-lim}_{t\to+\infty} W_t^* U_t F = \operatorname*{s-lim}_{t\to+\infty} W_t^* U_t \Omega\Omega^* = E\Omega^* = (\Omega E)^* = \Omega^*.$$

(b) Upon taking the adjoint of (5.42) and replacing in it $t$ by $-t$ (one has for example $U_t^* = U_{-t}$) one finds

$$\Omega^* U_t = W_t \Omega^* \qquad \forall t \in \mathbb{R}. \tag{5.52}$$

By using first (5.42) and then (5.52) one finds that $U_t F = U_t \Omega\Omega^* = \Omega W_t \Omega^* = \Omega\Omega^* U_t = F U_t$.

(c) The argument is similar to that in (b). By (5.44) we have $\Omega^*\varphi(A) = \varphi(B)\Omega^*$ if $\varphi$ is continuous and bounded [take the adjoint of (5.44) and take into account (4.50)]. Then $\varphi(A)F = \varphi(A)\Omega\Omega^* = \Omega\varphi(B)\Omega^* = \Omega\Omega^*\varphi(A) = F\varphi(A)$.

(d) Let $g \in \mathcal{N}$. There exists $f \in \mathcal{M}$ such that $g = \Omega f$. Then, by using (5.43) in the second step:

$$\langle g, E_\lambda^A g\rangle = \langle \Omega f, E_\lambda^A \Omega f\rangle = \langle \Omega f, \Omega E_\lambda^B f\rangle = \langle \Omega^*\Omega f, E_\lambda^B f\rangle = \langle f, E_\lambda^B f\rangle.$$

Thus, if $f \in \mathcal{H}_{\mathrm{ac}}(B)$, i.e. if the measure determined by $\langle f, E_\lambda^B f\rangle$ is absolutely continuous, then $g \in \mathcal{H}_{\mathrm{ac}}(A)$, i.e. the measure determined by $\langle g, E_\lambda^A g\rangle$ is also absolutely continuous (because these two measures are identical). This gives the first implication in (d). The second one is obtained in the same way. For the third one, assume that $f \in \mathcal{M} \cap \mathcal{M}_\infty^+(B)$. Observe that

$$\begin{aligned}\|F_R U_t g\| &= \|F_R[U_t g - W_t f] + F_R W_t f\| \leq \|F_R[U_t g - W_t f]\| + \|F_R W_t f\| \\ &\leq \|U_t g - W_t f\| + \|F_R W_t f\| = \|g - U_t^* W_t f\| + \|F_R W_t f\|. \end{aligned} \tag{5.53}$$

Now $\|g - U_t^* W_t f\|$ converges to zero as $t \to +\infty$; since $f \in \mathcal{M}_\infty^+(B)$, $\|F_R W_t f\|$ also converges to zero, and then the inequality (5.53) implies that $\|F_R U_t g\|$ also converges to zero as $t \to +\infty$, i.e. that $g \in \mathcal{M}_\infty^+(A)$. □

The existence of the limits defining the wave operators is sometimes easy to obtain by using the result of the following proposition (see §5.4.3 for examples and Remark 5.22 for comments).

**Proposition 5.16 (Cook Criterion).** *Assume that there exists a subset $\mathcal{D}$ of the subspace $\mathcal{M}$ having the following properties:*

*(1) $\mathcal{D}$ is total in $\mathcal{M}$,*

*(2) $W_t f \in \mathcal{D}(A) \cap \mathcal{D}(B)$ for each $f \in \mathcal{D}$ and each $t \in \mathbb{R}$,*

*(3) $\int_1^\infty \|(A - B)W_t f\|\, dt < \infty$ for each $f \in \mathcal{D}$.*

*Then $\Omega(A, B; E)$ exists (where $E = P_{\mathcal{M}}$).*

PROOF. (i) Let $f \in \mathcal{D}$. Then, by using the hypothesis (2) and the Leibniz rule:

$$\begin{aligned}\frac{d}{dt} U_t^* W_t f &= [\frac{d}{dt} U_t^*] W_t f + U_t^* [\frac{d}{dt} W_t f] \\ &= U_t^* i A W_t f + U_t^*(-iBW_t f) = iU_t^*(A - B)W_t f.\end{aligned}$$

By virtue of Proposition 1.5(c) one then has (assuming $t > s > 1$)

$$U_t^* W_t f - U_s^* W_s f = \int_s^t \frac{d}{d\tau} U_\tau^* W_\tau f\, d\tau = i \int_s^t U_\tau^*(A - B) W_\tau f\, d\tau. \tag{5.54}$$

Thus, by taking into account (1.28):

$$\begin{aligned}\|U_t^* W_t f - U_s^* W_s f\| &= \Big\| \int_s^t U_\tau^*(A - B) W_\tau f\, d\tau \Big\| \\ &\leq \int_s^t \|U_\tau^*(A - B) W_\tau f\|\, d\tau = \int_s^t \|(A - B) W_\tau f\|\, d\tau.\end{aligned}$$

Combined with the hypothesis (3) this implies that $\{U_t^* W_t f\}$ is strongly Cauchy as $t \to \infty$ (the right-hand side of the preceding inequality is arbitrarily small provided that $s$ and $t$ are sufficiently large).

(ii) We have shown that $\{U_t^* W_t f\}$ is strongly Cauchy for all $f$ in a subset of $\mathcal{M}$ which, by the hypothesis (1), is total in $\mathcal{M}$. As $\|U_t^* W_t\| = 1$ for each $t$, we can conclude that $\{U_t^* W_t f\}$ converges for each $f \in \mathcal{M}$ by applying Proposition 2.2. □

## 5.4 Simple scattering systems. Scattering operator

We discuss here the simplest situation in which, in physical terms, there is only elastic scattering (for example the scattering of a particle by a potential converging to zero at infinity, or equivalently the scattering between two particles described in their centre-of-mass frame). The description of scattering systems with inelastic scattering is similar but mathematically more complex; in this case there are several scattering channels (sometimes an infinite number of them) and consequently a family of wave operators (for each sign of time), in fact, for each sign of time, one wave operator for each channel. A simple example of a two-channel system is given by a particle on the real line subject to an infinitely long potential barrier, i.e. to a potential $V(x)$ which tends to different limits as $x \to -\infty$ and as $x \to +\infty$. Here reflection at the barrier is an elastic process whereas transmission is inelastic (non-conservation of the kinetic energy). Another simple example is that of scattering of a deuteron by a fixed centre of force. A deuteron can either be scattered elastically or it can be broken up into its constituent parts (a proton and a neutron) through the interaction, so again there are two scattering channels. More generally, in atomic collisions one can have creation of various new systems or particles, for example $\alpha$ particles.

**5.4.1.** We consider again the situation described in §5.3.1: $\mathcal{H} = L^2(\mathbb{R}^n)$, $B = H_0 = \vec{P}^2$ and $A = H = H_0 + V(\vec{Q})$, where $V\colon \mathbb{R}^n \to \mathbb{R}$ is a potential assumed to be "small" at infinity. We write $U_t = e^{-iHt}$ for the evolution group associated with the total Hamiltonian $H$. This group is often called the ***total evolution group***. We use the notation $U_t^0$ for the ***free evolution group***, i.e. for the evolution group determined by the free Hamiltonian $H_0$: $U_t^0 = e^{-iH_0 t}$.

Let us first consider the physical situation before the interaction takes place, i.e. the asymptotic condition at $t = -\infty$. A free particle[5] is directed towards the scattering centre. When the particle approaches the scattering centre the interaction will begin to become effective. Thus initially (as $t$ is very close to $-\infty$) the evolution of the state of the particle will be described (at least to a very good approximation) by the free evolution group $\{U_t^0\}$, but at later times by the total evolution group $\{U_t\}$. Let us first describe the free initial state. It is determined at all times if it is given at one time $t_0$. The preceding description suggests taking $t_0 = -\infty$, but this does not make

[5] In practice it is not a single particle but rather a beam of particles. We describe here the scattering of a single particle, and in Section 5.6 we treat the more realistic case of an initial beam so as to arrive at expressions for the scattering cross section.

sense mathematically. *One therefore adopts the convention that the state vectors are specified at time* $t = 0$. Thus, if a vector $f_- \in \mathcal{H}$ is taken as an initial state[6], it is understood that the prepared state at large negative time $t$ is $U_t^0 f_-$. So the particle will be (to a very good approximation) in the state $U_t^0 f_-$ if $t$ is close to $-\infty$, but in principle it will never be, in the course of its evolution, exactly in the state $U_t^0 f_-$ (because its evolution is not governed by the group $\{U_t^0\}$ but rather by the group $\{U_t\}$). To describe the real evolution of the particle one must find a vector $g$ which, under the total evolution $U_t$, approaches $U_t^0 f_-$ as $t \to -\infty$. So $g$ describes the real state (i.e. taking into account the interaction with the scatterer) of the particle at time $t = 0$ under the assumption that the initial state (taken at time $t = 0$ according to the above convention) is $f_-$. In mathematical terms, given $f_-$, the vector $g$ must satisfy the condition $\lim_{t\to-\infty} \|U_t g - U_t^0 f_-\| = 0$.

We see that $g$ is determined from $f_-$ by the relation $g = \Omega_-(H,H_0)f_-$, where $\Omega_-(H,H_0) = \text{s-lim}_{t\to-\infty} U_t^* U_t^0$ is the wave operator at $t = -\infty$ associated with the couple $\{H, H_0\}$. We have set $E = I$ which, in the considered physical situation, is reasonable because each vector in $L^2(\mathbb{R}^n)$ represents in principle a possible initial state. In other situations it would be natural to take for $E$ the projection onto the subspace $\mathcal{M}_\infty^-(H_0)$ formed by the scattering states of the free Hamiltonian $H_0$ at $t = -\infty$; in the present case, where $H_0 = \vec{P}^2$, we know that $\mathcal{M}_\infty^-(H_0) = \mathcal{H}$.

Let us consider the situation after the interaction has taken place (in a reasonable case in which there is no absorption), i.e. for $t \to +\infty$. The state leaving the scattering region will be given by $U_t g$, with $t \to +\infty$. Again it will behave to a very good approximation like a freely evolving state: there should exist a vector $f_+$ such that $U_t g$ approaches the free evolution $U_t^0 f_+$ of $f_+$ as $t \to +\infty$. $f_+$ is interpreted as the final state at time $t = 0$ (given the initial state $f_-$). Thus $f_+$ must obey the condition $\lim_{t\to+\infty} \|U_t g - U_t^0 f_+\| = 0$, i.e. $f_+ = \Omega_+(H_0, H; E_\infty^-(H))g$. Here we have inserted the projection $E_\infty^-(H)$ onto the subspace $\mathcal{M}_\infty^-(H)$ in the definition of the wave operator, because $g$ belongs to $\mathcal{M}_\infty^-(H)$ by Proposition 5.15(d) (for negative times).

In the preceding description of the scattering of a particle, the effect of the interaction is contained in the association $f_- \mapsto f_+$: with each possible initial state $f_-$ (here with each vector $f_- \in \mathcal{H}$) there is associated another vector $f_+$ interpreted as the final state corresponding to $f_-$, all states being given at time $t = 0$. In principle, if the correspondence $f_- \mapsto f_+$ is known, one must be able to calculate all physical observations that can be made after the scattering has taken place. The correspondence $f_- \mapsto f_+$ defines a linear operator called the ***scattering operator*** $S$ (sometimes also the $S$-matrix in the physics literature). Thus one writes $f_+ = Sf_-$, and the operator $S$ should determine the physical quantities measured after the scattering if the initial conditions are known.

The preceding description makes sense provided that the wave operators $\Omega_-(H,H_0)$ and $\Omega_+(H_0,H;E_\infty^-(H))$ exist. It is usual to apply a more symmetrical treatment to the sign of time and to define

$$\Omega_- \equiv \Omega_-(H,H_0;I) = \underset{t\to-\infty}{\text{s-lim}}\, U_t^* U_t^0 \tag{5.55}$$

[6] In the physics literature the state $f_-$ is frequently called $f_{\text{in}}$ or $\psi_{\text{in}}$.

and

$$\Omega_+ \equiv \Omega_+(H, H_0; I) = \underset{t\to+\infty}{\text{s-lim}}\, U_t^* U_t^0. \tag{5.56}$$

Then, in the above description, the relation between $g$ and $f_+$ can be written as $g = \Omega_+ f_+$ or $f_+ = \Omega_+^* g$ [see Proposition 2.7(e)]. Thus the scattering operator $S$ is given as follows: $Sf_- = f_+ = \Omega_+^* g = \Omega_+^* \Omega_- f_-$. Hence

$$S = \Omega_+^* \Omega_-. \tag{5.57}$$

This reformulation makes sense provided that one knows that the vector $g = \Omega_- f_-$ belongs to the final set of $\Omega_+$ [i.e. that $g \in \mathcal{R}(\Omega_+)$], otherwise the relation $g = \Omega_+ f_+$ has no sense. So one must know that $\mathcal{R}(\Omega_-) \subseteq \mathcal{R}(\Omega_+)$. On the other hand, if this condition is not satisfied (when one has to use the first formulation to clearly depict the physical situation), one can still define an operator $S$ by the formula (5.57) because $\Omega_+^*$ and $\Omega_-$ are well defined operators in $\mathcal{B}(\mathcal{H})$ (in fact of norm 1), assuming of course that the limits in (5.55) and (5.56) exist.

Figure 5.1 symbolically represents the situation. In this representation the points in the plane correspond to vectors of the Hilbert space, a straight line to the free evolution of a state vector and a curved line to the total evolution of a vector.

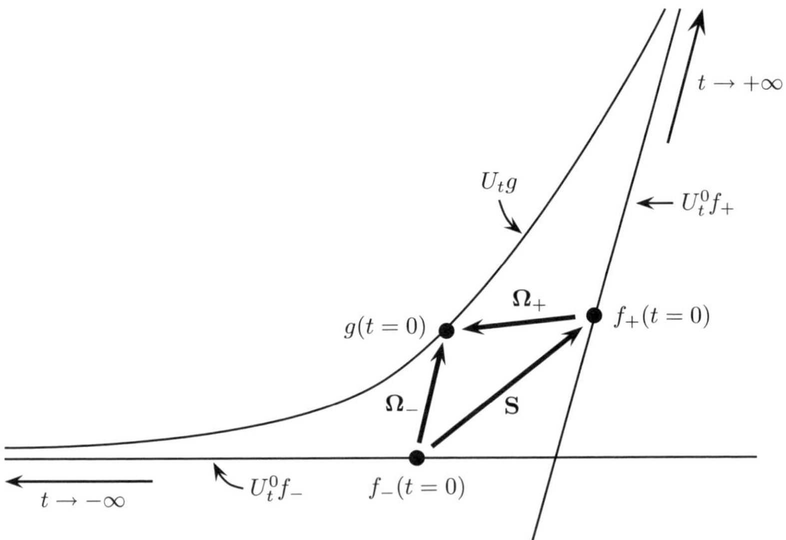


**Figure 5.1**

**5.4.2.** Before establishing the existence of the wave operators for a certain class of potentials $V$ and discussing the properties of the scattering operator $S$, we give some useful results concerning the free evolution group. These explicit characterisations of $U_t^0$ are possible because the operator $U_t^0 \equiv e^{-iH_0 t}$ assumes a simple form in momentum space, namely $U_t^0$ is the multiplication operator by $e^{-i\vec{k}^2 t}$:

$$[\mathcal{F} U_t^0 f](\vec{k}) = e^{-i\vec{k}^2 t} \tilde{f}(\vec{k}). \tag{5.58}$$

Indeed, for $f$ belonging to the dense subset $\widetilde{C}_0^\infty(\mathbb{R}^n)$ of infinitely differentiable functions of compact support in momentum space [i.e. $\tilde{f}(\vec{k}) = 0$ for $\vec{k}$ outside some bounded subset of $\mathbb{R}^n$], $e^{-iH_0t}f$ is given as follows (see Examples 4.12 and 4.17):

$$[\mathcal{F}e^{-iH_0t}f](\vec{k}) = \sum_{j=0}^{\infty} \frac{(-it)^j}{j!}[\mathcal{F}(\vec{P}^2)^j f](\vec{k}) = \sum_{j=0}^{\infty} \frac{(-it)^j}{j!}(\vec{k}^2)^j \tilde{f}(\vec{k}) = e^{-i\vec{k}^2 t}\tilde{f}(\vec{k}).$$

Eq. (5.58) expresses the free evolution in the momentum representation. By an inverse Fourier transformation one can obtain a formula for $U_t^0$ in configuration space:

**Proposition 5.17.** *In $L^2(\mathbb{R}^n)$, $U_t^0$ is an integral operator given by*

$$(U_t^0 f)(\vec{x}) = \frac{1}{(4\pi it)^{n/2}} \int_{\mathbb{R}^n} \exp\Big\{\frac{i|\vec{x}-\vec{y}|^2}{4t}\Big\} f(\vec{y})\, d^n y. \tag{5.59}$$

*The square root in* (5.59) *is defined such that*

$$\Big(\frac{4\pi it}{|4\pi it|}\Big)^{-n/2} = \begin{cases} \exp[-in\pi/4] & \text{if } t > 0 \\ \exp[+in\pi/4] & \text{if } t < 0. \end{cases}$$

REMARK. The integral in (5.59) has a meaning if $f \in L^1(\mathbb{R}^n)$. So $(U_t^0 f)(\vec{x})$ is given by (5.59) if $f \in L^2(\mathbb{R}^n) \cap L^1(\mathbb{R}^n)$. For vectors $f \in L^2(\mathbb{R}^n)$ not belonging to $L^1(\mathbb{R}^n)$, this integral must be interpreted as a limit in the mean, as in §2.5.2.

PROOF OF PROPOSITION 5.17. In order to simplify the notations we consider the case $n = 1$.

(i) One first establishes (5.59) for particular functions $f$, namely for Gaussian functions $f_a(x) = \exp[-(x-a)^2]$, $a \in \mathbb{R}$, by using the following explicit expression for Gaussian integrals: If $\rho > 0$, $\gamma \in \mathbb{R}$ and $\xi \in \mathbb{R}$, then

$$\frac{1}{\sqrt{2\pi}} \int_{-\infty}^{\infty} e^{iz\xi} e^{-i\gamma z^2 - \rho z^2}\, dz = \frac{1}{\sqrt{2(\rho + i\gamma)}} \exp\Big[-\frac{\xi^2}{4(\rho+i\gamma)}\Big], \tag{5.60}$$

where the determination of the square root is such that $\Re\sqrt{\rho + i\gamma} > 0$. In particular the Fourier transform of $f_a$ is $\widetilde{f_a}(k) = 2^{-1/2} e^{-ika} e^{-k^2/4}$, and

$$\begin{aligned}(U_t^0 f_a)(x) &= \frac{1}{\sqrt{2\pi}} \int_{-\infty}^{\infty} e^{ixk} e^{-itk^2} \widetilde{f_a}(k)\, dk = \frac{1}{\sqrt{4\pi}} \int_{-\infty}^{\infty} e^{i(x-a)k} e^{-itk^2} e^{-k^2/4} dk \\ &= \frac{1}{\sqrt{1+4it}} \exp\Big[-\frac{(x-a)^2}{1+4it}\Big].\end{aligned}$$

The same expression is obtained by using (5.60) to calculate the right-hand side of (5.59) for $f = f_a$:

$$\begin{aligned}\frac{1}{(4\pi it)^{1/2}} \int_{-\infty}^{\infty} \exp\Big\{\frac{i|x-y|^2}{4t}\Big\} f_a(y)\, dy &= \frac{1}{(2it)^{1/2}} \exp\Big[\frac{i(x-a)^2}{4t}\Big] \\ &\times \frac{1}{\sqrt{2\pi}} \int_{-\infty}^{\infty} \exp\Big[\frac{-i(y-a)(x-a)}{2t}\Big] \exp\Big[-(y-a)^2(1 - i/4t)\Big] dy \\ &= \frac{1}{\sqrt{1+4it}} \exp\Big[-\frac{(x-a)^2}{1+4it}\Big].\end{aligned}$$

(ii) The set of functions $\mathcal{N} := \{f_a\}_{a\in\mathbb{R}}$ is total in $L^2(\mathbb{R})$. Indeed, suppose that $g \perp \mathcal{N}$, i.e. that $\langle \widetilde{f}_a, \tilde{g}\rangle = 0$ for each $a \in \mathbb{R}$, so that $\int_{-\infty}^{\infty} e^{ika}e^{-k^2/4}\tilde{g}(k)dk = 0$ for each $a \in \mathbb{R}$. This means that the inverse Fourier transform of the function $\exp[-k^2/4]\tilde{g}(k)$ at the point $a$ is zero, and this for each $a \in \mathbb{R}$. Thus the inverse Fourier transform of this function is the zero vector $\mathbf{0}$, and by the unitarity of the Fourier transformation $\exp[-k^2/4]\tilde{g}(k) = 0$ almost everywhere, hence $\tilde{g}(k) = 0$ a.e., i.e. $g = \mathbf{0}$. By Proposition 1.6 the set $\mathcal{N}$ is total in $L^2(\mathbb{R})$.

(iii) If $f$ is an arbitrary vector in $L^2(\mathbb{R}) \cap L^1(\mathbb{R})$, it can be approximated by a sequence $\{g_j\}_{j\in\mathbb{N}}$ belonging to the dense linear manifold $\mathcal{D}$ spanned by $\mathcal{N}$. $(U_t^0 g_j)(x)$ is given by (5.59), and one obtains the validity of (5.59) for $f$ by a limiting argument. We omit the details because the result is not essential in what follows (it will suffice to know (5.59) for a dense set of vectors $f$, for example for the set $\mathcal{D}$ used here). □

**Proposition 5.18.** *Let $\mathcal{H} = L^2(\mathbb{R}^n)$. For $t \neq 0$ let $Z_t$ be the following unitary operator:*

$$Z_t = e^{i\vec{Q}^2/(4t)}, \qquad or \qquad (Z_t f)(x) = \exp\left[\frac{i\vec{x}^2}{4t}\right] f(x). \tag{5.61}$$

*Then one has* $\text{s-}\lim_{t\to\pm\infty} Z_t = I$ *and, if $\varphi \in L^\infty(\mathbb{R}^n)$:*

$$U_t^{0*}\varphi(\vec{Q})U_t^0 = Z_t^*\varphi(2t\vec{P})Z_t, \tag{5.62}$$

*where $\varphi(\vec{Q})$ and $\varphi(2t\vec{P})$ denote the multiplication operators by $\varphi(\vec{x})$ and $\varphi(2t\vec{k})$ in $L^2(\mathbb{R}^n)$ and $\tilde{L}^2(\mathbb{R}^n)$ respectively.*

PROOF. One has

$$\|[Z_t - I]f\|^2 = \int_{\mathbb{R}^n} \left| \exp\left[\frac{i\vec{x}^2}{4t}\right] - 1\right|^2 |f(\vec{x})|^2 d^n x.$$

By applying the Dominated Convergence Theorem (Proposition 1.9) one finds that this expression converges to zero as $t \to \pm\infty$. Next, since all operators in (5.62) belong to $\mathcal{B}(\mathcal{H})$, it suffices to obtain this equation on a total set of vectors $f$. We take $f \in \mathcal{N}^n$, where $\mathcal{N}^n$ is the $n$-dimensional analogue of the set $\mathcal{N}$ introduced in the preceding proof: $\mathcal{N}^n := \{f_{\vec{a}}\}_{\vec{a}\in\mathbb{R}^n}$, where $f_{\vec{a}}(\vec{x}) = \exp[-(\vec{x} - \vec{a})^2]$. So it suffices to show that $\langle g, U_t^{0*}\varphi(\vec{Q})U_t^0 f\rangle = \langle g, Z_t^*\varphi(2t\vec{P})Z_t f\rangle$ for all $f, g \in \mathcal{N}^n$. This is easily obtained by using (5.59) and making the change of variables $\vec{z} \mapsto \vec{k} = \vec{z}/2t$:

$$\begin{aligned}
&\langle U_t^0 g, \varphi(\vec{Q})U_t^0 f\rangle \\
&= \frac{1}{(4\pi|t|)^n}\int_{\mathbb{R}^n} d^n z\, \varphi(\vec{z}) \int_{\mathbb{R}^n} d^n x \exp\left[\frac{-i|\vec{z}-\vec{x}|^2}{4t}\right]\overline{g(\vec{x})} \int_{\mathbb{R}^n} d^n y \exp\left[\frac{i|\vec{z}-\vec{y}|^2}{4t}\right] f(\vec{y}) \\
&= \frac{1}{(2\pi)^n}\int_{\mathbb{R}^n} d^n k\, \varphi(2t\vec{k}) \int_{\mathbb{R}^n} d^n x e^{i\vec{k}\cdot\vec{x}} \exp\left[\frac{-i\vec{x}^2}{4t}\right]\overline{g(\vec{x})} \int_{\mathbb{R}^n} d^n y e^{-i\vec{k}\cdot\vec{y}} \exp\left[\frac{i\vec{y}^2}{4t}\right] f(\vec{y}).
\end{aligned}$$

Clearly the last expression is equal to $\langle Z_t g, \varphi(2t\vec{P})Z_t f\rangle$. □

**Remark 5.19.** Since $Z_t \approx I$ at large values of $|t|$, Eq. (5.62) shows that, under the free evolution and in the Heisenberg picture, the operator $\varphi(\vec{Q})$ is essentially the same as $\varphi(2t\vec{P})$ at large $|t|$, hence $\vec{Q}(t) \approx 2t\vec{P}$. This is a quantum-mechanical analogue of the fact that $\vec{x}(t) = x_0 + \vec{p}t/m$ for a freely moving classical particle (note that, with our choice $H_0 = \vec{P}^2$, one has $2m = 1$, hence $m = 1/2$ for the mass).

**5.4.3.** Let us consider some applications of Eqs. (5.59) and (5.62).
**(A)** Scattering states of $H_0$. The formula (5.59) implies that

$$|(U_t^0 f)(\vec{x})| \leq \frac{1}{(4\pi|t|)^{n/2}} \int_{\mathbb{R}^n} |f(\vec{y})| d^n y = \frac{1}{(4\pi|t|)^{n/2}} \|f\|_1. \tag{5.63}$$

It follows that, if $\theta$ is a function in $L^2(\mathbb{R}^n)$, then

$$\|\theta(\vec{Q})U_t^0 f\|^2 = \int_{\mathbb{R}^n} |\theta(\vec{x})|^2 |(U_t^0 f)(\vec{x})|^2 d^n x \leq \frac{1}{(4\pi|t|)^n} \|\theta\|_2^2 \, \|f\|_1^2. \tag{5.64}$$

Let us take for $\theta$ the characteristic function $\chi_R$ of the ball $B_R$ [see (5.22)]. Then (5.64) becomes

$$\|F_R U_t^0 f\|^2 \leq \frac{1}{(4\pi|t|)^n} \, |B_R| \, \|f\|_1^2.$$

where $|B_R|$ denotes the volume of this ball. Thus, for $f \in L^2(\mathbb{R}^n) \cap L^1(\mathbb{R}^n)$, the probability of presence in $B_R$ tends to zero like $|t|^{-n}$ as $|t| \to \infty$. So each vector in $L^2(\mathbb{R}^n) \cap L^1(\mathbb{R}^n)$ represents a scattering state of $H_0$. Since $L^2(\mathbb{R}^n) \cap L^1(\mathbb{R}^n)$ is dense in $L^2(\mathbb{R}^n)$, this gives a new proof of the fact that $\mathcal{M}_\infty^\pm(H_0) = \mathcal{H}$. [7]

**(B)** Existence of the wave operators, I. Let $H = H_0 + V(\vec{Q}) \equiv \vec{P}^2 + V(\vec{Q})$. Assume that $V$ is a (real) function belonging to $L^2(\mathbb{R}^n)$ and take $\theta = V$ in (5.64):

$$\|(H - H_0)U_t^0 f\| = \|V(\vec{Q})U_t^0 f\| \leq \frac{1}{(4\pi|t|)^{n/2}} \|V\|_2 \|f\|_1. \tag{5.65}$$

If $n \geq 3$, this function of the variable $t$ is integrable at $t = \pm\infty$. One can then apply the Cook Criterion (Proposition 5.16) to obtain the existence of the wave operators $\Omega_\pm$. Let us take $A = H$, $B = H_0$ (hence $W_t = U_t^0$) and $\mathcal{D} = \mathcal{N}^n$ in the Cook Criterion and consider the conditions (1) - (3) that have to be satisfied in this criterion:
(1) $\mathcal{D} \equiv \mathcal{N}^n$ is total in $L^2(\mathbb{R}^n)$.
(2) Observe that $\mathcal{N}^n \subset \mathcal{S}(\mathbb{R}^n)$. If $f \in \mathcal{S}(\mathbb{R}^n)$, then $U_t^0 f \in \mathcal{S}(\mathbb{R}^n)$ for each $t$ [this is evident in the momentum representation, see (5.58)], hence $U_t^0 f \in \mathcal{D}(H_0)$. One must also know that $U_t^0 f \in \mathcal{D}(H)$, which holds in particular if $\mathcal{D}(H) = \mathcal{D}(H_0)$. We know from Proposition 2.49 that this latter condition is satisfied if $V \in L^2(\mathbb{R}^n)$ in the case $n = 3$ and $V \in L^p(\mathbb{R}^n)$ for some $p > n/2$ if $n > 3$.
(3) We have seen that this integrability condition is satisfied if $V \in L^2(\mathbb{R}^n)$ and $n \geq 3$.

[7] We have verified (5.59) only for $f \in \mathcal{N}^n$. By replacing $L^1(\mathbb{R}^n)$ in the preceding argument by $\mathcal{N}^n$ one arrives at the same conclusion, because $\mathcal{N}^n$ spans a dense linear manifold in $L^2(\mathbb{R}^n)$.

In conclusion:

- *If* $n = 3$ *and* $V \in L^2(\mathbb{R}^3)$, *then the wave operators* $\Omega_\pm$ *exist.*
- *If* $n \geq 4$ *and* $V \in L^2(\mathbb{R}^3) \cap L^p(\mathbb{R}^n)$ *for some* $p > n/2$, *then the wave operators* $\Omega_\pm$ *exist.*

**(C)** Existence of the wave operators, II. The above proof of the existence of the wave operators in dimension $n \geq 3$ involves the assumption that $V \in L^2(\mathbb{R}^n)$. Let us see what this assumption implies for the asymptotic behaviour of the potential $V(\vec{x})$. Suppose that $|V(\vec{x})| = |\vec{x}|^{-\beta}$ ($\beta > 0$) in a neighbourhood of infinity. The function $\varphi(\vec{x}) = |\vec{x}|^{-\beta}$ is square-integrable in a neighbourhood of infinity in $\mathbb{R}^n$ if $\int^\infty r^{-2\beta} r^{n-1} dr < \infty$, i.e. if $\beta > n/2$. This is not an optimal decay assumption for the existence of the wave operators. Indeed, since $\| |\vec{Q}|^{-\beta} U_t^0 f \| = \| |2\vec{P}t|^{-\beta} Z_t f \|$, which behaves approximately like $(2|t|)^{-\beta} \| |\vec{P}|^{-\beta} f \|$ at large times (see Remark 5.19), it can be expected that $\|V U_t^0 f\|$ is integrable near $t = \pm\infty$ if $\beta > 1$ but not so if $\beta \leq 1$. We verify below that the wave operators do exist if $\beta > 1$ and more generally for potentials converging to zero at infinity at least as fast as $|\vec{x}|^{-\beta}$ for some $\beta > 1$ (called ***short range potentials***). On the other hand, if $\beta \leq 1$ (in particular for the Coulomb potential), the wave operators do not exist; one then speaks of ***long range potentials***. For such potentials it is not possible to approximate asymptotically the evolution $e^{-iHt} f$ of a scattering state $f$ by a free evolution [in the sense indicated in (5.39)]. We refer to Section 5.8 for some details on time-dependent scattering theory for long range potentials.

So let us now consider the case $\beta > 1$. If we assume $V$ to be bounded on $\mathbb{R}^n$ and $|V(\vec{x})| \leq c|\vec{x}|^{-\beta}$ in some neighbourhood of infinity, with $\beta > 1$, we have $V \in L^q(\mathbb{R}^n)$ for some $q = n - \varepsilon$ with $\varepsilon > 0$. If $\beta$ is close to 1, $\varepsilon$ will be very small; in any case we may assume that $\varepsilon < 1$, so that $n - \varepsilon > 2$ if $n \geq 3$. We now prove the existence of the wave operators if $V \in L^\infty(\mathbb{R}^n) \cap L^q(\mathbb{R}^n)$ for some $q \in (2, n)$ (with $n \geq 3$), by invoking the Cook Criterion. Let $f \in \mathcal{S}(\mathbb{R}^n)$. Then $(I + \vec{Q}^2)^n f \in \mathcal{S}(\mathbb{R}^n)$. By taking into account (5.62) one sees that

$$\begin{aligned} &\|V(\vec{Q}) U_t^0 f\| = \|U_t^{0*} V(\vec{Q}) U_t^0 f\| = \|Z_t^* V(2t\vec{P}) Z_t f\| \\ &= \|V(2t\vec{P})(I + \vec{Q}^2)^{-n} Z_t (I + \vec{Q}^2)^n f\| \leq \|V(2t\vec{P})(I + \vec{Q}^2)^{-n}\| \|(I + \vec{Q}^2)^n f\|. \end{aligned}$$

Thus it suffices to show that $\|V(2t\vec{P})(I + \vec{Q}^2)^{-n}\|$ is integrable with respect to $dt$ near $t = \pm\infty$. This norm can be estimated by using (2.87):

$$\|V(2t\vec{P})(I + \vec{Q}^2)^{-n}\| \leq \|V(2t\vec{k})\|_q \|(1 + \vec{x}^{\,2})^{-n}\|_q .$$

One has $\|(1 + \vec{x}^{\,2})^{-n}\|_q < \infty$ and (setting $\vec{z} = 2t\vec{k}$):

$$\|V(2t\vec{k})\|_q = \Big[ \int_{\mathbb{R}^n} |V(2t\vec{k})|^q d^n k \Big]^{1/q} = \Big[ \int_{\mathbb{R}^n} |V(\vec{z})|^q \frac{d^n z}{|2t|^n} \Big]^{1/q} = \|V\|_q |2t|^{-n/q}. \tag{5.66}$$

Since $n/q > 1$, the function $t \mapsto \|V(2t\vec{P})(I + \vec{Q}^2)^{-n}\|$ is integrable at $t = \pm\infty$.

By combining the results of I and II we have obtained:

**Proposition 5.20.** *Let $H = H_0 + V(\vec{Q})$ in $L^2(\mathbb{R}^n)$, with $V$ real, and $H_0 = \vec{P}^2$.*
*(a) If $n = 3$, let $V = V_1 + V_2$ with $V_1 \in L^2(\mathbb{R}^3)$ and $V_2 \in L^\infty(\mathbb{R}^3) \cap L^{3-\varepsilon}(\mathbb{R}^3)$ for some $\varepsilon \in (0,1)$. Then the wave operators $\Omega_\pm \equiv \Omega_\pm(H, H_0; I)$ exist.*
*(b) If $n > 3$, assume that $V = V_1 + V_2$ with $V_1 \in L^2(\mathbb{R}^n) \cap L^p(\mathbb{R}^n)$ for some $p > n/2$ and $V_2 \in L^\infty(\mathbb{R}^n) \cap L^{n-\varepsilon}(\mathbb{R}^n)$ for some $\varepsilon \in (0,1)$. Then the wave operators $\Omega_\pm \equiv \Omega_\pm(H, H_0; I)$ exist.*

REMARKS. (1) In the preceding proposition, the part $V_1$ of the potential characterises the local behaviour of $V$, and the part $V_2$ the behaviour of $V$ in a neighbourhood of infinity: $V$ may have local singularities of class $L^q$ with $q = 2$ if $n = 3$ and $q > n/2$ if $n > 3$, and it should be bounded and of class $L^{n-\varepsilon}$ in some neighbourhood of infinity $(0 < \varepsilon < 1)$.

(2) The proof of the existence of the wave operators given above cannot be applied in dimension $n = 1$ and in dimension $n = 2$. In these cases different methods are necessary. We present one such method in the following chapters and refer to Proposition 7.3 for results on wave operators in dimension 1 and 2.

(3) Let us mention an extension of the preceding considerations. The Cook method requires an estimate of $\|VU_t^0 f\|$ at large times. If $f$ is any state vector, then $U_t^0 f$ is essentially localised in a neighbourhood of infinity if $|t|$ is very large (each $f \in \mathcal{H}$ is a scattering state of $H_0$). Therefore only the behaviour of $V$ near infinity should be important in the estimate (5.65). Indeed one can prove the existence of the wave operators by imposing the conditions of Proposition 5.20 only in a neighbourhood of infinity ($V$ can be quite arbitrary in a finite part of configuration space!). Let us explain this **Theorem of Kupsch and Sandhas** in the case $n = 3$. Let $R$ be a fixed positive number and assume that, for $|\vec{x}| > R$, one may write $V(\vec{x}) = V_1(\vec{x}) + V_2(\vec{x})$ with $V_1 \in L^2(\mathbb{R}^3 \setminus B_R)$ and $V_2 \in L^\infty(\mathbb{R}^3 \setminus B_R) \cap L^q(\mathbb{R}^3 \setminus B_R)$ for some $q \in (2,3)$. We impose no condition on $V$ in the ball $B_R$ [except that $\mathcal{D}(H_0) \cap \mathcal{D}(V(\vec{Q}))$ must be dense in $L^2(\mathbb{R}^3)$]. One must first give a meaning to the Hamiltonian $H$. We admit that $H$ is an arbitrary self-adjoint extension of the symmetric operator $H_0 + V(\vec{Q})$ [with $H_0 + V(\vec{Q})$ defined on $\mathcal{D}(H_0) \cap \mathcal{D}(V(\vec{Q}))$], and we shall show that under these conditions the wave operators $\Omega_\pm(H, H_0; I)$ exist.

Let us fix a function $\phi \in C_0^\infty(\mathbb{R}^3)$ satisfying $\phi(\vec{x}) = 1$ on $B_R$ and set $\psi = 1 - \phi$. Then

$$U_t^* U_t^0 = U_t^* \phi(\vec{Q}) U_t^0 + U_t^* \psi(\vec{Q}) U_t^0.$$

As $|t| \to \infty$, the first term on the right-hand side converges strongly to zero [use for example (5.64) with $\theta = \phi$]. Thus $\Omega_\pm = \text{s-}\lim_{t \to \pm\infty} U_t^* \psi(\vec{Q}) U_t^0$. To obtain the existence of these limits one proceeds in analogy with the Cook method. The operator $\psi(\vec{Q})$ maps $\mathcal{D}(H_0)$ into $\mathcal{D}(H)$ (since $\mathcal{D}(H)$ and $\mathcal{D}(H_0)$ coincide locally in regions of configuration space where the potential is only moderately singular, see page 210; observe that $\psi(\vec{x}) \neq 0$ only for $|\vec{x}| > R$). By applying the Leibniz rule [Proposition 5.5(a)] one obtains for $f \in \mathcal{D}(H_0)$:

$$\frac{d}{dt} U_t^* \psi(\vec{Q}) U_t^0 f = i U_t^* \big[ H \psi(\vec{Q}) - \psi(\vec{Q}) H_0 \big] U_t^0 f.$$

Now $H\psi(\vec{Q}) - \psi(\vec{Q})H_0 = V\psi(\vec{Q}) + [H_0, \psi(\vec{Q})]$, hence

$$\|\frac{d}{dt}U_t^*\psi(\vec{Q})U_t^0 f\| \leq \|V\psi(\vec{Q})U_t^0 f\| + \|[H_0, \psi(\vec{Q})]U_t^0 f\|.$$

Since $V\psi$ is a function satisfying the conditions of Proposition 5.20, $\|V\psi(\vec{Q})U_t^0 f\|$ is integrable at $t = \pm\infty$ if $f \in \mathcal{S}(\mathbb{R}^n)$ by the arguments given in the derivation of Proposition 5.20. By using the Leibniz rule one finds that, for $g \in \mathcal{S}(\mathbb{R}^3)$:

$$\begin{aligned}([H_0, \psi(\vec{Q})]g)(\vec{x}) = ([-\Delta, \psi(\vec{Q})]g)(\vec{x}) &= -\Delta\{\psi(\vec{x})g(\vec{x})\} + \psi(\vec{x})\Delta g(\vec{x}) \\ &= -\{\Delta\psi(\vec{x})\}g(\vec{x}) - 2[\vec{\nabla}\psi(\vec{x})] \cdot [\vec{\nabla}g(\vec{x})],\end{aligned}$$

or in operatorial language: $[H_0, \psi(\vec{Q})] = -[\Delta\psi](\vec{Q}) - 2i[\vec{\nabla}\psi](\vec{Q}) \cdot \vec{P}$. Here for example $[\Delta\psi](\vec{Q})$ denotes the multiplication operator by the function $\Delta\psi$. This function belongs to $C_0^\infty(\mathbb{R}^3)$ because $\psi(\vec{x}) = 1$ in some neighbourhood of infinity. Hence $\|[\Delta\psi](\vec{Q})U_t^0 f\|$ behaves like $|t|^{-3/2}$ at infinity [using (5.64) with $\theta = \Delta\psi$] and is therefore integrable with respect to $dt$ at $t = \pm\infty$. Similarly each component of $\vec{\nabla}\psi$ belongs to $C_0^\infty(\mathbb{R}^3)$, so that $\|[\vec{\nabla}\psi](\vec{Q}) \cdot \vec{P}U_t^0 f\| \leq \sum_{k=1}^3 \|[\vec{\nabla}\psi]_k(\vec{Q})U_t^0 P_k f\|$ also behaves like $|t|^{-3/2}$ at infinity.

## 5.5 Scattering operator and $S$-matrix

**5.5.1.** We again consider the simple situation of §5.4.1. We recall the definition of the scattering operator: $S = \Omega_+^*\Omega_-$. Let us first indicate some important simple properties of this operator.

**Proposition 5.21.** *(a) The scattering operator commutes with the free Hamiltonian $H_0$. More precisely one has*

$$[S, U_t^0] = O \qquad \forall t \in \mathbb{R} \tag{5.67}$$

*and, if $f \in \mathcal{D}(H_0)$, then $Sf \in \mathcal{D}(H_0)$ and $SH_0 f = H_0 Sf$.*
*(b) $S$ is an isometric operator if and only if $\mathcal{R}(\Omega_-) \subseteq \mathcal{R}(\Omega_+)$.*
*(c) $S$ is a unitary operator if and only if $\mathcal{R}(\Omega_-) = \mathcal{R}(\Omega_+)$.*

PROOF. (a) Eq. (5.67) is an immediate consequence of the intertwining relation (5.42) and its adjoint (5.52): in the situation considered here ($W_t \equiv U_t^0$) one has

$$SU_t^0 = \Omega_+^*\Omega_- U_t^0 = \Omega_+^* U_t \Omega_- = U_t^0 \Omega_+^* \Omega_- = U_t^0 S.$$

Next let $f \in \mathcal{D}(H_0)$. By (5.67) one has then $it^{-1}S[U_t^0 - I]f = it^{-1}[U_t^0 - I]Sf$. As $t \to 0$ the left-hand side converges strongly to $SH_0 f$ by Stone's Theorem. Consequently the right-hand side is strongly convergent as $t \to 0$, so that (again by Stone's Theorem) $Sf$ belongs to $\mathcal{D}(H_0)$ and one has $SH_0 f = H_0 Sf$.

(b) We denote by $F_+$ the projection onto the subspace $\mathcal{R}(\Omega_+)$ and by $F_-$ that onto

$\mathcal{R}(\Omega_-)$. We recall that $F_\pm = \Omega_\pm \Omega_\pm^*$ [see (5.46)]. The hypothesis $\mathcal{R}(\Omega_-) \subseteq \mathcal{R}(\Omega_+)$ implies that $F_+\Omega_- = \Omega_-$. Under this hypothesis one will have

$$S^*S = \Omega_-^*\Omega_+\Omega_+^*\Omega_- = \Omega_-^* F_+\Omega_- = \Omega_-^*\Omega_- = I,$$

i.e. $S$ will be isometric. If the inclusion $\mathcal{R}(\Omega_-) \subseteq \mathcal{R}(\Omega_+)$ is not satisfied, there is a vector $g \in \mathcal{R}(\Omega_-)$ such that $g \notin \mathcal{R}(\Omega_+) \equiv \mathcal{R}(F_+)$. If $f = \Omega_-^* g$ (so that $g = \Omega_- f$), then $\|F_+g\| < \|g\| = \|\Omega_- f\| = \|f\|$ (because $\Omega_-$ is isometric) and consequently $\|Sf\| = \|\Omega_+^*\Omega_- f\| = \|\Omega_+^* g\| = \|\Omega_+\Omega_+^* g\| = \|F_+ g\| < \|f\|$ (using the isometry of $\Omega_+$ for the third equality). In this case $S$ cannot be isometric [see (2.37)].

(c) $S$ is unitary if and only if $S$ and $S^*$ are isometric. By (b) $S$ is isometric if and only if $\mathcal{R}(\Omega_-) \subseteq \mathcal{R}(\Omega_+)$. Since $S^* = \Omega_-^*\Omega_+$ one obtains [by exchanging in (b) the roles of the signs $+$ and $-$] that $S^*$ is isometric if and only if $\mathcal{R}(\Omega_+) \subseteq \mathcal{R}(\Omega_-)$. □

In relation with parts (b) and (c) of the preceding proposition, let us briefly discuss the notion of asymptotic completeness. It follows from Proposition 5.15(d) that $\mathcal{R}(\Omega_\pm) \subseteq \mathcal{H}_{\mathrm{ac}}(H)$. One says that the wave operator $\Omega_+$ (or $\Omega_-$) is ***complete*** if $\mathcal{R}(\Omega_+) = \mathcal{H}_{\mathrm{ac}}(H)$ [respectively if $\mathcal{R}(\Omega_-) = \mathcal{H}_{\mathrm{ac}}(H)$], and that one has ***asymptotic completeness*** if $\mathcal{R}(\Omega_+) = \mathcal{R}(\Omega_-) = \mathcal{H}_{\mathrm{ac}}(H)$, i.e. if both $\Omega_+$ and $\Omega_-$ are complete. This terminology is justified by the fact that, under the assumption of asymptotic completeness, one has a description of the large time behaviour of all scattering states of $H$; more precisely, if one assumes that the set of scattering states $\mathcal{M}_\infty(H)$ coincides with $\mathcal{H}_{\mathrm{ac}}(H)$, and if $g$ is an arbitrary vector in $\mathcal{H}_{\mathrm{ac}}(H)$, there exist vectors $f_+$ and $f_-$ in $\mathcal{H}$ such that $U_t g$ behaves asymptotically (i.e. as $t \to \pm\infty$) as $U_t^0 f_\pm$ (if $f_\pm = \Omega_\pm^* g$, then $\|U_t g - U_t^0 f_\pm\| \to 0$ as $t \to \pm\infty$).

One can consider other notions of asymptotic completeness, for example:
***Strong asymptotic completeness***: $\Omega_+$ and $\Omega_-$ are complete and $H$ has no singular continuous spectrum, i.e.

$$\mathcal{R}(\Omega_+) = \mathcal{R}(\Omega_-) = \mathcal{H}_{\mathrm{p}}(H)^\perp.$$

***Asymptotic completeness in the geometric sense***: we have seen that one always has $\mathcal{R}(\Omega_\pm) \subseteq \mathcal{M}_\infty^\pm(H)$ [Proposition 5.15(d)]. One speaks of asymptotic completeness in the geometric sense if $\mathcal{R}(\Omega_+) = \mathcal{M}_\infty^+(H)$ and $\mathcal{R}(\Omega_-) = \mathcal{M}_\infty^-(H)$ [without assuming a relation between $\mathcal{M}_\infty^\pm(H)$ and $\mathcal{H}_{\mathrm{ac}}(H)$].
***Generalised asymptotic completeness***: this notion has been used in the particular context where absorption of states by local singularities of a potential may occur, as explained at the end of Section 5.2. In such a situation one would say that generalised asymptotic completeness holds if $\mathcal{R}(\Omega_+) = \mathcal{M}_\infty^+(H)$, $\mathcal{R}(\Omega_-) = \mathcal{M}_\infty^-(H)$ and (5.38) is satisfied. As explained after (5.38), one again has a complete description of the large time behaviour of states in $\mathcal{H}_{\mathrm{c}}(H)$ (decomposition into an absorbed part and a part propagating to infinity and becoming free).

Asymptotic completeness (and a fortiori strong asymptotic completeness) implies the ***unitarity*** of the scattering operator $S$. In various physical theories the unitarity of $S$ is assumed essentially in an axiomatic way. Observe that the isometry of $S$ implies the "conservation of probability": if $f_-$ is an initial state and $S$ is isometric, then

$f_+ = Sf_-$ has the same norm as $f_-$, i.e. $\|f_+\| = \|f_-\|$; in the simple situation considered here, if the $S$-operator is isometric, then the scattering does not lead to a loss of probability. Evidently, in cases where the potential is absorptive, a part of the incident state could be absorbed at its singularities and the norm of the out-coming state $f_+$ could be less than $\|f_-\|$; nevertheless one should not speak of non-conservation of probability because the evolution is unitary and one knows [for example if (5.38) is satisfied] what happens with the part of the state that is not leaving the scattering region (if $f_-$ is the initial state, then the associated scattering state of $H$, i.e. the vector $\Omega_- f$, has a unique decomposition into a part that will be absorbed and a part that will be scattered as $t \to +\infty$); so it is only for an experimentalist (situated far from the scattering region) that an apparent loss of probability occurs.

**Remark 5.22.** We have seen that in the case in which $H_0$ is the operator $\vec{P}^2$ in dimension $n \geq 3$ and the interaction is given by a potential, the existence of the wave operators $\Omega_\pm \equiv \Omega_\pm(H, H_0; I)$ is easy to prove by using the Cook Criterion. To obtain asymptotic completeness one has to show that $\Omega_\pm(H_0, H; E_{ac}(H))$ exist[8]. For this the Cook Criterion is not useful because one would have to know for example that $\int_1^\infty \|V(\vec{Q})U_t f\| dt < \infty$ for sufficiently many vectors $f$; but the total evolution group $\{U_t\}$ cannot be written explicitly (as was the case for the free evolution group), hence one has no simple estimate of the asymptotic behaviour of $\|V(\vec{Q})U_t f\|$ as a function of the parameter $t$. To prove asymptotic completeness one has to have recourse to more elaborate mathematical methods. One such method, involving the theory of relatively smooth operators, is presented in Section 7.1.

**Remark 5.23.** Since $\mathcal{R}(\Omega_\pm) \subseteq \mathcal{H}_{ac}(H)$ it is clear that, in all cases in which the spectrum of the Hamiltonian $H$ is not purely absolutely continuous, the wave operators are not unitary (they are only isometric). In particular, if $H$ has eigenvalues, then $\Omega_+$ and $\Omega_-$ are not unitary, the eigenvectors being orthogonal to $\mathcal{R}(\Omega_+)$ and to $\mathcal{R}(\Omega_-)$.

Let us now discuss part (a) of Proposition 5.21, i.e. the fact that $S$ commutes with the free Hamiltonian $H_0$. In physical terms this roughly expresses the ***conservation of the kinetic energy*** by the scattering process. Indeed Eq. (5.67) implies that $S$ commutes with the spectral projections $E_0(\mathcal{V})$ of the free Hamiltonian $H_0$ (as explained in Remark 5.14), in particular one has $E_0(J)S = SE_0(J)$ for each interval $J$. Assume that the initial state $f_-$ satisfies $E_0(J)f_- = f_-$ for some interval $J$, in other terms that the kinetic energy of this state is localised with certainty in $J$. Then, however small the interval $J$, the outgoing state $f_+ = Sf_-$ has the same property, i.e. one has $E_0(J)f_+ = f_+$; indeed

$$E_0(J)f_+ = E_0(J)Sf_- = SE_0(J)f_- = Sf_- = f_+.$$

The following two observations are interesting in this context:
(1) If $S$ is not isometric, the amount of kinetic energy is not conserved for an observer

[8] For example if $g \in \mathcal{R}(\Omega_+)$, so that $g = \Omega_+ f$ for some $f \in \mathcal{H}$, then $\|f - U_t^{0*} U_t g\| \to 0$ as $t \to +\infty$, i.e. one has $f = \Omega_+(H_0, H; E_{ac}(H))g$ [see also Proposition 5.15(a)]. Thus, in order to have $\mathcal{R}(\Omega_+) = \mathcal{H}_{ac}(H)$, one must know the existence of $\text{s-lim}_{t \to +\infty} U_t^{0*} U_t$ on $\mathcal{H}_{ac}(H)$.

far from the scattering centre. The amount of kinetic energy of the initial state $f_-$ is given by $\|H_0 f_-\| = \{\int_0^\infty \lambda^2 \langle f_-, E_0(d\lambda) f_- \rangle\}^{1/2}$. The associated amount of kinetic energy of the out-coming state $f_+ = Sf_-$ will be $\|H_0 f_+\| = \|H_0 S f_-\| = \|S H_0 f_-\|$ and this number can be less that $\|H_0 f_-\|$ if $S$ is not isometric (this is by no means surprising: for example if a part of the scattering state remains absorbed in the potential, then this part will retain a certain amount of the total energy).

(2) The relation $E_0(J)S = SE_0(J)$ implies that the subspace $\mathcal{M}(J) \equiv \mathcal{R}(E_0(J))$ gets mapped into itself by the operator $S$. Thus the restriction of $S$ to this subspace defines a bounded operator in $\mathcal{M}(J)$, and it is natural to denote this operator by $S(J)$. If for example $S$ is unitary (as an operator in $\mathcal{H}$), then $S(J)$ is unitary as an operator in $\mathcal{M}(J)$. One may take for $J$ an arbitrarily small interval, and one may be tempted to consider the limit in which $J$ consists of a single point $\lambda$. One would then have an operator $S(\lambda)$ (unitary if $S$ is unitary) acting in the space $\mathcal{M}(\{\lambda\})$. However, since the spectrum of $H_0$ is purely continuous, the functions having support in a single point $\lambda$ with respect to the spectral measure of $H_0$ are equivalent to the zero vector **0** (one has $E_0(\{\lambda\}) = O$), so the notation $\mathcal{M}(\{\lambda\})$ does not correspond to a subspace of $\mathcal{H}$. Nevertheless, since the operator $H_0$ is very explicitly known, one can give a meaning to this notation by introducing the ***spectral representation of $H_0$***. This means that one chooses a representation of the Hilbert space $\mathcal{H} = L^2(\mathbb{R}^n)$ in which $H_0$ is a multiplication operator, in other terms one "diagonalises" $H_0$ by mapping $\mathcal{H}$ by means of a unitary operator $\mathcal{U}_0$ onto the space $L^2((0,\infty); L^2(S^{n-1}), d\lambda)$ [9] in such a way that, after transformation by $\mathcal{U}_0$, the operator $H_0$ becomes multiplication by the variable $\lambda$ in $L^2((0,\infty); L^2(S^{n-1}), d\lambda)$ [the interval $(0,\infty)$ appearing here corresponds to the spectrum of $H_0$, with the exception of the point $\lambda = 0$ which we omit because spherical polar co-ordinates are singular at this point, and of course $\{\lambda = 0\}$ is a null set].

Let us first give an explicit formula for $\mathcal{U}_0$. The construction is similar to that indicated at the beginning of §3.3.3, but it is carried through in the momentum representation $\tilde{L}^2(\mathbb{R}^n)$ of the Hilbert space. One denotes by $S^{n-1}$ the unit sphere in $\mathbb{R}^n$, i.e. the set of vectors $\vec{\omega} \in \mathbb{R}^n$ satisfying $|\vec{\omega}| = 1$, and one writes the wave vectors $\vec{k} \in \mathbb{R}^n$ in spherical polar co-ordinates: $\vec{k} = \sqrt{\lambda}\vec{\omega}$ (so $\lambda = \vec{k}^2$). $L^2(S^{n-1})$ is the set of (equivalence classes of) complex functions defined on $S^{n-1}$ that are measurable and square-integrable with respect to Lebesgue measure $d\omega$ on $S^{n-1}$ [for example for $n = 3$ with respect to $d\omega = \sin\theta\, d\theta\, d\phi$ as in (3.40)]. For $f \in L^2(\mathbb{R}^n)$ we set

$$(\mathcal{U}_0 f)_\lambda(\vec{\omega}) = 2^{-1/2} \lambda^{(n-2)/4} \tilde{f}(\sqrt{\lambda}\vec{\omega}). \tag{5.68}$$

For each $f$, Eq. (5.68) defines for almost every $\lambda > 0$ an element of $L^2(S^{n-1})$ (as a function of $\vec{\omega}$). By varying $\lambda$, one associates by Eq. (5.68) with each $f \in L^2(\mathbb{R}^n)$ an element of $L^2((0,\infty); L^2(S^{n-1}), d\lambda)$ [10]. The operator $\mathcal{U}_0$ is isometric as a mapping

[9] See (1.53) and below for the notations. Further comments on spectral representations can be found in the footnote on page 152.

[10] In order to be in agreement with the notations of Chapter 1, we should write $(\mathcal{U}_0 f)(\lambda)(\vec{\omega})$ rather than $(\mathcal{U}_0 f)_\lambda(\vec{\omega})$: one considers a vector-valued function [with values in $L^2(S^{n-1})$] defined on $(0,\infty)$, and in Chapter 1 we used the notation $(\mathcal{U}_0 f)(\lambda)$ for the value of such a function at the point $\lambda \in (0,\infty)$. Since $(\mathcal{U}_0 f)(\lambda)$ belongs to $L^2(S^{n-1})$, it is a function of the angular variable $\vec{\omega}$, and it is more convenient to write $(\mathcal{U}_0 f)_\lambda$ rather than $(\mathcal{U}_0 f)(\lambda)$.

from $L^2(\mathbb{R}^n)$ to $L^2((0,\infty);L^2(S^{n-1}),d\lambda)$ [see (1.53); observe that if one sets $k = |\vec{k}|$, then $d\lambda = 2kdk$ and consequently $d^nk = k^{n-1}dkd\omega = (1/2)\lambda^{(n-2)/2}d\lambda d\omega$]:

$$\int_0^\infty d\lambda \|(\mathcal{U}_0 f)_\lambda\|^2_{L^2(S^{n-1})} = \frac{1}{2}\int_0^\infty \lambda^{(n-2)/2}d\lambda \int_{S^{n-1}} d\omega |\tilde{f}(\sqrt{\lambda}\vec{\omega})|^2$$
$$= \int_{R^n} |\tilde{f}(\vec{k})|^2 d^nk = \|f\|^2.$$

It is also clear that $\mathcal{U}_0$ is surjective, hence a unitary operator. Its inverse is given as follows: if $g = \{g_\lambda\} \in L^2((0,\infty);L^2(S^{n-1}),d\lambda)$, then [11]

$$[\mathcal{F}\mathcal{U}_0^{-1}g](\vec{k}) = 2^{1/2}|\vec{k}|^{(2-n)/2}g_{\vec{k}^2}(\vec{\omega}), \qquad \text{with } \vec{\omega} = \vec{k}/|k|. \tag{5.69}$$

If $f \in \mathcal{D}(H_0)$, then $(\mathcal{U}_0 H_0 f)_\lambda(\vec{\omega}) = \lambda(\mathcal{U}_0 f)_\lambda(\vec{\omega})$, because in momentum space $H_0$ is the operator of multiplication by $\vec{k}^2$. More generally, if $\varphi\colon \mathbb{R}^n \to \mathbb{C}$ is a bounded continuous function, then [12]

$$[\mathcal{U}_0\varphi(H_0)f]_\lambda(\vec{\omega}) = \varphi(\lambda)(\mathcal{U}_0 f)_\lambda(\vec{\omega}) \qquad \forall f \in L^2(\mathbb{R}^n). \tag{5.70}$$

Since $S$ commutes with $H_0$, it cannot mix the different values of $\lambda$ in the spectral representation of $H_0$. This is expressed in the following manner: for each $\lambda > 0$ there exists a bounded operator $S(\lambda)$ acting in $L^2(S^{n-1})$ such that the component at $\lambda$ of the vector $Sf$ [i.e. the element $(\mathcal{U}_0 Sf)_\lambda$ of $L^2(S^{n-1})$] is obtained by acting with $S(\lambda)$ on the component at $\lambda$ of the vector $f$:

$$(\mathcal{U}_0 Sf)_\lambda = S(\lambda)(\mathcal{U}_0 f)_\lambda. \tag{5.71}$$

Equation (5.71) is (for each $\lambda > 0$) an equation in $L^2(S^{n-1})$. If $S$ is isometric or unitary as an operator in $L^2(\mathbb{R}^n)$, then each $S(\lambda)$ is an isometric or unitary operator, respectively [in $L^2(S^{n-1})$]. A priori $S(\lambda)$ is defined only for almost all $\lambda > 0$ (with respect to Lebesgue measure). In most situations one can show that one may choose $S(\lambda)$ to be continuous as a function of $\lambda$, and with such a choice $S(\lambda)$ is defined for all $\lambda > 0$ (physically this seems plausible, since physical quantities measured in a scattering experiment should depend continuously on the kinetic energy). The operator $S(\lambda)$ is frequently called the ***S-matrix at energy*** $\boldsymbol{\lambda}$. The space $L^2(S^{n-1})$ associated by (5.68) with a fixed value of $\lambda$ is sometimes called an ***energy shell***.

Above we have given a description of the mathematical context in relation with the concept of the $S$-matrix. We still have to verify these assertions, i.e. to prove the existence of $S(\lambda)$ and its properties. This will be done in §5.5.2 in an abstract setting (the situation considered above is a particular case of that treated in §5.5.2). Before this we indicate briefly how the scattering phase shifts, introduced in quantum mechanics courses, arise in the present formalism (in the case $n = 3$).

---

[11] We denote by $g_\lambda \in L^2(S^{n-1})$ the value of $g$ at the point $\lambda$.
[12] Eq. (5.70) will hold for almost all $\lambda \in (0,\infty)$ and almost all $\vec{\omega} \in S^{n-1}$.

Below we use the letter $\mathcal{R}$ to denote rotations in $\mathbb{R}^3$ around axes passing through the origin $\vec{x} = 0$. The set of all such rotations forms a group, the rotation group. A unitary representation of this group in $L^2(\mathbb{R}^3)$ is given by the following formula:

$$[U(\mathcal{R})f](\vec{x}) = f(\mathcal{R}^{-1}\vec{x}), \tag{5.72}$$

where $\mathcal{R}^{-1}$ is the inverse of the rotation $\mathcal{R}$. This formula associates with each rotation $\mathcal{R}$ a unitary operator $U(\mathcal{R})$ in $L^2(\mathbb{R}^3)$, and the collection of these unitary operators gives a strongly continuous representation of the rotation group. It is easy to calculate the Fourier transform of (5.72) to obtain the action of $U(\mathcal{R})$ in the momentum representation, i.e. on the Fourier transforms of the wave functions: $[\mathcal{F}U(\mathcal{R})f](\vec{k}) = \tilde{f}(\mathcal{R}^{-1}\vec{k})$. Thus $U(\mathcal{R})$ acts in $\tilde{L}^2(\mathbb{R}^3)$ exactly in the same way as in $L^2(\mathbb{R}^3)$. It is clear that each $U(\mathcal{R})$ commutes with the free Hamiltonian $H_0$, because

$$\begin{aligned}[\mathcal{F}U(\mathcal{R})H_0f](\vec{k}) &= [\mathcal{F}H_0f](\mathcal{R}^{-1}\vec{k}) = |\mathcal{R}^{-1}\vec{k}|^2\tilde{f}(\mathcal{R}^{-1}\vec{k})\\ &= |\vec{k}|^2\tilde{f}(\mathcal{R}^{-1}\vec{k}) = [\mathcal{F}H_0U(\mathcal{R})f](\vec{k})\end{aligned}$$

(the length of a vector of $\mathbb{R}^3$ remains unchanged under a rotation: $|\mathcal{R}^{-1}\vec{k}| = |\vec{k}|$). If $H = H_0 + V(\vec{Q})$ and if the potential $V$ is spherically symmetric (i.e. if $V(\vec{x})$ depends only on $r \equiv |\vec{x}|$), then $U(\mathcal{R})$ also commutes with $V(\vec{Q})$ and hence with $H$ [13]. In such a case $U(\mathcal{R})$ commutes with $U_t^0 = e^{-iH_0t}$ as well as with $U_t = e^{-iHt}$, hence also with $\Omega_+$ and with $\Omega_-$. The relation $U(\mathcal{R})\Omega_+ = \Omega_+U(\mathcal{R})$ implies that $\Omega_+^*U(\mathcal{R})^* = U(\mathcal{R})^*\Omega_+^*$; since $U(\mathcal{R})^* = U(\mathcal{R}^{-1})$, this shows that $\Omega_+^*$ commutes with $U(\mathcal{R}^{-1})$ for each rotation $\mathcal{R}$, hence with each $U(\mathcal{R})$ (when $\mathcal{R}$ varies over the rotation group, $\mathcal{R}^{-1}$ also varies over this group, each rotation being the inverse of another rotation). Thus each $U(\mathcal{R})$ commutes with $\Omega_-$ and with $\Omega_+^*$ and hence with $S \equiv \Omega_+^*\Omega_-$.

Let us now consider the subgroup of rotations around one of the three Cartesian co-ordinate axes. This subgroup is a one-parameter group (the parameter being the angle of rotation), and the representation of this subgroup (given by the operators $U(\mathcal{R})$ for $\mathcal{R}$ belonging to the subgroup) is a strongly continuous unitary one-parameter group (an evolution group in our terminology). Its infinitesimal generator is the angular momentum operator relative to the considered co-ordinate axis. By choosing successively the co-ordinate axis of $x_1$, then that of $x_2$ and finally that of $x_3$, one obtains the three operators $L_1$, $L_2$ and $L_3$ as infinitesimal generators of these three subgroups. Since the scattering operator $S$ commutes with the representation of each of these subgroups, it also commutes with their infinitesimal generators, i.e. with $L_1$, $L_2$ and $L_3$, and consequently also with $\vec{L}^2 := L_1^2 + L_2^2 + L_3^2$. Thus $S$ leaves invariant the subspace associated with an eigenvalue of any one of these operators (the subspace spanned by all eigenvectors associated with an eigenvalue of one of these operators). In particular $S$ leaves invariant the subspace $\mathcal{H}_{\ell m}$ formed by the common eigenvectors of $\vec{L}^2$ with eigenvalue $\ell(\ell+1)$ and of $L_3$ with eigenvalue $m$ ($\ell = 0, 1, 2, \ldots; m = \ell, \ell-1, \ldots, -\ell$), because $\vec{L}^2$ and $L_3$ commute.

[13] One assumes here that $V$ is only moderately singular, so that the operator $H_0 + V(\vec{Q})$ is self-adjoint on $\mathcal{D}(H_0) \cap \mathcal{D}(V(\vec{Q}))$. In more general situations $H$ could be a self-adjoint extension of $H_0 + V(\vec{Q})$ which does not commute with $U(\mathcal{R})$ (as indicated on page 126), and then the argument presented below does not apply.

Now in the representation (5.68) of $L^2(\mathbb{R}^3)$, i.e. in $L^2((0,\infty); L^2(S^2), d\lambda)$, the common eigenvectors of $\vec{L}^2$ with eigenvalue $\ell(\ell+1)$ and of $L_3$ with eigenvalue $m$ have the form $h(\lambda)Y_\ell^m(\vec{\omega})$, where $h$ is a complex function defined on $(0,\infty)$ satisfying $\int_0^\infty |h(\lambda)|^2 d\lambda < \infty$ (the dependence on the angular variables is fixed, of the form $Y_\ell^m(\vec{\omega})$, the dependence on the radial variable is free; the situation is essentially the same as that discussed in §3.3.3, but here the decomposition into partial waves is written in the variable $\vec{k}$ rather than in the variable $\vec{x}$, as discussed in Appendix A.3.3). For fixed $\lambda$ [i.e. in $L^2(S^2)$], the subspace formed by the common eigenvectors of $\vec{L}^2$ with eigenvalue $\ell(\ell+1)$ and of $L_3$ with eigenvalue $m$ is one-dimensional, given by the scalar multiples of $Y_\ell^m$. The operator $S(\lambda)$ (for fixed $\lambda$) maps this one-dimensional subspace into itself, so $S(\lambda)$ simply multiplies $Y_\ell^m$ by a number $b_{\ell m}$; this number may depend on $\lambda$:

$$S(\lambda)Y_\ell^m = b_{\ell m}(\lambda)Y_\ell^m.$$

If $S$ is unitary (which is the case for moderately singular potentials), then so is $S(\lambda)$. In this situation the absolute value of $b_{\ell m}(\lambda)$ must be 1: $|b_{\ell m}(\lambda)| = 1$. One may always write $b_{\ell m}(\lambda) = \exp[2i\delta_\ell^m(\lambda)]$, which defines $\delta_\ell^m(\lambda)$ (mod $\pi$). If $S$ is unitary, $\delta_\ell^m(\lambda)$ must be real. We shall see below that $\delta_\ell^m(\lambda)$ does not depend on $m$: $\delta_\ell^m(\lambda) = \delta_\ell(\lambda)$. Thus, in each partial wave (of angular momentum $\ell$), the operator $S$ is given by a function $\delta_\ell(\lambda)$ of the kinetic energy $\lambda$. This function $\delta_\ell$ is real if $S$ is unitary, and it is just the ***scattering phase shift*** at angular momentum $\ell$ introduced in quantum mechanics texts for scattering by a spherically symmetric potential.

Finally let us verify that $\delta_\ell^{m_1}(\lambda) = \delta_\ell^{m_2}(\lambda)$ for each $m_1, m_2 = \ell, \ell-1, \ldots, -\ell$. For this one uses the fact that $S$ also commutes with $L_1$ and $L_2$, hence with $L_\pm := L_1 \pm iL_2$, and that for example $L_+Y_\ell^m = c_{\ell m}Y_\ell^{m+1}$ for some constant $c_{\ell m}$ ($c_{\ell m} \neq 0$ if $m \neq \ell$). One then has for $m \neq \ell$:

$$\begin{aligned} e^{2i\delta_\ell^{m+1}(\lambda)}Y_\ell^{m+1} &= S(\lambda)Y_\ell^{m+1} = \frac{1}{c_{\ell m}}S(\lambda)L_+Y_\ell^m = \frac{1}{c_{\ell m}}L_+S(\lambda)Y_\ell^m \\ &= \frac{1}{c_{\ell m}}L_+e^{2i\delta_\ell^m(\lambda)}Y_\ell^m = e^{2i\delta_\ell^m(\lambda)}\frac{1}{c_{\ell m}}L_+Y_\ell^m = e^{2i\delta_\ell^m(\lambda)}Y_\ell^{m+1}, \end{aligned}$$

which shows that $e^{2i\delta_\ell^{m+1}(\lambda)} = e^{2i\delta_\ell^m(\lambda)}$.

**5.5.2.** Consider a Hilbert space $\mathcal{H} = L^2(\mathbb{R}; \mathcal{K}, m)$ as in (1.53), where $m$ is a $\sigma$-finite measure on $(\mathbb{R}, \mathcal{A}_{\mathrm{B}})$. We use the notation $x$ for the variable in $\mathbb{R}$, and we denote by $X$ the operator of multiplication by the variable $x$ in $\mathcal{H}$ (defined on the maximal domain for a multiplication operator): $(Xf)(x) = xf(x)$ if $f = \{f(x)\}$ is an element of $L^2(\mathbb{R}; \mathcal{K}, m)$ [thus $f(x) \in \mathcal{K}$ for each $x \in \mathbb{R}$] [14].

[14] In the application considered in §5.5.1, one has $\mathcal{K} = L^2(S^{n-1})$ and $X = \mathcal{U}_0 H_0 \mathcal{U}_0^{-1}$, and $m$ is identically zero on $(-\infty, 0]$ and coincides with the Lebesgue measure on $(0,\infty)$: $m(dx) = dx$ on $(0,\infty)$.

An operator $A$ in $\mathcal{H}$ is said to be ***decomposable*** if it has the following form: for each $x \in \mathbb{R}$ there exists an operator $A(x)$ in $\mathcal{B}(\mathcal{K})$ such that [15]

$$\mathcal{D}(A) = \{f \in \mathcal{H} \mid \int_{\mathbb{R}} \|A(x)f(x)\|_{\mathcal{K}}^2 m(dx) < \infty\} \quad \text{and} \quad (Af)(x) = A(x)f(x). \tag{5.73}$$

If $A$ is a decomposable operator we write $A = \{A(x)\}$, and we denote by $\|A(x)\|_{\mathcal{K}}$ the norm of the operator $A(x)$. The following properties of decomposable operators are easy to verify by mimicking the arguments given for multiplication operators in Section 2.5:

**Proposition 5.24.** *(a) Let $A$ be a decomposable operator. Then $A$ is bounded, with $\mathcal{D}(A) = \mathcal{H}$, if and only if the function $\psi(x) := \|A(x)\|_{\mathcal{K}}$ belongs to $L^\infty(\mathbb{R}, m)$, and one has $\|A\| = \|\psi\|_\infty$.*
*(b) If $A = \{A(x)\}$ and $B = \{B(x)\}$ are decomposable and belong to $\mathcal{B}(\mathcal{H})$, and if $\alpha \in \mathbb{C}$, then $A + \alpha B$, $AB$ and $A^*$ are also decomposable and given by*

$$A + \alpha B = \{A(x) + \alpha B(x)\}, \tag{5.74}$$

$$AB = \{A(x)B(x)\}, \tag{5.75}$$

$$A^* = \{A(x)^*\}. \tag{5.76}$$

**Corollary 5.25.** *Let $A$ be a decomposable operator. $A$ is unitary if and only if $A(x)$ is unitary m-almost everywhere. $A$ is isometric if and only if $A(x)$ is isometric m-a.e.*

A particular class of decomposable operators are the ***diagonalisable operators*** which have the form $A = \{\varphi(x)I\}$, where $\varphi$ is a complex measurable function defined on $\mathbb{R}$ and $I$ denotes the identity operator in $\mathcal{K}$. In this case $A$ is the multiplication operator by $\varphi(x)$ in $L^2(\mathbb{R}; \mathcal{K}, m)$, which is just the operator $\varphi(X)$. If $\varphi = \chi_{\mathcal{V}}$ is the characteristic function of a Borel set $\mathcal{V}$, then $\varphi(X) = \chi_{\mathcal{V}}(X)$ is the spectral projection $E(\mathcal{V})$ of $X$ associated with $\mathcal{V}$. It is clear that each decomposable operator commutes with all diagonalisable operators, in particular with $X$. In the next proposition we prove the converse: a bounded operator in $L^2(\mathbb{R}; \mathcal{K}, m)$ that commutes with $X$ is decomposable:

**Proposition 5.26.** *Let $\mathcal{H} = L^2(\mathbb{R}; \mathcal{K}, m)$. If $A \in \mathcal{B}(\mathcal{H})$ is such that $[A, e^{-iXt}] = O$ for each $t \in \mathbb{R}$, then $A$ is decomposable.*

PROOF. The hypothesis $[A, e^{-iXt}] = O$ for all $t \in \mathbb{R}$ implies that $AE(\mathcal{V}) = E(\mathcal{V})A$ for each Borel set $\mathcal{V}$ (see Remark 5.14), hence $E(\mathcal{V})A^* = A^*E(\mathcal{V})$ for all $\mathcal{V} \in \mathcal{A}_{\mathrm{B}}$.

(i) We fix an orthonormal basis $\{e_k^\circ\}$ of $\mathcal{K}$ and a function $\theta\colon \mathbb{R} \to \mathbb{R}$ such that $\int_{-\infty}^{\infty} |\theta(x)|^2 m(dx) < \infty$ and $\theta(x) \neq 0$ for each $x$ (for example $\theta(x) = e^{-|x|}$ if $m$ is the Lebesgue measure). We set $e_k(x) = \theta(x)e_k^\circ$. For each fixed $x$, $e_k(x)$ is a non-zero vector in $\mathcal{K}$. Furthermore, since $\int_{-\infty}^{\infty} \|e_k(x)\|_{\mathcal{K}}^2 m(dx) = \int_{-\infty}^{\infty} |\theta(x)|^2 \|e_k^\circ\|_{\mathcal{K}}^2 m(dx) = \int_{-\infty}^{\infty} |\theta(x)|^2 m(dx) < \infty$, the family $\{e_k(x)\}_{x \in X}$ (for fixed $k$) defines an element of the Hilbert space $\mathcal{H}$. We write $e_k$ for this element of $\mathcal{H}$, i.e. $e_k = \{e_k(x)\}$. We

[15] To be precise, the family $\{A(x)\}$ must also be *measurable*, i.e. for all $f, g \in \mathcal{K}$ the function $x \mapsto \langle f, A(x)g\rangle$ must be Borel measurable.

denote by $\mathcal{D}_0 \subseteq \mathcal{K}$ the set of finite linear combinations with *rational coefficients* of the vectors $\{e_k^{\rm o}\}$ and by $\mathcal{D} \subseteq \mathcal{H}$ the set of finite linear combinations, also with rational coefficients, of the vectors $\{e_k\}$. We observe that if $f = \{f(x)\}$, then $f \in \mathcal{D}$ if and only if $f(x) = \theta(x) f_0$ for some $f_0 \in \mathcal{D}_0$ (modulo equivalence).

For each $x \in \mathbb{R}$ we define an operator $B(x)$ in $\mathcal{K}$, with domain $\mathcal{D}_0$, by

$$B(x)e_k^{\rm o} = \frac{1}{\theta(x)}\,[A^*e_k](x) \tag{5.77}$$

(here $[A^*e_k](x)$ denotes the component at $x$ of the vector $A^*e_k$ of $\mathcal{H}$, i.e. $A^*e_k = \{[A^*e_k](x)\}$). Eq. (5.77) specifies the operator $B(x)$ on each vector $e_k^{\rm o}$, and its action on the vectors in $\mathcal{D}_0$ is defined from (5.77) by linearity. In (5.77) we have chosen for each $k$ a representative in the equivalence class of the vector-valued function $[A^*e_k](x)$; a different choice would change the definition of $B(x)$ at most on a countable union (over $k \in \mathbb{N}$) of null sets $\Gamma_k$, hence on a set of measure zero. Thus $B(x)$ is defined $m$-almost everywhere.

(ii) Let $f = \{\theta(x) f_0\} \in \mathcal{D}$. By using the relations $B(x)e_k(x) = [A^*e_k](x)$ and $E(\mathcal{V})A^* = A^*E(\mathcal{V})$ for $\mathcal{V} \in \mathcal{A}_{\rm B}$, one finds that

$$\begin{aligned}\int_{\mathcal{V}} \|B(x)f(x)\|_{\mathcal{K}}^2\, m(dx) &= \int_{\mathcal{V}} \|[A^*f](x)\|_{\mathcal{K}}^2\, m(dx) = \|E(\mathcal{V})A^*f\|^2 \\ &\leq \|A^*\|^2 \|E(\mathcal{V})f\|^2 = \|A\|^2 \int_{\mathcal{V}} \|f(x)\|_{\mathcal{K}}^2\, m(dx).\end{aligned}$$

Thus

$$\int_{\mathcal{V}} \big[\|A\|^2 \|f(x)\|_{\mathcal{K}}^2 - \|B(x)f(x)\|_{\mathcal{K}}^2\big] m(dx) \geq 0 \qquad \forall \mathcal{V} \in \mathcal{A}_{\rm B},$$

which implies that $\|B(x)f(x)\|_{\mathcal{K}} \leq \|A\| \|f(x)\|_{\mathcal{K}}$ $m$-almost everywhere. Consequently, for each $f \in \mathcal{D}$ there exists a Borel set $\Gamma(f)$ of measure zero such that $\|B(x)f(x)\|_{\mathcal{K}} \leq \|A\| \|f(x)\|_{\mathcal{K}}$ for each $x \notin \Gamma(f)$. Since $\mathcal{D}$ is a countable set, the union $\Gamma$ of all $\Gamma(f)$, as $f$ varies over $\mathcal{D}$, is a Borel set of measure zero, and for each $x \notin \Gamma$, $B(x)$ is a bounded operator (on $\mathcal{D}_0$). Thus the closure of $B(x)$ belongs to $\mathcal{B}(\mathcal{K})$ (see page 48), and the adjoint of the closure of $B(x)$ is equal to the adjoint of $B(x)$ (Proposition 2.20). Thus, if we set $A(x) = B(x)^*$ for $x \notin \Gamma$, we have $A(x) \in \mathcal{B}(\mathcal{K})$ and $\|A(x)\| \leq \|A\|$ (for each $x \notin \Gamma$). By finally defining $A(x) = O$ if $x \in \Gamma$ we have constructed a family $\{A(x)\}_{x \in \mathbb{R}}$ of operators in $\mathcal{B}(\mathcal{K})$ satisfying $\|A(x)\| \leq \|A\|$ for each $x \in \mathbb{R}$. This family $\{A(x)\}$ defines a decomposable operator belonging to $\mathcal{B}(\mathcal{H})$ [Proposition 5.24(a)]. We denote this operator by $A'$ (so $A' = \{A(x)\}$), and it suffices to show that $A = A'$.

Let us fix a number $k \in \mathbb{N}$. Let $h = \{h(x)\} \in \mathcal{H}$. By using the hypothesis that $E(\mathcal{V})A^* = A^*E(\mathcal{V})$ for each Borel set $\mathcal{V}$, one sees that

$$\begin{aligned}\int_{\mathcal{V}} \langle A(x)h(x), e_k(x)\rangle_{\mathcal{K}}\, m(dx) &= \int_{\mathcal{V}} \langle h(x), B(x)e_k(x)\rangle_{\mathcal{K}}\, m(dx) \\ &= \int_{\mathcal{V}} \langle h(x), [A^*e_k](x)\rangle_{\mathcal{K}}\, m(dx) = \langle h, E(\mathcal{V})A^*e_k\rangle \\ &= \langle h, A^*E(\mathcal{V})e_k\rangle = \langle Ah, E(\mathcal{V})e_k\rangle = \int_{\mathcal{V}} \langle [Ah](x), e_k(x)\rangle_{\mathcal{K}}\, m(dx).\end{aligned}$$

This implies that $\langle A(x)h(x), e_k(x)\rangle_{\mathcal{K}} = \langle [Ah](x), e_k(x)\rangle_{\mathcal{K}}$ $m$-a.e.[16] Since $e_k(x) = \theta(x)e_k^o$ and $\theta(x) \neq 0$, it follows that $\langle A(x)h(x), e_k^o\rangle_{\mathcal{K}} = \langle [Ah](x), e_k^o\rangle_{\mathcal{K}}$ $m$-a.e., i.e. for all $x \in \mathbb{R}$ not belonging to a certain Borel set $\Sigma_k(h)$ of measure zero. If one denotes by $\Sigma(h)$ the union of all sets $\Sigma_k(h)$, then $\Sigma(h)$ is a Borel set of measure zero, and one has $\langle A(x)h(x), e_k^o\rangle_{\mathcal{K}} = \langle [Ah](x), e_k^o\rangle_{\mathcal{K}}$ for all $x \notin \Sigma(h)$ and for all $k$. Since the family $\{e_k^o\}$ is total in $\mathcal{K}$, it follows (Proposition 1.6) that $[Ah](x) = A(x)h(x) = [A'h](x)$ for $x \notin \Sigma(h)$. Thus one has $Ah = A'h$ for each $h \in \mathcal{H}$, i.e. $A = A'$. □

## 5.6 Scattering cross sections

We begin with a mathematical model of a scattering experiment in which a beam of independent particles is scattered by a fixed scattering centre. We shall study the temporal behaviour of a single particle of this beam and then deduce (by using the scattering operator describing the scattering of this particle by the scattering centre) a formula for the probability that this particle will be observed, after the scattering has taken place ($t \to +\infty$), in a cone $\mathcal{C}$ with vertex at the co-ordinate origin. By adding the probabilities for all particles of the beam, one obtains the scattering cross section for the cone $\mathcal{C}$ (here one uses the hypothesis that the particles of the beam are scattered independently, i.e. one after another, because one adds the probabilities rather than the probability amplitudes – no super-positions of the states of the different particles). Let us recall the usual definition of the scattering cross section for a cone $\mathcal{C}$:

$$\sigma(\mathcal{C}) = \frac{\text{number of particles scattered into } \mathcal{C} \text{ per unit time}}{\text{number of particles incident per unit time and unit surface area}},$$

where the surface area is determined in the plane orthogonal to the direction of propagation of the beam.

We assume that $\mathcal{H} = L^2(\mathbb{R}^n)$ with $n \geq 2$ and that the free evolution is described by the group $U_t^0$ which was discussed in §5.4.2. We fix a direction $\vec{\omega}_0$ (a vector in $\mathbb{R}^n$ of length 1) which will give (approximately) the direction of the velocity of the particles of the incident beam, and we denote by $\Pi_0$ the hyperplane of dimension $n-1$ orthogonal to $\vec{\omega}_0$ and containing the origin. The beam will be described by an ensemble of vectors $\{f_{\vec{b}}\}_{\vec{b}\in\Pi_0}$ in $\mathcal{H}$, obtained from a fixed vector $f$ (with $\|f\| = 1$) by translating $f$ by $\vec{b}$ in the hyperplane $\Pi_0$ [more precisely by acting with the operator $T(\vec{b})$ on $f$, where $\{T(\cdot)\}$ is the usual unitary representation of the translation group of $\mathbb{R}^n$ in the Hilbert space $L^2(\mathbb{R}^n)$], and $f$ is such that the support of its Fourier transform $\tilde{f}$ is contained in a (small) neighbourhood of $\sqrt{\lambda_0}\,\vec{\omega}_0$. Here $\lambda_0$ is a fixed positive

[16] If $\psi$ is a function in $L^1(\mathbb{R}, m)$ satisfying $\int_{\mathcal{V}} \psi(x)m(dx) = 0$ for each Borel set $\mathcal{V}$, then $\psi(x) = 0$ $m$-a.e. This is clear if one knows that $\psi(x) \geq 0$ $m$-almost everywhere. If $\psi$ is real, one is reduced to the preceding situation by setting $\mathcal{W}_+ = \{x \mid \psi(x) \geq 0\}$ and $\mathcal{W}_- = \{x \mid \psi(x) < 0\}$; by observing that $\int_{\mathcal{V}} \psi(x)\chi_{\mathcal{W}_+}(x)m(dx) = \int_{\mathcal{V}\cap\mathcal{W}_+} \psi(x)m(dx) = 0$, one obtains $\psi(x)\chi_{\mathcal{W}_+}(x) = 0$ $m$-a.e. Similarly one finds that $\psi(x)\chi_{\mathcal{W}_-}(x) = 0$ $m$-a.e. If $\psi$ is complex, one applies the preceding argument to the real part and to the imaginary part of $\psi$ (observe that the relation $\int_{\mathcal{V}} \psi(x)m(dx) = 0$ implies that $\int_{\mathcal{V}} \Re[\psi(x)]m(dx) = \Re\int_{\mathcal{V}} \psi(x)m(dx) = 0$, and similarly for the imaginary part).

number, corresponding to the kinetic energy of a particle of the beam. Thus, to a good approximation, the momentum of each particle of the beam is $\vec{k}_0 \equiv \sqrt{\lambda_0}\,\vec{\omega}_0$.

So let $f$ be a vector in $L^2(\mathbb{R}^n)$ such that $\|f\| = 1$. We assume that its Fourier transform $\tilde{f}$ vanishes outside a compact subset (bounded, and also small) of the half-space $\{\vec{k} \in \mathbb{R}^n \mid \vec{k} \cdot \vec{\omega}_0 > 0\}$, i.e. that $\operatorname{supp}\tilde{f} \subset \{\vec{k} \in \mathbb{R}^n \mid \vec{k} \cdot \vec{\omega}_0 > 0\}$, where $\operatorname{supp}\tilde{f}$ stands for the support of the function $\tilde{f}$. We denote by $\mathcal{C}_{\text{in}}$ the cone in $\mathbb{R}^n$ (with vertex at the origin) that is spanned by $\operatorname{supp}\tilde{f}$, i.e. $\mathcal{C}_{\text{in}} = \{\vec{k} \in \mathbb{R}^n \mid \alpha\vec{k} \in \operatorname{supp}\tilde{f} \text{ for some } \alpha > 0\}$, and we assume that $\vec{\omega}_0 \in \mathcal{C}_{\text{in}}$. Finally we assume that $\tilde{f} \in L^\infty(\mathbb{R}^n)$. For each $\vec{b} \in \Pi_0$ we define the vector $f_{\vec{b}}$ by

$$f_{\vec{b}}(\vec{x}) = f(\vec{x} - \vec{b}) \qquad \text{or} \qquad [\mathcal{F}f_{\vec{b}}](\vec{k}) = e^{-i\vec{k}\cdot\vec{b}}\tilde{f}(\vec{k}). \tag{5.78}$$

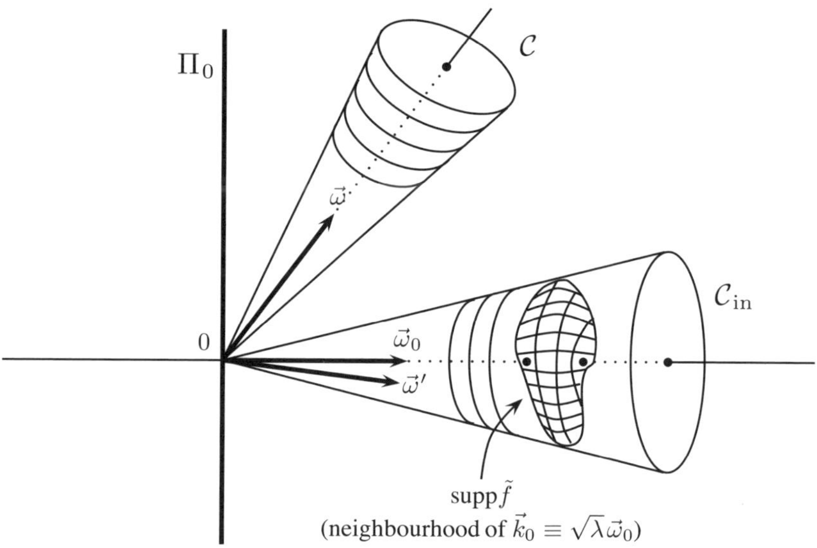


**Figure 5.2**

If $\mathcal{V}$ is a Borel set in $\mathbb{R}^n$ and $h \in L^2(\mathbb{R}^n)$, with $\|h\| = 1$, we denote by $\mathcal{P}(h;\mathcal{V})$ the probability that the scattering state $\Omega_- h$ associated with the initial state $h$ be localised in the region $\mathcal{V}$ of configuration space at time $t = +\infty$, i.e. (assuming that the limit exists):

$$\mathcal{P}(h;\mathcal{V}) = \lim_{t\to+\infty} \int_{\mathcal{V}} |(U_t\Omega_- h)(\vec{x})|^2\, d^n x. \tag{5.79}$$

If one adds the probabilities $\mathcal{P}(f_{\vec{b}};\mathcal{V})$ of all states in the ensemble $\{f_{\vec{b}}\}$, one obtains a quantity representing the number $n(f;\mathcal{V})$ of scattered particles localised asymptotically in $\mathcal{V}$:

$$n(f;\mathcal{V}) = \int_{\Pi_0} \mathcal{P}(f_{\vec{b}};\mathcal{V})\, d^{n-1}b. \tag{5.80}$$

We are mostly interested in the case in which $\mathcal{V}$ is a truncated cone $\mathcal{C}_\rho$, defined as $\mathcal{C}_\rho = \{\vec{x} \in \mathcal{C} \mid |\vec{x}| > \rho\}$, where $\mathcal{C}$ is a fixed cone with vertex at the origin and $\rho$ is a positive number (the observation of the scattered particles will be made at distances greater than $\rho$).

It should be mentioned here that in most texts on quantum mechanics, the scattering cross section is introduced in terms of probability currents across distant surfaces. The current density associated with a (differentiable) state vector $g \in L^2(\mathbb{R}^n)$ is given as $\vec{j}(\vec{x}) = 2\,\Im[\overline{g(\vec{x})}\vec{\nabla}g(\vec{x})]$, and the flux $\Phi_{\mathcal{O}}(g)$ across a hypersurface $\mathcal{O}$ is then

$$\Phi_{\mathcal{O}}(g) = 2\,\Im \int_{\mathcal{O}} \overline{g(\vec{x})}\,\vec{n}(\vec{x}) \cdot \vec{\nabla}g(\vec{x})\,d\sigma(\vec{x}), \tag{5.81}$$

where $\vec{n}(\vec{x}) \in \mathbb{R}^n$ is a unit vector orthogonal to $\mathcal{O}$ at the point $\vec{x} \in \mathcal{O}$ and $d\sigma(\vec{x})$ denotes the surface element at $\vec{x}$. Let $\mathcal{O} = S_d \cap \mathcal{C}$, where $S_d = \{\vec{x} \in \mathbb{R}^n \mid |\vec{x}| = d\}$, and let $\vec{n}(\vec{x})$ be the outward unit vector at the point $\vec{x} \in \mathcal{O}$. Then the integrated flux (counted positively in the outward direction) associated with $U_t\Omega_- h$, after some initial time $T$, across this hypersurface is given by

$$\int_T^\infty \Phi_{S_d \cap \mathcal{C}}(U_t\Omega_- h)\,dt.$$

It is of considerable interest to know that, for very large $d$ (and any finite $T$ and $\rho$), this quantity is the same as the probability $\mathcal{P}(h;\mathcal{C}_\rho)$ that the scattering state associated with $h$ is localised in the truncated cone $C_\rho$ at $t = +\infty$. More precisely one has, under suitable conditions:

$$\mathcal{P}(h;\mathcal{C}_\rho) = \lim_{d\to\infty} \int_T^\infty \Phi_{S_d \cap \mathcal{C}}(U_t\Omega_- h)\,dt. \tag{5.82}$$

A proof of Eq. (5.82) is presented in Section 7.2. The quantity on its right-hand side seems closer to an experimental situation, where one counts the particles passing (at some random time) through a detector located far from the scattering centre. Of course, in order to interpret the expression (5.82) as a scattering cross section, one would have to know that at large positive times the probability current through $S_d \cap \mathcal{C}$ is essentially outgoing. This point will be discussed at the end of Section 7.2. Here we work with the expression (5.79) for $\mathcal{P}(h;\mathcal{C}_\rho)$ in terms of the probability of presence in the truncated cone $\mathcal{C}_\rho$.

The ***scattering cross section*** $\sigma(f;\mathcal{C})$ for scattering into the cone $\mathcal{C}$, for a beam defined in terms of a vector $f$ as described above, is then simply $\sigma(f;\mathcal{C}) = n(f;\mathcal{C}_\rho)$, because the number of incident particles per unit surface in $\Pi_0$ is equal to 1 [it is given by $\int_\Sigma d^{n-1}b$, where $\Sigma$ is a subset of $\Pi_0$ of measure 1, the measure being the $(n-1)$-dimensional Lebesgue measure] [17]. We shall see that, under reasonable assumptions,

[17] Until now we have considered the scattering of a beam of particles by a *single* scattering centre. Often the target consists of a large number $N$ of scattering centres ($N$ of the order of $10^{20}$). If the spatial distribution of these scattering centres is random (and if multiple scattering is neglected), then the scattering cross section for the cone $\mathcal{C}$ will simply be $Nn(f;\mathcal{C}_\rho)$, because the probabilities for scattering into $\mathcal{C}$

the quantity $n(f;\mathcal{C}_\rho)$ is independent of $\rho$. We begin with a fundamental result, the ***Scattering-Into-Cones Formula***; this formula expresses a fact that is intuitively evident: if $h$ is an initial state, then the probability $\mathcal{P}(h;\mathcal{C}_\rho)$ of finding the out-coming particle in the truncated cone $\mathcal{C}_\rho$ is just the probability that the momentum of the corresponding final state $Sh$ lies in the cone $\mathcal{C}$, i.e. the probability that the out-coming particle propagates in a direction in which it will eventually penetrate into the cone $\mathcal{C}$.

**Proposition 5.27.** *Assume that the wave operators $\Omega_\pm = \Omega_\pm(H,H_0)$ exist and that $\mathcal{R}(\Omega_-)\subseteq\mathcal{R}(\Omega_+)$. Then one has for each $h\in L^2(\mathbb{R}^n)$ and for each $\rho>0$:*

$$\mathcal{P}(h;\mathcal{C}_\rho) = \int_{\mathcal{C}} \left|(\widetilde{Sh})(\vec{k})\right|^2 d^n k. \tag{5.83}$$

PROOF. If $\mathcal{V}$ is a Borel set in $\mathbb{R}^n$, we denote by $F_\mathcal{V}$ the projection onto the subspace of vectors with support in $\mathcal{V}$, i.e. $F_\mathcal{V} = \chi_\mathcal{V}(\vec{Q})$, where $\chi_\mathcal{V}$ is the characteristic function of $\mathcal{V}$. As before, $F_+$ is the projection onto the subspace $\mathcal{R}(\Omega_+)$.

(i) We first show that, under the assumptions of the proposition, one has for each Borel set $\mathcal{V}$ (again assuming that the limit exists):

$$\mathcal{P}(h;\mathcal{V}) = \lim_{t\to+\infty}\int_\mathcal{V}\left|[U_t^0 Sh](\vec{x})\right|^2 d^n x \equiv \lim_{t\to+\infty}\|F_\mathcal{V}U_t^0 Sh\|^2. \tag{5.84}$$

Indeed: $\mathcal{P}(h;\mathcal{V}) = \lim_{t\to+\infty}\|F_\mathcal{V}U_t\Omega_-h\|^2$, and it suffices to see that

$$\lim_{t\to+\infty}\{\|F_\mathcal{V}U_t\Omega_-h\|^2 - \|F_\mathcal{V}U_t^0Sh\|^2\} = 0. \tag{5.85}$$

Now [using (1.15)]:

$$\begin{aligned} &\big|\|F_\mathcal{V}U_t\Omega_-h\|^2 - \|F_\mathcal{V}U_t^0Sh\|^2\big| \\ &= \big[\|F_\mathcal{V}U_t\Omega_-h\| + \|F_\mathcal{V}U_t^0Sh\|\big]\cdot\big|\|F_\mathcal{V}U_t\Omega_-h\| - \|F_\mathcal{V}U_t^0Sh\|\big| \\ &\le 2\|h\|\|F_\mathcal{V}U_t\Omega_-h - F_\mathcal{V}U_t^0Sh\| \le 2\|h\|\|U_t\Omega_-h - U_t^0Sh\|. \end{aligned} \tag{5.86}$$

As $t\to+\infty$ one has

$$\|U_t\Omega_-h - U_t^0Sh\| \to \|\Omega_-h-\Omega_+Sh\| = \|\Omega_-h-\Omega_+\Omega_+^*\Omega_-h\| = \|\Omega_-h - F_+\Omega_-h\|.$$

Under the assumption that $\mathcal{R}(\Omega_-)\subseteq\mathcal{R}(\Omega_+)$ one has $F_+\Omega_-h = \Omega_-h$. Thus (5.84) is proved.

(ii) We now show that $\mathcal{P}(h;\mathcal{C}_\rho)$ does not depend on $\rho$, more precisely that one has $\mathcal{P}(h;\mathcal{C}_\rho) = \mathcal{P}(h;\mathcal{C})$ for each $\rho>0$. For this we decompose the cone $\mathcal{C}$ into two disjoint parts $\mathcal{C} = \mathcal{C}_\rho\cup[\mathcal{C}\backslash\mathcal{C}_\rho]$, which implies that

$$\mathcal{P}(h;\mathcal{C}) = \mathcal{P}(h;\mathcal{C}_\rho) + \mathcal{P}(h;\mathcal{C}\backslash\mathcal{C}_\rho),$$

from the different scattering centres are added. On the other hand if the spatial distribution of the scattering centres presents some regularity (for example if the target is a crystal), the theory discussed here cannot be directly applied (one must superpose the wave functions scattered by the different centres rather than add the probabilities).

and it suffices to show that $\mathcal{P}(h;\mathcal{C}\backslash\mathcal{C}_\rho) = 0$. Now (5.84) implies that (we use the notations of page 205)

$$\mathcal{P}(h;\mathcal{C}\backslash\mathcal{C}_\rho) = \lim_{t\to+\infty} \|F_{\mathcal{C}\backslash\mathcal{C}_\rho} U_t^0 Sh\|^2 = \lim_{t\to+\infty} \int_{\mathcal{C}\backslash\mathcal{C}_\rho} \left|(U_t^0 Sh)(\vec{x})\right|^2 d^n x$$
$$\leq \lim_{t\to+\infty} \int_{B_\rho} \left|(U_t^0 Sh)(\vec{x})\right|^2 d^n x = \lim_{t\to+\infty} \|\chi_\rho(\vec{Q}) U_t^0 Sh\|^2,$$

and this limit is zero because $\mathcal{M}_\infty(H_0) = \mathcal{H}$.

(iii) Let us finally calculate $\mathcal{P}(h;\mathcal{C}) \equiv \lim_{t\to+\infty} \|F_\mathcal{C} U_t^0 Sh\|^2$. We denote by $\chi_\mathcal{C}$ the characteristic function of the cone $\mathcal{C}$ [so that $F_\mathcal{C} = \chi_\mathcal{C}(\vec{Q})$], and we use (5.62):

$$\|F_\mathcal{C} U_t^0 Sh\| = \|U_t^{0*} F_\mathcal{C} U_t^0 Sh\| = \|Z_t^* \chi_\mathcal{C}(2t\vec{P}) Z_t Sh\| = \|\chi_\mathcal{C}(2t\vec{P}) Z_t Sh\|$$
$$= \|\chi_\mathcal{C}(2t\vec{P})[Z_t - I]Sh + \chi_\mathcal{C}(2t\vec{P})Sh\|.$$

Now $\|\chi_\mathcal{C}(2t\vec{P})[Z_t - I]Sh\| \leq \|[Z_t - I]Sh\| \to 0$ as $t \to +\infty$ (Proposition 5.18). Hence $\mathcal{P}(h;\mathcal{C}) \equiv \lim_{t\to+\infty} \|F_\mathcal{C} U_t^0 Sh\|^2 = \lim_{t\to+\infty} \|\chi_\mathcal{C}(2t\vec{P})Sh\|^2$. If $t > 0$ one has $\chi_\mathcal{C}(2t\vec{k}) = \chi_\mathcal{C}(\vec{k})$ for each $\vec{k} \in \mathbb{R}^n$, because for $t > 0$: $2t\vec{k} \in \mathcal{C}$ if and only if $\vec{k} \in \mathcal{C}$. Thus, writing the last norm in the momentum representation $\tilde{L}^2(\mathbb{R}^n)$:

$$\mathcal{P}(h;\mathcal{C}) = \lim_{t\to+\infty} \|\chi_\mathcal{C}(2t\vec{P})Sh\|^2 = \lim_{t\to+\infty} \|\chi_\mathcal{C}(\vec{P})Sh\|^2 = \int_\mathcal{C} \left|(\widetilde{Sh})(\vec{k})\right|^2 d^n k. \quad \square$$

It is useful to introduce an operator $R$ as follows

$$R = S - I. \tag{5.87}$$

Clearly $R$ is decomposable in the spectral representation of $H_0$:

$$(\mathcal{U}_0 Rh)_\lambda = R(\lambda)(\mathcal{U}_0 h)_\lambda, \qquad \text{with} \qquad R(\lambda) = S(\lambda) - I_\mathcal{K}, \tag{5.88}$$

where $I_\mathcal{K}$ denotes the identity operator in the Hilbert space $\mathcal{K} = L^2(S^{n-1})$. If $f$ is the state vector defining the incident beam (as described on page 236, with $\operatorname{supp}\tilde{f} \subset \mathcal{C}_{\text{in}}$), and if the cone $\mathcal{C}$ is disjoint from $\mathcal{C}_{\text{in}}$, one can express the formula for $\mathcal{P}(f;\mathcal{C}_\rho)$ given in (5.83) in terms of the operator $R$ (assuming the assumptions of Proposition 5.27 to be satisfied):

$$\mathcal{P}(f;\mathcal{C}_\rho) = \int_\mathcal{C} \left|(\widetilde{Rf})(\vec{k})\right|^2 d^n k \qquad \text{if } \mathcal{C} \cap \mathcal{C}_{\text{in}} = \varnothing. \tag{5.89}$$

Indeed, by writing $S = R + I$ and setting $h = f$ in (5.83), one has

$$\mathcal{P}(f;\mathcal{C}_\rho) = \int_\mathcal{C} \left|(\widetilde{Rf})(\vec{k})\right|^2 d^n k + \int_\mathcal{C} \left|\tilde{f}(\vec{k})\right|^2 d^n k + 2\Re \int_\mathcal{C} \overline{\tilde{f}(\vec{k})} (\widetilde{Rf})(\vec{k})\, d^n k,$$

and the last two terms are zero because $\tilde{f} = 0$ for $\vec{k} \in \mathcal{C}$ if $\operatorname{supp}\tilde{f} \subset \mathcal{C}_{\text{in}}$ and $\mathcal{C} \cap \mathcal{C}_{\text{in}} = \varnothing$.

To obtain an expression for $\sigma(f;\mathcal{C})$ having a simple interpretation, one must suppose that the ***R-matrix*** $R(\lambda)$ is an integral operator in $L^2(S^{n-1})$ (i.e. an integral operator with respect to the angular variables in momentum space, $\lambda$ being fixed). We know a criterion for this to be the case: it suffices to know that $R(\lambda)$ is a Hilbert-Schmidt operator (see Proposition 2.15). Further on we specify conditions implying that $R(\lambda) \in \mathcal{B}_2(L^2(S^{n-1}))$. We denote by $\|R(\lambda)\|_{\text{HS}}$ the Hilbert-Schmidt norm [in $L^2(S^{n-1})$] of the operator $R(\lambda)$ and by $r(\lambda;\vec{\omega},\vec{\omega}')$ its kernel. Thus one has

$$\|R(\lambda)\|_{\text{HS}}^2 = \int_{S^{n-1}} d\omega \int_{S^{n-1}} d\omega' \, |r(\lambda;\vec{\omega},\vec{\omega}')|^2. \tag{5.90}$$

**Proposition 5.28.** *Assume that $R(\lambda) \in \mathcal{B}_2(L^2(S^{n-1}))$ for each $\lambda > 0$. With the notations introduced above and under the hypotheses made (in particular $\mathcal{C} \cap \mathcal{C}_{\text{in}} = \varnothing$), the scattering cross section $\sigma(f;\mathcal{C})$ for scattering into the cone $\mathcal{C}$, for a beam defined in terms of a vector $f$ ($\|f\| = 1$), is given by the following expression:*

$$\sigma(f;\mathcal{C}) = \int_{S^{n-1}\cap\mathcal{C}} d\omega \int_0^\infty d\lambda \int_{S^{n-1}} d\omega' \left[\frac{2\pi}{\sqrt{\lambda}}\right]^{n-1} \frac{1}{\vec{\omega}_0\cdot\vec{\omega}'} \left|r(\lambda;\vec{\omega},\vec{\omega}')\right|^2 \left|(\mathcal{U}_0 f)_\lambda(\vec{\omega}')\right|^2. \tag{5.91}$$

PROOF. (i) We know from the remarks preceding Proposition 5.27 that

$$\sigma(f;\mathcal{C}) = n(f;\mathcal{C}_\rho) = \int_{\Pi_0} \mathcal{P}(f_{\vec{b}};\mathcal{C}_\rho)\, d^{n-1}b.$$

Now the support of the Fourier transform of $f_{\vec{b}}$ is identical with that of $f$ [see (5.78)], hence contained in $\mathcal{C}_{\text{in}}$. Since $\mathcal{C} \cap \mathcal{C}_{\text{in}} = \varnothing$, we can use (5.89) to find an expression for $\mathcal{P}(f_{\vec{b}};\mathcal{C}_\rho)$. By using the spectral representation of $H_0$, introduced in §5.5.1, one obtains (see also Figure 5.2):

$$\begin{aligned}
\sigma(f;\mathcal{C}) &= \int_{\Pi_0} d^{n-1}b \int_0^\infty d\lambda \int_{S^{n-1}\cap\mathcal{C}} d\omega \left|(\mathcal{U}_0 R f_{\vec{b}})_\lambda(\vec{\omega})\right|^2 \\
&= \int_{\Pi_0} d^{n-1}b \int_0^\infty d\lambda \int_{S^{n-1}\cap\mathcal{C}} d\omega \left|[R(\lambda)(\mathcal{U}_0 f_{\vec{b}})_\lambda](\vec{\omega})\right|^2 \\
&= \int_{\Pi_0} d^{n-1}b \int_0^\infty d\lambda \int_{S^{n-1}\cap\mathcal{C}} d\omega \left|\int_{S^{n-1}\cap\mathcal{C}_{\text{in}}} r(\lambda;\vec{\omega},\vec{\omega}')(\mathcal{U}_0 f_{\vec{b}})_\lambda(\vec{\omega}')\, d\omega'\right|^2 \\
&= \int_{\Pi_0} d^{n-1}b \int_0^\infty d\lambda \int_{S^{n-1}\cap\mathcal{C}} d\omega \left|\int_{S^{n-1}\cap\mathcal{C}_{\text{in}}} r(\lambda;\vec{\omega},\vec{\omega}') e^{-i\lambda^{1/2}\vec{\omega}'\cdot\vec{b}} (\mathcal{U}_0 f)_\lambda(\vec{\omega}')\, d\omega'\right|^2.
\end{aligned}$$

We interchange the order of integration over $d^{n-1}b$ with those over $d\lambda$ and $d\omega$. To arrive at (5.91) it then is enough to verify that

$$\begin{aligned}
&\int_{\Pi_0} d^{n-1}b \left|\int_{S^{n-1}\cap\mathcal{C}_{\text{in}}} r(\lambda;\vec{\omega},\vec{\omega}') e^{-i\lambda^{1/2}\vec{\omega}'\cdot\vec{b}} (\mathcal{U}_0 f)_\lambda(\vec{\omega}')\, d\omega'\right|^2 \\
&\qquad = \left[\frac{2\pi}{\sqrt{\lambda}}\right]^{n-1} \int_{S^{n-1}} \left|r(\lambda;\vec{\omega},\vec{\omega}')\right|^2 \left|(\mathcal{U}_0 f)_\lambda(\vec{\omega}')\right|^2 \frac{1}{\vec{\omega}_0\cdot\vec{\omega}'}\, d\omega'.
\end{aligned} \tag{5.92}$$

(ii) To prove (5.92), let us set $\varphi_{\lambda,\vec{\omega}}(\vec{\omega}') = r(\lambda;\vec{\omega},\vec{\omega}')(\mathcal{U}_0 f)_\lambda(\vec{\omega}')\chi_{\mathcal{C}_{\text{in}}\cap S^{n-1}}(\vec{\omega}')$, where $\chi_{\mathcal{C}_{\text{in}}\cap S^{n-1}}$ denotes the characteristic function of $C_{\text{in}} \cap S^{n-1}$. By virtue of the hypotheses made $[R(\lambda) \in \mathcal{B}_2(L^2(S^{n-1})),\ \tilde{f} \in L^\infty(\mathbb{R}^n)]$, the function $\varphi_{\lambda,\vec{\omega}}(\cdot)$ belongs to $L^2(S^{n-1})$ for each $\lambda$ and almost all $\vec{\omega}$. Its support is contained in the hemisphere $\{\vec{\omega}' \mid \vec{\omega}_0\cdot\vec{\omega}' > 0\}$. This allows us to parametrise the points $\vec{\omega}'$ in the support of $\varphi_{\lambda,\vec{\omega}}$ by their orthogonal projection onto the hyperplane $\Pi_0$; in other terms we write $\vec{\omega}' = \vec{z} + (\vec{\omega}_0\cdot\vec{\omega}')\vec{\omega}_0$ with $\vec{z}\in\Pi_0$, and we make the change of variables $\vec{\omega}' \mapsto \vec{z}$. We then have $d\omega' = [\vec{\omega}_0\cdot\vec{\omega}']^{-1}d^{n-1}z$ and $\vec{\omega}'\cdot\vec{b} = \vec{z}\cdot\vec{b}$ for $\vec{b}\in\Pi_0$. If we set $\vec{b}' = \sqrt{\lambda}\,\vec{b}$ and use the unitarity of the Fourier transformation in $L^2(\mathbb{R}^{n-1})$, we find that

$$\begin{aligned}
&\int_{\Pi_0} d^{n-1}b \left|\int_{S^{n-1}\cap\mathcal{C}_{\text{in}}} r(\lambda;\vec{\omega},\vec{\omega}')e^{-i\lambda^{1/2}\vec{\omega}'\cdot\vec{b}}(\mathcal{U}_0 f)_\lambda(\vec{\omega}')d\omega'\right|^2\\
&\quad\equiv \int_{\Pi_0} d^{n-1}b \left|\int_{S^{n-1}} e^{-i\lambda^{1/2}\vec{\omega}'\cdot\vec{b}}\varphi_{\lambda,\vec{\omega}}(\vec{\omega}')d\omega'\right|^2\\
&\quad= \int_{\Pi_0} \frac{d^{n-1}b'}{\lambda^{(n-1)/2}} \left|\int_{\Pi_0} e^{-i\vec{b}'\cdot\vec{z}}\varphi_{\lambda,\vec{\omega}}(\vec{\omega}'(\vec{z}))\,\frac{d^{n-1}z}{\vec{\omega}_0\cdot\vec{\omega}'(\vec{z})}\right|^2\\
&\quad= \left[\frac{2\pi}{\sqrt{\lambda}}\right]^{n-1}\int_{\Pi_0} \left|\varphi_{\lambda,\vec{\omega}}(\vec{\omega}'(\vec{z}))\right|^2\left[\vec{\omega}_0\cdot\vec{\omega}'(\vec{z})\right]^{-2}d^{n-1}z\\
&\quad= \left[\frac{2\pi}{\sqrt{\lambda}}\right]^{n-1}\int_{\Pi_0} \left|\varphi_{\lambda,\vec{\omega}}(\vec{\omega}')\right|^2\left[\vec{\omega}_0\cdot\vec{\omega}'\right]^{-1}d\omega'.
\end{aligned}$$

Upon inserting the definition of $\varphi_{\lambda,\vec{\omega}}(\vec{\omega}')$, one obtains (5.92). □

DISCUSSION OF EQUATION (5.91). Let us recall this equation:

$$\sigma(f;\mathcal{C}) = \int_{S^{n-1}\cap\mathcal{C}} d\omega \int_0^\infty d\lambda \int_{S^{n-1}} d\omega' \left[\frac{2\pi}{\sqrt{\lambda}}\right]^{n-1}\frac{1}{\vec{\omega}_0\cdot\vec{\omega}'}\left|r(\lambda;\vec{\omega},\vec{\omega}')\right|^2\left|(\mathcal{U}_0 f)_\lambda(\vec{\omega}')\right|^2. \tag{5.93}$$

We have assumed that the support of $\tilde{f}$ is contained in a small neighbourhood of a vector $\vec{k}_0 = \sqrt{\lambda_0}\,\vec{\omega}_0$. Thus the integral with respect to $d\omega'$ in (5.93) extends only over a small neighbourhood of $\vec{\omega}_0$, i.e. one has $\cos(\vec{\omega}_0\cdot\vec{\omega}') \approx 1$ for each $\vec{\omega}'$ belonging to this set. Moreover, if (for the vectors $\vec{\omega}$ in $\mathcal{C}$), $r(\lambda;\vec{\omega},\vec{\omega}')$ is a continuous function of $\vec{\omega}'$ and of $\lambda$ as $\vec{\omega}'$ varies over $\mathcal{C}_{\text{in}} \cap S^{n-1}$ and $\sqrt{\lambda}$ varies over a neighbourhood of $|\vec{k}_0|$ (more precisely as $\lambda$ varies over the support of the vector $f$ with respect to the spectral family of $H_0$), we may replace in Eq. (5.93) the quantity $|r(\lambda;\vec{\omega},\vec{\omega}')|^2$ by $|r(\lambda_0;\vec{\omega},\vec{\omega}_0)|^2$, where $\lambda_0 = |\vec{k}_0|^2$. Thus, if the support of $\tilde{f}$ is sufficiently small, one will have to a very good approximation [18]

$$\begin{aligned}
\sigma(f;\mathcal{C}) &= \int_{S^{n-1}\cap\mathcal{C}} d\omega \int_0^\infty d\lambda \int_{S^{n-1}} d\omega' \left[\frac{2\pi}{\sqrt{\lambda_0}}\right]^{n-1}\left|r(\lambda_0;\vec{\omega},\vec{\omega}_0)\right|^2\left|(\mathcal{U}_0 f)_\lambda(\vec{\omega}')\right|^2\\
&= \left[\frac{2\pi}{\sqrt{\lambda_0}}\right]^{n-1}\int_{S^{n-1}\cap\mathcal{C}}\left|r(\lambda_0;\vec{\omega},\vec{\omega}_0)\right|^2 d\omega.
\end{aligned} \tag{5.94}$$

[18] A rigourous justification can be obtained for example by considering a sequence $\{f_j\}$ of initial states such that $|(\mathcal{U}_0 f_j)_\lambda(\vec{\omega}')|^2$ is a Dirac sequence, i.e. such that $|(\mathcal{U}_0 f_j)_\lambda(\vec{\omega}')|^2 \to \delta(\lambda-\lambda_0)\delta(\vec{\omega}'-\vec{\omega}_0)$ as $j \to \infty$, see page 287 of [AJS].

Then the right-hand side of Eq. (5.94) can be interpreted as the scattering cross section $\sigma(\lambda_0, \vec{\omega}_0; \mathcal{C})$, for an initial beam of particles of momentum $\vec{k}_0 = \sqrt{\lambda_0}\,\vec{\omega}_0$, for scattering into a cone $\mathcal{C}$ satisfying $\mathcal{C} \cap \mathcal{C}_{\text{in}} = \varnothing$ for some open cone $\mathcal{C}_{\text{in}}$ with $\vec{\omega}_0 \in \mathcal{C}_{\text{in}}$. One sees that $\sigma(\lambda_0, \vec{\omega}_0; \cdot)$ defines a measure on $S^{n-1}\backslash\{\vec{\omega}_0\}$ which is absolutely continuous with respect to Lebesgue measure $d\omega$. The Radon-Nikodym derivative of this measure is called the ***differential scattering cross section*** $d\sigma/d\omega$:

$$\frac{d\sigma}{d\omega}(\lambda_0, \vec{\omega}_0; \vec{\omega}) = \left[\frac{2\pi}{\sqrt{\lambda_0}}\right]^{n-1} \left|r(\lambda_0; \vec{\omega}, \vec{\omega}_0)\right|^2, \tag{5.95}$$

and the quantity

$$f(\lambda_0; \vec{\omega}_0 \to \vec{\omega}) := \left[\frac{-2\pi i}{\sqrt{\lambda_0}}\right]^{(n-1)/2} r(\lambda_0; \vec{\omega}, \vec{\omega}_0) \tag{5.96}$$

may be interpreted as the ***scattering amplitude*** for scattering at energy $\lambda_0$ from the initial direction $\vec{\omega}_0$ into the direction $\vec{\omega}$ (the factor $-i$ has been inserted in order to obtain the usual result for the scattering amplitude calculated by considering the Schrödinger equation in $n = 3$ dimensions; we recall that, by definition, the differential scattering cross section is the square of the absolute value of the scattering amplitude). One sees that the kernel of the integral operator $R(\lambda) = S(\lambda) - I_{\mathcal{K}}$ coincides, apart from a factor, with the scattering amplitude!

Another important quantity is the ***total scattering cross section*** $\sigma_{\text{tot}}(\lambda_0, \vec{\omega}_0)$, defined as the integral of the differential scattering cross section $(d\sigma/d\omega)(\lambda_0, \vec{\omega}_0; \vec{\omega})$ over all final directions $\vec{\omega}$ ($\vec{\omega} \neq \vec{\omega}_0$):

$$\sigma_{\text{tot}}(\lambda_0, \vec{\omega}_0) = \int_{S^{n-1}} \frac{d\sigma}{d\omega}(\lambda_0, \vec{\omega}_0; \vec{\omega})\, d\omega. \tag{5.97}$$

It is also useful to introduce the ***averaged total scattering cross section*** $\overline{\sigma}(\lambda_0)$ obtained by averaging the quantity (5.97) over all initial directions $\vec{\omega}_0$. If we denote by $\Theta_n$ the surface area of the sphere $S^{n-1}$ [$\Theta_n = 4\pi$ if $n = 3$], then (we write $\lambda$ in place of $\lambda_0$):

$$\begin{aligned}\overline{\sigma}(\lambda) &= \frac{1}{\Theta_n}\int_{S^{n-1}} \sigma_{\text{tot}}(\lambda, \vec{\omega}_0)\, d\omega_0 = \frac{1}{\Theta_n}\int_{S^{n-1}} d\omega \int_{S^{n-1}} d\omega_0\, \frac{d\sigma}{d\omega}(\lambda, \vec{\omega}_0; \vec{\omega}) \\ &= \frac{1}{\Theta_n}\left[\frac{2\pi}{\sqrt{\lambda}}\right]^{n-1} \int_{S^{n-1}} d\omega \int_{S^{n-1}} d\omega_0 \left|r(\lambda; \vec{\omega}, \vec{\omega}_0)\right|^2 = \frac{1}{\Theta_n}\left[\frac{2\pi}{\sqrt{\lambda}}\right]^{n-1} \|R(\lambda)\|_{\text{HS}}^2.\end{aligned} \tag{5.98}$$

For example for $n = 3$: $\overline{\sigma}(\lambda) = [\pi/\lambda]\|R(\lambda)\|_{\text{HS}}^2$.

One notices that the total scattering cross section $\sigma_{\text{tot}}(\lambda, \vec{\omega}_0)$ at energy $\lambda$ is finite for almost all initial directions $\vec{\omega}_0$ when $R(\lambda) \in \mathcal{B}_2(L^2(S^{n-1}))$. This question will be further discussed in Section 5.7. If the potential $V$ is spherically symmetric, $\sigma_{\text{tot}}(\lambda, \vec{\omega}_0)$ is independent of the initial direction $\vec{\omega}_0$, and then one has $\overline{\sigma}(\lambda) = \sigma_{\text{tot}}(\lambda, \vec{\omega}_0)$. It is instructive to verify that in this case (with $n = 3$) one obtains the usual expression for the total scattering cross section in terms of the scattering phase shifts. We recall (see §5.5.1) that $S(\lambda)Y_\ell^m = e^{2i\delta_\ell(\lambda)}Y_\ell^m$. Hence

$$R(\lambda)Y_\ell^m = \left[e^{2i\delta_\ell(\lambda)} - 1\right]Y_\ell^m = 2ie^{i\delta_\ell(\lambda)}\sin[\delta_\ell(\lambda)]Y_\ell^m$$

and

$$\sigma_{\text{tot}}(\lambda, \vec{\omega}_0) = \overline{\sigma}(\lambda) = \frac{\pi}{\lambda}\|R(\lambda)\|_{\text{HS}}^2 = \frac{\pi}{\lambda}\sum_{\ell=0}^{\infty}\sum_{m=-\ell}^{\ell}\|R(\lambda)Y_\ell^m\|^2$$

$$= \frac{4\pi}{\lambda}\sum_{\ell=0}^{\infty}\sum_{m=-\ell}^{\ell}\sin^2[\delta_\ell(\lambda)] = \frac{4\pi}{\lambda}\sum_{\ell=0}^{\infty}(2\ell+1)\sin^2[\delta_\ell(\lambda)].$$

## 5.7 Bounds on scattering cross sections

**5.7.1.** We have seen above that the total scattering cross section at energy $\lambda$ is finite when $R(\lambda)$ is a Hilbert-Schmidt operator in $L^2(S^{n-1})$. We describe below the type of potentials $V$ for which this condition is satisfied, but we begin with some general remarks.

One may be surprised that the scattering cross section could be infinite. In fact this possibility is inherent in the definition of the cross section in which one considers the idealisation of an infinitely extended beam (in the directions orthogonal to its direction of propagation, i.e. in the hyperplane $\Pi_0$). This idealisation is equivalent to considering an incident plane wave (which is infinitely extended) in the usual description of scattering by means of the stationary Schrödinger equation. Let us recall the situation in classical mechanics (in $n = 3$ dimensions). If the potential $V$ is spherically symmetric, then the total scattering cross section at each energy $\lambda$ is equal to $\pi R^2$, where $R$ denotes the range of the potential [i.e. the smallest positive number such that $V(r) = 0$ for $r > R$]. In this case the total scattering cross section is identical with the (two-dimensional) area of the obstacle encountered by the particles of the incident beam [if a classical particle is incident with an impact parameter $\vec{b}$ satisfying $|\vec{b}| > R$, it will not penetrate into the region where $V(r) \neq 0$, hence it will not be scattered]. This explains the terminology 'scattering cross section'. If $R = \infty$, i.e. if the potential is non-zero in each neighbourhood of infinity, then the total scattering cross section is infinite.

In quantum mechanics the situation is somewhat different because of the uncertainty relations. If $V(r)$ is not of finite range but converges to zero sufficiently rapidly as $r \to \infty$, then the particles with large impact parameter will be only very weakly scattered (in some sense they are subject to very small forces), and the total scattering cross section is finite. On the other hand, if $V(r)$ converges to zero very slowly, the particles with large impact parameter will be more strongly scattered, which leads to an infinite total scattering cross section (see further on for details). We shall also see at the end of this section that the finiteness or infiniteness of the total scattering cross section depends only on the behaviour of the potential near infinity (the potential may be essentially arbitrary in a finite part of configuration space).

The fact that the scattering cross section is determined by $R(\lambda)$ rather than directly by $S(\lambda)$ is interesting. If the interaction is zero ($V = O$), then there is no scattering and $S = I$. Thus the effect of scattering is described by $R = S - I$ (one has $R = O$ in the absence of interaction). In the formalism of the time-independent Schrödinger equation, this corresponds to splitting the eigenfunction of the Hamiltonian (the outgoing wave function in the stationary formalism) into two parts, the incoming plane wave (the analogue of the operator $I$) and the scattered wave (the analogue of the operator $R$).

To discuss the scattering cross section at a fixed energy $\lambda$, one thus has to study the operator $R(\lambda)$. Now $R(\lambda)$ acts in the Hilbert space $L^2(S^{n-1})$ rather than in the Hilbert space $L^2(\mathbb{R}^n)$ in which the theory is initially defined. It is possible to prove formulas expressing $R(\lambda)$ in terms of the potential, of the resolvent $(H - \lambda - i\varepsilon)^{-1}$ for $\varepsilon \to 0$ [these operators act in $L^2(\mathbb{R}^n)$] and of certain operators from $L^2(\mathbb{R}^n)$ to $L^2(S^{n-1})$ which we call transition operators. These formulas allow one to obtain certain mathematical properties of $R(\lambda)$ [for example continuity or differentiability in $\lambda$, Hilbert-Schmidt property, behaviour of its kernel $r(\lambda; \vec{\omega}, \vec{\omega}')$ as a function of the angular variables] by using properties of the transition operators and the behaviour of the resolvent close to the real axis. An expression for $R(\lambda)$ of the indicated type will be derived in Section 7.4. In the present section we present a relatively simple method for estimating total scattering cross sections in terms of only operators acting in $L^2(\mathbb{R}^n)$. More precisely, we consider averages of the scattering cross section over an interval of energies and obtain bounds for these averages directly in terms of the potential. Let $\varphi\colon (0,\infty) \to [0,\infty)$ be a non-negative weight function, typically a smooth function vanishing outside some finite interval. We consider integrals of the type $\int_0^\infty \varphi(\lambda)\overline{\sigma}(\lambda)d\lambda$, and we shall obtain upper bounds for such integrals in the form of a product of a certain norm of the weight function $\varphi$ and of a certain norm of the potential $V$. In particular, if each of these two norms is finite, one will have $\overline{\sigma}(\lambda) < \infty$ for almost all $\lambda$ in the support of $\varphi$, hence also $R(\lambda) \in \mathcal{B}_2(L^2(S^{n-1}))$ for almost all of these $\lambda$.

**5.7.2.** We consider the spectral representation $L^2((0,\infty); L^2(S^{n-1}), d\lambda)$ of the free Hamiltonian $H_0$ given by means of the mapping $\mathcal{U}_0$ defined in (5.68). If $\eta\colon (0,\infty) \to \mathbb{R}$ is a continuous function, we denote by $P(\eta)$ the following operator in $L^2(\mathbb{R}^n)$ (defined by its action in the spectral representation of $H_0$):

$$\big[\mathcal{U}_0 P(\eta) f\big]_\lambda = \eta(\lambda)\Big[\int_0^\infty \eta(\mu)\big[\mathcal{U}_0 f\big]_\mu d\mu\Big]. \tag{5.99}$$

The integral in (5.99) is an integral of a vector-valued function in $L^2(S^{n-1})$; one may assume $[\mathcal{U}_0 f]_\mu$ to be continuous, because it suffices to define the bounded operator $P(\eta)$ on a dense subset of the underlying Hilbert space.

**Lemma 5.29.** *Assume that $\int_0^\infty |\eta(\lambda)|^2 d\lambda = 1$. Then*
*(a) $P(\eta)$ is a projection, more precisely the projection onto the subspace*

$$\mathcal{M}(\eta) = \{f \in L^2(\mathbb{R}^n) \mid (\mathcal{U}_0 f)_\lambda = \eta(\lambda) g \text{ for some } g \in L^2(S^{n-1})\}.^{19} \tag{5.100}$$

*(b) If $A \in \mathcal{B}(L^2(\mathbb{R}^n))$ is a decomposable operator in the spectral representation of $H_0$, i.e. such that $[\mathcal{U}_0 A f]_\lambda = A(\lambda)(\mathcal{U}_0 f)_\lambda$ with $A(\lambda) \in \mathcal{B}(L^2(S^{n-1}))$, then*

[19] In spherical polar co-ordinates, the Fourier transforms of the elements $f$ of $\mathcal{M}(\eta)$ have the form $\tilde{f}(\vec{k}) = \sqrt{2}|\vec{k}|^{(2-n)/2}\eta(\vec{k}^2)g(\vec{\omega})$, see (5.69).

$$\|AP(\eta)\|_{\mathrm{HS}}^2 = \int_0^\infty |\eta(\lambda)|^2 \|A(\lambda)\|_{\mathrm{HS}}^2 \, d\lambda, \tag{5.101}$$

*where the first Hilbert-Schmidt norm is in* $L^2(\mathbb{R}^n)$ *and the second one in* $L^2(S^{n-1})$.

REMARK. The importance of this result can be seen by taking for $A$ the operator $R \equiv S - I$. Indeed: $\|R(\lambda)\|_{\mathrm{HS}}^2$ is proportional to the total scattering cross section $\overline{\sigma}(\lambda)$ at energy $\lambda$ averaged over the angular variables, and in this case Eq. (5.101), combined with (5.98), allows one to obtain bounds on the scattering cross section by estimating an operator in $L^2(\mathbb{R}^n)$, i.e. without having to consider directly the $S$-matrix (without having to consider restrictions to energy shells). Of course these bounds are not pointwise bounds, i.e. they do not refer to $\overline{\sigma}(\lambda)$ at a fixed energy $\lambda$, but rather some sort of average of $\overline{\sigma}(\lambda)$ over an interval of energy (depending on the choice of $\eta$):

$$\|RP(\eta)\|_{\mathrm{HS}}^2 = (2\pi)^{1-n}\Theta_n \int_0^\infty \lambda^{(n-1)/2} |\eta(\lambda)|^2 \, \overline{\sigma}(\lambda) \, d\lambda, \tag{5.102}$$

where $\Theta_n$ is the surface area of the unit sphere $S^{n-1}$ in $\mathbb{R}^n$.

PROOF OF LEMMA 5.29. We set $\mathcal{K} = L^2(S^{n-1})$.

(a) (i) Let us first check that $P(\eta) \in \mathcal{B}(\mathcal{H})$. By using the continuous triangle inequality (1.28) and then the Schwarz inequality in $L^2(\mathbb{R})$, we obtain:

$$\begin{aligned}
\|P(\eta)f\|^2 &= \int_0^\infty d\lambda \, \eta(\lambda)^2 \Big\| \int_0^\infty \eta(\mu)(\mathcal{U}_0 f)_\mu d\mu \Big\|_{\mathcal{K}}^2 = \Big\| \int_0^\infty \eta(\mu)(\mathcal{U}_0 f)_\mu d\mu \Big\|_{\mathcal{K}}^2 \\
&\le \Big[ \int_0^\infty |\eta(\mu)| \, \|(\mathcal{U}_0 f)_\mu\|_{\mathcal{K}} d\mu \Big]^2 \le \Big[ \int_0^\infty |\eta(\mu)|^2 d\mu \Big] \Big[ \int_0^\infty \|(\mathcal{U}_0 f)_\mu\|_{\mathcal{K}}^2 d\mu \Big] \\
&= \int_0^\infty \|(\mathcal{U}_0 f)_\mu\|_{\mathcal{K}}^2 d\mu = \|f\|^2.
\end{aligned}$$

Hence $\|P(\eta)\| \le 1$.

(ii) Let us show that $P(\eta)^2 = P(\eta)$:

$$\begin{aligned}
\big[\mathcal{U}_0 P(\eta)^2 f\big]_\lambda &= \eta(\lambda) \Big[ \int_0^\infty \eta(\mu) \big[\mathcal{U}_0 P(\eta) f\big]_\mu d\mu \Big] \\
&= \eta(\lambda) \Big[ \int_0^\infty d\mu \, \eta(\mu) \cdot \eta(\mu) \int_0^\infty d\mu' \, \eta(\mu') \big(\mathcal{U}_0 f\big)_{\mu'} \Big] \\
&= \eta(\lambda) \Big[ \int_0^\infty d\mu' \, \eta(\mu') \big(\mathcal{U}_0 f\big)_{\mu'} \Big] = \big[\mathcal{U}_0 P(\eta) f\big]_\lambda,
\end{aligned}$$

where we have used the relation $\int_0^\infty |\eta(\mu)|^2 d\mu = 1$. Let us also see that $P(\eta)^* = P(\eta)$:

$$\begin{aligned}
\langle g, P(\eta) f \rangle = \langle \mathcal{U}_0 g, \mathcal{U}_0 P(\eta) f \rangle &= \int_0^\infty d\lambda \langle (\mathcal{U}_0 g)_\lambda, (\mathcal{U}_0 P(\eta) f)_\lambda \rangle_{\mathcal{K}} \\
&= \int_0^\infty d\lambda \, \eta(\lambda) \int_0^\infty d\mu \, \eta(\mu) \langle (\mathcal{U}_0 g)_\lambda, (\mathcal{U}_0 f)_\mu \rangle_{\mathcal{K}}.
\end{aligned}$$

After interchanging the order of integration and by using the assumption that $\eta$ is real, one finds that the last expression is equal to $\langle P(\eta) g, f \rangle$.

(iii) Clearly, in the spectral representation of $H_0$, the vectors in the range of $P(\eta)$ have the following form: $(\mathcal{U}_0 f)_\lambda = \eta(\lambda)g$ for some $g \in L^2(S^{n-1})$. Consequently $\mathcal{R}(P(\eta)) \subseteq \mathcal{M}(\eta)$. On the other hand, if $f \in \mathcal{M}(\eta)$, then (5.99) implies that

$$\begin{aligned}\left[\mathcal{U}_0 P(\eta) f\right]_\lambda &= \eta(\lambda) \int_0^\infty \eta(\mu) \left[\mathcal{U}_0 f\right]_\mu d\mu = \eta(\lambda) \int_0^\infty \eta(\mu)[\eta(\mu) g]\, d\mu \\ &= \int_0^\infty |\eta(\mu)|^2 d\mu \cdot \eta(\lambda) g = \eta(\lambda) g = (\mathcal{U}_0 f)_\lambda,\end{aligned}$$

which shows that $\mathcal{M}(\eta) \subseteq \mathcal{R}(P(\eta))$.

(b) One has $\|AP(\eta)\|_{\text{HS}}^2 = \sum_k \|Ae_k\|^2$, where $\{e_k\}$ is an arbitrary orthonormal basis of $\mathcal{M}(\eta)$. The vectors of such a basis are necessarily of the form $(\mathcal{U}_0 e_k)_\lambda = \eta(\lambda) e_k^{\text{o}}$, with $\{e_k^{\text{o}}\}$ an orthonormal basis of $\mathcal{K} \equiv L^2(S^{n-1})$. Hence

$$\begin{aligned}\|AP(\eta)\|_{\text{HS}}^2 &= \sum_k \|Ae_k\|^2 = \sum_k \|\mathcal{U}_0 A e_k\|^2 = \sum_k \int_0^\infty \|(\mathcal{U}_0 A e_k)_\lambda\|_{\mathcal{K}}^2 d\lambda \\ &= \sum_k \int_0^\infty \|A(\lambda) \cdot \eta(\lambda) e_k^{\text{o}}\|_{\mathcal{K}}^2 d\lambda = \int_0^\infty d\lambda\, |\eta(\lambda)|^2 \sum_k \|A(\lambda) e_k^{\text{o}}\|_{\mathcal{K}}^2 \\ &= \int_0^\infty d\lambda\, |\eta(\lambda)|^2 \|A(\lambda)\|_{\text{HS}}^2,\end{aligned}$$

which proves Eq. (5.101). □

By taking into account Eq. (5.102), we now have to estimate the Hilbert-Schmidt norm of the operator $RP(\eta)$. For this we observe that

$$\begin{aligned}R = S - I = \Omega_+^* \Omega_- - \Omega_+^* \Omega_+ &= \Omega_+^*(\Omega_- - \Omega_+) = -\Omega_+^* \int_{-\infty}^\infty \frac{d}{dt} U_t^* U_t^0\, dt \\ = -\Omega_+^* \int_{-\infty}^\infty U_t^*(iH - iH_0) U_t^0\, dt &= -i\Omega_+^* \int_{-\infty}^\infty U_t^* V U_t^0\, dt \end{aligned} \tag{5.103}$$

[this equation holds on $\mathcal{D}(H_0)$ if one assumes that $\mathcal{D}(H) \doteq \mathcal{D}(H_0)$]. Consequently

$$\left\|RP(\eta)\right\|_{\text{HS}} \leq \|\Omega_+^*\| \left\| \int_{-\infty}^\infty U_t^* V U_t^0 P(\eta)\, dt \right\|_{\text{HS}} \leq \int_{-\infty}^\infty \left\|V U_t^0 P(\eta)\right\|_{\text{HS}} dt. \tag{5.104}$$

It is seen that one can estimate $\|RP(\eta)\|_{\text{HS}}$ in terms of $\|VU_t^0 P(\eta)\|_{\text{HS}}$, and, as in the proof of the existence of the wave operators, this last expression involves only the free evolution group, which is known rather explicitly.

It turns out that, if the operator $V \equiv H - H_0$ is a potential, i.e. if $V = V(\vec{Q})$, then one can explicitly calculate integrals of the type occurring in the last expression of (5.104) if one replaces the integrand by its square. In the next proposition we denote by $\|\cdot\|$ the norm in $L^2(\mathbb{R}^n)$ (if the argument is a function defined on $\mathbb{R}^n$) as well as the norm in $L^2((0,\infty))$ [if the argument is a function defined on $(0,\infty)$]. If $\eta$ is a function defined on $(0,\infty)$ and $\alpha \in \mathbb{R}$, we denote by $\eta_\alpha$ the function given as $\eta_\alpha(\lambda) = \lambda^{\alpha/4}\eta(\lambda)$ and by $\eta_\alpha'$ its derivative. So we have for example:

$$\|\eta_\alpha\|^2 = \int_0^\infty \lambda^{\alpha/2}|\eta(\lambda)|^2\, d\lambda, \qquad \|\eta'_\alpha\|^2 = \int_0^\infty \left|\frac{d}{d\lambda}\left[\lambda^{\alpha/4}\eta(\lambda)\right]\right|^2 d\lambda,$$

$$\||\vec{Q}|w\|^2 = \int_{\mathbb{R}^n} |\vec{x}|^2 |w(\vec{x})|^2\, d^n x.$$

**Proposition 5.30.** *Let $W = W(\vec{Q})$ be the multiplication operator in $L^2(\mathbb{R}^n)$ by a real function $W(\vec{x})$. Let $\eta\colon (0,\infty) \to \mathbb{R}$ be continuously differentiable, $\eta(\lambda) = 0$ in a neighbourhood of $0$ and in a neighbourhood of $\infty$, and $\|\eta\| = 1$. If $W \in L^2(\mathbb{R}^n)$, one has*

$$\int_{-\infty}^{\infty} \|WU_t^0 P(\eta)\|_{\mathrm{HS}}^2\, dt = \frac{1}{2}(2\pi)^{1-n}\Theta_n \|\eta_{n-2}\|^2 \|W\|^2. \tag{5.105}$$

*If $(1 + |\cdot|)W(\cdot)$ belongs to $L^2(\mathbb{R}^n)$, then*

$$\begin{aligned}\int_{-\infty}^{\infty} t^2 \|WU_t^0 P(\eta)\|_{\mathrm{HS}}^2\, dt &= \frac{1}{2}(2\pi)^{1-n}\Theta_n \|\eta'_{n-2}\|^2 \|W\|^2 \\ &\quad + \frac{1}{8n}(2\pi)^{1-n}\Theta_n \|\eta_{n-4}\|^2 \||\vec{Q}|W\|^2. \end{aligned}\tag{5.106}$$

REMARK. It is interesting to observe that the terms on the right-hand sides of the equations (5.105) and (5.106) have the form of a product of a factor depending only on $\eta$ and a factor involving only $W$; apart from these factors there are different constants depending on $n$.

PROOF OF PROPOSITION 5.30. (i) By the relation $[WU_t^0 P(\eta)]^* = P(\eta)U_{-t}^0 W$, one has $\|WU_t^0 P(\eta)\|_{\mathrm{HS}} = \|P(\eta)U_{-t}^0 W\|_{\mathrm{HS}}$. To calculate this Hilbert-Schmidt norm, we express the operator $P(\eta)U_{-t}^0 W$ as an integral operator in momentum space and then apply Proposition 2.15. We observe that $W$ is a convolution operator in $\tilde{L}^2(\mathbb{R}^n)$, more precisely the integral operator with kernel $(2\pi)^{-n/2}\widetilde{W}(\vec{k} - \vec{k}')$, see (2.84). By setting $\vec{k} = \sqrt{\lambda}\vec{\omega}$ and using first (5.69), then the definition (5.99) of $P(\eta)$ and finally (5.68), one obtains

$$\begin{aligned}\left[\mathcal{F}P(\eta)U_{-t}^0 Wf\right](\vec{k}) &= 2^{1/2}\lambda^{(2-n)/4}\left[\mathcal{U}_0 P(\eta)U_{-t}^0 Wf\right]_\lambda(\vec{\omega}) \\ &= 2^{1/2}\lambda^{(2-n)/4}\eta(\lambda)\left[\int_0^\infty \eta(\mu)\left[\mathcal{U}_0 U_{-t}^0 Wf\right]_\mu d\mu\right](\vec{\omega}) \\ &= 2^{1/2}\lambda^{(2-n)/4}\eta(\lambda)\left[\int_0^\infty \eta(\mu)e^{i\mu t}(\mathcal{U}_0 Wf)_\mu\, d\mu\right](\vec{\omega}) \\ &= 2^{1/2}\lambda^{(2-n)/4}\eta(\lambda)\int_0^\infty d\mu\, \eta(\mu)e^{i\mu t}\left[2^{-1/2}\mu^{(n-2)/4}(\mathcal{F}Wf)\right](\sqrt{\mu}\vec{\omega}) \\ &= \lambda^{(2-n)/4}\eta(\lambda)\int_0^\infty d\mu\, \eta(\mu)e^{i\mu t}\mu^{(n-2)/4} \\ &\quad \times (2\pi)^{-n/2}\int_{\mathbb{R}^n} d^n\vec{k}'\, \widetilde{W}(\sqrt{\mu}\vec{\omega} - \vec{k}')\tilde{f}(\vec{k}'). \end{aligned}$$

This shows that $P(\eta)U^0_{-t}W$ is indeed an integral operator in $\tilde{L}^2(\mathbb{R}^n)$ and that its kernel $N_t(\vec{k},\vec{k}')$ is given by

$$N_t(\vec{k},\vec{k}') = (2\pi)^{-n/2}\lambda^{(2-n)/4}\eta(\lambda)\int_0^\infty d\mu\,\mu^{(n-2)/4}\eta(\mu)e^{i\mu t}\widetilde{W}(\sqrt{\mu}\vec{\omega}-\vec{k}'),$$

with $\vec{k}=\sqrt{\lambda}\vec{\omega}$. Thus, by observing that $d^nk=(1/2)\lambda^{(n-2)/2}d\lambda d\omega$, one obtains from Proposition 2.15 that

$$\begin{aligned}\|WU^0_tP(\eta)\|^2_{\text{HS}} &= \int_{\mathbb{R}^n} d^nk\int_{\mathbb{R}^n} d^nk'\left|N_t(\vec{k},\vec{k}')\right|^2\\ &= \frac{1}{2}(2\pi)^{-n}\|\eta\|^2\int_{S^{n-1}} d\omega\int_{\mathbb{R}^n} d^nk'\left|\int_0^\infty d\mu\,\eta_{n-2}(\mu)e^{i\mu t}\widetilde{W}(\sqrt{\mu}\vec{\omega}-\vec{k}')\right|^2. \quad (5.107)\end{aligned}$$

We recall that $\|\eta\|^2=1$. The integration with respect to $d\mu$ may be considered [apart from a factor $(2\pi)^{1/2}$] as a one-dimensional Fourier transformation (extend $\eta$ to $\mathbb{R}$ by setting $\eta(\lambda)=0$ for $\lambda\le 0$). Thus, upon applying $\int_{\mathbb{R}} dt$ to the expression (5.107), we can use the unitarity of the Fourier transformation to evaluate the integral with respect to $dt$:

$$\begin{aligned}\int_{-\infty}^\infty \|WU^0_tP(\eta)\|^2_{\text{HS}}\,dt &= \frac{1}{2}(2\pi)^{-n+1}\int d\omega\int d^nk'\int_0^\infty d\mu\left|\eta_{n-2}(\mu)\widetilde{W}(\sqrt{\mu}\vec{\omega}-\vec{k}')\right|^2\\ &= \frac{1}{2}(2\pi)^{-n+1}\int_{S^{n-1}} d\omega\int_0^\infty d\mu\,|\eta_{n-2}(\mu)|^2\|\widetilde{W}\|^2,\end{aligned}$$

which is identical with (5.105).

(ii) To prove (5.106) we set $W_j(\vec{x})=x_jW(\vec{x})$, $j=1,\ldots,n$. The kernel of the integral operator $tP(\eta)U^0_{-t}W$ is given as follows [writing $te^{i\mu t}=-i(d/d\mu)e^{i\mu t}$]:

$$-i(2\pi)^{-n/2}\lambda^{(2-n)/4}\eta(\lambda)\int_0^\infty d\mu\,\mu^{(n-2)/4}\eta(\mu)\widetilde{W}(\sqrt{\mu}\vec{\omega}-\vec{k}')\frac{d}{d\mu}e^{i\mu t},\ \text{with } \vec{k}=\sqrt{\lambda}\vec{\omega}.$$

In this expression one first integrates by parts with respect to the variable $\mu$, by observing that the boundary terms vanish because $\eta(0)=\eta(\infty)=0$. Then, as in (i) above, one can use the unitarity of the Fourier transformation to obtain that

$$\begin{aligned}\int_{-\infty}^\infty t^2\|WU^0_tP(\eta)\|^2_{\text{HS}}\,dt = \frac{1}{2}(2\pi)^{-n+1}\int d\omega\int d^nk'\int_0^\infty d\mu\Big|\eta'_{n-2}(\mu)\widetilde{W}(\sqrt{\mu}\vec{\omega}-\vec{k}')\\ +\eta_{n-2}(\mu)\frac{\partial}{\partial\mu}\widetilde{W}(\sqrt{\mu}\vec{\omega}-\vec{k}')\Big|^2. \quad (5.108)\end{aligned}$$

If one denotes by $\omega_j$ ($j=1,\ldots,n$) the components of the vector $\vec{\omega}$, one has

$$\begin{aligned}\frac{\partial}{\partial\mu}\widetilde{W}(\sqrt{\mu}\vec{\omega}-\vec{k}') &= \frac{1}{2}\mu^{-1/2}\frac{\partial}{\partial\sqrt{\mu}}\widetilde{W}(\sqrt{\mu}\vec{\omega}-\vec{k}')\\ &= \frac{1}{2}\mu^{-1/2}\frac{\partial}{\partial\sqrt{\mu}}(2\pi)^{-n/2}\int_{\mathbb{R}^n} e^{-i\mu^{1/2}\vec{\omega}\cdot\vec{x}}e^{i\vec{k}'\cdot\vec{x}}W(\vec{x})\,d^nx\\ &= -\frac{i}{2}\mu^{-1/2}\sum_{j=1}^n \omega_j\widetilde{W}_j(\sqrt{\mu}\vec{\omega}-\vec{k}').\end{aligned}$$

We insert this expression into (5.108) in which we make the change of variables $\vec{k}' \mapsto \vec{p} = \sqrt{\mu}\vec{\omega} - \vec{k}'$:

$$\int_{-\infty}^{\infty} t^2 \|WU_t^0 P(\eta)\|_{\mathrm{HS}}^2\, dt = \frac{1}{2}(2\pi)^{-n+1} \int_{S^{n-1}} d\omega \int_0^{\infty} d\mu \Big\| \eta_{n-2}'(\mu)\widetilde{W}(\cdot) - \frac{i}{2}\eta_{n-4}(\mu) \sum_{j=1}^{n} \omega_j \widetilde{W_j}(\cdot) \Big\|^2, \tag{5.109}$$

where the norm is in $\tilde{L}^2(\mathbb{R}^n)$ with respect to the variable $\cdot = \vec{p}$. Next one calculates the integral with respect to $d\omega$ by writing the norm occurring in (5.109) as a sum of scalar products. One has to know the values of the following integrals [20]:

$$\int_{S^{n-1}} d\omega = \Theta_n, \quad \int_{S^{n-1}} \omega_j\, d\omega = 0, \quad \int_{S^{n-1}} \omega_j \omega_k\, d\omega = \frac{\Theta_n}{n}\delta_{jk}.$$

One finds that the right-hand side of Eq. (5.109) is equal to

$$\frac{1}{2}(2\pi)^{-n+1}\Theta_n \Big[ \|\eta_{n-2}'\|^2 \|W\|^2 + \frac{1}{n}\Big\|\frac{1}{2}\eta_{n-4}\Big\|^2 \sum_{j=1}^{n} \|W_j\|^2 \Big],$$

which is identical with the right-hand side of Eq. (5.106), since

$$\sum_{j=1}^{n} \|W_j\|^2 = \sum_{j=1}^{n} \int_{\mathbb{R}^n} x_j^2 |W(\vec{x})|^2 d^n x = \int_{\mathbb{R}^n} |\vec{x}|^2 |W(\vec{x})|^2 d^n x = \||\vec{Q}|W\|^2. \quad \square$$

Let us apply the preceding proposition to the question of the finiteness of the total scattering cross section. Let $\eta$ be as in Proposition 5.30. By using Eq. (5.102), then the inequality (5.104) and the Schwarz inequality (for the integral with respect to $dt$), and finally (5.105) and (5.106) with $W = V$, one finds that

$$\begin{aligned}
\int_0^{\infty} \lambda^{(n-1)/2} |\eta(\lambda)|^2 \overline{\sigma}(\lambda)\, d\lambda &= \frac{(2\pi)^{n-1}}{\Theta_n} \|RP(\eta)\|_{\mathrm{HS}}^2 \\
&\leq \frac{(2\pi)^{n-1}}{\Theta_n} \Big[ \int_{-\infty}^{\infty} \|VU_t^0 P(\eta)\|_{\mathrm{HS}}\, dt \Big]^2 \\
&= \frac{(2\pi)^{n-1}}{\Theta_n} \Big[ \int_{-\infty}^{\infty} (1+t^2)^{-1/2} \cdot (1+t^2)^{1/2} \|VU_t^0 P(\eta)\|_{\mathrm{HS}}\, dt \Big]^2 \\
&\leq \frac{(2\pi)^{n-1}}{\Theta_n} \Big[ \int_{-\infty}^{\infty} (1+t^2)^{-1} dt \Big] \cdot \Big[ \int_{-\infty}^{\infty} (1+t^2) \|VU_t^0 P(\eta)\|_{\mathrm{HS}}^2\, dt \Big] \\
&= \frac{\pi}{2} \Big[ \|\eta_{n-2}\|^2 \|V\|^2 + \|\eta_{n-2}'\|^2 \|V\|^2 + \frac{1}{4n} \|\eta_{n-4}\|^2 \||\vec{Q}|V\|^2 \Big].
\end{aligned} \tag{5.110}$$

Since $\eta$ is differentiable and vanishes in a neighbourhood of 0 and of $\infty$, all norms in (5.110) containing the function $\eta$ are finite. Thus the right-hand side of (5.110) is finite

[20] For the last integral with $j = k$, observe that its value is independent of $j$ and that $\sum_{j=1}^{n} \omega_j^2 = 1$.

[and consequently the total scattering cross section $\sigma_{\text{tot}}(\lambda, \vec{\omega})$ is finite for almost all $\lambda > 0$ and $\vec{\omega} \in S^{n-1}$] provided that $(1+|\vec{x}|)V(\vec{x})$ is a function belonging to $L^2(\mathbb{R}^n)$. Assume for instance that $V$ is locally square-integrable and that $|V(\vec{x})| = c|\vec{x}|^{-\beta}$ in a neighbourhood of infinity. Then the preceding condition is satisfied if and only if $\beta > (n+2)/2$.

Consider the case $n = 3$. The condition $\beta > (n+2)/2$ becomes $\beta > 5/2$. This result is not optimal, it is known that the total scattering cross section is finite if $\beta > 2$ and infinite if $\beta \leq 2$ (for $n = 3$).

In order to improve the preceding result [i.e. the condition $\beta > (n+2)/2$], we observe that the integral $\int_{-\infty}^{\infty}(1+t^2)^{-1}dt$ in (5.110) converges too rapidly at infinity: for large $t$ the integrand behaves like $t^{-2}$, whereas a decay like $t^{-1-\varepsilon}$, with $\varepsilon > 0$, would be sufficient; in other terms it would suffice to insert in the calculation in (5.110), before applying the Schwarz inequality, the factor $(1+|t|)^{-1/2-\nu} \cdot (1+|t|)^{1/2+\nu}$ in place of $(1+t^2)^{-1/2} \cdot (1+t^2)^{1/2}$, with $\nu > 0$. One can see from (5.105) and (5.106) that the introduction of a factor $t$ in $\|WU_t^0P(\eta)\|_{\text{HS}}$ [more precisely the replacement of $\|WU_t^0P(\eta)\|_{\text{HS}}$ in (5.105) by $\|tWU_t^0P(\eta)\|_{\text{HS}}$] is essentially equivalent to the replacement of $W(\vec{x})$ by $|\vec{x}|W(\vec{x})$ in the norm appearing on the right-hand side of (5.105).

One can therefore expect that the replacement in Eq. (5.105) of $\|WU_t^0P(\eta)\|_{\text{HS}}$ by $\|t^\alpha WU_t^0P(\eta)\|_{\text{HS}}$ should amount to replacing $W$ on the right-hand side by $|\vec{Q}|^\alpha W$. Assuming this and taking $|W(\vec{x})| = |\vec{x}|^{-\beta}$, one has $|\vec{x}|^\alpha|W(\vec{x})| = |\vec{x}|^{\alpha-\beta}$, and this latter function is square-integrable in a neighbourhood of infinity if $\beta - \alpha > n/2$, i.e. $\beta > (n+1)/2$ if $\alpha$ is sufficiently close to $1/2$ (recall that one needs $\alpha = 1/2+\nu$, with $\nu > 0$, for the integral $\int_{-\infty}^{\infty}(1+|t|)^{-2\alpha}dt$ to be finite). For $n = 3$, the condition $\beta > (n+1)/2$ is equivalent to $\beta > 2$, so one can expect to arrive in this way at an essentially optimal result on the finiteness of total scattering cross sections.

The fact that the replacement of $\|WU_t^0P(\eta)\|_{\text{HS}}$ by $\|t^\alpha WU_t^0P(\eta)\|_{\text{HS}}$ in the left-hand side of (5.105) is essentially equivalent to the replacement of the operator $W(\vec{Q})$ by $|\vec{Q}|^\alpha W(\vec{Q})$ in the norm on the right-hand side is not really surprising. In a classical approximation one has $\vec{Q} \approx 2\vec{P}t$ at large times (see Remark 5.19). Also, by (5.62): $\|\varphi(\vec{Q})U_t^0P(\eta)\|_{\text{HS}} = \|\varphi(2\vec{P}t)Z_tP(\eta)\|_{\text{HS}}$, hence

$$\begin{aligned}\|t^\alpha W(\vec{Q})U_t^0P(\eta)\|_{\text{HS}} &= \|t^\alpha W(2\vec{P}t)Z_tP(\eta)\|_{\text{HS}}\\ &= \|[(2|\vec{P}|t)^\alpha W(2\vec{P}t)](2|\vec{P}|)^{-\alpha}Z_tP(\eta)\|_{\text{HS}}.\end{aligned}$$

Since $\eta$ is assumed to be zero near the origin, there exists a number $b > 0$ such that $\eta(\lambda) = 0$ if $0 < \lambda < b$, and hence state vectors in the range of $P(\eta)$ have a minimal momentum $b^{1/2}$. So, for large $|t|$ ($Z_t \approx I$), the last expression should be less than

$$(2b)^{-\alpha}\|[(2|\vec{P}|t)^\alpha W(2\vec{P}t)]Z_tP(\eta)\|_{\text{HS}} = (2b)^{-\alpha}\||\vec{Q}|^\alpha W(\vec{Q})U_t^0P(\eta)\|_{\text{HS}}.$$

This suggests that one has indeed $\int_{-\infty}^{\infty}\|t^\alpha WU_t^0P(\eta)\|_{\text{HS}}^2dt \leq C_n(\eta)\||\vec{Q}|^\alpha W(\vec{Q})\|^2$ for some constant $C_n(\eta)$ depending on $n$ and $\eta$. The next proposition gives a rigourous meaning to this.

**Proposition 5.31.** *Let $\eta$ be as in Proposition 5.30, $\alpha \in (0,1)$ and $V: \mathbb{R}^n \to \mathbb{R}$ such that $(1+|\cdot|)^\alpha V(\cdot) \in L^2(\mathbb{R}^n)$. Then there exists a finite constant $C = C(\eta)$ such that*

$$\int_{-\infty}^{\infty} (1+|t|)^{2\alpha} \|V(\vec{Q})U_t^0 P(\eta)\|_{\mathrm{HS}}^2\, dt \le C\|(I+|\vec{Q}|)^\alpha V\|^2. \tag{5.111}$$

PROOF. Denote the kernel of the Hilbert-Schmidt operator $V(\vec{Q})U_t^0P(\eta)$ by $M_t(\vec{x},\vec{y})$. Then one has

$$\int_{-\infty}^{\infty} (1+|t|)^{2\alpha}\|V(\vec{Q})U_t^0P(\eta)\|_{\mathrm{HS}}^2\, dt = \int_{-\infty}^{\infty} dt \int_{\mathbb{R}^n} d^n x \int_{\mathbb{R}^n} d^n y (1+|t|)^{2\alpha}|M_t(\vec{x},\vec{y})|^2.$$

For fixed $t$, one divides the domain of integration with respect to $d^n x$ into two parts $\mathcal{O}_1$ and $\mathcal{O}_2$, with $\mathcal{O}_1 = \{\vec{x}\in\mathbb{R}^n \mid |\vec{x}| > |t|\}$ and $\mathcal{O}_2 = \{\vec{x}\in\mathbb{R}^n \mid |\vec{x}| \le |t|\}$. In $\mathcal{O}_1$ we majorise $(1+|t|)^{2\alpha}$ by $(1+|\vec{x}|)^{2\alpha}$, then we extend the integral over $\mathcal{O}_1$ again to all of $\mathbb{R}^n$. In $\mathcal{O}_2$ we use the following estimate:

$$\begin{aligned}(1+|t|)^{2\alpha} &= \frac{(1+|t|)^{2\alpha}}{(1+|\vec{x}|)^{2\alpha}}\cdot(1+|\vec{x}|)^{2\alpha} \le \left\{\frac{(1+|t|)}{(1+|\vec{x}|)}\right\}^2\cdot(1+|\vec{x}|)^{2\alpha}\\ &\le \frac{2(1+t^2)}{(1+|\vec{x}|)^2}\cdot(1+|\vec{x}|)^{2\alpha} \le 2(1+|\vec{x}|)^{2\alpha} + 2t^2(1+|\vec{x}|)^{2\alpha-2}.\end{aligned}$$

Again we extend the integral over $\mathcal{O}_2$ to $\mathbb{R}^n$. We obtain

$$\begin{aligned}&\int_{-\infty}^{\infty}(1+|t|)^{2\alpha}\|V(\vec{Q})U_t^0P(\eta)\|_{\mathrm{HS}}^2\,dt\\ &\qquad\le 3\int_{-\infty}^{\infty} dt\int_{\mathbb{R}^n} d^n x\int_{\mathbb{R}^n} d^n y(1+|\vec{x}|)^{2\alpha}|M_t(\vec{x},\vec{y})|^2\\ &\qquad\quad + 2\int_{-\infty}^{\infty} t^2 dt\int_{\mathbb{R}^n} d^n x\int_{\mathbb{R}^n} d^n y(1+|\vec{x}|)^{2\alpha-2}|M_t(\vec{x},\vec{y})|^2\\ &\qquad= 3\int_{-\infty}^{\infty}\|(I+|\vec{Q}|)^\alpha V(\vec{Q})U_t^0P(\eta)\|_{\mathrm{HS}}^2\,dt\\ &\qquad\quad + 2\int_{-\infty}^{\infty} t^2\|(I+|\vec{Q}|)^{\alpha-1}V(\vec{Q})U_t^0P(\eta)\|_{\mathrm{HS}}^2\,dt.\end{aligned}$$

Hence, by using Proposition 5.30, one arrives at

$$\begin{aligned}\int_{-\infty}^{\infty}(1+|t|)^{2\alpha}\|V(\vec{Q})U_t^0P(\eta)\|_{\mathrm{HS}}^2\,dt &\le 3C_1(\eta)\|(I+|\vec{Q}|)^\alpha V\|^2\\ &\quad + 2C_2(\eta)\|(I+|\vec{Q}|)^{\alpha-1}V\|^2 + 2C_3(\eta)\||\vec{Q}|(I+|\vec{Q}|^{\alpha-1})V\|^2\\ &\le C(\eta)\|(I+|\vec{Q}|)^\alpha V\|^2,\end{aligned}$$

because $\|(I+|\vec{Q}|)^{-1}\| = \||\vec{Q}|(I+|\vec{Q}|)^{-1}\| = 1$. □

It is very interesting to observe that the local behaviour of the potential has no influence on the finiteness of the total scattering cross section. More precisely, if $\int_{|\vec{x}|\geq R}|(1+|\vec{x}|)^{\alpha}V(\vec{x})|^2 d^n x < \infty$ for some $R \in [0,\infty)$, then $\overline{\sigma}(\lambda) < \infty$ for almost all values of $\lambda$ [the Hamiltonian $H$ can be an arbitrary self-adjoint extension of the symmetric operator $H_0 + V(\vec{Q})$, as in point (3) on page 225]. To see this one uses the fact, proved on page 225, that $\Omega_{\pm} = \text{s-lim}_{t\to\pm\infty} U_t^* \psi(\vec{Q}) U_t^0$, where $\psi$ is a function of class $C^\infty$ such that $\psi(\vec{x}) = 0$ if $|\vec{x}| \leq R$ and $\psi(\vec{x}) = 1$ if $|\vec{x}| > 2R$. In place of (5.103) one can then write (by using the calculations and notations of page 226):

$$R = S - I = \Omega_+^*(\Omega_- - \Omega_+) = -\Omega_+^* \int_{-\infty}^{\infty} \frac{d}{dt} U_t^* \psi(\vec{Q}) U_t^0 \, dt$$

$$= -i\Omega_+^* \int_{-\infty}^{\infty} U_t^* \{V(\vec{Q})\psi(\vec{Q}) - [\Delta\psi](\vec{Q}) - 2i[\vec{\nabla}\psi](\vec{Q}) \cdot \vec{P}\} U_t^0 \, dt.$$

Consequently [compare with (5.104)]:

$$\|RP(\eta)\|_{\text{HS}} \leq \int_{-\infty}^{\infty} \|V(\vec{Q})\psi(\vec{Q})U_t^0 P(\eta)\|_{\text{HS}} dt + \int_{-\infty}^{\infty} \big\|[\Delta\psi](\vec{Q})U_t^0 P(\eta)\big\|_{\text{HS}} dt$$
$$+ 2\sum_{j=1}^{n} \int_{-\infty}^{\infty} \big\|[\vec{\nabla}\psi]_j(\vec{Q})U_t^0 P_j P(\eta)\big\|_{\text{HS}} dt. \qquad (5.112)$$

The first integral is finite under the specified condition on the potential $V$, and the remaining integrals are finite because the derivatives of $\psi$ have compact support. One can use Proposition 5.30 to estimate these integrals and obtain bounds in terms of certain norms on the function $\eta$, of the parameter $R$ and of $\int_{|\vec{x}|\geq R} |(1+|\vec{x}|)^{\alpha} V(\vec{x})|^2 d^n x$. We refer to [A] for details [see also Problem 5.17 for the last integral in (5.112)].

## 5.8 Coulomb scattering

In §5.4.3 it was pointed out that for a Coulomb potential, and more generally for long range potentials, the wave operators do not exist, namely that $U_t U_t^0$ will not be strongly convergent as $t \to \pm\infty$, and that this property is due to the slow decay of the potential at large distances from the scattering centre. It turns out that one can still introduce some kind of wave operators generalising the notion considered before. We shall first describe the definition and some properties of such modified wave operators for long range potentials, and then we shall prove their existence for Coulomb potentials.

**5.8.1.** The modified wave operators are given as strong limits, as $t \to \pm\infty$, of operators of the form $U_t U_t^0 e^{-iX_t}$, where $\{X_t\}_{t\in\mathbb{R}}$ is a time-dependent family of self-adjoint operators commuting with the free Hamiltonian $H_0$ (for spherically symmetric potentials the operators $X_t$ are functions of $H_0$, and more generally they are functions of the momentum operator $\vec{P}$). Thus the free evolution $U_t^0 = e^{-iH_0 t}$ occurring in the standard definition of wave operators is replaced here by unitary operators of the form $U_t^0 e^{-iX_t} = e^{-i[H_0 t + X_t]}$:

$$\Omega_\pm = \operatorname*{s-lim}_{t\to\pm\infty} U_t^* U_t^0 e^{-iX_t} = \operatorname*{s-lim}_{t\to\pm\infty} e^{iHt} e^{-i[H_0 t + X_t]}. \tag{5.113}$$

As a function of $t$, the operators $e^{-i[H_0 t + X_t]}$ will not form a group: the dependence of $X_t$ on the parameter $t$ is not linear but rather such that $X_t$ is small compared to $H_0 t$ at large values of $t$, viz. such that $X_t/t$ converges to zero as $t \to \pm\infty$ [see Proposition 5.32(a) for an alternative and more precise formulation of this property]. In this sense the modified free evolution used to define the (modified) wave operators for long range potentials is close to the free evolution $e^{-iH_0 t}$ at large times.

A justification for the introduction of these modified wave operators can be found through the algebraic scattering theory mentioned on page 213. Let us consider first a situation in which the usual wave operators $\Omega_\pm = \text{s-lim}_{t\to\pm\infty} U_t^* U_t^0$ exist. Let $A$ be a free observable, i.e. a bounded self-adjoint operator commuting with the free Hamiltonian $H_0$ (i.e. with each $U_t^0$). If $f$ is a vector in the range of $\Omega_+$, then [see Proposition 5.15(a)]

$$\langle U_t f, AU_t f\rangle = \langle U_t^{0*} U_t f, AU_t^{0*} U_t f\rangle \to \langle \Omega_+^* f, A\Omega_+^* f\rangle \qquad \text{as } t \to +\infty. \tag{5.114}$$

One sees that the asymptotic condition is satisfied in the algebraic sense, i.e. the expectation values $\langle U_t f, AU_t f\rangle$ have limits as $t \to +\infty$ for each observable $A$ commuting with $H_0$ and for each state vector $f$ in the range of $\Omega_+$. In many situations the range of $\Omega_+$ coincides with the set of all scattering states of the total Hamiltonian $H$, and then the asymptotic values of $\langle U_t f, AU_t f\rangle$ at $t = +\infty$ exist for each free observable $A$ and each scattering state of $H$. Analogous statements can be made for the limit $t \to -\infty$.

A similar reasoning applies for the modified wave operators. If $A$ is an observable commuting with each $U_t^0$ and also with each $X_t$, then we have as in (5.114), with $f \in \mathcal{R}(\Omega_+)$: [21]

$$\langle U_t f, AU_t f\rangle = \langle e^{iX_t} U_t^{0*} U_t f, Ae^{iX_t} U_t^{0*} U_t f\rangle \to \langle \Omega_+^* f, A\Omega_+^* f\rangle \qquad \text{as } t \to +\infty. \tag{5.115}$$

Again the ranges of $\Omega_\pm$ frequently coincide with the set of all scattering states of $H$, and in many situations (in particular for Coulomb potentials), the operators $X_t$ are functions of the free Hamiltonian $H_0$. Then the asymptotic condition in the algebraic sense is again satisfied for all scattering states of $H$ and all observables $A$ that commute with the free Hamiltonian (if the operators $X_t$ – which commute with $H_0$ – are not functions of $H_0$, then asymptotic convergence of $\langle U_t f, AU_t f\rangle$ will hold only for a more restricted class of free observables $A$ than that indicated above).

The equation (5.113) means that, in a long range potential, the evolution of a scattering state of $H$ will not be asymptotically comparable to a free evolution but rather to a somewhat distorted free evolution. If $f$ is a scattering state of $H$ then, as $t \to +\infty$, the state vector $e^{-iHt} f$ will not be close to $e^{-iH_0 t} f_+$ (for some scattering state $f_+$

[21] As in the proof of Proposition 5.15(a) one finds that the operators $e^{iX_t} U_t^{0*} U_t$ converge strongly to $\Omega_+^*$ on the subspace $\mathcal{R}(\Omega_+)$.

of $H_0$) as in Eq. (5.39) but rather to $e^{-i[H_0 t+X_t]}f_+$. The distortion of the free evolution (i.e. the operator $e^{-iX_t}$) takes into account the effect of the potential at very large distances, in particular it depends on the potential.

It should be said that the operators $X_t$ are highly non-unique. Only their asymptotic behaviour (as $t \to \pm\infty$) is essential for obtaining convergence of $U_t^* U_t^0 e^{-iX_t}$. For example the addition of a fixed self-adjoint operator $Z$ to each $X_t$ will lead to new wave operators $\Omega_\pm e^{-iZ}$ if $\Omega_\pm = \text{s-lim}_{t\to\pm\infty} U_t^* U_t^0 e^{-iX_t}$. However, provided that $Z$ is a function of $H_0$, the asymptotic values of $\langle U_t f, AU_t f\rangle$ in (5.115) will not depend on $Z$. More generally, without changing the asymptotic values of $\langle U_t f, AU_t f\rangle$, one has the freedom of adding to each $X_t$ an operator $Z_t$ depending on $H_0$ and on the value of $t$ such that the family $\{Z_t\}_{t\in\mathbb{R}}$ satisfies $\text{s-lim}_{t\to+\infty} e^{-iZ_t} - e^{iZ} = O$ for some fixed self-adjoint operator $Z$.

The determination of a suitable family $\{X_t\}$ for a given potential $V$ may be motivated by the use of an adaptation of the Cook Criterion (Proposition 5.16) for proving asymptotic convergence. By applying the Leibniz rule one finds, as in the proof of Proposition 5.16, that

$$\frac{d}{dt}U_t^* U_t^0 e^{-iX_t}f = iU_t^*\big[H - H_0 - \frac{d}{dt}X_t\big]U_t^0 e^{-iX_t}f = iU_t^*\big[V - \frac{d}{dt}X_t\big]U_t^0 e^{-iX_t}f, \tag{5.116}$$

and the existence of $\text{s-lim}_{t\to+\infty} U_t^* U_t^0 e^{-iX_t}$ is guaranteed if one knows that, for a dense set of vectors $f$, the function $t \mapsto \|[V - (d/dt)X_t]U_t^0 e^{-iX_t}f\|$ belongs to $L^1([M,\infty))$ for some $M \in [0,\infty)$. For a short range potential [when $X_t = O$, hence $(d/dt)X_t = O$], this integrability condition is satisfied as a consequence of the sufficiently rapid decay of the function $V$ at infinity, see the discussion under point (C) on page 224. For a long range potential the function $t \mapsto \|VU_t^0 f\|$ is not integrable at infinity, but the subtraction of a suitably chosen term of the form $(d/dt)X_t$ from $V$ will improve the decay of the occurring $L^2$ norm at large values of $t$. By taking into account that $\vec{Q}(t) \approx 2\vec{P}t$ at large times, one would take $X_t \equiv X_t(\vec{P})$ such that $(d/dt)X_t(\vec{P}) = V(2\vec{P}t)$. It turns out that, with this choice of $X_t$, the modified wave operators exist for long range potentials decaying like $|\vec{x}|^{-\beta}$, with $1/2 < \beta \le 1$ (in particular for Coulomb potentials).[22] For more slowly decaying potentials a suitable choice for $X_t$ is given by the solution of the non-linear equation

$$\frac{d}{dt}X_t(\vec{P}) = V\big(2\vec{P}t + (\vec{\nabla}X_t)(\vec{P})\big)$$

or by some approximation of this solution. In §5.8.2 below we discuss the modified wave operators for Coulomb potentials. More comprehensive treatments of long range scattering and references to the literature can be found for example in Chapter 13 of [AJS], in Section XI.9, Volume III of [RS] and in [P].

[22] Of course this construction of a modified free evolution can also be applied to short range potentials, for example for potentials decaying like $|\vec{x}|^{-\beta}$ with $\beta > 1$. In this case both the modified wave operators $\text{s-lim}_{t\to\pm\infty} U_t^* U_t^0 e^{-iX_t}$ and the usual wave operators $\text{s-lim}_{t\to\pm\infty} U_t^* U_t^0$ will exist, and they will be identical (for each sign of time) except for the non-uniqueness pointed out before.

To end this subsection, let us point out some general properties of the modified wave operators.

**Proposition 5.32.** *Assume that there are constants $c > 0$, $R > 0$ and $\beta > 0$ such that $|V(\vec{x})| \leq c|\vec{x}|^{-\beta}$ for $|\vec{x}| \geq R$. Let $X_t = X_t(\vec{P})$, where the function $X_t \colon \mathbb{R}^n \to \mathbb{R}$ satisfies $(\partial/\partial t)X_t(\vec{k}) = V(2\vec{k}t)$. Then*
*(a) The family of unitary operators $\{e^{-iX_t}\}$ is feebly oscillating in the following sense: $\text{s-lim}_{t\to\pm\infty}[e^{-iX_{t+\tau}} - e^{-iX_t}] = O$ for each fixed $\tau \in \mathbb{R}$.*
*(b) Assume that the wave operators $\Omega_\pm = \text{s-lim}_{t\to\pm\infty} U_t^* U_t^0 e^{-iX_t}$ exist. Then $\Omega_\pm$ are isometric operators which intertwine $H$ and $H_0$, i.e. one has $U_t\Omega_\pm = \Omega_\pm U_t^0$ for each $t \in \mathbb{R}$.*

PROOF. (a) We have for example for $t > 0$ and $\tau > 0$:

$$\left|e^{-iX_{t+\tau}(\vec{k})} - e^{-iX_t(\vec{k})}\right| \leq \int_t^{t+\tau} \left|\frac{\partial}{\partial s} e^{-iX_s(\vec{k})}\right| ds = \int_t^{t+\tau} |V(2\vec{k}s)|\, ds.$$

If $\vec{k} \neq 0$ and $t > 0$ is so large that $2|\vec{k}|t > R$, we obtain

$$\left|e^{-iX_{t+\tau}(\vec{k})} - e^{-iX_t(\vec{k})}\right| \leq c\int_t^{t+\tau} \frac{ds}{[2|\vec{k}|s]^\beta} \leq \frac{c\tau}{[2|\vec{k}|t]^\beta}.$$

This shows that, as $t \to +\infty$, $e^{-iX_{t+\tau}(\vec{k})} - e^{-iX_t(\vec{k})}$ converges to zero for almost all $\vec{k} \in \mathbb{R}^n$. Since $|e^{-iX_{t+\tau}(\vec{k})} - e^{-iX_t(\vec{k})}| \leq 2$, one finds by applying the Dominated Convergence Theorem in $\tilde{L}^2(\mathbb{R}^n)$ that $\|[e^{-iX_{t+\tau}} - e^{-iX_t}]f\|^2$ converges to zero for each vector $f$.

(b) The operators $\Omega_\pm$ are strong limits of a family of unitary operators, hence they are isometric [as in the proof of Proposition 5.13(a)]. To verify the intertwining property we write for $\tau \in \mathbb{R}$:

$$U_\tau\Omega_\pm = \operatorname*{s-lim}_{t\to\pm\infty} U_{t-\tau}^* U_t^0 e^{-iX_t} = \operatorname*{s-lim}_{s\to\pm\infty} U_s^* U_s^0 e^{-iX_s} \cdot e^{iX_s} e^{-iX_{s+\tau}} U_\tau^0. \tag{5.117}$$

Since $\text{s-lim}_{s\to\pm\infty} e^{iX_s} e^{-iX_{s+\tau}} = I$ as a consequence of the result of (a), (5.117) shows together with Proposition 2.1 that $U_\tau\Omega_\pm = \Omega_\pm U_\tau^0$. □

**5.8.2.** Now we consider Coulomb potentials $V(\vec{x}) = \gamma|\vec{x}|^{-1}$ in $L^2(\mathbb{R}^n)$, with $n \geq 3$. It is of course assumed that $\gamma \in \mathbb{R}$ and $\gamma \neq 0$. For $\vec{k} \in \mathbb{R}^n$ we set $k = |\vec{k}|$. A solution of the equation $(\partial/\partial t)X_t(\vec{k}) = V(2\vec{k}t) = \gamma(2kt)^{-1}$ is given by $X_t(\vec{k}) = \gamma(2k)^{-1}\log t$ for $t > 0$ and by $X_t(\vec{k}) = -\gamma(2k)^{-1}\log|t|$ for $t < 0$. We treat the limit $t \to +\infty$. To ascertain the existence of the wave operator $\Omega_+ = \text{s-lim}_{t\to+\infty} U_t^* U_t^0 e^{-iX_t}$, we show for a dense set $\mathcal{D}$ of vectors $f$ that $\|[V - (d/dt)X_t]U_t^0 e^{-iX_t} f\|$ is integrable as a function of $t$ in some neighbourhood of $t = +\infty$. Here $U_t = e^{-iH_C t}$, where $H_C = H_0 + \gamma|\vec{Q}|^{-1}$ is the Coulomb Hamiltonian, and $X_t(\vec{k}) = \gamma(2k)^{-1}\log t$.

We take $\mathcal{D} = \{f \in L^2(\mathbb{R}^n) \mid \tilde{f} \in C_0^\infty(\mathbb{R}^n \backslash \{0\})\}$. Clearly $\mathcal{D} \subset \mathcal{S}(\mathbb{R}^n)$, and furthermore $\mathcal{D} \in \mathcal{D}(H_0^m)$ for any $m \in \mathbb{R}$ because of the support properties of the Fourier transforms of the functions in $\mathcal{D}$. Also, if $f \in \mathcal{D}$, then $\mathcal{F}U_t^0 e^{-iX_t} f \in C_0^\infty(\mathbb{R}^n \setminus \{0\})$, hence $U_t^0 e^{-iX_t} f \in \mathcal{D}$, and $t \mapsto e^{-iX_t} f$ is strongly differentiable.

From now on we consider a fixed vector $f$ belonging to $\mathcal{D}$. Let $a > 0$ be such that $\tilde{f}(\vec{k}) = 0$ for all $\vec{k} \in \mathbb{R}^n$ satisfying $|\vec{k}| \leq a$. We choose a function $\phi\colon \mathbb{R}^n \to [0,1]$ of class $C^\infty$ satisfying $\phi(\vec{k}) = 0$ if $|\vec{k}| \leq a/2$ and $\phi(\vec{k}) = 1$ for $|\vec{k}| \geq a$. We set $\mathcal{V}(\vec{x}) = V(\vec{x})\phi(\vec{x}) = \gamma|\vec{x}|^{-1}\phi(\vec{x})$ and write $V = \mathcal{V} + W$, with $W(\vec{x}) = V(\vec{x})[1-\phi(\vec{x})]$. This gives a splitting of the Coulomb potential into a short range part $W$ (the support of the function $W$ is contained in the ball $\{\vec{x} \in \mathbb{R}^n \mid |\vec{x}| \leq a\}$) and a bounded long range potential $\mathcal{V}$ satisfying

$$|\nabla \mathcal{V}(\vec{x})| \leq c(1+|\vec{x}|)^{-2} \quad \text{and} \quad |\Delta \mathcal{V}(\vec{x})| \leq c(1+|\vec{x}|)^{-3} \tag{5.118}$$

for some constant $c$ and all $\vec{x} \in \mathbb{R}^n$ (notice that this decomposition of $V$ and the constant $c$ depend on the vector $f$).

We have

$$\begin{aligned} \big\|[V - \tfrac{d}{dt}X_t]U_t^0 e^{-iX_t} f\big\| &= \big\|[V(\vec{Q}) - V(2\vec{P}t)]U_t^0 e^{-iX_t} f\big\| \\ &\leq \big\|[\mathcal{V}(\vec{Q}) - \mathcal{V}(2\vec{P}t)]U_t^0 e^{-iX_t} f\big\| + \big\|W(\vec{Q})U_t^0 e^{-iX_t} f\big\| + \big\|W(2\vec{P}t) f\big\|. \end{aligned} \tag{5.119}$$

We now show that each of the three terms on the last line belongs $L^1([e,\infty))$ as a function of $t$. The lower endpoint of the involved interval is taken to be $e$ so that we can use the inequality $1 \leq \log t$, and from now on we assume that $t \geq e$.

We first discuss the long range part in (5.119). With $Z_t = e^{i\vec{Q}^2/(4t)}$ we have by (5.62):

$$\begin{aligned} \big\|[\mathcal{V}(\vec{Q}) - \mathcal{V}(2\vec{P}t)]U_t^0 e^{-iX_t} f\big\| &= \big\|[U_t^{0*}\mathcal{V}(\vec{Q})U_t^0 - \mathcal{V}(2\vec{P}t)]e^{-iX_t} f\big\| \\ &= \big\|[Z_t^* \mathcal{V}(2\vec{P}t) Z_t - \mathcal{V}(2\vec{P}t)]e^{-iX_t} f\big\| \\ &= \Big\| \int_0^1 d\varepsilon \frac{\partial}{\partial \varepsilon} e^{-i\varepsilon \vec{Q}^2/4t} \mathcal{V}(2\vec{P}t) e^{i\varepsilon \vec{Q}^2/4t} e^{-iX_t} f \Big\| \\ &= \frac{1}{4t} \Big\| \int_0^1 d\varepsilon\, Z_{t/\varepsilon}^* [\vec{Q}^2, \mathcal{V}(2\vec{P}t)] Z_{t/\varepsilon} e^{-iX_t} f \Big\| \\ &\leq \frac{1}{4t} \int_0^1 d\varepsilon \big\|[\vec{Q}^2, \mathcal{V}(2\vec{P}t)] Z_{t/\varepsilon} e^{-iX_t} f\big\|. \end{aligned}$$

The commutator is easily calculated by remembering that $\vec{Q}^2$ acts as $-\Delta$ in momentum space: $-[\vec{Q}^2, \mathcal{V}(2\vec{P}t)] = (2t)^2(\Delta\mathcal{V})(2\vec{P}t) - 4it(\vec{\nabla}\mathcal{V})(2\vec{P}t) \cdot \vec{Q}$. By using also the equation $Z_{t/\varepsilon}^* \vartheta(2\vec{P}t) Z_{t/\varepsilon} = U_{t/\varepsilon}^{0*} \vartheta(\varepsilon\vec{Q}) U_{t/\varepsilon}^0$, which is a simple consequence of (5.62), one finds

$$\begin{aligned} &\big\|[\mathcal{V}(\vec{Q}) - \mathcal{V}(2\vec{P}t)]U_t^0 e^{-iX_t} f\big\| \\ &\qquad \leq \int_0^1 d\varepsilon \big\| \big[t(\Delta\mathcal{V})(\varepsilon\vec{Q}) U_{t/\varepsilon}^0 e^{-iX_t} f - i(\vec{\nabla}\mathcal{V})(\varepsilon\vec{Q}) U_{t/\varepsilon}^0 \cdot \vec{Q} e^{-iX_t} f\big]\big\|. \end{aligned}$$

It is useful to introduce the notation $\langle \varepsilon Q\rangle = (I+\varepsilon^2\vec{Q}^2)^{1/2}$. Then the preceding inequality implies together with (5.118) that

$$\begin{aligned}&\|[\mathcal{V}(\vec{Q})-\mathcal{V}(2\vec{P}t)]U_t^0 e^{-iX_t}f\|\\&\quad\leq c\int_0^1 d\varepsilon\left[t\|\langle\varepsilon Q\rangle^{-3}U^0_{t/\varepsilon}e^{-iX_t}f\|+\sum_{j=1}^n\|\langle\varepsilon Q\rangle^{-2}U^0_{t/\varepsilon}Q_j e^{-iX_t}f\|\right].\end{aligned}$$

So it suffices to estimate $\|\langle\varepsilon Q\rangle^{-3}U^0_{t/\varepsilon}e^{-iX_t}f\|$ and $\|\langle\varepsilon Q\rangle^{-2}U^0_{t/\varepsilon}Q_j e^{-iX_t}f\|$ as functions of $\varepsilon$ and $t$. The results of the following lemma show that the integrand in the preceding inequality is bounded, uniformly in $\varepsilon\in(0,1]$, by $c_f(\log t)^3/t^2$ for some finite constant $c_f$ depending on $f$ (using the assumption that $\log t\geq 1$). Consequently $\|[\mathcal{V}(\vec{Q})-\mathcal{V}(2\vec{P}t)]U_t^0 e^{-iX_t}f\|$ belongs to $L^1([e,\infty))$.

**Lemma 5.33.** *Let $f\in L^2(\mathbb{R}^n)$ be such that $\tilde{f}\in C_0^\infty(\mathbb{R}^n\setminus\{0\})$, with $n\geq 3$, and let $k\in\{0,1,2,3,\dots\}$. Then there exist finite constants $c_k(f)$ and $\tilde{c}_k(f)$ such that for all $t\geq e$ and all $\varepsilon\in(0,1]$:*

$$\|\langle\varepsilon Q\rangle^{-k}U^0_{t/\varepsilon}e^{-iX_t}f\|\leq c_k(f)(\log t)^k t^{-k} \tag{5.120}$$

*and*

$$\|\langle\varepsilon Q\rangle^{-k}U^0_{t/\varepsilon}Q_j e^{-iX_t}f\|\leq \tilde{c}_k(f)(\log t)^{k+1}t^{-k}. \tag{5.121}$$

Before proving the lemma, let us consider the short range terms in (5.119). By the support properties of $\tilde{f}$ and $W$ we have $W(2\vec{k}t)\tilde{f}(\vec{k})=0$ for all $\vec{k}\in\mathbb{R}^n$ if $t>1/2$, so that $\|W(2\vec{P}t)f\|\equiv 0$ on $[e,\infty)$. For the term $\|W(\vec{Q})U_t^0 e^{-iX_t}f\|$ the argument is similar to that applied in the proof of existence of short range wave operators given in (C) on page 224. For $h\in\mathcal{S}(\mathbb{R}^n)$ we have $W(\vec{Q})h=\text{s-lim}_{N\to\infty}W_N(\vec{Q})h$ with $W_N(\vec{x})=\chi_{[1/N,\infty)}(|\vec{x}|)W(\vec{x})$. Since $W_N\in L^\infty(\mathbb{R}^n)$, we obtain by using (5.62):

$$\begin{aligned}\|W(\vec{Q})U_t^0e^{-iX_t}f\|&=\lim_{N\to\infty}\|W_N(\vec{Q})U_t^0e^{-iX_t}f\|=\lim_{N\to\infty}\|W_N(2\vec{P}t)Z_te^{-iX_t}f\|\\&\leq\lim_{N\to\infty}\|W_N(2\vec{P}t)\langle Q\rangle^{-2n}\|\|\langle Q\rangle^{2n}e^{-iX_t}f\|.\end{aligned}\tag{5.122}$$

Now [see (5.66)]

$$\begin{aligned}\|W_N(2\vec{P}t)\langle Q\rangle^{-2n}\|&\leq\|W_N(2\vec{k}t)\|_q\|(1+\vec{x}^2)^{-n}\|_q\leq\|W(2\vec{k}t)\|_q\|(1+\vec{x}^2)^{-n}\|_q\\&\leq t^{-n/q}\|W\|_q\|(1+\vec{x}^2)^{-n}\|_q.\end{aligned}\tag{5.123}$$

By choosing $q=n-1/4$, so that $\|W\|_q<\infty$, we have $\|W_N(2\vec{P}t)\langle Q\rangle^{-2n}\|\leq c_nt^{-1-\delta}$ for some constant $c_n$ and all $N\in\mathbb{N}$, with $\delta=1/(4n-1)>0$. Furthermore, since $Q_j$ acts as $\partial/\partial k_j$ in momentum space and $\log t\geq 1$, one has $\|\langle Q\rangle^{2n}e^{-iX_t}f\|\leq\tilde{c}_f(\log t)^{2n}$ for some constant $\tilde{c}_f$. So (5.122) and (5.123) imply that $\|W(\vec{Q})U_t^0e^{-iX_t}f\|$ belongs to $L^1([e,\infty))$.

PROOF OF LEMMA 5.33. We shall use the fact that, for any $m\in\mathbb{R}$, the vectors $H_0^{-m}f$, $P_jH_0^{-m}f$, $Q_jf$ and $P_jH_0^{-m}Q_jf$ all belong to $\mathcal{D}$.

(i) We first prove (5.120) by induction. Clearly (5.120) is satisfied for $k = 0$, with $c_0(f) = \|f\|$. As a preparation for the induction step we observe that, by the commutation relation $[Q_j, U_s^0] = 2P_j s U_s^0$, one has for $g \in \mathcal{S}(\mathbb{R}^n)$:

$$\varepsilon Q_j U_{t/\varepsilon}^0 P_j g = U_{t/\varepsilon}^0 (\varepsilon Q_j P_j + 2P_j^2 t) g.$$

After summing over $j$ and then setting $g = (2\vec{P}^2 t)^{-1} e^{-iX_t} f$, one obtains the following identity:

$$U_{t/\varepsilon}^0 e^{-iX_t} f = \frac{\varepsilon}{2t} \vec{Q} \cdot \vec{P} H_0^{-1} U_{t/\varepsilon}^0 e^{-iX_t} f - \frac{\varepsilon}{2t} U_{t/\varepsilon}^0 \vec{Q} \cdot \vec{P} H_0^{-1} e^{-iX_t} f. \tag{5.124}$$

Now assume that (5.120) holds for $k = N$, and consider $\|\langle \varepsilon Q \rangle^{-N-1} U_{t/\varepsilon}^0 e^{-iX_t} f\|$. By using (5.124) and observing that $\|\langle \varepsilon Q \rangle^{-1}\| = 1$ , one finds for $\varepsilon \in (0, 1]$:

$$\begin{aligned} \|\langle \varepsilon Q \rangle^{-N-1} U_{t/\varepsilon}^0 e^{-iX_t} f\| \leq & \frac{1}{2t} \sum_{j=1}^n \|\langle \varepsilon Q \rangle^{-1} \varepsilon Q_j\| \|\langle \varepsilon Q \rangle^{-N} U_{t/\varepsilon}^0 e^{-iX_t} P_j H_0^{-1} f\| \\ & + \frac{1}{2t} \|\langle \varepsilon Q \rangle^{-N} U_{t/\varepsilon}^0 \vec{Q} \cdot \vec{P} H_0^{-1} e^{-iX_t} f\|. \end{aligned} \tag{5.125}$$

Since $\|\langle \varepsilon Q \rangle^{-1} \varepsilon Q_j\| = 1$, the first term on the right-hand side of this inequality is bounded by $[c_N(f)/2](\log t)^N t^{-N-1}$, which is less than $[c_N(f)/2](\log t)^{N+1} t^{-N-1}$ if $t \geq e$. To handle the second term on the right-hand side we commute the operators $Q_j$ through $P_j H_0^{-1} e^{-iX_t}$; remembering that $Q_j$ acts as $\partial/\partial k_j$ in momentum space, this gives

$$\vec{Q} \cdot \vec{P} H_0^{-1} e^{-iX_t} f = e^{-iX_t} [(n-2) i H_0^{-1} f + \sum_{j=1}^n P_j H_0^{-1} Q_j f - \frac{\gamma}{2} (\log t) H_0^{-3/2} f. \tag{5.126}$$

By inserting this expression into the last term in (5.125) and by using the induction hypothesis, one finds that this term is bounded by $\kappa_N(f)(\log t)^{N+1} t^{-N-1}$ with

$$\kappa_N(f) = (n-2) c_N(H_0^{-1} f) + c_N(H_0^{-1} \vec{P} \cdot \vec{Q} f) + \frac{\gamma}{2} c_N(H_0^{-3/2} f).$$

We conclude that (5.120) holds for $k = N + 1$, with $c_{N+1}(f) = c_N(f)/2 + \kappa_N(f)$.

(ii) The inequality (5.121) follows easily from (5.120) by observing that $Q_j e^{-iX_t} f = e^{-iX_t} Q_j f - (\gamma/2)(\log t) e^{-iX_t} P_j H_0^{-3/2} f$. □

REMARKS. (a) By an adjustment of the arguments presented before the proof of Lemma 5.33 one easily sees that one may add a short range potential $V_S$ to $\gamma |\vec{x}|^{-1}$ and still get strong convergence of $e^{iHt} U_t^0 e^{-iX_t}$, where $H = H_0 + \gamma |\vec{x}|^{-1} + V_S$ and $X_t$ is as before.

(b) It is rather clear that essentially the same proof as that presented for Coulomb potentials can show the existence of modified wave operators for other long range potentials [for details see the references mentioned before Proposition 5.32].

(c) Unlike the asymptotic expectation values of observables, the scattering operator for long range potentials will not be unique. More precisely, if one requires the distortion $e^{-iX_t}$ of the free evolution to be a function of $H_0$, then $\lim_{t\to\pm\infty}\langle U_t f, AU_t f\rangle$ for a free observable $A$ are uniquely determined, whereas the scattering operator $S = \Omega_+^*\Omega_-$ is determined only up to a unitary factor $e^{iZ}$, where $Z$ is a self-adjoint function of $H_0$. So the $S$-matrices $S(\lambda)$ at energy $\lambda$ for different admissible distortions of the free evolution may differ by a factor $e^{i\alpha(\lambda)}$, with $\alpha(\lambda)\in\mathbb{R}$. However the probability of scattering into a cone $\mathcal{C}$ is not effected by this indeterminacy: the Scattering-Into-Cones Formula (5.83) is valid for Coulomb potentials (Problem 5.18), and the value of the integral in (5.83) does not change if $S$ is multiplied by a unitary operator of the form $e^{iZ(H_0)}$.

**5.8.3.** To complete our discussion of Coulomb scattering we show that the usual wave operators do not exist for a Coulomb potential, i.e. that $U_t^*U_t^0$ cannot have strong limits as $t\to\pm\infty$ if $U_t = e^{-iH_C t}$. Specifically we shall prove that $U_t^*U_t^0$ converge weakly to zero as $t\to\pm\infty$. Then, assuming the existence of $W_\pm := \text{s-}\lim_{t\to\pm\infty} U_t^*U_t^0$, one would have $O = \text{w-}\lim_{t\to\pm\infty} U_t^*U_t^0 = W_\pm$, a contradiction because the limit operators $W_\pm$ must be isometries as the strong limits of a family of unitary operators (see Proposition 5.13).

We begin by establishing an extension of the result of Lemma 5.33. Let $f$ be as in Lemma 5.33, and let $c_k(f)$ and $\tilde{c}_k(f)$ be the constants constructed in the proof of Lemma 5.33, $k\in\{0,1,2,3,\dots\}$. Then one has for $1<s<t$:

$$\|\langle\varepsilon Q\rangle^{-k}U^0_{t/\varepsilon}e^{-iX_t}e^{iX_s}f\| \le c_k(f)(\log t)^k t^{-k} \tag{5.127}$$

and

$$\|\langle\varepsilon Q\rangle^{-k}U^0_{t/\varepsilon}Q_j e^{-iX_t}e^{iX_s}f\| \le \tilde{c}_k(f)(\log t)^{k+1}t^{-k}. \tag{5.128}$$

PROOF. As in the proof of Lemma 5.33 one can proceed by induction. It is clear that (5.127) is satisfied for $k=0$. So assume that it holds for $k=N$ and consider the quantity $\|\langle\varepsilon Q\rangle^{-N-1}U^0_{t/\varepsilon}e^{-iX_t}e^{iX_s}f\|$. One can repeat the arguments given after the equation (5.124) by replacing everywhere $e^{-iX_t}$ by $e^{-i(X_t-X_s)}$. After this replacement all the expressions in the proof of Lemma 5.33 remain valid except that the last term in (5.126) now becomes $(\gamma/2)(\log t-\log s)H_0^{-3/2}f$. By observing that $|\log t-\log s|\le \log t$ if $1<s<t$, one finds as before that (5.127) holds for $k=N+1$, with the same constant $c_{N+1}(f)$. Finally (5.128) follows from (5.127) as in the proof of Lemma 5.33. □

Next let $f$ belong to the dense $\mathcal{D}$ set introduced at the beginning of §5.8.2. Then

$$\|\Omega_+e^{iX_s}f - U_s^*U_s^0 f\| = \|[\Omega_+ - U_s^*U_s^0e^{-iX_s}]e^{iX_s}f\|$$
$$\le \int_s^\infty \|\frac{d}{dt}(U_t^*U_t^0e^{-iX_t})e^{iX_s}f\|\,dt = \int_s^\infty \|[V(\vec{Q}) - V(2\vec{P}t)]U_t^0e^{-iX_t}e^{iX_s}f\|\,dt.$$

By following the proof of the existence of the modified wave operators given in §5.8.2, using (5.127) and (5.128) in the place of (5.120) and (5.121), one finds that the integrand in the last expression is majorized by an $s$-independent function of $t$ belonging to $L^1([s,\infty))$ if $s>e$. It follows that $\text{s-}\lim_{s\to\infty}[\Omega_+e^{iX_s} - U_s^*U_s^0] = O$.

Now, taking again $f \in \mathcal{D}$, we have for any $g \in \mathcal{H}$:

$$\langle g, U_s^* U_s^0 f\rangle = \langle \Omega_+^* g, e^{iX_s} f\rangle - \langle g, [\Omega_+ e^{iX_s} - U_s^* U_s^0] f\rangle.$$

As $s \to \infty$, the second term on the right-hand side converges to zero by the preceding result, and so does the first one because, as shown below, $e^{iX_s}$ converges weakly to zero as $s \to \infty$. Hence $\text{w-lim}_{s\to\infty} U_s^* U_s^0 f = \mathbf{0}$ for each $f \in \mathcal{D}$ and hence for each $f \in \mathcal{H}$ (see Problem 2.5).

It remains to prove that $\text{w-lim}_{s\to\infty} e^{iX_s} = O$, or equivalently (again by Problem 2.5) that $\lim_{s\to\infty} \langle g, e^{iX_s} f\rangle = 0$ for all $f, g \in \mathcal{D}$. So let $f, g \in \mathcal{D}$. Then, setting $h(\vec{k}) = \overline{\tilde{g}(\vec{k})}\tilde{f}(\vec{k})$ and $\theta(k) = k^{n-1}\int_{S^{n-1}} h(k\vec{\omega})d\omega$, with $k \equiv |\vec{k}| \in (0, \infty)$, we have

$$\langle g, e^{iX_s} f\rangle = \int_{\mathbb{R}^n} e^{i\gamma(\log s)/(2k)} h(\vec{k}) d^n k = \int_0^\infty e^{i\gamma(\log s)/(2k)} \theta(k) dk.$$

By making the change of variables $k \mapsto u = 1/(2k)$ this becomes

$$\langle g, e^{iX_s} f\rangle = \int_0^\infty e^{i\gamma u(\log s)} \theta(\frac{1}{2u}) \frac{du}{2u^2}. \tag{5.129}$$

Since $h \in \mathcal{D}$, the function $u \mapsto u^{-2}\theta(1/(2u))$ is bounded and has compact support in $(0, \infty)$, hence it belongs to $L^1((0, \infty))$. By the Riemann-Lebesgue Lemma (Proposition 5.6) the integral in (5.129) converges to zero as $s \to \infty$.

## Bibliographical Notes

We have essentially followed [A] for the discussion of evolution groups and scattering cross sections. For scattering theory we cite [AJS], [Ne], [P], Volume III of [RS] and [BW]. Extensive bibliographies concerning spectral and scattering theory are given in [BW] and in Chapter 14 of [P]. Alternative proofs of Proposition 5.9(a), based on the Mean Ergodic Theorem, can be found for example in [A] and in [14].

## Problems

**5.1.** Show that a one-parameter group $\{U_t\}$ of unitary operators is strongly continuous if and only if it is weakly continuous.

**5.2.** Prove the uniqueness of the infinitesimal generator of an evolution group: If $A$ and $B$ are self-adjoint operators such that $U_t = e^{-iAt} = e^{-iBt}$ for all $t \in \mathbb{R}$, then $A = B$.

**5.3.** Let $\{U_t\}$ be an evolution group and $A$ its infinitesimal generator. Define $\hat{U}_t$ by $\hat{U}_t = U_t^*$. Verify that the family $\{\hat{U}_t\}$ is also an evolution group and determine its infinitesimal generator.

**5.4.** Let $\{U_t\}$ be the following family of operators in $L^2(\mathbb{R}^n)$:

$$(U_t f)(\vec{x}) = e^{nt/2} f(e^t \vec{x}). \tag{5.130}$$

(a) Show that $t \mapsto U_t$ defines a strongly continuous unitary one-parameter group (an evolution group), which gives a unitary representation of the dilation group mentioned in Example 5.4.
(b) Determine the infinitesimal generator of this group.
(c) The space $C_0^\infty(\mathbb{R}^n \setminus \{0\})$ (the set of functions in $C_0^\infty(\mathbb{R}^n)$ that vanish near $\vec{x} = 0$) is dense in $L^2(\mathbb{R}^n)$. By using Nelson's Criterion (Proposition 5.3), show that the restriction of the infinitesimal generator to $C_0^\infty(\mathbb{R}^n \setminus \{0\})$ is an essentially self-adjoint operator.

**5.5.** Let $\{U_t\}$ be an evolution group satisfying $U(2\pi) = I$. Show that its infinitesimal generator has no continuous spectrum and that all its eigenvalues are integers, i.e. $\sigma(A) = \sigma_{\mathrm{p}}(A) \subseteq \mathbb{Z}$. What can be said about $\sigma_{\mathrm{d}}(A)$?

**5.6.** Prove the Mean Ergodic Theorem (Proposition 5.8).

**5.7.** Verify that the statement of Proposition 5.9(a) has a converse: if $f$ is a vector in $\mathcal{H}$ such that $\lim_{T\to\infty} T^{-1} \int_0^T |\langle h, U_t f\rangle|^2 dt = 0$ for each $h \in \mathcal{H}$, then $f \in \mathcal{H}_{\mathrm{c}}(A)$.

**5.8.** Prove that $\overline{\mathcal{M}_\infty^+}(H)$, $\overline{\mathcal{M}_\infty^-}(H)$ and $\overline{\mathcal{M}}_\infty(H) \equiv \overline{\mathcal{M}_\infty^+}(H) \cap \overline{\mathcal{M}_\infty^-}(H)$, defined in (5.30), are subspaces and that each of these subspaces is invariant under the operators $U_t$.

**5.9.** (a) Let $A$ be a self-adjoint operator in a Hilbert space $\mathcal{H}$, $\lambda$ a real number and $f \in \mathcal{H}$. Show that $e^{-i(A-\lambda)t} f$ is strongly convergent as $t \to +\infty$ if and only if $f$ is an eigenvector of $A$ associated with the eigenvalue $\lambda$.
(b) Let $B$ be a second self-adjoint operator. Assume that $\text{s-}\lim_{t\to+\infty} e^{iAt}e^{-iBt} g$ exists and that $Bg = \lambda g$ for some $\lambda \in \mathbb{R}$. Then $Ag = \lambda g$.

**5.10.** Let $h \in L^1(\mathbb{R}^3) \cap L^2(\mathbb{R}^3)$ such that $\|h\| \equiv \|h\|_2 = 1$. Let $V_h$ be the operator defined by

$$V_h f = \langle h, f\rangle h,$$

or in Dirac notation $V_h = |h\rangle\langle h|$. Prove the existence of the wave operators for $H = \vec{P}^2 + \gamma V_h$, where $\gamma$ is a real constant.

REMARK. An interaction of the form $\gamma V_h$ is called a ***separable potential*** and is used for example in models of nuclear physics. Mathematically such an operator is an operator of rank 1.

**5.11.** In $\mathcal{H} = L^2(\mathbb{R})$, let $P$ be the self-adjoint momentum operator [formally given by $-i(d/dx)$]. Let $A = P + V$, where $V$ is the multiplication operator by a bounded real function belonging to $L^1(\mathbb{R})$.
(a) Prove that the wave operators $\Omega_\pm = \text{s-}\lim_{t\to\pm\infty} e^{iAt}e^{-iPt}$ exist and are unitary operators.
(b) Give explicit expressions for $\Omega_\pm$ and for the scattering operator $S = \Omega_+^*\Omega_-$ in terms of the potential $V$.
[Hint: Show first that $[e^{-iPt}f](x) = f(x-t)$ and that there is a function $W$ such that $[e^{-iAt}f](x) = e^{-iW(x)+iW(x-t)} f(x-t)$.]

**5.12.** Let $\mathcal{H}$ be an infinite-dimensional Hilbert space and $\{e_{j,k}\}$ an orthonormal basis of $\mathcal{H}$, where $j = 1, 2, \ldots$ and $k = 0, \pm1, \pm2, \ldots$. Let $X$ and $Y$ be the following operators defined by their action on each $e_{j,k}$:

$$\begin{aligned} Xe_{j,k} &= e_{j,k+1} \\ Ye_{j,k} &= e_{j,k+1} \qquad \text{if } k \neq -j, -j+1 \\ Ye_{j,-j} &= e_{j+1,-j+1} \\ Ye_{j,-j+1} &= e_{j-1,-j+2} \qquad \text{if } j \neq 1 \\ Ye_{1,0} &= e_{1,1} \end{aligned}$$

(a) Make a pictorial representation of the action of $X$ and $Y$ by representing each $e_{j,k}$ by its co-ordinates $(j,k)$ in the plane $\mathbb{R}^2$ and verify that $X$ and $Y$ are unitary operators.
(b) For each integer $n$, let $U_n^0 = X^n$ and $U_n = Y^n$. Show that the wave operators $\Omega_\pm = \text{s-lim}_{n\to\pm\infty} U_n^* U_n^0$ exist.
(c) Determine $\Omega_\pm e_{j,k}$ and show that $\mathcal{R}(\Omega_+) = \mathcal{H}$ and $\mathcal{R}(\Omega_-) \neq \mathcal{H}$. Verify explicitly that $S = \Omega_+^*\Omega_-$ is isometric but not unitary and that $\text{w-lim}_{n\to-\infty} U_n^{0*} U_n f = \mathbf{0}$ for each $f$ orthogonal to $\mathcal{R}(\Omega_-)$.

**5.13.** Let $A_1$, $A_2$ and $A_3$ be self-adjoint operators in a Hilbert space $\mathcal{H}$. For $j = 1, 2, 3$ denote by $E_j$ the projection onto the absolutely continuous subspace $\mathcal{H}_{\text{ac}}(A_j)$ of $A_j$ and set $\Omega_\pm^{jk} = \text{s-lim}_{t\to\pm\infty} e^{iA_jt}e^{-iA_kt}E_k$ if the limit exists.
(a) Assume that $\Omega_\pm^{12}$ and $\Omega_\pm^{23}$ exist. Show that then $\Omega_\pm^{13}$ also exist and that $\Omega_\pm^{13} = \Omega_\pm^{12}\Omega_\pm^{23}$ (***chain rule*** for wave operators).
(b) Assume that $\Omega_\pm^{12}$ and $\Omega_\pm^{21}$ exist. Show that $\mathcal{R}(\Omega_\pm^{21}) = E_2\mathcal{H}$. In particular, if the spectrum of $A_1$ is purely absolutely continuous, the scattering operator $(\Omega_+^{21})^*\Omega_-^{21}$ for the couple $\{A_2, A_1\}$ is unitary.

**5.14.** In §5.3.1 the asymptotic condition was introduced on the basis of strong convergence, for example it was required that, for given $f \in \mathcal{M}_\infty^+(H)$, there should exist a vector $f_+$ such that $\text{s-lim}_{t\to+\infty}[U_t f - e^{-iH_0t}f_+] = 0$. Show that if this hypothesis is weakened to the requirement that $\text{w-lim}_{t\to+\infty}[U_t f - e^{-iH_0t}f_+] = 0$, then the vector $f_+$ will in general not be unique.

**5.15.** Let $\{S(t)\}_{t\geq0}$ be a strongly continuous family of bounded self-adjoint operators satisfying $S(t)S(\tau) = S(t+\tau)$, $S(0) = I$ and $\|S(t)\| \leq e^{mt}$ for some real constant $m$ and all $t, \tau \geq 0$. Prove that there is a self-adjoint operator $A$ such that $S(t) = e^{-At}$. [Hint: Follow the proof of Stone's Theorem.]

**5.16.** Prove Proposition 5.24 and Corollary 5.25.

**5.17.** Let $W$, $\eta$ and $\eta_\alpha$ be as in Proposition 5.30. Show that, for $j = 1, \ldots, n$:

$$\int_{-\infty}^{\infty} \|WP_jU_t^0P(\eta)\|_{\text{HS}}^2\, dt = \frac{1}{2n}(2\pi)^{1-n}\Theta_n\|\eta_n\|^2\|W\|^2.$$

**5.18.** Prove the validity of the Scattering-Into-Cones Formula (5.83) for scattering by a Coulomb potential.

# CHAPTER 6

# The Conjugate Operator Method

We have seen in Section 4.4 that, if $H$ is a self-adjoint operator and $\lambda \in \mathbb{R}$ a point in its spectrum, then $(H - \lambda - i\mu)^{-1}$ cannot converge [in $\mathcal{B}(\mathcal{H})$] as $\mu \to 0$, but that $\langle f, (H - \lambda - i\mu)^{-1} f\rangle$ may have a limit or remain bounded for certain vectors $f$. We have also seen that these properties of convergence or boundedness are linked to details of the spectrum of $H$ in a neighbourhood of $\lambda$. In particular, the positivity of the commutator of $iH$ with some other self-adjoint operator $A$ leads to absolute continuity of part of the spectrum of $H$ (see Proposition 4.32 and the remarks following its statement). We discuss here a method that was introduced by E. Mourre for proving the existence of the limits of $\langle f, (H - \lambda - i\mu)^{-1} f\rangle$ under certain conditions on $H$, $\lambda$ and $f$. It exploits in a subtle way the idea of obtaining absolute continuity from the positivity of a commutator. Important consequences are the absence of singular continuous spectrum of $H$ (see Corollary 6.9) and asymptotic completeness in scattering theory (see Section 7.1). The Mourre method is particularly interesting through the fact that it also allows the treatment of $N$-body Hamiltonians, which have a much more complex structure.

Let us give a brief description of the idea of this method. It relies on the introduction of two auxiliary quantities: a parameter $\varepsilon$ and a self-adjoint operator $A$ having suitable commutation properties with $H$. The function $\langle f, (H - \lambda - i\mu)^{-1} f\rangle$ is extended to an $\varepsilon$-dependent function $\Phi_\varepsilon(\lambda + i\mu)$, $0 \le \varepsilon \le 1$, with $\Phi_0(\lambda + i\mu) = \langle f, (H - \lambda - i\mu)^{-1} f\rangle$. The definition of $\Phi_\varepsilon(\lambda + i\mu)$ involves the commutator between $iH$ and the auxiliary operator $A$, and it is essential that this commutator be positive in some sense. One then establishes a ***differential inequality*** for $\Phi_\varepsilon$ (as a function of $\varepsilon$), i.e. an inequality of the form

$$|\Phi'_\varepsilon(z)| \equiv \Big|\frac{\partial}{\partial\varepsilon}\Phi_\varepsilon(z)\Big| \le h(\varepsilon, \Phi_\varepsilon(z)) \qquad (z = \lambda + i\mu, \mu > 0),$$

for some function $h$ of two variables (not depending explicitly on $z$). By integrating this differential inequality, one rather easily arrives at the existence of the limit $\lim_{\mu\to+0} \langle f, (H - \lambda - i\mu)^{-1} f\rangle$. The existence of $\lim_{\mu\to+0} \langle f, (H - \lambda + i\mu)^{-1} f\rangle$ can be obtained in a similar way. In general these two limits will be different.

As in the case of the Spectral Theorem (Appendix to Chapter 4), the arguments are simpler for self-adjoint operators $H$ that are bounded. For this reason we elaborate in Sections 6.2 and 6.3 the essentials of the Mourre method, as well as its consequences for the spectral properties of $H$, by assuming $H$ to be bounded. In Section 6.4 we extend the results to a large class of unbounded operators $H$ and apply them to Schrödinger operators in $L^2(\mathbb{R}^n)$. As a motivating introduction, we discuss in Section 6.1 a simple example that already exhibits many of the aspects of the method. In Section 6.5 we introduce the notion of smooth operators relative to $H$; this concept is related to the behaviour of $(H-\lambda-i\mu)^{-1}$ for small values of $\mu$ and has applications in particular in scattering theory. Section 6.6 is devoted to some resolvent estimates involving higher order commutators of $H$ with the auxiliary operator $A$; these results will be needed for some developments in Chapter 7.

Evaluation or estimation of commutators is often an important tool in proving mathematical results in quantum mechanics. Commutator calculations play a very important role in the present context as well as further on, especially in Section 7.2. Interesting examples can be found in §6.4.1 and, further on, in the proof of Lemma 7.5 and in the estimates of §7.2.5. These calculations are complicated through the fact that they involve unbounded operators. In order not to lengthen the arguments, we often determine the occurring commutators by formal manipulations. In each instance one would have to give a rigourous meaning to these manipulations by showing that they can be justified on a suitable domain of vectors in the Hilbert space. The final Section 6.7 of the present chapter is devoted to some comments on this and to the derivation of some commutator estimates that are used in the text.

## 6.1 A simple example

In this introductory section we calculate certain commutators between unbounded operators without taking into account domain questions. The aim is rather to illustrate the Mourre method, as outlined above, by means of a simple example. A more careful treatment will be presented in the subsequent sections of this chapter.

**6.1.1.** We begin by considering a very simple differential inequality involving two variables $\varepsilon$ and $\mu$. We fix some number $\varepsilon_0$ in $(0,1)$ and let $\mathfrak{D}$ be the following subset of $\mathbb{R}^2$: $\mathfrak{D}=\{(\varepsilon,\mu)\in\mathbb{R}^2 \mid 0\le\varepsilon\le\varepsilon_0,\ 0\le\mu\le 1,\ \varepsilon+\mu>0\}$, Thus $\mathfrak{D}$ is a closed rectangle with one vertex [the point $(0,0)$] removed. Let $F\colon\mathfrak{D}\to\mathbb{C}$ be a continuous function assumed to be everywhere non-zero and continuously differentiable with respect to the first variable $\varepsilon$ on $\mathfrak{D}_0=\{(\varepsilon,\mu)\in\mathfrak{D}\mid 0<\varepsilon<\varepsilon_0\}$ [1]. Assume that the following differential inequality is satisfied on $\mathfrak{D}_0$:

$$\Big|\frac{\partial}{\partial\varepsilon}F(\varepsilon,\mu)\Big|\le\vartheta(\varepsilon)|F(\varepsilon,\mu)|^{1/2}, \tag{6.1}$$

where $\vartheta$ is a non-negative function on $(0,\varepsilon_0]$ satisfying $\Theta:=\int_0^{\varepsilon_0}\vartheta(\varepsilon)d\varepsilon<\infty$. We shall show that, under these conditions, $F$ can be extended to a continuous function

[1] In the context explained at the beginning of this chapter, one would take $F(\varepsilon,\mu)=\Phi_\varepsilon(\lambda+i\mu)$, with a fixed value of $\lambda$.

on the closed rectangle $\overline{\mathfrak{D}}$, i.e. that there exists a complex number $F(0,0)$ such that $F(\varepsilon,\mu) \to F(0,0)$ as the point $(\varepsilon,\mu) \in \mathfrak{D}$ tends to $(0,0)$ in an arbitrary way. In particular $F(0,\mu)$ will converge as $\mu \to +0$, which is the consequence that is important in our context.

To obtain the claimed result, we rewrite (6.1) as follows:

$$\Big|\frac{\partial}{\partial\varepsilon}F(\varepsilon,\mu)^{1/2}\Big| \leq \frac{1}{2}\vartheta(\varepsilon) \leq \vartheta(\varepsilon). \tag{6.2}$$

For $0 < \varepsilon \leq \varepsilon_0$ we define a number $\varphi(\varepsilon)$ by $\varphi(\varepsilon) = \sup_{0\leq\mu\leq1}|F(\varepsilon,\mu)|$ and observe that $\varphi(\varepsilon)$ is finite as a consequence of the continuity of $F$ on $\mathfrak{D}$. It then follows from (6.2) that, for all $(\varepsilon,\mu) \in \mathfrak{D}$:

$$\begin{aligned} \big|F(\varepsilon,\mu)\big|^{1/2} = \big|F(\varepsilon,\mu)^{1/2}\big| &= \Big|F(\varepsilon_0,\mu)^{1/2} - \int_\varepsilon^{\varepsilon_0}\frac{\partial}{\partial\tau}F(\tau,\mu)^{1/2}\,d\tau\Big| \\ &\leq \sqrt{\varphi(\varepsilon_0)} + \int_0^{\varepsilon_0}\vartheta(\tau)\,d\tau. \end{aligned} \tag{6.3}$$

By inserting this estimate into (6.1), one obtains the inequality

$$\Big|\frac{\partial}{\partial\varepsilon}F(\varepsilon,\mu)\Big| \leq \kappa\vartheta(\varepsilon) \tag{6.4}$$

with the constant $\kappa = \sqrt{\varphi(\varepsilon_0)} + \Theta$. This inequality holds on $\mathfrak{D}_0$.

The inequality (6.4) implies that, for $0 < \eta < \varepsilon < \varepsilon_0$ and all $\mu \in [0,1]$:

$$|F(\varepsilon,\mu) - F(\eta,\mu)| = \Big|\int_\eta^\varepsilon \frac{\partial}{\partial\tau}F(\tau,\mu)\,d\tau\Big| \leq \kappa\int_\eta^\varepsilon \vartheta(\tau)\,d\tau. \tag{6.5}$$

Since $\int_0^{\varepsilon_0}\vartheta(\tau)d\tau < \infty$, this shows that, for each $\mu \in [0,1]$, $F(\varepsilon,\mu)$ is convergent as $\varepsilon \to 0$. By taking $\mu = 0$, one can define $F(0,0)$ as $F(0,0) = \lim_{\varepsilon\to0} F(\varepsilon,0)$.

We have thus extended the function $F$ to the set $\overline{\mathfrak{D}}$, and it remains to verify its continuity at the point $(0,0)$. We must show that, given $\delta > 0$, there exist $\varepsilon_1 \in (0,\varepsilon_0)$ and $\mu_0 \in (0,1)$ such that $|F(\varepsilon,\mu) - F(0,0)| < \delta$ for all $\varepsilon \in [0,\varepsilon_1]$ and all $\mu \in [0,\mu_0]$. For this we write

$$|F(\varepsilon,\mu)-F(0,0)| \leq |F(\varepsilon,\mu)-F(\varepsilon_1,\mu)|+|F(\varepsilon_1,\mu)-F(\varepsilon_1,0)|+|F(\varepsilon_1,0)-F(0,0)|. \tag{6.6}$$

The inequality (6.5) implies that we can choose $\varepsilon_1 \in (0,\varepsilon_0)$ such that the first and the third term on the right-hand side of (6.6) are less than $\delta/3$ for all $\varepsilon \in [0,\varepsilon_1]$ and any value of $\mu$. Then, by the continuity on $[0,1]$ of $\mu \mapsto F(\varepsilon_1,\mu)$, there exists $\mu_0 > 0$ such that $|F(\varepsilon_1,\mu) - F(\varepsilon_1,0)| < \delta/3$ for all $\mu \in [0,\mu_0]$. This completes the proof of the continuity of $F$ on $\overline{\mathfrak{D}}$.

**6.1.2.** We now discuss an example illustrating the usefulness of a differential inequality in studying boundary values of a resolvent. Let $H$ be a self-adjoint operator in a Hilbert space $\mathcal{H}$. We denote by $R(z) \equiv (H-z)^{-1}$ its resolvent at the point $z \in \rho(H)$

(see Section 2.6). We recall from (2.103) that $R(z)^* = R(\overline{z})$. Thus $\langle f, R(\overline{z})f\rangle$ is the complex conjugate of $\langle f, R(z)f\rangle$. Hence, by also using the first resolvent equation (2.93), we have

$$\|R(z)f\|^2 = \langle R(z)f, R(z)f\rangle = \langle f, R(\overline{z})R(z)f\rangle = \frac{1}{z-\overline{z}}[\langle f, R(z)f\rangle - \langle f, R(\overline{z})f\rangle]$$
$$= \frac{1}{2i\Im z}\left[2i\Im\langle f, R(z)f\rangle\right] = \frac{1}{\Im z}\Im\langle f, R(z)f\rangle. \tag{6.7}$$

Since $\Im\overline{z} = -\Im z$ and $\Im\langle f, R(\overline{z})f\rangle = \Im\langle R(z)f, f\rangle = -\Im\langle f, R(z)f\rangle$, the preceding equation implies that one also has [2]

$$\|R(\overline{z})f\|^2 = \|R(z)^*f\|^2 = \frac{1}{\Im z}\Im\langle f, R(z)f\rangle. \tag{6.8}$$

We know from (2.99) that $(d/dz)R(z) = R(z)^2$. Furthermore, if $A$ is an arbitrary operator, one has

$$\begin{aligned} R(z)[H, A]R(z) &= R(z)[H - z, A]R(z) \\ &= R(z)(H - z)AR(z) - R(z)A(H - z)R(z) \\ &= AR(z) - R(z)A, \end{aligned}$$

i.e.

$$[A, R(z)] = R(z)[H, A]R(z). \tag{6.9}$$

In particular, if $A$ is a self-adjoint operator such that $[H, iA] = H$, then one obtains from Eq. (6.9) that

$$[iA, R(z)] = R(z)HR(z) = R(z)(H - z)R(z) + zR(z)^2 = R(z) + z\frac{d}{dz}R(z). \tag{6.10}$$

Let $f$ be a vector in $\mathcal{H}$ and set $\Phi(z) = \langle f, R(z)f\rangle$. One would like to know if $\Phi(\lambda \pm i\mu)$ converges as $\mu \to +0$. As already pointed out, the Mourre method consists in somewhat modifying the resolvent. One considers a family $\{G_\varepsilon(z)\}$ of operators depending on a parameter $\varepsilon$ ($0 \le \varepsilon \le \varepsilon_0$) such that $G_0(z) = R(z)$ and such that, for $\varepsilon \ne 0$, the quantity $\Phi_\varepsilon(z) := \langle f, G_\varepsilon(z)f\rangle$ has a limit as $z$ tends to the real axis and satisfies a suitable differential inequality.

To illustrate this, we assume as above that $[H, iA] = H$ and choose $f$ in the domain $\mathcal{D}(A)$ of $A$. We take

$$G_\varepsilon(z) = (e^{-i\varepsilon}H - z)^{-1} = [e^{-i\varepsilon}(H - ze^{i\varepsilon})]^{-1} = e^{i\varepsilon}R(ze^{i\varepsilon}). \tag{6.11}$$

The following property of these operators follows from (6.7) and (6.8):

$$\|G_\varepsilon(z)f\|^2 = \|G_\varepsilon(z)^*f\|^2 = \frac{1}{\Im(ze^{i\varepsilon})}\Im\langle f, R(ze^{i\varepsilon})f\rangle = \frac{1}{\Im(ze^{i\varepsilon})}\Im\left[e^{-i\varepsilon}\langle f, G_\varepsilon(z)f\rangle\right]$$
$$= \left|\frac{1}{\Im(ze^{i\varepsilon})}\Im\left[e^{-i\varepsilon}\langle f, G_\varepsilon(z)f\rangle\right]\right| \le \frac{1}{|\Im(ze^{i\varepsilon})|}|\langle f, G_\varepsilon(z)f\rangle|. \tag{6.12}$$

[2] The identity $\|R(z)f\|^2 = \|R(\overline{z})f\|^2$ also follows from (4.42) by taking $\varphi(\lambda) = (\lambda - z)^{-1}$ or $\varphi(\lambda) = (\lambda - \overline{z})^{-1}$ in that equation.

Let us now calculate the derivative $(\partial/\partial\varepsilon)G_\varepsilon(z) \equiv G'_\varepsilon(z)$ of $G_\varepsilon(z)$ (in the sequel a prime will denote derivatives with respect to the parameter $\varepsilon$). Let us set $\zeta = ze^{i\varepsilon}$ and observe that, by using (6.10):

$$\begin{aligned}\frac{\partial}{\partial\varepsilon}R(ze^{i\varepsilon}) &= i\zeta\frac{d}{d\zeta}R(\zeta)\big|_{\zeta=ze^{i\varepsilon}} = i\{-R(\zeta)+[iA,R(\zeta)]\}\big|_{\zeta=ze^{i\varepsilon}}\\ &= -iR(ze^{i\varepsilon}) - [A,R(ze^{i\varepsilon})].\end{aligned}$$

Thus, by applying the Leibniz rule:

$$\begin{aligned}\frac{\partial}{\partial\varepsilon}G_\varepsilon(z) &= \frac{\partial}{\partial\varepsilon}\{e^{i\varepsilon}R(ze^{i\varepsilon})\} = ie^{i\varepsilon}R(ze^{i\varepsilon}) + e^{i\varepsilon}\frac{\partial}{\partial\varepsilon}R(ze^{i\varepsilon})\\ &= iG_\varepsilon(z) - ie^{i\varepsilon}R(ze^{i\varepsilon}) - e^{i\varepsilon}[A,R(ze^{i\varepsilon})]\\ &= -e^{i\varepsilon}[A,R(ze^{i\varepsilon})] = -[A,G_\varepsilon(z)].\end{aligned} \tag{6.13}$$

It is clear that, if $z = \lambda + i\mu$ with either $\lambda > 0$ and $\mu \geq 0$ or $\lambda < 0$ and $\mu \leq 0$, and if $\varepsilon$ is strictly positive and sufficiently small, then $ze^{i\varepsilon}$ belongs to the resolvent set of $H$, hence, by (6.11), $G_\varepsilon(\lambda \pm i\mu)$ is a well-defined operator in $\mathcal{B}(\mathcal{H})$ (in particular $\mu = 0$ is admitted). It remains to see how to treat the limit $\varepsilon \to 0$. Let us set $\Phi_\varepsilon(z) = \langle f, G_\varepsilon(z)f\rangle$ and use Eq. (6.13) to express its derivative with respect to $\varepsilon$:

$$|\Phi'_\varepsilon(z)| = |\langle G_\varepsilon(z)^*f, Af\rangle - \langle Af, G_\varepsilon(z)f\rangle| \leq \|Af\|\Big[\|G_\varepsilon(z)^*f\| + \|G_\varepsilon(z)f\|\Big].$$

By using (6.12) one finds that

$$|\Phi'_\varepsilon(z)| \leq 2\|Af\|\Big[\frac{1}{|\Im(ze^{i\varepsilon})|}|\langle f, G_\varepsilon(z)f\rangle|\Big]^{1/2} = 2\|Af\|\Big[\frac{1}{|\Im(ze^{i\varepsilon})|}|\Phi_\varepsilon(z)|\Big]^{1/2}.$$

Thus, for $z = \lambda + i\mu$ [hence $\Im(ze^{i\varepsilon}) = \lambda\sin\varepsilon + \mu\cos\varepsilon$], one has the following inequality:

$$|\Phi'_\varepsilon(\lambda+i\mu)| \leq 2\|Af\|\frac{1}{|\lambda\sin\varepsilon + \mu\cos\varepsilon|^{1/2}}|\Phi_\varepsilon(\lambda+i\mu)|^{1/2}. \tag{6.14}$$

If $\lambda > 0$, then for $\mu \geq 0$ and $\varepsilon \in (0, \pi/2]$: $\lambda\sin\varepsilon + \mu\cos\varepsilon \geq \lambda\sin\varepsilon \geq \lambda\cdot\varepsilon(2/\pi) = 2\pi^{-1}\lambda\varepsilon$, hence for these values of $\lambda$, $\mu$ and $\varepsilon$:

$$|\Phi'_\varepsilon(\lambda+i\mu)| \leq \sqrt{\frac{2\pi}{\lambda\varepsilon}}\|Af\||\Phi_\varepsilon(\lambda+i\mu)|^{1/2}. \tag{6.15}$$

We have arrived at a differential inequality for the function $\varepsilon \mapsto \Phi_\varepsilon(\lambda+i\mu)$. This differential inequality is of the form (6.1), with $F(\varepsilon,\mu) = \Phi_\varepsilon(\lambda+i\mu)$ (for a fixed value of $\lambda$) and with $\vartheta(\varepsilon) = \sqrt{2\pi/\lambda}\|Af\|\varepsilon^{-1/2}$. By the results of §6.1.1 we have the following conclusion[3]:

[3] It is easy to see from the definitions of $G_\varepsilon(z)$ and $\Phi_\varepsilon(z)$ that $F$ has all the properties required in §6.1.1.

*If there exists a self-adjoint operator $A$ such that $[H, iA] = H$, then for each $f$ in the dense linear manifold $\mathcal{D}(A)$ and each $\lambda > 0$, $\langle f, (H - \lambda - i\mu)^{-1} f \rangle$ converges as $\mu \to +0$.*

A simple adjustment of the above arguments shows that $\langle f, (H - \lambda - i\mu)^{-1} f \rangle$ converges as $\mu \to -0$ if $f \in \mathcal{D}(A)$ and $\lambda < 0$. One can also prove the existence of $\langle f, (H - \lambda - i\mu)^{-1} f \rangle$ for $\lambda > 0$ and $\mu \to -0$ and for $\lambda < 0$ and $\mu \to +0$ by taking for $G_\varepsilon(z)$ the operator $(e^{+i\varepsilon} H - z)^{-1}$ in place of the choice made in (6.11).

COMMENTS. (a) The method leads to a control on $\langle f, R(\lambda \pm i0) f \rangle$ for $\lambda \neq 0$ but gives no result for $\lambda = 0$.

(b) It can be shown that the limits $\lim_{\mu \to +0} \langle f, (H - \lambda - i\mu)^{-1} f \rangle$ exist uniformly in $\lambda$ on any interval $[\lambda_0, \infty)$ with $\lambda_0 > 0$ (Problem 6.1).

(c) It is clear that the given arguments involve only small values of the parameter $\varepsilon$. For small $\varepsilon$ one may write $e^{-i\varepsilon} \approx 1 - i\varepsilon$, hence $e^{-i\varepsilon} H \approx H - i\varepsilon H \equiv H - i\varepsilon[H, iA]$ and $G_\varepsilon(\lambda + i\mu) \approx \{H - \lambda - i(\mu + \varepsilon[H, iA])\}^{-1}$. Thus the replacement of $R(\lambda + i\mu) \equiv (H - \lambda - i\mu)^{-1}$ by $G_\varepsilon(\lambda + i\mu)$ essentially amounts to adding to $\mu$ a term $\varepsilon[H, iA]$. For this to be useful, it is necessary that the operator $[H, iA]$ be positive (we assume here that $\mu > 0$); this increases the imaginary part of the denominator (for $\mu < 0$ one would add a term $-\varepsilon[H, iA]$).

E. Mourre realised that the choice $G_\varepsilon(\lambda + i\mu) = \{H - \lambda - i(\mu + \varepsilon[H, iA])\}^{-1}$ allows one to obtain the existence of the limits of $R(\lambda \pm i\mu)$ as $\mu \to +0$ if $[H, iA]$ is a positive operator (in a neighbourhood of $\lambda$)[4] without assuming that $[H, iA] = H$. This will be discussed in the following sections.

(d) In the literature one says that a ***limiting absorption principle*** is satisfied at $\lambda$ if the limit of $\langle f, (H - \lambda - i\mu)^{-1} f \rangle$ as $\mu \to +0$ exists for each vector $f$ in some dense subset of $\mathcal{H}$.

(e) The validity of the stated results can be extended in several ways by introducing the parameter $\varepsilon$ also for other quantities (this complicates the calculations and the differential inequality but does not require new techniques):

(1) One can approximate $H$ by a family $\{H_\varepsilon\}_{0 < \varepsilon \leq \varepsilon_0}$ such that $H_\varepsilon \to H$ as $\varepsilon \to 0$ and such that $H_\varepsilon$ has better properties relative to $A$ than the operator $H$. One then takes $G_\varepsilon = \{H_\varepsilon - \lambda - i(\mu + \varepsilon[H_\varepsilon, iA])\}^{-1}$. This allows one to treat a larger class of operators $H$.

(2) One can obtain the limits $\langle f, R(\lambda \pm i0) f \rangle$ for a set of vectors $f$ that is larger than $\mathcal{D}(A)$ by approximating a vector $f$ in this larger set in an appropriate way by a family $\{f_\varepsilon\}_{0 < \varepsilon \leq \varepsilon_0}$, with $f_\varepsilon \in \mathcal{D}(A)$.

We shall not take into account the possibility (1) but introduce (2), because this has implications for scattering theory.

[4] The precise definition will be given in §6.2.2. In the example treated here, the operator $[H, iA] = H$ is positive on a subspace, the spectral subspace $E^H((0, \infty))\mathcal{H}$. Indeed we have been able to control the limit of $R(\lambda + i\mu)$ as $\mu \to +0$ for the spectral values $\lambda > 0$. On $E^H((-\infty, 0))\mathcal{H}$, the operator $[H, iA] = H$ is negative, and for the values $\lambda < 0$ we have obtained the existence of the limit of $R(\lambda + i\mu)$ for $\mu \to -0$.

**6.1.3.** We introduce here the ***dilation group*** in the Hilbert space $L^2(\mathbb{R}^n)$ and its infinitesimal generator which we denote by $\widehat{D}$. For numerous Schrödinger Hamiltonians this operator $\widehat{D}$ can assume the role played by the operator $A$ in §6.1.2, and the dilation group will be an important tool in settling certain technical points in this chapter.

For each $\theta \in \mathbb{R}$ we define an operator $\mathrm{U}_\theta$ in $L^2(\mathbb{R}^n)$ by

$$[\mathrm{U}_\theta f](\vec{x}) = e^{n\theta/2} f(e^\theta \vec{x}). \tag{6.16}$$

It is easy to check (Problem 5.4) that the family $\{\mathrm{U}_\theta\}_{\theta\in\mathbb{R}}$ forms a strongly continuous unitary group of operators (an evolution group in the terminology of Chapter 5). The action of $\mathrm{U}_\theta$ in momentum space is also easy to calculate (using the change of variables $\vec{x} \mapsto \vec{y} = e^\theta \vec{x}$, $d^n x = e^{-n\theta} d^n y$):

$$\begin{aligned}[\mathcal{F}\mathrm{U}_\theta f](\vec{k}) &= (2\pi)^{-n/2} \int_{\mathbb{R}^n} e^{-i\vec{k}\cdot\vec{x}} e^{n\theta/2} f(e^\theta \vec{x})\, d^n x \\ &= (2\pi)^{-n/2} e^{-n\theta/2} \int_{\mathbb{R}^n} e^{-ie^{-\theta}\vec{k}\cdot\vec{y}} f(\vec{y})\, d^n y = e^{-n\theta/2} \tilde{f}(e^{-\theta}\vec{k}). \qquad (6.17)\end{aligned}$$

To determine the infinitesimal generator $\widehat{D}$ of the group $\{\mathrm{U}_\theta\}_{\theta\in\mathbb{R}}$, we observe that $\mathcal{S}(\mathbb{R}^n)$ is invariant under $\mathrm{U}_\theta$: $f \in \mathcal{S}(\mathbb{R}^n) \implies \mathrm{U}_\theta f \in \mathcal{S}(\mathbb{R}^n)$. It is rather evident that $\mathcal{S}(\mathbb{R}^n)$ belongs to the domain of $\widehat{D}$ [for each fixed $\vec{x}$ the expression (6.16) is differentiable with respect to $\theta$, and the derivative still belongs to $\mathcal{S}(\mathbb{R}^n)$ and thus is square–integrable]. By Nelson's Criterion (Proposition 5.3), the restriction of $\widehat{D}$ to $\mathcal{S}(\mathbb{R}^n)$ is essentially self-adjoint. So let $f \in \mathcal{S}(\mathbb{R}^n)$. Then [5]

$$\frac{\partial}{\partial\theta}[\mathrm{U}_\theta f](\vec{x}) = \frac{\partial}{\partial\theta}[e^{n\theta/2} f(e^\theta \vec{x})] = \frac{n}{2} e^{n\theta/2} f(e^\theta \vec{x}) + e^{n\theta/2} \sum_{j=1}^n e^\theta x_j \left[\frac{\partial f}{\partial x_j}\right](e^\theta \vec{x}).$$

Hence

$$\begin{aligned}[\widehat{D}f](\vec{x}) = i\frac{\partial}{\partial\theta}[\mathrm{U}_\theta f](\vec{x})\big|_{\theta=0} &= \frac{in}{2} f(\vec{x}) + i\sum_{j=1}^n x_j \left[\frac{\partial f}{\partial x_j}\right](\vec{x}) \\ &= \big[(\frac{in}{2} I - \sum_{j=1}^n Q_j P_j) f\big](\vec{x}).\end{aligned}$$

By writing

$$Q_j P_j = \frac{1}{2} Q_j P_j + \frac{1}{2} P_j Q_j + \frac{1}{2}[Q_j, P_j] = \frac{1}{2} Q_j P_j + \frac{1}{2} P_j Q_j + \frac{1}{2} iI,$$

one finds that, for $f \in \mathcal{S}(\mathbb{R}^n)$:

$$[\widehat{D}f](\vec{x}) = -\frac{1}{2}\big[\sum_{j=1}^n (Q_j P_j + P_j Q_j) f\big](\vec{x}).$$

[5] A somewhat more careful way of arriving at the expression for $\widehat{D}f$, for $f \in \mathcal{S}(\mathbb{R}^n)$, would involve the use of the Dominated Convergence Theorem in analogy with what was done in the derivation of (5.10).

Thus: the infinitesimal generator $\widehat{D}$ of the dilation group is the closure of the following operator on $\mathcal{S}(\mathbb{R}^n)$:

$$\begin{aligned}
-\frac{1}{2}\sum_{j=1}^{n}(Q_jP_j + P_jQ_j) &\equiv -\frac{1}{2}(\vec{Q}\cdot\vec{P} + \vec{P}\cdot\vec{Q}) \\
&= -\frac{1}{2}(2\vec{Q}\cdot\vec{P} - inI) = -\frac{1}{2}(2\vec{P}\cdot\vec{Q} + inI). \qquad (6.18)
\end{aligned}$$

For later applications it is useful to define $D = -\widehat{D}/2$, hence formally

$$D = \frac{1}{4}(\vec{Q}\cdot\vec{P} + \vec{P}\cdot\vec{Q}) = \frac{1}{4}(2\vec{Q}\cdot\vec{P} - inI) = \frac{1}{4}(2\vec{P}\cdot\vec{Q} + inI). \qquad (6.19)$$

Let $H_0 = \vec{P}^2$. Since $[Q_j, P_k] = i\delta_{jk}I$ [on $\mathcal{S}(\mathbb{R}^n)$], one has $[Q_j, \vec{P}^2] = 2iP_j$, hence [for example on $\mathcal{S}(\mathbb{R}^n)$];

$$[H_0, iD] = \frac{i}{2}\sum_{j=1}^{n}[\vec{P}^2, Q_jP_j] = \frac{i}{2}\sum_{j=1}^{n}[\vec{P}^2, Q_j]P_j = \sum_{j=1}^{n}P_j^2 = \vec{P}^2 = H_0. \qquad (6.20)$$

Thus the example treated in §6.1.2 covers the case $H = \vec{P}^2$ (take $A = D$).

A similar calculation gives $[1/\vec{Q}^2, iD] = 1/\vec{Q}^2$. So, if $H = \vec{P}^2 + \gamma/\vec{Q}^2$ (with $\gamma \in \mathbb{R}$), then $[H, iD] = H$. However, here one should be careful if one wants to apply the results of §6.1.2, since for example if $n = 3$ and $\gamma < 3/4$, one has to impose boundary conditions at the co-ordinate origin to define $H$ as a self-adjoint operator (see Proposition 3.30), and the domain defined by these boundary conditions is not necessarily invariant under $\mathrm{U}_\theta$.

It is clear from Eq. (6.17) that the operators $\mathrm{U}_\theta$ map $\mathcal{D}(H_0)$ into $\mathcal{D}(H_0)$ [apply twice Eq. (6.17)]:

$$\begin{aligned}
[\mathcal{F}\mathrm{U}_\theta H_0 \mathrm{U}_{-\theta} f](\vec{k}) &= e^{-n\theta/2}[\mathcal{F}H_0\mathrm{U}_{-\theta}f](e^{-\theta}\vec{k}) = e^{-n\theta/2}e^{-2\theta}\vec{k}^2[\mathcal{F}\mathrm{U}_{-\theta}f](e^{-\theta}\vec{k}) \\
&= e^{-n\theta/2}e^{-2\theta}\vec{k}^2 e^{n\theta/2}\tilde{f}(\vec{k}) = e^{-2\theta}\vec{k}^2\tilde{f}(\vec{k}),
\end{aligned}$$

hence

$$\mathrm{U}_\theta H_0 \mathrm{U}_{-\theta} = e^{-2\theta}H_0. \qquad (6.21)$$

By differentiating this equation one obtains again Eq. (6.20) (recall that $\mathrm{U}_\theta = e^{-i\widehat{D}\theta} = e^{2iD\theta}$):

$$\frac{d}{d\theta}\mathrm{U}_\theta H_0 \mathrm{U}_{-\theta} f\big|_{\theta=0} = \mathrm{U}_\theta[2iDH_0 - 2iH_0D]\mathrm{U}_{-\theta}f\big|_{\theta=0} = -2[H_0, iD]f$$

and

$$\frac{d}{d\theta}e^{-2\theta}H_0 f\big|_{\theta=0} = -2e^{-2\theta}H_0 f\big|_{\theta=0} = -2H_0 f.$$

In the next chapter it will be useful to know how the operator $D$ acts in the spectral representation of $H_0$. Since in momentum space the operators $\mathrm{U}_\theta$ correspond to radial dilations, it is rather evident that their generator $D$ must be a first order differential operator in the variable $\lambda$. To find its explicit form, observe that $[\mathcal{F}\vec{P}\cdot\vec{Q}f](\vec{k}) = i\vec{k}\cdot\nabla\tilde{f}(\vec{k}) = ik\partial\tilde{f}(\vec{k})/\partial k$, where $k = |\vec{k}|$. Now, with $\lambda = k^2$, we have $k(d/dk) = 2\lambda(d/d\lambda)$, and by (5.68): $\tilde{f}(\sqrt{\lambda}\vec{\omega}) = 2^{1/2}\lambda^{-(n-2)/4}(\mathcal{U}_0 f)_\lambda$. Thus

$$\begin{aligned}(\mathcal{U}_0\vec{P}\cdot\vec{Q}f)_\lambda(\vec{\omega}) &= 2^{-1/2}\lambda^{(n-2)/4}2i\lambda\frac{\partial}{\partial\lambda}\tilde{f}(\sqrt{\lambda}\vec{\omega}) \\ &= -i\frac{n-2}{2}(\mathcal{U}_0 f)_\lambda(\vec{\omega}) + 2i\lambda\frac{\partial}{\partial\lambda}(\mathcal{U}_0 f)_\lambda(\vec{\omega}). \qquad (6.22)\end{aligned}$$

It follows that

$$(\mathcal{U}_0 Df)_\lambda = \frac{1}{4}\big[2(\mathcal{U}_0\vec{P}\cdot\vec{Q}f)_\lambda + in(\mathcal{U}_0 f)_\lambda\big] = \frac{1}{4}\big[2i(\mathcal{U}_0 f)_\lambda + 4i\lambda\frac{d}{d\lambda}(\mathcal{U}_0 f)_\lambda\big].$$

This shows that, in the spectral representation of $H_0$, $D$ is the following differential operator:

$$D = i\lambda\frac{d}{d\lambda} + \frac{i}{2} = \frac{i}{2}\big(\lambda\frac{d}{d\lambda} + \frac{d}{d\lambda}\lambda\big). \qquad (6.23)$$

## 6.2 The method of differential inequalities

In this and the next section we develop the essential technical aspects of the Mourre method. The proof of Theorem 6.3 contains the principal estimates leading to the existence of the boundary values of $\langle f, (H-\lambda \pm i\mu)^{-1}f\rangle$ as $\mu \to +0$ for suitable vectors $f$, under the assumption of positivity of the commutator $[H, iA]$ between the self-adjoint operator $H$ under consideration and an auxiliary self-adjoint operator $A$. Implications for the spectral properties of $H$ are presented in Theorem 6.4 and in Corollary 6.9.

**6.2.1.** We first discuss the differential inequality that one encounters in the general setting indicated in Comment (c) on page 268. It is similar to that given in the preceding section but involves two additional terms on the right-hand side. Also it is useful to admit operator-valued functions rather than just numerical functions. We let $\mathfrak{D}$ and $\mathfrak{D}_0$ be as in §6.1.1 and consider a norm-continuous function $F\colon \mathfrak{D} \to \mathcal{B}(\mathcal{H})$ that is continuously differentiable with respect to the first variable $\varepsilon$ on $(0, \varepsilon_0)$ and satisfies the following differential inequality on $\mathfrak{D}_0$:

$$\Big\|\frac{\partial}{\partial\varepsilon}F(\varepsilon,\mu)\Big\| \le \vartheta_1(\varepsilon) + \vartheta_2(\varepsilon)\|F(\varepsilon,\mu)\|^{1/2} + \gamma\|F(\varepsilon,\mu)\|, \qquad (6.24)$$

where the functions $\vartheta_k\colon (0,\varepsilon_0] \to [0,\infty)$ satisfy $\Theta_k := \int_0^{\varepsilon_0}\vartheta_k(\varepsilon)d\varepsilon < \infty$ $(k = 1, 2)$ and where $\gamma \ge 0$ is a constant. The aim is again to prove that $F(0,\mu)$ converges as $\mu \to +0$. We show below that the differential inequality (6.24) also leads to the estimate (6.4), namely that

$$\Big\|\frac{\partial}{\partial\varepsilon}F(\varepsilon,\mu)\Big\| \le \vartheta(\varepsilon) \qquad (6.25)$$

for some function $\vartheta$ with $\int_0^{\varepsilon_0}\vartheta(\varepsilon)d\varepsilon < \infty$. Once this estimate is established, the existence of the limit $\text{u-lim}_{\mu\to+0} F(0,\mu)$ is obtained as in §6.1.1 [see the reasoning after (6.4), in which $|\cdot|$ should be replaced by $\|\cdot\|$]. For the applications it will be important to have an explicit upper bound for $\|F(\varepsilon,\mu)\|$ in terms of $\Theta_1, \Theta_2, \gamma$ and $\varphi(\varepsilon_0)$, where $\varphi(\varepsilon)$ is again defined as $\varphi(\varepsilon) = \sup_{0\le\mu\le1}\|F(\varepsilon,\mu)\|$. Hence we must carefully treat these constants in the derivation of the inequality (6.25).

REMARK. It is important that the coefficients of the powers of $\|F(\varepsilon,\mu)\|$ in (6.24) belong to $L^1((0,\varepsilon_0))$ (as functions of $\varepsilon$) and also that they do not depend on the variable $\mu$. The last property could be weakened, but this is not necessary for our applications. Also one could replace the constant $\gamma$ in (6.24) by a function $\vartheta_3$ of finite integral and require continuity of $F$ only on the subset $\mathfrak{D}_0$ of $\mathfrak{D}$.

PROOF OF (6.25). The inequality (6.24) implies that one has on $\mathfrak{D}_0$:

$$\begin{aligned}\|F(\varepsilon,\mu)\| &= \|F(\varepsilon_0,\mu) - \int_\varepsilon^{\varepsilon_0}\frac{\partial}{\partial\tau}F(\tau,\mu)\,d\tau\| \le \|F(\varepsilon_0,\mu)\| + \int_\varepsilon^{\varepsilon_0}\|\frac{\partial}{\partial\tau}F(\tau,\mu)\|d\tau \\ &\le \varphi(\varepsilon_0) + \Theta_1 + \int_\varepsilon^{\varepsilon_0}\vartheta_2(\tau)\|F(\tau,\mu)\|^{1/2}d\tau + \gamma\int_\varepsilon^{\varepsilon_0}\|F(\tau,\mu)\|\,d\tau.\end{aligned} \tag{6.26}$$

We set $\kappa_1 = \varphi(\varepsilon_0) + \Theta_1$ and

$$\psi(\varepsilon,\mu) = \kappa_1 + \int_\varepsilon^{\varepsilon_0}\vartheta_2(\tau)\|F(\tau,\mu)\|^{1/2}d\tau + \gamma\int_\varepsilon^{\varepsilon_0}\|F(\tau,\mu)\|\,d\tau.$$

We notice that $\psi(\varepsilon,\mu) > 0$ and that $\psi(\varepsilon_0,\mu) = \kappa_1$ for all $\mu$. We may now write (6.26) in the short form

$$\|F(\varepsilon,\mu)\| \le \psi(\varepsilon,\mu). \tag{6.27}$$

Finally we set $\eta(\varepsilon,\mu) = \psi(\varepsilon,\mu)e^{\gamma(\varepsilon-\varepsilon_0)}$. Denoting again by a prime the differentiation with respect to $\varepsilon$ and omitting the arguments $\varepsilon$ and $\mu$, we have for $0 < \varepsilon < \varepsilon_0$:

$$\begin{aligned}\eta' &= (\psi' + \gamma\psi)e^{\gamma(\varepsilon-\varepsilon_0)} \ge (\psi' + \gamma\|F\|)e^{\gamma(\varepsilon-\varepsilon_0)} \\ &= (-\vartheta_2\|F\|^{1/2} - \gamma\|F\| + \gamma\|F\|)e^{\gamma(\varepsilon-\varepsilon_0)} = -\vartheta_2\|F\|^{1/2}e^{\gamma(\varepsilon-\varepsilon_0)} \\ &\ge -\vartheta_2\psi^{1/2}e^{\gamma(\varepsilon-\varepsilon_0)} = -\vartheta_2\eta^{1/2}e^{\gamma(\varepsilon-\varepsilon_0)/2} \ge -\vartheta_2\eta^{1/2}.\end{aligned}$$

It follows that [6]

$$\frac{\partial}{\partial\varepsilon}\eta(\varepsilon,\mu)^{1/2} \ge -\frac{1}{2}\vartheta_2(\varepsilon) \ge -\vartheta_2(\varepsilon). \tag{6.28}$$

Consequently, for $0 < \varepsilon < \varepsilon_0$ and all $\mu \in [0,1]$:

$$\begin{aligned}\eta(\varepsilon,\mu)^{1/2} &= \eta(\varepsilon_0,\mu)^{1/2} - \int_\varepsilon^{\varepsilon_0}\frac{\partial}{\partial\tau}\left[\eta(\tau,\mu)^{1/2}\right]d\tau \le \eta(\varepsilon_0,\mu)^{1/2} + \int_0^{\varepsilon_0}\vartheta_2(\tau)\,d\tau \\ &= \psi(\varepsilon_0,\mu)^{1/2} + \Theta_2 = \sqrt{\kappa_1} + \Theta_2.\end{aligned}$$

[6] Since $\eta(\varepsilon,\mu) > 0$, division by $\eta(\varepsilon,\mu)^{1/2}$ causes no problem.

This leads to the following estimate for $\|F(\varepsilon,\mu)\|$:

$$\|F(\varepsilon,\mu)\| \le \psi(\varepsilon,\mu) = \eta(\varepsilon,\mu)e^{\gamma(\varepsilon_0-\varepsilon)} \le \big[\sqrt{\kappa_1}+\Theta_2\big]^2 e^{\gamma\varepsilon_0}. \tag{6.29}$$

This bound is valid on $\mathfrak{O}$, it is independent of $\varepsilon$ and $\mu$. Insertion of this estimate into (6.24) gives the required inequality (6.25), with

$$\vartheta(\varepsilon) = e^{\gamma\varepsilon_0}\big\{\vartheta_1(\varepsilon) + \big[\sqrt{\kappa_1}+\Theta_2\big]\vartheta_2(\varepsilon) + \gamma\big[\sqrt{\kappa_1}+\Theta_2\big]^2\big\}. \qquad \square$$

**Remark 6.1.** The functions $F$ appearing in our applications will also depend on an additional argument $\lambda$. For each $\lambda$ in some interval $J_0$, let $F^{(\lambda)}: \mathfrak{O} \to \mathcal{B}(\mathcal{H})$ be norm continuous and continuously differentiable with respect to $\varepsilon$ for $0 < \varepsilon < \varepsilon_0$ (with $\varepsilon_0 \le 1$). Assume that each $F^{(\lambda)}$ satisfies the differential inequality (6.24), with $\vartheta_1, \vartheta_2$ and $\gamma$ independent of $\lambda$ and $\mu$. Then the norm limit of $F^{(\lambda)}(0,\mu)$ as $\mu \to +0$ exists for each fixed $\lambda \in J_0$. Moreover, setting $\phi(\varepsilon) := \sup_{\lambda\in J_0}\sup_{\mu\in[0,1]}\|F^{(\lambda)}(\varepsilon,\mu)\|$, the following bound for $F^{(\lambda)}(\varepsilon,\mu)$ is satisfied for all $\lambda\in J_0$, $\varepsilon\in[0,\varepsilon_0]$ and $\mu\in[0,1]$:

$$\|F^{(\lambda)}(\varepsilon,\mu)\| \le \big[\sqrt{\tilde\kappa_1}+\Theta_2\big]^2 e^{\gamma\varepsilon_0} \qquad \text{with } \tilde\kappa_1 = \phi(\varepsilon_0)+\Theta_1. \tag{6.30}$$

**6.2.2.** We now turn to the question of resolvent estimates and begin with a simple preliminary result which will be very useful in what follows.

**Lemma 6.2.** *Let $A$ be a self-adjoint operator in a Hilbert space $\mathcal{H}$ and $C \in \mathcal{B}(\mathcal{H})$. Assume that there exists an operator $Z \in \mathcal{B}(\mathcal{H})$ such that $\langle C^*f, Ag\rangle - \langle Af, Cg\rangle = \langle f, Zg\rangle$ for all $f, g \in \mathcal{D}(A)$. Then $C$ maps $\mathcal{D}(A)$ into $\mathcal{D}(A)$; hence $CA - AC$ is defined on $\mathcal{D}(A)$ and has an extension to $\mathcal{B}(\mathcal{H})$ (given by $Z$, which is identical with the closure of the operator $CA - AC$).*

PROOF. Let $f, g \in \mathcal{D}(A)$. By assumption we have $\langle Af, Cg\rangle = \langle f, CAg\rangle - \langle f, Zg\rangle = \langle f, g^*\rangle$, with $g^* = CAg - Zg \in \mathcal{H}$. So $Cg \in \mathcal{D}(A^*) \equiv \mathcal{D}(A)$ and $A^*Cg \equiv ACg = g^* = CAg - Zg$. $\square$

Now let $H$ and $A$ be self-adjoint operators in a Hilbert space $\mathcal{H}$. We denote by $\{E(\cdot)\}$ the spectral measure of $H$. One says that a ***strict Mourre inequality*** is satisfied on an open interval $J$ if there exists a number $a > 0$ such that

$$E(J)[H, iA]E(J) \ge aE(J). \tag{6.31}$$

This means that on the subspace $\mathcal{M}(J) \equiv E(J)\mathcal{H}$, the operator $[H, iA]$ is strictly positive, namely larger that $a$. [7]

In order for (6.31) to make sense, it is necessary to define a scalar product, corresponding to the commutator $[H, iA]$, for vectors in the range of $E(J)$. We first treat the situation where $H$ is a bounded operator which we denote by $T$, whereas the case

[7] Recall that $S \ge T$ means that $\langle f, Sf\rangle \ge \langle f, Tf\rangle$ for each $f \in \mathcal{H}$ [we assume here that S and T belong to $\mathcal{B}(\mathcal{H})$].

of unbounded $H$ will be discussed further on (see page 286ff.). So let $T \in \mathcal{B}(\mathcal{H})$. For $f, g \in \mathcal{D}(A)$ the number $\langle f, [T, iA]g\rangle$ is then well defined if one sets

$$\langle f, [T, iA]g\rangle = i\langle Tf, Ag\rangle - i\langle Af, Tg\rangle.$$

We make the assumption that there exists an operator $B \in \mathcal{B}(\mathcal{H})$ such that

$$\langle f, [T, iA]g\rangle = \langle f, Bg\rangle \qquad \forall f, g \in \mathcal{D}(A), \tag{6.32}$$

and we then say that the closure of $[T, iA]$ belongs to $\mathcal{B}(\mathcal{H})$. Indeed, by Lemma 6.2, Eq. (6.32) implies that $T$ maps $\mathcal{D}(A)$ into $\mathcal{D}(A)$. Then $[T, iA]g = iTAg - iATg$ is well defined for each $g \in \mathcal{D}(A)$: $[T, iA]$ defines a symmetric operator on the dense domain $\mathcal{D}(A)$, and by virtue of (6.32) this symmetric operator has an extension $B$ (its closure) to $\mathcal{B}(\mathcal{H})$. It is clear that $B$ is self-adjoint (closure of a bounded symmetric operator).

We shall also need the second commutator of $T$ with $A$, more precisely the operator $[[T, iA], iA] \equiv [B, iA]$. By analogy with the preceding developments, we assume that there exists an operator $Z \in \mathcal{B}(\mathcal{H})$ such that

$$\langle f, [B, iA]g\rangle \equiv i\langle Bf, Ag\rangle - i\langle Af, Bg\rangle = \langle f, Zg\rangle \qquad \forall f, g \in \mathcal{D}(A), \tag{6.33}$$

hence that $[B, iA]$ has an extension $Z$ to $\mathcal{B}(\mathcal{H})$. However we prefer to use the notation $[B, iA]$, rather than $Z$, for this extension [so we use the same notation $[B, iA]$ for the symmetric operator on $\mathcal{D}(A)$ defined by the left-hand side of (6.33) and for its closure in $\mathcal{B}(\mathcal{H})$].

The next theorem involves the operator $|A|^\nu$ for $\nu > 1/2$. This operator is defined by the functional calculus: $|A|^\nu = \int_{-\infty}^{\infty} |\lambda|^\nu E^A(d\lambda)$, where $\{E^A(\cdot)\}$ denotes the spectral measure of $A$. Let us observe that $\|Af\| = \||A|f\|$ if $f \in \mathcal{D}(A)$, because each of these norms is equal to $\left[\int_{-\infty}^{\infty} \lambda^2 m_f^A(d\lambda)\right]^{1/2}$.

**Theorem 6.3.** *Let $T \in \mathcal{B}(\mathcal{H})$ be self-adjoint. Assume that there exists a self-adjoint operator $A$ such that the closure $B$ of $[T, iA]$ and the closure of $[B, iA] \equiv i[[T, iA], A]$ belong to $\mathcal{B}(\mathcal{H})$ and such that a strict Mourre inequality is satisfied on some open interval $J$, i.e.*

$$E(J)BE(J) \geq aE(J) \qquad \text{with } a > 0, \tag{6.34}$$

*where $\{E(\cdot)\}$ is the spectral measure of $T$. Let $\nu > 1/2$. Then the norm limits $\text{u-lim}_{\mu\to+0}(I + |A|)^{-\nu}(T - \lambda \pm i\mu)^{-1}(I + |A|)^{-\nu}$ exist for each $\lambda \in J$. In particular, $\langle f, (T - \lambda \pm i\mu)^{-1}g\rangle$ converges as $\mu \to +0$ for all $f, g \in \mathcal{D}(|A|^\nu)$ and each $\lambda \in J$. Moreover, if $J_0$ is a bounded closed subinterval of $J$, then*

$$\sup_{\lambda\in J_0, \mu>0} \|(I + |A|)^{-\nu}(T - \lambda \pm i\mu)^{-1}(I + |A|)^{-\nu}\| < \infty. \tag{6.35}$$

REMARKS. (a) We shall see later that the limits in Theorem 6.3 exist in fact uniformly in $\lambda \in J_0$ if $J_0$ is any bounded closed subinterval of $J$.

(b) If $\lambda$ belongs to the spectrum of $T$, then the limits of $\langle f, (T - \lambda + i\mu)^{-1}f\rangle$ and

$\langle f, (T-\lambda-i\mu)^{-1}f\rangle$ as $\mu \to +0$, denoted somewhat formally by $\langle f, (T-\lambda \pm i0)^{-1}f\rangle$, are not the same.

(c) The hypotheses of boundedness of $[T, A]$ and $[[T, A], A]$ are not trivial. Indeed it is not difficult to show that, if $\sigma(T) \cap J \neq \varnothing$ and $A$ is bounded, then the inequality (6.32) can be satisfied only for $a \leq 0$.[8] Thus the validity of (6.32) for some $a > 0$ (and $J$ non-disjoint from the spectrum of $T$) implies that $A$ is an unbounded operator [if $\sigma(T) \cap J = \varnothing$, then $E(J) = O$ and (6.32) is trivially satisfied for each $a \in \mathbb{R}$; but this is uninteresting in the context of the Mourre method, since then $(T - \lambda)^{-1} \in \mathcal{B}(\mathcal{H})$ for $\lambda \in J$].

(d) It is possible to obtain the conclusions of the theorem with the assumption of boundedness of $[B, iA]$ replaced by a weaker condition.

PROOF OF THEOREM 6.3. The proof is carried out in three parts: (1) definition and properties of $G_\varepsilon$, (2) establishment of a differential inequality of the form (6.24), (3) boundary values of the resolvent as an application of the results of §6.2.1. We shall only treat the limit of $(T - \lambda - i\mu)^{-1}$ as $\mu \to +0$. The limit of $(T - \lambda + i\mu)^{-1}$ ($\mu \to +0$) is obtained in an analogous way. We fix a closed subinterval $J_0 = [\alpha, \beta]$ of $J$, and we assume throughout the proof that $z = \lambda + i\mu$ with $\lambda \in J_0$ and $0 \leq \mu \leq 1$. In calculations in which $z$ is fixed, we sometimes omit the argument $z$.

We shall denote by $c_k (k = 1, 2, 3, ...)$ various constants occurring in the proof. Their value is of no importance for arriving at the conclusions of the theorem. What matters is the fact that all these constants will be independent of $\lambda$ and $\mu$, of the parameter $\varepsilon$ and of the considered vectors $f, g, h$. For the moment we define $c_1$ by

$$c_1 = \|T\| + \|B\| + \|B\|^2 + \frac{4}{a}\|B\|^2 + \|[B, A]\| + a + |\alpha| + |\beta| + 1. \tag{6.36}$$

Finally, without loss of generality, we can assume in the proof that $\nu \leq 1$ (because $(I + |A|)^{-\nu} \leq (I + |A|)^{-1}$ if $\nu > 1$). The hypothesis that $\nu > 1/2$ is necessary in order for $|\Phi'_\varepsilon|$ to be integrable at $\varepsilon = 0$, see (6.56).

(1) Definition and properties of $G_\varepsilon$

As pointed out in Comment (c) on page 268, we wish to define $G_\varepsilon(z)$, for $0 \leq \varepsilon \leq 1$ and $z = \lambda + i\mu$ ($\lambda \in J_0$), by

$$G_\varepsilon(z) \equiv G_\varepsilon(\lambda + i\mu) = (T - \lambda - i\mu - i\varepsilon B)^{-1}. \tag{6.37}$$

We must show that the operator $(T - \lambda - i\mu - i\varepsilon B)$ is invertible in $\mathcal{B}(\mathcal{H})$.

(i) For $\varepsilon \in [0, 1]$ we define $N_\varepsilon^\pm \in \mathcal{B}(\mathcal{H})$ by $N_\varepsilon^\pm = T - \lambda \pm i(\mu + \varepsilon B)$. Clearly $\|N_\varepsilon^\pm\| \leq 4c_1$. We fix a number $\delta > 0$ chosen sufficiently small so that the interval $[\alpha - \delta, \beta + \delta]$ is contained in $J$. We write $I = E + E^\perp$ with $E = E([\alpha - \delta, \beta + \delta])$

[8] Let $\lambda \in \sigma(T) \cap J$. By Proposition 4.20(b) there exists a sequence of vectors $\{f_k\}$ in $E(J)\mathcal{H}$ satisfying $\|f_k\| = 1$ and $\|(T - \lambda)f_k\| \to 0$. Assume $A$ to be bounded. Then (6.32) implies that

$$\begin{aligned} a = a\|f_k\|^2 &\leq \langle f_k, [T, iA]f_k\rangle = \langle f_k, [T - \lambda, iA]f_k\rangle = i\langle (T - \lambda)f_k, Af_k\rangle - i\langle Af_k, (T - \lambda)f_k\rangle \\ &\leq 2\|(T - \lambda)f_k\|\|Af_k\| \leq 2\|A\|\|(T - \lambda)f_k\| \to 0 \quad \text{as } k \to \infty, \end{aligned}$$

hence $a \leq 0$.

and $E^{\perp} = E(\mathbb{R}\setminus[\alpha-\delta,\beta+\delta])$. Let us observe immediately that, if $\lambda \in J_0 = [\alpha,\beta]$ and $\mu \geq 0$ [using (4.51)]:

$$\|(T-\lambda \pm i\mu)^{-1}E^{\perp}\| \leq \sup_{\tau\in\mathbb{R}\setminus[\alpha-\delta,\beta+\delta]} \frac{1}{|\tau-\lambda\pm i\mu|} \leq \frac{1}{\sqrt{\delta^2+\mu^2}} \leq \frac{1}{\delta}. \quad (6.38)$$

Consequently one has for each $g \in \mathcal{H}$:

$$\begin{aligned}\|E^{\perp}g\|^2 &= \|(T-\lambda\pm i\mu)^{-1}E^{\perp}(N_\varepsilon^{\pm} \mp i\varepsilon B)g\|^2 \\ &\leq \|(T-\lambda\pm i\mu)^{-1}E^{\perp}\|^2\|(N_\varepsilon^{\pm}\mp i\varepsilon B)g\|^2 \\ &\leq \frac{2}{\delta^2}\left[\|N_\varepsilon^{\pm}g\|^2 + \varepsilon^2 c_1\|g\|^2\right]. \end{aligned} \quad (6.39)$$

(ii) The assumption (6.34) implies that $aE \leq EBE$ [9]. Thus, for each $g \in \mathcal{H}$:

$$a\|g\|^2 = a\|Eg\|^2 + a\|E^{\perp}g\|^2 = a\langle g, Eg\rangle + a\|E^{\perp}g\|^2 \leq \langle g, EBEg\rangle + a\|E^{\perp}g\|^2. \quad (6.40)$$

By using the inequality $|\langle f,h\rangle + \langle h,f\rangle| \leq \|f\|^2 + \|h\|^2$ (see page 3) one finds that

$$\begin{aligned}\langle g, EBEg\rangle &= \langle g,Bg\rangle - \langle g, BE^{\perp}g\rangle - \langle g, E^{\perp}Bg\rangle + \langle E^{\perp}g, BE^{\perp}g\rangle \\ &= \langle g,Bg\rangle - \langle \frac{\sqrt{a}}{2}g, \frac{2}{\sqrt{a}}BE^{\perp}g\rangle - \langle \frac{2}{\sqrt{a}}BE^{\perp}g, \frac{\sqrt{a}}{2}g\rangle + \langle E^{\perp}g, BE^{\perp}g\rangle \\ &\leq \langle g,Bg\rangle + \left[\frac{a}{4}\|g\|^2 + \frac{4}{a}\|B\|^2\|E^{\perp}g\|^2\right] + \|B\|\|E^{\perp}g\|^2 \\ &\leq \langle g,Bg\rangle + \frac{a}{4}\|g\|^2 + c_1\|E^{\perp}g\|^2.\end{aligned}$$

Insertion of this inequality into (6.40) leads to (using $a + c_1 \leq 2c_1$)

$$\frac{3a}{4}\|g\|^2 \leq \langle g,Bg\rangle + (a+c_1)\|E^{\perp}g\|^2 \leq \langle g,Bg\rangle + \frac{4c_1}{\delta^2}\|N_\varepsilon^{\pm}g\|^2 + \frac{4\varepsilon^2c_1^2}{\delta^2}\|g\|^2.$$

Let us choose $\varepsilon_0 \in (0,1)$ such that $4\varepsilon_0^2c_1^2/\delta^2 \leq a/4$. We find that, for $0 \leq \varepsilon \leq \varepsilon_0$, $\lambda \in J_0$ and $\mu \geq 0$:

$$\boxed{\frac{a}{2}\|g\|^2 \leq \langle g,Bg\rangle + \frac{4c_1}{\delta^2}\|N_\varepsilon^{\pm}g\|^2} \qquad \forall g \in \mathcal{H}. \quad (6.41)$$

This is the basic inequality for the proof.

(iii) Let us give a first application of the basic inequality (6.41). Since $\mu \geq 0$, this inequality implies that

$$\frac{a\varepsilon+\mu}{2}\|g\|^2 \leq \langle g, (\varepsilon B+\mu)g\rangle + \frac{4c_1\varepsilon}{\delta^2}\|N_\varepsilon^{\pm}g\|^2. \quad (6.42)$$

[9] One has $E = EE(J) = E(J)E$; hence, if $h \in \mathcal{H}$:

$$a\langle h, Eh\rangle = a\langle Eh, E(J)Eh\rangle \leq \langle Eh, E(J)BE(J)Eh\rangle = \langle h, EBEh\rangle.$$

Because $\langle g, (T-\lambda)g\rangle$ is real, one has $\langle g, (\varepsilon B+\mu)g\rangle = \pm\Im\langle g, N_\varepsilon^\pm g\rangle \leq |\langle g, N_\varepsilon^\pm g\rangle| \leq \|g\|\|N_\varepsilon^\pm g\|$. Consequently (using the bound $\|N_\varepsilon^\pm\| \leq 4c_1$)

$$\frac{a\varepsilon+\mu}{2}\|g\|^2 \leq \|g\|\|N_\varepsilon^\pm g\|\big[1+\frac{16c_1^2\varepsilon}{\delta^2}\big]. \tag{6.43}$$

Assume that $\varepsilon \neq 0$ or $\mu \neq 0$. After division by $\|g\|$ the inequality (6.43) has the form $\|N_\varepsilon^\pm g\| \geq \omega\|g\|$ for some $\omega > 0$ and for each $g \in \mathcal{H}$. Hence $N_\varepsilon^\pm g = \mathbf{0}$ implies that $g = \mathbf{0}$, i.e. the operators $N_\varepsilon^\pm$ are invertible. Moreover $\mathcal{R}(N_\varepsilon^\pm)$ are closed and one has $\|(N_\varepsilon^\pm)^{-1}\| \leq \omega^{-1}$ (see Lemma 3.1).

Now by Proposition 2.19(b): $\mathcal{R}(N_\varepsilon^\pm)^\perp = \mathcal{N}((N_\varepsilon^\pm)^*) = \mathcal{N}(N_\varepsilon^\mp) = \{\mathbf{0}\}$. Since $\mathcal{R}(N_\varepsilon^\pm)$ are closed, this implies that $\mathcal{R}(N_\varepsilon^\pm) = \mathcal{H}$. Thus $G_\varepsilon = (N_\varepsilon^-)^{-1}$ is defined on $\mathcal{H}$ and hence belongs to $\mathcal{B}(\mathcal{H})$, with

$$\|G_\varepsilon(\lambda+i\mu)\| \leq \frac{1}{\omega} \equiv \frac{2}{a\varepsilon+\mu}\big[1+\frac{16c_1^2\varepsilon}{\delta^2}\big].$$

One sees that $G_\varepsilon(\lambda+i\mu) \in \mathcal{B}(\mathcal{H})$ if $\varepsilon \neq 0$ or if $\mu \neq 0$. If one assumes in addition that $\varepsilon_0$ is chosen so small that $16c_1^2\varepsilon_0/\delta^2 \leq 1$, the preceding estimate is simplified as follows:

$$\|G_\varepsilon(\lambda+i\mu)\| \leq \frac{4}{a\varepsilon+\mu} \qquad (\text{for } 0 \leq \varepsilon \leq \varepsilon_0, \mu \geq 0). \tag{6.44}$$

In the remainder of the proof it is always understood that $\varepsilon \leq \varepsilon_0$.

(iv) It will be useful to know that, if $\varepsilon > 0$, then $G_\varepsilon(\lambda+i\mu)$ depends continuously on $\lambda$ and $\mu$. Setting $z = \lambda+i\mu$ and $z_0 = \lambda_0+i\mu_0$, one has

$$G_\varepsilon(z) - G_\varepsilon(z_0) = G_\varepsilon(z)[N_\varepsilon^-(z_0) - N_\varepsilon^-(z)]G_\varepsilon(z_0) = G_\varepsilon(z)[z-z_0]G_\varepsilon(z_0).$$

Consequently, by virtue of (6.44):

$$\|G_\varepsilon(\lambda+i\mu) - G_\varepsilon(\lambda_0+i\mu_0)\| \leq \frac{16}{a^2\varepsilon^2}\big[|\lambda-\lambda_0| + |\mu-\mu_0|\big]. \tag{6.45}$$

(v) Finally we deduce a useful estimate on $G_\varepsilon$. Let $h \in \mathcal{H}$ and set $g = G_\varepsilon h$ in (6.42) (remember that $\mu \geq 0$):

$$\begin{aligned}\frac{\varepsilon a}{2}\|G_\varepsilon h\|^2 &\leq \langle G_\varepsilon h, (\varepsilon B+\mu)G_\varepsilon h\rangle + \frac{4c_1\varepsilon}{\delta^2}\|h\|^2\\ &= \frac{1}{2i}\langle G_\varepsilon h, (N_\varepsilon^+ - N_\varepsilon^-)G_\varepsilon h\rangle + \frac{4c_1\varepsilon}{\delta^2}\|h\|^2.\end{aligned} \tag{6.46}$$

Now observe that $N_\varepsilon^- G_\varepsilon = I$ and $(G_\varepsilon^* N_\varepsilon^+)^* = N_\varepsilon^{+*}G_\varepsilon = N_\varepsilon^- G_\varepsilon = I$, hence $G_\varepsilon^* N_\varepsilon^+ = I$ and consequently $G_\varepsilon^*(N_\varepsilon^+ - N_\varepsilon^-)G_\varepsilon = G_\varepsilon - G_\varepsilon^*$. Thus

$$\langle G_\varepsilon h, (N_\varepsilon^+ - N_\varepsilon^-)G_\varepsilon h\rangle = \langle h, G_\varepsilon h\rangle - \langle h, G_\varepsilon^* h\rangle = \langle h, G_\varepsilon h\rangle - \langle G_\varepsilon h, h\rangle = 2i\Im\langle h, G_\varepsilon h\rangle,$$

and (6.46) implies that

$$\frac{\varepsilon a}{2}\|G_\varepsilon h\|^2 \leq |\Im\langle h, G_\varepsilon h\rangle| + \frac{4c_1\varepsilon}{\delta^2}\|h\|^2.$$

Thus

$$\|G_\varepsilon h\| \leq \sqrt{\frac{2}{a\varepsilon}}\,|\Im\langle h, G_\varepsilon h\rangle|^{1/2} + \frac{1}{\delta}\sqrt{\frac{8c_1}{a}}\,\|h\|. \tag{6.47}$$

Let $Y \in \mathcal{B}(\mathcal{H})$ be a self-adjoint operator satisfying $\|Y\| \leq 1$. By taking $h = Yf$ in (6.47) and setting $c_2 = \sqrt{2/a} + \delta^{-1}\sqrt{8c_1/a}$, one gets the following inequality:

$$\|G_\varepsilon Y\| \leq \frac{c_2}{\sqrt{\varepsilon}}\|YG_\varepsilon Y\|^{1/2} + c_2. \tag{6.48}$$

The same inequality can be obtained for $\|G_\varepsilon^* Y\|$ by setting $g = G_\varepsilon^* h$ in (6.42) and observing that $G_\varepsilon^* = (N_\varepsilon^+)^{-1}$. So

$$\|YG_\varepsilon\| = \|G_\varepsilon^* Y\| \leq \frac{c_2}{\sqrt{\varepsilon}}\|YG_\varepsilon Y\|^{1/2} + c_2. \tag{6.49}$$

(2) The differential inequality

(i) We first calculate the derivative $G_\varepsilon'(z) \equiv (\partial/\partial\varepsilon)G_\varepsilon(z)$. If $X$ and $Y$ are invertible operators in $\mathcal{B}(\mathcal{H})$, with $X^{-1}, Y^{-1} \in \mathcal{B}(\mathcal{H})$, then one has the identity $X^{-1} - Y^{-1} = X^{-1}(Y - X)Y^{-1}$. Hence

$$\begin{aligned} G_\tau - G_\varepsilon &= (T - \lambda - i\mu - i\tau B)^{-1} - (T - \lambda - i\mu - i\varepsilon B)^{-1} \\ &= i(\tau - \varepsilon)(T - \lambda - i\mu - i\tau B)^{-1}B(T - \lambda - i\mu - i\varepsilon B)^{-1}. \end{aligned} \tag{6.50}$$

Since $\|(T - \lambda - i\mu - i\tau B)^{-1}\| \leq 4/(a\tau)$ by (6.44), one deduces from Eq. (6.50) that $\|G_\tau - G_\varepsilon\| \to 0$ as $\tau \to \varepsilon$ ($\varepsilon \neq 0$). Thus $\varepsilon \mapsto G_\varepsilon$ is continuous in norm. By dividing (6.50) by $\tau - \varepsilon$ and then taking the limit $\tau \to \varepsilon$ (which exists in norm), one finds that $(\partial/\partial\varepsilon)G_\varepsilon = iG_\varepsilon BG_\varepsilon$. Hence $\varepsilon \mapsto G_\varepsilon$ is of class $C^1$ in the norm of $\mathcal{B}(\mathcal{H})$. The same reasoning as that leading to (6.9) gives the following result (use the formal identity $[A, X^{-1}] = X^{-1}[X, A]X^{-1}$):

$$\begin{aligned} [A, G_\varepsilon] &= G_\varepsilon[T - \lambda - i\mu - i\varepsilon B, A]G_\varepsilon = G_\varepsilon[T, A]G_\varepsilon - i\varepsilon G_\varepsilon[B, A]G_\varepsilon \\ &= -iG_\varepsilon BG_\varepsilon - i\varepsilon G_\varepsilon[B, A]G_\varepsilon = -G_\varepsilon' - i\varepsilon G_\varepsilon[B, A]G_\varepsilon. \end{aligned} \tag{6.51}$$

As in §6.1.2 this argument is somewhat formal, we have not considered the domain questions. The right-hand side of (6.51) belongs to $\mathcal{B}(\mathcal{H})$ as a consequence of the hypotheses of the theorem. The left-hand side has a meaning between vectors $f$ and $g$ belonging to $\mathcal{D}(A)$ (i.e. $\langle f, [A, G_\varepsilon]g\rangle$ is well defined for such vectors). Thus (6.51) has a meaning between $f, g \in \mathcal{D}(A)$, and it is in this form that it will be used [see Eq. (6.55)]. We indicate a more careful proof of (6.51) in §6.2.3. Here we rewrite (6.51) in the following form [which has a meaning between $f, g \in \mathcal{D}(A)$]:

$$G_\varepsilon'(\lambda + i\mu) = [G_\varepsilon(\lambda + i\mu), A] - i\varepsilon G_\varepsilon(\lambda + i\mu)[B, A]G_\varepsilon(\lambda + i\mu). \tag{6.52}$$

(ii) For $\varepsilon > 0$ we set $X_\varepsilon = (\varepsilon|A| + I)^{-1}$. Clearly $\mathcal{R}(X_\varepsilon) = \mathcal{D}(|A|) \equiv \mathcal{D}(A)$. Moreover $\|X_\varepsilon\| \leq 1$, $\text{s-lim}_{\varepsilon\to 0} X_\varepsilon = I$ and, if $0 < \beta \leq 1$:

$$\||A|^\beta X_\varepsilon\| = \sup_{x\geq 0}\frac{x^\beta}{\varepsilon x + 1} = \sup_{y\geq 0}\frac{(y/\varepsilon)^\beta}{y + 1} = \varepsilon^{-\beta}\sup_{y\geq 0}\frac{y^\beta}{y + 1} \leq \varepsilon^{-\beta}. \tag{6.53}$$

We also set $\mathcal{Z} = (I + |A|)^{-\nu}$ and define [10]

$$\Phi_\varepsilon(\lambda + i\mu) = (I + |A|)^{-\nu} X_\varepsilon G_\varepsilon(\lambda + i\mu) X_\varepsilon (I + |A|)^{-\nu} \equiv X_\varepsilon \mathcal{Z} G_\varepsilon \mathcal{Z} X_\varepsilon. \quad (6.54)$$

By using the Leibniz rule, the relation $X_\varepsilon' = -|A|(\varepsilon|A|+I)^{-2} = -|A|X_\varepsilon^2$ and finally (6.52), one obtains that

$$\begin{aligned}\Phi_\varepsilon' &= -|A|X_\varepsilon^2 \mathcal{Z} G_\varepsilon \mathcal{Z} X_\varepsilon - X_\varepsilon \mathcal{Z} G_\varepsilon \mathcal{Z} X_\varepsilon^2 |A| + X_\varepsilon \mathcal{Z} G_\varepsilon' \mathcal{Z} X_\varepsilon \\ &= -|A|X_\varepsilon^2 \mathcal{Z} G_\varepsilon \mathcal{Z} X_\varepsilon - X_\varepsilon \mathcal{Z} G_\varepsilon \mathcal{Z} X_\varepsilon^2 |A| + X_\varepsilon \mathcal{Z} G_\varepsilon A \mathcal{Z} X_\varepsilon - X_\varepsilon A \mathcal{Z} G_\varepsilon \mathcal{Z} X_\varepsilon \\ &\quad - i\varepsilon X_\varepsilon \mathcal{Z} G_\varepsilon [B,A] G_\varepsilon \mathcal{Z} X_\varepsilon. \end{aligned} \quad (6.55)$$

To estimate the norms of the individual terms in the last expression for $\Phi_\varepsilon'$, we write $|A|\mathcal{Z}X_\varepsilon$ as $|A|^\nu \mathcal{Z} \cdot |A|^{1-\nu} X_\varepsilon$. For the first term we then obtain:

$$\big\| |A| X_\varepsilon^2 \mathcal{Z} G_\varepsilon \mathcal{Z} X_\varepsilon \big\| \le \|X_\varepsilon\| \big\| |A|^\nu \mathcal{Z} \big\| \big\| |A|^{1-\nu} X_\varepsilon \big\| \|G_\varepsilon \mathcal{Z} X_\varepsilon\| \le \varepsilon^{\nu-1} \|G_\varepsilon \mathcal{Z} X_\varepsilon\|,$$

where the last inequality is obtained by using $\|X_\varepsilon\| \le 1$, $\||A|^\nu \mathcal{Z}\| \le 1$ and (6.53). Similarly the norm of the second term is bounded by $\varepsilon^{\nu-1}\|X_\varepsilon \mathcal{Z} G_\varepsilon\|$. For the third term one obtains in the same way:

$$\|X_\varepsilon \mathcal{Z} G_\varepsilon A \mathcal{Z} X_\varepsilon\| \le \|X_\varepsilon \mathcal{Z} G_\varepsilon\| \big\| |A|^\nu \mathcal{Z} \big\| \big\| |A|^{1-\nu} X_\varepsilon \big\| \le \varepsilon^{\nu-1} \|X_\varepsilon \mathcal{Z} G_\varepsilon\|,$$

and the norm of the fourth term is bounded by $\varepsilon^{\nu-1}\|G_\varepsilon \mathcal{Z} X_\varepsilon\|$. So (6.55) leads to the inequality

$$\|\Phi_\varepsilon'\| \le 2\varepsilon^{\nu-1}\big[\|G_\varepsilon \mathcal{Z} X_\varepsilon\| + \|X_\varepsilon \mathcal{Z} G_\varepsilon\|\big] + \varepsilon \|[B,A]\| \|X_\varepsilon \mathcal{Z} G_\varepsilon\| \|G_\varepsilon \mathcal{Z} X_\varepsilon\|.$$

From (6.48) and (6.49), with $Y = \mathcal{Z} X_\varepsilon \equiv X_\varepsilon \mathcal{Z}$, we see that each of the norms $\|G_\varepsilon \mathcal{Z} X_\varepsilon\|$ and $\|X_\varepsilon \mathcal{Z} G_\varepsilon\|$ is bounded by $c_2(1 + \varepsilon^{-1/2}\|\Phi_\varepsilon\|^{1/2})$. So, by taking into account the inequality $\|[B,A]\| \le c_1$ and remembering that $\nu \le 1$, hence $1 \le \varepsilon^{\nu-1}$, one finally obtains the estimate

$$\|\Phi_\varepsilon'\| \le c_3 \varepsilon^{\nu-3/2} \|\Phi_\varepsilon\|^{1/2} + c_3 \varepsilon^{\nu-1} + c_3 \|\Phi_\varepsilon\|, \quad (6.56)$$

with $c_3 = 4c_2 + 2c_1c_2^2$.

(3) Boundary values of the resolvent

The differential inequality (6.56) is of the form (6.24), with $F^{(\lambda)}(\varepsilon, \mu) = \Phi_\varepsilon(\lambda + i\mu)$, $\vartheta_1(\varepsilon) = c_3\varepsilon^{\nu-1}$, $\vartheta_2(\varepsilon) = c_3\varepsilon^{\nu-3/2}$ and $\gamma = c_3$ (see also Remark 6.1). Since $\nu > 1/2$, the integrals $\Theta_k \equiv \int_0^{\varepsilon_0} \vartheta_k(\varepsilon) d\varepsilon$ are finite. So, by the results of §6.2.1, $\Phi_0(\lambda + i\mu) \equiv (I + |A|)^{-\nu}(T - \lambda - i\mu)^{-1}(I + |A|)^{-\nu}$ converges as $\mu \to +0$. Furthermore

$$\begin{aligned}\phi(\varepsilon_0) &\equiv \sup_{\lambda \in J_0} \sup_{0 \le \mu \le 1} \|\Phi_{\varepsilon_0}(\lambda + i\mu)\| \\ &= \sup_{\lambda \in J_0} \sup_{0 \le \mu \le 1} \|(I + |A|)^{-\nu} X_{\varepsilon_0} G_{\varepsilon_0}(\lambda + i\mu) X_{\varepsilon_0} (I + |A|)^{-\nu}\| \\ &\le \sup_{\lambda \in J_0} \sup_{0 \le \mu \le 1} \|G_{\varepsilon_0}(\lambda + i\mu)\| \le \frac{4}{a\varepsilon_0}.\end{aligned}$$

[10] If $\nu = 1$ the arguments leading to (6.56) can be somewhat simplified: in this case one may set $X_\varepsilon = I$ and use $\|A\mathcal{Z}\| = \||A|\mathcal{Z}\| \le 1$.

So (6.30) shows that $\|F^{(\lambda)}(\varepsilon,\mu)\| \equiv \|\Phi_\varepsilon(\lambda+i\mu)\| \leq c_4$ for some constant $c_4$ which we do not write down explicitly [we recall that $\lambda \in J_0$, $0 \leq \mu \leq 1$ and $\varepsilon \in [0,\varepsilon_0)$]. Since one has $\|(I+|A|)^{-\nu}(T-\lambda-i\mu)^{-1}(I+|A|)^{-\nu}\| \leq 1$ if $|\mu| \geq 1$, the proof of (6.35) is complete. The convergence of $\langle f,(T-\lambda-i\mu)^{-1}g\rangle$ as $\mu \to +0$, for $f,g \in \mathcal{D}(|A|^\nu)$, is now obtained by writing $f=(I+|A|)^{-\nu}\hat{f}$, $g=(I+|A|)^{-\nu}\hat{g}$ with $\hat{f},\hat{g} \in \mathcal{H}$. □

**Theorem 6.4.** *Assume that the hypotheses of Theorem 6.3 are satisfied. Then*
*(a) The spectrum of $T$ in $J$ is purely absolutely continuous. In other terms $\sigma_s(T)\cap J = \varnothing$, or equivalently $E(J)\mathcal{H} \subseteq \mathcal{H}_{ac}(T)$.*
*(b) The convergence of $(I+|A|)^{-\nu}(T-\lambda \pm i\mu)(I+|A|)^{-\nu}$ as $\mu \to +0$ is uniform in $\lambda$ on each bounded closed subinterval $J_0$ of $J$, and the limit defines a norm continuous function of $\lambda$: if one sets $X(\lambda) = \text{u-lim}_{\mu\to+0}(I+|A|)^{-\nu}(T-\lambda-i\mu)^{-1}(I+|A|)^{-\nu}$, then $\|X(\lambda)-X(\lambda_0)\| \to 0$ as $\lambda \to \lambda_0$ (with $\lambda,\lambda_0 \in J$).*

PROOF. (a) In view of Theorem 6.3, the result of (a) follows from Proposition 4.31(b) by taking $\mathcal{D}=\mathcal{D}(|A|^\nu)$ in that proposition; the required condition of uniformity in $\lambda$ of $\lim_{\mu\to+0}\langle f,(T-\lambda-i\mu)^{-1}f\rangle$ is satisfied by part (b) of the present theorem.

(b) Let $J_0$ be a bounded closed subinterval of $J$. We let $\varepsilon_0>0$ be as in the preceding proof and use some estimates obtained there, valid for $0<\varepsilon\leq\varepsilon_0$. Let $\lambda,\lambda_0 \in J_0$ and $\mu,\mu_0 \in [0,1]$. We have

$$\begin{aligned}\big\|\Phi_0(\lambda+i\mu)-\Phi_0(\lambda_0+i\mu_0)\big\| &\leq \big\|\Phi_0(\lambda+i\mu)-\Phi_\varepsilon(\lambda+i\mu)\big\| \\ &+\big\|\Phi_\varepsilon(\lambda+i\mu)-\Phi_\varepsilon(\lambda_0+i\mu_0)\big\| + \big\|\Phi_0(\lambda_0+i\mu_0)-\Phi_\varepsilon(\lambda_0+i\mu_0)\big\|. \quad (6.57)\end{aligned}$$

Now

$$\big\|\Phi_0(\lambda+i\mu)-\Phi_\varepsilon(\lambda+i\mu)\big\| \leq \int_0^\varepsilon \Big\|\frac{\partial}{\partial\rho}\Phi_\rho(\lambda+i\mu)\Big\|\,d\rho.$$

Since $\|\Phi_\varepsilon(\lambda+i\mu)\| \leq c_4$, we see from (6.56) that the integrand is majorised by an integrable function which does not depend on $\lambda$ and $\mu$. Thus, given $\delta>0$, we can choose a number $\varepsilon>0$ such that the first and the third term on the right-hand side of (6.57) are less than $\delta/3$ for all $\lambda,\lambda_0 \in J_0$ and all $\mu,\mu_0 \in [0,1]$. For the second term we have by (6.45):

$$\begin{aligned}\big\|\Phi_\varepsilon(\lambda+i\mu)-\Phi_\varepsilon(\lambda_0+i\mu_0)\big\| &\leq \big\|G_\varepsilon(\lambda+i\mu)-G_\varepsilon(\lambda_0+i\mu_0)\big\| \\ &\leq \frac{16}{a^2\varepsilon^2}\big[|\lambda-\lambda_0|+|\mu-\mu_0|\big].\end{aligned}$$

For $\lambda=\lambda_0$ this is less than $\delta/3$ if $\mu,\mu_0 < a^2\varepsilon^2\delta/48$. Then the expression on the left-hand side of (6.57) is less than $\delta$ for all $\mu,\mu_0$ satisfying the preceding condition and for all $\lambda=\lambda_0 \in J_0$, which proves that the convergence of $\Phi_0(\lambda+i\mu)$ to $\Phi_0(\lambda+i0)$ is uniform in $\lambda$ on $J_0$.

Next, taking $\mu=\mu_0=0$ in (6.57), we see that its left-hand side is less than $\delta$ if $\lambda-\lambda_0 < a^2\varepsilon^2\delta/48$, which establishes the norm continuity of $X(\lambda)\equiv\Phi_0(\lambda+i0)$ on $J_0$, hence on $J$ since each $\lambda \in J$ belongs to some bounded closed subinterval $J_0$ of the interval $J$. □

**6.2.3.** We now present an extension of Lemma 6.2 and some applications of this result.

**Proposition 6.5.** *Let $A$ be a self-adjoint operator in a Hilbert space $\mathcal{H}$ and let $C \in \mathcal{B}(\mathcal{H})$. For $\tau \in \mathbb{R}$, set $C(\tau) = e^{iA\tau}Ce^{-iA\tau}$. Assume that there exists an operator $Z \in \mathcal{B}(\mathcal{H})$ such that $\langle C^*f, Ag\rangle - \langle Af, Cg\rangle = \langle f, Zg\rangle$ for all vectors $f, g \in \mathcal{D}(A)$. Then $\tau \mapsto C(\tau)$ is strongly of class $C^1$, and $(d/d\tau)C(\tau)|_{\tau=0} = -iZ$. If furthermore $C$ is invertible with $C^{-1} \in \mathcal{B}(\mathcal{H})$, then*

$$\langle C^{-1*}f, Ag\rangle - \langle Af, C^{-1}g\rangle = -\langle f, C^{-1}ZC^{-1}g\rangle \qquad \forall f, g \in \mathcal{D}(A), \tag{6.58}$$

*and $C^{-1}$ maps $\mathcal{D}(A)$ into $\mathcal{D}(A)$.*

REMARK. Formally these results are evident. For example:

$$\begin{aligned}\frac{d}{d\tau}C(\tau)\big|_{\tau=0} = \frac{d}{d\tau}e^{iA\tau}Ce^{-iA\tau}\big|_{\tau=0} &= e^{iA\tau}[iAC - iCA]e^{-iA\tau}\big|_{\tau=0}\\ &= i[A, C] = -iZ.\end{aligned}$$

The purpose of the proposition is to give a rigourous meaning to this formal calculation.

PROOF OF PROPOSITION 6.5. (i) Let $g \in \mathcal{D}(A)$. Let us write

$$\frac{i}{\varepsilon}\big(e^{iA\varepsilon}Ce^{-iA\varepsilon} - C\big)g = \Big(e^{iA\varepsilon}C\frac{I - e^{-iA\varepsilon}}{i\varepsilon}g - \frac{e^{iA\varepsilon} - I}{i\varepsilon}\Big)Cg.$$

Since $g$ and $Cg$ belong to $\mathcal{D}(A)$ (see Lemma 6.2), the right-hand side converges strongly as $\varepsilon \to 0$, the limit being $CAg - ACg$. Thus

$$i\frac{d}{d\varepsilon}(e^{iA\varepsilon}Ce^{-iA\varepsilon} - C)g\big|_{\varepsilon=0} = CAg - ACg = Zg \qquad \text{for } g \in \mathcal{D}(A)$$

(the last identity follows also from Lemma 6.2). Now observe that

$$\frac{i}{\varepsilon}\big[e^{iA(\tau+\varepsilon)}Ce^{-iA(\tau+\varepsilon)} - (e^{iA\tau}Ce^{-iA\tau}\big]g = e^{iA\tau}\big[\frac{i}{\varepsilon}(e^{iA\varepsilon}Ce^{-iA\varepsilon} - C)\big]e^{-iA\tau}g. \tag{6.59}$$

From Proposition 5.1 we have $e^{-iA\tau}g \in \mathcal{D}(A)$. By the preceding result the right-hand side of (6.59) is strongly convergent as $\varepsilon \to 0$, and

$$i\frac{d}{d\tau}e^{iA\tau}Ce^{-iA\tau}g = e^{iA\tau}Ze^{-iA\tau}g.$$

Thus we have for all $g \in \mathcal{D}(A)$ and all $\sigma \in \mathbb{R}$:

$$e^{iA\sigma}Ce^{-iA\sigma}g - Cg = \int_0^\sigma \frac{d}{d\tau}e^{iA\tau}Ce^{-iA\tau}g\,d\tau = -i\int_0^\sigma e^{iA\tau}Ze^{-iA\tau}g\,d\tau. \tag{6.60}$$

Since all operators occurring in this identity belong to $\mathcal{B}(\mathcal{H})$ and since $\mathcal{D}(A)$ is dense in $\mathcal{H}$, Eq. (6.60) can be extended to each $g \in \mathcal{H}$ [approximate $g$ by a sequence $\{g_k\} \in \mathcal{D}(A)$, write (6.60) for $g_k$ and then take the limit $k \to \infty$]. Now the right-hand side of

(6.60) is strongly differentiable with respect to $\sigma$ (for each $g \in \mathcal{H}$), consequently this is also the case for the left-hand side, and

$$\frac{d}{d\sigma}e^{iA\sigma}Ce^{-iA\sigma}g\big|_{\sigma=0} = -i\frac{d}{d\sigma}\int_0^\sigma e^{iA\tau}Ze^{-iA\tau}g\,d\tau\big|_{\sigma=0} = -iZg.$$

(ii) Now assume in addition that $C$ is invertible with $C^{-1} \in \mathcal{B}(\mathcal{H})$. Set $C^{-1}(\tau) = e^{iA\tau}C^{-1}e^{-iA\tau}$. Then $C^{-1}(\tau)C(\tau) = I$, and the following identity is evident:

$$C^{-1}(\tau) - C^{-1} = -C^{-1}(\tau)[C(\tau) - C]C^{-1}. \tag{6.61}$$

Now $\text{s-lim}_{\tau\to 0} C^{-1}(\tau) = C^{-1}$ and $\text{s-lim}_{\tau\to 0} \tau^{-1}[C(\tau) - C] = -iZ$. Thus (6.61) implies that $C^{-1}(\tau)$ is strongly differentiable at $\tau = 0$, and

$$\frac{d}{d\tau}C^{-1}(\tau)\big|_{\tau=0} = -C^{-1}\big[\frac{d}{d\tau}C(\tau)\big]\big|_{\tau=0}C^{-1} = iC^{-1}ZC^{-1}. \tag{6.62}$$

Eq. (6.58) follows easily by calculating the derivative, using the Leibniz rule: if $f$ and $g$ belong to $\mathcal{D}(A)$, then

$$\begin{aligned}\frac{d}{d\tau}\langle f, C^{-1}(\tau)g\rangle\big|_{\tau=0} &= \frac{d}{d\tau}\langle e^{-iA\tau}f, C^{-1}e^{-iA\tau}g\rangle\big|_{\tau=0}\\ &= i\langle Af, C^{-1}g\rangle - i\langle f, C^{-1}Ag\rangle = i\langle Af, C^{-1}g\rangle - i\langle C^{-1*}f, Ag\rangle.\end{aligned}$$

The fact that $C^{-1}$ maps $\mathcal{D}(A)$ into itself follows from Lemma 6.2. □

As a first application of Proposition 6.5, we give a rigourous derivation of the formula (6.51). We take $C = N_\varepsilon^-$, hence $C^{-1} = G_\varepsilon \equiv G_\varepsilon(\lambda + i\mu)$. To obtain (6.51) from Proposition 6.5, one has to know that

$$\langle (T - \lambda - i\mu - i\varepsilon B)^* f, Ag\rangle - \langle Af, (T - \lambda - i\mu - i\varepsilon B)g\rangle = \langle f, Zg\rangle$$

for some operator $Z \in \mathcal{B}(\mathcal{H})$ and all $f, g \in \mathcal{D}(A)$. The terms involving the constants $\lambda$ and $\mu$ on the left-hand side cancel, whereas for $T$ and $B$ this condition is satisfied as a consequence of the hypotheses (6.32) and (6.33) of Theorem 6.3. We thus have by Eq. (6.58):

$$\langle G_\varepsilon^* f, Ag\rangle - \langle Af, G_\varepsilon g\rangle = -\langle f, G_\varepsilon Z G_\varepsilon g\rangle \qquad \forall f, g \in \mathcal{D}(A),$$

where

$$Z = i\frac{d}{d\tau}\big[e^{iA\tau}N_\varepsilon^- e^{-iA\tau}\big]\big|_{\tau=0} = i\frac{d}{d\tau}\big[e^{iA\tau}Te^{-iA\tau} - \lambda - i\mu - i\varepsilon e^{iA\tau}Be^{-iA\tau}\big]\big|_{\tau=0}.$$

The derivatives can be obtained by using Proposition 6.5 (with $C = T$ and with $C = B$):

$$i\frac{d}{d\tau}e^{iA\tau}Te^{-iA\tau}\big|_{\tau=0} = -iB, \qquad \frac{d}{d\tau}e^{iA\tau}Be^{-iA\tau}\big|_{\tau=0} = -[B, iA]. \tag{6.63}$$

Thus one has for all $f, g \in \mathcal{D}(A)$:

$$\langle G_\varepsilon^* f, Ag\rangle - \langle Af, G_\varepsilon g\rangle = \langle f, iG_\varepsilon BG_\varepsilon g\rangle + \varepsilon\langle f, G_\varepsilon[B, iA]G_\varepsilon g\rangle,$$

which is identical with Eq. (6.51). Proposition 6.5 also implies that $G_\varepsilon$ maps $\mathcal{D}(A)$ into itself. □

As a second application of Proposition 6.5 we give a result that will be of great importance in the next section.

**Proposition 6.6 (Virial Theorem).** *Let $T$ and $A$ be self-adjoint operators, with $T \in \mathcal{B}(\mathcal{H})$, such that the closure $B$ of $[T, iA]$ also belongs to $\mathcal{B}(\mathcal{H})$. Let $f$ and $g$ be eigenvectors of $T$ with the same eigenvalue $\lambda$: $Tf = \lambda f$, $Tg = \lambda g$. Then $\langle f, Bg\rangle = 0$.*

PROOF. Formally the conclusion is clear, it suffices to open the commutator:

$$\langle f, Bg\rangle = i\langle Tf, Ag\rangle - i\langle Af, Tg\rangle = i\lambda\big[\langle f, Ag\rangle - \langle Af, g\rangle\big] = 0.$$

This argument is only formal because one does not know that $f$ and $g$ belong to $\mathcal{D}(A)$. A rigourous argument is as follows. We set $T(\tau) = e^{iA\tau}Te^{-iA\tau}$. Then $T(\tau)$ is strongly of class $C^1$ by Proposition 6.5, and $(d/d\tau)T(\tau)|_{\tau=0} = -B$. Writing $f$ as $f = \text{s-}\lim_{\tau\to 0} e^{iA\tau}f$, we obtain by virtue of Proposition 1.1(b):

$$\begin{aligned}\langle f, Bg\rangle = -\lim_{\tau\to 0}\frac{1}{\tau}\langle f, \{T(\tau) - T\}g\rangle &= -\lim_{\tau\to 0}\frac{1}{\tau}\langle e^{iA\tau}f, \{T(\tau) - T\}g\rangle \\ &\equiv -\lim_{\tau\to 0}\frac{1}{\tau}\big\{\langle Tf, e^{-iA\tau}g\rangle - \langle e^{iA\tau}f, Tg\rangle\big\} \\ &= -\lim_{\tau\to 0}\frac{\lambda}{\tau}\big\{\langle f, e^{-iA\tau}g\rangle - \langle e^{iA\tau}f, g\rangle\big\}.\end{aligned}$$

Since $\langle f, e^{-iA\tau}g\rangle = \langle e^{iA\tau}f, g\rangle$, we have $\langle f, Bg\rangle = 0$. □

## 6.3 The Mourre inequality

The results obtained so far were consequences of the hypothesis of the validity of a strict Mourre inequality $E(J)[H, iA]E(J) \geq aE(J)$ with $a > 0$, where $J$ is an open interval. A weaker condition, which is more easy to verify in concrete situations, is the validity of such an inequality modulo a compact operator. More precisely, if $H$ is a self-adjoint operator and $J$ an open interval, one says that a self-adjoint operator $A$ is ***conjugate to H*** on $J$ if there exist a number $a > 0$ and a self-adjoint compact operator $K$ such that

$$E(J)[H, iA]E(J) \geq aE(J) + K. \tag{6.64}$$

This is called a ***Mourre inequality***. We have seen that a strict Mourre inequality (6.31) implies that the part of the spectrum of $H$ in $J$ is absolutely continuous. We shall show that the validity of a Mourre inequality (6.64) leads to almost the same conclusion: the continuous spectrum of $H$ in $J$ is absolutely continuous, i.e. $H$ has no singular continuous spectrum in $J$; however $H$ may have eigenvalues in $J$, but each of these eigenvalues is of finite multiplicity and their number is finite. These statements will be proved in §6.3.1, again in the case in which $H$ is bounded (we again use the letter $T$ for such an operator).

**6.3.1.** Let $T \in \mathcal{B}(\mathcal{H})$ be a self-adjoint operator and $A$ an operator that is conjugate to $T$ on some open interval $J$. So $A$ is self-adjoint and there exist a number $a > 0$ and a compact operator $K$ such that

$$E(J)BE(J) \geq aE(J) + K, \tag{6.65}$$

where $B \equiv [T, iA]$ is assumed to belong to $\mathcal{B}(\mathcal{H})$ and $\{E(\cdot)\}$ denotes again the spectral measure of $T$.

**Proposition 6.7.** *Assume that the closure $B$ of $[T, iA]$ belongs to $\mathcal{B}(\mathcal{H})$ and that $A$ is conjugate to $T$ on an open interval $J$. Then the number of eigenvalues of $T$ (if any) in $J$ is finite, and each of these eigenvalues is of finite multiplicity.*

PROOF. We denote by $\mathcal{M}_{\mathrm{p}}(J)$ the subspace of $\mathcal{H}$ spanned by all eigenvectors of $T$ associated with the eigenvalues in $J$. We must show that $\dim \mathcal{M}_{\mathrm{p}}(J) < \infty$. Assume that this is not the case, i.e. that $\dim \mathcal{M}_{\mathrm{p}}(J) = \infty$. Then we can choose an infinite orthonormal sequence $\{f_j\}$ of vectors in $\mathcal{M}_{\mathrm{p}}(J)$ such that $\|f_j\| = 1$ and $Tf_j = \lambda_j f_j$, where $\lambda_j$ is an eigenvalue of $T$ in $J$. The inequality (6.65) implies that

$$\langle f_j, Bf_j\rangle \geq a\|f_j\|^2 + \langle f_j, Kf_j\rangle = a + \langle f_j, Kf_j\rangle.$$

But $\langle f_j, Bf_j\rangle = 0$ by the Virial Theorem, hence

$$0 \geq a + \langle f_j, Kf_j\rangle \qquad \forall j \in \mathbb{N}. \tag{6.66}$$

Now $|\langle f_j, Kf_j\rangle| \leq \|f_j\|\|Kf_j\| = \|Kf_j\|$, and $\|Kf_j\| \to 0$ as $j \to \infty$ [the sequence $\{f_j\}$ converges weakly to zero by Example 1.2, hence $\{Kf_j\}$ converges strongly to zero by Proposition 2.12(a) since $K$ is a compact operator]. By taking the limit $j \to \infty$ in (6.66), one obtains $0 \geq a$, which contradicts the hypothesis that $a > 0$. □

**Theorem 6.8.** *Let $T \in \mathcal{B}(\mathcal{H})$ be self-adjoint. Assume that there exists a self-adjoint operator $A$ such that the closure $B$ of $[T, iA]$ and $[B, A] \equiv i[[T, A], A]$ belong to $\mathcal{B}(\mathcal{H})$ and such that the Mourre inequality is satisfied on an open interval $J$. i.e.*

$$E(J)BE(J) \geq aE(J) + K \qquad \text{with } a > 0,\ K \in \mathcal{K}(\mathcal{H}). \tag{6.67}$$

*Let $\nu > 1/2$. Then*
*(a) If $\lambda \in J$ is not an eigenvalue of $T$, then $(I + |A|)^{-\nu}(T - \lambda \pm i\mu)^{-1}(I + |A|)^{-\nu}$ converge in norm as $\mu \to +0$. The convergence is uniform in $\lambda \in J_0$ on each bounded closed subinterval $J_0$ of $J \setminus \sigma_{\mathrm{p}}(T)$, and the limit operators depend continuously (in norm) on $\lambda$ in $J \setminus \sigma_{\mathrm{p}}(T)$. Furthermore, for $J_0$ as above:*

$$\sup_{\lambda \in J_0, \mu > 0} \|(I + |A|)^{-\nu}(T - \lambda \pm i\mu)^{-1}(I + |A|)^{-\nu}\| < \infty. \tag{6.68}$$

*(b) If $J_0$ is as in (a), one has $E(J_0)\mathcal{H} \subseteq \mathcal{H}_{\mathrm{ac}}(T)$.*

PROOF. (i) We shall prove in (ii) below the following fact: If $\lambda$ belongs to $J\setminus\sigma_{\mathrm{p}}(T)$, there exists a number $\delta_\lambda > 0$ such that a strict Mourre inequality is satisfied on the interval $(\lambda-\delta_\lambda, \lambda+\delta_\lambda)$, more precisely such that

$$E\big((\lambda-\delta_\lambda,\lambda+\delta_\lambda)\big)BE\big((\lambda-\delta_\lambda,\lambda+\delta_\lambda)\big) \geq \frac{a}{2}E\big((\lambda-\delta_\lambda,\lambda+\delta_\lambda)\big), \quad (6.69)$$

where $a$ is the constant in (6.67). So the first convergence result stated in (a) follows from Theorem 6.3. Then, by Theorems 6.3 and 6.4, the statements involving a bounded closed subinterval $J_0$ of $J\setminus\sigma_{\mathrm{p}}(T)$ are true on each interval $J_0$ of the form $[\lambda-\varepsilon_\lambda, \lambda+\varepsilon_\lambda]$ if $0<\varepsilon_\lambda<\delta_\lambda$, hence also on each open interval $(\lambda-\varepsilon_\lambda,\lambda+\varepsilon_\lambda)$ with $\varepsilon_\lambda\in(0,\delta_\lambda)$. For an arbitrary subinterval $J_0=[\alpha,\beta]$ of $J\setminus\sigma_{\mathrm{p}}(T)$ these results follow easily by observing that $J_0$ can be covered by a finite number of intervals of the form $(\lambda-\varepsilon_\lambda,\lambda+\varepsilon_\lambda)$ with $\lambda\in[\alpha,\beta]$ and $\varepsilon_\lambda$ as above.

(ii) We now prove (6.69). We use the following fact: If $C, M\in\mathcal{B}(\mathcal{H})$ and $C\geq O$, then $M^*CM\geq O$ [see Lemma 4.33(b)].

Let $\lambda\in J\setminus\sigma_{\mathrm{p}}(T)$. We choose $\varepsilon>0$ such that $(\lambda-\varepsilon,\lambda+\varepsilon)\subseteq J$. Upon multiplying (6.65) on the left and on the right by $E((\lambda-\delta,\lambda+\delta))$, where $0<\delta\leq\varepsilon$, one obtains by observing that $E(J)E((\lambda-\delta,\lambda+\delta)) = E(J\cap(\lambda-\delta,\lambda+\delta)) = E((\lambda-\delta,\lambda+\delta))$:

$$\begin{aligned} &E\big((\lambda-\delta,\lambda+\delta)\big)BE\big((\lambda-\delta,\lambda+\delta)\big)\\ &\qquad\geq aE\big((\lambda-\delta,\lambda+\delta)\big)+E\big((\lambda-\delta,\lambda+\delta)\big)KE\big((\lambda-\delta,\lambda+\delta)\big). \quad (6.70)\end{aligned}$$

Now $\text{s-lim}_{\delta\to 0}E((\lambda-\delta,\lambda+\delta)) = E(\{\lambda\}) = O$ ($\lambda$ is a point of continuity of the spectral family $\{E_\mu\}$). Consequently $\|KE((\lambda-\delta,\lambda+\delta))\|\to 0$ as $\delta\to 0$ [Proposition 2.12(b)]. Hence there exists a number $\delta_\lambda\in(0,\varepsilon)$ such that [11]

$$\frac{a}{2}I \geq E\big((\lambda-\delta_\lambda,\lambda+\delta_\lambda)\big)KE\big((\lambda-\delta_\lambda,\lambda+\delta_\lambda)\big) \geq -\frac{a}{2}I.$$

Upon multiplying the second of these inequalities from the left and from the right by $E((\lambda-\delta_\lambda,\lambda+\delta_\lambda))$, one obtains

$$E\big((\lambda-\delta_\lambda,\lambda+\delta_\lambda)\big)KE\big((\lambda-\delta_\lambda,\lambda+\delta_\lambda)\big) \geq -\frac{a}{2}E\big((\lambda-\delta_\lambda,\lambda+\delta_\lambda)\big). \quad (6.71)$$

By taking $\delta=\delta_\lambda$ in (6.70) and taking into account (6.71), one obtains (6.69). □

The following consequences of Theorem 6.8 and Proposition 6.7 are easy to verify (Problem 6.5):

**Corollary 6.9.** *Under the hypotheses of Theorem 6.8, the operator $T$ has no singular continuous spectrum in $J$, i.e. $\sigma_{\mathrm{sc}}(T)\cap J=\varnothing$, the number of eigenvalues of $T$ in $J$ is finite (or zero), and each of these eigenvalues is of finite multiplicity.*

---

[11] If $C=C^*$ and $\|C\|\leq\kappa$, and if $\{E^C(\cdot)\}$ denotes the spectral measure of $C$, one has $E^C(\mathcal{V}) = E^C(\mathcal{V}\cap[-\kappa,\kappa])$ (the spectrum of $C$ is contained in the interval $[-\kappa,\kappa]$). Hence, by setting $m_f(\mathcal{W}) = \|E^C(\mathcal{W})f\|^2$:

$$\kappa\langle f,f\rangle = \kappa\int_{-\kappa}^{\kappa} m_f(d\lambda) \geq \int_{-\kappa}^{\kappa}\lambda m_f(d\lambda) = \langle f, Cf\rangle \geq -\kappa\int_{-\kappa}^{\kappa} m_f(d\lambda) = -\kappa\langle f,f\rangle.$$

**6.3.2.** The principal results of Mourre theory obtained so far are valid if the operator $T$ is replaced by a general self-adjoint operator which will be denoted by $H$. The idea of their proof is the same, but for unbounded $H$ one has to take into account domain properties, which lengthens the arguments. For example one has to give a meaning to the operators $[H, iA]$ and $[[H, A], A]$ which are not bounded (this is seen already in the simple case in which $H$ is the free Hamiltonian $\vec{P}^2$ in $L^2(\mathbb{R}^n)$; a conjugate operator to $\vec{P}^2$ is the operator $D$ introduced in (6.19), and we have seen that $[\vec{P}^2, iD] = \vec{P}^2$, so this commutator is unbounded). Thus one will have to impose conditions relating the domains of $H$ and $A$. For example, it is required in many approaches that the group $\{e^{-iAs}\}_{s\in\mathbb{R}}$ should leave the domain $\mathcal{D}(H)$ of $H$ invariant, i.e. that $e^{-iAs}f \in \mathcal{D}(H)$ for each $f \in \mathcal{D}(H)$ and each $s \in \mathbb{R}$.

If the spectrum of $H$ does not cover the entire real axis, i.e. if $\rho(H) \cap \mathbb{R}$ is non-empty, one can make use of the results obtained so far by introducing the bounded self-adjoint operator $T := (H - \lambda_0)^{-1}$, where $\lambda_0$ is a fixed real number in the resolvent set $\rho(H)$ of $H$. The mapping $\lambda \mapsto 1/(\lambda - \lambda_0)$ gives a one-to-one correspondence between the spectrum of $H$ and that of $T$, which allows one to relate the spectral properties of $H$ to those of $T$ and to transform a Mourre inequality for $H$ into a Mourre inequality for $T$. One then arrives rather easily at the following theorem:

**Theorem 6.10.** *Let $H$ be a self-adjoint operator in a Hilbert space $\mathcal{H}$ such that $\lambda_0 \in \rho(H)$ for some $\lambda_0 \in \mathbb{R}$. Assume that there exists a self-adjoint operator $A$ such that the closures of $[(H - \lambda_0)^{-1}, iA]$ and $[[(H - \lambda_0)^{-1}, A], A]$ belong to $\mathcal{B}(\mathcal{H})$ and such that, for a bounded open interval $J$:*

$$E(J)[H, iA]E(J) \geq aE(J) + K \qquad \text{with } a > 0,\ K \in \mathcal{K}(\mathcal{H}), \tag{6.72}$$

*where $\{E(\cdot)\}$ denotes the spectral measure of $H$. Then all conclusions of Theorem 6.8 (with $T$ replaced by $H$) are true, and the spectrum of $H$ in $J$ is as described in Corollary 6.9.*

In §6.4.2 we give the proof of this theorem when $H$ is a Schrödinger operator. Except for some domain considerations [12], the proof of Theorem 6.10 in the general case is exactly the same.

## 6.4 Application to Schrödinger operators

The main purpose of this section is to apply the general theory to two-body Schrödinger Hamiltonians in the centre-of-mass frame with scattering-type potentials. Such Hamiltonians are unbounded self-adjoint operators. However they are bounded from below and therefore covered by Theorem 6.10. In §6.4.1 we discuss the necessary commutators and state sufficient conditions on the potential $V$ so that the the Mourre method is applicable. Certain more technical aspects are presented in §6.4.2 and §6.7.2.

[12] The necessary domain properties, all of which are consequences of the assumption that the commutator $[(H - \lambda_0)^{-1}, iA]$ belongs to $\mathcal{B}(\mathcal{H})$, are discussed in Theorem 6.2.10 of [ABG].

**6.4.1.** In order to make use of the results of Sections 6.2 and 6.3, one has to know that $[(H-\lambda_0)^{-1}, iA]$ and $[[(H-\lambda_0)^{-1}, A], A]$ belong to $\mathcal{B}(\mathcal{H})$ (in the sense specified at the beginning of §6.2.2) for some $\lambda_0 \in \rho(H)$. We make a few comments about this point, remaining on a formal level for the moment. By using (6.9) and the formal identity $[XY, Z] = [X, Z]Y + X[Y, Z]$, one sees that for any $z \in \rho(H)$:

$$[(H-z)^{-1}, A] = (H-z)^{-1}[A, H](H-z)^{-1} \tag{6.73}$$

and

$$\begin{aligned}[[(H-z)^{-1}, A], A] &= [(H-z)^{-1}[A, H](H-z)^{-1}, A] \\ &= [(H-z)^{-1}, A][A, H](H-z)^{-1} + (H-z)^{-1}[[A, H], A](H-z)^{-1} \\ &\quad + (H-z)^{-1}[A, H][(H-z)^{-1}, A]. \end{aligned} \tag{6.74}$$

This shows that the conditions $[(H-z)^{-1}, iA] \in \mathcal{B}(\mathcal{H})$ and $[[(H-z)^{-1}, A], A] \in \mathcal{B}(\mathcal{H})$ can be expressed in terms of certain conditions on the first and second commutator of $H$ with $A$.

Let us consider Hamiltonians $H$ in $L^2(\mathbb{R}^n)$ of the form $H = H_0 + V(\vec{Q})$, with $H_0 = \vec{P}^2$. We denote by $\{E(\cdot)\}$ the spectral measure of $H$, and we assume that $V(\vec{Q})$ is $H_0$-compact, so that $\sigma_{\text{ess}}(H) = \sigma_{\text{ess}}(H_0) = [0, \infty)$ by Proposition 4.26. Thus the spectrum of $H$ in $(-\infty, 0)$ is discrete and $H$ is bounded from below, and we shall apply the Mourre method for intervals $J$ contained in $(0, \infty)$ to obtain details of the nature of the essential spectrum of $H$. We know that a conjugate operator in the case $V \equiv 0$ is the operator $D = -\widehat{D}/2 = (\vec{Q}\cdot\vec{P}+\vec{P}\cdot\vec{Q})/4$ given in (6.19), where $\widehat{D}$ is the infinitesimal generator of the dilation group in $L^2(\mathbb{R}^n)$. We shall show that the same operator $D$ can serve as a conjugate operator to $H$ if the potential $V$ is sufficiently well behaved. We observe that the domain of $H$ is invariant under the dilation group; indeed $\mathcal{D}(H) = \mathcal{D}(H_0)$ by the Rellich-Kato Theorem (Proposition 2.44), and we have seen in §6.1.3 that $\mathcal{D}(H_0)$ is invariant under the dilation group.

To calculate $[H, iD]$, we again proceed somewhat formally. Recall from (6.20) that $[H_0, iD] = H_0$ [at least on $\mathcal{S}(\mathbb{R}^n)$]. It follows that

$$[H, iD] = [H_0, iD] + [V(\vec{Q}), iD] = H_0 + [V(\vec{Q}), iD] = H - V(\vec{Q}) + [V(\vec{Q}), iD].$$

The commutator $[V(\vec{Q}), iD]$ is easily calculated by using the following expression for $D$ in $L^2(\mathbb{R}^n)$ [see (6.19)]:

$$D = \frac{1}{4}(2\vec{Q}\cdot\vec{P} - inI) = -\frac{i}{4}(2\vec{Q}\cdot\vec{\nabla} + nI).$$

It follows that

$$[V(\vec{Q}), iD] = -\frac{1}{2}(\vec{Q}\cdot\vec{\nabla}V)(\vec{Q}) = -\frac{1}{2}\left(r\frac{\partial V}{\partial r}\right)(\vec{Q}). \tag{6.75}$$

The last relation follows from the fact that the projection of the gradient operator in the direction $\vec{x}$ is the radial derivative (in spherical polar co-ordinates): $\sum_{k=1}^{n} x_k(\partial/\partial x_k) = r(\partial/\partial r)$.

In Eq. (6.75), $[r(\partial V/\partial r)](\vec{Q})$ denotes the multiplication operator in $L^2(\mathbb{R}^n)$ by $[r(\partial V/\partial r)](\vec{x}) \equiv |\vec{x}|(\partial V/\partial r)(\vec{x})$. One must assume that the radial derivative $\partial V/\partial r$ of $V$ is defined in some sense. The simplest situation is that where the function $V$ (expressed in spherical polar co-ordinates) is differentiable in the usual sense, except possibly at the origin $r = 0$. Another possibility is to define $\vec{x} \cdot \vec{\nabla} V$ in the sense of distributions. If the function $V$ belongs to $L^1_{\text{loc}}(\mathbb{R}^n)$, then $\int_{\mathbb{R}^n} V(\vec{x})\phi(\vec{x})\, d^n x$ is well defined for each $\phi \in C_0^\infty(\mathbb{R}^n)$, so that $V$ defines a distribution on the set of test functions $C_0^\infty(\mathbb{R}^n)$, and its partial derivatives $\partial V/\partial x_k$ exist as distributions. Assume that these distributions are given by functions, i.e. that there exist functions $\theta_k \in L^1_{\text{loc}}(\mathbb{R}^n)$ such that

$$\int_{\mathbb{R}^n} V(\vec{x}) \frac{\partial \phi(\vec{x})}{\partial x_k}\, d^n x = -\int_{\mathbb{R}^n} \theta_k(\vec{x})\phi(\vec{x})\, d^n x \qquad \forall \phi \in C_0^\infty(\mathbb{R}^n).$$

We shall use the natural notation $(\partial V/\partial x_k)$ for these functions $\theta_k$, and we write $[r(\partial V/\partial r)](\vec{x})$ for $\sum_{k=1}^n x_k (\partial V/\partial x_k)(\vec{x})$. Under the stated integrability condition on $\theta_k$, the scalar products

$$\langle f, \Big(r\frac{\partial V}{\partial r}\Big)(\vec{Q})g\rangle \equiv \int_{\mathbb{R}^n} \overline{f(\vec{x})} \Big(r\frac{\partial V}{\partial r}\Big)(\vec{x}) g(\vec{x})\, d^n x$$

are well defined for all $f, g \in C_0^\infty(\mathbb{R}^n)$, and one has [in the sense of scalar products between vectors from $C_0^\infty(\mathbb{R}^n)$]

$$[H, iD] = H - V(\vec{Q}) - \frac{1}{2}\Big(r\frac{\partial V}{\partial r}\Big)(\vec{Q}) \equiv H - Z(\vec{Q}), \tag{6.76}$$

$$\text{with } Z(\vec{x}) = V(\vec{x}) + \frac{1}{2}\Big(r\frac{\partial V}{\partial r}\Big)(\vec{x}). \tag{6.77}$$

If the functions $\theta_k$ are locally square-integrable, then the operator of multiplication by the function $r(\partial V/\partial r)$ is defined on $C_0^\infty(\mathbb{R}^n)$, and Eq. (6.76) holds as an operator equation on $C_0^\infty(\mathbb{R}^n)$ (provided that $V$ is also locally square-integrable).

Proceeding in the same way one finds [using (6.75) for $Z$ in place of $V$] that

$$\begin{aligned} [[H, iD], iD] &= [H, iD] - [Z(\vec{Q}), iD] \\ &= H - Z(\vec{Q}) + \frac{1}{2}\big(r\frac{\partial V}{\partial r}\big)(\vec{Q}) + \frac{1}{4}\big(r\frac{\partial}{\partial r} r\frac{\partial V}{\partial r}\big)(\vec{Q}) \equiv H - Z_1(\vec{Q}), \end{aligned} \tag{6.78}$$

$$\text{with } Z_1(\vec{x}) = V(\vec{x}) - \frac{1}{4}\big(r\frac{\partial V}{\partial r}\big)(\vec{x}) - \frac{1}{4}\big(r^2\frac{\partial^2 V}{\partial r^2}\big)(\vec{x}). \tag{6.79}$$

We shall consider a rather simple class of (real-valued) potentials defined by the condition (C) stated below. Further on, when discussing properties of Schrödinger operators that rely on results obtained by the Mourre method, we shall usually assume that the potentials satisfy this condition.

(C) *$V$ and $r(\partial V/\partial r)$ belong to $L^p(\mathbb{R}^n) + L^\infty_\varepsilon(\mathbb{R}^n)$, and $r^2(\partial^2 V/\partial r^2)$ belongs to $L^p(\mathbb{R}^n) + L^\infty(\mathbb{R}^n)$, with $p = 2$ if $n = 1, 2, 3$ and $p \in (n/2, \infty)$ if $n \geq 4$.*

If (C) is satisfied, then $V(\vec{Q})$, $Z(\vec{Q})$ and $Z_1(\vec{Q})$ are defined not only on $C_0^\infty(\mathbb{R}^n)$ but on $\mathcal{D}(H_0)$, hence also on $\mathcal{D}(H)$. The operators $V(\vec{Q})$ and $Z(\vec{Q})$ are $H$-compact (see Propositions 2.49 and 2.47); $Z_1(\vec{Q})$ is $H_0$-bounded with relative bound 0, hence also $H$-bounded with relative bound 0 [see Proposition 2.49 and the proof of Proposition 2.44(b)]. Hence, if $z \in \rho(H)$, the operators $V(\vec{Q})(H-z)^{-1}$ and $Z(\vec{Q})(H-z)^{-1}$ are defined on $\mathcal{H}$ and are in fact compact, whereas $Z_1(\vec{Q})(H-z)^{-1} \in \mathcal{B}(\mathcal{H})$. In particular:

(1) $[H, D](H-z)^{-1}$ and $[[H, D], D](H-z)^{-1}$ belong to $\mathcal{B}(\mathcal{H})$ [this is evident by taking into account (6.76) and (6.78); see also §6.7.2 for a more careful calculation of the commutators in (6.76) and (6.78)];

(2) if $J$ is a finite interval, then $HE(J) \in \mathcal{B}(\mathcal{H})$ and $V(\vec{Q})E(J) = V(\vec{Q})(H-z)^{-1} \cdot (H-z)E(J) \in \mathcal{K}(\mathcal{H})$; similarly $Z(\vec{Q})E(J) \in \mathcal{K}(\mathcal{H})$ and $Z_1(\vec{Q})E(J) \in \mathcal{B}(\mathcal{H})$.

Assume now that $J$ is a bounded interval in $(\varepsilon, \infty)$ for some $\varepsilon > 0$. Then, for each $f \in \mathcal{H}$: $\langle E(J)f, HE(J)f\rangle = \int_J \lambda m_f(d\lambda) \geq \varepsilon \int_J m_f(d\lambda) = \varepsilon\langle f, E(J)f\rangle$, hence $E(J)HE(J) \geq \varepsilon E(J)$. By taking into account (6.76) we see that

$$E(J)[H, iD]E(J) = E(J)HE(J) - E(J)Z(\vec{Q})E(J) \geq \varepsilon E(J) - E(J)Z(\vec{Q})E(J).$$

Since $E(J)Z(\vec{Q})E(J)$ is a compact operator, this shows that $D$ *is conjugate to* $H$ *on each bounded interval* $J$ *in* $(0, \infty)$ *provided that* $0$ *does not belong to the closure of* $J$. [13] By the extension of the Mourre method to unbounded operators (see Theorem 6.10), one has the following results:

(a) $H$ has no singular continuous spectrum: $\sigma_{\text{sc}}(H) = \varnothing$.

(b) $\sigma_{\text{ac}}(H) = [0, \infty)$. [14]

(c) In each interval $J = [\alpha, \beta]$ with $0 < \alpha < \beta < \infty$, $H$ has at most a finite number of eigenvalues, each of them of finite multiplicity. [15]

(d) If $\lambda \in (0, \infty)$ is not an eigenvalue of $T$, then $(I+|A|)^{-\nu}(T-\lambda \pm i\mu)^{-1}(I+|A|)^{-\nu}$ converge in norm as $\mu \to +0$, for any $\nu > 1/2$. The convergence is uniform in $\lambda$ on each bounded closed subinterval $J$ of $(0, \infty)$ disjoint from the eigenvalues of $H$, and if $J$ is such an interval, then

$$\sup_{\lambda \in J,\, \mu > 0} \|(I+|A|)^{-\nu}(H-\lambda \pm i\mu)^{-1}(I+|A|)^{-\nu}\| < \infty. \tag{6.80}$$

---

[13] It is clear that the condition (C) is not optimal for arriving at this conclusion (we have not used at all the first factor $E(J)$ occurring in the Mourre inequality; for example $E(J)W(\vec{Q})E(J)$ may be compact without $W(\vec{Q})E(J)$ being compact).
Simple examples of potentials satisfying (C) for $n = 3$ are the Yukawa potential $\gamma e^{-\mu r}/r$ and the Coulomb potential $\gamma/r$ [$\gamma \in \mathbb{R}$, $\mu > 0$].

[14] We know that $\sigma_{\text{ess}}(H) = [0, \infty)$, and the assertion of (b) follows by taking into account (a) and (c).

[15] One can show that if $V$ satisfies the condition (C), then $H$ has no positive eigenvalues. On the other hand one knows spherically symmetric bounded potentials converging to zero at infinity for which the associated Hamiltonian has strictly positive eigenvalues (see Section 4.4 of [CFKS] and Section XIII.13, Volume IV of [RS]).

**6.4.2.** As explained in §6.3.2, we give here a proof of Theorem 6.10 when $H$ is a Schrödinger Hamiltonian with a potential $V$ satisfying the condition (C) of page 288. The idea of the proof is to obtain the results of the theorem from the corresponding statements for bounded operators. We fix a real point $\lambda_0$ in the resolvent set of $H$ (so $\lambda_0 < 0$) and denote by $\{E(\cdot)\}$ the spectral measure of $H$. By the results of §6.4.1 the boundedness assumptions of Theorem 6.10 are satisfied (with $A = D$), and one has a Mourre estimate for $H$ on each bounded open interval $J$ in $(0, \infty)$ the closure of which does not contain the point $\lambda = 0$. In the sequel we consider a fixed interval $J$ having these properties.

Since $\rho(H)$ is open, there exists a number $\delta > 0$ such that the interval $(\lambda_0 - \delta, \lambda_0 + \delta)$ belongs to $\rho(H)$, and we can assume without loss of generality that $\lambda_0 + \delta < 0$. One then has $\text{dist}(\lambda_0, J) > \delta > 0$.

(i) We set $T = -(H - \lambda_0)^{-1} = (\lambda_0 - H)^{-1}$. Clearly $T$ is a self-adjoint operator belonging to $\mathcal{B}(\mathcal{H})$, and we denote its spectral measure by $\{E^T(\cdot)\}$ [as we also use the spectral measure $\{E(\cdot)\}$ of $H$]. These two spectral measures are not independent, because $T$ is a function of $H$: $T = \varphi(H)$, with $\varphi(\lambda) = (\lambda_0 - \lambda)^{-1}$. We know (see Remark 4.14) that $E^T(\mathcal{V}) = \chi_{\mathcal{V}}(T) = [\chi_{\mathcal{V}} \circ \varphi](H)$. Let us calculate the function $\chi_{\mathcal{V}} \circ \varphi$:

$$[\chi_{\mathcal{V}} \circ \varphi](\lambda) = \begin{cases} 1 & \text{if } \varphi(\lambda) \in \mathcal{V} \\ 0 & \text{if } \varphi(\lambda) \notin \mathcal{V}. \end{cases}$$

Let us denote by $\mathcal{W}$ the pre-image of $\mathcal{V}$ under $\varphi$, i.e. $\mathcal{W} = \{\lambda \in \mathbb{R} \mid \varphi(\lambda) \in \mathcal{V}\}$. Thus $[\chi_{\mathcal{V}} \circ \varphi](\lambda) = \chi_{\mathcal{W}}(\lambda)$. In other terms $E^T(\mathcal{V}) = E(\mathcal{W})$: the spectral projection of $T$ for $\mathcal{V}$ is identical with the spectral projection of $H$ for $\mathcal{W} = \varphi^{-1}(\mathcal{V})$! In particular, if $\mathcal{W} = J$, let us set $\tilde{J} = \varphi(J) = \{\sigma \in \mathbb{R} \mid \sigma = (\lambda_0 - \lambda)^{-1} \text{ for some } \lambda \in J\}$. Then $E^T(\tilde{J}) = E(J)$.

(ii) One has $[T, iA] = -[(H - \lambda_0)^{-1}, iA] = (H - \lambda_0)^{-1}[H, iA](H - \lambda_0)^{-1} = T[H, iA]T$, hence (from the Mourre inequality on $J$) [16]:

$$\begin{aligned} E^T(\tilde{J})[T, iA]E^T(\tilde{J}) &= E(J)T[H, iA]TE(J) \\ &= TE(J)[H, iA]E(J)T \geq aTE(J)T + TKT. \end{aligned}$$

We set $\tilde{K} = TKT$. Clearly $\tilde{K}$ is a compact operator. Also there exists a number $\kappa > 0$ such that $TE(J)T \geq \kappa E(J)$: indeed, if $[\alpha, \beta]$ denotes the closure of the open interval $J$ (with $0 < \alpha < \beta$) and if one sets $\kappa = (\beta - \lambda_0)^{-2}$, then

$$\langle f, TE(J)Tf \rangle = \int_J \frac{1}{(\lambda - \lambda_0)^2} m_f(d\lambda) \geq \kappa \int_J m_f(d\lambda) = \kappa \langle f, E(J)f \rangle.$$

[16] We observe that for the Schrödinger Hamiltonians considered here:

$$\begin{aligned} E(J)T[H, iA]TE(J) &= E(J)(H - \lambda_0)^{-1}[H - Z(\vec{Q})](H - \lambda_0)^{-1}E(J) \\ &= (H - \lambda_0)^{-1}E(J)[H - Z(\vec{Q})]E(J)(H - \lambda_0)^{-1} \equiv TE(J)[H, iA]E(J)T, \end{aligned}$$

since $[H - Z(\vec{Q})]E(J)$ belongs to $\mathcal{B}(\mathcal{H})$. For a more general self-adjoint operator $H$ the establishment of this identity may require some care (see the footnote on page 286).

As a consequence we have

$$E^T(\tilde{J})[T, iA]E^T(\tilde{J}) \geq a\kappa E(J) + \tilde{K} = a\kappa E^T(\tilde{J}) + \tilde{K},$$

with $a\kappa > 0$ and $\tilde{K} \in \mathcal{K}(\mathcal{H})$. By Theorem 6.8 and Corollary 6.9, $T$ has at most a finite number of eigenvalues in $\tilde{J}$, each of finite multiplicity. Moreover, if $\tilde{J}_1$ is a closed subinterval of $\tilde{J} \setminus \sigma_{\rm p}(T)$, then the limits of $(I + |A|)^{-\nu}(T - \lambda \pm i\mu)^{-1}(I + |A|)^{-\nu}$ as $\mu \to +0$ exist uniformly in $\lambda \in \tilde{J}_1$ for any $\nu > 1/2$ [in fact the $\mathcal{B}(\mathcal{H})$-valued function $\Phi_0(z) = (I + |A|)^{-\nu}(T - z)^{-1}(I + |A|)^{-\nu}$ is uniformly norm continuous on $\mathcal{O}_1 := \{z = \lambda + i\mu \mid \lambda \in \tilde{J}_1, 0 \leq \mu \leq 1\}$]. Furthermore one has

$$\sup_{\lambda \in \tilde{J}_1, \mu > 0} \|(I + |A|)^{-\nu}(T - \lambda \pm i\mu)^{-1}(I + |A|)^{-\nu}\| < \infty. \tag{6.81}$$

(iii) The relation $E(\{\lambda\}) = E^T(\{\varphi(\lambda)\})$ shows that the eigenvectors of $H$ and of $T$ are the same. A real number $\lambda$ is an eigenvalue of $H$ if and only if $\varphi(\lambda)$ is an eigenvalue of $T$, with the same multiplicity [one can see explicitly that if $Hf = \lambda f$, then $Tf = (\lambda_0 - H)^{-1}f = (\lambda_0 - \lambda)^{-1}f = \varphi(\lambda)f$]. Since $T$ has at most a finite number of eigenvalues in $\tilde{J}$, each of them of finite multiplicity, it follows that $H$ has at most a finite number of eigenvalues in $J$, all of them of finite multiplicity.

(iv) Now let $J_0$ be a closed subinterval of $J \setminus \sigma_{\rm p}(H)$. Define $\tilde{J}_0$ by $\tilde{J}_0 = \varphi(J_0)$. $\tilde{J}_0$ is a closed subinterval of $\tilde{J} \setminus \sigma_{\rm p}(T)$. Observe that one has the following identity (it is evident if one replaces $H$ by $\lambda$):

$$\begin{aligned}(H - z)^{-1} &= (\lambda_0 - z)^{-1}\big[(\lambda_0 - H)^{-1} - (\lambda_0 - z)^{-1}\big]^{-1}(\lambda_0 - H)^{-1} \\ &= (\lambda_0 - z)^{-1}\big[(T - (\lambda_0 - z)^{-1}\big]^{-1}T.\end{aligned}$$

Consequently

$$\begin{aligned}&(I + |A|)^{-\nu}(H - \lambda - i\mu)^{-1}(I + |A|)^{-\nu} = (\lambda_0 - \lambda - i\mu)^{-1} \\ &\quad \times (I + |A|)^{-\nu}\big[T - (\lambda_0 - \lambda - i\mu)^{-1}\big]^{-1}(I + |A|)^{-\nu} \times (I + |A|)^{\nu}T(I + |A|)^{-\nu}.\end{aligned} \tag{6.82}$$

Consider the set $\mathcal{O} = \{z = \lambda + i\mu \mid \lambda \in J_0, 0 \leq \mu \leq \mu_0\}$, where $\mu_0 \in (0, 1]$ is fixed. The mapping $z \mapsto (\lambda_0 - z)^{-1}$ maps the interval $J_0$ onto $\tilde{J}_0$ and $\mathcal{O}$ in a continuous manner onto a complex set $\tilde{\mathcal{O}}$ containing $\tilde{J}_0$; if $\mu_0$ is taken to be small enough, $\tilde{\mathcal{O}}$ is contained in the rectangle $\{z = \sigma + i\varepsilon \mid 0 \leq \varepsilon \leq 1, \sigma \in \tilde{J}_1\}$, where $\tilde{J}_1$ is a closed subinterval of $\tilde{J}$ disjoint from $\sigma_{\rm p}(T)$ and containing $\tilde{J}_0$ in its interior. As $\mu \to +0$, $(\lambda_0 - \lambda - i\mu)^{-1}$ converges [uniformly in $\lambda \in J$] to $(\lambda_0 - \lambda)^{-1}$. So, since $(I + |A|)^{-\nu}(T - z)^{-1}(I + |A|)^{-\nu}$ is (uniformly) continuous on the set $\{z = \sigma + i\varepsilon \mid 0 \leq \varepsilon \leq 1, \sigma \in \tilde{J}_1\}$ (assuming again that $\nu > 1/2$), we find from Eq. (6.82) that $(I + |A|)^{-\nu}(H - \lambda - i\mu)^{-1}(I + |A|)^{-\nu}$ has limits in norm as $\mu \to +0$, uniformly in $\lambda \in J_0$, provided that one knows that $(I + |A|)^{\nu}T(I + |A|)^{-\nu}$ belongs to $\mathcal{B}(\mathcal{H})$. If $\nu = 1$ we know that this condition is satisfied, because $T \equiv -(H - \lambda_0)^{-1}$ maps $\mathcal{D}(A)$

into $\mathcal{D}(A)$ under the hypotheses of the theorem (see Lemma 6.2)[17]. If $\nu < 1$ one can show by an interpolation method that this is also the case [see the Appendix to this chapter, in particular Example 6.31].

(v) The results obtained so far imply the spectral properties of $H$ (the validity of Corollary 6.9 for $H$). The bound (6.80) for $H$ follows from (6.81) and (6.82). With this the proof of Theorem 6.10 is complete.

## 6.5 Relatively smooth operators

The concept of relatively smooth operators with respect to a self-adjoint operator $H$ fits naturally into the context of the present chapter, since it can be defined in terms of the resolvent of $H$ close to the real axis. In §6.5.1 we give the definition and some results that are important for our subsequent considerations, in particular a theorem that will permit us in Chapter 7 to prove asymptotic completeness in scattering theory. In §6.5.2 we indicate a class of smooth operators relative to the Hamiltonians that were considered in §6.4.1.

**6.5.1.** Let $H$ be a self-adjoint operator in a Hilbert space $\mathcal{H}$ and $\{E(\cdot)\}$ its spectral measure. Let $C$ be a closed operator in $\mathcal{H}$. One says that $C$ is ***smooth relative to H*** (or simply ***H-smooth***) if $\mathcal{D}(C) \supseteq \mathcal{D}(H)$ and if there exists a constant $c_0 \in (0,\infty)$ such that for each $\mu \in (0,1)$ and each $f \in \mathcal{H}$:

$$\int_{-\infty}^{\infty} \left\| C[(H-\lambda-i\mu)^{-1} - (H-\lambda+i\mu)^{-1}]f \right\|^2 d\lambda \le c_0 \|f\|^2. \tag{6.83}$$

We immediately give a result whose relevance for scattering theory is rather evident (see Chapter 7 for details):

**Proposition 6.11.** *Let $H$ be self-adjoint and let $C$ be a closed operator which is $H$-smooth, i.e. satisfying* (6.83)*. Then one has*

$$\int_{-\infty}^{\infty} \|Ce^{-iHt}f\|^2 dt \le \frac{c_0}{2\pi}\|f\|^2 \qquad \forall f \in \mathcal{D}(H). \tag{6.84}$$

PROOF. We set $R(z) = (H-z)^{-1}$, and we shall always assume that $f \in \mathcal{D}(H)$. If $z \in \rho(H)$, the operator $C(H-z)^{-1} \equiv CR(z)$ is defined on each vector $g \in \mathcal{H}$ and is closed, hence it belongs to $\mathcal{B}(\mathcal{H})$ by virtue of the Closed Graph Theorem (see page

[17] Observe that, for $\nu = 1$, one does have $\|(I+|A|)^{\nu}T(I+|A|)^{-\nu}f\| \le c\|f\|$ with $c < \infty$. Indeed:

$$\begin{aligned}\|(I+|A|)T(I+|A|)^{-1}f\| &\le \|T(I+|A|)^{-1}f\| + \|AT(I+|A|)^{-1}f\| \\ &\le \|T(I+|A|)^{-1}f\| + \|TA(I+|A|)^{-1}f\| + \|[A,T](I+|A|)^{-1}f\| \\ &\le \{\|T(I+|A|)^{-1}\| + \|T\|\|A(I+|A|)^{-1}\| + \|[A,T]\|\}\|f\| = c\|f\|.\end{aligned}$$

For $\nu = 0$ one has $(I+|A|)^{\nu}T(I+|A|)^{-\nu} = T \in \mathcal{B}(\mathcal{H})$, and for $0 < \nu < 1$ one must interpolate between these two cases, as explained in the Appendix.

68; in our applications it is not necessary to invoke this theorem, because the operators $C$ that will be considered are $H$-bounded as a consequence of results proved in earlier chapters). We then have $\|Ce^{-iHt}f\| \leq \|CR(i)\|\|(H-i)f\| < \infty$ if $f \in \mathcal{D}(H)$, i.e. the integrand in (6.84) is bounded on $\mathbb{R}$. As a consequence, if $\mu > 0$, then the function $\eta(t) := e^{-\mu|t|}\|Ce^{-iHt}f\|$ belongs to $L^2(\mathbb{R})$, in other terms the vector-valued function $t \mapsto e^{-\mu|t|}Ce^{-iHt}f$ belongs to $L^2(\mathbb{R};\mathcal{H})$ (see page 20 for this type of spaces).

By Eqs. (5.7) and (5.8) one has for $\mu > 0$:

$$R(\lambda + i\mu) - R(\lambda - i\mu) = i\int_{-\infty}^{\infty} e^{i\lambda t}e^{-\mu|t|}e^{-iHt}\,dt. \tag{6.85}$$

So $\lambda \mapsto C[R(\lambda+i\mu) - R(\lambda - i\mu)]f$ is the inverse Fourier transform of the vector-valued function $t \mapsto i\sqrt{2\pi}e^{-\mu|t|}Ce^{-iHt}f$. By using the unitarity of the Fourier transformation in $L^2(\mathbb{R};\mathcal{H})$ (i.e. for vector-valued functions defined on $\mathbb{R}$), one obtains

$$\int_{-\infty}^{\infty} e^{-2\mu|t|}\|Ce^{-iHt}f\|^2\,dt = \frac{1}{2\pi}\int_{-\infty}^{\infty}\|C\big[R(\lambda+i\mu)-R(\lambda-i\mu)\big]f\|^2\,d\lambda \leq \frac{c_0}{2\pi}\|f\|^2.$$

In particular one has for each $M \in \mathbb{N}$ and each $\mu > 0$:

$$\int_{-M}^{M} e^{-2\mu|t|}\|Ce^{-iHt}f\|^2\,dt \leq \frac{c_0}{2\pi}\|f\|^2.$$

This inequality is still true for $\mu = 0$ (use for example the Dominated Convergence Theorem by taking into account the fact that $\|Ce^{-iHt}f\|^2$ is bounded as a function of $t$, hence in $L^1([-M,M])$, or a more explicit argument):

$$\int_{-M}^{M}\|Ce^{-iHt}f\|^2\,dt = \lim_{\mu\to+0}\int_{-M}^{M} e^{-2\mu|t|}\|Ce^{-iHt}f\|^2\,dt \leq \frac{c_0}{2\pi}\|f\|^2 \quad \forall M \in \mathbb{N}.$$

As $M \to \infty$, the left-hand side (a non-decreasing and bounded sequence of real numbers) converges, the limit being $\int_{-\infty}^{\infty}\|Ce^{-iHt}f\|^2dt$, and this limit is still bounded by the same number $(2\pi)^{-1}c_0\|f\|^2$. □

We next introduce a local version of the concept of relatively smooth operators. Let again $C$ be a closed operator with $\mathcal{D}(C) \supseteq \mathcal{D}(H)$, and let $J$ be an interval (or more generally a Borel subset of $\mathbb{R}$). One says that $C$ is ***H*-smooth on *J*** if $CE(J)$ is $H$-smooth in the sense used before, i.e. if there exists a constant $c_0 \in (0,\infty)$ such that for each $\mu \in (0,1)$ and each $f \in \mathcal{H}$:

$$\int_{-\infty}^{\infty}\big\|C\big[(H-\lambda-i\mu)^{-1} - (H-\lambda+i\mu)^{-1}\big]E(J)f\big\|^2\,d\lambda \leq c_0\|f\|^2. \tag{6.86}$$

**Proposition 6.12.** *Let $H$ be self-adjoint and let $C$ be a closed operator with $\mathcal{D}(C) \supseteq \mathcal{D}(H)$. Let $J$ be a closed interval such that*

$$\sup_{\lambda\in J,\,0<\mu<1}\|C(H-\lambda-i\mu)^{-1}C^*\| < \infty. \tag{6.87}$$

*Then (6.86) is satisfied with $c_0 = 20\pi\sup_{\lambda\in J,0<\mu<1}\|C(H-\lambda-i\mu)^{-1}C^*\|$, and $C$ is $H$-smooth on $J$.*

REMARK. The operator $C(H-\lambda-i\mu)^{-1}C^*$ is well defined on a dense domain: indeed $\mathcal{D}(C^*)$ is dense [see the Comment (iii) on page 52; in our applications $C$ will even be self-adjoint], and $C(H-\lambda-i\mu)^{-1}$ belongs to $\mathcal{B}(\mathcal{H})$. Thus in condition (6.87) it is assumed in particular that the closure of $C(H-\lambda-i\mu)^{-1}C^*$ belongs to $\mathcal{B}(\mathcal{H})$.

PROOF OF PROPOSITION 6.12. Let again $R(z) = (H-z)^{-1}$. We assume throughout the proof that $\mu > 0$. We can then use the following identity [take into account (4.42)]:

$$\begin{aligned}\mu\int_{-\infty}^{\infty}\|R(\lambda-i\mu)f\|^2 d\lambda &= \int_{-\infty}^{\infty}d\lambda\int_{-\infty}^{\infty}\frac{\mu}{(\sigma-\lambda)^2+\mu^2}\,m_f(d\sigma)\\ &= \int_{-\infty}^{\infty}m_f(d\sigma)\int_{-\infty}^{\infty}\frac{\mu}{(\sigma-\lambda)^2+\mu^2}\,d\lambda\\ &= \int_{-\infty}^{\infty}m_f(d\sigma)\int_{-\infty}^{\infty}\frac{\mu}{x^2+\mu^2}\,dx = \pi\int_{-\infty}^{\infty}m_f(d\sigma) = \pi\|f\|^2. \end{aligned}\tag{6.88}$$

By the first resolvent equation (2.93) we have

$$R(\lambda+i\mu) - R(\lambda-i\mu) = 2i\mu R(\lambda+i\mu)R(\lambda-i\mu),$$

hence [taking into account (6.88) in the last step]

$$\begin{aligned}&\int_{-\infty}^{\infty}\|CE(J)\big[R(\lambda+i\mu)-R(\lambda-i\mu)\big]f\|^2 d\lambda\\ &\qquad= 4\mu^2\int_{-\infty}^{\infty}\|CR(\lambda+i\mu)E(J)R(\lambda-i\mu)f\|^2 d\lambda\\ &\qquad\le 4\mu^2\sup_{\sigma\in\mathbb{R}}\|CR(\sigma+i\mu)E(J)\|^2\int_{-\infty}^{\infty}\|R(\lambda-i\mu)f\|^2 d\lambda\\ &\qquad= 4\pi\mu\sup_{\sigma\in\mathbb{R}}\|CR(\sigma+i\mu)E(J)\|^2\|f\|^2. \end{aligned}\tag{6.89}$$

To estimate the supremum occurring in this formula, we consider separately the case in which $\sigma\in J$ and that in which $\sigma\notin J$.

(i) If $\sigma\in J$, we write (using $\|T^*T\| = \|T\|^2 = \|T^*\|^2$)

$$\begin{aligned}\|CR(\sigma+i\mu)E(J)\|^2 &\le \|CR(\sigma+i\mu)\|^2 \le \sup_{\lambda\in J}\|CR(\lambda+i\mu)\|^2\\ &= \sup_{\lambda\in J}\|[CR(\lambda+i\mu)][CR(\lambda+i\mu)]^*\| = \sup_{\lambda\in J}\|CR(\lambda+i\mu)R(\lambda-i\mu)C^*\|\\ &= \sup_{\lambda\in J}\frac{1}{|2i\mu|}\|C\big[R(\lambda+i\mu)-R(\lambda-i\mu)\big]C^*\|\\ &\le \frac{1}{2\mu}\sup_{\lambda\in J}\big[\|CR(\lambda+i\mu)C^*\| + \|CR(\lambda-i\mu)C^*\|\big]\\ &= \frac{1}{\mu}\sup_{\lambda\in J}\|CR(\lambda+i\mu)C^*\|, \end{aligned}\tag{6.90}$$

because $\|CR(\lambda-i\mu)C^*\| = \|[CR(\lambda-i\mu)C^*]^*\| = \|CR(\lambda+i\mu)C^*\|$.

(ii) If $\sigma \notin J$, let $\sigma_0 \in J$ be such that $\operatorname{dist}(\sigma, J) = |\sigma_0 - \sigma|$ (i.e. $\sigma_0$ is the endpoint of $J$ on the side of $\sigma$). One has

$$\begin{aligned}\|CR(\sigma+i\mu)E(J)\| &\leq \|CR(\sigma_0+i\mu)\|\|(H-\sigma_0-i\mu)R(\sigma+i\mu)E(J)\| \\ &= \|CR(\sigma_0+i\mu)\|\|[I+(\sigma-\sigma_0)R(\sigma+i\mu)]E(J)\|.\end{aligned}$$

As $\sigma_0 \in J$, $\|CR(\sigma_0+i\mu)\|^2$ is majorised by (6.90). On the other hand

$$\|[I+(\sigma-\sigma_0)R(\sigma+i\mu)]E(J)\| = \sup_{\lambda\in J}\left[1+\frac{|\sigma-\sigma_0|}{|\lambda-\sigma-i\mu|}\right] \leq 2.$$

Summing up we have

$$\begin{aligned}&\sup_{\sigma\in\mathbb{R}}\|CR(\sigma+i\mu)E(J)\|^2 \\ &\qquad \leq \sup_{\sigma\in J}\|CR(\sigma+i\mu)E(J)\|^2 + \sup_{\sigma\notin J}\|CR(\sigma+i\mu)E(J)\|^2 \\ &\qquad \leq \frac{1}{\mu}\sup_{\lambda\in J}\|CR(\lambda+i\mu)C^*\|\left(1+2^2\right) = \frac{5}{\mu}\sup_{\lambda\in J}\|CR(\lambda+i\mu)C^*\|.\end{aligned}$$

Insertion into (6.89) gives (note that the factor $20\pi$ is not optimal):

$$\int_{-\infty}^{\infty}\left\|CE(J)\left[R(\lambda+i\mu)-R(\lambda-i\mu)\right]f\right\|^2 d\lambda \leq 20\pi\sup_{\lambda\in J}\|CR(\lambda+i\mu)C^*\|\|f\|^2.$$

By taking into account the hypothesis (6.87), one obtains the validity of (6.86). □

**6.5.2.** Proposition 6.12, combined with Theorem 6.10, gives the following result:

**Proposition 6.13.** *If $H$ and $A$ are two self-adjoint operators satisfying the hypotheses of Theorem 6.10 on a bounded open interval $J$, then for each $\nu > 1/2$ the operator $(I+|A|)^{-\nu}$ is $H$-smooth on each bounded closed subinterval $J_0$ of $J\setminus\sigma_{\mathrm{p}}(H)$.*

If $H = H_0 + V(\vec{Q})$ is a Hamiltonian of the class considered in §6.4.1, we find that $(I+|D|)^{-\nu}$ is $H$-smooth on each interval $[\alpha,\beta] \subset (0,\infty)\setminus\sigma_{\mathrm{p}}(H)$. This can be used to show that certain functions of the position operator are also $H$-smooth [for example $(I+|\vec{Q}|)^{-\nu}$ if $\nu > 1/2$], a fact that will be very useful in scattering theory. Again the case $\nu = 1$ is simple, whereas for $\nu < 1$ one needs an interpolation result. In the following lemma we present the general idea permitting the replacement of $(I+|A|)^{-\nu}$ by another operator C.

**Lemma 6.14.** *Let $H$ be a lower semibounded self-adjoint operator. Let $\nu > 1/2$ and let $J$ be a bounded interval. Let $A$ be a self-adjoint operator and let $C_1, C_2$ be closed operators such that, for some real number $\zeta$ with $-\zeta < \inf \sigma(H)$, one has $C_k(H+\zeta)^{-1/2} \in \mathcal{B}(\mathcal{H})$ [18] and such that the closures of $C_k(H+\zeta)^{-1}(I+|A|)^{\nu}$ belong to $\mathcal{B}(\mathcal{H})$ $(k=1,2)$.*
*(a) Assume that*

$$\sup_{\lambda \in J, 0<\mu<1} \|(I+|A|)^{-\nu}(H-\lambda \pm i\mu)^{-1}(I+|A|)^{-\nu}\| < \infty. \tag{6.91}$$

*Then one has*

$$\sup_{\lambda \in J, 0<\mu<1} \|C_1(H-\lambda \pm i\mu)^{-1}C_2^*\| < \infty. \tag{6.92}$$

*In particular the operators $C_k$ are $H$-smooth on $J$.*
*(b) Assume that* $\text{u-lim}_{\mu \to +0}(I+|A|)^{-\nu}(H-\lambda-i\mu)^{-1}(I+|A|)^{-\nu}$ *exists for each $\lambda \in J$, the convergence being uniform in $\lambda$. Then* $\text{u-lim}_{\mu \to +0} C_1(H-\lambda-i\mu)^{-1}C_2^*$ *exists for each $\lambda \in J$; the convergence is uniform in $\lambda$ and the limit operators depend continuously on $\lambda$ (in norm).*

PROOF. (a) One has $-\zeta \in \rho(H)$ and $H+\zeta \equiv H-(-\zeta) > O$. Hence $(H+\zeta)^{-1/2}$ is a well defined bounded and positive operator. By iterating the first resolvent equation (2.93) one obtains, for $z \in \rho(H)$:

$$\begin{aligned}(H-z)^{-1} &= (H+\zeta)^{-1} + (z+\zeta)(H+\zeta)^{-1}(H-z)^{-1}\\ &= (H+\zeta)^{-1} + (z+\zeta)(H+\zeta)^{-2} + (z+\zeta)^2(H+\zeta)^{-1}(H-z)^{-1}(H+\zeta)^{-1}.\end{aligned}$$

Thus

$$\begin{aligned}C_1(H-\lambda \pm i\mu)^{-1}C_2^* &= C_1(H+\zeta)^{-1}C_2^* + (\lambda \mp i\mu + \zeta)C_1(H+\zeta)^{-2}C_2^*\\ &\quad + (\lambda \mp i\mu+\zeta)^2 C_1(H+\zeta)^{-1}(H-\lambda \pm i\mu)^{-1}(H+\zeta)^{-1}C_2^*.\end{aligned} \tag{6.93}$$

Now $C_1(H+\zeta)^{-1}C_2^* = C_1(H+\zeta)^{-1/2} \cdot [C_2(H+\zeta)^{-1/2}]^* \in \mathcal{B}(\mathcal{H})$, and similarly $C_1(H+\zeta)^{-2}C_2^* = C_1(H+\zeta)^{-1/2} \cdot (H+\zeta)^{-1} \cdot [C_2(H+\zeta)^{-1/2}]^* \in \mathcal{B}(\mathcal{H})$. Consequently insertion of the first two terms of (6.93) into (6.92) gives a finite contribution to the supremum, because $J$ is bounded so that $|\lambda \mp i\mu + \zeta| \le c < \infty$ for $\lambda \in J$ and all $|\mu| \le 1$. Thus it suffices to show that

$$\sup_{\lambda \in J, 0<\mu<1} \|C_1(H+\zeta)^{-1}(H-\lambda \pm i\mu)^{-1}(H+\zeta)^{-1}C_2^*\| < \infty.$$

Let us write

$$\begin{aligned}&C_1(H+\zeta)^{-1}(H-\lambda \pm i\mu)^{-1}(H+\zeta)^{-1}C_2^*\\ &\qquad = C_1(H+\zeta)^{-1}(I+|A|)^{\nu} \times (I+|A|)^{-\nu}(H-\lambda \pm i\mu)^{-1}(I+|A|)^{-\nu}\\ &\qquad\quad \times (I+|A|)^{\nu}(H+\zeta)^{-1}C_2^*.\end{aligned} \tag{6.94}$$

[18] It suffices to know that $\mathcal{D}(C_k) \supseteq \mathcal{D}(|H|^{1/2})$ [by the Closed Graph Theorem]. One has $\inf \sigma(H) \equiv \inf\{\lambda \mid \lambda \in \sigma(H)\} > -\infty$.

The first and the third factor on the right-hand side are bounded by assumption [the third one is the adjoint of $C_2(H+\zeta)^{-1}(I+|A|)^\nu$ because $\zeta$ is real]. Hence

$$\begin{aligned}\sup_{\lambda\in J, 0<\mu<1} &\|C_1(H+\zeta)^{-1}(H-\lambda\pm i\mu)^{-1}(H+\zeta)^{-1}C_2^*\| \\ &\le \|C_1(H+\zeta)^{-1}(I+|A|)^\nu\|\|C_2(H+\zeta)^{-1}(I+|A|)^\nu\| \\ &\quad\times \sup_{\lambda\in J,\mu>0}\|(I+|A|)^{-\nu}(H-\lambda\pm i\mu)^{-1}(I+|A|)^{-\nu}\| < \infty.\end{aligned}$$

(b) This is easily seen by taking into account Eqs. (6.93) and (6.94). □

**Remark 6.15.** The preceding proof shows that one has the following inequality:

$$\begin{aligned}\sup_{\lambda\in J,0<\mu<1} &\|C_1(H-\lambda\pm i\mu)^{-1}C_2^*\| \le c\|C_1(H+\zeta)^{-1/2}\|\|C_2(H+\zeta)^{-1/2}\| \\ &+ c^2\|C_1(H+\zeta)^{-1}(I+|A|)^\nu\|\|C_2(H+\zeta)^{-1}(I+|A|)^\nu\| \\ &\times \sup_{\lambda\in J,\mu>0}\|(I+|A|)^{-\nu}(H-\lambda\pm i\mu)^{-1}(I+|A|)^{-\nu}\|, \end{aligned}\tag{6.95}$$

where $c$ is a constant depending only on $J$ and $\zeta$.

We now apply the preceding lemma to the case in which $H = H_0 + V(\vec{Q})$, where the function $V$ satisfies the condition (C) of page 288. We shall take $C_1 = C_2 = W(\vec{Q})$, where $W$ is a function defined on $\mathbb{R}^n$, and $A = D = (2\vec{Q}\cdot\vec{P} - inI)/4$.

**Proposition 6.16.** *Let $H = H_0 + V(\vec{Q})$, where $V$ satisfies condition (C) of page 288. (a) Let $W\colon \mathbb{R}^n \to \mathbb{C}$ be a function of the form $W(\vec{x}) = [w_1(\vec{x}) + w_2(\vec{x})](1+|\vec{x}|)^{-\nu}$ with $w_1 \in L^\infty(\mathbb{R}^n)$ and $w_2 \in L^p(\mathbb{R}^n)$ for some $p$ satisfying $p\in(n,\infty)$ and $p\ge 2$, and with $\nu > 1/2$. Then $W(\vec{Q})$ is $H$-smooth on each bounded closed interval $J = [\alpha,\beta]$ in $(0,\infty)\setminus\sigma_{\rm p}(H)$. Moreover, for each such interval there exists a constant $c_J$ such that for all $f\in\mathcal{H}$:*

$$\begin{aligned}\int_{-\infty}^{\infty} &\|W(\vec{Q})\big[(H-\lambda-i\mu)^{-1} - (H-\lambda+i\mu)^{-1}\big]E(J)f\|^2 d\lambda \\ &\le c_J\big[\|w_1\|_\infty + \|w_2\|_p\big]^2\|f\|^2.\end{aligned}\tag{6.96}$$

*(b) Let $U\colon\mathbb{R}^n\to\mathbb{C}$ be a second function satisfying the hypotheses imposed on $W$ in (a), and let $J$ be a bounded closed interval as in (a). Then, for each $\lambda\in J$, the limit $\text{u-lim}_{\mu\to+0} W(\vec{Q})(H-\lambda-i\mu)^{-1}U(\vec{Q})^*$ exists; the convergence is uniform in $\lambda\in J$, and the limit operators depend continuously on $\lambda$ (in norm).*

PROOF. We treat the case $\nu = 1$. This covers all cases in which $\nu\ge 1$, whereas for $\nu\in(1/2,1)$ one has to have recourse to interpolation theory (this will be explained after the present proof and in the Appendix to this chapter). We fix $\zeta > 1$ so large that the interval $(-\infty,-\zeta]$ belongs to $\rho(H)$ (recall that $H$ is lower semibounded).

(a) We use the fact that $[w_1(\vec{Q}) + w_2(\vec{Q})](H_0+\zeta)^{-1/2}$ belongs to $\mathcal{B}(\mathcal{H})$. For the term with $w_1$ this is evident; for that with $w_2$ one applies Proposition 2.33 by observing that the function $\vec{k} \mapsto (\vec{k}^2+\zeta)^{-1/2}$ belongs to $L^p(\mathbb{R}^n)$ if $p>n$. Thus, by observing that $\|(H_0+\zeta)^{-1/2}\|<1$ because $\zeta>1$, we have for a certain constant $c_1$:

$$\begin{aligned}\|[w_1(\vec{Q}) + w_2(\vec{Q})](H_0+\zeta)^{-1/2}\| &\leq \|w_1(\vec{Q})(H_0+\zeta)^{-1/2}\| \\ &+ \|w_2(\vec{Q})(H_0+\zeta)^{-1/2}\| \leq \|w_1\|_\infty + c_1\|w_2\|_p. \end{aligned} \tag{6.97}$$

(i) The inequality (6.97) implies that

$$\begin{aligned}&\|[w_1(\vec{Q})+w_2(\vec{Q})](H+\zeta)^{-1/2}\| \leq \|[w_1(\vec{Q})+w_2(\vec{Q})](H_0+\zeta)^{-1/2}\| \\ &\times \|(H_0+\zeta)^{1/2}(H+\zeta)^{-1/2}\| \leq \big[\|w_1\|_\infty + c_1\|w_2\|_p\big]\|(H_0+\zeta)^{1/2}(H+\zeta)^{-1/2}\|.\end{aligned}$$

By using the fact that $(H_0+\zeta)^{1/2}(H+\zeta)^{-1/2}\in\mathcal{B}(\mathcal{H})$ (see below), one deduces from the preceding inequality that $(I+|\vec{Q}|)^\nu W(\vec{Q})(H+\zeta)^{-1/2}\in\mathcal{B}(\mathcal{H})$ and that, for some constant $c_2$:

$$\|(I+|\vec{Q}|)^\nu W(\vec{Q})(H+\zeta)^{-1/2}\| \leq c_2\big[\|w_1\|_\infty + \|w_2\|_p\big]. \tag{6.98}$$

The inclusion $(H_0+\zeta)^{1/2}(H+\zeta)^{-1/2}\in\mathcal{B}(\mathcal{H})$ [or $\mathcal{D}(|H|^{1/2})\subseteq\mathcal{D}(H_0^{1/2})$] is obtained by interpolation from the fact that $\mathcal{D}(H)=\mathcal{D}(H_0)$, see Example 6.30. If $w_2=0$, the validity of (6.98) is evident; in this case interpolation theory is not necessary.

(ii) From now on we assume that $\nu=1$. Let us first show that the closure of the operator $W(\vec{Q})(H+\zeta)^{-1}D$ belongs to $\mathcal{B}(\mathcal{H})$. In fact the unbounded factors $Q_j$ occurring in $D$ can be compensated by the decay of $W$, and the unbounded factors $P_j$ by the resolvent of $H$. To make this precise we observe that, by (6.19), we have on $\mathcal{S}(\mathbb{R}^n)$:

$$2W(\vec{Q})(H+\zeta)^{-1}D = \sum_{j=1}^n W(\vec{Q})(H+\zeta)^{-1}Q_jP_j - \frac{in}{2}W(\vec{Q})(H+\zeta)^{-1}. \tag{6.99}$$

By virtue of (6.98) one has

$$\|W(\vec{Q})(H+\zeta)^{-1}\| \leq c_2\big[\|w_1\|_\infty + \|w_2\|_p\big]\|(H+\zeta)^{-1/2}\| \leq c_3\big[\|w_1\|_\infty + \|w_2\|_p\big].$$

So the last term in (6.99) is bounded. The same conclusion holds for each term in the sum on the right-hand side of (6.99). Indeed we may write [using (6.117)]

$$\begin{aligned}W(\vec{Q})(H+\zeta)^{-1}Q_jP_j &= Q_jW(\vec{Q})(H+\zeta)^{-1}P_j \\ &+ 2iW(\vec{Q})(H+\zeta)^{-1}P_j(H+\zeta)^{-1}P_j.\end{aligned} \tag{6.100}$$

The operator $W(\vec{Q})(H+\zeta)^{-1}$ belongs to $\mathcal{B}(\mathcal{H})$ because $W(\vec{Q})$ is $H$-compact, and the closure of $P_j(H+\zeta)^{-1}P_j = [P_j(H+\zeta)^{-1/2}][P_j(H+\zeta)^{-1/2}]^*$ belongs to $\mathcal{B}(\mathcal{H})$ because $\|P_j(H+\zeta)^{-1/2}\| \leq \|(H_0+\zeta)^{1/2}(H+\zeta)^{-1/2}\| < \infty$. This establishes the boundedness of the last term in (6.100). That of the first term on the right-hand side of (6.100) is obtained by noticing that $Q_jW(\vec{Q})(H+\zeta)^{-1/2}\in\mathcal{B}(\mathcal{H})$ [by (6.98) with $\nu=1$] and by remembering that $P_j(H+\zeta)^{-1/2}\in\mathcal{B}(\mathcal{H})$. By using the bound

(6.98) for $\|Q_j W(\vec{Q})(H+\zeta)^{-1/2}\|$ and $\|W(\vec{Q})(H+\zeta)^{-1/2}\|$, we get the following inequality from (6.100): $\|W(\vec{Q})(H+\zeta)^{-1}Q_j P_j\| \leq c_4\big[\|w_1\|_\infty + \|w_2\|_p\big]$. We have thus shown that

$$\|W(\vec{Q})(H+\zeta)^{-1}D\| \leq c_5\big[\|w_1\|_\infty + \|w_2\|_p\big] \qquad \text{for some constant } c_5.$$

(iii) Let $f \in \mathcal{D}(|D|) \equiv \mathcal{D}(D)$ and write $f = f_+ + f_-$ with $f_+ = E^D([0,\infty))f$ and $f_- = E^D((-\infty,0))f$. Then $f_\pm \in \mathcal{D}(D)$, $|D|f = Df_+ - Df_-$ and

$$\begin{aligned}
&\|W(\vec{Q})(H+\zeta)^{-1}(I+|D|)f\| \\
&\leq \|W(\vec{Q})(H+\zeta)^{-1}f\| + \|W(\vec{Q})(H+\zeta)^{-1}Df_+\| + \|W(\vec{Q})(H+\zeta)^{-1}Df_-\| \\
&\leq \Big[\|W(\vec{Q})(H+\zeta)^{-1}\| + 2\|W(\vec{Q})(H+\zeta)^{-1}D\|\Big]\|f\|.
\end{aligned}$$

Hence the closure of $W(\vec{Q})(H+\zeta)^{-1}(I+|D|)$ belongs to $\mathcal{B}(\mathcal{H})$, with

$$\|W(\vec{Q})(H+\zeta)^{-1}(I+|D|)\| \leq c_6\big[\|w_1\|_\infty + \|w_2\|_p\big]. \tag{6.101}$$

(iv) The above results show that the operator $C = W(\vec{Q})$ satisfies the hypotheses of Lemma 6.14(a) [with $\nu = 1$; take $C_1 = C_2 = W(\vec{Q})$ in Lemma 6.14]. Thus $W(\vec{Q})$ is $H$-smooth on each bounded closed subinterval of $(0,\infty)\backslash\sigma_{\mathrm{p}}(H)$ [because (6.91) is satisfied on each interval of this type, with $A = D$]. To obtain (6.96) one introduces the estimates (6.98) and (6.101) into (6.95), which gives

$$\sup_{\lambda\in J, 0<\mu<1} \|W(\vec{Q})(H-\lambda\pm i\mu)^{-1}W(\vec{Q})^*\| \leq c_7\big[\|w_1\|_\infty + \|w_2\|_p\big]^2.$$

By Proposition 6.12 the inequality (6.96) is satisfied with $c_J = 20\pi c_7$.

(b) This follows from Lemma 6.14(b) by taking into account the observations made above [recall that the hypothesis of Lemma 6.14(b), with $A = D$, is satisfied for each bounded closed subinterval of $(0,\infty)\backslash\sigma_{\mathrm{p}}(H)$]. □

We have thus proved Proposition 6.16 for $\nu = 1$, with the exception of the fact that $(H_0+\zeta)^{1/2}(H+\zeta)^{-1/2} \in \mathcal{B}(\mathcal{H})$. The verification of this fact and the proof of the proposition for $\nu < 1$ involve an interesting interpolation theorem that we now state; its proof and various applications (Examples 6.30 - 6.33) are presented in the Appendix to this chapter.

**Proposition 6.17.** *Let $R$ and $S$ be positive invertible self-adjoint operators with $R \in \mathcal{B}(\mathcal{H})$, and let $X \in \mathcal{B}(\mathcal{H})$. Assume that the closure of $RXS$ belongs to $\mathcal{B}(\mathcal{H})$. Then, for each $\nu \in [0,1]$, the operator $X^*$ maps $\mathcal{R}(R^\nu)$ into $\mathcal{D}(S^\nu)$. One has $S^\nu X^* R^\nu \in \mathcal{B}(\mathcal{H})$, and the closure of the operator $R^\nu X S^\nu$ also belongs to $\mathcal{B}(\mathcal{H})$. Moreover*

$$\|R^\nu X S^\nu\| \leq \|X\|^{1-\nu}\|RXS\|^\nu. \tag{6.102}$$

We use this proposition to prove the result of Proposition 6.16 for $\nu \in (1/2, 1)$. So we assume that $W(\vec{x}) = [w_1(\vec{x}) + w_2(\vec{x})](1+|\vec{x}|)^{-\nu}$ with $\nu \in (1/2, 1)$. We apply

Proposition 6.17 with $X = [w_1(\vec{Q}) + w_2(\vec{Q})](H+\zeta)^{-1}$, $R = (I + |\vec{Q}|)^{-1}$ and $S = I + |D|$. We have seen in the proof of Proposition 6.16 that $X \in \mathcal{B}(\mathcal{H})$ and $(I+|\vec{Q}|)^{-1}X(I+|D|) \in \mathcal{B}(\mathcal{H})$, and also that $\|X\| + \|(I+|\vec{Q}|)^{-1}X(I+|D|)\| \leq c[\|w_1\|_\infty + \|w_2\|_p]$. Thus (6.102) shows that

$$\|R^\nu X S^\nu\| \equiv \|W(\vec{Q})(H+\zeta)^{-1}(I+|D|)^\nu\| \leq c\big[\|w_1\|_\infty + \|w_2\|_p\big].$$

This inequality generalises (6.101) for values of $\nu$ belonging to $(1/2, 1)$, and the reasoning that follows (6.101) is again applicable.

## 6.6 Higher order resolvent estimates

We present here a generalisation of Theorem 6.3 and of its consequences for powers of the resolvent $(T - \lambda - i\mu)^{-1}$. This has important applications in scattering theory, as seen for example in Section 7.2. To obtain a bound on $\langle f, (T-\lambda-i\mu)^{-N} f\rangle$ for suitable vectors $f$, one assumes that the multiple commutators (up to order $N$) of $T$ with $iA$ belong to $\mathcal{B}(\mathcal{H})$ (in the sense explained before Theorem 6.3). We set $B_k = [[\cdots[T, iA], iA], \ldots, iA]$ (k commutators). Observe that $B_1$ coincides with the operator $B$ used in Section 6.2 and that each $B_k$ is formally self-adjoint. We then have the following generalisation of Theorem 6.3:

**Theorem 6.18.** *Let $T$ and $A$ be self-adjoint operators with $T \in \mathcal{B}(\mathcal{H})$, and let $N \in \{2, 3, 4, \ldots\}$. Assume that each of the operators $B_k, k = 1, \ldots, N+1$, belongs to $\mathcal{B}(\mathcal{H})$ and that a strict Mourre inequality* (6.34) *is satisfied on some open interval $J$. Let $J_0$ be a bounded closed subinterval of $J$ and $\nu > N - (1/2)$. Then the norm limits $\text{u-lim}_{\mu\to+0}(I+|A|)^{-\nu}(T-\lambda\pm i\mu)^{-N}(I+|A|)^{-\nu}$ exist uniformly in $\lambda \in J_0$, and*

$$\sup_{\lambda\in J_0, \mu>0} \|(I+|A|)^{-\nu}(T-\lambda\pm i\mu)^{-N}(I+|A|)^{-\nu}\| < \infty. \tag{6.103}$$

PROOF. The proof follows the lines of that of Theorem 6.3. We just indicate the adjustments that are necessary. First the term $\|[B, A]\|$ in the constant $c_1$ in (6.36) should be replaced by $\sum_{k=2}^N \|B_k\| + \big[\sum_{k=1}^N \|B_k\|\big]^2$.

(1) Definition and properties of $G_\varepsilon$

- Points (i) and (ii). Here the appropriate definition of $N_\varepsilon^\pm$ is as follows:

$$N_\varepsilon^\pm = T - \lambda \pm i\mu + Z^\pm(\varepsilon) \quad \text{with} \quad Z^\pm(\varepsilon) = \sum_{k=1}^N \big[(\pm i\varepsilon)^k B_k/k!\big].$$

Hence $G_\varepsilon \equiv G_\varepsilon(\lambda + i\mu) = [T - \lambda - i\mu + Z^-(\varepsilon)]^{-1}$, with $\varepsilon \in [0, 1]$. The basic inequality (6.41) remains true, the only change in its verification occurs in the derivation of (6.39) where $\mp i\varepsilon B$ must be replaced by $-Z^\pm(\varepsilon)$.

• Point (iii). Here we have

$$\langle g, (\varepsilon B + \mu) g\rangle = \pm\Im\langle g, N_\varepsilon^\pm g\rangle - \sum_{k=3,5,7,\ldots}^{N} (-1)^{(k+1)/2} \frac{\varepsilon^k}{k!}\langle g, B_k g\rangle$$

(only odd values of $k$ contribute to the preceding sum, because the terms in $Z^\pm(\varepsilon)$ containing even powers of $\varepsilon$ are self-adjoint and thus do not occur in $\Im N_\varepsilon^\pm$). If $N > 2$ this leads to an additional term on the right-hand side of (6.43) of the form $c\varepsilon^3\|g\|^2$ for some constant $c < c_1$. By absorbing this additional term in the left-hand side, it follows that, for sufficiently small $\varepsilon$, the inequality (6.43) remains true provided that its left-hand side is replaced by $(a\varepsilon + \mu)\|g\|^2/4$. Then the bound on $\|G_\varepsilon(\lambda + i\mu)\|$ in (6.44) obtains an additional factor 2 (and this unimportant factor should be taken into account whenever this bound is used further on). Also the value of $\varepsilon_0$ may be smaller than in the earlier proof.

• Point (iv). This remains unchanged, in particular (6.45) holds.

• Point (v). This is handled somewhat differently. We set $G_\varepsilon^{(1)} = T - \lambda - i\mu - i\varepsilon B_1$ and then have [using the equation $X^{-1} - Y^{-1} = X^{-1}(Y - X)Y^{-1}$]:

$$G_\varepsilon = G_\varepsilon^{(1)} - \sum_{k=2}^{N} \frac{(-i\varepsilon)^k}{k!} G_\varepsilon B_k G_\varepsilon^{(1)} = G_\varepsilon^{(1)} - \sum_{k=2}^{N} \frac{(-i\varepsilon)^k}{k!} G_\varepsilon^{(1)} B_k G_\varepsilon. \tag{6.104}$$

Now $G_\varepsilon^{(1)}$ is just the operator $G_\varepsilon$ used in Section 6.2, so that by (6.48), (6.49) and (6.35): $\|G_\varepsilon^{(1)}(I + |A|)^{-1}\| + \|(I + |A|)^{-1}G_\varepsilon^{(1)}\| \le c\varepsilon^{-1/2}$ for some constant $c$ and all $\varepsilon \in (0, \varepsilon_0)$ (with $c$ independent of $\lambda$ and $\mu$). Furthermore $\varepsilon^k\|G_\varepsilon\| \le 8/a$ if $k \ge 1$ and $0 < \varepsilon < \varepsilon_0$ [see (6.44)], and $\|B_k\| < c_1$. By using these inequalities one obtains from (6.104) that

$$\|G_\varepsilon(I + |A|)^{-1}\| + \|(I + |A|)^{-1}G_\varepsilon\| \le c\varepsilon^{-1/2} \tag{6.105}$$

for some constant $c$, all $\varepsilon \in (0, \varepsilon_0)$, $\lambda \in J_0$ and $\mu \in (0, 1)$.

(2) <u>The differential inequality</u>

• Point (i). Using the identity $a^k - b^k = (a - b)(a^{k-1} + a^{k-2}b + \cdots + ab^{k-2} + b^{k-1})$, with $a = -i\tau, b = -i\varepsilon$, a calculation as in (6.50) shows that

$$G_\varepsilon' \equiv \frac{d}{d\varepsilon} G_\varepsilon = i \sum_{k=1}^{N} \frac{(-i\varepsilon)^{k-1}}{(k-1)!} G_\varepsilon B_k G_\varepsilon. \tag{6.106}$$

On the other hand, the analogue of Eq. (6.51) is here as follows: [19]

$$[A, G_\varepsilon] = -iG_\varepsilon B_1 G_\varepsilon - i \sum_{k=1}^{N} \frac{(-i\varepsilon)^k}{k!} G_\varepsilon B_{k+1} G_\varepsilon = -i \sum_{k=0}^{N} \frac{(-i\varepsilon)^k}{k!} G_\varepsilon B_{k+1} G_\varepsilon.$$

Comparison with (6.106) leads to

$$G_\varepsilon'(\lambda + i\mu) = [G_\varepsilon(\lambda + i\mu), A] - i\frac{(-i\varepsilon)^N}{N!} G_\varepsilon(\lambda + i\mu) B_{N+1} G_\varepsilon(\lambda + i\mu). \tag{6.107}$$

[19] We observe that, as in §6.3.2, the operators $G_\varepsilon$ map $\mathcal{D}(A)$ into $\mathcal{D}(A)$.

We also need a formula for $(d/d\varepsilon)[G_\varepsilon]^N$. By the Leibniz rule this is as follows:

$$\begin{aligned}\frac{d}{d\varepsilon}(G_\varepsilon)^N &= \sum_{k=0}^{N-1}(G_\varepsilon)^k G_\varepsilon'(G_\varepsilon)^{N-k-1}\\ &= \sum_{k=0}^{N-1}(G_\varepsilon)^k[G_\varepsilon, A](G_\varepsilon)^{N-k-1} - i\frac{(-i\varepsilon)^N}{N!}\sum_{k=0}^{N-1}(G_\varepsilon)^{k+1}B_{N+1}(G_\varepsilon)^{N-k}\\ &= [(G_\varepsilon)^N, A] - i\frac{(-i\varepsilon)^N}{N!}\sum_{k=0}^{N-1}(G_\varepsilon)^{k+1}B_{N+1}(G_\varepsilon)^{N-k}. \qquad (6.108)\end{aligned}$$

• Point (ii). We replace $X_\varepsilon$ simply by $I$ and set $\mathcal{Z} = (I + |A|)^{-1}$. We deduce a differential inequality for the $\mathcal{B}(\mathcal{H})$-valued function $\varepsilon \mapsto \Phi_\varepsilon := \mathcal{Z}^\nu(G_\varepsilon)^N\mathcal{Z}^\nu$. For this we observe that

$$\begin{aligned}\Phi_\varepsilon' &= \mathcal{Z}^\nu\Big[\frac{d}{d\varepsilon}(G_\varepsilon)^N\Big]\mathcal{Z}^\nu\\ &= \mathcal{Z}^\nu[(G_\varepsilon)^N, A]\mathcal{Z}^\nu - i\frac{(-i\varepsilon)^N}{N!}\sum_{k=0}^{N-1}\mathcal{Z}^\nu(G_\varepsilon)^{k+1}B_{N+1}(G_\varepsilon)^{N-k}\mathcal{Z}^\nu. \qquad (6.109)\end{aligned}$$

Now, since $\|A\mathcal{Z}\| \le 1$:

$$\begin{aligned}\|\mathcal{Z}^\nu[(G_\varepsilon)^N, A]\mathcal{Z}^\nu\| &\le \|\mathcal{Z}^\nu(G_\varepsilon)^N\mathcal{Z}^{\nu-1}\cdot A\mathcal{Z}\| + \|A\mathcal{Z}\cdot\mathcal{Z}^{\nu-1}(G_\varepsilon)^N\mathcal{Z}^\nu\|\\ &\le \|\mathcal{Z}^\nu(G_\varepsilon)^N\mathcal{Z}^{\nu-1}\| + \|\mathcal{Z}^{\nu-1}(G_\varepsilon)^N\mathcal{Z}^\nu\|. \qquad (6.110)\end{aligned}$$

The norms on the last line are treated as follows by an interpolation argument. In Example 6.33 set $B = \mathcal{Z}^{-1}$ and $Y = (G_\varepsilon)^N$. Then (6.134), for $\kappa = \nu$, leads to the following bound on the first norm in (6.110):

$$\|\mathcal{Z}^\nu(G_\varepsilon)^N\mathcal{Z}^{\nu-1}\| \le \|\mathcal{Z}^\nu(G_\varepsilon)^N\mathcal{Z}^\nu\|^{1-(1/\nu)}\|\mathcal{Z}^\nu(G_\varepsilon)^N\|^{1/\nu}.$$

Now, by (6.105) and (6.44), we have for $j \ge 1$:

$$\|\mathcal{Z}^\nu(G_\varepsilon)^j\| \le \|\mathcal{Z}^\nu G_\varepsilon\|\|G_\varepsilon\|^{j-1} \le c_j\varepsilon^{-1/2}\varepsilon^{-(j-1)} = c_j\varepsilon^{-[j-(1/2)]}$$

for some constant $c_j$. Consequently, taking $j = N$:

$$\|\mathcal{Z}^\nu(G_\varepsilon)^N\mathcal{Z}^{\nu-1}\| \le c_N\varepsilon^{-[N-(1/2)]/\nu}\|\mathcal{Z}^\nu(G_\varepsilon)^N\mathcal{Z}^\nu\|^{1-(1/\nu)} \equiv c_N\varepsilon^{-\beta}\|\Phi_\varepsilon\|^\delta,$$

with $\beta = [N - (1/2)]/\nu$ and $\delta = 1 - (1/\nu)$; observe that $\beta, \delta \in (0, 1)$ because $\nu > N - (1/2)$. The same bound holds for the second norm on the last line of (6.110) (use the same argument with $G_\varepsilon$ replaced by $G_\varepsilon^*$).

We next estimate the last term in (6.109). Its norm is majorised by

$$\frac{\varepsilon^N}{N!}\sum_{k=0}^{N-1}\|\mathcal{Z}^\nu G_\varepsilon\|\|G_\varepsilon\|^k\|B_{N+1}\|\|G_\varepsilon\|^{N-k-1}\|G_\varepsilon\mathcal{Z}^\nu\|$$

$$\le c\frac{\varepsilon^N}{N!}\sum_{k=0}^{N-1}\varepsilon^{-1/2}\varepsilon^{-k}\varepsilon^{-N+k+1}\varepsilon^{-1/2} \le c$$

for some constant $c$. Thus (6.109) leads to the following differential inequality:

$$\|\Phi_\varepsilon'\| \le c\big[\varepsilon^{-\beta}\|\Phi_\varepsilon\|^\delta + 1\big] \tag{6.111}$$

for some constant $c$, with $\beta \in (0,1)$ and $\delta \in (0,1)$.

(3) Boundary values

The preceding differential inequality is of the form (6.24), except that the exponent $\delta$ of the fractional power of $\|\Phi_\varepsilon\|$ may be different from $1/2$ (but less than 1). By a straightforward adjustment of the proof of (6.25) given in §6.2.1 one finds that (Problem 6.3)

$$\|\Phi_\varepsilon'\| \le c\varepsilon^{-\beta}$$

for some constant $c$ that is independent of $\lambda$ and $\mu$. As before, the preceding inequality implies the existence of $\text{u-lim}_{\mu\to+0}\,\mathcal{Z}^\nu(T-\lambda-i\mu)^{-N}\mathcal{Z}^\nu$, and one obtains a bound of the form $\|\mathcal{Z}^\nu(T-\lambda-i\mu)^{-N}\mathcal{Z}^\nu\| \le const.$ for all $\lambda \in J_0$ and all $\mu \in [0,1]$. □

**Corollary 6.19.** *Under the hypothesis of Proposition 6.18 [in particular $\nu > N - (1/2)$], the operator-valued function $\lambda \mapsto (I+|A|)^{-\nu}(T-\lambda-i0)^{-1}(I+|A|)^{-\nu}$ is $N-1$ times continuously differentiable in norm on the interval $J$, and for $k = 1,2,\dots,N-1$ one has*

$$\begin{aligned}\Big(\frac{d}{d\lambda}\Big)^k &(I+|A|)^{-\nu}(T-\lambda-i0)^{-1}(I+|A|)^{-\nu}\\ &= k!\,(I+|A|)^{-\nu}(T-\lambda-i0)^{-(k+1)}(I+|A|)^{-\nu}.\end{aligned} \tag{6.112}$$

*Here $(I+|A|)^{-\nu}(T-\lambda-i0)^{-\kappa}(I+|A|)^{-\nu}$ is a formal notation for the operator $\text{u-lim}_{\varepsilon\to+0}(I+|A|)^{-\nu}G_\varepsilon(\lambda+i0)^\kappa(I+|A|)^{-\nu}$, with $G_\varepsilon(z)$ defined in the preceding proof.*

PROOF. For $\varepsilon \ne 0$, the operators $\mathcal{Z}^\nu G_\varepsilon(\lambda+i0)\mathcal{Z}^\nu$ are infinitely differentiable (in norm) with respect to $\lambda$, with $(d/d\lambda)^k\mathcal{Z}^\nu G_\varepsilon(\lambda+i0)\mathcal{Z}^\nu = k!\,\mathcal{Z}^\nu G_\varepsilon(\lambda+i0)^{k+1}\mathcal{Z}^\nu$. As $\varepsilon \to +0$, the derivative $k\mathcal{Z}^\nu G_\varepsilon(\lambda+i0)^{k+1}\mathcal{Z}^\nu$ of $\mathcal{Z}^\nu G_\varepsilon(\lambda+i0)^k\mathcal{Z}^\nu$ converges in norm to $k\mathcal{Z}^\nu(T-\lambda-i0)^{-(k+1)}\mathcal{Z}^\nu$ on each closed bounded subinterval $J_0$ of $J$, provided that $k \in \{1,2,\dots,N-1\}$. By a standard result from analysis[20], this implies that $\mathcal{Z}^\nu(T-\lambda-i0)^{-k}\mathcal{Z}^\nu$ is differentiable with respect to $\lambda$ (in the interior of $J_0$) and that its derivative is $k\mathcal{Z}^\nu(T-\lambda-i0)^{-(k+1)}\mathcal{Z}^\nu$. Since $J_0$ can be any bounded closed subinterval of $J$, one obtains in particular (6.112). □

By proceeding as in Section 6.4, one can extend the results of Theorem 6.18 and Corollary 6.19 to unbounded operators $H$ satisfying $\rho(H) \cap \mathbb{R} \ne \varnothing$. As in §6.4.2, this is done by setting $T = (\lambda_0 - H)^{-1}$, with $\lambda_0 \in \rho(H)$. One has to know that the multiple commutators of $(\lambda_0 - H)^{-1}$ with $A$ (up to order $N+1$) belong to $\mathcal{B}(\mathcal{H})$. For $H$ of the form $H_0 + V(\vec{Q})$, this is ensured by suitably generalising the condition (C) of page 288. A convenient condition, sufficient for our applications in Chapter 7, is to assume that $V$ is of class $C^{N+1}$ and that $r^k(\partial^k V/\partial r^k)$ belongs to $L^\infty_\varepsilon(\mathbb{R}^n)$ for

[20] Interchange of limits with derivatives; see for example §9 in Chapter V of [L].

$k = 0, 1, 2, \ldots, N+1$. When repeating the arguments of §6.4.2, one arrives at the following generalisation of (6.82):

$$(I+|A|)^{-\nu}(H-z)^{-k}(I+|A|)^{-\nu} = (\lambda_0 - z)^{-k}(I+|A|)^{-\nu}\big[(T-(\lambda_0-z)^{-1}\big]^{-k}(I+|A|)^{-\nu} \times (I+|A|)^{\nu}T^k(I+|A|)^{-\nu}.$$

Here one has to know that $(I+|A|)^{\nu}T^k(I+|A|)^{-\nu} \in \mathcal{B}(\mathcal{H})$ for $k \le N$ and $\nu > N-(1/2)$. This is easily obtained by interpolation (Example 6.32) by using the fact that $T$ maps $\mathcal{D}(A^N)$ into itself (see Problem 6.4).

Under suitable assumptions it is again possible to replace $(I+|A|)^{-\nu}$ on the left-hand side of (6.112) by some other operator. In particular, if $H$ is a Schrödinger operator, one can control the derivatives of $(I+|Q|)^{-\nu}(H-\lambda \pm i0)^{-1}(I+|Q|)^{-\nu}$ and similar expressions if the potential decays sufficiently rapidly, as explained in the following two propositions.

**Proposition 6.20.** *Let $k \in \mathbb{N}$. Let $U, W\colon \mathbb{R}^n \to \mathbb{C}$ be two functions of the form $[w_1(\vec{x}) + w_2(\vec{x})](1+|\vec{x}|)^{-\nu}$ with $w_1 \in L^\infty(\mathbb{R}^n)$ and $w_2 \in L^p(\mathbb{R}^n)$ for some $p$ satisfying $p \in (n,\infty)$ and $p \ge 2$, and with $\nu > k+1/2$. Then the operator-valued function $\lambda \mapsto W(\vec{Q})(H_0 - \lambda - i0)^{-1}U(\vec{Q})^*$ is $k$ times continuously norm differentiable on $(0,\infty)$.*

PROOF. Given an interval $J = (\alpha, \beta)$ with $0 < \alpha < \beta < \infty$, we choose a function $\phi \in C_0^\infty(\mathbb{R})$ such that $\phi(\mu) = 1$ for all $\mu$ in $(\alpha - \delta, \beta + \delta)$ for some $\delta > 0$. If $\Im z \neq 0$, we have [setting $U = U(\vec{Q})$ and $W = W(\vec{Q})$]

$$W(H_0 - z)^{-1}U^* = W\phi(H_0)(H_0-z)^{-1}\phi(H_0)U^* + W[I-\phi(H_0)^2](H_0-z)^{-1}U^*. \tag{6.113}$$

The second term on the right-hand side may be written as

$$W(H_0+1)^{-1/2}\cdot[I-\phi(H_0)^2][I+(1+z)(H_0-z)^{-1}]\cdot(H_0+1)^{-1/2}U^*.$$

Since $W(H_0+1)^{-1/2}$ and $U(H_0+1)^{-1/2}$ are bounded operators (see the proof of Proposition 6.16) and since $1-\phi(\mu)^2 = 0$ for $\mu \in (\alpha-\delta, \beta+\delta)$, this term is infinitely differentiable in norm in the domain $\{z \in \mathbb{C} \mid \Re z \in J\}$.

To treat the first term on the right-hand side of (6.113), we express it in the form

$$W\phi(H_0)(I+D^2)^{\nu/2}\cdot(I+D^2)^{-\nu/2}(H_0-z)^{-1}(I+D^2)^{-\nu/2}\cdot(I+D^2)^{\nu/2}\phi(H_0)U^*.$$

The function $\lambda \mapsto (I+D^2)^{-\nu/2}(H_0-\lambda-i0)^{-1}(I+D^2)^{-\nu/2}$ is $k$ times continuously norm differentiable on $J$ by Corollary 6.19, and the closures of $(I+D^2)^{\nu/2}\phi(H_0)U^*$ and $W\phi(H_0)(I+D^2)^{\nu/2}$ are bounded.[21] □

[21] We have $(I+D^2)^{\nu/2}\phi(H_0)U^* = (I+D^2)^{\nu/2}\varphi(H_0)(I+|Q|)^{-\nu}\cdot(I+|Q|)^{\nu}(H_0+1)^{-1}U^*$, with $\varphi$ given by $\varphi(\mu) = (\mu+1)\phi(\mu)$. Now $(I+|Q|)^{\nu}(H_0+1)^{-1}U^*$ is bounded (see Example 6.23). For $N \in \mathbb{N}$, the operator $D^N\varphi(H_0)(I+|Q|)^{-N}$ is bounded, which is easily seen by commuting all of the operators $Q_j$ occurring in $D^N$ through $\varphi(H_0)$ and noticing that $P^m\psi(H_0) \in \mathcal{B}(\mathcal{H})$ if $\psi \in C_0^\infty(\mathbb{R})$ (Problem 6.9). By interpolation one finds that $(I+D^2)^{\nu/2}\varphi(H_0)(I+|Q|)^{-\nu}$ is bounded for each $\nu > 0$ [take $R = (I+|Q|)^{-2N}$, $X = \varphi(H_0)$ and $S = (I+D^2)^N$ in Proposition 6.17, with $N > \nu/2$].

**Proposition 6.21.** *Let $k \in \mathbb{N}$. Assume that $V\colon \mathbb{R}^n \to \mathbb{R}$ satisfies condition (C) of page 288 and may be written in the form $V(\vec{x}) = [v_1(\vec{x}) + v_2(\vec{x})](1+|\vec{x}|)^{-\kappa}$ with $v_1 \in L^\infty(\mathbb{R}^n)$, $v_2 \in L^q(\mathbb{R}^n)$ for some $q$ satisfying $q \in (n/2, \infty)$ and $q \geq 1$, and with $\kappa > k+1$. Let $H = H_0 + V(\vec{Q})$ and let $U, W\colon \mathbb{R}^n \to \mathbb{C}$ be as in Proposition 6.20. Then the operator-valued function $\lambda \mapsto W(\vec{Q})(H-\lambda-i0)^{-1}U(\vec{Q})^*$ is $k$ times continuously differentiable in norm on $(0,\infty)\backslash\sigma_{\rm p}(H)$.*

PROOF. The proof can be made inductively in $k$ by using the result of the preceding proposition. We discuss the cases $k=1$ and $k=2$. We put $R_z = (H-z)^{-1}$ and $R_z^0 = (H_0 - z)^{-1}$.

(i) Let $k=1$. By differentiating the second resolvent equation $R_z = R_z^0 - R_z^0 V R_z$ we get [$' = d/dz$]:

$$R_z' = R_z^{0\prime} - R_z^{0\prime} V R_z - R_z^0 V R_z'.$$

Thus $(I + R_z^0 V)R_z' = R_z^{0\prime}(I - VR_z)$. Since $(I - R_z V)(I + R_z^0 V) = I$ (by the second resolvent equation), we then obtain

$$R_z' = (I - R_z V)(I + R_z^0 V)R_z' = (I - R_z V)R_z^{0\prime}(I - VR_z). \tag{6.114}$$

We choose $\kappa_1 > 1/2$, $\kappa_2 > 3/2$ with $\kappa_1 + \kappa_2 = \kappa$, set $v(\vec{x}) = v_1(\vec{x}) + v_2(\vec{x})$ and factor the function $V$ into $V = V_1 V_2$, with $V_1(\vec{x}) = |v(\vec{x})|^{1/2}(1+|\vec{x}|)^{-\kappa_1}$ and $V_2(\vec{x}) = |v(\vec{x})|^{1/2}[\text{sign}\, v(\vec{x})](1+|\vec{x}|)^{-\kappa_2}$. Then we have by (6.114):

$$\begin{aligned} WR_z'U^* = {} & WR_z^{0\prime}U^* - WR_z^{0\prime}V_2 \cdot V_1 R_z U^* - WR_z V_1 \cdot V_2 R_z^{0\prime} U^* \\ & + WR_z V_1 \cdot V_2 R_z^{0\prime} V_2 \cdot V_1 R_z U^*. \end{aligned} \tag{6.115}$$

Let $J$ be a bounded closed interval in $(0,\infty)\backslash\sigma_{\rm p}(H)$ and let $z = \lambda + i\varepsilon$ in Eq. (6.115), with $\lambda \in J$ and $\varepsilon > 0$. By taking into account Proposition 6.16(b) [for the operators of the form $X_1(\vec{Q})R_{\lambda+i\varepsilon}X_2(\vec{Q})$] as well as Proposition 6.20 [for those of the form $X_1(\vec{Q})R^{0\prime}_{\lambda+i\varepsilon}X_2(\vec{Q})$], one sees that the right-hand-side of (6.115), for $z = \lambda + i\varepsilon$, is norm convergent as $\varepsilon \to +0$, and that the convergence is uniform in $\lambda$ on $J$. So both $WR_{\lambda+i\varepsilon}U^*$ and its derivative with respect to $\lambda$ are convergent as $\varepsilon \to +0$, uniformly in $\lambda \in J$. This implies that the limit $WR_{\lambda+i0}U^*$ is differentiable in norm, and that its derivative is given by the right-hand side of (6.115) with $z = \lambda + i0$ (see for example Theorem 12 in §9, Chapter V of [L]). So $\lambda \mapsto WR_{\lambda+i0}U^*$ is continuously norm differentiable on $J$.

(ii) Let $k=2$. By differentiating (6.115) one gets

$$\begin{aligned} WR_z''U^* = {} & WR_z^{0\prime\prime}U^* - WR_z^{0\prime\prime}VR_zU^* - WR_z^{0\prime}VR_z'U^* \\ & - WR_z'VR_z^{0\prime}U^* - WR_zVR_z^{0\prime\prime}U^* + WR_z'VR_z^{0\prime}VR_zU^* \\ & + WR_zVR_z^{0\prime\prime}VR_zU^* + WR_zVR_z^{0\prime}VR_z'U^*. \end{aligned} \tag{6.116}$$

Each operator $V$ occurring on the right-hand side is factored into $V_1V_2$, with $V_1(\vec{x})$ and $V_2(\vec{x})$ of the same form as in (i) above but with the exponents $\kappa_1$ and$\kappa_2$ depending on the position of $V$. If $V$ stands next to an operator $R_z^{0\prime}$, we take $\kappa_1 = \kappa_2 = \kappa/2 > 3/2$

(so that for example $WR_z^{0\prime}V_1$ and $V_2R_z^{\prime}U^*$ converge as $z \to J$). If $V$ stands on the left of an operator $R_z^{0\prime\prime}$, we take $\kappa_1 > 1/2$ and $\kappa_2 > 5/2$ (so that for example $V_2R_z^{0\prime\prime}U^*$ has boundary values on $J$), and if $V$ stands on the right of an operator $R_z^{0\prime\prime}$, we take $\kappa_1 > 5/2$ and $\kappa_2 > 1/2$. Then, as in (i), the right-hand side of (6.116), for $z = \lambda + i\varepsilon$, is norm convergent as $\varepsilon \to +0$, uniformly in $\lambda$ on each bounded closed subinterval $J$ of $(0,\infty)\backslash\sigma_{\rm p}(H)$. Hence $\lambda \mapsto WR_{\lambda+i0}^{\prime}U^*$ is differentiable in norm, and its derivative is given by the right-hand side of (6.116) with $z = \lambda + i0$. So $\lambda \mapsto WR_{\lambda+i0}U^*$ is twice continuously norm differentiable on each $J$. □

## 6.7 Some commutators

**6.7.1.** As mentioned in the introduction to this chapter, the evaluation of commutators involving unbounded operators requires care. In the present section we illustrate this by means of examples the results of which are quoted in the text. The Hilbert space is $\mathcal{H} = L^2(\mathbb{R}^n)$, and we shall use the notations $\langle P\rangle = (I + \vec{P}^2)^{1/2}$, $\langle Q\rangle = (I + \vec{Q}^2)^{1/2}$ and $\partial_j = \partial/\partial x_j$ $(j = 1, \ldots, n)$.

**Example 6.22.** In various situations one can proceed by using scalar products with vectors belonging to a suitable dense set $\mathcal{D}$, chosen such that the commutator makes sense on $\mathcal{D}$, and then taking into account domain properties to transfer the operators to the other argument in the scalar product. As a simple example, let us prove that $[P_j, \varphi(\vec{Q})] = -i(\partial_j\varphi)(\vec{Q})$ on $\mathcal{D}(P_j)$ if $\varphi$ is of class $C^1$ with $\varphi$ and its first order derivatives $\partial_j\varphi$ bounded. Note that we must show in particular that if $f \in \mathcal{D}(P_j)$, then $\varphi(\vec{Q})f \in \mathcal{D}(P_j)$.

(i) We first let $g, h \in \mathcal{S}(\mathbb{R}^n)$. Then

$$\begin{aligned}\langle P_jh, \varphi(\vec{Q})g\rangle &= i\langle\partial_jh, \varphi g\rangle = -i\langle h, \varphi\partial_jg\rangle - i\langle h, (\partial_j\varphi)g\rangle \\ &= \langle h, \varphi(\vec{Q})P_jg\rangle - i\langle h, (\partial_j\varphi)(\vec{Q})g\rangle,\end{aligned}$$

where the second equality is obtained upon integrating by parts. Since $\varphi(\vec{Q})P_jg$ and $(\partial_j\varphi)(\vec{Q})g$ belong to the Hilbert space, this implies that $\varphi(\vec{Q})g \in \mathcal{D}(P_j)$ [remember that $P_j$ is essentially self-adjoint on $\mathcal{S}(\mathbb{R}^n)$] and that $P_j\varphi(\vec{Q})g = \varphi(\vec{Q})P_jg - i(\partial_j\varphi)(\vec{Q})g$.

(ii) Now let $f \in \mathcal{D}(P_j)$ and take $\mathcal{D} = \mathcal{S}(\mathbb{R}^n)$. For $g \in \mathcal{S}(\mathbb{R}^n)$ we find by using the final result of (i) (with $\overline{\varphi}$ in place of $\varphi$) that

$$\begin{aligned}\langle P_jg, \varphi(\vec{Q})f\rangle &= \langle\overline{\varphi}(\vec{Q})P_jg, f\rangle = \langle P_j\overline{\varphi}(\vec{Q})g, f\rangle - i\langle(\partial_j\overline{\varphi})(\vec{Q})g, f\rangle \\ &= \langle g, \varphi(\vec{Q})P_jf\rangle - i\langle g, (\partial_j\varphi)(\vec{Q})f\rangle.\end{aligned}$$

As in (i), this implies that $\varphi(\vec{Q})f \in \mathcal{D}(P_j)$ and $P_j\varphi(\vec{Q})f = \varphi(\vec{Q})P_jf - i(\partial_j\varphi)(\vec{Q})f$, which completes the proof.

By setting $f = (I + |P_j|)^{-1}h$ in the last relation above, one finds that the operator $P_j\varphi(\vec{Q})(I + |P_j|)^{-1}$ belongs to $\mathcal{B}(\mathcal{H})$. In the same way as above one can obtain the result that $(P_j)^N\varphi(\vec{Q})(I+|\vec{P}|)^{-N}\in\mathcal{B}(\mathcal{H})$ if $\varphi$ is a bounded function of class $C^N$ such that all its partial derivatives up to order $N$ are bounded ($N\in\mathbb{N}$). This implies that $\langle P\rangle^N\varphi(\vec{Q})\langle P\rangle^{-N}\in\mathcal{B}(\mathcal{H})$ for even $N$, hence also $\langle P\rangle^\nu\varphi(\vec{Q})\langle P\rangle^{-\nu}\in\mathcal{B}(\mathcal{H})$ for each $\nu\in[0,N]$, by interpolation (applying Proposition 6.17 as in Example 6.31).

Let us point out a different way of getting the result of (ii) above. For $f,g\in\mathcal{D}(P_j)$ we have

$$\langle P_jf,\varphi(\vec{Q})g\rangle-\langle f,\varphi(\vec{Q})P_jg\rangle=i\frac{d}{dt}\langle e^{iP_jt}f,\varphi(\vec{Q})e^{iP_jt}g\rangle\big|_{t=0}=i\frac{d}{dt}\langle f,\varphi_t(\vec{Q})g\rangle\big|_{t=0},$$

with $\varphi_t(\vec{Q})=e^{-iP_jt}\varphi(\vec{Q})e^{iP_jt}$. Now, since $\{e^{-iP_jt}\}$ is just the translation group along the $j$-th co-ordinate axis (see Example 5.4), $\varphi_t(\vec{Q})$ is the multiplication operator by $\varphi(x_1,\dots,x_{j-1},x_j-t,x_{j+1},\dots,x_n)$. So $(\partial/\partial t)\varphi_t(\vec{x})|_{t=0}=-\partial_j\varphi(\vec{x})$ and we get (by using the Dominated Convergence Theorem):

$$\langle P_jf,\varphi(\vec{Q})g\rangle=\langle f,\varphi(\vec{Q})P_jg\rangle-i\langle f,(\partial_j\varphi)(\vec{Q})g\rangle.$$

This again implies that $\varphi(\vec{Q})g\in\mathcal{D}(P_j)$ and that $P_j\varphi(\vec{Q})g=\varphi(\vec{Q})P_jg-i(\partial_j\varphi)(\vec{Q})g$.

**Example 6.23.** By interchanging the roles of $\vec{P}$ and $\vec{Q}$ in the arguments of the preceding example, one sees that $\langle Q\rangle^\nu\varphi(\vec{P})\langle Q\rangle^{-\nu}\in\mathcal{B}(\mathcal{H})$ for each $\nu\in[0,N]$ if $\varphi$ is a bounded function of class $C^N$ such that all its partial derivatives up to order $N$ are bounded. Particularly important examples in our context are the following functions: $\varphi(\vec{k})=(\vec{k}^2-z)^{-1}$ and $\varphi(\vec{k})=k_j(\vec{k}^2-z)^{-1}$ $[j=1,\dots,n;\ z\in\mathbb{C}\backslash\mathbb{R}]$. These functions are of class $C^\infty$ and their partial derivatives of any order are bounded. Therefore $\langle Q\rangle^\nu(H_0-z)^{-1}\langle Q\rangle^{-\nu}\in\mathcal{B}(\mathcal{H})$ and $\langle Q\rangle^\nu P_j(H_0-z)^{-1}\langle Q\rangle^{-\nu}\in\mathcal{B}(\mathcal{H})$ for each $\nu\geq 0$ and each $z\in\rho(H_0)$.

Furthermore, by observing that $(Q_j)^N(H_0-z)^{-1}$ is given by a sum of terms of the form $\theta_k(\vec{P})(H_0-z)^{-1}(Q_j)^k$ with $\theta_k\in L^\infty(\mathbb{R}^n)$ and $k=0,\dots,N$, one finds that the closure of $\langle Q\rangle^N(H_0-z)^{-1}W(\vec{Q})\langle Q\rangle^{-N}$ belongs to $\mathcal{B}(\mathcal{H})$ if $W\colon\mathbb{R}^n\to\mathbb{C}$ is such that $W(\vec{Q})$ is a $H_0$-bounded operator. By interpolation one finds that the closure of $\langle Q\rangle^\nu(H_0-z)^{-1}W(\vec{Q})\langle Q\rangle^{-\nu}$ belongs to $\mathcal{B}(\mathcal{H})$ for all $\nu\geq 0$.

**Example 6.24.** Consider a Hamiltonian of the form $H=H_0+V(\vec{Q})$, assumed to be self-adjoint on $\mathcal{D}(H_0)$. Let us show that, for $z\in\rho(H)$, the resolvent $(H-z)^{-1}$ maps $\mathcal{D}(Q_j)$ into $\mathcal{D}(Q_j)$ and that for each $g\in\mathcal{D}(Q_j)$:

$$Q_j(H-z)^{-1}g=(H-z)^{-1}Q_jg-2i(H-z)^{-1}P_j(H-z)^{-1}g. \tag{6.117}$$

For the proof, let $f,g\in\mathcal{D}(Q_j)$. Then we have:

$$\langle Q_jf,(H-z)^{-1}g\rangle-\langle f,(H-z)^{-1}Q_jg\rangle=i\frac{d}{dt}\langle e^{iQ_jt}f,(H-z)^{-1}e^{iQ_jt}g\rangle\big|_{t=0}.$$

We set $H(t)=e^{-iQ_jt}He^{iQ_jt}$, which is well defined on $\mathcal{D}(H)$ since this domain is invariant under $e^{iQ_jt}$.[22] Then $[H(t)-z]^{-1}=e^{-iQ_jt}(H-z)^{-1}e^{iQ_jt}$ and hence

$$\langle e^{iQ_jt}f,(H-z)^{-1}e^{iQ_jt}g\rangle=\langle f,[H(t)-z]^{-1}g\rangle.$$

[22] We have $\mathcal{D}(H)=\mathcal{D}(H_0)$, and $e^{iQ_jt}\mathcal{D}(H_0)\subseteq\mathcal{D}(H_0)$ (Problem 6.7).

To calculate $(d/dt)[H(t)-z]^{-1}|_{t=0}$ we observe that

$$\frac{1}{t}\{[H(t)-z]^{-1}-(H-z)^{-1}\} = -[H(t)-z]^{-1}\frac{H(t)-H}{t}(H-z)^{-1}.$$

Now $H(t) = H_0(t) + V(\vec{Q})$, with $H_0(t) = e^{-iQ_jt}\vec{P}^2 e^{iQ_jt} = \widehat{\sum}_{k=1}^{n} P_k^2 + (P_j+t)^2$, where $\widehat{\sum}_{k=1}^{n}$ means that the term with $k=j$ has been omitted in the sum $\sum_{k=1}^{n}$. So $(d/dt)H(t)\big|_{t=0} = (d/dt)H_0(t)\big|_{t=0} = 2P_j$ [on $\mathcal{D}(H_0)$]. Hence we have

$$\langle Q_j f, (H-z)^{-1}g\rangle = \langle f, (H-z)^{-1}Q_j g\rangle - 2i\langle f, (H-z)^{-1}P_j(H-z)^{-1}g\rangle,$$

from which the claimed results follow.

**Example 6.25.** Consider again a Hamiltonian of the form $H = H_0 + V(\vec{Q})$, where $V\colon \mathbb{R}^n \to \mathbb{R}$ is such that $V(\vec{Q})$ is $H_0$-compact and $\langle Q\rangle V(\vec{Q})$ is $H_0$-bounded. Then $H$ is self-adjoint on $\mathcal{D}(H_0)$. Let us show that the operator $\langle Q\rangle^{\nu}(H-z)^{-1}\langle Q\rangle^{-\nu}$ belongs to $\mathcal{B}(\mathcal{H})$ for every $z\in\rho(H)$ and each $\nu\geq 0$. By the second resolvent equation (2.115) we have

$$\begin{aligned}\langle Q\rangle^{\nu}(H-z)^{-1}\langle Q\rangle^{-\nu} &= \left[\langle Q\rangle^{\nu}(H_0-z)^{-1}\langle Q\rangle^{-\nu}\right]\\ &\quad - \left[\langle Q\rangle^{\nu}(H_0-z)^{-1}\langle Q\rangle V(\vec{Q})\langle Q\rangle^{-\nu}\right]\left[\langle Q\rangle^{\nu-1}(H-z)^{-1}\langle Q\rangle^{-\nu}\right]. \qquad (6.118)\end{aligned}$$

By Example 6.23, the first term on the right-hand side belongs to $\mathcal{B}(\mathcal{H})$ for any $\nu\geq 0$. To handle the second term on the right-hand side, we first observe that its first factor also belongs to $\mathcal{B}(\mathcal{H})$ for each $\nu\geq 0$, again by Example 6.23. Its second factor [i.e. $\langle Q\rangle^{\nu-1}(H-z)^{-1}\langle Q\rangle^{-\nu}$] trivially belongs to $\mathcal{B}(\mathcal{H})$ if $0\leq\nu\leq 1$. Consequently $\langle Q\rangle^{\nu}(H-z)^{-1}\langle Q\rangle^{-\nu}\in\mathcal{B}(\mathcal{H})$ for $\nu\in[0,1]$. Now let $\nu\in(1,2]$ in (6.118). Then, since $\langle Q\rangle^{\nu-1}(H-z)^{-1}\langle Q\rangle^{-\nu}\in\mathcal{B}(\mathcal{H})$ by the preceding result, we find from (6.118) that $\langle Q\rangle^{\nu}(H-z)^{-1}\langle Q\rangle^{-\nu}\in\mathcal{B}(\mathcal{H})$ for $\nu\in(1,2]$. By continuing this argument (taking $\nu\in(2,3]$, then $\nu\in(3,4]$ etc.), one obtains that $\langle Q\rangle^{\nu}(H-z)^{-1}\langle Q\rangle^{-\nu}$ belongs to $\mathcal{B}(\mathcal{H})$ for all $\nu\geq 0$.

**Example 6.26.** Here we give two boundedness results that will be needed in Section 7.2. We consider the class of Hamiltonians $H$ of Example 6.25 and assume in addition that $V$ is a bounded function. By the second resolvent equation we have for any $\nu\geq 0$:

$$\begin{aligned}\langle Q\rangle^{\nu}(H-z)^{-1}P_j\langle Q\rangle^{-\nu} &= \left[I - \langle Q\rangle^{\nu}(H-z)^{-1}\langle Q\rangle^{-\nu}V\right]\\ &\quad\times\left[\langle Q\rangle^{\nu}(H_0-z)^{-1}P_j\langle Q\rangle^{-\nu}\right], \qquad (6.119)\end{aligned}$$

and by Examples 6.23 and 6.25 above, each factor on the right-hand side belongs to $\mathcal{B}(\mathcal{H})$. Thus, taking $\nu=2$, we have for $f,g\in\mathcal{S}(\mathbb{R}^n)$:

$$\begin{aligned}\left|\langle f, \langle Q\rangle^{-2}P_j(H-i)^{-1}\langle Q\rangle^{2}g\rangle\right| &= \left|\langle\langle Q\rangle^{2}(H+i)^{-1}P_j\langle Q\rangle^{-2}f, g\rangle\right|\\ &\leq \|f\|\|\langle Q\rangle^{2}(H+i)^{-1}P_j\langle Q\rangle^{-2}\|\|g\|,\end{aligned}$$

and (2.4) shows that the closure of $\langle Q\rangle^{-2}P_j(H-i)^{-1}\langle Q\rangle^{2}$ belongs to $\mathcal{B}(\mathcal{H})$.

Now let $W\colon \mathbb{R}^n \to \mathbb{C}$ be such that $W(\vec{Q})$ is a $H_0$-bounded operator. Upon replacing in (6.119) the operator $P_j$ by $W(\vec{Q})$, one finds as above that the closure of $\langle Q\rangle^{-\nu} W(\vec{Q})(H \pm i)^{-1}\langle Q\rangle^{\nu}$ belongs to $\mathcal{B}(\mathcal{H})$ for each $\nu \geq 0$.

**6.7.2.** We add some remarks concerning (6.73) and (6.74), in the framework considered in §6.4.1, hence with $A = D$ and by assuming condition (C) on page 288 to be satisfied. To be able to use the conventions made at the beginning of §6.2.2, we would have to show that the identities (6.73) and (6.74) hold between arbitrary vectors $f, g$ belonging to $\mathcal{D}(A) \equiv \mathcal{D}(D)$, for example that, for some $z \in \rho(H) \cap \mathbb{R}$ and all $f, g \in \mathcal{D}(D)$:

$$\langle f, (H-z)^{-1} Dg\rangle - \langle Df, (H-z)^{-1} g\rangle = i\langle f, (H-z)^{-1}[H - Z(\vec{Q})](H-z)^{-1} g\rangle. \tag{6.120}$$

To obtain this equation, we follow the method of Example 6.24. We observe that, if $f, g \in \mathcal{D}(D)$:

$$\langle f, (H-z)^{-1} Dg\rangle - \langle Df, (H-z)^{-1} g\rangle = -i\frac{d}{d\theta}\langle \mathrm{U}_{\theta/2} f, (H-z)^{-1}\mathrm{U}_{\theta/2} g\rangle\big|_{\theta=0},$$

where $\{\mathrm{U}_\kappa\}_{\kappa\in\mathbb{R}}$ is the dilation group defined in (6.16) [we had $\mathrm{U}_\kappa = e^{-i\widehat{D}\kappa} \equiv e^{2iD\kappa}$]. Let us set $H(\theta) = \mathrm{U}^*_{\theta/2} H \mathrm{U}_{\theta/2}$, which is well defined on $\mathcal{D}(H)$ since this domain is invariant under the dilation group. Then $[H(\theta) - z]^{-1} = \mathrm{U}^*_{\theta/2}(H-z)^{-1}\mathrm{U}_{\theta/2}$ and consequently

$$\langle \mathrm{U}_{\theta/2} f, (H-z)^{-1}\mathrm{U}_{\theta/2} g\rangle = \langle f, \mathrm{U}^*_{\theta/2}(H-z)^{-1}\mathrm{U}_{\theta/2} g\rangle = \langle f, [H(\theta) - z]^{-1} g\rangle.$$

It remains to calculate $(d/d\theta)[H(\theta) - z]^{-1}|_{\theta=0}$ and to check that this operator is identical with $-(H-z)^{-1}[H - Z(\vec{Q})](H-z)^{-1}$, where $Z(\vec{x})$ is given by (6.77). Now:

$$\frac{1}{\theta}\{[H(\theta) - z]^{-1} - (H-z)^{-1}\} = -[H(\theta) - z]^{-1}\frac{H(\theta) - H}{\theta}(H-z)^{-1}.$$

By taking the strong limit $\theta \to 0$ we get

$$\frac{d}{d\theta}[H(\theta) - z]^{-1}\big|_{\theta=0} = -(H-z)^{-1}\frac{d}{d\theta}H(\theta)\big|_{\theta=0}(H-z)^{-1}.$$

By Eq. (6.21) we have $H(\theta) = e^\theta H_0 + V_\theta$, with $V_\theta \equiv V_\theta(\vec{Q}) = \mathrm{U}^*_{\theta/2} V(\vec{Q})\mathrm{U}_{\theta/2} = V(e^{-\theta/2}\vec{Q})$. Since $(d/d\theta)e^\theta H_0|_{\theta=0} = H_0 = H - V(\vec{Q})$, it remains to show that

$$\frac{d}{d\theta}(H-z)^{-1} V_\theta (H-z)^{-1}\big|_{\theta=0} = -\frac{1}{2}(H-z)^{-1}\left[r\frac{\partial V}{\partial r}\right](\vec{Q})(H-z)^{-1}. \tag{6.121}$$

To obtain this last identity, let us set $\mathcal{D} = (H-z)C_0^\infty(\mathbb{R}^n)$. Then $\mathcal{D}$ is a dense linear manifold in $\mathcal{H}$, [23] and one has $(H-z)^{-1}\mathcal{D} = C_0^\infty(\mathbb{R}^n)$. If $f, g \in \mathcal{D}$, we obtain

[23] This follows because $H$ is essentially self-adjoint on $C_0^\infty(\mathbb{R}^n)$. To see this, remember that $H_0$ is essentially self-adjoint on $C_0^\infty(\mathbb{R}^n)$ and apply the following generalisation of the Rellich-Kato Theorem: *Proposition 2.44 remains true if one replaces everywhere in its statement the term "self-adjoint" by "essentially self-adjoint"*. The proof of this variant is practically identical with that of Proposition 2.44.

by using Eq. (6.75) and the invariance of $C_0^\infty(\mathbb{R}^n)$ under $\mathrm{U}_{\theta/2}$:

$$\frac{d}{d\theta}\langle f,(H-z)^{-1}V(\theta)(H-z)^{-1}g\rangle = -\frac{1}{2}\langle f,(H-z)^{-1}\mathrm{U}_{\theta/2}^*\Big[r\frac{\partial V}{\partial r}\Big](\vec{Q})\mathrm{U}_{\theta/2}(H-z)^{-1}g\rangle. \tag{6.122}$$

In particular (set $\theta=0$): Eq. (6.121) is satisfied between vectors $f,g\in\mathcal{D}$. Let us now integrate the identity (6.122) on $[0,\varepsilon]$:

$$\langle f,(H-z)^{-1}[V(\varepsilon)-V](H-z)^{-1}g\rangle = -\frac{1}{2}\int_0^\varepsilon\langle f,(H-z)^{-1}\mathrm{U}_{\theta/2}^*\Big[r\frac{\partial V}{\partial r}\Big](\vec{Q})\mathrm{U}_{\theta/2}(H-z)^{-1}g\rangle d\theta. \tag{6.123}$$

Now

$$\Big[r\frac{\partial V}{\partial r}\Big](\vec{Q})\mathrm{U}_{\theta/2}(H-z)^{-1} = \Big[r\frac{\partial V}{\partial r}\Big](\vec{Q})(H_0-z)^{-1}\cdot(H_0-z)\mathrm{U}_{\theta/2}(H-z)^{-1}$$

and

$$(H_0-z)\mathrm{U}_{\theta/2}(H-z)^{-1} = \mathrm{U}_{\theta/2}(e^\theta H_0-z)(H-z)^{-1}.$$

Since $\big[r(\partial V/\partial r)\big](\vec{Q})(H_0-z)^{-1}\in\mathcal{B}(\mathcal{H})$ by assumption, this shows that the correspondence $\theta\mapsto\big[r(\partial V/\partial r)\big](\vec{Q})\mathrm{U}_{\theta/2}(H-z)^{-1}$ determines a strongly continuous $\mathcal{B}(\mathcal{H})$-valued function. Thus the expression

$$\int_0^\varepsilon(H-z)^{-1}\mathrm{U}_{\theta/2}^*\Big[r\frac{\partial V}{\partial r}\Big](\vec{Q})\mathrm{U}_{\theta/2}(H-z)^{-1}d\theta$$

defines an operator in $\mathcal{B}(\mathcal{H})$. Since $(H-z)^{-1}[V(\varepsilon)-V](H-z)^{-1}$ also belongs to $\mathcal{B}(\mathcal{H})$, Eq. (6.123) is satisfied for all $f,g\in\mathcal{H}$ (approximate these vectors by vectors belonging to $\mathcal{D}$, as in the proof of Proposition 6.5). By differentiating (6.123) with respect to $\varepsilon$ (at $\varepsilon=0$, and with arbitrary $f,g\in\mathcal{H}$), one obtains the validity of (6.121).

Above we have shown how to obtain (6.73) for arbitrary $f,g$ in $\mathcal{D}(A)$. An analogous argument [using also the assumption made on $r^2(\partial^2V/\partial r^2)$ in the condition (C)] allows one to obtain the validity of (6.74) between arbitrary vectors $f,g\in\mathcal{D}(A)$.

## Appendix to Chapter 6: Interpolation of operators

The theory of interpolation for Banach spaces and for operators is an important topic in functional analysis (see for example [BL], [BS]). One distinguishes between the *real method of interpolation* and the *complex method of interpolation*. We give here a simple result in the framework of complex interpolation. This result is based on the following lemma characterising certain holomorphic functions in a strip of the form

$$\mathcal{O}=\{z\in\mathbb{C}\,|\,0<\Re z<1\}. \tag{6.124}$$

**Lemma 6.27 (Hadamard Three Line Theorem).** *Let $F: \mathcal{O} \to \mathbb{C}$ be a holomorphic function. Assume that $F$ has a continuous bounded extension to the closed strip $\overline{\mathcal{O}} = \{z \in \mathbb{C} \mid 0 \le \Re z \le 1\}$. Let $M_0 = \sup_{y\in\mathbb{R}} |F(iy)|$ and $M_1 = \sup_{y\in\mathbb{R}} |F(1+iy)|$. Then*

$$|F(x+iy)| \le (M_0)^{1-x}(M_1)^x \qquad \forall x \in [0,1], \forall y \in \mathbb{R}. \tag{6.125}$$

PROOF. For $z \in \overline{\mathcal{O}}$ and $\varepsilon > 0$ we set $F_\varepsilon(z) = e^{\varepsilon z^2 + \kappa z} F(z)$, where $\kappa$ is a real constant that will be fixed further on. One has ($z = x + iy$ with $0 \le x \le 1$):

$$|F_\varepsilon(z)| = e^{\varepsilon(x^2-y^2)} e^{\kappa x} |F(x+iy)| \le e^{\varepsilon + \kappa} e^{-\varepsilon y^2} \sup_{z \in \overline{\mathcal{O}}} |F(z)|, \tag{6.126}$$

hence $|F_\varepsilon(z)| \to 0$ as $y \equiv \Im z \to \pm\infty$. Let us also observe that

$$|F_\varepsilon(iy)| \le M_0 \quad , \qquad |F_\varepsilon(1+iy)| \le e^{\varepsilon+\kappa} M_1.$$

Since $F_\varepsilon$ is holomorphic in $\mathcal{O}$ and continuous on $\overline{\mathcal{O}}$, the function $|F_\varepsilon|$ attains its maximum on the boundary of $\mathcal{O}$ (by the maximum principle for analytic functions), hence in our situation on one of the two lines $\{z = iy \mid y \in \mathbb{R}\}$ and $\{z = 1+iy \mid y \in \mathbb{R}\}$ (because $|F_\varepsilon(z)| = 0$ at infinity):

$$|F_\varepsilon(z)| \le \max\big\{\sup_{y\in\mathbb{R}} |F_\varepsilon(iy)|, \sup_{y\in\mathbb{R}} |F_\varepsilon(1+iy)|\big\} \qquad \forall z \in \overline{\mathcal{O}}.$$

In other terms $|F_\varepsilon(z)| \le \max\{M_0, e^{\varepsilon+\kappa} M_1\}$ for each $z \in \overline{\mathcal{O}}$. By taking into account the first equation in (6.126), this can be rewritten as

$$\begin{aligned} |F(x+iy)| &\le e^{-\varepsilon(x^2-y^2)} e^{-\kappa x} \max\{M_0, e^{\varepsilon+\kappa} M_1\} \\ &\le e^{\varepsilon} e^{-\varepsilon(x^2-y^2)} e^{-\kappa x} \max\{M_0, e^{\kappa} M_1\}. \end{aligned}$$

This inequality holds for each $\varepsilon > 0$, hence also for $\varepsilon = 0$:

$$|F(x+iy)| \le e^{-\kappa x} \max\{M_0, e^{\kappa} M_1\}.$$

Let us set $\kappa = \log[M_0/M_1]$ (one may assume that $M_0, M_1 > 0$, otherwise $F \equiv 0$), so that $e^\kappa M_1 = M_0$. Then

$$|F(x+iy)| \le M_0 e^{-\log[M_0/M_1]x} = M_0 \big(\frac{M_1}{M_0}\big)^x = M_0^{1-x} M_1^x. \qquad \square$$

**Lemma 6.28.** *Let $C$ be a positive invertible self-adjoint operator and let $f \in \mathcal{D}(C)$. Then the vector-valued function $z \mapsto C^z f$ is strongly differentiable in the strip $\mathcal{O}$ defined in* (6.124) *(hence this function is strongly holomorphic in $\mathcal{O}$) and strongly continuous on $\overline{\mathcal{O}}$.*

PROOF. We denote by $\{E(\cdot)\}$ the spectral measure of $C$. The operator $C^z$ is given by the functional calculus: $C^z = \int_0^\infty \lambda^z E(d\lambda)$. Formally one has $(d/dz)C^z = C^z \log C$.

Since $\Re z \in (0,1)$ and $C > O$, the vector $C^z \log C f \equiv \int_0^\infty \lambda^{x+iy}(\log\lambda)E(d\lambda)f$ is well defined for $f \in \mathcal{D}(C)$. We must verify that

$$\int_0^\infty \Big|\frac{\lambda^{z+h}-\lambda^z}{h} - \lambda^z\log\lambda\Big|^2 m_f(d\lambda) \to 0 \quad \text{as } |h|\to 0 \qquad (h\in\mathbb{C}). \tag{6.127}$$

Let us show that this follows from the Dominated Convergence Theorem (Proposition 1.9). For each fixed $\lambda > 0$, the integrand converges to zero as $|h| \to 0$. So it remains to find an $h$-independent upper bound of the integrand by a function belonging to $L^1((0,\infty), m_f)$. For this we use Taylor's formula in second order (which is easily verified by partial integration):

$$\varphi(z+h) = \varphi(z) + h\varphi'(z) + h^2\int_0^1 (1-\tau)\varphi''(z+\tau h)\,d\tau \qquad ('\equiv \frac{d}{dz}).$$

We take $\varphi(z) = \lambda^z$, hence $\varphi'(z) = \lambda^z\log\lambda$ and $\varphi''(z) = (\log\lambda)^2\lambda^z$. Then

$$\begin{aligned}\Big|\frac{\lambda^{z+h}-\lambda^z}{h} - \lambda^z\log\lambda\Big| &\le |h|\int_0^1 (1-\tau)(\log\lambda)^2|\lambda^{z+\tau h}|\,d\tau \\ &\le |h|\int_0^1 (\log\lambda)^2\lambda^{\Re(z+\tau h)}\,d\tau.\end{aligned}$$

We assume that $|h| < \min\{x/2, (1-x)/2\}$. Since $x \in (0,1)$, we then have $0 < x/2 < \Re(z+\tau h) < (1+x)/2 < 1$ for all $\tau\in[0,1]$, so that $|h|(\log\lambda)^2\lambda^{\Re(z+\tau h)} \le c_x \max\{1,\lambda\}$ for some constant $c_x$ depending on $x$ and for all $\tau \in [0,1]$. The function $\lambda \mapsto \max\{1,\lambda\}$ belongs to $L^1((0,\infty), m_f)$ for $f\in\mathcal{D}(C)$.

We have obtained the complex differentiability of $C^z f$ in $\mathcal{O}$. Its strong continuity on $\overline{\mathcal{O}}$ is proved in the same way. □

In what follows we use several times the formula (2.4): If $\mathcal{D}_1$ and $\mathcal{D}_2$ are dense linear manifolds in $\mathcal{H}$ and $Z$ is an operator with $\mathcal{D}_2 \subseteq \mathcal{D}(Z)$, then

$$\|Z\| = \sup_{f\in\mathcal{D}_1, \|f\|=1}\ \sup_{g\in\mathcal{D}_2,\|g\|=1} |\langle f, Zg\rangle| \tag{6.128}$$

[the right-hand side is finite if and only if the closure of $Z$ belongs to $\mathcal{B}(\mathcal{H})$].

**Lemma 6.29.** *Let $C_1, C_2$ be self-adjoint operators with $C_1 \in \mathcal{B}(\mathcal{H})$, and let $X$ be an operator in $\mathcal{B}(\mathcal{H})$. Then the following two statements are equivalent:*
*(a) $X^*$ maps $\mathcal{R}(C_1)$ into $\mathcal{D}(C_2)$ and $C_2X^*C_1 \in \mathcal{B}(\mathcal{H})$.*
*(b) The closure of $C_1XC_2$ [defined on $\mathcal{D}(C_2)$] belongs to $\mathcal{B}(\mathcal{H})$.*
*If one of these two conditions is satisfied, one has $C_2X^*C_1 = [C_1XC_2]^*$, in particular $\|C_2X^*C_1\| = \|C_1XC_2\|$.*

PROOF. For $f\in\mathcal{H}$ and $g \in \mathcal{D}(C_2)$ one has

$$\langle f, C_1XC_2 g\rangle = \langle X^*C_1 f, C_2 g\rangle. \tag{6.129}$$

(i) Under the hypothesis of (b), (6.129) implies that

$$|\langle X^* C_1 f, C_2 g\rangle| \leq \|C_1 X C_2\| \|f\| \|g\|. \tag{6.130}$$

For fixed $f$, (6.130) shows that the scalar product in (6.129) defines a bounded linear functional $\varphi_f$ on $\mathcal{D}(C_2)$. One can extend $\varphi_f$ to a bounded linear functional on all of $\mathcal{H}$ [if $g \in \mathcal{H}$ and $\{g_k\}$ is a sequence in $\mathcal{D}(C_2)$ such that $\|g - g_k\| \to 0$, set $\varphi_f(g) = \lim_{k\to\infty} \varphi_f(g_k)$; the limit exists because $\varphi_f(g_j) - \varphi_f(g_k) \to 0$ as $j, k \to \infty$, by (6.130)]. By the Riesz Lemma (Proposition 1.8) there exists a vector $h \in \mathcal{H}$ such that $\varphi_f(g) = \langle h, g\rangle$ for each $g \in \mathcal{H}$. For $g \in \mathcal{D}(C_2)$ this means that

$$\langle X^* C_1 f, C_2 g\rangle = \langle h, g\rangle.$$

Hence $X^* C_1 f$ belongs to the domain of $C_2^* \equiv C_2$, and $C_2 X^* C_1 f = h$. Thus the operator $C_2 X^* C_1$ is defined on all of $\mathcal{H}$; in particular $X^*$ maps $\mathcal{R}(C_1)$ into $\mathcal{D}(C_2)$. (6.130) shows [combined with (6.128)] that $\|C_2 X^* C_1\| \leq \|C_1 X C_2\|$.

(ii) Under the hypotheses of (a) one finds from (6.129) that

$$|\langle f, C_1 X C_2 g\rangle| = |\langle C_2 X^* C_1 f, g\rangle| \leq \|C_2 X^* C_1\| \|f\| \|g\|$$

for all $f \in \mathcal{H}$ and all $g \in \mathcal{D}(C_2)$. Hence

$$\|C_1 X C_2\|_{\mathcal{D}(C_2)} = \sup_{f\in\mathcal{H}, \|f\|=1} \ \sup_{g\in\mathcal{D}(C_2), \|g\|=1} |\langle f, C_1 X C_2 g\rangle| \leq \|C_2 X^* C_1\|.$$

This shows that the closure of the operator $C_1 X C_2$ [defined on $\mathcal{D}(C_2)$] belongs to $\mathcal{B}(\mathcal{H})$. □

**Proof of Proposition 6.17**. We use the notations and the result of Lemma 6.27. Let $g \in \mathcal{D}(S)$. Then the vector-valued function $z \mapsto R^z X S^z g$ is strongly holomorphic in $\mathcal{O}$ and strongly continuous on $\overline{\mathcal{O}}$ [Leibniz rule, by taking into account Lemma 6.28 and the boundedness of $R$, see (2.16)]. Let $f \in \mathcal{H}$ and define

$$F(z) = \langle f, R^z X S^z g\rangle. \tag{6.131}$$

One has

$$M_0 = \sup_{y\in\mathbb{R}} |F(iy)| \leq \|X\| \|R^{iy} f\| \|S^{iy} g\| = \|X\| \|f\| \|g\|$$

(because $R^{iy}$ and $S^{iy}$ are a unitary operators), and

$$M_1 = \sup_{y\in\mathbb{R}} |F(1+iy)| = \sup_{y\in\mathbb{R}} |\langle R^{iy} f, RXSS^{iy} g\rangle| \leq \|RXS\| \|f\| \|g\|.$$

Thus Lemma 6.27 implies that

$$|F(x+iy)| \leq c_x \|f\| \|g\|, \quad \text{with } c_x = \|X\|^{1-x} \|RXS\|^x. \tag{6.132}$$

Hence, taking $x = \nu$ and $y = 0$:

$$\|R^\nu X S^\nu\| = \sup_{f\in\mathcal{H}, \|f\|=1} \ \sup_{g\in\mathcal{D}(S), \|g\|=1} |\langle f, R^\nu X S^\nu g\rangle| \leq c_\nu.$$

This shows that the closure of the operator $R^\nu X S^\nu$ [defined on $\mathcal{D}(S)$] belongs to $\mathcal{B}(\mathcal{H})$, and all assertions follow from Lemma 6.29 (take $C_1 = R^\nu$ and $C_2 = S^\nu$ in Lemma 6.29). □

We present some applications of Proposition 6.17 that are used in the text. Below the operators $R, S$ and $X$ refer to Proposition 6.17, identified in each example with operators of some application in the text.

**Example 6.30.** In the situation of Proposition 6.16, let us take $R = (H+\zeta)^{-1}$, $S = H_0 + \zeta$ and $X = I$, where $\zeta > 0$ is chosen such that $H + \zeta > 0$. The operator $(H_0+\zeta)(H+\zeta)^{-1} \equiv I - V(H+\zeta)^{-1}$ belongs to $\mathcal{B}(\mathcal{H})$. Hence the closure of $(H+\zeta)^{-1}(H_0+\zeta)$ belongs to $\mathcal{B}(\mathcal{H})$ (use for example Lemma 6.29). So Proposition 6.17 implies that the closure of $(H+\zeta)^{-\nu}(H_0+\zeta)^\nu$ belongs to $\mathcal{B}(\mathcal{H})$ for each $\nu \in [0,1]$. Applying again Lemma 6.29, we then also have $(H_0+\zeta)^\nu(H+\zeta)^{-\nu} \in \mathcal{B}(\mathcal{H})$. This also implies that $(H_0+\zeta)^\nu(H+\zeta)^{-1}(H_0+\zeta)^{-\nu} \in \mathcal{B}(\mathcal{H})$ for $\nu \in [0,1]$.

**Example 6.31.** In the situation of Theorem 6.10, let us take $R = (I+|A|)^{-1}$, $S = I + |A|$ and $X = T$. One has $(I+|A|)T(I+|A|)^{-1} \in \mathcal{B}(\mathcal{H})$ by the footnote on page 292. Since $T$ maps $\mathcal{D}(|A|)$ into itself (page 274), this implies that the closure of $RXS \equiv (I+|A|)^{-1}T(I+|A|)$ belongs to $\mathcal{B}(\mathcal{H})$ (using Lemma 6.29). Hence for each $\nu \in [0,1]$, the closure of $(I+|A|)^{-\nu}T(I+|A|)^\nu$ belongs to $\mathcal{B}(\mathcal{H})$, and $T$ maps $\mathcal{D}(|A|^\nu)$ into $\mathcal{D}(|A|^\nu)$. In particular $(I+|A|)^\nu T(I+|A|)^{-\nu} \in \mathcal{B}(\mathcal{H})$.

**Example 6.32.** Let $B$ and $C$ be positive invertible self-adjoint operators in a Hilbert space $\mathcal{H}$, with $B^{-1} \in \mathcal{B}(\mathcal{H})$, and let $X \in \mathcal{B}(\mathcal{H})$. In Proposition 6.17, take $R = B^{-\kappa}$ and $S = C^{-\kappa}$ for some $\kappa > 0$. Then, for $\nu \in (0,1)$:

$$\|B^{-\nu\kappa}XC^{-\nu\kappa}\| \le \|X\|^{1-\nu}\|B^{-\kappa}XC^{-\kappa}\|^\nu.$$

Setting $\theta = \nu\kappa$, this reads as follows:

$$\|B^{-\theta}XC^{-\theta}\| \le \|X\|^{1-(\theta/\kappa)}\|B^{-\kappa}XC^{-\kappa}\|^{\theta/\kappa}, \qquad 0 < \theta < \kappa. \tag{6.133}$$

**Example 6.33.** Let $B$ be a positive self-adjoint operator in a Hilbert space $\mathcal{H}$, with inverse in $\mathcal{B}(\mathcal{H})$. Let $Y$ be some operator in $\mathcal{B}(\mathcal{H})$ and $X = B^{-\kappa}YB^{-\kappa}$ for some $\kappa > 0$. In Proposition 6.17, take $R = I$ and $S = B^\kappa$. Then, for $\nu \in (0,1)$:

$$\|XB^{\nu\kappa}\| \equiv \|B^{-\kappa}YB^{-(1-\nu)\kappa}\| \le \|B^{-\kappa}YB^{-\kappa}\|^{1-\nu}\|B^{-\kappa}Y\|^\nu.$$

In particular, for $\nu = 1/\kappa$ (assuming that $\kappa > 1$):

$$\|B^{-\kappa}YB^{-\kappa+1}\| \le \|B^{-\kappa}YB^{-\kappa}\|^{1-(1/\kappa)}\|B^{-\kappa}Y\|^{1/\kappa}. \tag{6.134}$$

## Bibliographical Notes

A very extensive exposition of Mourre theory is given in the monograph [ABG]. Various other methods for determining spectra of quantum-mechanical operators are discussed in [DK]. A recent review of spectral results for Schrödinger operators is given in [6], and a presentation of spectral and scattering theory for non-relativistic N-body Hamiltonians can be found in [9]. For a more complete discussion of the theory of relatively smooth operators we refer to Volume IV of [RS].

## Problems

**6.1.** In the example considered in §6.1.2 ($[H, iA] = H$, $f \in \mathcal{D}(A)$) it was shown that $\langle f, (H - \lambda - i\mu)^{-1} f\rangle$ converges for each $\lambda > 0$ as $\mu \to +0$. Show that this convergence is uniform in $\lambda$ on any interval $[\lambda_0, \infty)$ with $\lambda_0 > 0$.

**6.2.** Let $H$ and $A$ be bounded self-adjoint operators such that $[H, iA] \geq C \geq O$ for some bounded self-adjoint operator $C$. Show that $C^{1/2}$ is $H$-smooth and that $\int_{\mathbb{R}} \|C^{1/2} e^{-iHt} f\|^2\, dt$ is finite for each $f \in \mathcal{H}$.
[Hint: Putnam's Theorem.]

**6.3.** Let $\mathfrak{D}$ and $\mathfrak{D}_0$ be as in §6.1.1 (page 264). Let $F\colon \mathfrak{D} \to \mathcal{B}(\mathcal{H})$ be continuous in norm and continuously differentiable with respect to the first variable $\varepsilon$ on $\mathfrak{D}_0$. Assume that $F$ satisfies the following differential inequality on $\mathfrak{D}_0$: $\|(\partial/\partial\varepsilon) F(\varepsilon, \mu)\| \leq c\,[\varepsilon^{-\beta} \|F(\varepsilon, \mu)\|^{\delta} + 1]$, with $\beta, \delta \in [0, 1)$. Show that $\text{u-}\lim_{\mu \to +0} F(0, \mu)$ exists.

**6.4.** Let $T \in \mathcal{B}(\mathcal{H})$ and let $A$ be self-adjoint.
(a) If $[T, A]$ and $[[T, A], A]$ belong to $\mathcal{B}(\mathcal{H})$ (in the sense of §6.2.2), then $T$ maps $\mathcal{D}(A^2)$ into $\mathcal{D}(A^2)$.
(b) If $[T, A]$, $[[T, A], A]$ and $[[[T, A], A], A]$ belong to $\mathcal{B}(\mathcal{H})$, then $T$ maps $\mathcal{D}(A^3)$ into $\mathcal{D}(A^3)$.

**6.5.** Prove Corollary 6.9.

**6.6.** Let $H$ be self-adjoint and let $C$ be a closed operator that is $H$-smooth on some interval $J$.
(a) Verify that if $\mathfrak{J} \subseteq J$, then $C$ is also $H$-smooth on $\mathfrak{J}$.
(b) Show that all eigenvectors of $H$ in $J$ belong to the null space of $C$: if $f$ satisfies $Hf = \lambda f$ with $\lambda \in J$, then $Cf = \mathbf{0}$.

**6.7.** In $L^2(\mathbb{R}^n)$, show that $f \in \mathcal{D}(H_0) \Longrightarrow e^{iQ_j t} f \in \mathcal{D}(H_0)$ for all real $t$.

**6.8.** In $L^2(\mathbb{R})$, consider the Stark Hamiltonian $H = P^2 + EQ$, where $E > 0$ is a constant.
(a) Verify that $H$ is essentially self-adjoint on $\mathcal{S}(\mathbb{R})$.
(b) By using the results of Section 6.2 (for unbounded $T$), show that the spectrum of $H$ is purely absolutely continuous: $\sigma(H) = \sigma_{\text{ac}}(H)$.
(c) Verify that $H$ is unitarily equivalent to the differential operator $\widehat{H} = iE(d/dx) + Q^2$ in $L^2(\mathbb{R})$.
(d) By using Weyl's criterion, show that $\sigma(H) \equiv \sigma(\widehat{H}) = \sigma_{\text{ess}}(\widehat{H}) = \mathbb{R}$ and conclude that $\sigma(H) = \sigma_{\text{ac}}(H) = \mathbb{R}$.
(e) Show that the operator $\widehat{H}$ is unitarily equivalent to $iE(d/dx)$ in $L^2(\mathbb{R})$. This again proves that $\sigma(H) = \sigma_{\text{ac}}(H) = \mathbb{R}$.
(f) Establish similar results for the $n$-dimensional Stark Hamiltonian $H = \vec{P}^2 + EQ_1$ in $L^2(\mathbb{R}^n)$, $n \geq 2$.
[Hints. For (b): the operator $P$ is conjugate to $H$. For (d): use functions $f_n$ of the form $f_n(x) = c_n e^{-ix^3/3} e^{i\lambda x} f(x/n)$, $\lambda \in \mathbb{R}$, $f \in L^2(\mathbb{R})$.]

**6.9** Let $N \in \mathbb{N}$, $\psi \in C_0^\infty(\mathbb{R})$ and $D = (\vec{Q} \cdot \vec{P} + \vec{P} \cdot \vec{Q})/4$ in $\mathcal{H} = L^2(\mathbb{R}^n)$. Verify that the operator $D^N \varphi(H_0) \langle Q\rangle^{-N}$ belongs to $\mathcal{B}(\mathcal{H})$.

**6.10.** In $L^2(\mathbb{R}^n)$, let $H = H_0 + V(\vec{Q})$, where $V: \mathbb{R}^n \to \mathbb{R}$ is a potential of class $\mathfrak{V}_\infty$ (see page 332).
(a) Show that $\mathcal{D}(H^k) = \mathcal{D}((H_0)^k)$ for each $k \in \mathbb{N}$.
(b) For $z \in \rho(H)$ and $j, k, \ell, m = 1, \ldots, n$, show that the following are bounded operators: $P_j Q_k (H-z)^{-1} \langle Q \rangle^{-1}$, $Q_j Q_k (H-z)^{-1} \langle Q \rangle^{-2}$, $P_j Q_k Q_\ell (H-z)^{-1} \langle Q \rangle^{-2}$ and $P_j P_k Q_\ell Q_m (H-z)^{-1} \langle Q \rangle^{-2}$.
(c) For $N \geq 1$, let $\mathcal{P}^{(N)}$ be a polynomial in $P_1, \ldots, P_n$ of degree $N$, and let $\mathcal{Q}^{(N)}$ be a polynomial in $Q_1, \ldots, Q_n$ of degree $N$. Let $M$ be an integer $\geq N/2$. Show that the operator $\mathcal{P}^{(N)} \mathcal{Q}^{(N)} (H-z)^{-M} \langle Q \rangle^{-N}$ is bounded $[z \in \rho(H)]$.
(d) For $N$, $M$ and $z$ as in (c), show that the operator $D^N (H-z)^{-M} \langle Q \rangle^{-N}$ is bounded.

**6.11.** Let $V, U, W: \mathbb{R}^n \to \mathbb{R}$ satisfy the conditions of Proposition 6.16 and assume in addition that $V(\vec{x}) = U(\vec{x}) W(\vec{x})$. Prove the following statements.
(a) For $z \in \mathbb{C} \setminus \mathbb{R}$, the operator $X_z := W(\vec{Q})(H_0 - z)^{-1} U(\vec{Q})$ is compact, $I + X_z$ is invertible in $\mathcal{B}(\mathcal{H})$ with $(I + X_z)^{-1} = I - W(\vec{Q})(H-z)^{-1} U(\vec{Q})$.
(b) For $z \in \mathbb{C} \setminus \mathbb{R}$, the operator $W(\vec{Q})(H-z)^{-1} U(\vec{Q})$ is compact.
(c) For $\lambda \in (0, \infty) \setminus \sigma_{\mathrm{p}}(H)$, the operators $X_{\lambda + i0}$ and $W(\vec{Q})(H - \lambda - i0)^{-1} U(\vec{Q})$ are compact, $I + X_{\lambda+i0}$ is invertible in $\mathcal{B}(\mathcal{H})$, and

$$(I + X_{\lambda+i0})^{-1} = I - W(\vec{Q})(H - \lambda - i0)^{-1} U(\vec{Q}).$$

(d) If $\lambda > 0$ is such that $I + X_{\lambda+i0}$ is not invertible, then $\lambda$ is an eigenvalue of $H$.
(e) If $\lambda > 0$ is an eigenvalue of $H$, then $I + X_{\lambda+i0}$ is not invertible: if $Hg = \lambda g$ with $g \in \mathcal{D}(H)$, then $f := W(\vec{Q}) g$ satisfies $(I + X_{\lambda+i0}) f = \mathbf{0}$, and $f \neq \mathbf{0}$.

CHAPTER 7

# Further Topics in Scattering Theory

This chapter contains some applications to scattering theory of the results obtained in Chapter 6. In the first section we obtain asymptotic completeness by using properties of relatively smooth operators. In Section 7.2 we prove the relation between scattering into cones and the probability current across distant surfaces. Section 7.3 is devoted to a derivation of the principal equations of stationary state scattering theory as a consequence of the asymptotic condition. These equations allow us to obtain an important stationary expression for the $S$-matrix $S(\lambda)$ in Section 7.4. In the final Section 7.5 we present some aspects of the concept of time delay in scattering theory.

## 7.1 Asymptotic completeness

**7.1.1.** Let us first consider the general situation of §5.3.2: $A$ and $B$ are two self-adjoint operators and $U_t = e^{-iAt}$, $W_t = e^{-iBt}$. We denote by $\{E^A(\cdot)\}$ the spectral measure of $A$, by $\{E^B(\cdot)\}$ that of $B$, by $E^A_{\mathrm{ac}}$ the projection onto $\mathcal{H}_{\mathrm{ac}}(A)$ and by $E^B_{\mathrm{ac}}$ that onto $\mathcal{H}_{\mathrm{ac}}(B)$. Assume that $\Omega \equiv \Omega(A,B;E^B_{\mathrm{ac}}) = \text{s-lim}_{t\to+\infty} U_t^* W_t E^B_{\mathrm{ac}}$ exists. In this general setting one would say that $\Omega$ is complete if $\mathcal{R}(\Omega) = \mathcal{H}_{\mathrm{ac}}(A)$ [we know from Proposition 5.15(d) that $\mathcal{R}(\Omega) \subseteq \mathcal{H}_{\mathrm{ac}}(A)$]. We have the following completeness criterion:

**Proposition 7.1.** *Assume that* $\Omega \equiv \Omega(A,B;E^B_{\mathrm{ac}}) = \text{s-lim}_{t\to+\infty} U_t^* W_t E^B_{\mathrm{ac}}$ *exists. Then* $\mathcal{R}(\Omega) = \mathcal{H}_{\mathrm{ac}}(A)$ *if and only if* $\Omega(B,A;E^A_{\mathrm{ac}}) = \text{s-lim}_{t\to+\infty} W_t^* U_t E^A_{\mathrm{ac}}$ *exists.*

PROOF. (i) Assume that $\mathcal{R}(\Omega) = \mathcal{H}_{\mathrm{ac}}(A)$. Then, given $g \in \mathcal{H}_{\mathrm{ac}}(A)$, there exists a vector $f \in \mathcal{H}_{\mathrm{ac}}(B)$ such that $g = \Omega f$, so that

$$\|W_t^* U_t g - f\| = \|g - U_t^* W_t f\| \to 0 \qquad \text{as } t \to +\infty.$$

This shows that $W_t^* U_t g$ is strongly convergent. Hence $\Omega(B,A;E^A_{\mathrm{ac}})$ exists.

(ii) Assume that $\Omega(B,A;E^A_{\mathrm{ac}})$ exists. Let $g \in \mathcal{H}_{\mathrm{ac}}(A)$ and $f = \Omega(B,A;E^A_{\mathrm{ac}})g$. One has $f \in \mathcal{H}_{\mathrm{ac}}(B)$ [see Proposition 5.15(d)] and $\|f - W_t^* U_t g\| \to 0$ as $t \to \infty$, hence $\|g - U_t^* W_t f\| \to 0$. In other terms, $g = \Omega(A,B;E^B_{\mathrm{ac}})f \equiv \Omega f$, hence $g \in \mathcal{R}(\Omega)$. □

The next theorem gives a criterion for the existence of wave operators. It is entirely symmetric in $A$ and $B$ (except for a sign, which is of no importance since, if $C$ is an $H$-smooth operator, then so is $-C$). Thus, in applications this theorem should allow one to prove at the same time the existence and the completeness of wave operators (let us recall that, for a class of non-relativistic Hamiltonians in dimension $n \geq 3$, the existence of the wave operators has already been obtained in Section 5.4, so in this case the theorem can be used to obtain their completeness).

**Proposition 7.2 (Lavine's Theorem).** *Let $A$ and $B$ be self-adjoint operators in a Hilbert space $\mathcal{H}$. Assume that $A - B$ can be factored into $A - B = C_1^* C_2$ in such a way that $C_1$ is A-bounded, $C_2$ is B-bounded, $C_1$ is A-smooth on some open interval $J$ and $C_2$ is B-smooth on $J$. More explicitly, assume that there exists an operator $C_1$ that is A-smooth on $J$ and an operator $C_2$ that is B-smooth on $J$ such that $C_1(A+i)^{-1} \in \mathcal{B}(\mathcal{H})$, $C_2(B+i)^{-1} \in \mathcal{B}(\mathcal{H})$ and*

$$\langle Af, g\rangle - \langle f, Bg\rangle = \langle C_1 f, C_2 g\rangle \qquad \forall f \in \mathcal{D}(A), \forall g \in \mathcal{D}(B). \tag{7.1}$$

*Then* $\operatorname{s-lim}_{t\to\pm\infty} e^{iAt}e^{-iBt}E^B(J)$ *and* $\operatorname{s-lim}_{t\to\pm\infty} e^{iBt}e^{-iAt}E^A(J)$ *exist.*

PROOF. If $C$ is $H$-smooth, then so is $-C$. Thus, due to the symmetry of the hypotheses in $A$ and $B$ (except for a sign), it is enough to prove the existence of the limits $\operatorname{s-lim}_{t\to\pm\infty} e^{iAt}e^{-iBt}E^B(J)$. We treat the case $t \to +\infty$. By virtue of Proposition 2.2 it suffices to show that $\operatorname{s-lim}_{t\to+\infty} e^{iAt}e^{-iBt}E^B(J)f$ exists for a set $\mathcal{D}$ of vectors $f$ which is dense in $\mathcal{R}(E^B(J))$. We take for $\mathcal{D}$ the set of vectors having compact support in $J$ with respect to the spectral measure of $B$; so $f \in \mathcal{D}$ if and only if there exists a bounded closed subinterval $\mathfrak{J} = [\alpha, \beta]$ of $J$ with $E^B(\mathfrak{J})f = f$. We shall show that, for such vectors $f$,

$$\operatorname*{s-lim}_{t\to+\infty} E^A(J)e^{iAt}e^{-iBt}f \quad \text{exists} \tag{7.2}$$

and

$$\operatorname*{s-lim}_{t\to+\infty} [I - E^A(J)]e^{iAt}e^{-iBt}f = \mathbf{0}, \tag{7.3}$$

which implies the existence of $\Omega(A, B; E^B(J))$. Throughout the proof we assume that $f$ is a fixed vector in $\mathcal{D}$.

(i) Let us first verify (7.2). We set $X(t) = e^{iAt}e^{-iBt}$. For $0 < s < t < \infty$ and $g \in \mathcal{D}(A)$ one has:

$$\begin{aligned}
&\langle g, E^A(J)[X(t) - X(s)]f\rangle = \langle E^A(J)g, [X(t) - X(s)]E^B(J)f\rangle \\
&= \int_s^t \frac{d}{d\tau}\langle E^A(J)g, X(\tau)E^B(J)f\rangle\, d\tau = \int_s^t \frac{d}{d\tau}\langle E^A(J)e^{-iA\tau}g, E^B(J)e^{-iB\tau}f\rangle\, d\tau \\
&= i\int_s^t \left[\langle AE^A(J)e^{-iA\tau}g, E^B(J)e^{-iB\tau}f\rangle - \langle E^A(J)e^{-iA\tau}g, BE^B(J)e^{-iB\tau}f\rangle\right] d\tau \\
&= i\int_s^t \langle C_1E^A(J)e^{-iA\tau}g, C_2E^B(J)e^{-iB\tau}f\rangle\, d\tau.
\end{aligned}$$

By using the Schwarz inequality in $\mathcal{H}$ and then in $L^2((s,t))$, one obtains

$$\begin{aligned}
&\big|\langle g, E^A(J)[X(t)-X(s)]f\rangle\big| \leq \int_s^t \|C_1E^A(J)e^{-iA\tau}g\|\|C_2E^B(J)e^{-iB\tau}f\|\,d\tau \\
&\leq \Big[\int_s^t \|C_1E^A(J)e^{-iA\tau}g\|^2\,d\tau\Big]^{1/2}\Big[\int_s^t \|C_2E^B(J)e^{-iB\tau}f\|^2\,d\tau\Big]^{1/2} \\
&\leq \Big[\int_{-\infty}^{\infty} \|C_1E^A(J)e^{-iA\tau}g\|^2\,d\tau\Big]^{1/2}\Big[\int_s^t \|C_2E^B(J)e^{-iB\tau}f\|^2\,d\tau\Big]^{1/2} \\
&\leq c\|g\|\Big[\int_s^t \|C_2E^B(J)e^{-iB\tau}f\|^2\,d\tau\Big]^{1/2},
\end{aligned}$$

where the last inequality follows from Proposition 6.11 since $C_1E^A(J)$ is $A$-smooth. Thus

$$\begin{aligned}
\|E^A(J)[X(t)-X(s)]f\| &= \sup_{g\in\mathcal{D}(A),\|g\|=1} \big|\langle g, E^A(J)[X(t)-X(s)]f\rangle\big| \\
&\leq c\Big[\int_s^t \|C_2E^B(J)e^{-iB\tau}f\|^2\,d\tau\Big]^{1/2}. \qquad (7.4)
\end{aligned}$$

Since $C_2$ is $B$-smooth on $J$, one has $\int_{-\infty}^{\infty} \|C_2E^B(J)e^{-iB\tau}f\|^2 d\tau < \infty$, hence (7.4) implies that $\{E^A(J)X(t)f\}$ is strongly Cauchy as $t \to +\infty$. This proves (7.2).

(ii) The proof of (7.3) is based on the fact that, for each $z \in \mathbb{C}\setminus\mathbb{R}$:

$$\operatorname*{s-lim}_{t\to+\infty} \big[(A-z)^{-1} - (B-z)^{-1}\big]e^{-iBt}f = \mathbf{0} \qquad (f \in \mathcal{D}). \tag{7.5}$$

In our applications, Eq. (7.5) will be satisfied since $f$ will belong to $\mathcal{H}_{\rm ac}(B)$, hence $e^{-iBt}f \to \mathbf{0}$ weakly as $t \to +\infty$ [see Proposition 5.7(a)], and because $(A-z)^{-1} - (B-z)^{-1} = (B-z)^{-1}(B-A)(A-z)^{-1}$ will be compact (we usually assume $B-A$ to be $A$-compact). However, these special assumptions are not necessary, because (7.5) is satisfied under the hypotheses of the proposition; this will be seen in part (iii) of the proof, but we first show that (7.5) implies (7.3).

We use the following observation. Let $\mathfrak{J} = [\alpha,\beta]$ be a closed subinterval of $J$ such that $E^B(\mathfrak{J})f = f$. Let $\Upsilon$ be a closed contour in the complex plane, oriented as in Figure 7.1(a), containing $\mathfrak{J}$ in its interior and intersecting the real axis in two points belonging to $J$ (say at distance $\delta > 0$ from $\alpha$ and from $\beta$). Then the operator $(B-z)^{-1}E^B(\mathfrak{J})$ belongs to $\mathcal{B}(\mathcal{H})$ for each $z$ on $\Upsilon$, with $\|(B-z)^{-1}E^B(\mathfrak{J})\| = [\mathrm{dist}(\Upsilon,\mathfrak{J})]^{-1}$. In particular the integral

$$\frac{1}{2\pi i}\int_\Upsilon (B-z)^{-1}E^B(\mathfrak{J})\,dz \tag{7.6}$$

exists in $\mathcal{B}(\mathcal{H})$. Thus, for each $g \in \mathcal{H}$, the following integral exists:

$$\frac{1}{2\pi i}\int_\Upsilon \langle g, (B-z)^{-1}f\rangle\,dz \tag{7.7}$$

[remember that $f = E^B(\mathfrak{J})f$], and the integrand is a holomorphic function in the domain $\mathbb{C}\backslash[\alpha,\beta]$. The contour of integration $\Upsilon$ can be deformed (by keeping its points of intersection with the real axis fixed) without changing the value of the integrals in (7.6) and (7.7). In particular

$$\frac{1}{2\pi i}\int_{\Upsilon}\langle g,(B-z)^{-1}f\rangle\,dz = \frac{1}{2\pi i}\int_{\Upsilon_\varepsilon}\langle g,(B-z)^{-1}f\rangle\,dz$$

if $\Upsilon_\varepsilon\,(\varepsilon>0)$ is as indicated in Figure 7.1(b). By letting $\varepsilon$ tend to 0, one finds by taking into account Stone's Formula (4.80) that

$$\frac{1}{2\pi i}\int_{\Upsilon}\langle g,(B-z)^{-1}f\rangle\,dz = \langle g, E^B((\alpha-\delta,\beta+\delta))f\rangle = \langle g,f\rangle.$$

Since this identity holds for each $g\in\mathcal{H}$, one has

$$\frac{1}{2\pi i}\int_{\Upsilon}(B-z)^{-1}f\,dz = f \equiv E^B(\mathfrak{J})f. \tag{7.8}$$

If $h$ is a vector such that $E^B(J)h = \mathbf{0}$, then the function $z\mapsto\langle g,(B-z)^{-1}h\rangle$ is holomorphic in $[\mathbb{C}\backslash\mathbb{R}]\cup J$, in particular in the interior of $\Upsilon$, and one obtains

$$\frac{1}{2\pi i}\int_{\Upsilon}(B-z)^{-1}h\,dz = \mathbf{0}. \tag{7.9}$$

The condition $E^B(J)h=\mathbf{0}$ is equivalent to $[I-E^B(J)]h = h$; hence (7.9) is satisfied for each $h\in\mathcal{R}(I-E^B(J))$, in other terms one has

$$\frac{1}{2\pi i}\int_{\Upsilon}(B-z)^{-1}[I-E^B(J)]\,dz = O. \tag{7.10}$$

By using (7.8) we may write

$$[I-E^A(J)]X(t)f = \frac{1}{2\pi i}\int_{\Upsilon}[I-E^A(J)]X(t)(B-z)^{-1}f\,dz. \tag{7.11}$$

By using (7.10) (for $A$ in place of $B$) we have

$$\frac{1}{2\pi i}\int_{\Upsilon}(A-z)^{-1}[I-E^A(J)]X(t)f\,dz = \mathbf{0}. \tag{7.12}$$

By combining (7.11) and (7.12) one finds that

$$[I-E^A(J)]X(t)f = \frac{1}{2\pi i}\int_{\Upsilon}[I-E^A(J)]e^{iAt}[(B-z)^{-1}-(A-z)^{-1}]e^{-iBt}f\,dz,$$

hence

$$\|[I-E^A(J)]X(t)f\| \le \frac{1}{2\pi}\int_{\Upsilon}\|[I-E^A(J)][(B-z)^{-1}-(A-z)^{-1}]e^{-iBt}f\|\,dz. \tag{7.13}$$

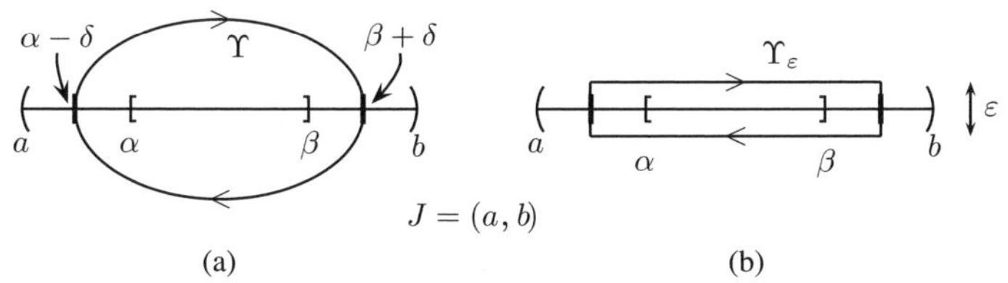


**Figure 7.1**

By (7.5), the integrand in (7.13) converges to zero as $t \to +\infty$ for almost all $z$ on $\Upsilon$. It is majorised by

$$\begin{aligned}&\Big\{\sup_{z\in\Upsilon}\|(B-z)^{-1}E^B(\mathfrak{J})\| + \sup_{z\in\Upsilon}\|[I-E^A(J)](A-z)^{-1}\|\Big\}\|f\| \\ &\quad = \Big\{[\mathrm{dist}(\Upsilon,\mathfrak{J})]^{-1} + [\mathrm{dist}(\Upsilon,\mathbb{R}\backslash J]^{-1}\Big\}\|f\|,\end{aligned}$$

which is a finite constant. The Dominated Convergence Theorem (Proposition 1.9), applied to the integral in (7.13), implies that $\|[I-E^A(J)]W(t)f\| \to 0$ as $t \to +\infty$, which proves (7.3).

(iii) Finally let us prove (7.5) from the hypotheses of the proposition. For any $g \in \mathcal{H}$ we may write by using the assumption (7.1):

$$\begin{aligned}&\big|\langle g,[(A-z)^{-1}-(B-z)^{-1}]e^{-iBt}f\rangle\big| \\ &\quad = \big|\langle (A-\overline{z})^{-1}g,(B-z)(B-z)^{-1}e^{-iBt}f\rangle \\ &\qquad - \langle (A-\overline{z})(A-\overline{z})^{-1}g,(B-z)^{-1}e^{-iBt}f\rangle\big| \\ &\quad = \big|\langle (A-\overline{z})^{-1}g,B(B-z)^{-1}e^{-iBt}f\rangle - \langle A(A-\overline{z})^{-1}g,(B-z)^{-1}e^{-iBt}f\rangle\big| \\ &\quad = \big|\langle C_1(A-\overline{z})^{-1}g,C_2(B-z)^{-1}e^{-iBt}f\rangle\big| \\ &\quad \le \|C_1(A-\overline{z})^{-1}\|\|g\|\|C_2(B-z)^{-1}e^{-iBt}f\|.\end{aligned}$$

Since $C_1$ is assumed to be $A$-bounded, one has $\|C_1(A-\overline{z})^{-1}\| \equiv c_1 < \infty$. Hence, remembering that $f = E^B(J)f$:

$$\begin{aligned}&\|[(A-z)^{-1}-(B-z)^{-1}]e^{-iBt}f\| \\ &\qquad = \sup_{g\in\mathcal{H},\|g\|=1}\big|\langle g,[(A-z)^{-1}-(B-z)^{-1}]e^{-iBt}f\rangle\big| \\ &\qquad \le c_1\|C_2(B-z)^{-1}e^{-iBt}f\| = c_1\|C_2E^B(J)e^{-iBt}(B-z)^{-1}f\|.\end{aligned}$$

Thus it suffices to show that $\lim_{t\to+\infty}\|C_2E^B(J)e^{-iBt}(B-z)^{-1}f\| = 0$. We have for $0 < s < t < \infty$:

$$\begin{aligned}&\|C_2E^B(J)e^{-iBt}(B-z)^{-1}f\|^2 - \|C_2E^B(J)e^{-iBs}(B-z)^{-1}f\|^2 \\ &\qquad = \int_s^t \frac{d}{d\tau}\|C_2E^B(J)e^{-iB\tau}(B-z)^{-1}f\|^2\,d\tau,\end{aligned}$$

hence it is enough to show that $(d/d\tau)\|C_2 E^B(J)e^{-iB\tau}(B-z)^{-1}f\|^2$ belongs to $L^1((0,\infty))$. Now

$$\begin{aligned}
&\frac{d}{d\tau}\|C_2 E^B(J)e^{-iB\tau}(B-z)^{-1}f\|^2 \\
&\quad = \frac{d}{d\tau}\langle C_2 E^B(J)e^{-iB\tau}(B-z)^{-1}f, C_2 E^B(J)e^{-iB\tau}(B-z)^{-1}f\rangle \\
&\quad = i\langle C_2 E^B(J)e^{-iB\tau}B(B-z)^{-1}f, C_2 E^B(J)e^{-iB\tau}(B-z)^{-1}f\rangle \\
&\qquad - i\langle C_2 E^B(J)e^{-iB\tau}(B-z)^{-1}f, C_2 E^B(J)e^{-iB\tau}B(B-z)^{-1}f\rangle.
\end{aligned}$$

Thus

$$\begin{aligned}
&\left|\frac{d}{d\tau}\|C_2 E^B(J)e^{-iB\tau}(B-z)^{-1}f\|^2\right| \\
&\quad \le 2\|C_2 E^B(J)e^{-iB\tau}(B-z)^{-1}f\|\|C_2 E^B(J)e^{-iB\tau}B(B-z)^{-1}f\|. \qquad (7.14)
\end{aligned}$$

Since $C_2$ is $B$-smooth on $J$, each of the norms on the right-hand side of (7.14) belongs to $L^2(\mathbb{R})$ (as functions of $\tau$), so by the Schwarz inequality the right-hand side of (7.14) belongs to $L^1(\mathbb{R})$. □

**7.1.2.** We now give an application of Lavine's Theorem to non-relativistic Hamiltonians. We again consider the simple situation of §5.4.1, i.e. $H_0 = \vec{P}^2$ is the free Hamiltonian in $L^2(\mathbb{R}^n)$ and $H = H_0 + V(\vec{Q})$ the total Hamiltonian. We have seen in Section 5.4 that, if $n \ge 3$, it is rather easy to prove the existence of the wave operators $\Omega_\pm = \text{s-lim}_{t\to\pm\infty} e^{iHt}e^{iH_0t}$. Sufficient conditions for the existence of these limits have been given in Proposition 5.20.

The concept of asymptotic completeness was introduced in §5.5.1. One speaks of ***strong asymptotic completeness*** if $\mathcal{R}(\Omega_-) = \mathcal{R}(\Omega_+) = \mathcal{H}_{\text{ac}}(H)$ and $\sigma_{\text{sc}}(H) = \varnothing$.

In what follows we denote by $\{E(\cdot)\}$ the spectral measure of $H$ and by $\{E_0(\cdot)\}$ that of $H_0$. We shall consider only potentials satisfying the condition (C) of §6.4.1 (page 288). Then the Hamiltonian has no singular continuous spectrum, $\sigma_{\text{ac}}(H) = [0,\infty)$ and the eigenvalues of $H$ in $(0,\infty)$ form a discrete set. Hence $(0,\infty)\backslash\sigma_{\text{p}}(H)$ is the union of at most a countable number of open intervals $\{J_k\}$ [1], and $E((0,\infty)\backslash\sigma_{\text{p}}(H)) \equiv \sum_k E(J_k)$ is the projection onto $\mathcal{H}_{\text{ac}}(H)$. Let $\mathcal{D}$ be the union of all subspaces of the form $E(J)\mathcal{H}$, where $J$ runs over all open subintervals of $(0,\infty)\backslash\sigma_{\text{p}}(H)$ such that $\overline{J}$ (the closure of $J$) still belongs to $(0,\infty)\backslash\sigma_{\text{p}}(H)$; then $\mathcal{D}$ is a total subset in $\mathcal{H}_{\text{ac}}(H)$. Similarly the union of all subspaces of the form $E_0(J)\mathcal{H}$, where $J$ runs over the same set of open intervals, is a total set in $\mathcal{H}_{\text{ac}}(H_0) \equiv \mathcal{H}$.

To obtain the existence and the completeness of the wave operators, it suffices to show that $\text{s-lim}_{t\to\pm\infty} e^{iHt}e^{-iH_0t}E_0(J)$ and $\text{s-lim}_{t\to\pm\infty} e^{iH_0t}e^{-iHt}E(J)$ exist for each open subinterval $J$ of $(0,\infty)\backslash\sigma_{\text{p}}(H)$ such that $\overline{J}$ also belongs to $(0,\infty)\backslash\sigma_{\text{p}}(H)$. For this we apply Lavine's Theorem (Proposition 7.2). We must factor the operator $V(\vec{Q})$ into $V(\vec{Q}) = C_1C_2$, where $C_1$ must be $H$-bounded and $H$-smooth on $J$, and $C_2$ must be $H_0$-bounded and $H_0$-smooth on $J$ (with $J$ an interval of the indicated

[1] We do not take into account the fact (not proved in this text) that $\sigma_{\text{p}}(H) \cap (0,\infty)$ is empty.

class). The simplest way is to use a symmetrical factorisation of $V$. Let us define $U_1, U_2 \colon \mathbb{R}^n \to \mathbb{R}$ by $U_1(\vec{x}) = |V(\vec{x})|^{1/2}$ and $U_2(\vec{x}) = |V(\vec{x})|^{1/2} \operatorname{sign} V(\vec{x})$. Then $V(\vec{Q}) = U_1(\vec{Q})U_2(\vec{Q})$. Let us take $C_1 = U_1(\vec{Q})$ and $C_2 = U_2(\vec{Q})$. We apply Proposition 6.16 (first with $W = U_1$ and $V$ equal to the considered potential, then with $W = U_2$ and $V$ equal to zero); it follows that $U_1$ is $H$-smooth on $\overline{J}$ (hence also on $J$) [2] and that $U_2$ is $H_0$-smooth on $\overline{J}$ (hence also on $J$) if $U_1$ is of the following form:

$$U_1(\vec{x}) = [w_1(\vec{x}) + w_2(\vec{x})](1 + |\vec{x}|)^{-\nu}$$

with $\nu > 1/2$, $w_1 \in L^\infty(\mathbb{R}^n)$ and $w_2 \in L^p(\mathbb{R}^n)$ for some $p > n$ and $\geq 2$. One obtains in this way the following (non-optimal, of course) result on the existence and completeness of wave operators:

**Proposition 7.3.** *Let $H_0 = \vec{P}^2$ and $H = H_0 + V(\vec{Q})$. Assume that $V$ is a real function of the form $V(\vec{x}) = (1 + |\vec{x}|)^{-1-\delta}[v_\infty(\vec{x}) + v_q(\vec{x})]$ with $\delta > 0$, $v_\infty \in L^\infty(\mathbb{R}^n)$ and $v_q \in L^q(\mathbb{R}^n)$ for some $q > n/2$ (and $q \geq 1$) and that*
*(i) if $n = 1, 2$ or $3$ : $V$ and $r(\partial V/\partial r)$ belong to $L^2(\mathbb{R}^n) + L^\infty_\varepsilon(\mathbb{R}^n)$, and $r^2(\partial^2 V/\partial r^2)$ belongs to $L^2(\mathbb{R}^n) + L^\infty(\mathbb{R}^n)$,*
*(ii) if $n \geq 4$ : $V$ and $r(\partial V/\partial r)$ belong to $L^p(\mathbb{R}^n) + L^\infty_\varepsilon(\mathbb{R}^n)$ and $r^2(\partial^2 V/\partial r^2) \in L^p(\mathbb{R}^n) + L^\infty(\mathbb{R}^n)$ for some $p \in (n/2, \infty)$.*
*Then the wave operators $\Omega_\pm = \text{s-lim}_{t\to\pm\infty} e^{iHt}e^{-iH_0t}$ exist and satisfy strong asymptotic completeness.*

PROOF. The conditions (i) and (ii) guarantee that (C) of §6.4.1 is satisfied. The assumption $V(\vec{x}) = (1 + |\vec{x}|)^{-(1-\delta)}[v_\infty(\vec{x}) + v_q(\vec{x})]$ ($q \geq 1, q > n/2$) implies that

$$|U_1(\vec{x})| \equiv |V(\vec{x})|^{1/2} \leq (1 + |\vec{x}|)^{-(1-\delta)/2}[w_1(\vec{x}) + w_2(\vec{x})],$$

where $w_1(\vec{x}) = |v_\infty(\vec{x})|^{1/2} \in L^\infty(\mathbb{R}^n)$ and $w_2(\vec{x}) = |v_q(\vec{x})|^{1/2} \in L^p(\mathbb{R}^n)$ with $p = 2q$ (hence $p > n$ and $p \geq 2$). The conclusions of the proposition follow from the remarks made before its statement; the assumptions of Proposition 6.16 are satisfied with $\nu = 1/2 + \delta/2 > 1/2$. □

EXAMPLES. We give some very simple examples of spherically symmetric potentials for which the hypotheses of Proposition 7.3 are satisfied, in the case $n = 3$ (it is clear that the spherical symmetry is not essential).
(a) The Yukawa potential $V(r) = \gamma e^{-\mu r}/r$ with $\gamma \in \mathbb{R}$, $\mu > 0$ (take for example $\delta = 1$, $v_\infty(r) = 0$ and $v_q(r) = \gamma(1 + r)^2 e^{-\mu r}/r$ with $q = 2$).
(b) The potentials $V(r) = \gamma(1 + r)^{-\beta}$ with $\gamma \in \mathbb{R}$ and $\beta > 1$ (take for example $\delta = \beta - 1$, $v_q(r) = 0$ and $v_\infty(r) = \gamma$).
(c) The potentials $\gamma r^{-\beta}$ with $\gamma \in \mathbb{R}$ and $1 < \beta < 3/2$ [take for example $\delta = \beta - 1$, $v_\infty(r) = \gamma(1 + r)^\beta r^{-\beta}$ for $r \geq 1$, $v_\infty(r) = 0$ for $r < 1$, $v_q(r) = 0$ for $r \geq 1$, $v_q(r) = \gamma(1 + r)^\beta r^{-\beta}$ for $r < 1$, with $q \in (3/2, 3/\beta)$].
(d) Let us briefly consider the Coulomb potential $V(r) = \gamma r^{-1}$. This is a limiting case of the potentials considered in (c). The condition (C), and hence the assumption

[2] If (6.86) is satisfied for an interval $J$, it is also satisfied if $J$ is replaced by any subset of $J$.

(i) of Proposition 7.3, is satisfied for the Coulomb potential (see page 78). On the other hand one knows that the wave operators do not exist for this potential. In fact it is not possible to write this potential in the form $(1+r)^{-2\nu}[v_\infty(r)+v_q(r)]$ with $\nu > 1/2$ (only with $\nu = 1/2$), because $V(r)$ does not decay sufficiently fast at infinity.

## 7.2 Flux and scattering into cones

When discussing scattering cross sections at the beginning of Section 5.6, we pointed out an important relation [Eq. (5.82)] between the probability of scattering into a cone $\mathcal{C}$ (with vertex at the origin) and the integrated flux over a distant spherical surface in this cone, viz. [3]

$$\mathcal{P}(h;\mathcal{C}) = \lim_{d\to\infty} \int_T^\infty \Phi_{S_d\cap\mathcal{C}}(U_t\Omega_- h)\,dt. \tag{7.15}$$

Here $S_d$ denotes the sphere of radius $d$ in $\mathbb{R}^n$ centred at the origin, and $U_t = e^{-iHt}$. The number $\mathcal{P}(h;\mathcal{C})$ is the probability that the scattering state $U_t\Omega_- h$ associated with an initial state $h$ is localised in the cone $\mathcal{C}$ at time $t = +\infty$:

$$\mathcal{P}(h;\mathcal{C}) = \lim_{t\to+\infty} \int_{\mathcal{C}} |(U_t\Omega_- h)(\vec{x})|^2\,d^n x. \tag{7.16}$$

The flux $\Phi_{\mathcal{O}}(f)$ of a function $f$ across a hypersurface $\mathcal{O}$ is given by the following expression, in which $\vec{n}(\vec{x}) \in \mathbb{R}^n$ is a - suitably oriented - unit vector orthogonal to $\mathcal{O}$ at the point $\vec{x} \in \mathcal{O}$ and $d\sigma(\vec{x})$ denotes the surface element at $\vec{x}$:

$$\Phi_{\mathcal{O}}(f) = 2\,\Im \int_{\mathcal{O}} \overline{f(\vec{x})}\,\vec{n}(\vec{x}) \cdot \vec{\nabla} f(\vec{x})\,d\sigma(\vec{x}). \tag{7.17}$$

In the present section we intend to give a proof of the relation (7.15) for some of the simple scattering systems considered in Section 5.4, under suitable assumptions on the cone $\mathcal{C}$, the state vector $g \equiv \Omega_- h$ and the potential $V$. We are not aware of a sufficiently simple proof in the literature. We shall use here results obtained in Chapter 6 by the conjugate operator method to establish propagation properties for certain observables. A ***propagation estimate*** for an observable, represented by a self-adjoint operator $A$, is an estimate on the large time behaviour (in the Heisenberg picture) of this operator in the subspace of scattering states of the Hamiltonian $H$. A scattering state will at large times be localised in a part of configuration space where the potential is very small. So for example, thinking in terms of classical mechanics, its position $\vec{x}(t)$ at time $t$ will be given as $\vec{x}(t) \approx \vec{x}_0 + (\vec{p}/m)(t-t_0)$ for some constants $\vec{x}_0$ and $t_0$ ($\vec{p}$ is the momentum and $m$ the mass). Hence, as $t \to +\infty$, one will have $|\vec{x}(t)|^{-\kappa} \approx [(|\vec{p}|/m)t]^{-\kappa}$ for any $\kappa > 0$. The quantum-mechanical analogue is an estimate of the form $||U_t^*|\vec{Q}|^{-\kappa}U_t f|| \leq c/t^\kappa$ for large $t$ and for some constant $c$; such an estimate

[3] We assume here that the probability of scattering into a truncated cone $\mathcal{C}_\rho$ is the same as that of scattering into $\mathcal{C}$, i.e. that $\mathcal{P}(h;\mathcal{C}_\rho) = \mathcal{P}(h;\mathcal{C})$ for any $\rho > 0$. This is frequently the case, see for example Proposition 5.27.

is expected to be (approximately) satisfied provided that the scattering state $f$ has a non-zero minimal velocity. In Lemma 7.6 we shall prove a precise inequality of this type for certain scattering states $f$.

In what follows we write $g$ for the vector $\Omega_- h$ occurring in (7.15) and (7.16) [with $g \in \mathcal{H}_{\rm ac}(H)$]. In fact the existence of the wave operators will not be used explicitly and is thus not required to hold.

**7.2.1.** We begin by fixing convenient notations (Figure 7.2) and then restate in these notations the result to be proved. For $n \geq 2$, let $\mathcal{C}$ be a fixed closed cone in $\mathbb{R}^n$ with vertex at the origin ($\mathcal{C} \neq \mathbb{R}^n$). We assume that $\mathcal{C}$ is convex and with smooth lateral surface.[4] If $0 < \rho < d < \infty$ we use the notations $\mathcal{C}_{\rho,d}$ for the doubly truncated cone $\mathcal{C}_{\rho,d} = \{\vec{x} \in \mathcal{C} \mid \rho < |\vec{x}| < d\}$ and $\mathcal{C}_\rho$ for the simply truncated cone corresponding to $d = \infty$, i.e. $\mathcal{C}_\rho = \{\vec{x} \in C \mid |\vec{x}| > \rho\}$. The boundary $\partial\mathcal{C}_{\rho,d}$ of the doubly truncated cone $\mathcal{C}_{\rho,d}$ is the union of two spherical surfaces $\Sigma_\rho = \{\vec{x} \in \mathcal{C} \mid |\vec{x}| = \rho\}$ and $\Sigma_d = \{\vec{x} \in \mathcal{C} \mid |\vec{x}| = d\}$ and of the lateral surface $\Lambda_{\rho,d} = \{\vec{x} \in \partial\mathcal{C} \mid \rho < |\vec{x}| < d\}$. We also set $\Lambda_\rho = \{\vec{x} \in \partial\mathcal{C} \mid |\vec{x}| > \rho\}$. For $\vec{x} \in \partial\mathcal{C}_{\rho,d}$, we let $\vec{n}(\vec{x})$ be the *outward* unit vector at the point $\vec{x}$. Finally we use the following notation for the integrated flux through a hypersurface $\mathcal{O}$:

$$\mathcal{I}_{\mathcal{O}}(g, T) = \int_T^\infty \Phi_{\mathcal{O}}(U_t g)\, dt. \tag{7.18}$$

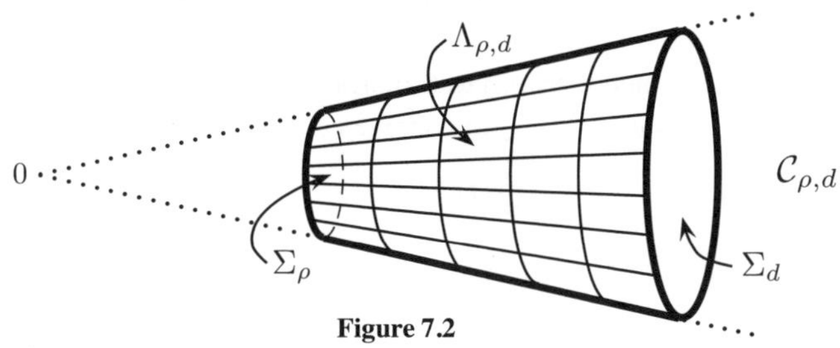


**Figure 7.2**

In the preceding notations the result to be established (sometimes called the **Flux-Across-Surfaces Theorem** in the literature) is as follows: if $g \in \mathcal{M}_\infty^+(H)$, then one has for each $T \in \mathbb{R}$:

$$\lim_{t\to+\infty} \|\chi_{\mathcal{C}}(\vec{Q})U_t g\|^2 = \lim_{d\to\infty} \int_T^\infty \Phi_{\Sigma_d}(U_\tau g)\, d\tau \equiv \lim_{d\to\infty} \mathcal{I}_{\Sigma_d}(g, T), \tag{7.19}$$

where $\chi_{\mathcal{C}}(\vec{Q})$ is the multiplication operator in $L^2(\mathbb{R}^n)$ by the characteristic function $\chi_{\mathcal{C}}$ of the cone $\mathcal{C}$.

[4] Our considerations can be applied under considerably weaker assumptions. For example in $n = 3$ dimensions, it suffices to assume that $\mathcal{C}$ is convex and such that its lateral surface can be divided by $N$ straight lines, each passing through the vertex at $\vec{x} = 0$, into $N$ smooth disjoint surfaces ($N < \infty$). By considering disjoint unions or differences of such cones, the final result of this section – the Flux-Across-Surfaces Theorem – is easily seen to hold for a quite general class of cones.

Conditions on the potential $V$ and on the state vector $g$ that are suitable for our proof of (7.19) will be given further on. For the moment we just assume that $H = H_0 + V(\vec{Q})$ is self-adjoint on $\mathcal{D}(H_0)$ and that all states in $\mathcal{H}_{\rm ac}(H)$ are scattering states under $U_t$, i.e. that $\mathcal{H}_{\rm ac}(H) \subseteq \mathcal{M}_\infty^+(H)$.

Let us give an outline of the proof of (7.19). Assume that $g \in \mathcal{D}(H) \cap \mathcal{H}_{\rm ac}(H)$. Then, since $g$ is a scattering state, we have for any $\rho > 0$:

$$\lim_{t\to+\infty} \|\chi_{\mathcal{C}}(\vec{Q})U_t g\|^2 = \lim_{t\to+\infty} \|\chi_{\mathcal{C}_\rho}(\vec{Q})U_t g\|^2.$$

Furthermore

$$\|\chi_{\mathcal{C}_\rho}(\vec{Q})U_t g\|^2 = \lim_{d\to\infty} \|\chi_{\mathcal{C}_{\rho,d}}(\vec{Q})U_t g\|^2 = -\lim_{d\to\infty} \int_t^\infty \frac{d}{d\tau}\|\chi_{\mathcal{C}_{\rho,d}}(\vec{Q})U_\tau g\|^2 \, d\tau. \tag{7.20}$$

Now

$$\begin{aligned} \frac{d}{d\tau}\|\chi_{\mathcal{C}_{\rho,d}}(\vec{Q})U_\tau g\|^2 &= \frac{d}{d\tau}\langle U_\tau g, \chi_{\mathcal{C}_{\rho,d}}(\vec{Q})U_\tau g\rangle \\ &= i\big[\langle HU_\tau g, \chi_{\mathcal{C}_{\rho,d}}(\vec{Q})U_\tau g\rangle - \langle U_\tau g, \chi_{\mathcal{C}_{\rho,d}}(\vec{Q})HU_\tau g\rangle\big] \\ &= i\big[\langle H_0U_\tau g, \chi_{\mathcal{C}_{\rho,d}}(\vec{Q})U_\tau g\rangle - \langle U_\tau g, \chi_{\mathcal{C}_{\rho,d}}(\vec{Q})H_0U_\tau g\rangle\big], \end{aligned} \tag{7.21}$$

since $\chi_{\mathcal{C}_{\rho,d}}(\vec{Q})$ commutes with the potential. The expression on the last line has the form $i[\langle H_0 f, \chi_{\mathcal{C}_{\rho,d}}(\vec{Q})f\rangle - \langle f, \chi_{\mathcal{C}_{\rho,d}}(\vec{Q})H_0 f\rangle]$, with $f = U_\tau g$. By writing $H_0 = -\Delta$ and then using the divergence theorem to transform the resulting volume integral into an integral over the boundary of $\mathcal{C}_{\rho,d}$, one arrives at

$$\begin{aligned} & i\big[\langle H_0 f, \chi_{\mathcal{C}_{\rho,d}}(\vec{Q})f\rangle - \langle f, \chi_{\mathcal{C}_{\rho,d}}(\vec{Q})H_0 f\rangle\big] \\ &\quad = -i\int_{\mathcal{C}_{\rho,d}} \big[f(\vec{x})\Delta\overline{f(\vec{x})} - \overline{f(\vec{x})}\Delta f(\vec{x})\big]\, d^n x \\ &\quad = -i\int_{\mathcal{C}_{\rho,d}} \vec{\nabla}\big[f(\vec{x})\vec{\nabla}\overline{f(\vec{x})} - \overline{f(\vec{x})}\vec{\nabla} f(\vec{x})\big]\, d^n x \\ &\quad = -i\int_{\partial\mathcal{C}_{\rho,d}} \big[f(\vec{x})\vec{\nabla}\overline{f(\vec{x})} - \overline{f(\vec{x})}\vec{\nabla} f(\vec{x})\big]\cdot\vec{n}(\vec{x})\, d\sigma(\vec{x}) = -\Phi_{\partial\mathcal{C}_{\rho,d}}(f). \end{aligned} \tag{7.22}$$

The application of the divergence theorem above is justified under the condition that $f$ is a function of class $C^2$. However, by an approximation argument, the validity of (7.22) can be shown to hold for all $f \in \mathcal{D}(H_0)$.[5] Then (7.21) leads to the relation

$$\frac{d}{d\tau}\|\chi_{\mathcal{C}_{\rho,d}}(\vec{Q})U_\tau g\|^2 = -\big[\Phi_{\Sigma_\rho}(U_\tau g) + \Phi_{\Sigma_d}(U_\tau g) + \Phi_{\Lambda_{\rho,d}}(U_\tau g)\big]. \tag{7.23}$$

---

[5] If $f \in \mathcal{D}(H_0)$, choose a sequence $\{g_k\}$ in $\mathcal{S}(\mathbb{R}^n)$ such that $\text{s-lim}_{k\to\infty} g_k = (H_0 + 1)f$. Then $f_k := (H_0+1)^{-1}g_k$ belongs to $\mathcal{S}(\mathbb{R}^n)$ and $\text{s-lim}_{k\to\infty} f_k = f$, hence also $\text{s-lim}_{k\to\infty} H_0 f_k = H_0 f$. By Eq. (7.22) we have $i[\langle H_0 f_k, \chi_{\mathcal{C}_{\rho,d}}(\vec{Q})f_k\rangle - \langle f_k, \chi_{\mathcal{C}_{\rho,d}}(\vec{Q})H_0 f_k\rangle] = -\Phi_{\partial\mathcal{C}_{\rho,d}}(f_k)$ for each $k$, and the validity of (7.22) for $f$ follows upon letting $k \to \infty$ in this equation, using the fact that $\Phi_{\partial\mathcal{C}_{\rho,d}}(f_k)$ converges to $\Phi_{\partial\mathcal{C}_{\rho,d}}(f)$ by (7.37) with $g = f_k$.

Assuming that, as a function of $\tau$, each of the fluxes occurring on the right-hand side is integrable on $[t, \infty)$, insertion of (7.23) into (7.20) shows that, for any $\rho > 0$:

$$\|\chi_{\mathcal{C}_\rho}(\vec{Q})U_t g\|^2 = \mathcal{I}_{\Sigma_\rho}(g,t) + \lim_{d\to\infty}\big[\mathcal{I}_{\Sigma_d}(g,t) + \mathcal{I}_{\Lambda_{\rho,d}}(g,t)\big]. \tag{7.24}$$

We shall prove in §7.2.5 that under convenient conditions on $V$ and $g$:

$$\Phi_{\Lambda_{\rho,d}}(U_\tau g) \in L^1([1,\infty)),\ \lim_{d\to\infty}\mathcal{I}_{\Lambda_{\rho,d}}(g,t) \text{ exists, and } \lim_{t\to+\infty}\lim_{d\to\infty}\mathcal{I}_{\Lambda_{\rho,d}}(g,t) = 0. \tag{7.25}$$

It then follows from Eq. (7.24) (since its left-hand side does not depend on $d$) that $\lim_{d\to\infty}\mathcal{I}_{\Sigma_d}(g,t)$ exists, and we get

$$\lim_{t\to+\infty}\|\chi_{\mathcal{C}}(\vec{Q})U_t g\|^2 = \lim_{t\to+\infty}\mathcal{I}_{\Sigma_\rho}(g,t) + \lim_{t\to+\infty}\big[\lim_{d\to\infty}\mathcal{I}_{\Sigma_d}(g,t)\big].$$

To arrive at (7.19), it remains to know the following results (they will be established in §7.2.4):

$$\Phi_{\Sigma_b}(U_\tau g) \in L^1([1,\infty)) \quad \text{and} \quad \lim_{t\to+\infty}\mathcal{I}_{\Sigma_b}(g,t) = 0 \quad \text{for each } b > 0, \tag{7.26}$$

$$\lim_{d\to\infty}\mathcal{I}_{\Sigma_d}(g,t) \text{ is independent of } t. \tag{7.27}$$

It is instructive to consider the interpretation of (7.25) - (7.27) in terms of propagation properties of scattering states. Here (7.25) essentially means that, at large times $t$, the flux through the lateral surface $\Lambda$ of a cone $\mathcal{C}$ is small (sufficiently small that its integral up to $t = \infty$ is finite). One can expect this to be the case because at large times the particle current is essentially radial, so that the probability of crossing the conical surface $\Lambda$ should be small. For (7.26) we observe that at large times the scattering states are localised at very large distances from the co-ordinate origin, which should imply that the flux through any bounded hypersurface will be small at very large times. Finally, during any finite time interval $[T_1, T_2]$, a scattering state will have a very small probability of being localised in a neighbourhood of $|\vec{x}| = \infty$, so that the flux through a very distant spherical surface, integrated over $[T_1, T_2]$, should be very small. This is expressed by (7.27).

**7.2.2.** The flux $\Phi_{\mathcal{O}}(f)$ is given in terms of the values of the function $f$ and its first order derivatives on a hypersurface of dimension $n-1$, which is a subset of $\mathbb{R}^n$ of Lebesgue measure zero. Thus $\Phi_{\mathcal{O}}(f)$ does not have an immediate Hilbert space meaning: given a vector in $L^2(\mathbb{R}^n)$, the various representative functions of its equivalence class will give different values for the flux, and for most (if not all) representatives, the quantity $\Phi_{\mathcal{O}}(f)$ in Eq. (7.17) will not even make sense. The expression for $\Phi_{\mathcal{O}}(f)$ has a well defined meaning for functions $f$ of class $C^1$. Thus it is possible to associate a flux $\Phi_{\mathcal{O}}$ to those elements of $L^2(\mathbb{R}^n)$ that contain a representative of class $C^1$ [this condition on $f$ is satisfied for example if the Fourier transforms of $f$ and of $P_j f$, $j = 1, \ldots, n$, belong to $L^1(\mathbb{R}^n)$, see Proposition 5.6].

A convenient description of the flux in terms of linear operators between Hilbert spaces can be obtained from the following result characterising the restriction of smooth functions to hyper-surfaces. We denote by $\gamma_{\mathcal{O}}$ the linear mapping $f \mapsto f\big|_{\mathcal{O}}$ defined on the set of continuous functions $f\colon \mathbb{R}^n \to \mathbb{C}$, where $f\big|_{\mathcal{O}}$ stands for the restriction of a function $f$ to the hypersurface $\mathcal{O}$. We shall write $\langle P\rangle$ for the operator $(I + \vec{P}^2)^{1/2}$. Clearly $\mathcal{D}(\langle P\rangle) = \mathcal{D}(|\vec{P}|)$ and $\langle P\rangle^{-1} \in \mathcal{B}(\mathcal{H})$.

**Proposition 7.4.** *Let $\mathcal{O}$ be one of the sets $\Sigma_b, \Lambda_{\rho,d}$ or $\Lambda_\rho$, with $b \in (0,\infty)$ and with $0 < \rho < d < \infty$. Then there is a constant $c_{\mathcal{O}} < \infty$ such that the following inequality holds for all functions $f, \varphi\colon \mathbb{R}^n \to \mathbb{C}$ of class $C^1$ satisfying*
*(i) $f$ defines a vector in $L^2(\mathbb{R}^n)$ belonging to the domain $D(\langle P\rangle)$ of the operator $\langle P\rangle$,*
*(ii) $\varphi$ and its first order derivatives are bounded, and $\varphi(\vec{x}) = 1$ for all $\vec{x} \in \mathcal{O}$:* [6]

$$\|\gamma_{\mathcal{O}} f\|_{L^2(\mathcal{O})} \equiv \left\{\int_{\mathcal{O}} |f(\vec{y})|^2\, d\sigma(\vec{y})\right\}^{1/2} \le c_{\mathcal{O}} \|\langle P\rangle \varphi(\vec{Q}) f\|_{L^2(\mathbb{R}^n)}. \tag{7.28}$$

*Furthermore there is a finite constant $\kappa$, depending on $n$, such that*

$$c_{\Sigma_b} < \kappa \qquad \forall b \ge 1. \tag{7.29}$$

The conditions imposed on $f$ and $\varphi$ in this proposition guarantee that the left-hand side of (7.28) is well defined and that the right-hand side is finite. A proof of the proposition can be found in the Appendix to the present chapter (page 374). We mention that more general results can be obtained by using the theory of relatively smooth operators (see Proposition 7.16 and the comments preceding it). For example $\gamma_{\mathcal{O}}$ can be defined for more general hyper-surfaces $\mathcal{O}$, and the condition $f \in \mathcal{D}(\langle P\rangle)$ can be weakened to $f \in \mathcal{D}(\langle P\rangle^\nu)$ for any $\nu > 1/2$ [the operator $\langle P\rangle$ can be replaced by $\langle P\rangle^\nu$ on the right-hand side of (7.28)]. An illustration is presented in Problem 7.3.

We now consider some consequences of Proposition 7.4. First it is clear that $c_{\Lambda_{\rho,d}}$ can be chosen such that

$$c_{\Lambda_{\rho,d}} \le c_{\Lambda_\rho} \qquad \forall d > \rho. \tag{7.30}$$

Next, if $h \in L^2(\mathbb{R}^n)$ is such that the equivalence class of $\langle P\rangle^{-1} h$ contains a function $f$ of class $C^1$, then (7.28) with $\varphi \equiv 1$ shows that

$$\|\gamma_{\mathcal{O}} f\|_{L^2(\mathcal{O})} \le c_{\mathcal{O}} \|\langle P\rangle f\| = c_{\mathcal{O}} \|h\|, \tag{7.31}$$

where $\|\cdot\|$ denotes the norm in $L^2(\mathbb{R}^n)$. This means that $\gamma_{\mathcal{O}}\langle P\rangle^{-1}$ may be interpreted as a bounded operator from the domain $\mathcal{D} := \{h \in L^2(\mathbb{R}^n) \mid h = \langle P\rangle f$ for some $f \in L^2(\mathbb{R}^n) \cap C^1(\mathbb{R}^n)\}$ to $L^2(\mathcal{O})$. It is easy to see that this operator is closable, and we denote its closure by $\Gamma_{\mathcal{O}}$. It turns out that $\Gamma_{\mathcal{O}} \in \mathcal{B}(L^2(\mathbb{R}^n), L^2(\mathcal{O}))$, i.e. $\Gamma_{\mathcal{O}}$ is defined on all of $L^2(\mathbb{R}^n)$. To see this, we must know (cf. §2.4.1) that the above domain $\mathcal{D}$ is dense in $L^2(\mathbb{R}^n)$, i.e. that $g \perp \mathcal{D}$ implies $g = \mathbf{0}$. Now, if $g \perp \mathcal{D}$, we have in particular $\langle g, \langle P\rangle f\rangle = 0$ for all $f \in \mathcal{S}(\mathbb{R}^n)$. This shows that $g$ belongs to the domain of the adjoint of $\langle P\rangle|_{\mathcal{S}(\mathbb{R}^n)}$ and that $[\langle P\rangle|_{\mathcal{S}(\mathbb{R}^n)}]^* g = \mathbf{0}$. Now the operator $\langle P\rangle$

[6] We write $L^2(\mathcal{O})$ for the $L^2$ space on $\mathcal{O}$ with respect to the measure determined by $d\sigma(\cdot)$.

is essentially self-adjoint on $\mathcal{S}(\mathbb{R}^n)$ [by Proposition 2.29 applied in $\tilde{L}^2(\mathbb{R}^n)$], hence $g \in \mathcal{D}(\langle P \rangle)$ and $\langle P \rangle g = \mathbf{0}$. Since $\langle P \rangle$ is invertible, it follows that $g = \mathbf{0}$. From (7.31) we obtain the following estimate for the norm of $\Gamma_{\mathcal{O}}$:

$$\|\Gamma_{\mathcal{O}}\| \leq c_{\mathcal{O}}. \tag{7.32}$$

In view of the preceding result, we may express the flux [Eq. (7.17)] as follows, where $\varphi$ is any function satisfying the conditions of Proposition 7.4 [observe that $\gamma_{\mathcal{O}} f = \gamma_{\mathcal{O}} \varphi f \equiv \gamma_{\mathcal{O}} \varphi(\vec{Q}) f$]:

$$\Phi_{\mathcal{O}}(f) = 2\,\Re \sum_{k=1}^{n} \int_{\mathcal{O}} \overline{[\gamma_{\mathcal{O}} \varphi f](\vec{x})}\, n_k(\vec{x}) [\gamma_{\mathcal{O}} P_k f](\vec{x})\, d\sigma(\vec{x}) \tag{7.33}$$

$$= 2\,\Re \sum_{k=1}^{n} \langle \gamma_{\mathcal{O}} \varphi(\vec{Q}) f, n_k(\cdot) \gamma_{\mathcal{O}} P_k f \rangle_{L^2(\mathcal{O})}$$

$$= 2\,\Re \sum_{k=1}^{n} \langle \Gamma_{\mathcal{O}} \langle P \rangle \varphi(\vec{Q}) f, n_k(\cdot) \Gamma_{\mathcal{O}} P_k \langle P \rangle f \rangle_{L^2(\mathcal{O})}. \tag{7.34}$$

The last expression for $\Phi_{\mathcal{O}}(f)$ is meaningful for all $f$ in $\mathcal{D}(H_0)$, and by the Schwarz inequality one has [7]

$$|\Phi_{\mathcal{O}}(f)| \leq 2\|\Gamma_{\mathcal{O}}\|^2 \|\langle P \rangle \varphi(\vec{Q}) f\| \sum_{k=1}^{n} \|P_k \langle P \rangle f\| \tag{7.35}$$

$$\leq 2nc_{\mathcal{O}}^2 \|(I + H_0)\varphi(\vec{Q}) f\| \|(I + H_0) f\|. \tag{7.36}$$

Another useful relation concerns the difference of fluxes: If $f$ and $g$ belong to $\mathcal{D}(H_0)$, then

$$\Phi_{\mathcal{O}}(f) - \Phi_{\mathcal{O}}(g) = 2\Re \sum_{k=1}^{n} \langle \Gamma_{\mathcal{O}} \langle P \rangle f, n_k(\cdot) \Gamma_{\mathcal{O}} P_k \langle P \rangle (f - g) \rangle$$
$$+ 2\Re \sum_{k=1}^{n} \langle \Gamma_{\mathcal{O}} \langle P \rangle (f - g), n_k(\cdot) \Gamma_{\mathcal{O}} P_k \langle P \rangle g \rangle.$$

As above, this leads to the following inequality:

$$|\Phi_{\mathcal{O}}(f) - \Phi_{\mathcal{O}}(g)| \leq 2nc_{\mathcal{O}}^2 \big[\|(I + H_0) f\| + \|(I + H_0) g\|\big] \|(I + H_0)(f - g)\|. \tag{7.37}$$

The following property of $\Gamma_{\Sigma_d}$ will be useful:

$$\operatorname*{s-lim}_{d \to \infty} \Gamma_{\Sigma_d} = O. \tag{7.38}$$

[7] For the second inequality, use (7.32) and observe that $\|\langle P \rangle h\| \leq \|\langle P \rangle^2 h\| = \|(I + H_0) h\|$ and $\|P_k \langle P \rangle h\| \leq \|\langle P \rangle^2 h\| = \|(I + H_0) h\|$.

Indeed, let us choose a function $\varphi$ on $\mathbb{R}^n$ of the form $\varphi(\vec{x}) = \eta(\vec{x}/d)$, where $\eta$ is a smooth function such that $\eta(\vec{x}) = 0$ for $|\vec{x}| < 1/4$ and $\eta(\vec{x}) = 1$ for $|\vec{x}| > 1/2$. By using (7.28) and (7.29), then (7.148) and commuting $P_k$ through $\varphi(\vec{Q})$, one gets for any $h \in \mathcal{H}$:

$$\|\Gamma_{\Sigma_d} h\| \leq \kappa \|\langle P\rangle \varphi(\vec{Q})\langle P\rangle^{-1} h\| \leq \kappa \|\varphi(\vec{Q})\langle P\rangle^{-1} h\| + \kappa \sum_{k=1}^{n} \|P_k \varphi(\vec{Q})\langle P\rangle^{-1} h\|$$

$$= \kappa \|\varphi(\vec{Q})\langle P\rangle^{-1} h\| + \kappa \sum_{k=1}^{n} \|\varphi(\vec{Q}) P_k \langle P\rangle^{-1} h - i[(\partial/\partial x_k)\varphi](\vec{Q})\langle P\rangle^{-1} h\|$$

$$= \kappa \big\|\eta(\tfrac{\vec{Q}}{d})\langle P\rangle^{-1} h\big\| + \kappa \sum_{k=1}^{n} \big\|\eta(\tfrac{\vec{Q}}{d}) P_k \langle P\rangle^{-1} h - \tfrac{i}{d}\,[(\partial/\partial x_k)\eta](\tfrac{\vec{Q}}{d})\langle P\rangle^{-1} h\big\|.$$

Now (7.38) follows because $\eta(\vec{Q}/d)$ converges strongly to zero as $d \to \infty$ and because $\|[(\partial/\partial x_k)\eta](\vec{Q}/d)\| \leq \|(\partial/\partial x_k)\eta\|_\infty < \infty$.

We finally note the following simple consequences of (7.28):

$$\|\Gamma_{\Lambda_{\rho,d}}\| \leq \|\Gamma_{\Lambda_\rho}\| \quad \forall d > \rho \qquad \text{and} \qquad \operatorname*{s-lim}_{d\to\infty} \Gamma_{\Lambda_{\rho,d}} = \Gamma_{\Lambda_\rho}. \tag{7.39}$$

**7.2.3.** We present here some preliminary results and some simple commutation relations that will be needed in the proof of the Flux-Across-Surfaces Theorem; some of these are also of interest by themselves. The following notations will be used: $\langle P\rangle = (I+\vec{P}^2)^{1/2}$, $\langle Q\rangle = (I+\vec{Q}^2)^{1/2}$, $\langle x\rangle = (I+|\vec{x}|^2)^{1/2}$ ($\vec{x} \in \mathbb{R}^n$) and $\partial_j = \partial/\partial x_j$ ($j = 1, \dots, n$).

For $\nu > 0$ we set

$$\mathcal{D}_\nu(H) = \big\{ g \in \mathcal{H} \mid g = \psi(H)\langle Q\rangle^{-\nu} h \text{ for some } h \in \mathcal{H} \text{ and some } \psi \in C_0^\infty(\mathbb{R}) \big\}.$$

Clearly $\mathcal{D}_\nu(H) \subset \mathcal{D}(H) = \mathcal{D}(H_0)$, and $\mathcal{D}_\nu(H)$ is dense in $\mathcal{H} = L^2(\mathbb{R}^n)$. Observe also that $\mathcal{D}_\nu(H) \subseteq \mathcal{D}_\sigma(H)$ if $\nu \geq \sigma$. The subset of functions in $\mathcal{D}_\nu(H)$ having support in $(0,\infty)$ (with respect to the spectral measure of $H$) will be denoted by $\mathcal{D}_\nu^+(H)$:

$$\mathcal{D}_\nu^+(H) = \big\{ g \in \mathcal{H} \mid g = \psi(H)\langle Q\rangle^{-\nu} h \text{ for some } h \in \mathcal{H} \text{ and some } \psi \in C_0^\infty((0,\infty)) \big\}.$$

For the usual Schrödinger Hamiltonians used in scattering theory, $\mathcal{D}_\nu^+(H)$ is a dense subset of $\mathcal{H}_{\mathrm{ac}}(H)$.[8] We shall prove the Flux-Across-Surfaces Theorem for $g \in \mathcal{D}_\nu^+(H)$ with $\nu \geq 2$.

**Lemma 7.5.** *Assume that the potential $V$ satisfies $|V(\vec{x})| \leq c\langle x\rangle^{-1}$ and $|\partial_j V(\vec{x})| \leq c$ for some constant $c$ and $j = 1, \dots, n$. Let $g \in \mathcal{D}_1^+(H)$. Then:*

[8] To see this, one uses the fact that $\sigma_{\mathrm{ac}}(H) = [0,\infty)$ and that $H$ has no singular continuous spectrum and no positive eigenvalues (see page 289). Without using the absence of positive eigenvalues, one may replace the condition $\psi \in C_0^\infty((0,\infty))$ in the definition of $\mathcal{D}_\nu^+(H)$ by $\psi \in C_0^\infty((0,\infty)\setminus\sigma_{\mathrm{p}}(H))$. See also Problem 7.4.

*(a) One has $U_t g \in \mathcal{D}(H_0) \cap \mathcal{D}(Q_j) \cap \mathcal{D}(Q_j P_k) \cap \mathcal{D}(P_k Q_j)$ for each $t \in \mathbb{R}$, and*

$$P_j U_t g = U_t P_j g - U_t \int_0^t U_{-\tau}[(\partial_j V)(\vec{Q})] U_\tau g \, d\tau, \tag{7.40}$$

$$Q_j U_t g = U_t Q_j g + 2U_t \int_0^t U_{-\tau} P_j U_\tau g \, d\tau. \tag{7.41}$$

*(b) For $j = 1, \ldots, n$ let $Y_t^{(j)} \in \mathcal{B}(\mathcal{H})$ be given by $Y_t^{(j)} = \int_0^t \tau U_{-\tau}[(\partial_j V)(\vec{Q})] U_\tau \, d\tau$. Then*

$$(P_j - \frac{Q_j}{2t}) U_t g = -\frac{1}{2t} U_t \big[Q_j + 2Y_t^{(j)}\big] g. \tag{7.42}$$

PROOF. (i) For fixed $t \in \mathbb{R}$ and $\psi \in C_0^\infty((0,\infty))$, we set $\psi_t(\lambda) = e^{-i\lambda t}\psi(\lambda)$. Then $\psi_t \in C_0^\infty((0,\infty))$, hence $U_t g \equiv \psi_t(H)\langle Q\rangle^{-1} h \in \mathcal{D}_1^+(H)$ if $g = \psi(H)\langle Q\rangle^{-1} h$. So, to know that $U_t g \in \mathcal{D}(Q_j)$, it is enough to show that $Q_j \eta(H)\langle Q\rangle^{-1}$ belongs to $\mathcal{B}(\mathcal{H})$ for each $\eta \in C_0^\infty((0,\infty))$. This follows from the fact (shown below) that the operators $Q_j \eta(H_0)\langle Q\rangle^{-1}$ and $Q_j[\eta(H) - \eta(H_0)]$ belong to $\mathcal{B}(\mathcal{H})$.

We have seen in Example 6.23 that $Q_j \eta(H_0)\langle Q\rangle^{-1} \in \mathcal{B}(\mathcal{H})$. Next let us set $L = (H+\omega)^{-1}$, $L_0 = (H_0+\omega)^{-1}$ and $\vartheta(\lambda) = \eta(\lambda^{-1} - \omega)$, where $\omega$ is a positive number such that $H + \omega > I$ [so the spectrum of $H$ is contained in $[-\omega+1, \infty)$]. We observe that $L_0 - L = L_0 V L$ (by the second resolvent equation) and that $\vartheta \in C_0^\infty(\mathbb{R})$ with support in $(0,\infty)$. Then, by using (4.57):

$$\begin{aligned} \eta(H) - \eta(H_0) = \vartheta(L) - \vartheta(L_0) &= \frac{1}{\sqrt{2\pi}} \int_{-\infty}^{\infty} \tilde{\vartheta}(t) \big[e^{iLt} - e^{iL_0 t}\big] dt \\ &= \frac{1}{\sqrt{2\pi}} \int_{-\infty}^{\infty} \tilde{\vartheta}(t) e^{iL_0 t} \big[e^{-iL_0 t} e^{iLt} - I\big] dt \\ &= \frac{-i}{\sqrt{2\pi}} \int_{-\infty}^{\infty} dt \, \tilde{\vartheta}(t) e^{iL_0 t} \int_0^t e^{-iL_0 s} L_0 V L e^{iLs} ds. \end{aligned} \tag{7.43}$$

Now $Q_j e^{iL_0(t-s)} = e^{iL_0(t-s)}[Q_j + 2(t-s)P_j L_0^2]$, with $P_j L_0^2 \in \mathcal{B}(\mathcal{H})$ [for a rigourous argument, consult Example 6.22]. By using the assumption that $|V(\vec{x})| \le c\langle x\rangle^{-1}$, one obtains

$$\|Q_j[\eta(H) - \eta(H_0)]\| \le \frac{c}{\sqrt{2\pi}} \int_{-\infty}^{\infty} dt \, |t\tilde{\vartheta}(t)| \big[\|Q_j L_0 \langle Q\rangle^{-1}\| + 2|t| \|P_j L_0^3\|\big] \|L\|.$$

The right-hand side is finite because $\tilde{\vartheta} \in \mathcal{S}(\mathbb{R})$ and $Q_j L_0 \langle Q\rangle^{-1} \in \mathcal{B}(\mathcal{H})$ (see Example 6.23). This completes the proof of the inclusion $U_t g \in \mathcal{D}(Q_j)$ in (a).

The inclusion $U_t g \in \mathcal{D}(Q_j P_k)$ is obtained similarly by showing that the operator $Q_j P_k \eta(H)\langle Q\rangle^{-1}$ belongs to $\mathcal{B}(\mathcal{H})$ for each $\eta \in C_0^\infty((0,\infty))$. First we get by simple commutator calculations that

$$\begin{aligned} Q_j P_k \eta(H_0)\langle Q\rangle^{-1} &= P_k Q_j \eta(H_0)\langle Q\rangle^{-1} + i\delta_{jk}\eta(H_0)\langle Q\rangle^{-1} \\ &= P_k \eta(H_0) Q_j \langle Q\rangle^{-1} + 2iP_k P_j[(\partial_j \eta)(H_0)]\langle Q\rangle^{-1} + i\delta_{jk}\eta(H_0)\langle Q\rangle^{-1}, \end{aligned}$$

which clearly belongs to $\mathcal{B}(\mathcal{H})$. To see that $Q_jP_k[\eta(H) - \eta(H_0] \in \mathcal{B}(\mathcal{H})$, we use (7.43) to write

$$P_k[\eta(H) - \eta(H_0)] = \frac{-i}{\sqrt{2\pi}} \int_{-\infty}^{\infty} dt\, \tilde{\vartheta}(t) e^{iL_0 t} e^{-iL_0 s} P_k L_0 V L e^{iLs} ds.$$

As before, on sees that $Q_jP_k[\eta(H)-\eta(H_0)] \in \mathcal{B}(\mathcal{H})$. The verification of the inclusion $U_t g \in \mathcal{D}(P_kQ_j)$ is left as an exercise.

(ii) We next indicate a formal proof of (7.40) and (7.41). These equations are particular cases of the following formal identity:

$$[A, U_t] = U_t[U_t^* A U_t - A] = U_t \int_0^t \frac{d}{d\tau} U_\tau^* A U_\tau \, d\tau = iU_t \int_0^t U_\tau^* [H, A] U_\tau \, d\tau.$$

It suffices to observe that, for $A = P_j$, one has $[H, A] = i(\partial_j V)(\vec{Q})$, whereas with $A = Q_j$: $[H, A] = -2iP_j$. We note that each term on the right-hand sides of (7.40) and (7.41) is well defined as a vector in $\mathcal{H}$, and we present a careful derivation of these two equations in the Appendix to this chapter (page 375).

(iii) By (7.40) and (7.41) we have

$$\begin{aligned}(P_j - \frac{Q_j}{2t})U_t g = U_t P_j g - U_t \int_0^t U_{-\tau}[(\partial_j V)(\vec{Q})] U_\tau g \, d\tau \\ - \frac{1}{2t} U_t Q_j g - \frac{1}{t} U_t \int_0^t U_{-\tau} P_j U_\tau g \, d\tau.\end{aligned} \tag{7.44}$$

Now, commuting $P_j$ through $U_\tau$ in the last integral [using (7.40)] and then integrating by parts, one obtains

$$\begin{aligned}\int_0^t U_{-\tau} P_j U_\tau g \, d\tau &= tP_j g - \int_0^t d\tau \int_0^\tau U_{-s}[(\partial_j V)(\vec{Q})] U_s g \, ds \\ &= tP_j g - \left[\tau \int_0^\tau U_{-s}[(\partial_j V)(\vec{Q})] U_s g \, ds\right]\Big|_{\tau=0}^{\tau=t} + \int_0^t \tau U_{-\tau}[(\partial_j V)(\vec{Q})] U_\tau g \, d\tau \\ &= tP_j g - t \int_0^t U_{-s}[(\partial_j V)(\vec{Q})] U_s g \, ds + Y_t^{(j)} g.\end{aligned}$$

Upon inserting this expression into (7.44), one arrives at (7.42). □

We say that a function $V : \mathbb{R}^n \to \mathbb{R}$ is of class $\mathfrak{V}_\infty$ if it is infinitely differentiable, satisfies $|V(\vec{x})| \leq c_0 \langle x \rangle^{-1}$ and if, for each integer $\ell \geq 1$, all partial derivatives of $V$ of order $\ell$ are bounded by $c_\ell \langle x \rangle^{-1-\ell}$ for some constant $c_\ell$ (so for example if $\ell = 2$: $|\partial_j \partial_k V(\vec{x})| \leq c_2 \langle x \rangle^{-3}$ for $j, k = 1, \ldots, n$). In §7.2.4 and §7.2.5 we always assume $V$ to belong to this class. In §7.2.6 we also consider potentials with local singularities, in particular the Coulomb potential.

**Lemma 7.6.** *Assume that the potential* $V$ *is of class* $\mathfrak{V}_\infty$. *Let* $\psi \in C_0^\infty((0,\infty))$, $\nu > 0$ *and* $\varepsilon > 0$. *Then there is a constant* $c$ *such that the following inequalities are satisfied for all* $t \in \mathbb{R}$:[9]

$$\|\langle Q\rangle^{-\nu}\psi(H)U_t\langle Q\rangle^{-\nu}\| \le c(1+|t|)^{-\nu+\varepsilon}, \tag{7.45}$$

$$\|\langle Q\rangle^{-\nu}\psi(H)U_t(I+D^2)^{-\nu/2}\| \le c(1+|t|)^{-\nu+\varepsilon}. \tag{7.46}$$

*In particular, if* $g = \psi(H)\langle Q\rangle^{-\nu}h \in \mathcal{D}_\nu^+(H)$ *and* $\mu \ge \nu$, *then*

$$\|\langle Q\rangle^{-\mu}U_t g\| \le c(1+|t|)^{-\nu+\varepsilon}\|h\|. \tag{7.47}$$

PROOF. We denote by $\{E(\cdot)\}$ the spectral measure of $H$ and set $R_z = (H-z)^{-1}$ and $Z = (I+D^2)^{-1/2}$. If $\kappa > 1/2$ and $\mathfrak{J}$ is a bounded closed interval in $(0,\infty)$, then by Stone's Formula (4.80):

$$\begin{aligned} Z^\kappa E(\mathfrak{J})Z^\kappa &= \frac{1}{2\pi i}\,\text{s-}\lim_{\varepsilon\to+0}\int_{\mathfrak{J}} Z^\kappa[R_{\lambda+i\varepsilon} - R_{\lambda-i\varepsilon}]Z^\kappa d\lambda \\ &= \frac{1}{2\pi i}\int_{\mathfrak{J}}\big[Z^\kappa(H-\lambda-i0)^{-1}Z^\kappa - Z^\kappa(H-\lambda+i0)^{-1}Z^\kappa\big]d\lambda. \end{aligned} \tag{7.48}$$

Here $Z^\kappa(H-\lambda\pm i0)^{-1}Z^\kappa$ denote the norm limits of $Z^\kappa R_{\lambda\mp i\varepsilon}Z^\kappa$ as $\varepsilon \to +0$. These limits exist, see point (d) on page 289, and the interchange of the limits with the integral can easily be justified since the preceding limits exist uniformly in $\lambda \in \mathfrak{J}$. If $\kappa > N + (1/2)$, then $Z^\kappa(H-\lambda\pm i0)^{-1}Z^\kappa$ are $N$ times continuously differentiable on $\mathfrak{J}$ with respect to $\lambda$, the derivative of order $\ell \le N$ being the strong limit of $\ell!\,Z^\kappa(H-\lambda\pm i\varepsilon)^{-(\ell+1)}Z^\kappa$ as $\varepsilon \to +0$, see Corollary 6.19 and the ensuing remarks.

Eq. (7.48) may be formally written as

$$Z^\kappa E(d\lambda)Z^\kappa = \frac{1}{2\pi i}\big[Z^\kappa R_{\lambda+i0}Z^\kappa - Z^\kappa R_{\lambda-i0}Z^\kappa\big]d\lambda \equiv \zeta(\lambda)d\lambda,$$

with the obvious definition of $\zeta(\lambda)$. Now let $\phi \in C_0^\infty((0,\infty))$ and denote its support by $J$. Then [10]

$$t^N Z^\kappa\phi(H)U_t Z^\kappa = t^N\int_J e^{-i\lambda t}\phi(\lambda)\zeta(\lambda)d\lambda = i^N\int_J\big[\frac{d^N}{d\lambda^N}e^{-i\lambda t}\big]\phi(\lambda)\zeta(\lambda)d\lambda.$$

Assume that $\kappa > N + (1/2)$, so that $\zeta$ is $N$ times continuously differentiable on $J$ and one may integrate $N$ times by parts in the last expression. The boundary terms arising from the integrations by parts are all zero because $\phi$ and all its derivatives vanish at the endpoints of $J$. In this way one finds, by taking norms, that $(1+|t|^N)\|Z^\kappa\phi(H)U_t Z^\kappa\|$ is majorised by a finite sum of terms of the following form

$$c\int_J |\phi^{(k)}(\lambda)|\,\|\zeta^{(\ell)}(\lambda)\|\,d\lambda$$

[9] See (6.19) for the definition of the operator $D$.

[10] A more careful derivation of this equation is achieved by introducing a sequence $\{\Pi_r\}$ of partitions of $J$, by writing $\phi(H)U_t$ as a limit of finite sums as in Proposition 4.10 and using (7.48) on each subinterval $(s_{k-1}, s_k]$ appearing in these partitions.

where $c$ is some constant (depending on $N$) and $k, \ell \in \{0, 1, 2, \ldots, N\}$. Since $J$ is bounded, these integrals are finite. Hence, by taking $\kappa = N + 1$ and by using the inequality $(1 + a)^N \leq 2^N(1 + a^N)$ for $a \geq 0$, one has

$$\|Z^{N+1}\phi(H)U_t Z^{N+1}\| \leq c_1(1 + |t|^N)^{-1} \leq 2^N c_1(1 + |t|)^{-N}$$

for some constant $c_1$.

Next let $\psi \in C_0^\infty((0, \infty))$ and set $\phi(\lambda) = (\lambda^2 + 1)^{N+1}\psi(\lambda)$. Then

$$\langle Q\rangle^{-(N+1)}\psi(H)U_t\langle Q\rangle^{-(N+1)} = X_N^* Z^{N+1}\phi(H)U_t Z^{N+1}X_N$$

with $X_N = (I + D^2)^{(N+1)/2}(H + i)^{-(N+1)}\langle Q\rangle^{-(N+1)}$. Since $X_N$ is a bounded operator (see Problem 6.10), it follows that

$$\|\langle Q\rangle^{-(N+1)}\psi(H)U_t\langle Q\rangle^{-(N+1)}\| \leq c_2(1 + |t|)^{-N}.$$

Now (7.45) is obtained by interpolation: in Example 6.32 take $B = C = \langle Q\rangle$ and $X = \psi(H)U_t$. Taking $N > \nu$, Eq. (6.133), with $\kappa = N + 1$ and $\theta = \nu$, gives

$$\begin{aligned}\|\langle Q\rangle^{-\nu}\psi(H)U_t\langle Q\rangle^{-\nu}\| &\leq c_3\|\langle Q\rangle^{-(N+1)}\psi(H)U_t\langle Q\rangle^{-(N+1)}\|^{\nu/(N+1)} \\ &\leq c_4(1 + |t|)^{-N\nu/(N+1)} = c_4(1 + |t|)^{-\nu+[\nu/(N+1)]}.\end{aligned}$$

Choosing $N$ so large that $\nu/(N + 1) < \varepsilon$, one obtains (7.45). The estimate (7.46) is obtained by the same interpolation argument but with $C = (I + D^2)^{1/2}$ instead of $C = \langle Q\rangle$. □

For later reference we add the following simple facts. (i) If $h \in \mathcal{D}(H)$, then one has for each $t \in \mathbb{R}$:

$$\|(I + H_0)U_t h\| \leq N_0(H; h) := \|(I + H_0)(H - i)^{-1}\|\|(H - i)h\|. \tag{7.49}$$

Note that $N_0(H; h)$ is finite since $(I + H_0)(H - i)^{-1} \in \mathcal{B}(\mathcal{H})$.
(ii) If $\varphi$ is a bounded function of class $C^2$, with bounded first and second order derivatives, then the following identities hold on $\mathcal{D}(H_0)$:

$$H_0\varphi(\vec{Q}) = \varphi(\vec{Q})H_0 - (\Delta\varphi)(\vec{Q}) - 2i(\vec{\nabla}\varphi)(\vec{Q}) \cdot \vec{P} \tag{7.50}$$

$$= \varphi(\vec{Q})H_0 + (\Delta\varphi)(\vec{Q}) - 2i\vec{P} \cdot (\vec{\nabla}\varphi)(\vec{Q}). \tag{7.51}$$

Formally this is easily verified by writing $H_0$ as $-\Delta$ [acting on functions $f(\vec{x})$] and using the Leibniz rule. See also Example 6.22.
(iii) As a consequence of (7.50) and (7.51) we have for $z$ in the resolvent set of $H$:

$$\begin{aligned}[\varphi(\vec{Q}), (H - z)^{-1}] &= (H - z)^{-1}[H, \varphi(\vec{Q})](H - z)^{-1} \\ &= -(H - z)^{-1}\{(\Delta\varphi)(\vec{Q}) + 2i(\vec{\nabla}\varphi)(\vec{Q}) \cdot \vec{P}\}(H - z)^{-1} \\ &= (H - z)^{-1}\{(\Delta\varphi)(\vec{Q}) - 2i\vec{P} \cdot (\vec{\nabla}\varphi)(\vec{Q})\}(H - z)^{-1}. \quad (7.52)\end{aligned}$$

**7.2.4.** We now prove (7.26) and (7.27) for $V$ of class $\mathfrak{V}_\infty$ and $g \in \mathcal{D}_2^+(H)$. We first show that $\mathcal{I}_{\Sigma_b}(U_t g)$ exists for each $b > 0$ and each $t \in \mathbb{R}$. For this we use (7.36) with $\varphi \in C_0^\infty(\mathbb{R}^n)$ such that $\varphi(\vec{x}) = 1$ if $|\vec{x}| < b+1$ and $\varphi(\vec{x}) = 0$ if $|\vec{x}| > b+2$. Then, taking into account (7.49):

$$\left|\Phi_{\Sigma_b}(U_\tau g)\right| \le 2nc_{\Sigma_b}^2 N_0(H;g)\|(I+H_0)\varphi(\vec{Q})U_\tau g\|.$$

By (7.50) we have

$$\begin{aligned}(I+H_0)\varphi(\vec{Q})U_\tau g &= \varphi(\vec{Q})U_\tau g + \varphi(\vec{Q})[H - V(\vec{Q})]U_\tau g \\ &\quad - [(\Delta\varphi)(\vec{Q})]U_\tau g - 2i[(\vec{\nabla}\varphi)(\vec{Q})]\cdot \vec{P}U_\tau g,\end{aligned}$$

and it suffices to show that the norm of each term on the right-hand side admits a bound of the form $const \cdot (1+|\tau|)^{-1-\varepsilon}$ for some constant that may depend on $b$, and for some $\varepsilon > 0$. For this we write $g = \psi(H)\langle Q\rangle^{-2}h$ and set $\psi_1(\lambda) = \lambda\psi(\lambda)$ and $\psi_2(\lambda) = (\lambda - i)\psi(\lambda)$. From the propagation estimate (7.45) we then get the following inequalities:

$$\begin{aligned}
&\|\varphi(\vec{Q})U_\tau g\| \le \|\langle Q\rangle^2\varphi(\vec{Q})\|\|\langle Q\rangle^{-2}U_\tau\psi(H)\langle Q\rangle^{-2}\|\|h\| \le const\cdot(1+|\tau|)^{-3/2},\\
&\|\varphi(\vec{Q})V(\vec{Q})U_\tau g\| \le \|V\|_\infty\|\varphi(\vec{Q})U_\tau g\| \le const\cdot(1+|\tau|)^{-3/2},\\
&\|\varphi(\vec{Q})HU_\tau g\| = \|\langle Q\rangle^2\varphi(\vec{Q})\|\|\langle Q\rangle^{-2}U_\tau\psi_1(H)\langle Q\rangle^{-2}\|\|h\| \le const\cdot(1+|\tau|)^{-3/2},\\
&\|[(\Delta\varphi)(\vec{Q})]U_\tau g\| \le \|\langle Q\rangle^2(\Delta\varphi)(\vec{Q})\|\|\langle Q\rangle^{-2}U_\tau\psi(H)\langle Q\rangle^{-2}\|\|h\|\\
&\qquad\qquad \le const\cdot(1+|\tau|)^{-3/2},\\
&\|[(\partial_j\varphi)(\vec{Q})]P_jU_\tau g\| \le \|\langle Q\rangle^2(\partial_j\varphi)(\vec{Q})\|\|\langle Q\rangle^{-2}P_j(H-i)^{-1}\langle Q\rangle^2\|\\
&\qquad\qquad \times\|\langle Q\rangle^{-2}U_\tau\psi_2(H)\langle Q\rangle^{-2}\|\|h\| \le const\cdot(1+|\tau|)^{-3/2},
\end{aligned}$$

since all occurring $\tau$-independent norms are finite (for $\|\langle Q\rangle^{-2}P_j(H-i)^{-1}\langle Q\rangle^2\|$ see Example 6.26). We have thus shown that for each $b \in (0,\infty)$, there is a constant $c(b)$ such that

$$\left|\Phi_{\Sigma_b}(U_\tau g)\right| \le c(b)(1+|\tau|)^{-3/2}.$$

This also implies that $\lim_{t\to\infty}\mathcal{I}_{\Sigma_b}(g,t) = 0$ and completes the proof of (7.26)

We now verify (7.27). It suffices to show that for any $t_1, t_2 \in \mathbb{R}$ and each $g \in \mathcal{D}(H)$:

$$\lim_{d\to\infty}\int_{t_1}^{t_2}\Phi_{\Sigma_d}(U_\tau g)\,d\tau = 0.$$

This is easily obtained by using the Dominated Convergence Theorem. (i) By (7.34) and (7.38), $\Phi_{\Sigma_d}(U_\tau g) \to 0$ as $d \to \infty$ for each fixed $t$. (ii) Since $c_{\Sigma_d} \le \kappa < \infty$ for all $d \ge 1$ [see (7.29)], we obtain from (7.36), with $\varphi \equiv 1$, the following $d$-independent bound for the integrand:

$$\left|\Phi_{\Sigma_d}(U_\tau g)\right| \le 2nc_{\Sigma_d}^2\|(I+H_0)U_\tau g\|^2 \le 2n\kappa^2[N_0(H;g)]^2,$$

which is integrable on each finite interval $[t_1, t_2]$.

**7.2.5.** Here we prove (7.25), again for $V$ of class $\mathfrak{V}_\infty$ and $g = \psi(H)\langle Q\rangle^{-2}h$ in $\mathcal{D}_2^+(H)$. Before producing the necessary decay estimates, we explain the idea of the proof. We observe that if $\mathcal{O} = \Lambda_{\rho,d}$, then $n(\vec{x})\cdot\vec{x} = 0$ for all $\vec{x}\in\mathcal{O}$. Hence, instead of (7.33), we may use the following expression for the flux across $\mathcal{O} = \Lambda_{\rho,d}$:

$$\Phi_{\mathcal{O}}(f) = 2\,\Re\sum_{k=1}^{n}\int_{\mathcal{O}}\overline{[\gamma_{\mathcal{O}}\varphi f](\vec{x})}\,n_k(\vec{x})\big[\gamma_{\mathcal{O}}(P_k - \tfrac{Q_k}{s})f\big](\vec{x})\,d\sigma(\vec{x}),$$

where $s$ is an arbitrary number. For $f = U_\tau g$ and $s = 2\tau$, the subtraction of $\vec{Q}/(2\tau)$ from $\vec{P}$ should improve the decay rate, since (in classical terms) the momentum $\vec{p}$ is expected to behave at large times $\tau$ approximately like $m\vec{x}/\tau$ and we have taken $m = 1/2$ for the mass.

Instead of (7.35) we now obtain the following bound for $\Phi_{\Lambda_{\rho,d}}(U_\tau g)$:

$$\big|\Phi_{\Lambda_{\rho,d}}(U_\tau g)\big| \le 2c_{\Lambda_\rho}^2\|\langle P\rangle\varphi(\vec{Q})U_\tau g\|\sum_{k=1}^{n}\|\langle P\rangle(P_k - \frac{Q_k}{2t})U_\tau g\|. \tag{7.53}$$

We shall choose the function $\varphi$ such that its support belongs to a small neighbourhood of the cone $\mathcal{C}$. Such a neighbourhood has a finite extent in directions orthogonal to $\Lambda_\rho$, so that one can expect that $\|\varphi(\vec{Q})U_\tau g\|$ decays as $|\tau|^{-1/2}$ due to the spreading effect for wave packets in one dimension [see for example (5.64) for the case $V = 0$]. Since (7.25) involves only positive times, we assume in the following estimates that $t \ge 1$ [in which case $1/t \le 2/(1+t)$]. We shall show that, given $\varepsilon > 0$, there is a constant $c$ (depending on $g$) such that for $t \ge 1$ and $j, k = 1,\dots,n$:

$$\|\varphi(\vec{Q})U_t g\| + \|P_k\varphi(\vec{Q})U_t g\| \le c(1+t)^{-1/2+\varepsilon} \tag{7.54}$$

and

$$\|(P_j - \frac{Q_j}{2t})U_t g\| + \|P_k(P_j - \frac{Q_j}{2t})U_t g\| \le c(1+t)^{-1+\varepsilon}. \tag{7.55}$$

These inequalities imply that $|\Phi_{\Lambda_{\rho,d}}(U_\tau g)| \le c'(1+\tau)^{-3/2+2\varepsilon}$ for some constant $c'$, all $\tau \ge 1$ and all $d \in (\rho,\infty]$ [see also (7.148)]. This gives the first statement in (7.25). The second one is obtained by means of the Dominated Convergence Theorem: by using the preceding bound on $|\Phi_{\Lambda_{\rho,d}}(U_\tau g)|$ and the fact that $\Phi_{\Lambda_{\rho,d}}(U_\tau g) - \Phi_{\Lambda_\rho}(U_\tau g) \equiv -\Phi_{\Lambda_d}(U_\tau g) \to 0$ as $d \to \infty$ for each fixed $\tau$ [see (7.39)], one finds that $\lim_{d\to\infty}\mathcal{I}_{\Lambda_{\rho,d}}(g,t) = \mathcal{I}_{\Lambda_\rho}(g,t)$. The third statement in (7.25) then follows since $\mathcal{I}_{\Lambda_\rho}(g,t) \le c''(1+t)^{-1/2+2\varepsilon}$.

In what follows we us the notation $c, c_j$ and *const* for various constants; these numbers may depend on some of the involved parameters or on a vector in $\mathcal{H}$ but they are independent of $t$ (they are constant with regard to the variable $t$).

We first verify the estimates in (7.55). Recalling that $Y_t^{(j)} = \int_0^t \tau U_{-\tau}(\partial_j V)U_\tau d\tau$ and that $|(\partial_j V)(\vec{x})| \le c_1\langle x\rangle^{-2}$, we obtain from the propagation estimate (7.47) that

$$\|Y_t^{(j)}g\| \le c\int_0^t(1+\tau)(1+\tau)^{-2+\varepsilon}d\tau\,\|h\| \le \mathit{const}\cdot(1+t)^{\varepsilon}\|h\|.$$

Hence [see (7.42)]

$$\|(P_j - \frac{Q_j}{2t})U_t g\| \le \frac{1}{2t}[\|Q_j g\| + 2\|Y_t^{(j)} g\|] \le \frac{1}{t}[\|Q_j g\| + const\cdot(1+t)^{\varepsilon}\|h\|],$$

which gives the inequality (7.55) for the first term on its left-hand side. To treat the second one we set $\hat{g} = (H+\omega)g \in \mathcal{D}_2^+(H)$, with $\omega$ as in the proof of Lemma 7.5, and then have [using again (7.42)]

$$\|P_k(P_j - \frac{Q_j}{2t})U_t g\| \le \frac{1}{2t}[\|P_k U_t Q_j (H+\omega)^{-1}\hat{g}\| + 2\|P_k U_t Y_t^{(j)}(H+\omega)^{-1}\hat{g}\|]. \tag{7.56}$$

To see that the norms on the right-hand side of this inequality are bounded by $c(1+t)^{\varepsilon}$ for some constant $c$, we commute $(H+\omega)^{-1}$ through $Q_j$ and $Y_t^{(j)}$ respectively. For the first term, using (6.117), this leads to

$$\|P_k U_t Q_j (H+\omega)^{-1}\hat{g}\| \le \|P_k (H+\omega)^{-1}\|[\|Q_j\hat{g}\| + 2\|P_j g\|] \equiv c_1, \tag{7.57}$$

with $c_1 < \infty$ because $\hat{g} \in \mathcal{D}(Q_j)$ and $g \in \mathcal{D}(P_j)$. For the second norm on the right-hand side of (7.56) we get

$$\begin{aligned}\|P_k U_t Y_t^{(j)}(H+\omega)^{-1}\hat{g}\| &\le \|P_k(H+\omega)^{-1}\|\|Y_t^{(j)}\hat{g}\| + \|P_k U_t [Y_t^{(j)}, (H+\omega)^{-1}]\hat{g}\| \\ &\le const\cdot(1+t)^{\varepsilon} + \|P_k U_t (H+\omega)^{-1}[H, Y_t^{(j)}]g\|,\end{aligned}$$

and it remains to check that $\|P_k U_t (H+\omega)^{-1}[H, Y_t^{(j)}]g\| \le const\cdot(1+t)^{\varepsilon}$. For this we observe that

$$\|P_k U_t (H+\omega)^{-1}[H, Y_t^{(j)}]g\| \le \|P_k (H+\omega)^{-1/2}\|\|(H+\omega)^{-1/2}[H, Y_t^{(j)}]g\|$$

and, by using (7.51) with $\varphi = \partial_j V$:

$$\begin{aligned}(H+\omega)^{-1/2}[H, Y_t^{(j)}]g &= \int_0^t \tau U_{-\tau}(H+\omega)^{-1/2}[H, (\partial_j V)(\vec{Q})]U_\tau g\, d\tau \\ &= \int_0^t \tau U_{-\tau}(H+\omega)^{-1/2}(\partial_j \Delta V)(\vec{Q})U_\tau g\, d\tau \\ &\quad - 2i\sum_{\ell=1}^{n}\int_0^t \tau U_{-\tau}(H+\omega)^{-1/2}P_\ell(\partial_j\partial_\ell V)(\vec{Q})U_\tau g\, d\tau.\end{aligned} \tag{7.58}$$

From the hypothesis on the decay of the derivatives of $V$ and the propagation estimate (7.47), we have $\|(\partial_j\Delta V)(\vec{Q})U_\tau g\| + \|(\partial_j\partial_\ell V)(\vec{Q})U_\tau g\| \le c_0(1+\tau)^{-2+\varepsilon}$ for some constant $c_0$ [remembering that $g \in \mathcal{D}_2^+(H)$]. Since the closure of $(H+\omega)^{-1/2}P_\ell$ belongs to $\mathcal{B}(\mathcal{H})$, one finds that the norm of each of the integrands in the final expression in (7.58) belongs to $L^1((0,\infty))$, so that $\|(H+\omega)^{-1/2}[H, Y_t^{(j)}]g\| \le c_1(1+t)^{\varepsilon}$ for all $t \geq 0$. Hence $\|P_k U_t (H+\omega)^{-1}[H, Y_t^{(j)}]g\| \leq c_2(1+t)^\varepsilon$ for all $t \geq 0$, and (7.55) is completely verified.

Now we prove (7.54). Let $\vec{\xi}$ be a unit vector pointing into the interior of the cone $\mathcal{C}$. For $\sigma > 0$ we denote by $\mathcal{C}(\sigma)$ the domain in $\mathbb{R}^n$ obtained by translating $\partial\mathcal{C}$ along $\vec{\xi}$ by distances less than $\sigma$, more precisely $\mathcal{C}(\sigma) = \{\vec{x} \in \mathbb{R}^n \mid \vec{x} = \vec{y} + \mu\vec{\xi}$ with $\vec{y} \in \partial\mathcal{C}$ and $-\sigma < \mu < \sigma\}$. Let $\theta$ be a bounded smooth function having support in $\mathcal{C}(\sigma)$. Then

$$\|\theta(\vec{Q})U_t g\| \leq \|\theta\|_\infty \|\chi_{\mathcal{C}(\sigma)}(\vec{Q})U_t^0(I+|\vec{\xi}\cdot\vec{Q}|)^{-1}\| \|(I+|\vec{\xi}\cdot\vec{Q}|)U_{-t}^0 U_t g\|. \quad (7.59)$$

By using the relation $[Q_k, U_{-t}^0] = -2tU_{-t}^0 P_k$ and then (7.55) we get, for some constant $c$ (with $t \geq 1$):

$$\|(I+|\vec{\xi}\cdot\vec{Q}|)U_{-t}^0 U_t g\| \leq \|g\| + 2t\sum_{k=1}^n \|(\frac{Q_k}{2t} - P_k)U_t g\| \leq c(1+t)^\varepsilon. \quad (7.60)$$

To estimate $\|\chi_{\mathcal{C}(\sigma)}(\vec{Q})U_t^0(I+|\vec{\xi}\cdot\vec{Q}|)^{-1}\|$, we choose a Cartesian co-ordinate system with the 1-axis is along $\vec{\xi}$, and we introduce the notation $\vec{x} = (x_1, \vec{x}\,')$ with $\vec{x}\,' \in \mathbb{R}^{n-1}$. Then $\vec{\xi}\cdot\vec{Q} = Q_1$. We use the following factorisation of $e^{-iP_1^2 t}$ which follows easily from (5.59) with $n=1$: $e^{-iP_1^2 t} = Z_t G_t Z_t$, where

$$(Z_t h)(\vec{x}) = e^{ix_1^2/4t}h(\vec{x}), \quad (G_t h)(\vec{x}) = \frac{1}{\sqrt{4\pi i t}}\int_{-\infty}^{\infty} e^{-iux_1/2t}h(u, x_2, \dots, x_n)\,du.$$

Since the operators $P_k$, $k = 2, \dots, n$, commute with $Q_1$, we then get

$$\begin{aligned}\|\chi_{\mathcal{C}(\sigma)}(\vec{Q})U_t^0(I+|\vec{\xi}\cdot\vec{Q}|)^{-1}\| &= \|\chi_{\mathcal{C}(\sigma)}(\vec{Q})e^{-iP_1^2 t}(I+|\vec{\xi}\cdot\vec{Q}|)^{-1}\| \\ &= \|\chi_{\mathcal{C}(\sigma)}(\vec{Q})G_t(I+|Q_1|)^{-1}\|. \quad (7.61)\end{aligned}$$

For fixed $\vec{x}\,' \in \mathbb{R}^{n-1}$, we denote by $S(\vec{x}\,')$ the intersection of $\mathcal{C}(\sigma)$ and the line $\vec{x}\,' + s\vec{\xi}$, where $s$ varies over $\mathbb{R}$: $S(\vec{x}\,') = \{\vec{x} \in \mathbb{R}^n \mid \vec{x} \equiv (x_1, \vec{x}\,') \in \mathcal{C}(\sigma)\}$. Since $\mathcal{C}$ is convex, $S(\vec{x}\,')$ is an interval of length $2\sigma$. Then, by the Schwarz inequality in $L^2(\mathbb{R})$, we have for $h \in \mathcal{S}(\mathbb{R}^n)$ and any $t \in \mathbb{R}$:

$$\begin{aligned}&\|\chi_{\mathcal{C}(\sigma)}(\vec{Q})G_t(I+|Q_1|)^{-1}h\|^2 \\ &\quad = \frac{1}{4|t|}\int_{\mathbb{R}^{n-1}} d^{n-1}x' \int_{S(\vec{x}\,')} dx_1 \left|\int_{-\infty}^{\infty} e^{-iux_1/2t}(1+|u|)^{-1}h(u,\vec{x}\,')\,du\right|^2 \\ &\quad \leq \frac{1}{4|t|}\int_{S(\vec{x}\,')} dx_1 \int_{\mathbb{R}^{n-1}} d^{n-1}x' \int_{-\infty}^{\infty} |h(u,\vec{x}\,')|^2 du \int_{-\infty}^{\infty} (1+|v|)^{-2}dv \\ &\quad = \frac{\sigma}{2|t|}\|h\|^2 \|(1+|\cdot|)^{-1}\|_{L^2(\mathbb{R})}^2.\end{aligned}$$

Thus, by taking into account (7.61), we have for $t \geq 1$:

$$\|\chi_{\mathcal{C}(\sigma)}(\vec{Q})U_t^0(I+|\vec{\xi}\cdot\vec{Q}|)^{-1}\| \leq \mathit{const}\cdot(1+t)^{-1/2}. \quad (7.62)$$

Inserting (7.62) and (7.60) into (7.59) leads to

$$\|\theta(\vec{Q})U_t g\| \leq c(1+t)^{-1/2+\varepsilon}, \tag{7.63}$$

where $c$ is some constant. In particular, for $\theta = \varphi$, with $\varphi$ as in (7.54) [choosing $\sigma$ such that the support of $\varphi$ belongs to $\mathcal{C}(\sigma)$], this gives the inequality (7.54) for the first term on its left-hand side.

Finally let us consider the second term on the left-hand side of (7.54). The argument here is similar to that applied in relation with (7.56). We again write $g = (H+\omega)^{-1}\hat{g}$ and commute $(H+\omega)^{-1}$ through $\varphi(\vec{Q})$ in the term under consideration, using (7.52):

$$\begin{aligned}\|P_k\varphi(\vec{Q})U_t g\| &\leq \|P_k(H+\omega)^{-1}\|\|\varphi(\vec{Q})U_t\hat{g}\| + \|P_k(H+\omega)^{-1}[H,\varphi(\vec{Q})]U_t g\|]\\ &\leq c(1+t)^{-1/2+\varepsilon} + \|P_k(H+\omega)^{-1}\big[(\Delta\varphi)(\vec{Q})U_t g - 2i\sum_{\ell=1}^{n} P_\ell(\partial_\ell\varphi)(\vec{Q})U_t g\big]\|.\end{aligned}$$

Since $\Delta\varphi$ and $\partial_\ell\varphi$ have support in $\mathcal{C}(\sigma)$, we find from (7.63) that $\|(\Delta\varphi)(\vec{Q})U_t g\| \leq c(1+t)^{-1/2+\varepsilon}$ and $\|(\partial_\ell\varphi)(\vec{Q})U_t g\| \leq c(1+t)^{-1/2+\varepsilon}$. Hence, by observing that the closure of the operator $P_k(H+\omega)^{-1}P_\ell$ belongs to $\mathcal{B}(\mathcal{H})$, we obtain the estimate (7.54) for the second term on its left-hand side.

**7.2.6.** So far we have proved the Flux-Across-Surfaces Theorem for smooth potentials decaying like $|\vec{x}|^{-1}$ or faster. We now show that this result remains true for potentials with similar decay but admitting local singularities. This class of potentials includes the Coulomb potential in $n = 3$ dimensions. We assume that $V = V_1 + V_2$, where $V_1$ is of class $\mathfrak{V}_\infty$ and $V_2(\vec{x}) = \langle x\rangle^{-\sigma}W(\vec{x})$ with $\sigma > 2$ and $W \in L^p(\mathbb{R}^n) + L^\infty(\mathbb{R}^n)$, where $p = 2$ for $n = 2, 3$ and $p > n/2$ for $n \geq 4$. We set $H_1 = H_0 + V_1$ and have $H = H_0 + V = H_1 + V_2$, and we shall use results from the preceding subsections for the operator $H_1$. As before, $\{U_t\}$ denotes the evolution group associated with $H$, i.e. $U_t = e^{-iHt}$, and we set $U_t^1 = e^{-iH_1 t}$.

We first discuss the basic decay estimate. Let $f \in \mathcal{D}_\nu^+(H_1)$ with $\nu \in (2,\sigma]$, and let $\varepsilon < (\nu-2)/2$. Then there is a constant $c$ such that for all $\tau \in \mathbb{R}$:

$$\|V_2 U_\tau^1 f\| \leq c(1+|\tau|)^{-2-\varepsilon}. \tag{7.64}$$

This is obtained by writing $f = \psi(H_1)\langle Q\rangle^{-\nu}h$ and

$$\|V_2 U_\tau^1 f\| \leq \|W\langle Q\rangle^{-\sigma}(H_1+i)^{-1}\langle Q\rangle^{\sigma}\|\|\langle Q\rangle^{-\sigma}\psi_0(H_1)U_\tau^1\langle Q\rangle^{-\nu}h\|, \tag{7.65}$$

where $\psi_0(\lambda) = (\lambda+i)\psi(\lambda)$. The operator in the first norm on the right-hand side is bounded: the unboundedness of $\langle Q\rangle^{\sigma}$ is compensated by $\langle Q\rangle^{-\sigma}$ and that of $W$ (if $W$ is not bounded) by the resolvent $(H_1+i)^{-1}$; we refer to Example 6.26 for details. By virtue of (7.45) in Lemma 7.6, the last norm in (7.65) admits a bound of the form $c(1+|\tau|)^{-2-\varepsilon}$ for $\varepsilon < (\nu-2)/2$.

The decay estimate (7.64) implies that, for $f$ and $\varepsilon$ as above:

$$\|\frac{d}{d\tau}U_\tau^* U_\tau^1 f\| = \|V_2 U_\tau^1 f\| \leq c(1+|\tau|)^{-2-\varepsilon}. \tag{7.66}$$

Thus, by Proposition 5.16, the wave operators $\Omega_\pm = \text{s-lim}_{t\to\pm\infty} U_t^* U_t^1 E_{\text{ac}}(H_1)$ exist. The fact that $\mathcal{R}(\Omega_+) = \mathcal{H}_{\text{ac}}(H)$ can be obtained as in §7.1.2 from Lavine's Theorem (cf. Problem 7.5).

We shall use the following consequence of (7.64): Let $f \in \mathcal{D}_\nu^+(H_1)$ with $\nu \in (2, \sigma]$, and let $\varepsilon < (\nu - 2)/2$. Then there is a constant $c$ such that for all $t > 0$:

$$\|(I + H_0)(U_t\Omega_+ f - U_t^1 f)\| \leq c(1+t)^{-1-\varepsilon}. \tag{7.67}$$

Let us prove this inequality. First we have for $t > 0$:

$$\|U_t\Omega_+ f - U_t^1 f\| = \|\Omega_+ f - U_t^* U_t^1 f\| = \Big\| \int_t^\infty \frac{d}{d\tau} U_\tau^* U_\tau^1 f \, d\tau \Big\| \leq \int_t^\infty \|V_2 U_\tau^1 f\| \, d\tau$$
$$\leq c \int_t^\infty (1+\tau)^{-2-\varepsilon} d\tau = c'(1+t)^{-1-\varepsilon}. \tag{7.68}$$

Next, setting $f_1 = (H_1 + i)f$ and using the intertwining relation $\Omega_+(H_1+i)^{-1} = (H+i)^{-1}\Omega_+$:

$$\|H_0(U_t\Omega_+ f - U_t^1 f)\| = \big\|H_0(H+i)^{-1}U_t\Omega_+ f_1 - H_0(H_1+i)^{-1}U_t^1 f_1\big\|$$
$$\leq \big\|H_0(H+i)^{-1}[U_t\Omega_+ f_1 - U_t^1 f_1]\big\| + \big\|H_0[(H+i)^{-1} - (H_1+i)^{-1}]U_t^1 f_1]\big\|.$$

By (7.68) the first term on the bottom line is bounded by $c(1+t)^{-1-\varepsilon}$ for some constant $c$. By the second resolvent equation the second term on this line is equal to $\|H_0(H+i)^{-1}V_2 U_t^1 f]\|$, which is bounded by $c(1+t)^{-2-\varepsilon}$, see (7.64). This completes the proof of (7.67).

We now verify the validity of the Flux-Across-Surfaces Theorem, Eq. (7.19), for the class of Hamiltonians $H$ considered here, with $g = \Omega_+ f$, where $f \in \mathcal{D}_\nu^+(H_1)$ for some $\nu > 2$. By using the fact that the distance between $U_t g$ and $U_t^1 f$ becomes very small at large values of $t$ [Eq. (7.67)], we shall find that $U_t g$ and $U_t^1 f$ have the same probability of being localised in a cone $\mathcal{C}$ at $t = +\infty$, and that the associated integrated fluxes across $\Sigma_d$ become identical as $d \to \infty$. Assuming these results for the moment, we then have

$$\lim_{t\to+\infty} \|\chi_{\mathcal{C}}(\vec{Q})U_t\Omega_+ f\|^2 = \lim_{t\to+\infty} \|\chi_{\mathcal{C}}(\vec{Q})U_t^1 f\|^2 = \lim_{d\to\infty} \int_T^\infty \Phi_{\Sigma_d}(U_t^1 f)\, dt$$
$$= \lim_{d\to\infty} \int_T^\infty \Phi_{\Sigma_d}(U_t\Omega_+ f)\, dt, \tag{7.69}$$

where the second equality is just the Flux-Across-Surfaces Theorem for $U_t^1 f$ obtained before. Equation (7.69) shows that the Flux-Across-Surfaces Theorem is satisfied for $U_t g \equiv U_t\Omega_+ f$.

It remains for one to check the two asymptotic relations stated before (7.69). First we have [as in (5.86)]

$$\big|\|\chi_{\mathcal{C}}(\vec{Q})U_t\Omega_+ f\|^2 - \|\chi_{\mathcal{C}}(\vec{Q})U_t^1 f\|^2\big| \leq 2\|f\|\big|\|\chi_{\mathcal{C}}(\vec{Q})U_t\Omega_+ f\| - \|\chi_{\mathcal{C}}(\vec{Q})U_t^1 f\|\big|$$
$$\leq 2\|f\|\|\chi_{\mathcal{C}}(\vec{Q})[U_t\Omega_+ f - U_t^1 f]\| \to 0 \quad \text{as } t \to \infty,$$

which implies the first identity in (7.69). The third one (asymptotic identity of the integrated fluxes) is based on the observation that, by (7.37) and (7.67):

$$\begin{aligned}\left|\Phi_{\Sigma_d}(U_t\Omega_+ f) - \Phi_{\Sigma_d}(U_t^1 f)\right| &\leq 2n\kappa^2[N_0(H;\Omega_+ f) + N_0(H_1; f)] \\ &\quad \times \|(I+H_0)(U_t\Omega_+ f - U_t^1 f)\| \leq c(1+t)^{-1-\varepsilon},\end{aligned} \tag{7.70}$$

with some constant $c$ that is independent of $d$ [see (7.49) for the definition of $N_0(\cdot\,;\cdot)$ and (7.29) for that of $\kappa$]. This inequality allows us to use the Dominated Convergence Theorem to obtain the third identity in (7.69), since by (7.38) both $\Phi_{\Sigma_d}(U_t\Omega_+ f)$ and $\Phi_{\Sigma_d}(U_t^1 f)$ converge to zero as $d \to \infty$.

REMARK. We have proved the Flux-Across-Surfaces Theorem under conditions on the scattering state $g$ [$g \in \mathcal{D}_2^+(H)$ if $V \in \mathfrak{V}_\infty$, and $g \in \Omega_+\mathcal{D}_\nu^+(H_1)$ in the situation of the present subsection, with $\nu > 2$]. In the short range case [11] one has $g = \Omega_- f$, and it is interesting to know conditions on the free initial state $f$ under which this result is true. For this one has to determine mapping properties of the wave operators $\Omega_\pm$ and of the scattering operator $S$. We refer to [11] for such results. In the first instance ($V \in \mathfrak{V}_\infty$) the wave operators map $\mathcal{D}_\nu^+(H_0)$ into $\mathcal{D}_{\nu'}^+(H)$ for any $\nu > \nu' > 0$, so that the Flux-Across-Surfaces Theorem holds for initial states $f \in \mathcal{D}_\nu^+(H_0)$ with $\nu > 2$. For the potentials considered in the present subsection, with $V_1 = 0$, the operator $S$ maps $\mathcal{D}_\nu^+(H_0)$ into $\mathcal{D}_\nu^+(H_0)$ for $\nu \in (0, \sigma - 1)$, and the wave operators map $\mathcal{D}_\nu^+(H_0)$ into $\mathcal{D}_\nu^+(H)$ for the same values of $\nu$. Since here $g = \Omega_+ Sf$, with $Sf \in \mathcal{D}_\nu^+(H_0)$ ($\nu > 2$), it is then seen that the Flux-Across-Surfaces Theorem is satisfied for initial states $f$ in $\mathcal{D}_\nu^+(H_0)$ with $\nu > 2$ (this requires that $\sigma > 3$ in the definition of $V_2$).

**7.2.7.** To end this section we comment on the condition, stated after (5.82), that at very large positive times the flux should be purely outgoing. This may be expressed for example in the following form:

$$\lim_{d\to\infty} \int_T^\infty \Phi_{\Sigma_d}(U_t\Omega_+ f)\,dt = \lim_{d\to\infty} \int_T^\infty |\Phi_{\Sigma_d}(U_t\Omega_+ f)|\,dt. \tag{7.71}$$

The proof of this relation is easily reduced to showing that

$$\lim_{d\to\infty} \int_T^\infty \Phi_{\Sigma_d}(U_t^1 f)\,dt = \lim_{d\to\infty} \int_T^\infty |\Phi_{\Sigma_d}(U_t^1 f)|\,dt. \tag{7.72}$$

Indeed, we have seen in (7.69) above that the left-hand sides of (7.71) and (7.72) are identical. The equality of their right-hand sides is obtained by essentially the same argument after noticing that, by (7.70):

$$\left||\Phi_{\Sigma_d}(U_t\Omega_+ f)| - |\Phi_{\Sigma_d}(U_t^1 f)|\right| \leq \left|\Phi_{\Sigma_d}(U_t\Omega_+ f) - \Phi_{\Sigma_d}(U_t^1 f)\right| \leq c(1+t)^{-1-\varepsilon}.$$

To discuss (7.72), it is convenient to express the flux across $\Sigma_d$ in a slightly different form. We notice that on $\Sigma_d$ one has $\vec{n}(\vec{x}) = \vec{x}/d$, so that [by using (7.33) and the

[11] Also in the long range case with an appropriate definition of the wave operators, see Section 5.8.

definition (6.19) of the infinitesimal generator $D$ of the dilation group]:

$$\begin{aligned}
\Phi_{\Sigma_d}(h) &= 2\,\Re \sum_{k=1}^{n} \int_{\Sigma_d} \overline{[\gamma_{\Sigma_d} h](\vec{x})}\, n_k(\vec{x}) [\gamma_{\Sigma_d} P_k h](\vec{x})\, d\sigma(\vec{x}) \\
&= \frac{2}{d}\,\Re \sum_{k=1}^{n} \int_{\Sigma_d} \overline{[\gamma_{\Sigma_d} h](\vec{x})} [\gamma_{\Sigma_d} Q_k P_k h](\vec{x})\, d\sigma(\vec{x}) \\
&= \frac{1}{d}\,\Re \sum_{k=1}^{n} \int_{\Sigma_d} \overline{[\gamma_{\Sigma_d} h](\vec{x})} \big[\gamma_{\Sigma_d}(4D + inI)h\big](\vec{x})\, d\sigma(\vec{x}) \\
&= \frac{4}{d}\,\Re \sum_{k=1}^{n} \int_{\Sigma_d} \overline{[\gamma_{\Sigma_d} h](\vec{x})} [\gamma_{\Sigma_d} D h](\vec{x})\, d\sigma(\vec{x}).
\end{aligned} \tag{7.73}$$

This shows that the flux across $\Sigma_d$ is determined by the self-adjoint operator $D$ (momentum in the radial directions).

Let us first consider potentials $V$ with $V_1 = 0$, in other terms short range potentials decaying more rapidly than $|\vec{x}|^{-2}$. Then $U_t^1 = e^{-iH_0 t} = U_t^0$, and one can check the validity of (7.72) (with $U_t^1$ replaced by $U_t^0$) by explicit calculations. This was done in [4] in terms of eigenfunctions of the operator $H_0$ and in [3] by using the expressions (7.73) for the flux and (5.59) for the free evolution.

For potentials $V$ with $V_1 \neq 0$, such explicit calculations are not possible because they would involve the evolution group $U_t^1$. The fact that the flux of a scattering state $g \in \mathcal{H}_{\mathrm{ac}}(H_1)$ is outgoing at very large positive times can be explained as follows. Assume that the spectral support of $g$ with respect to the spectral measure of $H_1$ is contained in some interval $J = [\lambda_1, \lambda_2]$, with $0 < \lambda_1 < \lambda_2$. Then $(d/dt)\langle U_t^1 g, DU_t^1 g\rangle = \langle U_t^1 g, [H, iD]U_t^1 g\rangle \geq a\|g\|^2$ for some $a > 0$, as a consequence of a strict Mourre inequality satisfied on $J$. This implies that $\langle U_t^1 g, DU_t^1 g\rangle$ will be positive after some time $t_0$.

The preceding conclusion can be obtained in a somewhat different way by considering the subspace $\mathcal{M}_-^D$ corresponding to the negative spectrum of the operator $D$ (i.e. $\mathcal{M}_-^D = E^D((-\infty, 0])\mathcal{H}$). Then, under the evolution group $\{U_t^1\}$, all scattering states of $H_1$ propagate away from this subspace at large positive times, more precisely $\|E^D((-\infty, 0])U_t^1 f\| \to 0$ as $t \to +\infty$ for each $f \in \mathcal{M}_\infty(H_1)$. The proof of this result is based on the following propagation estimate (combined with a density argument): if $\nu, \varepsilon > 0$ and $\psi \in C_0^\infty((0, \infty))$, there is a constant $c$ such that

$$\|E^D((-\infty, 0])U_t^1 \psi(H_1)(I + |D|)^{-\nu}\| \leq c(1 + t)^{-\nu+\varepsilon} \qquad \forall t > 0.$$

This estimate can be obtained by using the Mourre method as in Chapter 6 to show that $\sup_{\lambda \in J_0, \mu > 0} \|E^D((-\infty, 0])(H_1 - \lambda - i\mu)^{-N}(I + |D|)^{-\nu}\| < \infty$, where the notations are as in Theorem 6.18. Details can be found for example in [10].

## 7.3 Time-independent scattering theory

In Chapter 5 we have developed the *time-dependent scattering theory*. This theory was formulated in terms of the two evolution groups associated with the free Hamiltonian $H_0$ and with the total Hamiltonian $H$; it involves as fundamental objects the wave operators $\Omega_+$ and $\Omega_-$ and the scattering operator $S = \Omega_+^*\Omega_-$. The *time-independent scattering theory* (also called *stationary state scattering theory*) gives formulas for these operators in terms of the spectral measures and the resolvents of $H_0$ and $H$. To obtain these equations, one first expresses the wave operators as integrals over the time variable $t$ (Proposition 7.8). Then one represents the evolution groups in the integrand in terms of the corresponding spectral measures (by using the functional calculus); the expressions obtained in this way contain the resolvents in the form indicated in Corollary 5.2. We present here some interesting and useful formulas, and we refer to Chapter 6 of [AJS] for additional material and more details.

Before entering into mathematical details, let us indicate formally how one arrives at a time-independent formula for the wave operator $\Omega_+$ by following the procedure indicated above. In the situation considered in §5.4.1 [$H = H_0 + V(\vec{Q})$, $H_0 = \vec{P}^2$, $U_t = e^{-iHt}$, $U_t^0 = e^{-iH_0t}$] we have $\Omega_+ = \text{s-lim}_{t\to+\infty} U_t^* U_t^0$. If this limit exists, it is also given as the limit of a temporal mean; we saw this in Chapter 5 where we considered temporal means of the type $T^{-1}\int_0^T \ldots dt$. Here it is more convenient to take temporal means of the form $\varepsilon\int_0^\infty e^{-\varepsilon t}\ldots dt$ ($\varepsilon > 0$). One then has $\Omega_+ = \text{s-lim}_{\varepsilon\to+0}\,\varepsilon\int_0^\infty e^{-\varepsilon t}U_t^*U_t^0 dt$. One replaces in this expression the operator $U_t^*$ by $U_t^* = \int_{\mathbb{R}} e^{i\lambda t}E(d\lambda)$, where $\{E(\cdot)\}$ denotes the spectral measure of $H$, then one interchanges the order of integration:

$$\begin{aligned}\Omega_+ = \underset{\varepsilon\to+0}{\text{s-lim}}\;\varepsilon\int_0^\infty e^{-\varepsilon t}U_t^*U_t^0\,dt &= \underset{\varepsilon\to+0}{\text{s-lim}}\;\varepsilon\int_0^\infty dt\int_{\mathbb{R}} E(d\lambda)e^{i\lambda t-\varepsilon t}U_t^0\\ &= \underset{\varepsilon\to+0}{\text{s-lim}}\;\varepsilon\int_{\mathbb{R}} E(d\lambda)\int_0^\infty dt\,e^{i\lambda t-\varepsilon t}U_t^0.\end{aligned}$$

By taking into account (5.7) one arrives at the following stationary expression for $\Omega_+$:

$$\Omega_+ = \underset{\varepsilon\to+0}{\text{s-lim}}\,(-i\varepsilon)\int_{\mathbb{R}} E(d\lambda)(H_0-\lambda-i\varepsilon)^{-1}. \tag{7.74}$$

The integral in (7.74) is a spectral integral in the sense of Riemann (strong limit of approximating sums), see the proof of the following lemma for a precise example. In (7.74) the spectral measure is on the left of the integrand $(H_0-\lambda-i\varepsilon)^{-1}$ [the operators $E(\cdot)$ and $(H_0-\lambda-i\varepsilon)^{-1}$ do not commute]. A different expression for $\Omega_+$ that can be obtained similarly is $\Omega_+ = \text{s-lim}_{\varepsilon\to+0}\, i\varepsilon\int_{\mathbb{R}}(H-\lambda-i\varepsilon)^{-1}E_0(d\lambda)$, see Corollary 7.10 (here $\{E_0(\cdot)\}$ denotes the spectral measure of the free Hamiltonian $H_0 = \vec{P}^2$).

One can see that the stationary expressions will involve spectral integrals with operator-valued integrands. In Chapter 4 we have not given any criteria for the existence of such integrals. It turns out that such criteria are not necessary here: each of the occurring spectral integrals is equal to an integral with respect to a numerical measure (in fact Lebesgue measure), and the integrals of the latter type exist as a consequence of the existence (and sometimes also the completeness) of the wave operators. This is already well illustrated by the formal derivation of a stationary expression for $\Omega_+$ given above, and it can be seen in more details from the proofs given further on in this section.

**Lemma 7.7.** *Let $A$ be a self-adjoint operator and $\{E^A(\cdot)\}$ its spectral measure.*
*(a) Let $(a,b]$ be a finite interval and let $\{C_\lambda\}_{\lambda\in(a,b]}$ be a family of operators in $\mathcal{B}(\mathcal{H})$ such that $\|C_\lambda\| \le M < \infty$ for all $\lambda \in (a,b]$. Then $\int_a^b C_\lambda(A-\lambda)E^A(d\lambda) = O$.*
*(b) Let $\{C_\lambda\}_{\lambda\in\mathbb{R}}$ be a family of operators in $\mathcal{B}(\mathcal{H})$ such that, for each finite interval $J$, there exists a constant $M_J \in (0,\infty)$ with $\|C_\lambda\| \le M_J$ for all $\lambda \in J$. Then $\int_{\mathbb{R}} C_\lambda(A-\lambda)E^A(d\lambda) = O$.*

PROOF. (a) Since $AE^A([\lambda,\lambda+d\lambda]) \approx \lambda E^A([\lambda,\lambda+d\lambda])$, hence $(A-\lambda)E^A(d\lambda) \approx O$, the stated result is formally evident. For a rigourous proof, let us consider partitions $\Pi_N = \{s_0, s_1, \dots, s_N; u_1, \dots, u_N\}$ of the interval $(a,b]$ (see page 6). We set $J_k = (s_{k-1}, s_k]$ and we have to show that, for each vector $f \in \mathcal{H}$, the sequence of vectors $\{\sum_{k=1}^N C_{u_k}(A-u_k)E^A(J_k)f\}$ converges strongly to **0** as $N \to \infty$ and $|\Pi_N| \to 0$. Let us first observe that [see (4.42)]

$$\|(A-u_k)E^A(J_k)f\|^2 = \int_{J_k} |\mu - u_k|^2 m_f(d\mu) \le |s_k - s_{k-1}|^2 \int_{J_k} m_f(d\mu)$$
$$= |s_k - s_{k-1}|^2 \|E^A(J_k)f\|^2.$$

By using this estimate, the triangle inequality (1.13) and the Schwarz inequality for a sum, we obtain that

$$\begin{aligned}\Big\|\sum_{k=1}^N C_{u_k}(A-u_k)E^A(J_k)f\Big\| &\le M\sum_{k=1}^N \|(A-u_k)E^A(J_k)f\| \\ &\le M\sum_{k=1}^N |\Pi_N|^{1/2}(s_k - s_{k-1})^{1/2}\|E^A(J_k)f\| \\ &\le M|\Pi_N|^{1/2}\Big\{\sum_{k=1}^N (s_k - s_{k-1})\Big\}^{1/2}\Big\{\sum_{k=1}^N \|E^A(J_k)f\|^2\Big\}^{1/2} \\ &= M|\Pi_N|^{1/2}(b-a)^{1/2}\|E^A((a,b])f\|.\end{aligned}$$

This shows that the spectral integral $\int_a^b C_\lambda(A-\lambda)E^A(d\lambda)$ exists and is equal to $O$.

(b) The spectral integral over an infinite interval is defined as a strong limit of a sequence of spectral integrals over finite subintervals. Hence, by taking into account the result of (a):

$$\int_{\mathbb{R}} C_\lambda(A-\lambda)E^A(d\lambda) \equiv \operatorname*{s-lim}_{\substack{a\to-\infty\\ b\to\infty}} \int_a^b C_\lambda(A-\lambda)E^A(d\lambda) = \operatorname*{s-lim}_{\substack{a\to-\infty\\ b\to\infty}} O = O. \qquad \square$$

**Proposition 7.8.** *Let $A$ and $B$ be self-adjoint operators and $U_t = e^{-iAt}$, $W_t = e^{-iBt}$. Assume that $\Omega_\pm \equiv \Omega_\pm(A, B; E^B_{\rm ac}) = \text{s-lim}_{t\to\pm\infty} U_t^* W_t E^B_{\rm ac}$ exist. Then one has*

$$\Omega_+ = \underset{\varepsilon\to+0}{\text{s-lim}}\, \varepsilon \int_0^\infty e^{-\varepsilon t} U_t^* W_t E^B_{\rm ac}\, dt \quad \textit{and} \quad \Omega_- = \underset{\varepsilon\to+0}{\text{s-lim}}\, \varepsilon \int_{-\infty}^0 e^{\varepsilon t} U_t^* W_t E^B_{\rm ac}\, dt. \tag{7.75}$$

PROOF. If $\varepsilon > 0$, the operator-valued function $t \mapsto e^{-\varepsilon t} U_t^* W_t$ is strongly continuous, and $t \mapsto \|e^{-\varepsilon t} U_t^* W_t\|$ is integrable on $[0, \infty)$. Hence $\int_0^\infty e^{-\varepsilon t} U_t^* W_t\, dt$ exists [Proposition 2.3(c)]. Since $\varepsilon \int_0^\infty e^{-\varepsilon t} dt = 1$, one has for each $f \in \mathcal{R}(E^B_{\rm ac})$:

$$\begin{aligned} \Big\| \varepsilon \int_0^\infty e^{-\varepsilon t} U_t^* W_t E^B_{\rm ac} f\, dt - \Omega_+ f \Big\| &= \Big\| \varepsilon \int_0^\infty e^{-\varepsilon t} [U_t^* W_t - \Omega_+] f\, dt \Big\| \\ &\le \varepsilon \int_0^\infty e^{-\varepsilon t} \| [U_t^* W_t - \Omega_+] f \|\, dt \\ = \varepsilon \int_0^T e^{-\varepsilon t} \| [U_t^* W_t - \Omega_+] f \|\, dt &+ \varepsilon \int_T^\infty e^{-\varepsilon t} \| [U_t^* W_t - \Omega_+] f \|\, dt. \end{aligned} \tag{7.76}$$

Let $\delta > 0$ be fixed. One can choose $T$ such that $\| [U_t^* W_t - \Omega_+] f \| < \delta/2$ for each $t \ge T$, so that the second integral on the last line above is less than $\delta/2$ for each $\varepsilon > 0$. On the other hand

$$\varepsilon \int_0^T e^{-\varepsilon t} \| [U_t^* W_t - \Omega_+] f \|\, dt \le \varepsilon \int_0^T \big[ \|U_t^* W_t f\| + \|\Omega_+ f\| \big] dt \le 2\varepsilon T \|f\|,$$

which is less than $\delta/2$ for each $\varepsilon \in (0, \varepsilon_0)$ if $\varepsilon_0 = \delta (4T\|f\|)^{-1}$. So the first identity in (7.75) follows from (7.76). The second one is obtained similarly. □

**Lemma 7.9.** *Let $A$ and $B$ be self-adjoint operators and $U_t = e^{-iAt}$, $W_t = e^{-iBt}$.*
*(a) For each $\varepsilon > 0$, the integrals below exist and belong to $\mathcal{B}(\mathcal{H})$, and one has the following identities:*

$$\int_0^\infty e^{-\varepsilon t} U_t^* W_t\, dt = i \int_{\mathbb{R}} (A - \lambda + i\varepsilon)^{-1} E^B(d\lambda), \tag{7.77}$$

$$\int_{-\infty}^0 e^{\varepsilon t} U_t^* W_t\, dt = -i \int_{\mathbb{R}} (A - \lambda - i\varepsilon)^{-1} E^B(d\lambda). \tag{7.78}$$

*(b) For all $\varepsilon > 0$ and $\delta > 0$, the integrals below exist and belong to $\mathcal{B}(\mathcal{H})$, and one has the following identities:*

$$\int_0^\infty dt\, e^{-\varepsilon t} W_t^* \int_{-\infty}^0 ds\, e^{\delta s} U_s^* W_{s+t} = \int_{\mathbb{R}} (B - \lambda + i\varepsilon)^{-1} (A - \lambda - i\delta)^{-1} E^B(d\lambda), \tag{7.79}$$

$$\int_{-\infty}^0 dt\, e^{\varepsilon t} W_t^* \int_{-\infty}^0 ds\, e^{\delta s} U_s^* W_{s+t} = -\int_{\mathbb{R}} (B - \lambda - i\varepsilon)^{-1} (A - \lambda - i\delta)^{-1} E^B(d\lambda). \tag{7.80}$$

PROOF. (i) The existence of the integrals and double integrals with respect to $dt$ and $ds$ and the fact that they belong to $\mathcal{B}(\mathcal{H})$ is obtained by the arguments used in the preceding proof. Formally the identities (7.77) - (7.80) are rather evident [follow the arguments in the derivation of (7.74)]. We give here a detailed proof of (7.77); for (7.78) - (7.80) one proceeds in the same way, but certain expressions are somewhat longer to write out.

(ii) We fix a number $M \in (0, \infty)$, set $E_M = E^B((-M, M])$ and consider partitions $\Pi_N = \{s_0, s_1, \ldots, s_N; u_1, \ldots, u_N\}$ of the interval $(-M, M]$ (see page 6). We set $J_k = (s_{k-1}, s_k]$ and define, for $f \in \mathcal{H}$:

$$X_{\Pi_N}(f) = \int_0^\infty dt\, e^{-\varepsilon t} U_t^* \sum_{k=1}^N e^{-iu_k t} E^B(J_k) f.$$

By using (1.28) it follows that:

$$\Big\|\int_0^\infty e^{-\varepsilon t} U_t^* W_t E_M f\, dt - X_{\Pi_N}(f)\Big\| \leq \int_0^\infty dt\, e^{-\varepsilon t} \Big\|W_t E_M f - \sum_{k=1}^N e^{-iu_k t} E^B(J_k) f\Big\|. \tag{7.81}$$

Consider a sequence of partitions $\{\Pi_N\}$ with $|\Pi_N| \to 0$ as $N \to \infty$. Then the integrand on the right-hand side of (7.81) converges to zero for each fixed $t$, and it is majorised for all $N$ by the following integrable function:

$$e^{-\varepsilon t}\Big[\|f\| + \Big\|\sum_{k=1}^N e^{-iu_k t} E^B(J_k) f\Big\|\Big] \leq 2e^{-\varepsilon t}\|f\|.$$ [12]

So, by applying the Dominated Convergence Theorem (Proposition 1.9), one obtains

$$\int_0^\infty e^{-\varepsilon t} U_t^* W_t E_M f\, dt = \operatorname*{s-lim}_{N\to\infty} X_{\Pi_N}(f).$$

(iii) By making the change of variables $t \mapsto -t$, one obtains with (5.8) that

$$X_{\Pi_N}(f) = \sum_{k=1}^N \int_{-\infty}^0 dt\, e^{i(u_k - i\varepsilon)t} U_t E^B(J_k) f = i \sum_{k=1}^N (A - u_k + i\varepsilon)^{-1} E^B(J_k) f.$$

Hence

$$\begin{aligned}\int_0^\infty e^{-\varepsilon t} U_t^* W_t E_M f\, dt &= \operatorname*{s-lim}_{N\to\infty} i \sum_{k=1}^N (A - u_k + i\varepsilon)^{-1} E^B(J_k) f \\ &\equiv i \int_{-M}^M (A - \lambda + i\varepsilon)^{-1} E^B(d\lambda) f.\end{aligned}$$

---

[12] By (4.44), with $\varphi(x) = e^{-ixt}$, one has $\|\sum_{k=1}^N e^{-iu_k t} E^B(J_k) f\|^2 =$
$= \sum_{k=1}^N |e^{-iu_k t}|^2 \langle f, E^B(J_k) f\rangle = \sum_{k=1}^N \langle f, E^B(J_k) f\rangle = \langle f, E_M f\rangle = \|E_M f\|^2 \leq \|f\|^2.$

By letting $M \to \infty$ in this equation, one obtains

$$\int_0^\infty e^{-\varepsilon t} U_t^* W_t f \, dt = \underset{M\to+\infty}{\text{s-lim}} \int_0^\infty e^{-\varepsilon t} U_t^* W_t E_M f \, dt$$

$$= \underset{M\to+\infty}{\text{s-lim}}\; i \int_{-M}^{M} (A - \lambda + i\varepsilon)^{-1} E^B(d\lambda) f \equiv i \int_{\mathbb{R}} (A - \lambda + i\varepsilon)^{-1} E^B(d\lambda) f. \quad \square$$

By combining Proposition 7.8 and Lemma 7.9, one obtains the following stationary expressions for $\Omega_\pm$ (similar formulas are valid for $\Omega_\pm^*$):

**Corollary 7.10.** *Let $A$ and $B$ be self-adjoint operators and $U_t = e^{-iAt}$, $W_t = e^{-iBt}$. Assume that $\Omega_\pm \equiv \Omega_\pm(A, B; E_{\text{ac}}^B) = \text{s-lim}_{t\to\pm\infty} U_t^* W_t E_{\text{ac}}^B$ exist. Then*

$$\Omega_+ = \underset{\varepsilon\to+0}{\text{s-lim}}\; i\varepsilon \int_{\mathbb{R}} (A - \lambda + i\varepsilon)^{-1} E^B(d\lambda) E_{\text{ac}}^B \tag{7.82}$$

$$\Omega_- = \underset{\varepsilon\to+0}{\text{s-lim}}\; (-i\varepsilon) \int_{\mathbb{R}} (A - \lambda - i\varepsilon)^{-1} E^B(d\lambda) E_{\text{ac}}^B. \tag{7.83}$$

In what follows we consider the case in which $B = H_0 = \vec{P}^2$, $A = H = H_0 + V$ with $\mathcal{D}(H) = \mathcal{D}(H_0)$, where $V$ is a self-adjoint operator [for example $V = V(\vec{Q})$]. It is clear that similar formulas are true in the general case. We use the notations of Section 5.4. In particular the operators $U_t^0 = e^{-iH_0 t}$ and $U_t = e^{-iHt}$ give the free and the total evolution group respectively, and $\{E_0(\cdot)\}$ denotes the spectral measure of $H_0$.

**Proposition 7.11 (Lippmann-Schwinger equations).** *Let $H_0 = \vec{P}^2$ and $H = H_0 + V$, where $V = V^*$ is $H_0$-bounded and $H$-bounded. If the wave operators $\Omega_\pm = \text{s-lim}_{t\to\pm\infty} e^{iHt} e^{-iH_0 t}$ exist, then*

$$\begin{aligned}\Omega_+ &= I - \underset{\varepsilon\to+0}{\text{s-lim}} \int_{\mathbb{R}} (H - \lambda + i\varepsilon)^{-1} V E_0(d\lambda) \\ &= I - \underset{\varepsilon\to+0}{\text{s-lim}} \int_{\mathbb{R}} (H_0 - \lambda + i\varepsilon)^{-1} V \Omega_+ E_0(d\lambda),\end{aligned} \tag{7.84}$$

$$\begin{aligned}\Omega_- &= I - \underset{\varepsilon\to+0}{\text{s-lim}} \int_{\mathbb{R}} (H - \lambda - i\varepsilon)^{-1} V E_0(d\lambda) \\ &= I - \underset{\varepsilon\to+0}{\text{s-lim}} \int_{\mathbb{R}} (H_0 - \lambda - i\varepsilon)^{-1} V \Omega_- E_0(d\lambda).\end{aligned} \tag{7.85}$$

PROOF. Let $\varepsilon > 0$. Since $\|(H - \lambda \pm i\varepsilon)^{-1}\| \leq 1/\varepsilon$, Lemma 7.7 implies that for each interval $(a, b]$:

$$\int_a^b (H - \lambda \pm i\varepsilon)^{-1} (H_0 - \lambda) E_0(d\lambda) = O. \tag{7.86}$$

On the other hand one has by Corollary 7.10:

$$\Omega_+ = \underset{\varepsilon\to+0}{\text{s-lim}}\; \Omega_{+,\varepsilon}\,, \quad \text{with} \quad \Omega_{+,\varepsilon} = i\varepsilon \int_{\mathbb{R}} (H - \lambda + i\varepsilon)^{-1} E_0(d\lambda) \in \mathcal{B}(\mathcal{H}).$$

For $M \in (0, \infty)$ we can write, by using (7.86):

$$\begin{aligned}
\Omega_{+,\varepsilon} E_0((0,M]) &= i\varepsilon \int_0^M (H - \lambda + i\varepsilon)^{-1} E_0(d\lambda) \\
&= \int_0^M (H - \lambda + i\varepsilon)^{-1}(H_0 - \lambda + i\varepsilon) E_0(d\lambda) = \\
&= \int_0^M (H - \lambda + i\varepsilon)^{-1}(H - \lambda + i\varepsilon - V) E_0(d\lambda) \\
&= \int_0^M \big[I - (H - \lambda + i\varepsilon)^{-1} V\big] E_0(d\lambda) \\
&= E_0((0,M]) - \int_0^M (H - \lambda + i\varepsilon)^{-1} V E_0(d\lambda). \qquad (7.87)
\end{aligned}$$

Observe that $VE_0(J) \in \mathcal{B}(\mathcal{H})$ if $J$ is a finite interval, because $V$ is an $H_0$-bounded operator. The assumption made on $V$ has also allowed us to write $H_0E_0(d\lambda) = HE_0(d\lambda) - VE_0(d\lambda)$ for $\lambda \in (0, M]$. Since $V$ is also $H$-bounded, the closure of $(H-\lambda+i\varepsilon)^{-1}V$ is an operator in $\mathcal{B}(\mathcal{H})$. As $M \to \infty$, $E_0((0,M])$ converges strongly to the identity operator $I$ and $\Omega_{+,\varepsilon}E_0((0,M])$ converges strongly to $\Omega_{+,\varepsilon}$. Thus (7.87) implies that $\int_0^M (H - \lambda + i\varepsilon)^{-1} V E_0(d\lambda)$ is strongly convergent as $M \to \infty$, i.e. the spectral integral $\int_0^\infty (H - \lambda + i\varepsilon)^{-1} V E_0(d\lambda)$ exists, and one has

$$\Omega_{+,\varepsilon} = I - \int_0^\infty (H - \lambda + i\varepsilon)^{-1} V E_0(d\lambda) \equiv I - \int_{\mathbb{R}} (H - \lambda + i\varepsilon)^{-1} V E_0(d\lambda)$$

[remember that $E_0(\mathcal{V}) = O$ if $\mathcal{V} \cap (0,\infty) = \varnothing$]. As $\Omega_+ = \text{s-lim}_{\varepsilon \to +0} \Omega_{+,\varepsilon}$, we have verified the first identity in (7.84).

The second relation in (7.84) is obtained similarly [using Proposition 5.15(a)]:

$$\begin{aligned}
I = \Omega_+^* \Omega_+ &= \operatorname*{s-lim}_{\varepsilon \to +0} \varepsilon \int_0^\infty dt\, e^{-\varepsilon t} U_t^{0*} U_t \Omega_+ \\
&= \operatorname*{s-lim}_{\varepsilon \to +0} i\varepsilon \int_{\mathbb{R}} (H_0 - \lambda + i\varepsilon)^{-1} E(d\lambda) \Omega_+ \\
&= \operatorname*{s-lim}_{\varepsilon \to +0} \int_{\mathbb{R}} (H_0 - \lambda + i\varepsilon)^{-1} (H - \lambda + i\varepsilon) E(d\lambda) \Omega_+ \\
&= \Omega_+ + \operatorname*{s-lim}_{\varepsilon \to +0} \int_{\mathbb{R}} (H_0 - \lambda + i\varepsilon)^{-1} V E(d\lambda) \Omega_+,
\end{aligned}$$

and it now suffices to take into account the intertwining relation $E(\mathcal{V})\Omega_+ = \Omega_+ E_0(\mathcal{V})$. □

In order to represent the operator $R \equiv S - I$ as a spectral integral, we write $R = \Omega_+^* \Omega_- - \Omega_-^* \Omega_-$ and use the following expressions for $\Omega_+^* \Omega_-$ and $\Omega_-^* \Omega_-$:

**Lemma 7.12.** *Assume that the wave operators exist and are complete. Then*

$$\begin{aligned}\Omega_+^*\Omega_- &= \operatorname*{s-lim}_{\varepsilon\to+0}\operatorname*{s-lim}_{\delta\to+0}(\varepsilon+\delta)\delta\int_0^\infty dt\, e^{-\varepsilon t}U_t^{0*}\int_{-\infty}^0 ds\, e^{\delta s}U_s^*U_{s+t}^0\\ &= \operatorname*{s-lim}_{\varepsilon\to+0}\operatorname*{s-lim}_{\delta\to+0}(\varepsilon+\delta)\delta\int_{\mathbb{R}}(H_0-\lambda+i\varepsilon)^{-1}(H-\lambda-i\delta)^{-1}E_0(d\lambda),\end{aligned} \tag{7.88}$$

$$\begin{aligned}\Omega_-^*\Omega_- &= \operatorname*{s-lim}_{\varepsilon\to+0}\operatorname*{s-lim}_{\delta\to+0}(\varepsilon-\delta)\delta\int_{-\infty}^0 dt\, e^{\varepsilon t}U_t^{0*}\int_{-\infty}^0 ds\, e^{\delta s}U_s^*U_{s+t}^0\\ &= \operatorname*{s-lim}_{\varepsilon\to+0}\operatorname*{s-lim}_{\delta\to+0}(\delta-\varepsilon)\delta\int_{\mathbb{R}}(H_0-\lambda-i\varepsilon)^{-1}(H-\lambda-i\delta)^{-1}E_0(d\lambda).\end{aligned} \tag{7.89}$$

PROOF. We give the proof of (7.88). By taking into account the intertwining relation $U_t\Omega_- = \Omega_- U_t^0$ one obtains

$$\begin{aligned}\Omega_+^*\Omega_- &= \operatorname*{s-lim}_{t\to+\infty} U_t^{0*}U_t\Omega_- = \operatorname*{s-lim}_{\varepsilon\to+0}\varepsilon\int_0^\infty dt\, e^{-\varepsilon t}U_t^{0*}U_t\Omega_-\\ &= \operatorname*{s-lim}_{\varepsilon\to+0}\varepsilon\int_0^\infty dt\, e^{-\varepsilon t}U_t^{0*}\Omega_-U_t^0.\end{aligned} \tag{7.90}$$

For $\delta>0$, we set $\Omega_{-,\delta}=\delta\int_{-\infty}^0 e^{\delta t}U_t^*U_t^0\,dt$. Then $\|\Omega_{-,\delta}\|\le\delta\int_{-\infty}^0 e^{\delta t}\|U_t^*U_t^0\|\,dt = \delta\int_{-\infty}^0 e^{\delta t}dt=1$. One has

$$\left\|\int_0^\infty dt\, e^{-\varepsilon t}U_t^{0*}(\Omega_- - \Omega_{-,\delta})U_t^0 f\right\| \le \int_0^\infty dt\, e^{-\varepsilon t}\|(\Omega_- - \Omega_{-,\delta})U_t^0 f\|. \tag{7.91}$$

For each fixed $t$ the integrand on the right-hand side of (7.91) converges to zero as $\delta\to0$, and it is majorised by $2e^{-\varepsilon t}\|f\|$, which is independent of $\delta$ and integrable on $[0,\infty)$. So the Dominated Convergence Theorem (Proposition1.9) implies that

$$\operatorname*{s-lim}_{\delta\to+0}\int_0^\infty dt\, e^{-\varepsilon t}U_t^{0*}(\Omega_- - \Omega_{-,\delta})U_t^0 f = \mathbf{0} \qquad \forall f\in\mathcal{H}.$$

Hence (7.90) may be written as

$$\Omega_+^*\Omega_- = \operatorname*{s-lim}_{\varepsilon\to+0}\operatorname*{s-lim}_{\delta\to+0}\delta\varepsilon\int_0^\infty dt e^{-\varepsilon t}U_t^{0*}\int_{-\infty}^0 ds e^{\delta s}U_s^*U_{s+t}^0.$$

The difference between this expression and the first line in (7.88) resides in the term

$$-\operatorname*{s-lim}_{\varepsilon\to+0}\operatorname*{s-lim}_{\delta\to+0}\delta\int_0^\infty e^{-\varepsilon t}U_t^{0*}\Omega_{-,\delta}U_t^0\,dt,$$

and we have to show that this expression is zero. For this we observe that, for fixed $\varepsilon>0$:

$$\|\delta\int_0^\infty e^{-\varepsilon t}U_t^{0*}\Omega_{-,\delta}U_t^0\,dt\| \le \delta\int_0^\infty e^{-\varepsilon t}\|U_t^{0*}\Omega_{-,\delta}U_t^0\|\,dt \le \delta\int_0^\infty e^{-\varepsilon t}dt = \frac{\delta}{\varepsilon},$$

which converges to zero as $\delta \to 0$. The second expression in (7.88) now follows by taking into account (7.79). □

REMARKS. (a) The existence of the spectral integral in (7.88) (for $\varepsilon \neq 0$ and $\delta \neq 0$) follows from Lemma 7.9. The existence of the limits as $\delta \to +0$ and $\varepsilon \to +0$ is a consequence of the assumption that the wave operators exist and are complete.
(b) The expressions (7.88) and (7.89) look rather complicated because of the multiple limits. The spectral integral on $\mathbb{R}$ itself is already a double strong limit (first on finite intervals, then as the length of these intervals tends to infinity); this is followed by the two limits $\delta \to +0$ and $\varepsilon \to +0$. It is important to note that the order of these limits cannot be altered. Nevertheless the expressions (7.88) and (7.89), or rather their consequence (7.92) proved below, lead to a relatively simple and important expression for the $S$-matrix $S(\lambda)$, as we shall see further on.

**Proposition 7.13.** *Let $H_0 = \vec{P}^2$ and $H = H_0 + V$, where $\mathcal{D}(H) = \mathcal{D}(H_0)$ and the self-adjoint operator $V$ is $H_0$-bounded and $H$-bounded. For $z \in \mathbb{C}\setminus\mathbb{R}$, set $R_z^0 = (H_0 - z)^{-1}$ and $R_z = (H - z)^{-1}$. Assume that the wave operators exist and are complete. Then*

$$R \equiv S - I = \operatorname*{s-lim}_{\varepsilon \to +0} \operatorname*{s-lim}_{\delta \to +0} \int_{\mathbb{R}} (R^0_{\lambda - i\varepsilon} - R^0_{\lambda + i\varepsilon})(V - VR_{\lambda + i\delta}V)E_0(d\lambda). \tag{7.92}$$

PROOF. As in the proof of Proposition 7.11 we use the fact that $VE_0(\mathcal{W}) \in \mathcal{B}(\mathcal{H})$ if $\mathcal{W}$ is a bounded interval. So the spectral integral in (7.92) is well defined on each bounded subinterval of $\mathbb{R}$. Also, in the special situation considered here, where $H_0 = \vec{P}^2$, the integral in (7.92) extends effectively only over $(0, \infty)$.

By using (7.88) and (7.89) one sees that

$$R = \Omega_+^*\Omega_- - \Omega_-^*\Omega_- = \operatorname*{s-lim}_{\varepsilon \to +0} \operatorname*{s-lim}_{\delta \to +0} \delta \int_0^\infty X_{\varepsilon,\delta}(\lambda)E_0(d\lambda), \tag{7.93}$$

with

$$X_{\varepsilon,\delta}(\lambda) = (\varepsilon + \delta)R^0_{\lambda - i\varepsilon}R_{\lambda + i\delta} + (\varepsilon - \delta)R^0_{\lambda + i\varepsilon}R_{\lambda + i\delta}.$$

One may re-express $X_{\varepsilon,\delta}(\lambda)$ by using the second resolvent equation (2.115) and the first resolvent equation (2.93): if $z, \zeta \in \mathbb{C}\setminus\mathbb{R}$, one has

$$\begin{aligned}(\zeta - z)R_\zeta^0 R_z &= (\zeta - z)R_\zeta^0 R_z^0[I - VR_z] = [R_\zeta^0 - R_z^0][I - VR_z] \\ &= R_\zeta^0 - R_\zeta^0 VR_z - R_z^0[I - VR_z] = R_\zeta^0 - R_\zeta^0 VR_z - R_z.\end{aligned}$$

By taking $z = \lambda + i\delta$, $\zeta = \lambda - i\varepsilon$ or $\zeta = \lambda + i\varepsilon$, one obtains

$$\begin{aligned}X_{\varepsilon,\delta}(\lambda) &= i\{R^0_{\lambda - i\varepsilon} - R^0_{\lambda - i\varepsilon}VR_{\lambda + i\delta} - R_{\lambda + i\delta}\} \\ &\quad - i\{R^0_{\lambda + i\varepsilon} - R^0_{\lambda + i\varepsilon}VR_{\lambda + i\delta} - R_{\lambda + i\delta}\} \\ &= i[R^0_{\lambda - i\varepsilon} - R^0_{\lambda + i\varepsilon}] - i[R^0_{\lambda - i\varepsilon} - R^0_{\lambda + i\varepsilon}]VR_{\lambda + i\delta}.\end{aligned} \tag{7.94}$$

We now observe that, by Lemma 7.9:

$$\int_{\mathbb{R}} R^0_{\lambda-i\varepsilon} E_0(d\lambda) = -i\int_0^{\infty} e^{-\varepsilon t} U_t^{0*} U_t^0 \, dt = \frac{-i}{\varepsilon} I,$$

$$\int_{\mathbb{R}} R^0_{\lambda+i\varepsilon} E_0(d\lambda) = i\int_{-\infty}^{0} e^{\varepsilon t} U_t^{0*} U_t^0 \, dt = \frac{i}{\varepsilon} I.$$

This implies that for each $\varepsilon > 0$:

$$\operatorname*{s-lim}_{\delta\to+0} \delta \int_{\mathbb{R}} [R^0_{\lambda-i\varepsilon} - R^0_{\lambda+i\varepsilon}] E_0(d\lambda) = \operatorname*{s-lim}_{\delta\to+0} \frac{-2i\delta}{\varepsilon} I = O.$$

Thus, by inserting (7.94) into (7.93), one arrives at

$$\begin{aligned} R &= \operatorname*{s-lim}_{\varepsilon\to+0} \operatorname*{s-lim}_{\delta\to+0} (-i\delta) \int_0^{\infty} [R^0_{\lambda-i\varepsilon} - R^0_{\lambda+i\varepsilon}] V R_{\lambda+i\delta} E_0(d\lambda) \\ &= \operatorname*{s-lim}_{\varepsilon\to+0} \operatorname*{s-lim}_{\delta\to+0} \int_0^{\infty} [R^0_{\lambda-i\varepsilon} - R^0_{\lambda+i\varepsilon}] V R_{\lambda+i\delta} (H_0 - \lambda - i\delta) E_0(d\lambda), \end{aligned} \tag{7.95}$$

where the second identity is obtained by taking into account Lemma 7.7 [13]. By writing $H_0 = H - V$, one has $VR_{\lambda+i\delta}(H_0 - \lambda - i\delta) = VR_{\lambda+i\delta}(H - \lambda - i\delta - V) = V - VR_{\lambda+i\delta}V$, and (7.95) becomes identical with (7.92). □

Equation (7.92) indicates that the operator $R(\lambda) = S(\lambda) - I$ must be determined in a certain sense by the operators $T_z := V - VR_zV$ as the complex number $z$ tends to the spectral point $\lambda$ of $H_0$. A rigourous derivation of this relation is presented in Section 7.4. Let us already give a formal argument here. Assume that $T_z$ is an integral operator in the spectral representation of $H_0$ [see (5.68)], i.e. that there exists a family $\{T_z(\lambda, \mu)\}$, with $\lambda, \mu \in (0, \infty)$, of operators acting in $L^2(S^{n-1})$ such that

$$(\mathcal{U}_0 T_z f)_\lambda = \int_0^{\infty} T_z(\lambda, \mu)(\mathcal{U}_0 f)_\mu \, d\mu.$$

By observing that $(\mathcal{U}_0 R^0_{\lambda-i\varepsilon} f - \mathcal{U}_0 R^0_{\lambda+i\varepsilon} f)_\mu = -2i\varepsilon[(\lambda-\mu)^2 + \varepsilon^2]^{-1}(\mathcal{U}_0 f)_\mu$, one can write (7.92) in the spectral representation of $H_0$ as

$$\begin{aligned} \langle f, Rg \rangle = -\lim_{\varepsilon\to+0} \lim_{\delta\to+0} \int_0^{\infty} d\mu \int_0^{\infty} d\lambda \, \frac{2i\varepsilon}{(\lambda-\mu)^2 + \varepsilon^2} \\ \times \langle (\mathcal{U}_0 f)_\mu, T_{\lambda+i\delta}(\mu, \lambda)(\mathcal{U}_0 g)_\lambda \rangle_{L^2(S^{n-1})}. \end{aligned}$$

By taking first the limit $\varepsilon \to +0$ [remembering that $\lim_{\varepsilon\to+0} \varepsilon[(\lambda-\mu)^2 + \varepsilon^2]^{-1} = \pi\delta(\lambda - \mu)$], one obtains

$$\langle f, Rg \rangle = -2\pi i \int_0^{\infty} d\lambda \, \langle (\mathcal{U}_0 f)_\lambda, T_{\lambda+i0}(\lambda, \lambda)(\mathcal{U}_0 g)_\lambda \rangle_{L^2(S^{n-1})}.$$

[13] One has $\|[R^0_{\lambda-i\varepsilon} - R^0_{\lambda+i\varepsilon}] V R_{\lambda+i\delta}\| = \|[R^0_{\lambda-i\varepsilon} - R^0_{\lambda+i\varepsilon}] V R_i (H - i) R_{\lambda+i\delta}\| \leq 2|\varepsilon|^{-1} \|V(H-i)^{-1}\| \|(H-i)(H-\lambda-i\delta)^{-1}\|$, and $\|(H-i)(H-\lambda-i\delta)^{-1}\| = \|I + (\lambda + i\delta - i)(H - \lambda - i\delta)^{-1}\| \leq 1 + (|\lambda| + |\delta| + 1)|\delta|^{-1}$, which is a bounded function of $\lambda$ when $\lambda$ varies on a finite interval $J$.

By comparing with the formula $\langle f, Rg\rangle = \int_0^\infty d\lambda \, \langle(\mathcal{U}_0 f)_\lambda, R(\lambda)(\mathcal{U}_0 g)_\lambda\rangle_{L^2(S^{n-1})}$, one concludes from this formal argument that $R(\lambda)$ should be given by the expression $R(\lambda) = -2\pi i T_{\lambda+i0}(\lambda, \lambda)$.

## 7.4 The scattering matrix

For a simple scattering system the scattering matrix at energy $\lambda$ is a bounded operator $S(\lambda)$ acting in $L^2(S^{n-1})$. We have already mentioned on page 244, without proof, that there exist stationary expressions for this operator $S(\lambda)$ in terms of the resolvent $(H - z)^{-1}$ of the total Hamiltonian $H$, of the interaction $V$ and of certain transition operators mapping $L^2(\mathbb{R}^n)$ to $L^2(S^{n-1})$. We first discuss a useful class of such transition operators and then derive a stationary formula for $S(\lambda)$. We assume in this section that $n \geq 2$.

**7.4.1.** The operator $S(\lambda)$ may be viewed as acting on the angular variables $\vec{\omega}$ in the Fourier transforms of the wave functions in spherical polar co-ordinates, with the radial variable $k = \sqrt{\lambda}$ fixed. If $f \in L^2(\mathbb{R}^n)$ is a wave function and $\tilde{f}$ its Fourier transform, the restriction of $\tilde{f}$ to the sphere of radius $\sqrt{\lambda}$ is not well defined in general, because this sphere is a set of Lebesgue measure zero in $\mathbb{R}^n$. For certain wave functions it is possible to give a meaning to this restriction, as already discussed in §7.2.2. If for example $f \in L^1(\mathbb{R}^n)$, then $\tilde{f}$ is (equivalent to) a continuous function (see Proposition 5.6), and its restriction to a sphere in $\mathbb{R}^n$ is well defined. One could try to define in this way an operator with domain $L^2(\mathbb{R}^n) \cap L^1(\mathbb{R}^n)$ and values in $L^2(S^{n-1})$; such an operator would certainly be unbounded, and we are interested in constructing bounded transition operators to find an expression for the bounded operator $S(\lambda)$.

A simple method is as follows. We fix a function $W$ in $L^2(\mathbb{R}^n)$. For $f \in L^2(\mathbb{R}^n)$, we set $(Wf)(\vec{x}) = W(\vec{x})f(\vec{x})$. Then $Wf$ belongs to $L^1(\mathbb{R}^n)$, so that the restriction of its Fourier transform to a sphere has a well defined meaning. This holds for each $f \in L^2(\mathbb{R}^n)$. In this way one obtains a linear operator defined on $L^2(\mathbb{R}^n)$ associating to each $f \in L^2(\mathbb{R}^n)$ a continuous function of $\vec{\omega}$ (for fixed $\sqrt{\lambda}$), hence an element of $L^2(S^{n-1})$. We denote this operator by $M_W(\lambda)$. Thus

$$\begin{aligned}[M_W(\lambda)f](\vec{\omega}) &= 2^{-1/2}\lambda^{(n-2)/4}[\mathcal{F}Wf](\sqrt{\lambda}\vec{\omega}) \\ &= 2^{-1/2}\lambda^{(n-2)/4}(2\pi)^{-n/2}\int_{\mathbb{R}^n} e^{-i\sqrt{\lambda}\vec{\omega}\cdot\vec{x}}W(\vec{x})f(\vec{x})\,d^n x. \qquad (7.96)\end{aligned}$$

The factor $2^{-1/2}\lambda^{(n-2)/4}$ has been introduced for convenience; one then has for all $f \in \mathcal{D}(W(\vec{Q}))$ [see (5.68)]:

$$M_W(\lambda)f = \left[\mathcal{U}_0 W(\vec{Q})f\right]_\lambda. \qquad (7.97)$$

The integral in (7.96) depends only on the equivalence class of $f$, hence Eq. (7.96) defines, for each fixed $\lambda$, a continuous function of $\vec{\omega}$. The operator $M_W(\lambda)$ is a bounded operator from $L^2(\mathbb{R}^n)$ to $L^2(S^{n-1})$, i.e. $M_W(\lambda) \in \mathcal{B}(L^2(\mathbb{R}^n), L^2(S^{n-1}))$ [see §2.1.7 for the notation]. Indeed:

$$\begin{aligned}\big|[M_W(\lambda)f](\vec{\omega})\big| &\le 2^{-1/2}(2\pi)^{-n/2}\lambda^{(n-2)/4}\int_{\mathbb{R}^n}\big|W(\vec{x})f(\vec{x})\big|\,d^nx \\ &\le 2^{-1/2}(2\pi)^{-n/2}\lambda^{(n-2)/4}\|W\|_{L^2(\mathbb{R}^n)}\|f\|_{L^2(\mathbb{R}^n)}.\end{aligned}$$

Hence, denoting by $\Theta_n$ the surface area of the sphere $S^{n-1}$:

$$\begin{aligned}\|M_W(\lambda)f\|^2_{L^2(S^{n-1})} &= \int_{S^{n-1}}\big|[M_W(\lambda)f](\vec{\omega})\big|^2\,d\omega \\ &\le \frac{\Theta_n}{2(2\pi)^n}\lambda^{(n-2)/2}\|W\|^2_{L^2(\mathbb{R}^n)}\|f\|^2,\end{aligned}$$

i.e.

$$\|M_W(\lambda)\| \le 2^{-1/2}(2\pi)^{-n/2}\sqrt{\Theta_n}\,\lambda^{(n-2)/4}\|W\|_{L^2(\mathbb{R}^n)}. \tag{7.98}$$

In fact $M_W(\lambda)$ is even a Hilbert-Schmidt operator. Here we are considering Hilbert-Schmidt operators between two different Hilbert spaces. If $\mathcal{H}$ and $\mathcal{K}$ are Hilbert spaces, an everywhere defined operator $A\colon \mathcal{H} \to \mathcal{K}$ is said to be a Hilbert-Schmidt operator if $\sum_k \|Ae_k\|^2_{\mathcal{K}} < \infty$, where $\{e_k\}$ is an arbitrary orthonormal basis of $\mathcal{H}$. If $\mathcal{H}$ and $\mathcal{K}$ are $L^2$ spaces, an operator $A\colon \mathcal{H} \to \mathcal{K}$ is a Hilbert-Schmidt operator if and only if $A$ is an integral operator with square-integrable kernel; the proof consists in an evident adaptation of that of Proposition 2.15.

In the case of interest here [$\mathcal{H} = L^2(\mathbb{R}^n)$, $\mathcal{K} = L^2(S^{n-1})$], Eq. (7.96) shows that $M_W(\lambda)$ is an integral operator with kernel

$$a_\lambda(\vec{\omega},\vec{x}) = 2^{-1/2}(2\pi)^{-n/2}\lambda^{(n-2)/4}e^{-i\sqrt{\lambda}\vec{\omega}\cdot\vec{x}}W(\vec{x}). \tag{7.99}$$

Hence

$$\begin{aligned}\|M_W(\lambda)\|^2_{\text{HS}} &= \int_{S^{n-1}}d\omega\int_{\mathbb{R}^n}d^nx\,|a_\lambda(\vec{\omega},\vec{x})|^2 \\ &= \frac{1}{2(2\pi)^n}\lambda^{(n-2)/2}\int_{S^{n-1}}d\omega\int_{\mathbb{R}^n}d^nx\,|W(\vec{x})|^2 = \frac{1}{2(2\pi)^n}\lambda^{(n-2)/2}\Theta_n\|W\|^2_{L^2(\mathbb{R}^n)}.\end{aligned} \tag{7.100}$$

Some important properties of the operators $M_W(\lambda)$ are given in the following proposition. The adjoint of an operator between different Hilbert spaces has been described in §2.1.7.

**Proposition 7.14.** *Let $W \in L^2(\mathbb{R}^n)$. For $\lambda > 0$, let $M_W(\lambda)\colon L^2(\mathbb{R}^n) \to L^2(S^{n-1})$ be the operator defined by* (7.97). *Then:*

*(a) $M_W(\lambda)$ is a Hilbert-Schmidt operator. The operator-valued function $\lambda \mapsto M_W(\lambda)$ is continuous in Hilbert-Schmidt norm on $(0,\infty)$: $\|M_W(\lambda) - M_W(\mu)\|_{\text{HS}} \to 0$ as $\mu \to \lambda$.*

*(b) The adjoint* $M_W(\lambda)^*$ *of* $M_W(\lambda)$ *is a Hilbert-Schmidt operator from* $L^2(S^{n-1})$ *to* $L^2(\mathbb{R}^n)$*; the operator-valued function* $\lambda \mapsto M_W(\lambda)^*$ *is continuous in Hilbert-Schmidt norm, and one has for each Borel set* $\mathcal{V}$ *in* $\mathbb{R}$*:*

$$W(\vec{Q})^* E_0(\mathcal{V}) f = \int_{\mathcal{V}} M_W(\lambda)^* (\mathcal{U}_0 f)_\lambda d\lambda \qquad \forall f \in \mathcal{S}(\mathbb{R}^n). \tag{7.101}$$

**Remark 7.15.** The integral in (7.101) is a vector-valued integral in $L^2(\mathbb{R}^n)$ [the range of $M_W(\lambda)^*$ belongs to $L^2(\mathbb{R}^n)$]. If $f \in \mathcal{S}(\mathbb{R}^n)$, then $\lambda \mapsto (\mathcal{U}_0 f)_\lambda$ is a continuous $L^2(S^{n-1})$-valued function, and $\|(\mathcal{U}_0 f)_\lambda\|$ is of rapid decay as $\lambda \to \infty$. On the other hand $M_W(\lambda)^*$ is continuous (in Hilbert-Schmidt norm), and $\|M_W(\lambda)^*\| \le c\lambda^{(n-2)/4}$. Thus the integral in (7.101) is well defined with our conventions of §1.2.

PROOF OF PROPOSITION 7.14. (a) In view of (7.100) it just remains to prove the continuity in Hilbert-Schmidt norm of $\lambda \mapsto M_W(\lambda)$. One has

$$\|M_W(\lambda) - M_W(\mu)\|_{\mathrm{HS}}^2 = \int_{S^{n-1}} d\omega \int_{\mathbb{R}^n} d^n x \left|a_\lambda(\vec{\omega}, \vec{x}) - a_\mu(\vec{\omega}, \vec{x})\right|^2. \tag{7.102}$$

It is clear from (7.99) that $a_\lambda(\vec{\omega}, \vec{x}) - a_\mu(\vec{\omega}, \vec{x}) \to 0$ as $\mu \to \lambda$. Also, for $\mu \in (\lambda/2, 2\lambda]$:

$$\begin{aligned}\left|a_\lambda(\vec{\omega}, \vec{x}) - a_\mu(\vec{\omega}, \vec{x})\right|^2 &\le 2\{|a_\lambda(\vec{\omega}, \vec{x})|^2 + |a_\mu(\vec{\omega}, \vec{x})|^2\} \\ &\le \frac{2}{(2\pi)^n}(2\lambda)^{(n-2)/2}|W(\vec{x})|^2,\end{aligned}$$

which is an integrable function on $S^{n-1} \times \mathbb{R}^n$. By virtue of the Dominated Convergence Theorem, the expression (7.102) converges to zero as $\mu \to \lambda$.

(b) The first two assertions in (b) follow from the corresponding statements in (a) by passing to the adjoint [taking into account (2.45)]. One could also repeat the arguments given in the proof of (a) by observing that $M_W(\lambda)^*$ is an integral operator from $L^2(S^{n-1})$ to $L^2(\mathbb{R}^n)$ with kernel

$$2^{-1/2}(2\pi)^{-n/2}\lambda^{(n-2)/4}\overline{W(\vec{x})}e^{i\sqrt{\lambda}\vec{x}\cdot\vec{\omega}}.$$

It remains to verify (7.101). Let $f \in \mathcal{S}(\mathbb{R}^n)$ and $g \in \mathcal{D}(W(\vec{Q}))$. Then

$$\begin{aligned}&\langle E_0(\mathcal{V})f, W(\vec{Q})g\rangle = \langle \mathcal{U}_0 E_0(\mathcal{V})f, \mathcal{U}_0 W(\vec{Q})g\rangle \\ &= \int_{\mathcal{V}} d\lambda \left\langle (\mathcal{U}_0 f)_\lambda, (\mathcal{U}_0 W(\vec{Q})g)_\lambda \right\rangle_{L^2(S^{n-1})} = \int_{\mathcal{V}} d\lambda \left\langle (\mathcal{U}_0 f)_\lambda, M_W(\lambda) g \right\rangle_{L^2(S^{n-1})} \\ &= \int_{\mathcal{V}} d\lambda \langle M_W(\lambda)^*(\mathcal{U}_0 f)_\lambda, g\rangle = \left\langle \int_{\mathcal{V}} d\lambda\, M_W(\lambda)^*(\mathcal{U}_0 f)_\lambda, g\right\rangle. \end{aligned}\tag{7.103}$$

We have seen in Remark 7.15 that the integral $\int_{\mathcal{V}} d\lambda\, M_W(\lambda)^*(\mathcal{U}_0 f)_\lambda$ defines a vector in $L^2(\mathbb{R}^n)$. Denoting this vector by $f^*$, Eq. (7.103) implies that $E_0(\mathcal{V})f$ belongs to $\mathcal{D}(W(\vec{Q})^*)$ and that $W(\vec{Q})^* E_0(\mathcal{V}) f = f^* = \int_{\mathcal{V}} d\lambda\, M_W(\lambda)^*(\mathcal{U}_0 f)_\lambda$. □

The functions $W$ considered so far are not sufficiently general for our later applications. Indeed, if one assumes that $W(\vec{x}) = |\vec{x}|^{-\beta}$ in a neighbourhood of infinity, then the condition that $W \in L^2(\mathbb{R}^n)$ implies that $\beta > n/2$, i.e. a relatively rapid decay of $W$ at infinity. It is possible to associate transition operators $M_A(\lambda)$ to a much larger class of operators $A$ acting in $L^2(\mathbb{R}^n)$. One would define in analogy with (7.97): $M_A(\lambda)f = (\mathcal{U}_0 Af)_\lambda$ for $f \in \mathcal{D}(A)$. If the operator $A^*$ is locally $H_0$-smooth, then $M_A(\lambda)$ belongs to $\mathcal{B}(L^2(\mathbb{R}^n), L^2(S^{n-1}))$ for almost all $\lambda > 0$. We refer to §5 of [12] or §17.1.2 of [BW] for details, and we treat here only the case in which $A$ is a multiplication operator, i.e. $A = W(\vec{Q})$, where $W$ is a function belonging to the class considered in Proposition 6.16 [hence $W(\vec{Q})^*$ is $H_0$-smooth on each closed subinterval of $(0, \infty)$]. For such a function $W$ one can define $M_W(\lambda)$ so that its closure is defined on each vector of $L^2(\mathbb{R}^n)$. We also denote this closure by $M_W(\lambda)$.

In more detail, let us first recall the condition on $W$ (Proposition 6.16):

(S) $W(\vec{x}) = [w_1(\vec{x}) + w_2(\vec{x})](1 + |\vec{x}|)^{-\nu}$ *with* $w_1 \in L^\infty(\mathbb{R}^n)$ *and* $w_2 \in L^p(\mathbb{R}^n)$ *for some* $p$ *satisfying* $n < p < \infty$ *and* $p \geq 2$, *and with* $\nu > 1/2$.

This allows us to define $M_W(\lambda)f$ for $f \in \mathcal{S}(\mathbb{R}^n)$ as in (7.96) or (7.97), because the function $W(\vec{x})f(\vec{x})$ belongs to $L^1(\mathbb{R}^n) \cap L^2(\mathbb{R}^n)$ if $f \in \mathcal{S}(\mathbb{R}^n)$ (see the proof below). In the next proposition we state the essential properties of these operators $M_W(\lambda)$.

**Proposition 7.16.** *Let $W$ be a function satisfying the condition (S). For $\lambda > 0$, let $M_W(\lambda)$ be the closure of the operator defined on $\mathcal{S}(\mathbb{R}^n)$ by $M_W(\lambda)f = [\mathcal{U}_0 W(\vec{Q})f]_\lambda$. Then:*

*(a) $M_W(\lambda)$ is a compact operator from $L^2(\mathbb{R}^n)$ to $L^2(S^{n-1}))$, the operator-valued function $\lambda \mapsto M_W(\lambda)$ is continuous in norm on $(0, \infty)$, and*

$$\|M_W(\lambda)\| \leq 2^{-1/2}\lambda^{(n-2)/4}c_\lambda\left[\|w_1\|_\infty + \|w_2\|_p\right], \tag{7.104}$$

*where $c_\lambda$ is some constant depending on $\lambda$.*

*(b) The adjoint $M_W(\lambda)^*$ of $M_W(\lambda)$ is a compact operator from $L^2(S^{n-1})$ to $L^2(\mathbb{R}^n)$; the operator-valued function $\lambda \mapsto M_W(\lambda)^*$ is continuous in norm, and one has for each bounded closed interval $J \subset (0, \infty)$:*

$$W(\vec{Q})^* E_0(J) f = \int_J M_W(\lambda)^* (\mathcal{U}_0 f)_\lambda d\lambda \qquad \forall f \in \mathcal{S}(\mathbb{R}^n). \tag{7.105}$$

*In particular*

$$W(\vec{Q})^* f = \int_0^\infty M_W(\lambda)^* (\mathcal{U}_0 f)_\lambda d\lambda \qquad \text{if} \quad \tilde{f} \in C_0^\infty(\mathbb{R}^n \setminus \{0\}). \tag{7.106}$$

**Remark 7.17.** In contrast to the case covered by Proposition 7.14, we do not derive here a bound on the behaviour of $\|M_W(\lambda)\|$ as a function of $\lambda$ as $\lambda \to \infty$ or $\lambda \to 0$.[14] To be sure that the integral in (7.105) exists, we have assumed that $\mathcal{V}$ (here called $J$) is a bounded set. We take into account that $\|M_W(\lambda)\|$ is a bounded function of $\lambda$ on each closed interval $[\alpha, \beta] \in (0, \infty)$ (a continuous function on a compact set is bounded). In (7.106) we have assumed that $(\mathcal{U}_0 f)_\lambda = 0$ for $\lambda$ in some neighbourhood of $\lambda = \infty$ and in some neighbourhood of $\lambda = 0$; thus the integral in (7.106) extends effectively only over some interval $J$ as in (7.105) [$J$ is such that $E_0(J)f = f$].

[14] If $\nu \geq 1$ in the condition (S), we get from (7.29) that $\sup_{\lambda \geq 1} c_\lambda < \infty$, see the proof that follows.

PROOF OF PROPOSITION 7.16. We prove the boundedness of $M_W(\lambda)$ and the estimate (7.104) for its norm only under the assumption that $\nu \geq 1$ in the condition (S), because this allows us to use results from Section 7.2. The proof for $1/2 < \nu < 1$ uses Stone's Formula (4.81) and the fact that $W(\vec{Q})^*$ is locally $H_0$-smooth (Proposition 6.16).

(i) Assume that $\nu \geq 1$ in the condition (S). We show here that $\|M_W(\lambda)f\| \leq c\|f\|$ for some constant $c$ (depending on $\lambda$) and all $f \in \mathcal{S}(\mathbb{R}^n)$. If $f \in \mathcal{S}(\mathbb{R}^n)$, then clearly $w_1(\vec{Q})f \in L^1(\mathbb{R}^n) \cap L^2(\mathbb{R}^n)$. By the Hölder inequality (2.121), $w_2(\vec{Q})f$ also belongs to $L^1(\mathbb{R}^n) \cap L^2(\mathbb{R}^n)$ (because $\mathcal{S}(\mathbb{R}^n) \subseteq L^q(\mathbb{R}^n)$ for all $q \in [1, \infty]$). It follows that $W(\vec{Q})f \in L^1(\mathbb{R}^n) \cap \mathcal{D}(\langle Q\rangle)$ for each $f \in \mathcal{S}(\mathbb{R}^n)$, so that $M_W(\lambda)f \equiv (\mathcal{U}_0 W(\vec{Q})f)_\lambda$ is well defined for $\lambda \in (0, \infty)$ [we recall the notation $\langle Q\rangle = (I + \vec{Q}^2)^{1/2}$].

As in §7.2.2 we denote by $\gamma_{\Sigma_b}$ the operation of restricting functions defined on $\mathbb{R}^n$ to the sphere $\Sigma_b$ of radius $b$ in $\mathbb{R}^n$: $\gamma_{\Sigma_b} f = f\big|_{\Sigma_b}$. We then obtain from the definition (5.68) of $\mathcal{U}_0$:

$$M_W(\lambda)f = (\mathcal{U}_0 W(\vec{Q})f)_\lambda = 2^{-1/2}\lambda^{(n-2)/4}\gamma_{\Sigma_{\sqrt{\lambda}}}\widetilde{Wf}. \tag{7.107}$$

Since $Wf$ belongs to $\mathcal{D}(\langle Q\rangle)$, we obtain from Proposition 7.4, with the roles of $P$ and $Q$ interchanged, that

$$\|M_W(\lambda)f\| \leq 2^{-1/2}\lambda^{(n-2)/4}c\|\langle Q\rangle\varphi(\vec{P})W(\vec{Q})f\|,$$

where $\varphi$ can be any function in $C_0^\infty(\mathbb{R}^n)$ satisfying $\varphi(\vec{k}) = 1$ in some neighbourhood of the sphere $\Sigma_{\sqrt{\lambda}}$ and $c \equiv c_{\Sigma_{\sqrt{\lambda}}}$ is a constant depending on $\lambda$. To estimate the norm of $\langle Q\rangle\varphi(\vec{P})W(\vec{Q})$, we make use of the commutation relation $[Q_j, \varphi(\vec{P})] = i(\partial_j\varphi)(\vec{P})$ to get [see also (7.148) and use (1.14)]:

$$\begin{aligned}\|\langle Q\rangle\varphi(\vec{P})W(\vec{Q})f\|^2 &= \|\varphi(\vec{P})W(\vec{Q})f\|^2 + \sum_{j=1}^n \|Q_j\varphi(\vec{P})W(\vec{Q})f\|^2 \\ &\leq \|\varphi(\vec{P})W(\vec{Q})f\|^2 + 2\sum_{j=1}^n \|\varphi(\vec{P})Q_jW(\vec{Q})f\|^2 + 2\sum_{j=1}^n \|(\partial_j\varphi)(\vec{P})W(\vec{Q})f\|^2.\end{aligned}$$

Each of the norms on the last line has the form $\|\phi(\vec{P})(H_0 + 1)^{-1/2}\mathcal{W}(\vec{Q})f\|$, with $\phi \in C_0^\infty(\mathbb{R}^n)$ and $\mathcal{W}(\vec{Q}) = W(\vec{Q})$ or $\mathcal{W}(\vec{Q}) = Q_jW(\vec{Q})$. From the estimate (6.97) we get that $\|(H_0 + 1)^{-1/2}\mathcal{W}(\vec{Q})\| \leq c_0[\|w_1\|_\infty + \|w_2\|_p]$ for some constant $c_0$. This implies the boundedness of $M_W(\lambda)$ on $\mathcal{S}(\mathbb{R}^n)$ and the inequality (7.104) for its norm.

(ii) The compactness of $M_W(\lambda)$ can be obtained (for any $\nu > 1/2$) by decomposing $W$ as $W = W_R + W_R^\perp$, where $W_R(\vec{x}) = W(\vec{x})\chi_{[0,R]}(|\vec{x}|)$ and $W_R^\perp(\vec{x}) = W(\vec{x}) - W_R(\vec{x})$, $R > 0$. Then $M_W(\lambda) = M_{W_R} + M_{W_R^\perp}$. By Proposition 7.14 and (1.49), $M_{W_R}(\lambda)$ is a Hilbert-Schmidt operator, and $\lambda \mapsto M_{W_R}(\lambda)$ is continuous in Hilbert-Schmidt norm. The compactness of $M_W(\lambda)$ and its norm-continuity as a function of $\lambda$ then follow from Proposition 2.11(d) provided that one knows that $\lim_{R\to\infty} \|M_{W_R^\perp}(\lambda)\| = 0$, uniformly in $\lambda$ on each closed subinterval $[\alpha, \beta]$ of $(0, \infty)$. For this we choose a number $\nu'$ satisfying $1/2 < \nu' < \nu$, and we set

$\hat{w}_j(\vec{x}) = (1+|\vec{x}|)^{-(\nu-\nu')} w_j(\vec{x})$ (j = 1,2). Then $W(\vec{x}) = (1+|\vec{x}|)^{-\nu'}[\hat{w}_1(\vec{x})+\hat{w}_2(\vec{x})]$, and by (7.104) we have:

$$\|M_{W_R^\perp}(\lambda)\| \le \lambda^{(n-2)/4} c_\lambda \{ \sup_{|\vec{x}|>R} |\hat{w}_1(\vec{x})| + \Big[\int_{|\vec{x}|>R} |\hat{w}_2(\vec{x})|^p \, d^n x\Big]^{1/p} \}. \quad (7.108)$$

Since $|\hat{w}_1(\vec{x})| \le (1+|\vec{x}|)^{-(\nu-\nu')} \|w_1\|_\infty$ and $\hat{w}_2 \in L^p(\mathbb{R}^n)$, the inequality (7.108) implies that $\|M_{W_R^\perp}(\lambda)\| \to 0$ as $R \to \infty$. The required uniformity in $\lambda$ follows from the fact that $\sup_{\lambda \in [\alpha,\beta]} c_\lambda < \infty$ [see (7.29) which is satisfied if $\nu \ge 1$, or else part (iii) of the proof of Proposition 7.4 given on page 375].

(iii) The first assertions of (b) follow from those of (a) by passing to the adjoint. Equation (7.105) is obtained in the same way as (7.101), see the proof of Proposition 7.14(b): Eq. (7.103) is still valid, then one takes into account the observations made in Remark 7.17. □

To arrive at the stationary formula for $S(\lambda)$ we shall also need the following results:

**Lemma 7.18.** *Let $W$ be a function satisfying the condition (S).*
*(a) Let $\lambda > 0$ and assume that $f\colon \mathbb{R}^n \to \mathbb{C}$ is such that $\tilde{f} \in C_0^\infty(\mathbb{R}^n \setminus \{0\})$. Then $W(\vec{Q})^*(H_0 - \lambda - i\varepsilon)^{-1} f - W(\vec{Q})^*(H_0 - \lambda + i\varepsilon)^{-1} f$ converges weakly in $L^2(\mathbb{R}^n)$ to $2\pi i M_W(\lambda)^*(\mathcal{U}_0 f)_\lambda$ as $\varepsilon \to +0$.*
*(b) Let $J = [a,b]$ be a finite closed interval in $(0,\infty)$ and $C\colon J \to \mathcal{B}(\mathcal{H})$ a strongly continuous operator-valued function. Then one has for each $g \in \mathcal{S}(\mathbb{R}^n)$:*

$$\int_J C(\lambda) M_W(\lambda)^* (\mathcal{U}_0 g)_\lambda \, d\lambda = \int_J C(\lambda) W(\vec{Q})^* E_0(d\lambda) g. \quad (7.109)$$

PROOF. (a) Let $\tilde{f} \in C_0^\infty(\mathbb{R}^n \setminus \{0\})$, $g \in \mathcal{H} \equiv L^2(\mathbb{R}^n)$ and $\lambda, \varepsilon > 0$. We observe that the Fourier transform of $[(H_0 - \lambda - i\varepsilon)^{-1} - (H_0 - \lambda + i\varepsilon)^{-1}] f$ also belongs to $C_0^\infty(\mathbb{R}^n \setminus \{0\})$. Hence we have by virtue of (7.106):

$$\begin{aligned} &\langle g, W(\vec{Q})^* [(H_0 - \lambda - i\varepsilon)^{-1} - (H_0 - \lambda + i\varepsilon)^{-1}] f \rangle \\ &\quad = \int_0^\infty \langle g, M_W(\mu)^* (\mathcal{U}_0 [(H_0 - \lambda - i\varepsilon)^{-1} - (H_0 - \lambda + i\varepsilon)^{-1}] f)_\mu \rangle \, d\mu \\ &\quad = \int_0^\infty [(\mu - \lambda - i\varepsilon)^{-1} - (\mu - \lambda + i\varepsilon)^{-1}] \langle g, M_W(\mu)^* (\mathcal{U}_0 f)_\mu \rangle \, d\mu. \end{aligned} \quad (7.110)$$

Now $\mu \mapsto \langle g, M_W(\mu)^* (\mathcal{U}_0 f)_\mu \rangle$ is a continuous function of compact support (see Remark 7.17), and $[(\mu - \lambda - i\varepsilon)^{-1} - (\mu - \lambda + i\varepsilon)^{-1}]$ is a Dirac sequence as $\varepsilon \to +0$, more precisely $\lim_{\varepsilon \to +0} [(\mu - \lambda - i\varepsilon)^{-1} - (\mu - \lambda + i\varepsilon)^{-1}] = 2\pi i \delta(\mu - \lambda)$. Thus, as $\varepsilon \to +0$, the last expression in (7.110) converges to $2\pi i \langle g, M_W(\lambda)^* (\mathcal{U}_0 f)_\lambda \rangle$. This proves the result of (a).

(b) The integrand on the left-hand side of (7.109) is a strongly continuous vector-valued function [with values in $L^2(\mathbb{R}^n)$], hence the integral exists by Proposition 1.5.

Let us consider partitions $\Pi_N = \{s_0, s_1, \dots, s_N; u_1, \dots, u_N\}$ of the interval $J$ (see page 6) and set $J_k = (s_{k-1}, s_k]$. We can write

$$\int_J C(\lambda)M_W(\lambda)^*(\mathcal{U}_0 g)_\lambda d\lambda = \sum_{k=1}^N \int_{J_k} C(\lambda)M_W(\lambda)^*(\mathcal{U}_0 g)_\lambda d\lambda.$$

Let $\varepsilon > 0$. If the partition $\Pi_N$ is sufficiently fine, one has $\|C(\lambda) - C(u_k)\| \le \varepsilon$ for all $\lambda \in J_k$ and for each $k = 1, \dots, N$. Consequently, setting $f(\lambda) = M_W(\lambda)^*(\mathcal{U}_0 g)_\lambda$:

$$\begin{aligned}\Big\|\int_J C(\lambda)f(\lambda)d\lambda - \sum_{k=1}^N \int_{J_k} C(u_k)f(\lambda)d\lambda\Big\| &= \Big\|\sum_{k=1}^N \int_{J_k} [C(\lambda) - C(u_k)]f(\lambda)d\lambda\Big\| \\ &\le \sum_{k=1}^N \varepsilon \int_{J_k} \|f(\lambda)\| d\lambda = \varepsilon \int_J \|f(\lambda)\| d\lambda.\end{aligned}$$

This implies that $\int_J C(\lambda)f(\lambda)d\lambda = \text{s-lim}_{|\Pi_N|\to 0} \sum_{k=1}^N C(u_k)\int_{J_k} f(\lambda)d\lambda$. Now, by (7.105): $\int_{J_k} f(\lambda)d\lambda \equiv \int_{J_k} M_W(\lambda)^*(\mathcal{U}_0 g)_\lambda d\lambda = W(\vec{Q})^* E_0(J_k)g$. So we have

$$\int_J C(\lambda)M_W(\lambda)^*(\mathcal{U}_0 g)_\lambda d\lambda = \underset{|\Pi_N|\to 0}{\text{s-lim}} \sum_{k=1}^N C(u_k)W(\vec{Q})^* E_0(J_k)g.$$

The right-hand side is by definition equal to $\int_J C(\lambda)W(\vec{Q})^* E_0(d\lambda)g$. This proves the existence of the spectral integral as well as Eq. (7.109). □

**Remark 7.19.** Without proof we mention the following interesting result: If $W_1$ and $W_2$ are two functions satisfying the condition (S), then one has for each $\lambda > 0$:

$$2\pi i M_{W_1}(\lambda)^* M_{W_2}(\lambda) = \underset{\varepsilon\to+0}{\text{u-lim}}\, W_1(\vec{Q})^*\big[(H_0 - \lambda - i\varepsilon)^{-1} - (H_0 - \lambda + i\varepsilon)^{-1}\big]W_2(\vec{Q}). \tag{7.111}$$

**7.4.2.** We now present the stationary formula for $R(\lambda)$. We have seen at the end of Section 7.3 that $R(\lambda)$ should be given by the diagonal part of the operator $-2\pi i\, T_{\lambda+i0}$ in the spectral representation of $H_0$. We recall that $T_z = V - V(H-z)^{-1}V$. The idea is to factor $V$ as $V = UW$, hence to write $T_z = U[I - W(H-z)^{-1}U]W$, and then make use of the first factor $U$ and the last factor $W$ in this expression to introduce the transition operators. One obtains in this way the representation

$$R(\lambda) = -2\pi i M_U(\lambda)\big[I - W(H - \lambda - i0)^{-1}U\big]M_W(\lambda)^*. \tag{7.112}$$

Here $M_W(\lambda)^*$ maps $L^2(S^{n-1})$ into $L^2(\mathbb{R}^n)$, the operator $I - W(H - \lambda - i0)^{-1}U$ acts in $L^2(\mathbb{R}^n)$, and $M_U(\lambda)$ maps back from $L^2(\mathbb{R}^n)$ to $L^2(S^{n-1})$. Evidently it is necessary to impose conditions so that $M_U(\lambda)$ and $M_W(\lambda)$ are well defined and that $W(H - \lambda - i\varepsilon)^{-1}U$ converges as $\varepsilon \to +0$ [the notation $W(H - \lambda - i0)^{-1}U$ is then used for the limit]. This is the content of the following proposition. Let us recall that a factorisation of $V$ has already been used in the proof of asymptotic completeness in Section 7.1. One can take the same factorisation here (with different notations): one can take for $U$ the operator of multiplication by $|V(\vec{x})|^{1/2}$ and for $W$ the operator of multiplication by $|V(\vec{x})|^{1/2}\operatorname{sign}V(\vec{x})$, as in Proposition 7.3. Other factorisations will also be useful.

**Proposition 7.20.** *Assume that the function $V: \mathbb{R}^n \to \mathbb{R}$ satisfies the assumptions of Proposition 7.3. Let $U(\vec{x})$ and $W(\vec{x})$ be real such that $V(\vec{x}) = U(\vec{x})W(\vec{x})$ and such that $U$ and $W$ satisfy the condition (S) on page 355. Then:*
*(a) the wave operators $\Omega_\pm = \Omega_\pm(H, H_0; I)$ exist and are complete,*
*(b) for $\lambda \in (0,\infty)\setminus\sigma_p(H)$, the operator $R(\lambda)$ is given by (7.112), with $U = U(\vec{Q})$ and $W = W(\vec{Q})$.*

PROOF. The existence of $\Omega_\pm$ and strong asymptotic completeness have been obtained in Proposition 7.3. Moreover we know from Proposition 6.16 that the operators $U \equiv U(\vec{Q})$ and $W \equiv W(\vec{Q})$ are $H$-smooth. We have also seen that $W(H-\lambda-i\delta)^{-1}U$ converges (in norm) as $\delta \to +0$ if $\lambda \notin \sigma_{\rm p}(H)$, and that the convergence is uniform on each bounded closed subinterval $J$ of $(0,\infty)\setminus\sigma_{\rm p}(H)$. Since the functions $U(\cdot)$ and $W(\cdot)$ satisfy the condition (S), the operators $M_U(\lambda)$ and $M_W(\lambda)$ are well defined by Proposition 7.16. These facts will be used for proving (7.112).

Let $J$ be a bounded closed subinterval of $(0,\infty)\setminus\sigma_{\rm p}(H)$. Let $f \in \mathcal{S}(\mathbb{R}^n)$. We use the notations of Proposition 7.13 and write $V - VR_{\lambda+i\delta}V = U(I - WR_{\lambda+i\delta}U)W$. By using first (7.92) and then (7.109) we obtain

$$\begin{aligned} RE_0(J)f &= \operatorname*{s-lim}_{\varepsilon\to+0}\operatorname*{s-lim}_{\delta\to+0}\int_J (R^0_{\lambda-i\varepsilon} - R^0_{\lambda+i\varepsilon})(V - VR_{\lambda+i\delta}V)E_0(d\lambda)f \\ &= \operatorname*{s-lim}_{\varepsilon\to+0}\operatorname*{s-lim}_{\delta\to+0}\int_J (R^0_{\lambda-i\varepsilon} - R^0_{\lambda+i\varepsilon})U(I - WR_{\lambda+i\delta}U)M_W(\lambda)^*(\mathcal{U}_0 f)_\lambda\, d\lambda. \end{aligned}$$

Hence, for any $g \in \mathcal{H}$:

$$\begin{aligned} &\langle g, RE_0(J)f\rangle = \\ &= \lim_{\varepsilon\to+0}\lim_{\delta\to+0}\int_J \langle g, (R^0_{\lambda-i\varepsilon} - R^0_{\lambda+i\varepsilon})U(I - WR_{\lambda+i\delta}U)M_W(\lambda)^*(\mathcal{U}_0 f)_\lambda\rangle\, d\lambda \\ &= \lim_{\varepsilon\to+0}\lim_{\delta\to+0}\int_J \langle U(R^0_{\lambda+i\varepsilon} - R^0_{\lambda-i\varepsilon})g, (I - WR_{\lambda+i\delta}U)M_W(\lambda)^*(\mathcal{U}_0 f)_\lambda\rangle\, d\lambda. \end{aligned} \tag{7.113}$$

Now assume that $g \in \mathcal{S}(\mathbb{R}^n)$. Then

$$\|U(R^0_{\lambda+i\varepsilon} - R^0_{\lambda-i\varepsilon})g\| \le \|U(R^0_{\lambda+i\varepsilon} - R^0_{\lambda-i\varepsilon})(I + |\vec{Q}|)^{-1}\|\|(I + |\vec{Q}|)g\|,$$

which is bounded by a finite constant for all $\lambda \in J$ [by Proposition 6.16(b)]. Furthermore $\|M_W(\lambda)^*(\mathcal{U}_0 f)_\lambda\| \le c < \infty$ for all $\lambda \in J$. As $WR_{\lambda+i\delta}U$ converges in norm when $\delta \to +0$ (uniformly in $\lambda \in J$), the Dominated Convergence Theorem allows us to take the limit $\delta \to +0$ inside the integral in (7.113), i.e. to replace $WR_{\lambda+i\delta}U$ in the integrand by $WR_{\lambda+i0}U$. By using the estimate $\|WR_{\lambda+i0}U\| \leq c < \infty$ for all $\lambda \in J$ and by writing again $U(R^0_{\lambda+i\varepsilon} - R^0_{\lambda-i\varepsilon})g = U(R^0_{\lambda+i\varepsilon} - R^0_{\lambda-i\varepsilon})(I+|\vec{Q}|)^{-1}(I+|\vec{Q}|)g$, one can once more apply the Dominated Convergence Theorem to take the limit $\varepsilon \to +0$ inside the integral. One finds in this way that

$$\langle g, RE_0(J)f\rangle = \int_J \lim_{\varepsilon\to+0} \langle U(R^0_{\lambda+i\varepsilon} - R^0_{\lambda-i\varepsilon})g, (I - WR_{\lambda+i0}U)M_W(\lambda)^*(\mathcal{U}_0 f)_\lambda\rangle d\lambda.$$

For each fixed $\lambda$, one can evaluate the limit under the integral by using Lemma 7.18(a), provided that we impose the additional condition that $\tilde{g} \in C_0^\infty(\mathbb{R}^n \setminus \{0\})$:

$$\begin{aligned}\langle g, RE_0)f\rangle &= \int_J \langle 2\pi i M_U(\lambda)^*(\mathcal{U}_0 g)_\lambda, (I - WR_{\lambda+i0}U)M_W(\lambda)^*(\mathcal{U}_0 f)_\lambda\rangle d\lambda \\ &= -2\pi i \int_J \langle (\mathcal{U}_0 g)_\lambda, M_U(\lambda)(I - WR_{\lambda+i0}U)M_W(\lambda)^*(\mathcal{U}_0 f)_\lambda\rangle d\lambda.\end{aligned}$$

Since $\langle g, RE_0(J)f\rangle = \int_J \langle (\mathcal{U}_0 g)_\lambda, R(\lambda)(\mathcal{U}_0 f)_\lambda\rangle d\lambda$, we now have

$$\begin{aligned}&\int_J \langle (\mathcal{U}_0 g)_\lambda, R(\lambda)(\mathcal{U}_0 f)_\lambda\rangle d\lambda \\ &\qquad = -2\pi i \int_J \langle (\mathcal{U}_0 g)_\lambda, M_U(\lambda)(I - WR_{\lambda+i0}U)M_W(\lambda)^*(\mathcal{U}_0 f)_\lambda\rangle d\lambda. \quad (7.114)\end{aligned}$$

We wish to deduce from (7.114) that $R(\lambda) = -2\pi i M_U(\lambda)(I - WR_{\lambda+i0}U)M_W(\lambda)^*$ for almost all $\lambda \in J$. For this, let us first observe that, if $J = [\alpha, \beta]$, one can repeat the preceding arguments on each subinterval of the form $[\alpha, x]$, with $\alpha < x \leq \beta$. This gives

$$\begin{aligned}&\int_\alpha^x \langle (\mathcal{U}_0 g)_\lambda, R(\lambda)(\mathcal{U}_0 f)_\lambda\rangle d\lambda \\ &\qquad = -2\pi i \int_\alpha^x \langle (\mathcal{U}_0 g)_\lambda, M_U(\lambda)(I - WR_{\lambda+i0}U)M_W(\lambda)^*(\mathcal{U}_0 f)_\lambda\rangle d\lambda,\end{aligned}$$

and after differentiation with respect to $x$:

$$\langle (\mathcal{U}_0 g)_\lambda, R(\lambda)(\mathcal{U}_0 f)_\lambda\rangle = -2\pi i \langle (\mathcal{U}_0 g)_\lambda, M_U(\lambda)(I - WR_{\lambda+i0}U)M_W(\lambda)^*(\mathcal{U}_0 f)_\lambda\rangle. \quad (7.115)$$

This identity holds for almost all $\lambda$ in $J$ (each side is in principle defined almost everywhere). In other terms there exists a subset $\mathcal{N}(f,g)$ of $J$ of Lebesgue measure zero (and depending on $f$ and $g$) such that (7.115) is satisfied for all $\lambda \in J \setminus \mathcal{N}(f,g)$.

We now choose particular vectors $f$ and $g$, motivated by the proof of Proposition 5.26. We fix an orthonormal basis $\{e_k^\circ\}$ of $L^2(S^{n-1})$ and a function $\theta$ in $C_0^\infty((0,\infty))$ such that $\theta(\lambda) \neq 0$ on $J$. For each $k = 1, 2, \ldots$ we define a vector $e_k$ in $L^2(\mathbb{R}^n)$ by setting $(\mathcal{U}_0 e_k)_\lambda = \theta(\lambda)e_k^\circ$. Then $\widetilde{e_k}$ belongs to $C_0^\infty(\mathbb{R}^n \setminus \{0\})$ provided that $e_k^\circ$ is chosen to be of class $C^\infty$. If we take in (7.115) $f = e_k$ and $g = e_j$ we find that

$$|\theta(\lambda)|^2 \langle e_j^\circ, R(\lambda)e_k^\circ\rangle = -2\pi i |\theta(\lambda)|^2 \langle e_j^\circ, M_U(\lambda)(I - WR_{\lambda+i0}U)M_W(\lambda)^* e_k^\circ\rangle,$$

or

$$\langle e_j^{\circ}, R(\lambda) e_k^{\circ}\rangle = -2\pi i \langle e_j^{\circ}, M_U(\lambda)(I - WR_{\lambda+i0}U)M_W(\lambda)^* e_k^{\circ}\rangle \qquad \forall \lambda \in J \setminus \mathcal{N}_{jk},$$

where $\mathcal{N}_{jk}$ is a subset of $J$ of Lebesgue measure zero. Let now $\mathcal{N} = \bigcup_{j,k} \mathcal{N}_{jk}$. Then $\mathcal{N}$ is a subset of $J$ of Lebesgue measure zero, and one has for all $j, k \in \mathbb{N}$:

$$\langle e_j^{\circ}, R(\lambda) e_k^{\circ}\rangle = -2\pi i \langle e_j^{\circ}, M_U(\lambda)(I - WR_{\lambda+i0}U)M_W(\lambda)^* e_k^{\circ}\rangle \qquad \forall \lambda \in J \setminus \mathcal{N}.$$

This implies that $R(\lambda)e_k^{\circ} = -2\pi i M_U(\lambda)(I - WR_{\lambda+i0}U)M_W(\lambda)^* e_k^{\circ}$ for $\lambda \in J \setminus \mathcal{N}$ and each $k$ [apply Proposition 1.6 by taking into account that the set $\{e_j^{\circ}\}$ is total in $L^2(S^{n-1})$]. We have thus obtained the validity of (7.112), for almost all $\lambda$, on the vectors of an orthonormal basis of $L^2(S^{n-1})$. By linearity, (7.112) holds on all vectors of $L^2(S^{n-1})$, for the same values of $\lambda$. □

REMARKS AND EXAMPLES. (a) By the results of §5.5.2, the operator $R(\lambda)$ is defined only for almost all $\lambda \in (0, \infty)$. The right-hand side of (7.112) however is defined for every $\lambda > 0$ that is not an eigenvalue of $H$. One can choose to identify $R(\lambda)$ with the right-hand side of (7.112) for each $\lambda \in (0, \infty) \setminus \sigma_{\mathrm{p}}(H)$. This choice is compatible with the scattering operator $S$.

(b) As a first example, let us assume that $V(\vec{x}) = (1 + |\vec{x}|)^{-1-\varepsilon}[v_\infty(\vec{x}) + v_q(\vec{x})]$, with $\varepsilon > 0$, $v_\infty \in L^\infty(\mathbb{R}^n)$ and $v_q \in L^q(\mathbb{R}^n)$ for some $q > n/2$ ($n \geq 2$), and that $V$ satisfies the assumption (i) or (ii) of Proposition 7.3 (depending on the value of $n$). In this case we take a symmetric factorisation of $V$, as in the proof of Proposition 7.3: $U(\vec{x}) = |V(\vec{x})|^{1/2}$ and $W(\vec{x}) = |V(\vec{x})|^{1/2} \operatorname{sign} V(\vec{x})$. The operators $M_U(\lambda)$ and $M_W(\lambda)^*$ in (7.112) are compact and norm-continuous as functions of $\lambda$ (Proposition 7.16). On the other hand we know that $W(\vec{Q})(H - \lambda - i0)^{-1}U(\vec{Q})$ is continuous in norm as a function of $\lambda$, away from the eigenvalues of $H$. By assuming the choice specified in (a) above, this implies that, for each $\lambda \in (0, \infty) \setminus \sigma_{\mathrm{p}}(H)$, $R(\lambda)$ is a compact operator in $L^2(S^{n-1})$, and the operator-valued function $\lambda \mapsto R(\lambda)$ is continuous in norm on $(0, \infty) \setminus \sigma_{\mathrm{p}}(H)$.

(c) Above the only assumption on $\varepsilon$ was that $\varepsilon > 0$. This covers potentials decaying like $|\vec{x}|^{-\beta}$ with $\beta > 1$ and which may have local singularities, hence essentially the class of potentials for which we have shown the existence (and completeness) of the wave operators. We know that, if $\beta > (n+1)/2$, then $R(\lambda)$ is a Hilbert-Schmidt operator for almost all $\lambda$ (see Section 5.7). So let us assume that $V(\vec{x}) = (1 + |\vec{x}|)^{-\beta}[v_\infty(\vec{x}) + v_q(\vec{x})]$ with $\beta > (n+1)/2$ and $q > n, q \geq 2$, and again that $V$ satisfies the condition (i) or (ii) of Proposition 7.3. Let us choose a number $\nu > 1/2$ such that $\beta - \nu > n/2$ and take $U(\vec{x}) = (1 + |\vec{x}|)^{-\nu}[v_\infty(\vec{x}) + v_q(\vec{x})]$ and $W(\vec{x}) = (1 + |\vec{x}|)^{-(\beta-\nu)}$. Both $U$ and $W$ satisfy the condition (S) on page 355. $M_U(\lambda)$ is a compact operator depending continuously (in norm) on $\lambda$ (Proposition 7.16), and $M_W(\lambda)^*$ is a Hilbert-Schmidt operator depending continuously on $\lambda$ in Hilbert-Schmidt norm. Thus (7.112) implies that $R(\lambda) \in \mathcal{B}_2(L^2(S^{n-1}))$. Furthermore the operator-valued function $\lambda \mapsto R(\lambda)$ is continuous in Hilbert-Schmidt norm on $(0,\infty)\setminus\sigma_{\mathrm{p}}(H)$[15] [the averaged total scattering cross section $\overline{\sigma}(\lambda)$ can be considered to be a continuous function of $\lambda$ if the choice specified in (a) is made].

(d) For later reference we mention the following expression for $R(\lambda)$. Assume that $V(\vec{x}) = V_1(\vec{x}) + V_2(\vec{x})$ and that $V_\ell(\vec{x}) = U_\ell(\vec{x})W_\ell(\vec{x})$, where $U_\ell$ and $W_\ell$ are real functions satisfying condition (S) ($\ell = 1, 2$). Then (Problem 7.10):

$$R(\lambda) = -\,2\pi i\sum_{\ell=1}^{2} M_{U_\ell}(\lambda)M_{W_\ell}(\lambda)^* + 2\pi i\sum_{j,\ell=1}^{2} M_{U_j}(\lambda)[W_j(\vec{Q})(H-\lambda-i0)^{-1}U_\ell(\vec{Q})]M_{W_\ell}(\lambda)^*. \tag{7.116}$$

**7.4.3.** We present here a result on the differentiability of the scattering matrix $S(\lambda) \equiv R(\lambda) + I$ with respect to the variable $\lambda$. We recall from (7.112) that

$$R(\lambda) = -2\pi i M_U(\lambda)\big[I - W(H-\lambda-i0)^{-1}U\big]M_W(\lambda)^*,$$

which has the form of a product of three $\lambda$-dependent operators. Differentiability properties of $W(H-\lambda-i0)^{-1}U$ have already been established in Proposition 6.21, so it is enough to know conditions under which operators of the form $M_W(\lambda)$ are differentiable with respect to $\lambda$. Now $M_W(\lambda)$ is an integral operator with kernel

$$a_\lambda(\vec{\omega},\vec{x}) = 2^{-1/2}(2\pi)^{-n/2}\lambda^{(n-2)/4}e^{-i\sqrt{\lambda}\vec{\omega}\cdot\vec{x}}W(\vec{x}).$$

Differentiation of $a_\lambda(\vec{\omega},\vec{x})$ with respect to $\lambda$ leads (at least formally) to the following expression for the derivative of $M_W(\lambda)$:

$$\frac{d}{d\lambda}M_W(\lambda) = \frac{n-2}{4\lambda}M_W(\lambda) - \frac{i}{2\sqrt{\lambda}}\sum_{j=1}^{n} N_j M_{Q_jW}(\lambda), \tag{7.117}$$

where $N_j$ denotes the multiplication operator by $\omega_j$ in $L^2(S^{n-1})$, with $\vec{\omega} \in S^{n-1}$. Note that, for $\lambda > 0$, the right-hand side of Eq. (7.117) represents a well defined compact operator from $L^2(\mathbb{R}^n)$ to $L^2(S^{n-1})$ if the functions $W$ and $x_jW$ satisfy condition (S) on page 355, and it is norm continuous as a function of $\lambda$. We refer to part A.7.3 of the Appendix for a proof of Eq. (7.117) [with differentiability in norm!] and note the following consequence of this equation:

*Assume that $(1+|\vec{x}|)^kW$ satisfies conditions (S) on page 355 for some $k\in\mathbb{N}$. Then $M_W(\lambda)$ is $k$ times continuously norm differentiable on $(0,\infty)$.*

The next proposition applies to Schrödinger operators for the class of potentials considered in Proposition 7.3, but with stronger assumptions on the decay of $V$.

---

[15] Let $\lambda \mapsto C(\lambda) \in \mathcal{B}(\mathcal{H})$ be continuous in norm and $\lambda \mapsto Z(\lambda) \in \mathcal{B}_2(\mathcal{H})$ continuous in Hilbert-Schmidt norm. Then [see (2.47)]:

$$\begin{aligned}\|C(\lambda)Z(\lambda) - C(\mu)Z(\mu)\|_{\mathrm{HS}} &= \|C(\lambda)[Z(\lambda)-Z(\mu)]\|_{\mathrm{HS}} + \|[C(\lambda)-C(\mu)]Z(\mu)\|_{\mathrm{HS}}\\ &\le \|C(\lambda)\|\|Z(\lambda)-Z(\mu)\|_{\mathrm{HS}} + \|C(\lambda)-C(\mu)\|\|Z(\mu)\|_{\mathrm{HS}},\end{aligned}$$

and each term converges to zero as $\lambda \to \mu$.

**Proposition 7.21.** *Let $H = H_0 + V(\vec{Q})$ in $L^2(\mathbb{R}^n)$, where $V : \mathbb{R}^n \to \mathbb{R}$ has the form $V(\vec{x}) = (1 + |\vec{x}|)^{-1-\delta}[v_\infty(\vec{x}) + v_q(\vec{x})]$ with $\delta > k$ for some $k \in \mathbb{N}$, $v_\infty \in L^\infty(\mathbb{R}^n)$, $v_q \in L^q(\mathbb{R}^n)$ for some $q > n/2$ (and $q \geq 1$) and*
*(i) if $n = 1, 2$ or $3$ : $V$ and $r(\partial V/\partial r)$ belong to $L^2(\mathbb{R}^n) + L^\infty_\varepsilon(\mathbb{R}^n)$, and $r^2(\partial^2 V/\partial r^2)$ belongs to $L^2(\mathbb{R}^n) + L^\infty(\mathbb{R}^n)$,*
*(ii) if $n \geq 4$ : $V$ and $r(\partial V/\partial r)$ belong to $L^p(\mathbb{R}^n) + L^\infty_\varepsilon(\mathbb{R}^n)$ and $r^2(\partial^2 V/\partial r^2) \in L^p(\mathbb{R}^n) + L^\infty(\mathbb{R}^n)$ for some $p \in (n/2, \infty)$.*
*Then $S(\lambda)$ is $k$ times continuously differentiable in norm on $(0, \infty)\backslash\sigma_{\mathrm{p}}(H)$.*

PROOF. (i) We first prove the differentiability of $R(\lambda)$ under a stronger decay assumption on $V$, namely we assume that $\delta > 2k$ in the proposition. We denote by $V_1$ and $V_2$ the multiplication operators by $v_\infty(\vec{x})(1 + |\vec{x}|)^{-1-\delta}$ and $v_q(\vec{x})(1 + |\vec{x}|)^{-1-\delta}$ respectively. Let $\nu = (1 + \delta)/2$ and observe that $\nu > k + 1/2$ if $\delta > 2k$, and write $V_\ell = U_\ell W_\ell$ with $U_1(\vec{x}) = |v_\infty(\vec{x})|^{1/2}(1 + |\vec{x}|)^{-\nu}$, $U_2(\vec{x}) = |v_q(\vec{x})|^{1/2}(1 + |\vec{x}|)^{-\nu}$ and $W_\ell(\vec{x}) = U_\ell(\vec{x})\,\mathrm{sign}\,V_\ell(\vec{x})$ ($\ell = 1, 2$). The functions $(1+|\vec{x}|)^k U_\ell$ and $(1+|\vec{x}|)^k W_\ell$ satisfy condition (S), hence $M_{U_\ell}(\lambda)$ and $M_{W_\ell}(\lambda)$ are $k$ times continuously norm differentiable. By Proposition 6.21 the operators $W_j(\vec{Q})(H-\lambda-i0)^{-1}U_\ell(\vec{Q})$ are $k$ times continuously norm differentiable away from $\sigma_{\mathrm{p}}(H)$. Hence, in view of Eq. (7.116), $R(\lambda)$ is $k$ times continuously differentiable.

(ii) We note that $R(\lambda)$ in Eq. (7.116) is a sum of terms each of which has the form of a product of two or three $\lambda$-dependent operators. Such a product is differentiable if each factor is differentiable [as used in (i) above]. However, due to the special form of these products, one can obtain their differentiability without assuming the individual factors to be themselves differentiable. This is achieved by shifting operators of the form $\langle Q\rangle^{-\kappa}$ between these factors when writing the derivatives, using the identity $M_{AB}(\lambda) = M_A(\lambda)B$ and its adjoint $M_{AB}(\lambda)^* = B^* M_A(\lambda)^*$. In this way the proposition can be proved under the weaker assumption that $\delta > k$ (rather than $\delta > 2k$). We leave the detailed verification as an exercise (Problem 7.11) and just explain the method by considering the derivative of a term of the form $M_U(\lambda)M_W(\lambda)^*$.

For $U$ and $W$ satisfying the condition (S) and $\alpha \geq 0$, $\beta \geq 0$, set $U_{(\alpha)} = \langle Q\rangle^{-\alpha}U$ and $W_{(\beta)} = \langle Q\rangle^{-\beta}W$. By virtue of the identities on $M$-operators given above, we have for all $\lambda, \mu > 0$:

$$M_{U_{(\alpha)}}(\lambda)M_{W_{(\beta)}}(\mu)^* = M_U(\lambda)\langle Q\rangle^{-(\alpha+\beta)}M_W(\mu)^*.$$

This implies that

$$M_{U_{(\alpha)}}(\lambda)M_{W_{(\beta)}}(\mu)^* = M_{U_{(\gamma)}}(\lambda)M_{W_{(\delta)}}(\mu)^* \qquad \text{whenever } \alpha + \beta = \gamma + \delta.$$

For $\alpha + \beta = 1$, we may thus write

$$\begin{aligned} M_{U_{(\alpha)}}(\lambda+\varepsilon)M_{W_{(\beta)}}(\lambda + \varepsilon)^* &- M_{U_{(\alpha)}}(\lambda)M_{W_{(\beta)}}(\lambda)^* \\ &= \big[M_{U_{(1)}}(\lambda + \varepsilon) - M_{U_{(1)}}(\lambda)\big]M_W(\lambda + \varepsilon)^* \\ &\quad + M_U(\lambda)\big[M_{W_{(1)}}(\lambda + \varepsilon)^* - M_{W_{(1)}}(\lambda)^*\big]. \end{aligned}$$

Upon division by $\varepsilon$ and then taking the limit $\varepsilon \to 0$, one finds that

$$\frac{d}{d\lambda} M_{U_{(\alpha)}}(\lambda) M_{W_{(\beta)}}(\mu)^* = \Big[\frac{d}{d\lambda} M_{U_{(1)}}(\lambda)\Big] M_W(\lambda)^* + M_U(\lambda)\Big[\frac{d}{d\lambda} M_{W_{(1)}}(\lambda)^*\Big]. \tag{7.118}$$

The limit $\varepsilon \to 0$ exists in norm and the derivative, given by the right-hand side of (7.118), is continuous in norm. If the exponent $\nu$ in condition (S) is very close to $1/2$ (and $\alpha + \beta = 1$), then the function $U_{(\alpha)} W_{(\beta)}$ decays like $|\vec{x}|^{-2-\epsilon}$ with very small $\epsilon > 0$. In such a situation at least one of the two operators $M_{U_{(\alpha)}}(\lambda)$ and $M_{W_{(\beta)}}(\mu)^*$ is not differentiable, but their product is continuously differentiable in norm! □

## 7.5 Time delay

Time delay is one of the fundamental concepts of scattering theory, and it is of much theoretical importance in various fields of quantum physics. In the simplest situations, time delay expresses the excess time that scattered particles spend in the scattering region when compared to free particles subject to the same initial conditions. A positive time delay means that particles take more time to pass through the region where they are influenced by the interaction than particles propagating freely through the same region. A negative time delay means that on average the scattered particles are accelerated by the effects of the interaction.

In this final section we present a Hilbert space approach to time delay. It will be shown in particular that the time delay can be expressed in terms of a self-adjoint operator, called the Eisenbud-Wigner time delay operator, which is entirely determined by the scattering operator $S$. References to some applications can be found in the bibliographical notes to this chapter.

**7.5.1.** As in Section 5.2 we consider a physical system described by the states in the Hilbert space $\mathcal{H} = L^2(\mathbb{R}^n)$. Let $H$ be its Hamiltonian and $\{U_t\}_{t\in\mathbb{R}}$ the associated unitary evolution group, with $U_t = e^{-iHt}$. We denote by $\chi \in L^\infty(\mathbb{R})$ the characteristic function of the interval $[0, 1]$. Then $F_R = \chi(|\vec{Q}|/R)$ is the projection operator describing localisation in the ball $B_R := \{\vec{x} \in \mathbb{R}^n \,|\, |\vec{x}| \leq R\}$ in $\mathbb{R}^n$ [see Eq. (5.22)]. To each non-zero $g \in \mathcal{H}$ and each $R > 0$ we associate a quantity $\mathcal{T}_R(g)$ by setting

$$\mathcal{T}_R(g) = \frac{1}{\|g\|^2}\int_{-\infty}^{\infty} \|F_R U_t g\|^2 dt \equiv \frac{1}{\|g\|^2}\int_{-\infty}^{\infty} dt \int_{|\vec{x}|\leq R} |(U_t g)(\vec{x})|^2 d^n x. \tag{7.119}$$

Since $\|F_R U_t g\|^2/\|g\|^2$ is the probability that the state represented by the vector $g$ is localised in the ball $B_R$ at time $t$, the positive number $\mathcal{T}_R(g)$ may be interpreted as the total time that the state $g$ spends in the ball $B_R$ during its evolution, if the time evolution is governed by the group $\{U_t\}$.

Scattering theory is based on the comparison of the evolution of states with a free evolution. If $H_0$ denotes the free Hamiltonian and if $U_t^0 = e^{-iH_0 t}$, then the total time spent in the ball $B_R$ by a state $f$ evolving under the free evolution group $\{U_t^0\}$ will be given by the number

$$\mathcal{T}_R^0(f) = \frac{1}{\|f\|^2}\int_{-\infty}^{\infty} \|F_R U_t^0 f\|^2 dt. \tag{7.120}$$

As said in the introduction to this section, the time delay in the ball $B_R$ for scattering with initial state $f$ is the difference of the time spent in $B_R$ by the associated scattering state $\Omega_- f$ and the time spent in $B_R$ by the freely evolving state $f$, i.e. the quantity $\mathcal{T}_R(\Omega_- f) - \mathcal{T}_R^0(f)$.[16] Omitting the normalisation factors $1/\|\cdot\|^2$ occurring in (7.119) and (7.120), we define

$$(\Delta\mathcal{T})_R(f) = \int_{-\infty}^{\infty} \|F_R U_t \Omega_- f\|^2 dt - \int_{-\infty}^{\infty} \|F_R U_t^0 f\|^2 dt \tag{7.121}$$

which, for normalised state vectors ($\|f\| = 1$), is the time delay for the ball $B_R$. In the framework of quantum mechanics in $\mathbb{R}^n$, the effective scattering region is all of $\mathbb{R}^n$, so that one must consider the limit of $(\Delta\mathcal{T})_R(f)$ as $R \to \infty$. This limit (if it exists) is called the ***global time delay*** $\Delta\mathcal{T}(f)$ for the initial state $f$:[17]

$$\Delta\mathcal{T}(f) = \lim_{R\to\infty} (\Delta\mathcal{T})_R(f). \tag{7.122}$$

In most situations both integrals appearing in (7.121) are finite. Since the integrands are bounded continuous functions of $t$, the quantity $(\Delta\mathcal{T})_R(f)$ may be interpreted as the improper Riemann integral of the function $t \mapsto \|F_R U_t \Omega_- f\|^2 - \|F_R U_t^0 f\|^2$. As $R \to \infty$, each of the two integrals in (7.121) tends to $+\infty$, but, for suitable vectors $f$, their difference may be convergent and thus define a finite global time delay $\Delta\mathcal{T}(f)$. Before considering these questions we give a result that allows one to express the global time delay in terms of the scattering operator. We assume that $\mathcal{M}_\infty^\pm(H_0) = \mathcal{H}$, and that the wave operators $\Omega_\pm = \text{s-lim}_{t\to\pm\infty} U_t^* U_t^0$ exist and satisfy asymptotic completeness.[18] The crucial condition is a sufficiently rapid convergence of $U_t^* U_t^0$ to the wave operators for suitable vectors (note that one has $\|\Omega_\pm f - U_t^* U_t^0 f\| = \|(\Omega_\pm - I)U_t^0 f\|$).

**Proposition 7.22.** *Let $f$ be such that, for every fixed $R < \infty$, each of the functions $t \mapsto \|F_R U_t^0 f\|$ and $t \mapsto \|F_R U_t^0 S f\|$ belongs to $L^2([0,\infty))$. Furthermore assume that the wave operators exist and are complete, that the function $t \mapsto \|(\Omega_- - U_t^* U_t^0) f\|$ belongs to $L^1((-\infty, 0])$ and that $t \mapsto \|(\Omega_+ - U_t^* U_t^0) S f\|$ belongs to $L^1([0,\infty))$. Then*

$$\Delta\mathcal{T}(f) = \lim_{R\to\infty} \int_0^\infty \left[\langle Sf, U_t^{0*} F_R U_t^0 S f\rangle - \langle f, U_t^{0*} F_R U_t^0 f\rangle\right] dt \tag{7.123}$$

---

[16] We assume here that each vector $f \in \mathcal{H}$ is a possible initial state, i.e. that $\mathcal{M}_\infty^-(H_0) = \mathcal{H}$, and that the wave operators exist on $\mathcal{H}$.

[17] As pointed out, the interpretation of $\Delta\mathcal{T}(f)$ as a time delay is justified for vectors satisfying $\|f\| = 1$. The advantage of not including the normalisation factors $1/\|f\|^2$ in Eq (7.121) is that the quantity $\Delta\mathcal{T}(f)$ will be the expectation value of a *linear* operator.

[18] The explicit form of $H_0$ is not essential in this proposition, and the result could also be formulated in terms of an abstract family of localising operators as in Section 5.2.

*in the sense that the limit in* (7.122) *exists if and only if that in* (7.123) *exists, with equality of these two limits.*

PROOF. For $g \in \mathcal{H}$ we set $J_R^{\pm}(g,t) = \|F_R U_t \Omega_\pm g\|^2 - \|F_R U_t^0 g\|^2$. Since $\Omega_+ S f = \Omega_- f$, we have

$$J_R^+(Sf,t) = J_R^-(f,t) + \|F_R U_t^0 f\|^2 - \|F_R U_t^0 S f\|^2.$$

Hence we may write

$$\begin{aligned}(\Delta T)_R(f) &\equiv \int_{-\infty}^{\infty} J_R^-(f,t)dt = \int_{-\infty}^{0} J_R^-(f,t)dt + \int_0^{\infty} J_R^-(f,t)dt \\ &= \int_{-\infty}^{0} J_R^-(f,t)dt + \int_0^{\infty} J_R^+(Sf,t)dt + \int_0^{\infty} \left[\|F_R U_t^0 S f\|^2 - \|F_R U_t^0 f\|^2\right]dt.\end{aligned}$$

We shall show that each integral on the last line exists (as an improper Riemann integral) and that the first two of these integrals converge to zero as $R \to \infty$. The result of the proposition then follows by observing that the integrand of the third integral is identical with the integrand in (7.123).

So let us consider each of the three integrals on the last line above. It is clear that the third one exists by the hypotheses on $\|F_R U_t^0 S f\|$ and $\|F_R U_t^0 f\|$. To deal with the other two integrals, we observe [using (1.15)] that

$$\begin{aligned}\left|J_R^{\pm}(g,t)\right| &= \left[\|F_R U_t \Omega_\pm g\| + \|F_R U_t^0 g\|\right]\left|\|F_R U_t \Omega_\pm g\| - \|F_R U_t^0 g\|\right| \\ &\leq 2\|g\|\|U_t \Omega_\pm g - U_t^0 g\| = 2\|g\|\|(\Omega_\pm - U_t^* U_t^0)g\|. \qquad (7.124)\end{aligned}$$

By taking $g = f$ and $g = Sf$ and using the remaining assumptions of the proposition, one obtains the existence of the first two integrals under consideration. Finally we observe that $J_R^{\pm}(g,t)$ converge to zero as $R \to \infty$ for each fixed $t \in \mathbb{R}$. As functions of $t$, $J_R^-(f,t)$ and $J_R^+(Sf,t)$ admit $L^1$ bounds on $(-\infty, 0]$ and $[0,\infty)$ respectively, uniformly in $R$, see (7.124). From the Dominated Convergence Theorem, one obtains the convergence to zero of their integrals as $R \to \infty$. □

Proposition 7.22 shows that the time delay $(\Delta T)_R$ for a finite ball $B_R$ can be expressed in terms of expectation values of the operator $\int_0^\infty U_t^{0*} F_R U_t^0 dt$. We next discuss the asymptotic behaviour ($R \to \infty$) of such expectation values in the case in which $H_0 = \vec{P}^2$. For this we denote by $\mathcal{D}_\nu^+$ ($\nu > 0$) the following dense subsets of $L^2(\mathbb{R}^n)$ [which are just the sets $\mathcal{D}_\nu^+(H_0)$ introduced before Lemma 7.5]:

$$\mathcal{D}_\nu^+ = \left\{f \in \mathcal{H} \mid f = \psi(H_0)\langle Q\rangle^{-\nu} h \text{ for some } h \in \mathcal{H} \text{ and some } \psi \in C_0^\infty((0,\infty))\right\}.$$

$\mathcal{D}_\nu^+$ is contained in $\mathcal{D}(\langle Q\rangle^\nu)$, see Example 6.23. We also introduce an operator $D_0$ as follows [$D$ being defined in (6.19)]:

$$D_0 = \frac{1}{2}\left[\frac{1}{|\vec{P}|^2}\vec{P}\cdot\vec{Q} + \vec{Q}\cdot\vec{P}\frac{1}{|\vec{P}|^2}\right] = \left[\frac{1}{|\vec{P}|^2}D + D\frac{1}{|\vec{P}|^2}\right], \quad \text{with } \mathcal{D}(D_0) = \mathcal{D}_1^+. \qquad (7.125)$$

It is easy to check that this operator is well defined and symmetric (Problem 7.12). We recall that $F_R = \chi(R^{-1}|\vec{Q}|)$.

**Proposition 7.23.** *In $L^2(\mathbb{R}^n)$, let $U_t^0 = e^{-iH_0 t}$ with $H_0 = \vec{P}^2$. If $f, g \in \mathcal{D}_2^+$, then*

$$\lim_{R\to\infty}\Big[\int_0^\infty \langle f, U_t^{0*}\chi\big(\frac{|\vec{Q}|}{R}\big)U_t^0 g\rangle dt - \frac{R}{2}\langle f, H_0^{-1/2} g\rangle\Big] = -\frac{1}{2}\langle f, D_0 g\rangle. \qquad (7.126)$$

PROOF. (i) By virtue of the polarisation identity (2.15), it suffices to prove (7.126) for $g = f$. Let us denote the left-hand side of (7.126), with $g = f$, by $\mathcal{Y}(f)$. We must show that $\mathcal{Y}(f)$ exists (as a limit) and is equal to $-\langle f, D_0 f\rangle/2$. We first derive a convenient expression for $\mathcal{Y}(f)$. We set $W_\tau = \exp(i\vec{Q}^2\tau)$ and then get from (5.62) that $U_t^{0*}\chi(|\vec{Q}|/R)U_t^0 = W_{1/4t}^*\chi(2t|\vec{P}|/R)W_{1/4t}$. By making the change of variables $t \mapsto s = (2t)^{-1}R$ and setting $\varepsilon = (2R)^{-1}$ one finds that

$$\begin{aligned}\int_0^\infty \langle f, U_t^{0*}\chi\big(\frac{|\vec{Q}|}{R}\big)U_t^0 f\rangle dt &= \int_0^\infty \langle f, W_{1/4t}^*\chi\big(\frac{2t|\vec{P}|}{R}\big)W_{1/4t} f\rangle dt\\ &= \frac{1}{4\varepsilon}\int_0^\infty \langle f, W_{\varepsilon s}^*\chi\big(\frac{|\vec{P}|}{s}\big)W_{\varepsilon s} f\rangle \frac{ds}{s^2}.\end{aligned}$$

Similarly, setting again $\varepsilon = (2R)^{-1}$ and making the change of variables $u \mapsto s = |\vec{k}|/u$ in the second expression below, one obtains

$$\begin{aligned}&\frac{R}{2}\langle f, H_0^{-1/2} f\rangle = \frac{1}{4\varepsilon}\int_{\mathbb{R}^n} d^n k \int_0^\infty du\, \frac{\chi(u)}{|\vec{k}|}|\tilde{f}(\vec{k})|^2\\ &= \frac{1}{4\varepsilon}\int_{\mathbb{R}^n} d^n k \int_0^\infty \frac{ds}{s^2}\chi\big(\frac{|\vec{k}|}{s}\big)|\tilde{f}(\vec{k})|^2 = \frac{1}{4\varepsilon}\int_0^\infty \frac{ds}{s^2}\langle f, \chi\big(\frac{|\vec{P}|}{s}\big) f\rangle.\end{aligned}$$

The preceding two equations show that

$$\begin{aligned}\mathcal{Y}(f) &= \lim_{\varepsilon\to+0}\int_0^\infty \frac{1}{4\varepsilon s^2}\Big[\langle f, W_{\varepsilon s}^*\chi\big(\frac{|\vec{P}|}{s}\big)W_{\varepsilon s} f\rangle - \langle f, \chi\big(\frac{|\vec{P}|}{s}\big) f\rangle\Big] ds \qquad (7.127)\\ &\equiv \lim_{\varepsilon\to+0}\int_0^\infty K_\varepsilon(s) ds,\end{aligned}$$

where $K_\varepsilon(s)$ denotes the integrand in (7.127).

(ii) By using the Dominated Convergence Theorem one can justify the interchange of the limit and the integral in Eq. (7.127). The details are elaborated in (iii) below. Assuming this result for the moment, we obtain from (7.127) that

$$\begin{aligned}\mathcal{Y}(f) &= \int_0^\infty \frac{ds}{4s^2}\frac{\partial}{\partial\varepsilon}\langle f, W_{\varepsilon s}^*\chi\big(\frac{|\vec{P}|}{s}\big)W_{\varepsilon s} f\rangle\big|_{\varepsilon=0}\\ &= \frac{i}{4}\int_0^\infty \frac{ds}{s}\Big[\langle f, \chi\big(\frac{|\vec{P}|}{s}\big)\vec{Q}^2 f\rangle - \langle \vec{Q}^2 f, \chi\big(\frac{|\vec{P}|}{s}\big) f\rangle\Big]\\ &= \frac{i}{4}\int_0^\infty \frac{ds}{s}\int_{|\vec{k}|\le s} d^n k\Big[\overline{\tilde{f}(\vec{k})}(\mathcal{F}\vec{Q}^2 f)(\vec{k}) - \overline{(\mathcal{F}\vec{Q}^2 f)(\vec{k})}\tilde{f}(\vec{k})\Big]. \qquad (7.128)\end{aligned}$$

The integral over $d^n k$ in (7.128) is zero for small $s$, because $\tilde{f}(\vec{k}) = 0$ near the origin.

For large values of $s$ this integral is equal to $\langle f, \vec{Q}^2 f\rangle - \langle \vec{Q}^2 f, f\rangle = 0$, because $\tilde{f}(\vec{k}) = 0$ near infinity and $\vec{Q}^2$ is a symmetric operator. It follows that the last expression for $\mathcal{Y}(f)$ is finite, which proves the existence of $\mathcal{Y}(f)$. In spherical polar co-ordinates $\vec{k} = k\vec{\omega}$, with $k = |\vec{k}|$ and $\vec{\omega} = \vec{k}/k \in S^{n-1}$, the integral over $d^n k$ in (7.128) is as follows:

$$\int_0^s k^{n-1} dk \int_{S^{n-1}} d\omega \Big[ \overline{\tilde{f}(k\vec{\omega})}(\mathcal{F}\vec{Q}^2 f)(k\vec{\omega}) - \overline{(\mathcal{F}\vec{Q}^2 f)(k\vec{\omega})}\tilde{f}(k\vec{\omega})\Big].$$

Upon inserting this expression into (7.128) and integrating by parts in the variable $s$, one finds (after observing that the boundary terms at $s = 0$ and $s = \infty$ vanish) that

$$\begin{aligned}\mathcal{Y}(f) &= -\frac{i}{4}\int_0^s (\log s) s^{n-1} ds \int_{S^{n-1}} d\omega \Big[ \overline{\tilde{f}(s\vec{\omega})}(\mathcal{F}\vec{Q}^2 f)(s\vec{\omega}) - \overline{(\mathcal{F}\vec{Q}^2 f)(s\vec{\omega})}\tilde{f}(s\vec{\omega})\Big] \\ &= -\frac{i}{4}\Big[\langle (\log|\vec{P}|) f, \vec{Q}^2 f\rangle - \langle \vec{Q}^2 f, (\log|\vec{P}|) f\rangle\Big].\end{aligned}$$

To complete the proof, it remains to check that $(\log|\vec{P}|)\vec{Q}^2 - \vec{Q}^2 \log|\vec{P}| = -2iD_0$ on $\mathcal{D}_2^+$. This is easily done by observing that $[Q_j, \log|\vec{P}|] = iP_j/|\vec{P}|^2$:

$$\begin{aligned}\big[\log|\vec{P}|, \vec{Q}^2\big] &= \sum_{j=1}^n \Big\{ \big[\log|\vec{P}|, Q_j\big] Q_j + Q_j \big[\log|\vec{P}|, Q_j\big] \Big\} \\ &= -i \sum_{j=1}^n \Big\{ \frac{P_j}{\vec{P}^2} Q_j + Q_j \frac{P_j}{\vec{P}^2} \Big\} = -i \Big\{ \frac{1}{\vec{P}^2} \vec{P}\cdot\vec{Q} + \vec{Q}\cdot\vec{P}\frac{1}{\vec{P}^2}\Big\} = -2iD_0.\end{aligned}$$

(iii) We finally exhibit an $\varepsilon$-independent integrable bound for the integrand $K_\varepsilon(s)$ in (7.127), thus justifying the application of the Dominated Convergence Theorem on the right-hand side of (7.127). It will be necessary to consider separately the domains $0 < s \le 1$ and $1 < s < \infty$. The following alternative expressions for $K_\varepsilon(s)$ are easily obtained by simple algebraic manipulations, using the unitarity of $W_{\varepsilon s}$ and defining $\chi^\perp : [0, \infty) \to \mathbb{R}$ by $\chi^\perp(u) = 1 - \chi(u)$:

$$K_\varepsilon(s) = \frac{1}{4s}\Big[\langle \frac{W_{\varepsilon s} - I}{\varepsilon s} f, \chi\big(\frac{|\vec{P}|}{s}\big) W_{\varepsilon s} f\rangle + \langle \chi\big(\frac{|\vec{P}|}{s}\big) f, \frac{W_{\varepsilon s} - I}{\varepsilon s} f\rangle\Big] \tag{7.129}$$

$$= -\frac{1}{4s}\Big[\langle \frac{W_{\varepsilon s} - I}{\varepsilon s} f, \chi^\perp\big(\frac{|\vec{P}|}{s}\big) W_{\varepsilon s} f\rangle + \langle \chi^\perp\big(\frac{|\vec{P}|}{s}\big) f, \frac{W_{\varepsilon s} - I}{\varepsilon s} f\rangle\Big]. \tag{7.130}$$

We shall use the following bound for $\|\tau^{-1}(W_\tau - I) f\|$:

$$\Big\| \frac{W_\tau - I}{\tau} f \Big\| \le \min\Big\{ \|\vec{Q}^2 f\|, \frac{2}{|\tau|}\|f\| \Big\} \qquad \forall \tau \ne 0. \tag{7.131}$$

Indeed, as $|e^{i\alpha} - 1| = |\int_0^\alpha e^{it} dt| \le |\alpha|$ for $\alpha \in \mathbb{R}$, we have $\|(W_\tau - I) f\| \le \|\tau \vec{Q}^2 f\|$, and it is clear that $\|(W_\tau - I) f\| \le 2\|f\|$.

Now let $s \in (0,1]$. By (7.129) and (7.131) we have for $s > 0$:

$$|K_\varepsilon(s)| \leq \frac{1}{4s}\|\vec{Q}^2 f\|\Big[\|\chi\big(\frac{|\vec{P}|}{s}\big)W_{\varepsilon s}f\| + \|\chi\big(\frac{|\vec{P}|}{s}\big)f\|\Big].$$

To arrive at an $\varepsilon$-independent integrable bound for $K_\varepsilon(s)$ on the interval $(0,1]$, it is enough to show that $\|\chi(|\vec{P}|/s)W_\tau f\| \leq cs^\delta$ for all $\tau \in \mathbb{R}$ and some $\delta > 0$. We take $\delta = 1/4$. By observing that $\chi(x/s) = \chi(x/s)\chi(x)$ if $s \in (0,1]$ and that $|\chi(u)| \leq u^{-1/4}$ for all $u > 0$, we get for all $\tau \in \mathbb{R}$:

$$\|\chi\big(\frac{|\vec{P}|}{s}\big)W_\tau f\| = \|\chi\big(\frac{|\vec{P}|}{s}\big)\chi(|\vec{P}|)W_\tau f\| \leq s^{1/4}\||\vec{P}|^{-1/4}\chi(|\vec{P}|)\langle Q\rangle^{-2}\|\|\langle Q\rangle^2 f\|,$$

and the required estimate $\|\chi(|\vec{P}|/s)W_\tau f\| \leq cs^{1/4}$ follows since $|\vec{P}|^{-1/4}\chi(|\vec{P}|)\langle Q\rangle^{-2}$ belongs to $\mathcal{B}(\mathcal{H})$ by Proposition 2.33.

Next let us consider the case $s \in [1,\infty)$. Since $|\chi^\perp(u)| \leq u$ for all $u \geq 0$, we find by taking into account (7.147) that

$$\|\chi^\perp\big(\frac{|\vec{P}|}{s}\big)W_\tau f\| \leq \Big\|\frac{|\vec{P}|}{s}W_\tau f\Big\| \leq \frac{1}{s}\Big[\sum_{j=1}^n \|P_j W_\tau f\|^2\Big]^{1/2}$$

$$\leq \frac{1}{s}\sum_{j=1}^n \|W_\tau P_j f + 2\tau Q_j W_\tau f\| \leq \frac{1}{s}\sum_{j=1}^n \Big[\|P_j f\| + 2|\tau|\|Q_j f\|\Big].$$

Hence, by using (7.130) and (7.131), we obtain

$$\begin{aligned}|K_\varepsilon(s)| &\leq \frac{1}{4s}\Big[\min\big\{\|\vec{Q}^2 f\|, \frac{2}{\varepsilon s}\|f\|\big\}\Big]\Big[\|\chi^\perp\big(\frac{|\vec{P}|}{s}\big)W_{\varepsilon s}f\| + \|\chi^\perp\big(\frac{|\vec{P}|}{s}\big)f\|\Big]\\
&\leq \frac{1}{4s^2}\Big[\min\big\{\|\vec{Q}^2 f\|, \frac{2}{\varepsilon s}\|f\|\big\}\Big]\sum_{j=1}^n\Big[2\|P_j f\| + 2\varepsilon s\|Q_j f\|\Big]\\
&\leq \frac{1}{s^2}\sum_{j=1}^n\Big[\|\vec{Q}^2 f\|\|P_j f\| + \|f\|\|Q_j f\|\Big],\end{aligned}$$

which gives an integrable $\varepsilon$-independent bound for $K_\varepsilon(s)$ on the interval $[1,\infty)$. □

We can now prove the existence of the global time delay $\Delta\mathcal{T}(f)$ for suitable vectors $f$ in $L^2(\mathbb{R}^n)$ and a class of self-adjoint Hamiltonians of the form $H = H_0 + V(\vec{Q})$, with $H_0 = \vec{P}^2$.

**Proposition 7.24.** *Let $H = H_0 + V(\vec{Q})$ in $L^2(\mathbb{R}^n)$, where $V\colon \mathbb{R}^n \to \mathbb{R}$ has the form $V(\vec{x}) = (1+|\vec{x}|)^{-1-\delta}[v_\infty(\vec{x}) + v_q(\vec{x})]$ with $\delta > 3$, $v_\infty \in L^\infty(\mathbb{R}^n)$ and $v_q \in L^q(\mathbb{R}^n)$ for some $q$ with $q > n/2$ and $q \geq 2$. Also assume that $V$ satisfies the condition (C) on page 288. Let $f \in \mathcal{D}_\rho^+$ for some $\rho > 7/2$, and assume that the support of $(\mathcal{U}_0 f)_\lambda$ is disjoint from the point spectrum of $H$. Then the limit in* (7.122) *exists and is given by the following expression:*

$$\Delta\mathcal{T}(f) = -\frac{1}{2}\langle f, S^*[D_0, S]f\rangle. \qquad (7.132)$$

PROOF. We know from Proposition 7.3 that the wave operators exist and that the scattering operator $S$ is unitary. We shall show in (ii) - (iii) that the assumptions of Proposition 7.22 are satisfied and that the vector $Sf$ belongs to $\mathcal{D}_2^+$. One can then apply (7.123) and (7.126). By adding and subtracting $(R/2)\langle f, H_0^{-1/2} f\rangle \equiv (R/2)\langle Sf, H_0^{-1/2} Sf\rangle$ in the first integral below (observe that $S$ commutes with $H_0^{-1/2}$), one obtains

$$\begin{aligned}
\Delta\mathcal{T}(f) &= \lim_{R\to\infty}\int_0^\infty \left[\langle Sf, U_t^{0*}F_R U_t^0 Sf\rangle - \langle f, U_t^{0*}F_R U_t^0 f\rangle\right] dt \\
&= \lim_{R\to\infty}\Big[\int_0^\infty \langle Sf, U_t^{0*}F_R U_t^0 Sf\rangle dt - \frac{R}{2}\langle Sf, H_0^{-1/2} Sf\rangle\Big] \\
&\quad - \lim_{R\to\infty}\Big[\int_0^\infty \langle f, U_t^{0*}F_R U_t^0 f\rangle dt - \frac{R}{2}\langle f, H_0^{-1/2} f\rangle\Big] \\
&= -\frac{1}{2}\langle Sf, D_0 Sf\rangle + \frac{1}{2}\langle f, D_0 f\rangle.
\end{aligned}$$

This then proves (7.132).

(i) We fix a closed interval $J$ in $(0,\infty)\setminus\sigma_p(H)$ such that $(\mathcal{U}_0 f)_\lambda$ [and hence also $(\mathcal{U}_0 Sf)_\lambda$] is zero for $\lambda$ outside $J$. We first observe that the vector-valued function $\lambda \mapsto (\mathcal{U}_0 Sf)_\lambda \equiv S(\lambda)(\mathcal{U}_0 f)_\lambda$ is three times continuously differentiable on $J$. Indeed, since $V$ decays as $|\vec{x}|^{-\kappa}$ with $\kappa > 4$, $S(\lambda)$ is three times continuously differentiable in norm on $J$ (Proposition 7.21), and $(\mathcal{U}_0 f)_\lambda \equiv M_{\langle Q\rangle^{-\rho}}(\lambda)\langle Q\rangle^\rho f$ is three times continuously differentiable since $\rho - 3 > 1/2$ (see page 362).

It follows in particular that, if $D$ denotes the operator defined in (6.19), then $Sf$ belongs to $\mathcal{D}(D^3)$ [in the spectral representation of $H_0$, the operator $D^3$ is a third order differential operator with coefficients that are bounded on $J$, see Eq. (6.23)].

(ii) We now show that $Sf \in \mathcal{D}_2^+$. Let $\theta$ be a function in $C_0^\infty((0,\infty))$ satisfying $\theta(\lambda) = 1$ for all $\lambda \in J$. Then one may write $Sf = \theta(H_0)\langle Q\rangle^{-2}\langle Q\rangle^2 Sf$, and the inclusion $Sf \in \mathcal{D}_2^+$ follows once we have shown that $Sf \in \mathcal{D}(\vec{Q}^2)$. For this we express the operator $\vec{Q}^2$ in spherical polar co-ordinates in momentum space, viz.

$$(\mathcal{F}\vec{Q}^2 g)(k\vec{\omega}) = -\Delta\tilde{g}(k\vec{\omega}) = -\Big[\frac{\partial^2}{\partial k^2} + \frac{n-1}{k}\frac{\partial}{\partial k} + \frac{\Delta_S}{k^2}\Big]\tilde{g}(k\vec{\omega}). \qquad (7.133)$$

Here $\Delta_S$ denotes the spherical Laplacian (the Laplace-Beltrami operator on the unit sphere $S^{n-1}$, containing the derivatives with respect to the angular variables; an explicit formula for this operator will not be needed). By transcribing the preceding expression into the spectral representation of $H_0$ as in (6.22), one obtains (with $k = |\vec{k}| = \sqrt{\lambda}$):

$$(\mathcal{U}_0\vec{Q}^2 g)_\lambda = -\Big[4\lambda\big(\frac{d}{d\lambda}\big)^2 + 4\frac{d}{d\lambda} - \frac{(n-2)^2}{4\lambda}\Big](\mathcal{U}_0 g)_\lambda + \frac{\Delta_S}{\lambda}(\mathcal{U}_0 g)_\lambda.$$

For $g = Sf$, the first term on the right-hand side defines an element of $L^2(\mathbb{R}^n)$ because $(\mathcal{U}_0 Sf)_\lambda$ is three times continuously differentiable with respect to $\lambda$ and vanishes near $\lambda = 0$ and near infinity. So it remains to show that $\Delta_S(\mathcal{U}_0 Sf)_\lambda \equiv \Delta_S(\mathcal{U}_0 f)_\lambda + \Delta_S(\mathcal{U}_0 Rf)_\lambda$ represents a vector in $L^2(\mathbb{R}^n)$.

To treat the term $\Delta_S(\mathcal{U}_0 f)_\lambda$ we write it as $\Delta_S(\mathcal{U}_0 f)_\lambda = \Delta_S M_{\langle Q\rangle^{-3}}(\lambda)\langle Q\rangle^3 f$, and we use Lemma 7.25 in the Appendix which shows that $\Delta_S M_{\langle Q\rangle^{-3}}(\lambda)$ is a bounded operator from $L^2(\mathbb{R}^n)$ to $L^2(S^{n-1})$ that depends continuously on $\lambda$ on $(0,\infty)$. Hence $\|\Delta_S(\mathcal{U}_0 f)_\lambda\|$ is a bounded function of $\lambda$ on the interval $J$, so that $\lambda^{-1}\Delta_S(\mathcal{U}_0 f)_\lambda$ defines an element of $L^2(\mathbb{R}^n)$.

To handle the term $\Delta_S(\mathcal{U}_0 Rf)_\lambda$ we write the potential as $V = UW$ with $U(\vec{x}) = (1+|\vec{x}|)^{-\nu}|v_\infty(\vec{x})+v_q(\vec{x})|^{1/2}$, $W(\vec{x}) = (1+|\vec{x}|)^{\nu-1-\delta}|v_\infty(\vec{x})+v_q(\vec{x})|^{1/2}\operatorname{sign}V(\vec{x})$, with $\nu$ satisfying $5/2 < \nu < 3$. The functions $U$ and $W$ satisfy condition (S), hence $M_U(\lambda)$ and $M_W(\lambda)$ are well defined and depend continuously on $\lambda$ (Proposition 7.16). Then we have by (7.112):

$$\Delta_S(\mathcal{U}_0 Rf)_\lambda = -2\pi i\Delta_S M_U(\lambda)[I - W(\vec{Q})(H-\lambda-i0)^{-1}U(\vec{Q})]M_W(\lambda)^*(\mathcal{U}_0 f)_\lambda.$$

Again by Lemma 7.25 in the Appendix, the operator $\Delta_S M_U(\lambda)$ is bounded from $L^2(\mathbb{R}^n)$ to $L^2(S^{n-1})$ and depends continuously on $\lambda$ on $(0,\infty)$ (here the condition $\nu > 5/2$ is crucial). It is then seen from the preceding equation that $\|\Delta_S(\mathcal{U}_0 Rf)_\lambda\|$ is a bounded function of $\lambda$ on the interval $J$, hence $\lambda^{-1}\Delta_S(\mathcal{U}_0 Rf)_\lambda$ defines an element of $L^2(\mathbb{R}^n)$.

(iii) We finally verify that the assumptions of Proposition 7.22 are satisfied. Let $J$ be as in (i) above. Then $F_R$ is $H_0$-smooth on $J$ (see Proposition 6.16), consequently each of the functions $t \mapsto \|F_R U_t^0 f\| \equiv \|F_R E_0(J) U_t^0 f\|$ and $t \mapsto \|F_R U_t^0 Sf\|$ belongs to $L^2(\mathbb{R})$ (Proposition 6.11).

Next we have

$$\|(\Omega_- - U_t^* U_t^0)f\| = \Big\|\int_{-\infty}^t \frac{d}{d\tau}U_\tau^* U_\tau^0 f\,d\tau\Big\| = \Big\|\int_{-\infty}^t U_\tau^* V U_\tau^0 f\,d\tau\Big\|$$
$$\leq \int_{-\infty}^t \|VU_\tau^0 f\|d\tau.$$

Now, assuming that $f = \psi(H_0)\langle\vec{Q}\rangle^{-\rho}h$ and setting $\phi(\lambda) = (\lambda+i)\psi(\lambda)$:

$$\|VU_\tau^0 f\| \leq \|V(H_0+i)^{-1}\langle\vec{Q}\rangle^\rho\|\|\langle\vec{Q}\rangle^{-\rho}\phi(H_0)U_\tau^0\langle\vec{Q}\rangle^{-\rho}\|\|\langle\vec{Q}\rangle^\rho f\|,$$

which is less than $c(1+|\tau|)^{-3}$ for some constant $c$ by (7.45), since $\rho > 3$ (the first norm on the right-hand side is finite by Example 6.26). Thus $\|(\Omega_- - U_t^* U_t^0)f\| \leq c(1+|t|)^{-2}$ for negative $t$, which is an integrable function of $t$ on $(-\infty, 0]$.

The verification of the integrability of $\|(\Omega_+ - U_t^* U_t^0)Sf\|$ on $[0,\infty)$ is similar. Let again $\theta$ be a function in $C_0^\infty((0,\infty))$ satisfying $\theta(\lambda) = 1$ for all $\lambda \in J$. Then one may write $Sf = \theta(H_0)(I+D^2)^{-3/2}h$, with $h = (I+D^2)^{3/2}Sf \in \mathcal{H}$ because $Sf \in \mathcal{D}(D^3)$ as seen in (i). Setting $\varphi(\lambda) = (\lambda+i)\theta(\lambda)$, we then have

$$\|VU_\tau^0 Sf\| \leq \|V(H_0+i)^{-1}\langle\vec{Q}\rangle^\rho\|\|\langle\vec{Q}\rangle^{-\rho}\varphi(H_0)U_\tau^0(I+D^2)^{-3/2}\|\|h\|,$$

which is less than $c_\varepsilon(1+|\tau|)^{-3+\varepsilon}$ for any $\varepsilon > 0$ by (7.46). So $\|(\Omega_+ - U_t^* U_t^0)Sf\| \leq \int_t^\infty \|VU_\tau^0 Sf\|d\tau \leq c_\varepsilon(1+t)^{-2+\varepsilon}$ for $t > 0$. □

**7.5.2.** By taking into account the intertwining relation $U_\tau\Omega_- = \Omega_- U_\tau^0$, one sees from the definition (7.121) of the local time delay that $(\Delta T)_R(U_\tau^0 f) = (\Delta T)_R(f)$ for each $\tau \in \mathbb{R}$. Consequently the global time delay, if it exists, will also be invariant under the free evolution. In fact this is easily verified by observing that $U_\tau^{0*} D_0 U_\tau^0 = D_0 + 2\tau I$ on $\mathcal{D}_1^+$, so that $U_\tau^{0*} S^*[D_0, S] U_\tau^0 \equiv S^*[U_\tau^{0*} D_0 U_\tau^0, S] = S^*[D_0, S]$. The last property also shows that the operator $S^*[D_0, S]$ commutes with each $U_\tau^0$, hence with $H_0$ (on a suitable domain).

In analogy with (5.71), we conclude from the commutativity of $S^*[D_0, S]$ and $H_0$ that there will be a family $\{T(\lambda)\}_{\lambda>0}$ of operators acting in $L^2(S^{n-1})$ such that $(-1/2)(\mathcal{U}_0 S^*[D_0, S]f)_\lambda = T(\lambda)(\mathcal{U}_0 f)_\lambda$ for $\lambda > 0$, in other terms that the (unbounded) operator $S^*[D_0, S]$ will be decomposable in the spectral representation of $H_0$ (see also Proposition 5.26). The operator $T(\lambda)$ represents the time delay at energy $\lambda$, and it will be a symmetric operator because $S^*[D_0, S]$ is symmetric. In fact it is easy to obtain an explicit expression for $T(\lambda)$ in terms of the $S$-matrix: by using the second expression in (7.125) for the operator $D_0$ and the fact that $D$ acts as $i\lambda(d/d\lambda) + i/2$ in the spectral representation of $H_0$ [see (6.23)], one finds that the action of $D_0$ in this representation is as follows:

$$(\mathcal{U}_0 D_0 f)_\lambda = (\mathcal{U}_0 H_0^{-1} D f)_\lambda + (\mathcal{U}_0 D H_0^{-1} f)_\lambda = 2i\frac{d}{d\lambda}(\mathcal{U}_0 f)_\lambda. \tag{7.134}$$

In other terms, in the spectral representation of $H_0$, the operator $D_0$ is the differential operator $2i(d/d\lambda)$. We have seen in Section 3.2 that this operator is maximal symmetric but not essentially self-adjoint (page 101). We now get

$$T(\lambda) = -iS(\lambda)^*\Big[\frac{d}{d\lambda}, S(\lambda)\Big] = -iS(\lambda)^* S'(\lambda). \tag{7.135}$$

Here $S'(\lambda)$ denotes the derivative of the $\mathcal{B}(L^2(S^{n-1}))$-valued function $\lambda \mapsto S(\lambda)$ at the point $\lambda$. Note that the definition of this derivative at $\lambda$ involves the $S$-matrix also at neighbouring values of $\lambda$, but, since $T(\lambda)$ is defined as a decomposable operator, the action of $S'(\lambda)$ in (7.135) is as follows: $[\mathcal{U}_0 S'(\lambda) f]_\lambda = S'(\lambda)(\mathcal{U}_0 f)_\lambda$.

In situations in which $\lambda \mapsto S(\lambda)$ is continuously differentiable on $(0, \infty)$, the operator $T(\lambda)$ will be bounded for each $\lambda > 0$, hence not only symmetric but self-adjoint, and $\lambda \mapsto T(\lambda)$ will be continuous in norm. The self-adjointness of $T(\lambda)$ can also be inferred from the unitarity of $S(\lambda)$: by differentiating the equation $S(\lambda)^* S(\lambda) = I$, one finds that $S'(\lambda)^* S(\lambda) + S(\lambda)^* S'(\lambda) = O$, i.e. $S'(\lambda)^* S(\lambda) = -S(\lambda)^* S'(\lambda)$. Thus

$$T(\lambda)^* = \big[-iS(\lambda)^* S'(\lambda)\big]^* = iS'(\lambda)^* S(\lambda) = -iS(\lambda)^* S'(\lambda) = T(\lambda).$$

The family $\{T(\lambda)\}_{\lambda>0}$ determines a self-adjoint operator $T$ in $L^2(\mathbb{R}^n)$, called the ***Eisenbud-Wigner time delay operator***, given as follows by its action in the spectral representation of $H_0$:

$$(\mathcal{U}_0 T f)_\lambda = T(\lambda)(\mathcal{U}_0 f)_\lambda, \tag{7.136}$$

$$\mathcal{D}(T) = \Big\{f \in L^2(\mathbb{R}^n) \,\Big|\, \int_0^\infty d\lambda \, \|T(\lambda)[\mathcal{U}_0 f]_\lambda\|^2_{L^2(S^{n-1})} < \infty\Big\}. \tag{7.137}$$

In general this operator $T$ will be unbounded, even though each $T(\lambda)$ is bounded. This is so because $\|T(\lambda)\|$ usually diverges as $\lambda \to 0$ [19]. As for the multiplication operators considered in Section 2.5, the domain (7.137) is the maximal domain on which $T$ can be defined in a natural way. The global time delay $(\Delta T)(f)$ for an initial state $f$ ($\|f\| = 1$) is then the expectation value of the observable $T$ in this state.

In the simple situation considered at the end of §5.5.1 (scattering by a spherically symmetric potential in $n = 3$ dimensions), the $S$-matrix is given in each partial wave subspace by a number $S(\lambda) = e^{2i\delta_\ell(\lambda)}$. In this situation the time delay operator $T$ leaves each partial wave subspace invariant, and in the partial wave of angular momentum $\ell$ the operator $T(\lambda)$ is determined by the derivative of the phase shift $\delta_\lambda$, viz. $T(\lambda)$ is just multiplication by $2\delta'_\ell(\lambda)$.

**7.5.3.** We mention another quantity of interest, viz. the trace of the operator $T(\lambda)$. Let us consider potentials of the form $V(\vec{x}) = U(\vec{x})W(\vec{x})$ with $U$ and $W$ satisfying the condition (S). If both functions $U$ and $W$ belong to $L^2(\mathbb{R}^n)$, then $M_U(\lambda)$ and $M_W(\lambda)$ are Hilbert-Schmidt operators, and it follows from (7.112) that $R(\lambda)$ is a trace class operator in $L^2(S^{n-1})$. One can similarly find conditions implying that $T(\lambda)$ belongs to the trace class. If $(1+|\vec{x}|)W(\vec{x})$ belongs to $L^2(\mathbb{R}^n)$, then $(d/d\lambda)M_W(\lambda)$ is a Hilbert-Schmidt operator, as is easily seen from (7.117). By using this property in the calculation of the derivative of $R(\lambda)$ indicated in part (ii) of the proof of Proposition 7.21, one finds that $T(\lambda)$ belongs to the trace class if one assumes that $(1 + |\vec{x}|)V(\vec{x})$ belongs to $L^1(\mathbb{R}^n)$ (Problem 7.11). This condition requires a decay of $V(\vec{x})$ faster than $|\vec{x}|^{-(n+1)}$ [this result is not optimal; it is known that a decay faster than $|\vec{x}|^{-n}$ is sufficient for the inclusion $T(\lambda) \in \mathcal{B}_1(L^2(S^{n-1}))$].

Similar to the Hilbert-Schmidt norm of $R(\lambda)$, which determines the total scattering cross section for a beam, averaged over all initial directions, the trace of $T(\lambda)$ has an interpretation in relation to the scattering of a beam. Consider a family $\{f_{\vec{b}}\}$ of vectors obtained as in (5.78) from a fixed vector $f$ by translations in the hyperplane $\Pi_0$ orthogonal to an initial direction $\vec{\omega}_0$ and describing an incoming beam as in Section 5.6. The total time delay for such a beam will be the quantity $\int_{\Pi_0}(\Delta T)(f_{\vec{b}})$. Assuming that the Hilbert-Schmidt kernels of the trace class operators $T(\lambda)$ are continuous functions, and considering wave functions $f$ such that the support of $\tilde{f}$ belongs to a very small neighbourhood of a vector $\vec{k}_0 = \sqrt{\lambda_0}\vec{\omega}_0$, one finds that, after averaging over all initial directions $\vec{\omega}_0 \in S^{n-1}$, the averaged time delay for such a beam is determined by $\operatorname{Tr} T(\lambda_0)$:

$$\overline{T}(\lambda_0) := \frac{1}{\Theta_n}\int_{S^{n-1}} d\vec{\omega}_0 \int_{\Pi_0 \perp \vec{\omega}_0} (\Delta T)(f_{\vec{b}})\, d^{n-1}b \approx \frac{(2\pi)^{n-1}}{\lambda_0^{(n-1)/2}\Theta_n} \operatorname{Tr} T(\lambda_0).$$

The verification uses arguments similar to those leading to the total scattering cross section in Section 5.6. The details can be found in [2].

[19] Due to the inverse powers of $\sqrt{\lambda}$ occurring in (7.117), $S'(\lambda)$ will become unbounded as $\lambda \to 0$.

## Appendix to Chapter 7

**A.7.1. Proof of Proposition 7.4.** (i) We embed $\mathcal{O}$ in a shell $G_\delta$ of thickness $2\delta > 0$: Let $\mathcal{O}_s$ be the hypersurface obtained from $\mathcal{O}$ by displacements of length $s$ orthogonal to $\mathcal{O}$, i.e. $\mathcal{O}_s = \{\vec{x} \in \mathbb{R}^n \mid \vec{x} = \vec{y} + s\vec{n}(\vec{y}), \vec{y} \in \mathcal{O}\}$, and let $G_\delta = \bigcup_{-\delta \le s \le \delta} \mathcal{O}_s$. If $\mathcal{O} = \Sigma_b$, we take $\delta = b/2$; if $\mathcal{O} = \Lambda_{\rho,d}$, we choose $\delta$ to be sufficiently small so that each vector in $\mathcal{O}_\delta$ has a unique decomposition into $\vec{x} = \vec{y} + s\vec{n}(\vec{y})$ with $\vec{y} \in \mathcal{O}$ and $s \in [-\delta, \delta]$. Throughout this proof it is understood that $\vec{x}$ and $\vec{y}$ are related in the preceding manner (so, as $\vec{x}$ varies over $G_\delta$, $\vec{y}$ and $s$ are functions of $\vec{x}$).

We denote by $d\sigma_s(\vec{y})$ the surface element on $\mathcal{O}_s$ at the point $\vec{y} + s\vec{n}(\vec{y})$ and by $d\sigma(\vec{y}) \equiv d\sigma_0(\vec{y})$ the corresponding quantity on $\mathcal{O}$. There is a constant $\beta \in (1, \infty)$ such that

$$\frac{1}{\beta}\, d\sigma(\vec{y}) \le d\sigma_s(\vec{y}) \le \beta\, d\sigma(\vec{y}) \qquad \forall\, \vec{y} \in \mathcal{O}, \forall\, s \in [-\delta, \delta]. \tag{7.138}$$

(ii) Let $f$ be as stated in the proposition and $\vec{x} \in G_\delta$. For $\tau \in [0, 1]$ and $k = 1, \ldots, n$ we set

$$f_\tau^{(k)}(\vec{x}) = \big(\partial_k f\big)\big(\vec{y} + \tau(\vec{x} - \vec{y})\big) = i\big(P_k f\big)\big(\vec{y} + \tau s\vec{n}(\vec{y})\big).$$

By Newton's formula one has

$$f(\vec{x}) = f(\vec{y}) + \sum_{k=1}^{n} (x_k - y_k) \int_0^1 f_\tau^{(k)}(\vec{x})\, d\tau. \tag{7.139}$$

With (1.58) this implies that

$$\begin{aligned} \int_{\mathcal{O}} |f(\vec{y})|^2 d\sigma(\vec{y}) &\equiv \frac{1}{2\delta} \int_{-\delta}^{\delta} ds \int_{\mathcal{O}} |f(\vec{y})|^2 d\sigma(\vec{y}) \le \frac{\beta}{2\delta} \int_{-\delta}^{\delta} ds \int_{\mathcal{O}} |f(\vec{y})|^2 d\sigma_s(\vec{y}) \\ &\le \frac{2\beta}{2\delta} \int_{G_\delta} |f(\vec{x})|^2 d^n x + \frac{2n\beta}{2\delta} \int_{G_\delta} d^n x \sum_{k=1}^{n} \delta^2 \left| \int_0^1 f_\tau^{(k)}(\vec{x})\, d\tau \right|^2 \\ &\le \frac{\beta}{\delta} \|f\|^2 + n\beta\delta \sum_{k=1}^{n} \Big\| \int_0^1 f_\tau^{(k)}(\cdot)\, d\tau \Big\|_{L^2(G_\delta)}^2. \end{aligned} \tag{7.140}$$

Now, by the triangle inequality:

$$\Big\| \int_0^1 f_\tau^{(k)}(\cdot)\, d\tau \Big\|_{L^2(G_\delta)} \le \int_0^1 \|f_\tau^{(k)}\|_{L^2(G_\delta)}\, d\tau, \tag{7.141}$$

and, for $0 < \tau \le 1$ [using (7.138)]:

$$
\begin{aligned}
\|f_\tau^{(k)}\|^2_{L^2(G_\delta)} &= \int_{-\delta}^{\delta} ds \int_{\mathcal{O}} d\sigma_s(\vec{y}) \big| f_\tau^{(k)}(\vec{y}+s\vec{n}(\vec{y}))\big|^2 \\
&= \int_{-\delta}^{\delta} ds \int_{\mathcal{O}} d\sigma_s(\vec{y}) \big| (P_k f)(\vec{y}+\tau s\vec{n}(\vec{y}))\big|^2 \\
&= \frac{1}{\tau}\int_{-\tau\delta}^{\tau\delta} du \int_{\mathcal{O}} d\sigma_{u/\tau}(\vec{y}) \big| (P_k f)(\vec{y}+u\vec{n}(\vec{y}))\big|^2 \\
&\le \frac{\beta^2}{\tau}\int_{-\delta}^{\delta} du \int_{\mathcal{O}} d\sigma_{u}(\vec{y}) \big| (P_k f)(\vec{y}+u\vec{n}(\vec{y}))\big|^2 = \frac{\beta^2}{\tau}\|P_k f\|^2_{L^2(G_\delta)}.
\end{aligned}
$$

Together with (7.140) and (7.141) this implies that

$$
\|f|_{\mathcal{O}}\|^2_{L^2(\mathcal{O})} \le \frac{\beta}{\delta}\|f\|^2 + n\beta^3\delta\Big[\int_0^1 \tau^{-1/2}d\tau\Big]^2 \sum_{k=1}^n \|P_k f\|^2.
$$

Thus (7.28) holds for $\varphi \equiv 1$, with $c_{\mathcal{O}} = \max\{\sqrt{\beta/\delta}, 2\beta\sqrt{n\beta\delta}\}$ [take into account Eq. (7.148)]. The result for general $\varphi$ follows immediately because $f|_{\mathcal{O}} = (\varphi f)|_{\mathcal{O}}$.

(iii) We finally verify the inequality (7.29). Here $\mathcal{O} = \Sigma_b$, and we have $c_{\Sigma_b} \le \max\{\sqrt{\beta_b/\delta}, 2\beta_b\sqrt{n\beta_b\delta}\}$, where by (7.138) $\beta_b > 1$ is such that

$$
\frac{1}{\beta_b}b^{n-1} \le (b+s)^{n-1} \le \beta_b b^{n-1} \qquad \forall s\in[-\delta,\delta] \tag{7.142}
$$

(because $d\sigma_s(\vec{y}) = (b+s)^{n-1}d\omega$ in spherical polar co-ordinates). Since we consider only values of $b$ that are $\ge 1$, we may take $\delta = 1/4$ in (i) above (rather than $\delta = b/2$). Then (7.142) is satisfied for all $b \ge 1$ and all $s \in [-1/4, 1/4]$ with $\beta_b = (4/3)^{n-1}$. So we have $\sup_{b\ge 1} c_{\Sigma_b} \le \kappa_n$ with a finite constant $\kappa_n$, depending on $n$, which we do not determine explicitly. □

**A.7.2. Proof of (7.40) and (7.41)**

(i) The operator $P_j$ is $H$-bounded and maps $\mathcal{D}(H^2)$ into $\mathcal{D}(H)$, for $P_j(H+i)^{-2} = (H+i)^{-1}P_j(H+i)^{-1} + i(H+i)^{-1}[(\partial_j V)(\vec{Q})](H+i)^{-2}$. If $g = \psi(H)\langle Q\rangle^{-1}h \in \mathcal{D}_1(H)$, then $U_\tau g \in \mathcal{D}(H^2)$ for each $\tau \in \mathbb{R}$. By Proposition 5.5(b) (Leibniz rule), $t \mapsto U_t^* P_j U_t g$ is strongly differentiable. Hence

$$
\begin{aligned}
U_t^* P_j U_t g - P_j g &= \int_0^t \frac{d}{d\tau} U_\tau^* P_j U_\tau g\, d\tau = i\int_0^t U_\tau^*[H, P_j]U_\tau g\, d\tau \\
&= -\int_0^t U_\tau^*[(\partial_j V)(\vec{Q})]U_\tau g\, d\tau.
\end{aligned}
$$

Upon multiplying this equation by $U_t$ one obtains (7.40).

(ii) For $\varepsilon > 0$, let us set $Q_{j,\varepsilon} = Q_j(I+\varepsilon\vec{Q}^2)^{-1/2}$. Then one has $Q_{j,\varepsilon} \in \mathcal{B}(\mathcal{H})$ and $\text{s-lim}_{\varepsilon\to 0} Q_{j,\varepsilon}f = Q_j f$ for each $f \in \mathcal{D}(Q_j)$. As in Example 6.22 one finds that $Q_{j,\varepsilon}\mathcal{D}(H) \subseteq \mathcal{D}(H) \equiv \mathcal{D}(H_0)$ and that on $\mathcal{D}(H)$ [see (7.50)]:

$$
[H, Q_{j,\varepsilon}] = A_{j,\varepsilon} + 2i\vec{B}_{j,\varepsilon}\cdot\vec{P} - 2i(I+\varepsilon\vec{Q}^2)^{-1/2}P_j
$$

with

$$A_{j,\varepsilon} = \varepsilon Q_{j,\varepsilon}\Big[\frac{n+2}{I+\varepsilon\vec{Q}^2} - 3\frac{\varepsilon\vec{Q}^2}{(I+\varepsilon\vec{Q}^2)^2}\Big], \qquad [B_{j,\varepsilon}]_k = \varepsilon Q_{j,\varepsilon}\frac{Q_k}{I+\varepsilon\vec{Q}^2}\,.$$

Note that $\|\varepsilon Q_{j,\varepsilon}\| \le \varepsilon^{1/2}$, hence $\|A_{j,\varepsilon}\| \le (n+5)\varepsilon^{1/2}$, and that $\|[B_{j,\varepsilon}]_k\| \le 1$.

We can apply Proposition 5.5(a) to see that $\tau \mapsto U_\tau^* Q_{j,\varepsilon} U_\tau g$ is strongly differentiable if $g \in \mathcal{D}(H)$, with derivative $iU_\tau^*[A_{j,\varepsilon} + 2i\vec{B}_{j,\varepsilon}\cdot\vec{P} - 2i(I+\varepsilon\vec{Q}^2)^{-1/2}P_j]U_\tau g$. Hence

$$U_t^* Q_{j,\varepsilon} U_t g - Q_{j,\varepsilon} g = \int_0^t U_\tau^*\big[iA_{j,\varepsilon} - 2\vec{B}_{j,\varepsilon}\cdot\vec{P} + 2(I+\varepsilon\vec{Q}^2)^{-1/2}P_j\big]U_\tau g\,d\tau. \tag{7.143}$$

To obtain (7.41), we assume that $g = \psi(H)\langle Q\rangle^{-1}h \in \mathcal{D}_1^+(H)$ and determine the limit of (7.143) as $\varepsilon \to 0$. Since $U_t g \in \mathcal{D}(Q_j)\cap\mathcal{D}(H)$ [by part (i) of the proof of Lemma 7.5], the left-hand side of (7.143) converges strongly to $U_t^* Q_j U_t g - Q_j g$ as $\varepsilon \to 0$. Let us then consider each of the three terms on the right-hand side of (7.143).

First we have $\|\int_0^t U_\tau^* A_{j,\varepsilon} U_\tau\,d\tau\| \le |t|\|A_{j,\varepsilon}\|$. Since $\|A_{j,\varepsilon}\| \to 0$ as $\varepsilon \to 0$, the operator $i\int_0^t U_\tau^* A_{j,\varepsilon} U_\tau\,d\tau$ converges to zero in norm as $\varepsilon \to 0$. For the remaining two terms we apply the Dominated Convergence Theorem. We observe that

$$\Big\|\int_0^t U_\tau^*[B_{j,\varepsilon}]_k P_k U_\tau g\,d\tau\Big\| \le \int_0^t \|[B_{j,\varepsilon}]_k P_k U_\tau g\|\,d\tau.$$

The integrand on the right-hand side is bounded by a $\tau$-independent constant, since $\|[B_{j,\varepsilon}]_k\| \le 1$ and $\|P_k U_\tau g\| = \|P_k\psi(H)U_\tau\langle Q\rangle^{-1}h\| \le \|P_k\psi(H)\|\|h\|$. Furthermore this integrand converges to zero for each fixed $\tau$:

$$\|[B_{j,\varepsilon}]_k P_k U_\tau g\| \le \|\varepsilon Q_{j,\varepsilon}\|\|Q_k P_k U_\tau g\| \le \varepsilon^{1/2}\|Q_k P_k U_\tau g\|,$$

and $U_\tau g$ belongs to $\mathcal{D}(Q_k P_k)$ by Lemma 7.5(a). So the second term on the right-hand side of (7.143) converges strongly to zero. It remains to be shown that the third term converges strongly to $2\int_0^t U_\tau^* P_j U_\tau g\,d\tau$, i.e. that

$$\lim_{\varepsilon\to 0}\Big\|\int_0^t U_\tau^*[(I+\varepsilon\vec{Q}^2)^{-1/2} - I]P_j U_\tau g\,d\tau\Big\| = 0.$$

This is obtained as for the second term above by taking into account the fact that $\text{s-lim}_{\varepsilon\to 0}[(I+\varepsilon\vec{Q}^2)^{-1/2} - I] = 0$. □

**A.7.3. Differentiability of $M_W(\lambda)$.** Under the assumption that $(1+|\vec{x}|)W$ satisfies condition (S) on page 355, let us prove that $M_W(\lambda)$ is differentiable in norm, with derivative given by (7.117). By (7.107) we have for $f \in \mathcal{S}(\mathbb{R}^n)$

$$M_W(\lambda)f = \big(\mathcal{U}_0 W(\vec{Q})f\big)_\lambda = 2^{-1/2}\lambda^{(n-2)/4}\gamma_{\Sigma_{\sqrt{\lambda}}}\mathcal{F}Wf.$$

Differentiation of the factor $\lambda^{(n-2)/4}$ gives rise to the first term on the right-hand side of (7.117). So it suffices to show that the norm derivative of $\gamma_{\Sigma_{\sqrt{\lambda}}}\mathcal{F}W$ with respect to $\lambda$ exists and is equal to $-i(2\sqrt{\lambda})^{-1}\sum_{j=1}^n N_j\gamma_{\Sigma_{\sqrt{\lambda}}}\mathcal{F}Q_jW$, or equivalently [setting $\sqrt{\lambda}=\mu$, hence $d/d\lambda = (2\sqrt{\lambda})^{-1}d/d\mu$] that the norm derivative with respect to $\mu$ of $\gamma_{\Sigma_\mu}\mathcal{F}W$ exists and is equal to $-i\sum_{j=1}^n N_j\gamma_{\Sigma_\mu}\mathcal{F}Q_jW$.

With the notation $X_\varepsilon(\mu) = \varepsilon^{-1}[\gamma_{\Sigma_{\mu+\varepsilon}}\mathcal{F}W - \gamma_{\Sigma_\mu}\mathcal{F}W] + i\sum_{j=1}^n N_j\gamma_{\Sigma_\mu}\mathcal{F}Q_jW$, we must show that

$$\lim_{\varepsilon\to 0}\sup_{\substack{f\in\mathcal{S}(\mathbb{R}^n)\\ \|f\|=1}}\|X_\varepsilon(\mu)f\|_{L^2(S^{n-1})} = 0. \tag{7.144}$$

Now, by Newton's formula (7.139), we have for $f \in \mathcal{S}(\mathbb{R}^n)$:

$$\begin{aligned}
[X_\varepsilon(\mu)f](\vec{\omega}) &= \frac{1}{\varepsilon}[(\gamma_{\Sigma_{\mu+\varepsilon}}\widetilde{Wf})(\vec{\omega}) - (\gamma_{\Sigma_\mu}\widetilde{Wf})(\vec{\omega})] + i\sum_{j=1}^n \omega_j(\gamma_{\Sigma_\mu}\widetilde{Q_jWf})(\vec{\omega})\\
&= \frac{1}{\varepsilon}\int_0^1\Big[\sum_{j=1}^n \varepsilon\omega_j\partial_j(\widetilde{Wf})(\mu\vec{\omega}+s\varepsilon\vec{\omega}) + i\varepsilon\sum_{j=1}^n\omega_j(\widetilde{Q_jWf})(\mu\vec{\omega})\Big]ds\\
&= \int_0^1\Big[\sum_{j=1}^n \omega_j(-i)(\widetilde{Q_jWf})(\mu\vec{\omega}+s\varepsilon\vec{\omega}) + i\sum_{j=1}^n\omega_j(\widetilde{Q_jWf})(\mu\vec{\omega})\Big]ds\\
&= -i\int_0^1\Big[\sum_{j=1}^n \omega_j(\gamma_{\Sigma_{\mu+s\varepsilon}}\widetilde{Q_jWf})(\vec{\omega}) - \sum_{j=1}^n\omega_j(\gamma_{\Sigma_\mu}\widetilde{Q_jWf})(\vec{\omega})\Big]ds\\
&= -i\sum_{j=1}^n\int_0^1\Big[(N_j\gamma_{\Sigma_{\mu+s\varepsilon}}\mathcal{F}Q_jWf)(\vec{\omega}) - (N_j\gamma_{\Sigma_\mu}\mathcal{F}Q_jWf)(\vec{\omega})\Big]ds.
\end{aligned}$$

Since $\gamma_{\Sigma_\mu}\widetilde{Uf} = \sqrt{2}\mu^{-(n-2)/2}M_U(\mu^2)$, the integrand of the last expression is a strongly continuous $L^2(S^{n-1})$-valued function of $s$, and by using results from Section 1.2 we then obtain the following inequality, valid for $f \in \mathcal{S}(\mathbb{R}^n)$:

$$\begin{aligned}
&\|X_\varepsilon(\mu)f\|_{L^2(S^{n-1})}\\
&\le \sqrt{2}\sum_{j=1}^n\int_0^1 ds\,\Big\|\frac{1}{(\mu+s\varepsilon)^{(n-2)/2}}M_{Q_jW}\big((\mu+s\varepsilon)^2\big) - \frac{1}{\mu^{(n-2)/2}}M_{Q_jW}(\mu^2)\Big\|\,\|f\|.
\end{aligned}$$

Since $\sigma \mapsto \sigma^{-(n-2)/2}M_{Q_jW}(\sigma^2)$ is norm continuous on $(0,\infty)$, this implies the validity of (7.144) (using for example the Dominated Convergence Theorem). $\square$

**A.7.4. An estimate for the spherical Laplacian $\Delta_S$.** We establish here a result that was used in the proof of Theorem 7.24.

**Lemma 7.25.** *Let $W$ satisfy the condition (S) with $\nu > 5/2$. Then, denoting by $N_\ell$ the multiplication operator in $\mathcal{B}(L^2(S^{n-1}))$ by the function $N_\ell(\vec{\omega}) = \omega_\ell$ ($\ell = 1, \ldots, n$):*

$$\Delta_S M_W(\lambda) = -\lambda M_{\vec{Q}^2 W}(\lambda) + \lambda \sum_{j,\ell=1}^{n} N_j N_\ell M_{Q_j Q_\ell W}(\lambda) + (n-1) i\sqrt{\lambda} \sum_{\ell=1}^{n} N_\ell M_{Q_\ell W}(\lambda). \tag{7.145}$$

*The operator $\Delta_S M_W(\lambda)$ is bounded from $L^2(\mathbb{R}^n)$ to $L^2(S^{n-1})$ for each $\lambda > 0$, and the mapping $\lambda \mapsto \Delta_S M_W(\lambda)$ is continuous in norm.*

PROOF. Since $Q_\ell W$ and $Q_j Q_\ell W$ satisfy the condition (S) [with $\nu > 1/2$], the right-hand side of (7.145) is a norm continuous $\mathcal{B}(L^2(\mathbb{R}^n), L^2(S^{n-1}))$-valued function of $\lambda$ on $(0, \infty)$, see Proposition 7.16. To calculate $\Delta_S M_W(\lambda) f$, for fixed $\lambda$ and with $f \in \mathcal{S}(\mathbb{R}^n)$, we extend the function $\vec{\omega} \mapsto [M_W(\lambda) f](\vec{\omega})$ to a function $F$ defined on some neighbourhood of $S^{n-1}$ by setting, for $\vec{p} \in \mathbb{R}^n \setminus \{0\}$: $F(\vec{p}) = [M_W(\lambda) f](\vec{p}/p)$, where $p = |\vec{p}|$. $F$ is a homogeneous function on $\mathbb{R}^n$, independent of the radial variable $p$, and as a function of $\vec{\omega}$ it coincides with $M_W(\lambda) f$. As $F$ is independent of $p$, we have [see (7.133)]

$$(\Delta F)(p\vec{\omega}) = \left[ \frac{\partial^2}{\partial p^2} + \frac{n-1}{p} \frac{\partial}{\partial p} + \frac{\Delta_S}{p^2} \right] F(p\vec{\omega}) = \frac{\Delta_S}{p^2} F(p\vec{\omega}),$$

in particular (for $p = 1$)

$$\Delta_S M_W(\lambda) f = (\Delta F)\big|_{S^{n-1}}. \tag{7.146}$$

Now

$$F(\vec{p}) = 2^{-1/2} \lambda^{(n-2)/4} (2\pi)^{-n/2} \int_{\mathbb{R}^n} e^{-i\sqrt{\lambda}\vec{x}\cdot\vec{p}/p} W(\vec{x}) f(\vec{x}) d^n x$$

and, for $\vec{y} \in \mathbb{R}^n$:

$$\Delta_{\vec{p}}\, e^{-i\vec{y}\cdot\vec{p}/p} = \left[ -\frac{\vec{y}^{\,2}}{p^2} + \frac{(\vec{y}\cdot\vec{p})^2}{p^4} + i\frac{n-1}{p^3}\, \vec{y}\cdot\vec{p} \right] e^{-i\vec{y}\cdot\vec{p}/p}.$$

Consequently (setting $\vec{y} = \sqrt{\lambda}\vec{x}$)

$$\Delta F(\vec{p})\big|_{p=1} = 2^{-1/2} \lambda^{(n-2)/4} (2\pi)^{-n/2} \times \int_{\mathbb{R}^n} e^{-i\sqrt{\lambda}\vec{x}\cdot\vec{\omega}} \{ -\lambda \vec{x}^{\,2} + \lambda(\vec{x}\cdot\vec{\omega})^2 + i(n-1)\sqrt{\lambda}\,\vec{x}\cdot\vec{\omega} \} W(\vec{x}) f(\vec{x}) d^n x,$$

which, in view of (7.146), is equivalent to (7.145). □

## Bibliographical Notes

A presentation of various methods that have been developed for proving the existence and completeness of wave operators can be found in Part IV of [BW]. For Schrödinger operators V. Enss [8] introduced a time-dependent method giving strong asymptotic completeness (see also Section XI.17, Volume III of [RS] for this method, and [9] for subsequent developments). A review of the literature concerning the Flux-Across-Surfaces Theorem can be found in [7]. As regards time delay, the reader can find a presentation of numerous physical aspects and applications of this concept in the review [5] and references to the more mathematical literature in [15]; the paper [15] also discusses generalisations of the definition given in Section 7.5.

## Problems

**7.1.** In $L^2(\mathbb{R}^n)$, show that $\mathcal{D}(|\vec{P}|) = \bigcap_{k=1,\dots,n}\mathcal{D}(P_k)$ and that the following identities hold for $f \in \mathcal{D}(|\vec{P}|)$:

$$\||P|f\|^2 = \sum_{k=1}^{n} \|P_k f\|^2, \tag{7.147}$$

$$\|\langle P\rangle f\|^2 = \|f\|^2 + \sum_{k=1}^{n} \|P_k f\|^2. \tag{7.148}$$

**7.2.** Show that $\text{s-lim}_{d\to\infty}\Gamma_{\Lambda_d} = O$.

**7.3.** Let $n \geq 2$ and use the following notation for vectors in $\mathbb{R}^n$: $\vec{x} = (\hat{x}, x_n)$ and $\vec{k} = (\hat{k}, k_n)$, where $\hat{x}, \hat{k} \in \mathbb{R}^{n-1}$ [e.g. $\hat{x} = (x_1, \dots, x_{n-1})$]. For $\nu > 1/2$ and $\mu \in \mathbb{R}$, let $M_\nu(\mu)$ be the following operator from $L^2(\mathbb{R}^n)$ to $L^2(\mathbb{R}^{n-1})$, with domain $\mathcal{S}(\mathbb{R}^n)$:

$$[M_\nu(\mu)f](\hat{k}) = \big[\mathcal{F}(I + |\vec{Q}|)^{-\nu} f\big](\hat{k}, \mu) \qquad (\hat{k} \in \mathbb{R}^{n-1})$$

[restriction of the Fourier transform of $(I + |\vec{Q}|)^{-\nu} f$ to the $(n-1)$-dimensional hyperplane $k_n = \mu$].

(a) Show that $M_\nu(\mu)$ is bounded [hence its closure belongs to $\mathcal{B}(L^2(\mathbb{R}^n), L^2(\mathbb{R}^{n-1}))$].

(b) Show that $\mu \mapsto M_\nu(\mu)$ is continuous in the operator norm.

Hint for (a): by using the Schwarz inequality in $L^2(\mathbb{R})$ (variable $x_n$), prove that

$$\Big|\big[\mathcal{F}(I + |\vec{Q}|)^{-\nu} f\big](\hat{k}, \mu)\Big|^2 \leq \frac{1}{(2\pi)^n}\|(1 + |x_n|)^{-\nu}\|^2_{L^2(\mathbb{R})}$$
$$\times \int_{\mathbb{R}} dx_n \Big|\int_{\mathbb{R}^{n-1}} e^{-i\hat{k}\cdot\hat{x}}\left[\frac{1+|x_n|}{1+|\vec{x}|}\right]^\nu f(\hat{x}, x_n)\, d^{n-1}\hat{x}\Big|^2.$$

Show that this implies that $\|M_\nu(\mu) f\|^2_{L^2(\mathbb{R}^{n-1})} \leq (2\pi)^{-n}\|(1 + |x_n|)^{-\nu}\|^2_{L^2(\mathbb{R})}\|f\|^2$.

**7.4.** (a) Verify that, for any self-adjoint operator $H$ in $L^2(\mathbb{R}^n)$, the set $\mathcal{D}_\nu(H)$ (see page 330) is dense in $\mathcal{H}$.

(b) Let $H = \vec{P}^2 + V$ in $L^2(\mathbb{R}^n)$. Assume that the part of the spectrum of $H$ on $(0, \infty)$ is purely absolutely continuous and that $\sigma_{\text{ac}}(H) = [0, \infty)$. Show that the set $\mathcal{D}_\nu^+(H)$ is dense in $\mathcal{H}_{\text{ac}}(H)$.

**7.5.** In $L^2(\mathbb{R}^3)$ consider a Coulomb Hamiltonian $H_{\rm C} = H_0 + V(|\vec{Q}|)$, with $H_0 = \vec{P}^2$ and $V(r) = \gamma/r$ ($\gamma \in \mathbb{R}$). Let $W$ be a real function defined on $\mathbb{R}^3$ of the form $W(\vec{x}) = (1+|\vec{x}|)^{-1-\delta}[w_\infty(\vec{x}) + w_2(\vec{x})]$ with $\delta > 0$, $w_\infty \in L^\infty(\mathbb{R}^3)$, $w_2 \in L^2(\mathbb{R}^3)$, $r(\partial W/\partial r) \in L^2(\mathbb{R}^3) + L^\infty_\varepsilon(\mathbb{R}^3)$ and $r^2(\partial^2 W/\partial r^2) \in L^2(\mathbb{R}^3) + L^\infty(\mathbb{R}^3)$, and let $H = H_C + W(\vec{Q})$. Show that $H_{\rm C}$ and $H$ have no singular continuous spectrum, and that the relative wave operators $\Omega_\pm = \text{s-lim}_{t\to\pm\infty} e^{iHt}e^{-iH_{\rm C}t}E_{\rm ac}(H_{\rm C})$ exist and satisfy strong asymptotic completeness [i.e. $\mathcal{R}(\Omega_\pm) = \mathcal{H}_{\rm ac}(H) = \mathcal{H}_{\rm p}(H)^\perp$].

**7.6.** For $V$ of class $\mathfrak{V}_\infty$, show that the operators $Y_t^{(j)}$ defined in Lemma 7.5 map $\mathcal{D}(H)$ into itself and satisfy the commutation relation (7.58).

**7.7.** Under the conditions of Lemma 7.5, verify that one has $g \in \mathcal{D}(P_k Q_j)$.

**7.8.** Let $W : \mathbb{R}^n \to \mathbb{C}$ satisfy $\int_{\mathbb{R}^n}(1+\vec{x}^2)|W(\vec{x})|^2 d^n x < \infty$. Show that, as a function of $\lambda$, the operator $M_W(\lambda)$ is differentiable in Hilbert-Schmidt norm for $\lambda \in (0,\infty)$, and determine its derivative.

**7.9.** In the context of Section 7.4, find conditions on the potential $V$ such that the averaged total scattering cross section $\overline{\sigma}(\lambda)$, Eq. (5.98), is continuously differentiable.

**7.10.** Prove Eq. (7.116).

**7.11.** (a) Complete the proof of Proposition 7.21 for the case in which it is only assumed that $\delta > k$.

(b) Show that $(d/d\lambda)R(\lambda)$ is a trace class operator if $\delta > n$ in Proposition 7.21.

**7.12.** Show that the operator $D_0$ in Eq. (7.125) is well defined on $\mathcal{D}_1^+(H_0)$ and symmetric on this domain.

# References

## Books and Monographs


[A] W. O. AMREIN: *Non-Relativistic Quantum Dynamics*. Reidel, Dordrecht (1981).

[ABG] W. O. AMREIN, A. BOUTET DE MONVEL, V. GEORGESCU: *$C_0$-Groups, Commutator Methods and Spectral Theory of $N$-Body Hamiltonians*. Birkhäuser, Basel (1996).

[Ad] R. A. ADAMS: *Sobolev Spaces*. Academic Press, New York (1975).

[AG] N. I. AKHIEZER, I. M. GLAZMAN: *Theory of Linear Operators in Hilbert Space*. Pitman, London (1981).

[AJS] W. O. AMREIN, J. M. JAUCH, K. B. SINHA: *Scattering Theory in Quantum Mechanics*. Benjamin, Reading (1977).

[B] H. S. BEAR: *A Primer of Lebesgue Integration*. Academic Press, San Diego (1995).

[BL] J. BERGH, J. LÖFSTRÖM: *Interpolation Spaces, An Introduction*. Springer, Berlin (1976).

[Br] M. BRAUN: *Differential Equations and Their Applications*. Springer, New York (1993).

[BS] C. BENNETT, R. SHARPLEY: *Interpolation of Operators*. Academic Press, Boston (1988).

[BW] H. BAUMGÄRTEL, M. WOLLENBERG: *Mathematical Scattering Theory*. Birkhäuser, Basel (1983).

[CFKS] H. L. CYCON, R. G. FROESE, W. KIRSCH, B. SIMON: *Schrödinger Operators, with Application to Quantum Mechanics and Global Geometry*. Springer, Berlin (1987).

[CK] M. Capiński, P. E. Kopp: *Measure, Integral and Probability*. Springer, London (2004).

[CL] E. A. Coddington, N. Levinson: *Theory of Ordinary Differential Equations*. McGraw-Hill, New York (1955).

[CV] M. Carter, B. van Brunt: *The Lebesgue-Stieltjes Integral*. Springer, New York (2000).

[D] G. De Barra: *Introduction to Measure Theory*. Van Nostrand, New York (1974).

[DK] M. Demuth, M. Krishna: *Determining Spectra in Quantum Theory*. Birkhäuser, Boston (2005).

[DS] N. Dunford, J. T. Schwartz: *Linear Operators, Part II: Spectral Theory*. Interscience Publishers, New York (1963).

[FJ] F. G. Friedlander, M. Joshi: *Introduction to the Theory of Distributions*. Cambridge Univ. Press (1982, 1998).

[GO] B. R. Gelbaum, J. M. H. Olmsted: *Counterexamples in Analysis*. Holden-Day, San Francisco (1964).

[GS] I. M. Gel'fand, G. E. Shilov: *Generalized Functions*, Vol. I. Academic Press, New York (1964).

[H] P. R. Halmos: *Measure Theory*. Van Nostrand, Princeton (1950).

[HS] E. Hewitt, K. Stromberg: *Real and Abstract Analysis*. Springer, New York (1965).

[K] T. Kato: *Perturbation Theory for Linear Operators*. Springer, New York (1976).

[L] S. Lang: *Real Analysis*. Addison-Wesley, Reading MA (1973).

[N] M. A. Naimark: *Linear Differential Operators*. Harrap, London (1968).

[Na] I. P. Natanson: *Theory of Functions of a Real Variable*, Volume I. Ungar, New York (1964).

[Ne] R. G. Newton: *Scattering Theory of Waves and Particles*. McGraw-Hill, New York (1966).

[P] D. B. Pearson: *Quantum Scattering and Spectral Theory*. Academic Press, London (1988).

[R] H. L. Royden: *Real Analysis*. Macmillan, New York (1968).

[Ra] I. K. RANA: *An Introduction to Measure and Integration*. American Math. Society, Providence RI (2002).

[RK] F. S. ROFE-BEKETOV, A. M. KHOLKIN: *Spectral Analysis of Differential Operators*. World Scientific, New Jersey (2005).

[RN] F. RIESZ, B. SZ.-NAGY: *Leçons d'analyse fonctionnelle*. Gauthier-Villars, Paris (1955).

[RS] M. REED, B. SIMON: *Methods of Modern Mathematical Physics*, Vols. I-IV. Academic Press, New York (1972-1979).

[ST] M. SAMUELIDES, L. TOUZILLIER: *Analyse fonctionnelle*. Cepadues-Editions, Toulouse (1989).

[W1] J. WEIDMANN: *Linear Operators in Hilbert Space*. Springer, New York (1980).

[W2] J. WEIDMANN: *Lineare Operatoren in Hilberträumen, Teil I: Grundlagen, Teil II: Anwendungen*. Teubner, Stuttgart (2000, 2003).

## Articles

[1] W. O. AMREIN, V. GEORGESCU: *On the Characterization of Bound States and Scattering States in Quantum Mechanics*, Helv. Phys. Acta **46**, 635-658 (1973).

[2] W. O. AMREIN, K. B. SINHA: *Time-Delay and Resonances in Potential Scattering*, J. Phys. A: Math. Gen. **39**, 9231-9254 (2006).

[3] W. O. AMREIN, L. ZULETA: *Flux and Scattering into Cones in Potential Scattering*, Helv. Phys. Acta **70**, 1-15 (1997).

[4] M. DAUMER, D. DÜRR, S. GOLDSTEIN, N. ZANGHI: *On the flux-across-surfaces theorem*, Lett. Math. Phys. **38**, 103-116 (1996).

[5] C. A. A. DE CARVALHO, H. M. NUSSENZVEIG: *Time Delay*, Phys. Rep. **364**, 83-174 (2002).

[6] S. A. DENISOV, A. KISELEV: *Spectral Properties of Schrödinger Operators with Decaying Potentials*, Proceedings of Symposia in Pure Mathematics, Vol. **76**, Part 2, 565-589, American Math. Society, Providence (2006).

[7] D. DÜRR, T. MOSER, P. PICKL: *The Flux-Across-Surfaces Theorem under conditions on the scattering state*, J. Phys. A: Math. Gen. **39**, 163-183 (2006).

[8] V. ENSS: *Asymptotic Completeness for Quantum Mechanical Potential Scattering*, Comm. Math. Phys. **61**, 285-291 (1978).

[9] W. HUNZIKER, I. M. SIGAL: *The quantum $N$-body problem*, J. Math. Phys. **41**, 3448-3510 (2000).

[10] A. JENSEN, E. MOURRE, P. PERRY: *Multiple commutator estimates and resolvent smoothness in quantum scattering theory*, Ann. Inst. Henri Poincaré **41**, 207-225 (1984).

[11] A. JENSEN, S. NAKAMURA: *Mapping Properties of Wave and Scattering Operators for Two-Body Schrödinger Operators*, Lett. Math. Phys. **24**, 295-305 (1992).

[12] T. KATO: *Smooth operators and commutators*, Studia Mathematica **31**, 535-546 (1968).

[13] D. B. PEARSON: *Singular Continuous Measures in Scattering Theory*, Comm. Math. Phys. **60**, 13-36 (1978).

[14] D. W. ROBINSON: *Propagation Properties in Scattering Theory*, J. Austral. Math. Soc. **21** (Series B), 474-485 (1980).

[15] R. TIEDRA DE ALDECOA: *Time delay for dispersive systems in quantum scattering theory*, preprint http://arxiv.org/abs/0810.1032 (2008), to appear in *Rev. Math. Phys.*

# Notation Index

# Subject Index